*Edited by
Richard Dronskowski,
Shinichi Kikkawa, and
Andreas Stein*

**Handbook of
Solid State Chemistry**

*Edited by
Richard Dronskowski,
Shinichi Kikkawa, and
Andreas Stein*

Handbook of Solid State Chemistry

Volume 3: Characterization

WILEY-VCH

WILEY-VCH Verlag GmbH & Co. KGaA

Editors

Richard Dronskowski
RWTH Aachen
Institute of Inorganic Chemistry
Landoltweg 1
52056 Aachen
Germany

Shinichi Kikkawa
Hokkaido University
Faculty of Engineering
N13 W8, Kita-ku
060-8628 Sapporo
Japan

Andreas Stein
University of Minnesota
Department of Chemistry
207 Pleasant St. SE
Minneapolis, MN 55455
USA

Cover Credit: Sven Lidin, Arndt Simon and Franck Tessier

All books published by **Wiley-VCH** are carefully produced. Nevertheless, authors, editors, and publisher do not warrant the information contained in these books, including this book, to be free of errors. Readers are advised to keep in mind that statements, data, illustrations, procedural details or other items may inadvertently be inaccurate.

Library of Congress Card No.: applied for

British Library Cataloguing-in-Publication Data
A catalogue record for this book is available from the British Library.

Bibliographic information published by the Deutsche Nationalbibliothek
The Deutsche Nationalbibliothek lists this publication in the Deutsche Nationalbibliografie; detailed bibliographic data are available on the Internet at http://dnb.d-nb.de.

© 2017 Wiley-VCH Verlag GmbH & Co. KGaA, Boschstr. 12, 69469 Weinheim, Germany

All rights reserved (including those of translation into other languages). No part of this book may be reproduced in any form – by photoprinting, microfilm, or any other means – nor transmitted or translated into a machine language without written permission from the publishers. Registered names, trademarks, etc. used in this book, even when not specifically marked as such, are not to be considered unprotected by law.

Print ISBN: 978-3-527-32587-0
oBook ISBN: 978-3-527-69103-6

Cover Design Formgeber
Typesetting Thomson Digital, Noida, India
Printing and Binding Markono Print Media Pte Ltd, Singapore

Printed on acid-free paper

Preface

When you do great science, you do not have to make a lot of fuss. This oft-forgotten saying from the twentieth century has served these editors pretty well, so the foreword to this definitive six-volume *Handbook of Solid-State Chemistry* in the early twenty-first century will be brief. After all, is there any real need to highlight the paramount successes of solid-state chemistry in the last half century? – Successes that have led to novel magnets, solid-state lighting, dielectrics, phase-change materials, batteries, superconducting compounds, and a lot more? Probably not, but we should stress that many of these exciting matters were derived from curiosity-driven research — work that many practitioners of our beloved branch of chemistry truly appreciate, and this is exactly why they do it. Our objects of study may be immensely important for various applications but, first of all, they are interesting to us; that is, how chemistry defines and challenges itself. Let us also not forget that solid-state chemistry is a neighbor to physics, crystallography, materials science, and other fields, so there is plenty of room at the border, to paraphrase another important quote from a courageous physicist.

Given the incredibly rich heritage of solid-state chemistry, it is probably hard for a newcomer (a young doctoral student, for example) to see the forest for all the trees. In other words, there is a real need to cover solid-state chemistry in its entirety, but only if it is conveniently grouped into digestible categories. Because such an endeavor is not possible in introductory textbooks, this is what we have tried to put together here. The compendium starts with an overview of materials and of the structure of solids. Not too surprisingly, the next volume deals with synthetic techniques, followed by another volume on various ways of (structural) characterization. Being a timely handbook, the fourth volume touches upon nano and hybrid materials, while volume V introduces the reader to the theoretical description of the solid state. Finally, the sixth volume reaches into the real world by focusing on functional materials. Should we have considered more volumes? Yes, probably, but life is short, dear friends.

This handbook would have been impossible to compile for three authors, let alone a single one. Instead, the editors take enormous pride in saying that they managed to motivate more than a hundred first-class scientists living across the globe, each of them specializing in (and sometimes even shaping) a subfield of solid-state chemistry or a related discipline, and all of these wonderful colleagues did their very best to make our dream come true. Thanks to all of you; we sincerely appreciate your contributions. Thank you, once again, on behalf of solid-state chemistry. The editors also would like to thank Wiley-VCH, in particular Dr. Waltraud Wüst and also Dr. Frank Otmar Weinreich, for spiritually (and practically) accompanying us over a few years, and for reminding us here and there that there must be a final deadline. That being said, it is up to the reader to judge whether the tremendous effort was justified. We sincerely hope that this is the case.

A toast to our wonderful science! Long live solid-state chemistry!

Richard Dronskowski
RWTH Aachen, Aachen, Germany

Shinichi Kikkawa
Hokkaido University, Sapporo, Japan

Andreas Stein
University of Minnesota, Minneapolis, USA

Contents

Volume 1: Materials and Structure of Solids

1   Intermetallic Compounds and Alloy Bonding Theory Derived from Quantum Mechanical One-Electron Models   *1*
Stephen Lee and Daniel C. Fredrickson

2   Quasicrystal Approximants   *73*
Sven Lidin

3   Medium-Range Order in Oxide Glasses   *93*
Hellmut Eckert

4   Suboxides and Other Low-Valent Species   *139*
Arndt Simon

5   Introduction to the Crystal Chemistry of Transition Metal Oxides   *161*
J.E. Greedan

6   Perovskite Structure Compounds   *221*
Yuichi Shimakawa

7   Nitrides of Non-Main Group Elements   *251*
P. Höhn and R. Niewa

8   Fluorite-Type Transition Metal Oxynitrides   *361*
Franck Tessier

9   Mechanochemical Synthesis, Vacancy-Ordered Structures and Low-Dimensional Properties of Transition Metal Chalcogenides   *383*
Yutaka Ueda and Tsukio Ohtani

10	**Metal Borides: Versatile Structures and Properties** *435* Barbara Albert and Kathrin Hofmann	
11	**Metal Pnictides: Structures and Thermoelectric Properties** *455* Abdeljalil Assoud and Holger Kleinke	
12	**Metal Hydrides** *477* Yaoqing Zhang, Maarten C. Verbraeken, Cédric Tassel, and Hiroshi Kageyama	
13	**Local Atomic Order in Intermetallics and Alloys** *521* Frank Haarmann	
14	**Layered Double Hydroxides: Structure–Property Relationships** *541* Shan He, Jingbin Han, Mingfei Shao, Ruizheng Liang, Min Wei, David G. Evans, and Xue Duan	
15	**Structural Diversity in Complex Layered Oxides** *571* S. Uma	
16	**Magnetoresistance Materials** *595* Ichiro Terasaki	
17	**Magnetic Frustration in Spinels, Spin Ice Compounds, $A_3B_5O_{12}$ Garnet, and Multiferroic Materials** *617* Hongyang Zhao, Hideo Kimura, Zhenxiang Cheng, and Tingting Jia	
18	**Structures and Properties of Dielectrics and Ferroelectrics** *643* Mitsuru Itoh	
19	**Defect Chemistry and Its Relevance for Ionic Conduction and Reactivity** *665* Joachim Maier	
20	**Molecular Magnets** *703* J.V. Yakhmi	
21	**Ge–Sb–Te Phase-Change Materials** *735* Volker L. Deringer and Matthias Wuttig	
	Index *751*	

Volume 2: Synthesis

1 **High-Temperature Methods** *1*
  Rainer Pöttgen and Oliver Janka

2 **High-Pressure Methods in Solid-State Chemistry** *23*
  Hubert Huppertz, Gunter Heymann, Ulrich Schwarz, and Marcus R. Schwarz

3 **High-Pressure Perovskite: Synthesis, Structure, and Phase Relation** *49*
  Yoshiyuki Inaguma

4 **Solvothermal Methods** *107*
  Nobuhiro Kumada

5 **High-Throughput Synthesis Under Hydrothermal Conditions** *123*
  Nobuaki Aoki, Gimyeong Seong, Tsutomu Aida, Daisuke Hojo,
  Seiichi Takami, and Tadafumi Adschiri

6 **Particle-Mediated Crystal Growth** *155*
  R. Lee Penn

7 **Sol–Gel Synthesis of Solid-State Materials** *179*
  Guido Kickelbick and Patrick Wenderoth

8 **Templated Synthesis for Nanostructured Materials** *201*
  Yoshiyuki Kuroda and Kazuyuki Kuroda

9 **Bio-Inspired Synthesis and Application of Functional Inorganic Materials by Polymer-Controlled Crystallization** *233*
  Lei Liu and Shu-Hong Yu

10 **Reactive Fluxes** *275*

11 **Glass Formation and Crystallization** *287*
  T. Komatsu

12 **Glass-Forming Ability, Recent Trends, and Synthesis Methods of Metallic Glasses** *319*
  Hidemi Kato, Takeshi Wada, Rui Yamada, and Junji Saida

13 **Crystal Growth Via the Gas Phase by Chemical Vapor Transport Reactions** *351*
  Michael Binnewies, Robert Glaum, Marcus Schmidt, and Peer Schmidt

14	**Thermodynamic and Kinetic Aspects of Crystal Growth** *375* *Detlef Klimm*
15	**Chemical Vapor Deposition** *399* *Takashi Goto and Hirokazu Katsui*
16	**Growth of Wide Bandgap Semiconductors by Halide Vapor Phase Epitaxy** *429* *Yuichi Oshima, Encarnación G. Víllora, and Kiyoshi Shimamura*
17	**Growth of Silicon Nanowires** *467* *Fengji Li and Sam Zhang*
18	**Chemical Patterning on Surfaces and in Bulk Gels** *539* *Olaf Karthaus*
19	**Microcontact Printing** *563* *Kiyoshi Yase*
20	**Nanolithography Based on Surface Plasmon** *573* *Kosei Ueno and Hiroaki Misawa*

Index *589*

Volume 3: Characterization

1	**Single-Crystal X-Ray Diffraction** *1*
	Ulli Englert
1.1	Introduction *1*
1.2	Symmetry *1*
1.3	Diffraction *6*
1.4	Instrumentation *11*
1.5	Solution of the Phase Problem *13*
1.5.1	The Patterson Method *13*
1.5.2	Difference Fourier Synthesis *14*
1.5.3	Direct Methods *15*
1.5.4	Charge Flipping *17*
1.5.5	Other Methods for Structure Solution *18*
1.6	Refinement *18*
1.6.1	Agreement Factors *18*
1.6.2	Constraints and Restraints *19*
1.6.3	Troubleshooting *20*
1.6.4	Twin Refinement *21*
1.6.5	Refinement of Modulated Structures *22*
1.7	Interpretation of Diffraction Data *22*

1.7.1	Geometry and Displacement Parameters	23
1.7.2	Intermolecular Interactions	23
1.7.3	CIF	24
1.7.4	Databases	24
1.7.5	Electron Density	25
	Acknowledgment	26
	References	26
2	**Laboratory and Synchrotron Powder Diffraction**	**29**
	R. E. Dinnebier, M. Etter, and T. Runcevski	
2.1	Introduction	29
2.2	History and Basics of Powder Diffraction	30
2.3	Modern Powder Diffractometers in the Laboratory and at the Synchrotron	41
2.4	The Rietveld Method	46
2.5	Sequential and Parametric Rietveld Refinements	49
2.6	Whole Powder Pattern Decomposition	49
2.7	Concept of Symmetry Modes	50
2.8	Structure Determination from Powders	52
2.9	Case Studies	56
2.9.1	Case Study 1: Crystal Structure Determination of Codeine Phosphate Sesquihydrate by a Direct Space Global Optimization Method of Simulated Annealing Using Rigid Bodies	56
2.9.2	Case Study 2: Application of *In Situ* X-Ray Powder Diffraction in Elucidating of Physical Phenomena: The Thermosalient (Jumping Crystal) Effect	60
2.9.3	Case Study 3: Crystal Structure Determination of Corrosion Product by a Dual-Space Charge Flipping Algorithm, Aided by Inspection of Difference Fourier Maps	64
2.9.4	Case Study 4: Utilizing Symmetry Modes for the Investigation of High-Pressure Phase Transitions in the Rare Earth Orthoferrite $LaFeO_3$	67
	References	70
3	**Neutron Diffraction**	**77**
	Martin Meven and Georg Roth	
3.1	Introduction	77
3.2	Neutrons as Structural Probes	78
3.2.1	Characteristic Properties of Neutrons	78
3.2.2	Generation of Neutrons and Neutron Sources	81
3.3	Neutron Diffraction	82
3.3.1	Diffracted Intensities and Corrections	82
3.3.2	Concepts and Components of Instrumentation	84
3.3.3	Powder Neutron Diffraction	86
3.3.4	Single-Crystal Diffraction	88

3.3.4.1	Instrumentation	*90*
3.3.4.2	Sample Environment	*90*
3.4	Examples	*91*
3.4.1	The Hydrogen Challenge	*92*
3.4.1.1	H/D Ordering in Ferroelectric RbH_2PO_4 (RDP)	*92*
3.4.2	Diffraction Contrast	*95*
3.4.2.1	Crystal Structure and Site Occupation of $(Mn_{1-x}Cr_x)_{1+\delta}Sb$	*96*
3.4.3	Static Disorder and Displacements in HT Superconductors	*96*
3.4.4	Crystal Structure and Magnetic Order in Co_2SiO_4	*98*
3.4.5	Magnetic Structures from Neutron Diffraction	*100*
3.4.6	Electron Densities from X-Rays and Neutrons	*102*
3.4.7	Magnetization Densities from Neutron Diffraction	*104*
3.4.8	Spatially Resolved In Situ Neutron Diffraction – ZEBRA® cell	*105*
3.5	Outlook	*107*
	Acknowledgments	*107*
	References	*107*

4	**Modulated Crystal Structures**	*109*
	Sander van Smaalen	
4.1	Order Without Translational Symmetry	*109*
4.2	Displacive Modulation of the Basic Structure	*112*
4.3	Modulation Functions and Fourier Maps	*116*
4.4	Crystal Chemical Analysis: *t*-Plots	*119*
4.5	Modulated Molecular Crystals	*122*
4.6	Charge Order and Related Phenomena	*124*
4.7	Outlook	*127*
	References	*127*

5	**Characterization of Quasicrystals**	*131*
	Walter Steurer	
5.1	Introduction	*131*
5.2	Structure Analysis of Quasicrystals	*134*
5.3	Transmission Electron Microscopy for Quasicrystal Structure Imaging	*135*
5.3.1	High-Resolution Transmission Electron Microscopy (HRTEM)	*137*
5.3.2	High-Angle Annular Dark-Field and Annular Bright-Field Scanning Transmission Electron Microscopy (HAADF- and ABF-STEM)	*139*
5.3.3	Surface Structure Analysis of Quasicrystals	*140*
5.4	Diffraction Methods for the Analysis of Quasicrystal Structures	*140*
5.4.1	Charge Flipping (CF) and Low-Density Elimination (LDE)	*144*
5.4.1.1	Charge Flipping (CF)	*144*
5.4.1.2	Low-density Elimination (LDE)	*145*
5.4.2	nD and 3D Model Structure Refinement, Respectively	*147*

5.5	Spectroscopic Methods	*149*
5.6	Summary	*150*
	References	*150*

6 Transmission Electron Microscopy *155*
Krumeich Frank

6.1	Introduction	*155*
6.2	Operation Principle	*156*
6.2.1	Properties of the Electron	*156*
6.2.2	The Electron Lens	*157*
6.2.3	Transmission Electron Microscope	*158*
6.2.4	Scanning Transmission Electron Microscope	*160*
6.2.5	Sample Preparation	*162*
6.3	Electron–Matter Interactions and Their Applications	*163*
6.3.1	Elastic Interactions	*164*
6.3.1.1	Incoherent Scattering	*164*
6.3.1.2	Mass Thickness Contrast	*166*
6.3.1.3	Coherent Scattering	*167*
6.3.1.4	Electron Diffraction	*168*
6.3.1.5	Diffraction Contrast	*169*
6.3.1.6	High-Resolution Transmission Electron Microscopy (HRTEM)	*170*
6.3.1.7	Examples of HRTEM Studies	*172*
6.3.2	Inelastic Interactions	*174*
6.3.2.1	Overview	*174*
6.3.2.2	X-Ray Spectroscopy	*175*
6.3.2.3	Electron Energy Loss Spectroscopy (EELS)	*175*
6.3.2.4	Analytical Electron Microscopy	*177*
6.4	The Impact of Aberration Correction	*179*
6.4.1	High Resolution	*179*
6.4.2	*In Situ* Studies	*179*
6.5	Conclusions	*180*
	Acknowledgments	*181*
	References	*181*

7 Scanning Probe Microscopy *183*
Marek Nowicki and Klaus Wandelt

7.1	Introduction	*183*
7.2	Scanning Tunneling Microscopy	*185*
7.2.1	Physical Background	*185*
7.2.2	Instrumental Aspects	*190*
7.2.2.1	Control of Tip Position	*190*
7.2.2.2	STM Setup in UHV	*193*
7.2.2.3	Electrochemical STM Setup	*193*
7.2.2.4	Tip Preparation	*195*
7.2.3	Interpretation of STM Images	*197*

7.3	Atomic Force Microscopy	*199*
7.3.1	Physical Background	*199*
7.3.1.1	AFM Contact Mode	*200*
7.3.1.2	AFM Noncontact Mode	*203*
7.3.1.3	Tuning Fork Sensor	*206*
7.3.1.4	Length-Extension Resonator	*207*
7.4	Other Types of Scanning Probes	*207*
7.5	Exemplary Results	*208*
7.5.1	Investigations in UHV	*208*
7.5.1.1	Surface Structure: Reconstruction	*209*
7.5.1.2	Surface Composition: Alloys	*210*
7.5.1.3	Surface Oxidation	*211*
7.5.1.4	Amorphous Structures	*215*
7.5.1.5	Adsorption	*216*
7.5.1.6	Atomic Manipulation	*223*
7.5.1.7	Tip-Induced Surface Reaction	*223*
7.5.2	Scanning Tunneling Microscopy at High Gas Pressure	*224*
7.5.3	Scanning Tunneling Microscopy in Solution	*226*
7.5.3.1	Surface Structure	*227*
7.5.3.2	Electrochemical Double-Layer	*228*
7.5.3.3	Anion Adsorption on Cu(111)	*229*
7.5.3.4	Underpotential Metal Deposition	*234*
7.5.3.5	Molecular Self-Assembly at Metal–Electrolyte Interfaces	*236*
7.6	Summary	*238*
	References	*239*
8	**Solid-State NMR Spectroscopy: Introduction for Solid-State Chemists**	**245**
	Christoph S. Zehe, Renée Siegel, and Jürgen Senker	
8.1	Introduction	*245*
8.2	Theory and Methods	*246*
8.2.1	Important Interactions for NMR Spectroscopy	*246*
8.2.1.1	Tensors and Rotations	*246*
8.2.1.2	Zeeman Interaction	*248*
8.2.1.3	Chemical Shift	*249*
8.2.1.4	Dipolar Interactions	*250*
8.2.1.5	Quadrupolar Interaction	*250*
8.2.1.6	Radio Frequency Pulses	*251*
8.2.2	Time Evolution of Spin Systems	*251*
8.2.2.1	Spin Density Formalism	*252*
8.2.2.2	Simulation of NMR Experiments	*253*
8.2.3	Lineshapes in Solid-State NMR Spectroscopy	*253*
8.2.3.1	Magic-Angle Spinning	*253*
8.2.3.2	Lineshapes	*254*
8.3	Applications	*256*

8.3.1	Wideline Experiments	*256*
8.3.2	High-Resolution Experiments	*260*
8.3.3	Dipolar Recoupling Experiments	*265*
8.3.3.1	Homonuclear Dipolar Recoupling	*266*
8.3.3.2	Heteronuclear Dipolar Recoupling	*268*
	References *271*	

9 Modern Electron Paramagnetic Resonance Techniques and Their Applications to Magnetic Systems *279*
Andrej Zorko, Matej Pregelj, and Denis Arčon

9.1	Introduction *279*	
9.1.1	Relevant Interactions and Their Typical Energy Scales	*280*
9.1.2	Information Extracted from an EPR Experiment	*281*
9.1.3	Prime Fields of Application	*282*
9.2	Advanced Electron Paramagnetic Resonance Techniques	*284*
9.2.1	Pulsed versus c.w. Electron Paramagnetic Resonance Techniques	*284*
9.2.2	Toward Single Spin Electron Paramagnetic Resonance Detection	*284*
9.2.3	High-Frequency EPR Spectroscopy	*285*
9.3	Electron Paramagnetic Resonance of Exchange Coupled Systems	*286*
9.3.1	Kubo–Tomita Theory of the EPR Line Width and Exchange Narrowing	*287*
9.3.2	Nagata–Tazuke Theory of the EPR Line Shift	*288*
9.3.3	Applications of the Kubo–Tomita and Nagata–Tazuke Theories to Real Materials	*289*
9.3.3.1	$SrCu_2(BO_3)_2$: An Orthogonal Dimer System	*289*
9.3.3.2	Magnetic Anisotropy on the Kagome Lattice	*290*
9.3.4	Breakdown of the Kubo–Tomita Model and Alternative Approaches	*292*
9.3.4.1	EPR and Exact Calculations	*293*
9.3.4.2	Oshikawa–Affleck Theory for 1D Antiferromagnets	*293*
9.3.4.3	EPR Moments Approach	*294*
9.4	(Anti)Ferromagnetic Resonance in Magnetically Ordered Phases	*295*
9.4.1	Molecular Field Approach	*296*
9.4.2	Application of the Molecular Field Approach to Real Materials	*298*
9.4.2.1	$CuSe_2O_5$: A Spin-1/2 Chain System	*298*
9.4.2.2	$Ba_3NbFe_3Si_2O_{14}$: A Two-Dimensional Triangular Magnetic Lattice	*299*
9.4.3	Linear Spin-Wave Theory	*300*
9.5	Summary and Outlook	*302*
	References *302*	

10 Photoelectron Spectroscopy *311*
Stephan Breuer and Klaus Wandelt

10.1	Introduction *311*	
10.2	The Photoemission Experiment	*313*
10.2.1	The Sample	*314*

10.2.2 The Photon Source 315
10.2.3 Electron Analyzer and Detector 315
10.2.4 Typical Spectral Features 318
10.3 The Photoemission Process 324
10.3.1 Excitation, Relaxation 324
10.3.2 Spin–Orbit and Multiplet Splitting 326
10.3.3 Intensity 327
10.3.4 Photoemission from Solids 328
10.3.5 Extrinsic Satellites 330
10.3.6 Auger Spectroscopy 331
10.4 Analytical Applications 335
10.4.1 Qualitative Analysis 335
10.4.2 Chemical Shift 337
10.4.2.1 Influence of Ionicity 337
10.4.2.2 Standard Spectra 339
10.4.2.3 Surface Core-Level Shifts 341
10.4.2.4 Valence Band Emission 343
10.4.3 Core-Level Angular Dependence 345
10.4.4 Quantitative Analysis 348
10.4.5 Depth Profiling 352
10.5 Possible Misinterpretations 356
References 357

11 Recent Developments in Soft X-Ray Absorption Spectroscopy 361
Alexander Moewes
11.1 Introduction 361
11.1.1 Different Measurement Techniques: Electron Yield and Fluorescence Yield 365
11.2 Recent Developments 369
11.2.1 Calculation of Soft X-Ray Absorption Spectra 369
11.2.1.1 Notation and Formulas 369
11.2.1.2 Linewidth and Broadening 371
11.2.1.3 Different Approaches, Codes, and Optimization 375
11.2.2 Recent Technical Developments 377
11.2.2.1 Inverse Partial Fluorescence Yield (IPFY) 377
11.2.2.2 Measuring Soft X-Ray Absorption in Transmission Mode 379
11.2.3 New Sources, Time, and Spatial Resolution 380
11.2.4 Detectors 382
11.2.5 Polarization Effects 382
11.2.6 Liquids and in Operando XAS 384
11.2.7 Other 386
11.3 Conclusions and Outlook 388
References 389

12	**Vibrational Spectroscopy** *393*	
	Götz Eckold and Helmut Schober	
12.1	Why Spectroscopy *393*	
12.2	Fundamentals of Lattice Dynamics *394*	
12.2.1	Phonons in the Harmonic Approximation *394*	
12.2.2	Amplitudes of Lattice Vibrations and Normal Coordinates *402*	
12.2.3	Phonon Density of States and Lattice Heat Capacity *403*	
12.3	Spectroscopic Studies of Selected Systems *404*	
12.3.1	IR Spectroscopy *406*	
12.3.1.1	Thin Films versus Bulk Samples *406*	
12.3.1.2	IR Reflectivity of Multiferroic Systems *407*	
12.3.2	Raman Spectroscopy *409*	
12.3.2.1	Lattice Anomalies in Multiferroics *411*	
12.3.2.2	Raman Spectra of Multidomain Samples *412*	
12.3.2.3	Local Distortions Revealed by Raman Scattering *412*	
12.3.3	Inelastic Neutron Scattering *414*	
12.3.3.1	Elastic Constants and Mechanical Anisotropy *418*	
12.3.3.2	Softmode Phase Transitions *419*	
12.3.3.3	Anharmonicity *423*	
12.3.3.4	Nonequilibrium Studies *426*	
12.3.3.5	Molecular Crystals *429*	
12.3.3.6	Guest–Host Systems *430*	
12.3.3.7	Phonon-Enhanced Ionic Transport *434*	
12.3.4	Inelastic X-Ray Scattering *434*	
12.3.4.1	INXS Probing Superconductivity *435*	
12.3.4.2	RIXS Probing Electron–Phonon Coupling in Semiconductors *436*	
12.4	Concluding Remarks *438*	
	References *440*	
13	**Mößbauer Spectroscopy** *443*	
	Hermann Raphael	
13.1	Introduction *443*	
13.2	Spectral Parameters *446*	
13.2.1	Area and Recoil-Free Fraction *447*	
13.2.2	Isomer Shift *448*	
13.2.3	Second-Order Doppler Shift *449*	
13.2.4	Quadrupole Splitting *451*	
13.2.5	Magnetic Hyperfine Splitting *454*	
13.2.6	Relaxation Phenomena *455*	
13.3	Practical Aspects *456*	
13.3.1	Experimental Implementation *456*	
13.3.2	Radiation Sources *456*	
13.3.3	Detectors *459*	
13.3.4	Sample Thickness and Sample Environment *460*	

13.3.5 Calibration and Data Treatment *461*
13.3.6 Synchrotron Radiation *462*
13.3.7 Phonon-Assisted Nuclear Resonance Absorption *464*
13.4 Examples and Applications *466*
13.4.1 Valence Determination *466*
13.4.2 Site Occupancy/Distribution *466*
13.4.3 Spin Crossover *468*
13.4.4 Charge Order and Relaxation *468*
13.4.5 Magnetic Phase Transitions *470*
13.4.6 Superparamagnetism *471*
13.4.7 Transferred Field *471*
13.4.8 Superconductivity *471*
13.4.9 *In Situ*, *Operando*, Remote Operation *472*
13.4.10 Nuclear Resonance Scattering *473*
13.5 Conclusion *474*
Acknowledgments *474*
Abbreviations *474*
References *475*

14 Macroscopic Magnetic Behavior: Spontaneous Magnetic Ordering *485*
Heiko Lueken and Manfred Speldrich
14.1 Introduction *485*
14.2 Basic Prinziples *487*
14.2.1 Magnetic Quantities and Units *487*
14.2.1.1 Magnetic Dipole Moment m *487*
14.2.1.2 Magnetic Field, Magnetization, and Magnetic Susceptibility *487*
14.2.1.3 Brillouin Function, Curie Law *489*
14.2.2 Classification of Magnetic Materials *490*
14.2.2.1 Magnetically Dilute Systems *490*
14.2.2.2 Local Magnetic Ordering within a Dinuclear Molecule *493*
14.2.2.3 Magnetically Condensed Systems *493*
14.3 Crystal Field Splittings *495*
14.3.1 Introduction of the Crystal Field *495*
14.3.2 Splitting of d Orbitals in Octahedral Symmetry *495*
14.3.3 Splitting of d Orbitals in Tetrahedral and Other Symmetries *498*
14.3.4 Hole Formalism: d^1 and d^9 Electron Configurations *500*
14.3.5 Multiple Transitions for $3d^2$ Electron Configurations *502*
14.3.6 The Spectrochemical Series *505*
14.4 Magnetic Materials *510*
14.4.1 Magnetic Properties of Ferromagnets *510*
14.4.1.1 Permeability *510*
14.4.1.2 Hysteresis *510*
14.4.1.3 Saturation Magnetization *511*
14.4.1.4 Remanence *512*

14.4.1.5 Coercivity *512*
14.4.1.6 Differential Permeability *513*
14.4.1.7 Curie Temperature *513*
14.4.1.8 Classification of Ferromagnetic Materials *513*
14.4.1.9 Permanent Magnets *514*
14.4.2   Paramagnetism and Diamagnetism *515*
14.4.2.1 Paramagnets *515*
14.4.2.2 Temperature Dependence of Paramagnetic Susceptibility *516*
14.4.2.3 Applications of Paramagnets *518*
14.4.2.4 Diamagnets *518*
14.4.2.5 Superconductors *519*
         References *520*

15   Dielectric Properties *523*
         Rainer Waser and Susanne Hoffmann-Eifert
15.1   Applications of Dielectrics *523*
15.2   Polarization of Condensed Matter *524*
15.2.1   Dielectrics in an Electrostatic Field *525*
15.2.2   Electrostatic Field Energy *526*
15.2.3   Dipoles and Polarization *527*
15.3   Mechanisms of Polarization *528*
15.3.1   Electronic Polarization *528*
15.3.2   Ionic Polarization *530*
15.3.3   Orientation Polarization *532*
15.3.4   Maxwell–Wagner Polarization *535*
15.3.5   Material Classification by the Polarization Mechanism *536*
15.4   Electric Fields in Matter *537*
15.4.1   Macroscopic Description *537*
15.4.2   Microscopic Description and the Lorentz Field *538*
15.4.3   Clausius–Mosotti Equation *540*
15.4.4   Temperature Coefficient of the Dielectric Permittivity *540*
15.5   The Frequency Dependence of the Dielectric Behavior of Matter *541*
15.5.1   Dielectrics in Alternating Electrical Fields *541*
15.5.2   The Complex Dielectric Function *543*
15.5.3   Resonance Phenomena *544*
15.5.4   Relaxation Phenomena *546*
15.5.4.1 Debye Relaxation *546*
15.5.4.2 Maxwell–Wagner Relaxation *550*
15.5.4.3 Dielectrics with a Distribution of Relaxation Times *551*
15.6   Polarization Waves in Ionic Crystals *553*
15.6.1   The Lyddane–Sachs–Teller Relation *554*
15.6.2   Softening of the Transverse Optical Phonon *554*
15.6.3   Characteristic Oscillations in Perovskite-Type Oxides *556*
15.6.4   Temperature Dependence of the Permittivity in Titanates *557*

	15.6.5	Voltage Dependence of the Permittivity in Ionic Crystals 558
	15.7	Summary 559
		References 560

16 Mechanical Properties *561*

Volker Schnabel, Moritz to Baben, Denis Music, William J. Clegg, and Jochen M. Schneider

16.1	Introduction into Mechanical Properties 561
16.2	Elastic Behavior 562
16.3	Nonelastic Behavior 565
16.3.1	Plasticity 565
16.3.2	Fracture 570
16.3.3	Friction and Wear 575
16.4	Characterization Techniques 578
16.4.1	Diamond Anvil Cell 578
16.4.2	Scattering Techniques for Elasticity Measurements 579
16.4.3	Tensile Test 580
16.4.4	Nanoindentation 581
16.4.5	Fracture Properties 582
16.4.6	Friction and Wear Test 583
	Acknowledgments 585
	References 585

17 Calorimetry *589*

Hitoshi Kawaji

17.1	Introduction 589
17.2	Nonreaction Calorimetry for Heat Capacity Measurements 591
17.2.1	Adiabatic Calorimetry for Heat Capacity Measurements 591
17.2.2	Relaxation Method for Heat Capacity Measurements 596
17.2.3	AC Calorimetry for Heat Capacity Measurements 598
17.2.4	Heat Pulse Method for Heat Capacity Measurements 600
17.2.5	Drop Calorimetry for Heat Capacity Measurements 602
17.2.6	Differential Scanning Calorimetry for Heat Capacity Measurements 603
17.2.7	Heat Capacity Spectroscopy 607
17.3	Reaction Calorimetry for Enthalpy Change Measurements 608
17.3.1	Calvet Calorimeter for High-Temperature Reaction Calorimetry 609
	References 611

Index *615*

Volume 4: Nano and Hybrid Materials

1 Self-Assembly of Molecular Metal Oxide Nanoclusters *1*
   Laia Vilà-Nadal and Leroy Cronin

2 Inorganic Nanotubes and Fullerene–Like Nanoparticles from Layered (2D) Compounds *21*
   L. Yadgarov, R. Popovitz-Biro, and R. Tenne

3 Layered Materials: Oxides and Hydroxides *53*
   Ida Shintaro

4 Organoclays and Polymer-Clay Nanocomposites *79*
   M.A. Vicente and A. Gil

5 Zeolite and Zeolite-Like Materials *97*
   Watcharop Chaikittisilp and Tatsuya Okubo

6 Ordered Mesoporous Materials *121*
   Michal Kruk

7 Porous Coordination Polymers/Metal–Organic Frameworks *141*
   Ohtani Ryo and Kitagawa Susumu

8 Metal–Organic Frameworks: An Emerging Class of Solid-State Materials *165*
   Joseph E. Mondloch, Rachel C. Klet, Ashlee J. Howarth, Joseph T. Hupp, and Omar K. Farha

9 Sol–Gel Processing of Porous Materials *195*
   Kazuki Nakanishi, Kazuyoshi Kanamori, Yasuaki Tokudome, George Hasegawa, and Yang Zhu

10 Macroporous Materials Synthesized by Colloidal Crystal Templating *243*
    Jinbo Hu and Andreas Stein

11 Optical Properties of Hybrid Organic–Inorganic Materials and their Applications – Part I: Luminescence and Photochromism *275*
    Stephane Parola, Beatriz Julián-López, Luís D. Carlos, and Clément Sanchez

12 Optical Properties of Hybrid Organic–inorganic Materials and their Applications – Part II: Nonlinear Optics and Plasmonics *317*
    Stephane Parola, Beatriz Julián-López, Luís D. Carlos, and Clément Sanchez

13	**Bioactive Glasses** *357* Hirotaka Maeda and Toshihiro Kasuga	
14	**Materials for Tissue Engineering** *383* María Vallet-Regí and Antonio J. Salinas	

Index *411*

Volume 5: Theoretical Description

1	**Density Functional Theory** *1* Michael Springborg and Yi Dong	
2	**Eliminating Core Electrons in Electronic Structure Calculations: Pseudopotentials and PAW Potentials** *29* Stefan Goedecker and Santanu Saha	
3	**Periodic Local Møller–Plesset Perturbation Theory of Second Order for Solids** *59* Denis Usvyat, Lorenzo Maschio, and Martin Schütz	
4	**Resonating Valence Bonds in Chemistry and Solid State** *87* Evgeny A. Plekhanov and Andrei L. Tchougréeff	
5	**Many Body Perturbation Theory, Dynamical Mean Field Theory and All That** *119* Silke Biermann and Alexander Lichtenstein	
6	**Semiempirical Molecular Orbital Methods** *159* Thomas Bredow and Karl Jug	
7	**Tight-Binding Density Functional Theory: DFTB** *203* Gotthard Seifert	
8	**DFT Calculations for Real Solids** *227* Karlheinz Schwarz and Peter Blaha	
9	**Spin Polarization** *261* Dong-Kyun Seo	
10	**Magnetic Properties from the Perspectives of Electronic Hamiltonian: Spin Exchange Parameters, Spin Orientation, and Spin-Half Misconception** *285* Myung-Hwan Whangbo and Hongjun Xiang	

11   Basic Properties of Well-Known Intermetallics and Some New Complex
      Magnetic Intermetallics   *345*
      Peter Entel

12   Chemical Bonding in Solids   *405*
      Gordon J. Miller, Yuemei Zhang, and Frank R. Wagner

13   Lattice Dynamics and Thermochemistry of Solid-State Materials from
      First-Principles Quantum-Chemical Calculations   *491*
      Ralf Peter Stoffel and Richard Dronskowski

14   Predicting the Structure and Chemistry of Low-Dimensional
      Materials   *527*
      Xiaohu Yu, Artem R. Oganov, Zhenhai Wang, Gabriele Saleh, Vinit Sharma,
      Qiang Zhu, Qinggao Wang, Xiang-Feng Zhou, Ivan A. Popov,
      Alexander I. Boldyrev, Vladimir S. Baturin, and Sergey V. Lepeshkin

15   The Pressing Role of Theory in Studies of Compressed Matter   *571*
      Eva Zurek

16   First-Principles Computation of NMR Parameters in
      Solid-State Chemistry   *607*
      Jérôme Cuny, Régis Gautier, and Jean-François Halet

17   Quantum Mechanical/Molecular Mechanical (QM/MM) Approaches   *647*
      C. Richard A. Catlow, John Buckeridge, Matthew R. Farrow,
      Andrew J. Logsdail, and Alexey A. Sokol

18   Modeling Crystal Nucleation and Growth and Polymorphic
      Transitions   *681*
      Dirk Zahn

      Index   *701*

Volume 6: Functional Materials

1   Electrical Energy Storage: Batteries   *1*
      Eric McCalla

2   Electrical Energy Storage: Supercapacitors   *25*
      Enbo Zhao, Wentian Gu, and Gleb Yushin

3   Dye-Sensitized Solar Cells   *61*
      Anna Nikolskaia and Oleg Shevaleevskiy

4	**Electronics and Bioelectronic Interfaces** *75* Seong-Min Kim, Sungjun Park, Won-June Lee, and Myung-Han Yoon	
5	**Designing Thermoelectric Materials Using 2D Layers** *93* Sage R. Bauers and David C. Johnson	
6	**Magnetically Responsive Photonic Nanostructures for Display Applications** *123* Mingsheng Wang and Yadong Yin	
7	**Functional Materials: For Sensing/Diagnostics** *151* Rujuta D. Munje, Shalini Prasad, and Edward Graef	
8	**Superhard Materials** *175* Ralf Riedel, Leonore Wiehl, Andreas Zerr, Pavel Zinin, and Peter Kroll	
9	**Self-healing Materials** *201* Martin D. Hager	
10	**Functional Surfaces for Biomaterials** *227* Akiko Nagai, Naohiro Horiuchi, Miho Nakamura, Norio Wada, and Kimihiro Yamashita	
11	**Functional Materials for Gas Storage. Part I: Carbon Dioxide and Toxic Compounds** *249* L. Reguera and E. Reguera	
12	**Functional Materials for Gas Storage. Part II: Hydrogen and Methane** *281* L. Reguera and E. Reguera	
13	**Supported Catalysts** *313* Isao Ogino, Pedro Serna, and Bruce C. Gates	
14	**Hydrogenation by Metals** *339* Xin Jin and Raghunath V. Chaudhari	
15	**Catalysis/Selective Oxidation by Metal Oxides** *393* Wataru Ueda	
16	**Activity of Zeolitic Catalysts** *417* Xiangju Meng, Liang Wang, and Feng-Shou Xiao	
17	**Nanocatalysis: Catalysis with Nanoscale Materials** *443* Tewodros Asefa and Xiaoxi Huang	

18	**Heterogeneous Asymmetric Catalysis** *479* *Ágnes Mastalir and Mihály Bartók*	
19	**Catalysis by Metal Carbides and Nitrides** *511* *Connor Nash, Matt Yung, Yuan Chen, Sarah Carl, Levi Thompson, and Josh Schaidle*	
20	**Combinatorial Approaches for Bulk Solid-State Synthesis of Oxides** *553* *Paul J. McGinn*	

Index *573*

1
Single-Crystal X-Ray Diffraction

Ulli Englert

RWTH Aachen University, Institute of Inorganic Chemistry, Landoltweg 1, 52074 Aachen, Germany

1.1 Introduction

Many of the figures that adorn our scientific publications and textbook cover result from diffraction experiments on single crystals. The vast majority of these studies are performed with X-rays; it is hardly necessary to stress the relevance of the method. This chapter is intended as a first and necessarily short overview, hopefully suitable to guide the reader toward the more specialized literature. Let us first address the two key concepts in the title, *single crystals* and *X-ray diffraction*. In this chapter, twinned crystals will be covered together with single crystals: they are tackled with the same instrumentation and largely the same methods. Often, twinning is only detected during or after data collection, and the question whether a crystal is single or twinned may well depend on external parameters such as temperature. The majority of X-ray diffraction experiments are performed with laboratory sources; a few of the benefits of synchrotron radiation will be briefly addressed. Powder diffraction and experiments with neutrons will be treated in separate chapters of this handbook.

We will follow the sequence of a classical diffraction experiment and start with crystal symmetry, move to diffraction and data collection, tackle the phase problem, and refine the structure model. We will finally address the interpretation of the derived structure and conclude the chapter with a selection of the advanced literature.

1.2 Symmetry

Crystals owe their simple diffraction pattern to their simple microstructure: the building blocks – atoms, ions, or molecules – in a crystalline solid are arranged

Handbook of Solid State Chemistry, First Edition. Edited by Richard Dronskowski, Shinichi Kikkawa, and Andreas Stein.
© 2017 Wiley-VCH Verlag GmbH & Co. KGaA. Published 2017 by Wiley-VCH Verlag GmbH & Co. KGaA.

in an ordered, "symmetric" fashion. We assume that the reader is familiar with point groups. that is, with the symmetry of finite objects such as molecules; in the case of crystals, the relevant additional symmetry operation is *translation*, the infinite repetition of a motif in three dimensions. A crystalline solid can be structurally characterized by describing the motif and the required translations by which it will fill 3D space. The regular arrangement of the motif is referred to as a lattice; each lattice point may be reached by vectorial addition of the translation vectors. The smallest part of a crystal that is repeated by translation only is called the *unit cell* of the crystal. This unit cell itself may exhibit some kind of symmetry; the smallest section that can generate the whole crystal combining translation with all other symmetry operations corresponds to the so-called *asymmetric unit*. Figure 1.1 shows the two-dimensional projection of a crystal structure: Translation of the unit cell fills space (in Figure 1.1: the plane of projection); each combination of translation vectors terminates in a lattice point (marked in blue in Figure 1.1). In our example, the unit cell contains two molecules, and the gray area within the unit cell comprising a single molecule represents the asymmetric unit. In 3D space, the three noncoplanar translation vectors correspond to the edges a, b, c of the cell; they subtend the angles α, β, γ. Each position in the unit cell can be addressed by three so-called fractional coordinates x, y, z along the cell parameters.

Translation can occur in combination with point-group operations: Rotation leaves all points on the rotation axis invariant, whereas its translation-coupled counterpart, screw rotation, translates the rotated image in a direction perpendicular to the rotation. Similarly, a mirror plane preserves all points in the mirror, whereas a glide plane combines reflection with a translation perpendicular to it (Figure 1.2).

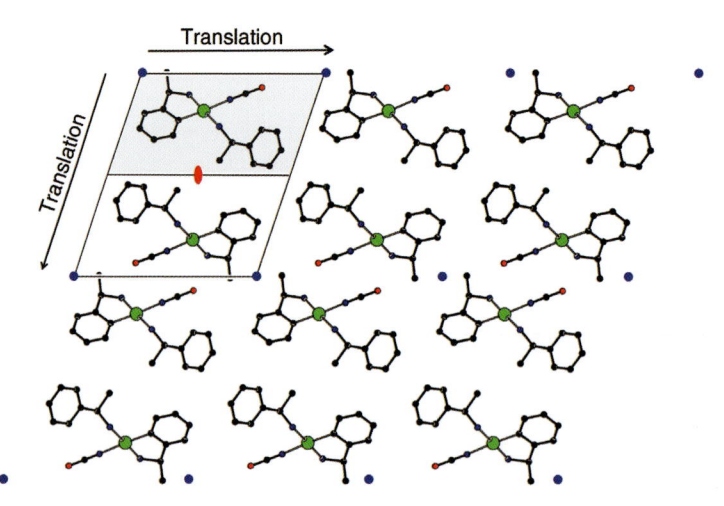

Figure 1.1 Unit cell, asymmetric unit, and translations in the two-dimensional projection of a crystal structure [1].

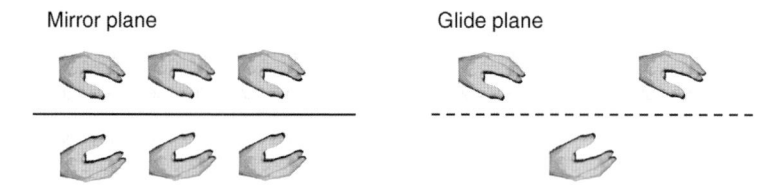

Figure 1.2 Mirror plane versus glide plane.

Symmetry in crystals does not only imply additional features but also brings about restrictions: Only twofold, threefold, fourfold, and sixfold rotations are compatible with translation, whereas the multiplicity of a rotation axis in a point group can in principle correspond to any integer number. Thus, only 32 of the infinite number of point groups are relevant for crystalline solids; they are referred to as *crystal classes* and correspond to the point groups of macroscopic crystals.

In order to exploit symmetry as efficiently as possible, crystal structures are described in symmetry-adapted coordinate systems (crystal systems). These crystal systems also decide about the most economic way of denoting symmetry in terms of view directions. The trigonal and hexagonal crystal system can be combined to the hexagonal crystal family, and both are often handled in the hexagonal coordinate system. Table 1.1 summarizes the crystal families and their coordinate systems with metric requirements and view directions.

In three-dimensional space, each of the above-mentioned coordinate systems can result in a primitive cell. Depending on the symmetry of the unit cell, additional lattice centerings may be possible; they are described with the help of fractional translation vectors. This description by *14 Bravais types*, either primitive or centered, is popular in crystallography. As an alternative to the centered unit cells, smaller primitive cells of a more complicated shape can be used; the latter approach is more common in solid-state physics.

A macroscopic crystal is a finite object and hence described by a crystallographic point group, but its microstructure can be approximated as an infinite repetition of unit cells: the combination of lattice translations with all possible

Table 1.1 Crystal families, metric restrictions, and view directions.

Crystal family	Metric restrictions	View directions		
Triclinic	None	—		
Monoclinic	$\alpha = \beta = 90°$	b		
Orthorhombic	$\alpha = \beta = \gamma = 90°$	a	b	c
Tetragonal	$a = b; \alpha = \beta = \gamma = 90°$	c	a	[110]
Hexagonal	$a = b; \alpha = \beta = 90°; \gamma = 120°$	c	a	[210]
Cubic	$a = b = c; \alpha = \beta = \gamma = 90°$	a	[111]	[110]

symmetry operations, including those with a translational component, results in the *space group* of the crystal. A total of 230 different space group types are possible in 3D space. They are designated by the *Hermann–Mauguin symbol*: An initial capital letter denotes the Bravais centering and is followed by up to three symbols in the view directions; a slash separates symmetry operations in the same view direction. The International Tables for Crystallography (IT) represent a convenient source of information for scientists involved in the structural characterization of solids and the understanding of their properties. Volume A of the IT is dedicated to symmetry and very frequently consulted during the structural characterization of solids; it is available both as hard copy [2] and web-based resource [3]. The following example (Figure 1.3) of the frequently encountered space group $P2_1/c$ summarizes the most important information provided in Vol. A of the IT.

The top row in Figure 1.3 gives the space group names according to Hermann–Mauguin (which is commonly used) and to Schoenflies. As the Schoenflies nomenclature does not explicitly denote translation, all space groups of the same crystal class share the same Schoenflies symbol; the distinction is made by a numeric superscript after the name ("5" in the example of Figure 1.3). The last two entries in the top row specify the crystal class and crystal system. The second row gives the space group number (14), the full symbol and Patterson symmetry. The projections of the unit cell have been rearranged in Figure 1.3: The first is oriented in the view direction of highest symmetry and shows the symbols for symmetry operations. These projections are followed by an enumeration of the operations selected as generators and a table of Wyckoff positions, site symmetry, coordinate triples for atoms and reflection conditions; the latter are useful for space group determination (see Section 1.5). The top entry in the table refers to general, the following rows to special positions with decreasing multiplicity.

The following paragraph in each space group entry provides information on group–subgroup relationships. We will give two reasons for doing so: (i) Different but chemically related solids can often be described by related space groups. This may apply for superstructures or for enantiomerically pure compounds and their racemic counterparts. (ii) Well-defined mathematical relationships can be observed for different phases of the same compound when temperature or pressure varies. Upon cooling, a high-temperature modification transforms into a phase of more favorable energy and lower symmetry: Either translational or point symmetry is lost and the corresponding phase transitions are referred to by the German adjectives *klassengleich* or *translationengleich*, respectively. Group–subgroup relationships [4,5] thus allow to organize families of structures and represent an important tool for the understanding of structural phase transitions; the IT compiles maximal subgroups and minimal supergroups.

We will shortly address twinned crystals. Twins consist of a small number, most often two, domains and form by systematic intergrowth or as the result of a (translationengleiche) structural phase transition. Overlap between the diffraction patterns of the contributing domains may be partial or complete. The latter can be due to randomly degenerate lattice parameters or can occur for reasons of symmetry: In merohedral twins, the relationship between the domains, the

$P2_1/c$ C_{2h}^5 $2/m$ Monoclinic

No. 14 $P12_1/c1$ Patterson symmetry $P12/m1$

UNIQUE AXIS b, CELL CHOICE 1

Generators selected (1); $t(1,0,0)$; $t(0,1,0)$; $t(0,0,1)$; (2); (3)

Positions

Multiplicity, Wyckoff letter, Site symmetry | | Coordinates | | | Reflection conditions

4	e	1	(1) x,y,z (2) $\bar{x}, y+\frac{1}{2}, \bar{z}+\frac{1}{2}$ (3) $\bar{x},\bar{y},\bar{z}$ (4) $x, \bar{y}+\frac{1}{2}, z+\frac{1}{2}$		General: $h0l : l = 2n$ $0k0 : k = 2n$ $00l : l = 2n$

Special: as above, plus

2	d	$\bar{1}$	$\frac{1}{2},0,\frac{1}{2}$	$\frac{1}{2},\frac{1}{2},0$	$hkl : k+l = 2n$
2	c	$\bar{1}$	$0,0,\frac{1}{2}$	$0,\frac{1}{2},0$	$hkl : k+l = 2n$
2	b	$\bar{1}$	$\frac{1}{2},0,0$	$\frac{1}{2},\frac{1}{2},\frac{1}{2}$	$hkl : k+l = 2n$
2	a	$\bar{1}$	$0,0,0$	$0,\frac{1}{2},\frac{1}{2}$	$hkl : k+l = 2n$

Symmetry of special projections

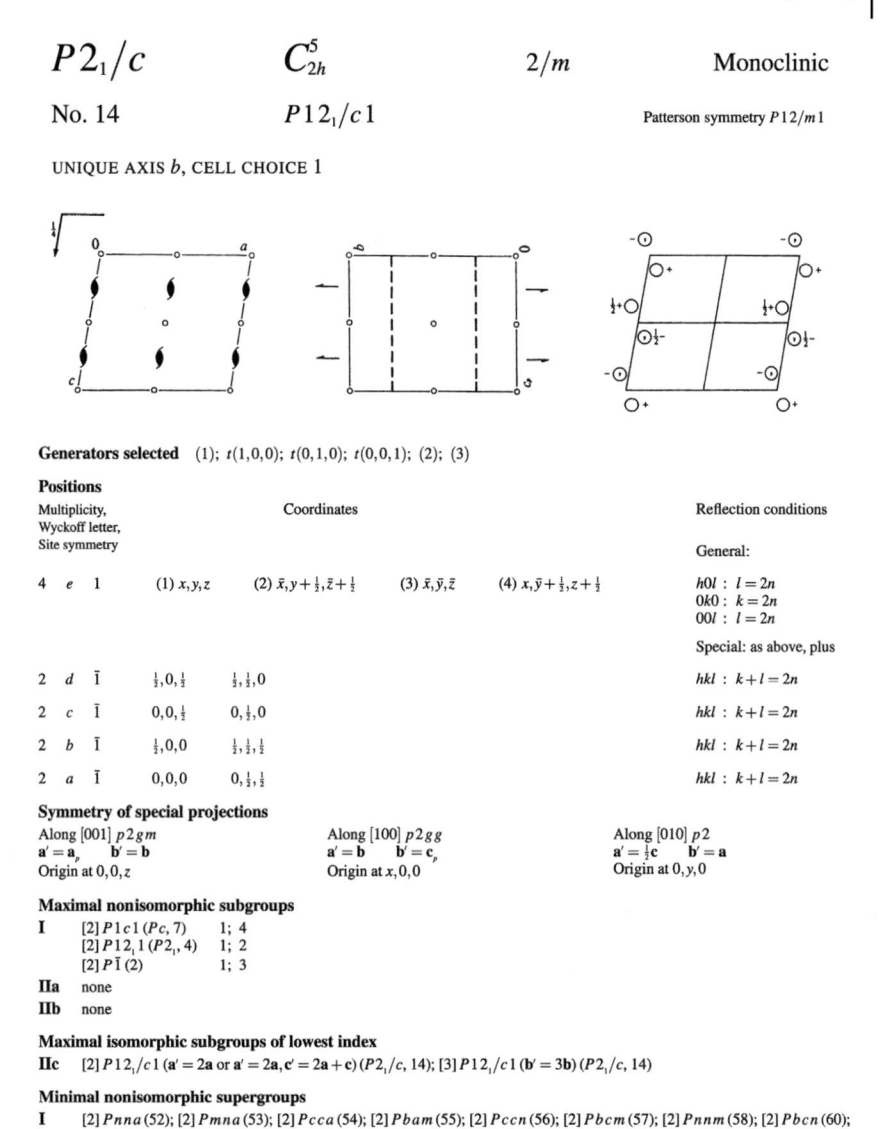

Along [001] $p2gm$ Along [100] $p2gg$ Along [010] $p2$
$a' = a_p$ $b' = b$ $a' = b$ $b' = c_p$ $a' = \frac{1}{2}c$ $b' = a$
Origin at $0,0,z$ Origin at $x,0,0$ Origin at $0,y,0$

Maximal nonisomorphic subgroups
I   [2] $P1c1$ $(Pc, 7)$ 1; 4
    [2] $P12_11$ $(P2_1, 4)$ 1; 2
    [2] $P\bar{1}$ (2)   1; 3
IIa none
IIb none

Maximal isomorphic subgroups of lowest index
IIc [2] $P12_1/c1$ ($a' = 2a$ or $a' = 2a, c' = 2a + c$) ($P2_1/c$, 14); [3] $P12_1/c1$ ($b' = 3b$) ($P2_1/c$, 14)

Minimal nonisomorphic supergroups
I   [2] $Pnna$ (52); [2] $Pmna$ (53); [2] $Pcca$ (54); [2] $Pbam$ (55); [2] $Pccn$ (56); [2] $Pbcm$ (57); [2] $Pnnm$ (58); [2] $Pbcn$ (60); [2] $Pbca$ (61); [2] $Pnma$ (62); [2] $Cmce$ (64)
II [2] $A12/m1$ ($C2/m$, 12); [2] $C12/c1$ ($C2/c$, 15); [2] $I12/c1$ ($C2/c$, 15); [2] $P12_1/m1$ ($c' = \frac{1}{2}c$) ($P2_1/m$, 11); [2] $P12/c1$ ($b' = \frac{1}{2}b$) ($P2/c$, 13)

Figure 1.3 Sample page of the IT; [2] projections of the unit cell have been rearranged with respect to the original page. (Reprinted with permission of the IUCr).

so-called twin law, is a symmetry operation of the crystal system, but no symmetry operation of the point group of the individual domains. As a consequence of twinning, determination of the correct symmetry – crystals system, point group, and space group – may be hampered. We will address further implications of twinning in later sections of this chapter.

We come back to the beginning of this section on symmetry and mention an additional aspect of the relationship between the concepts *order* and *3D translational symmetry*: Objects may have a long-range order (and thus give rise to a sharp diffraction pattern) without respecting periodicity in three dimensions. These aperiodic or quasiperiodic objects may be handled in a higher dimensional superspace [6]; they comprise modulated structures and quasicrystals. Symmetry of the former is restricted to the crystallographic point groups whereas the latter may show noncrystallographic point group symmetry, for example, fivefold or eightfold rotation. In 2011, Dan Shechtman was awarded the Nobel Prize in Chemistry for the discovery of quasicrystals.

1.3
Diffraction

A complete picture of the interaction between radiation and matter has to take many aspects into account. Fortunately, simplifications are justified in an introductory text about X-ray diffraction. Ionization, inelastic scattering, and multiple scattering can to a first approximation be neglected. X-rays interact with the electron clouds of the scattering atoms, hence each X-ray diffraction experiment in principle probes the *electron density*, albeit at only moderate resolution in standard experiments. The fact that electron density is responsible for scattering has implications for the accuracy of an X-ray diffraction experiment: Atoms associated with low electron density, in particular hydrogen, are detected less reliably, and distinguishing atoms with very similar electron count and radius may prove a challenge. Figure 1.4 summarizes both aspects of this contrast

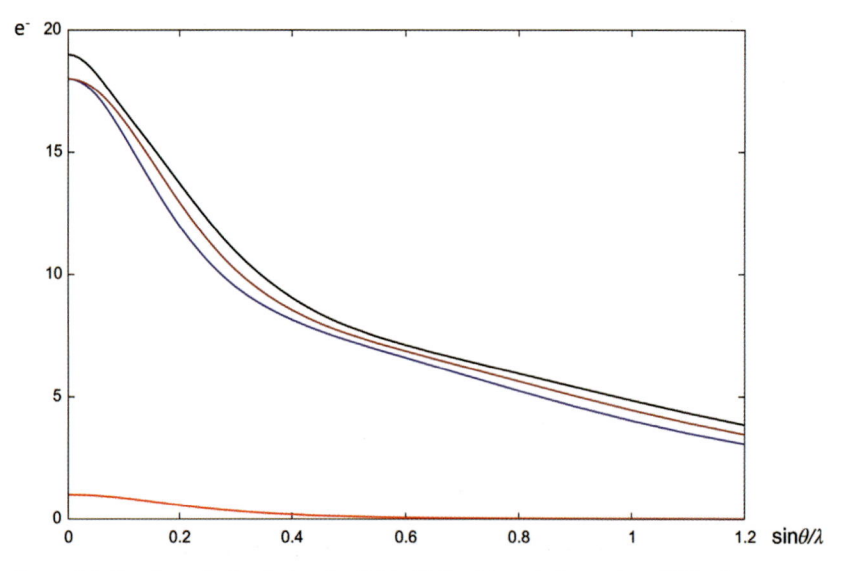

Figure 1.4 Atomic scattering factors for K (black), K+ (blue), Ar (brown), and H (red).

problem: The so-called atomic scattering factor for H is much smaller than that for the many-electron species K, K$^+$, or Ar. These 18- or 19-electron scatterers are clearly different from hydrogen but might be difficult to distinguish from each other based on their scattering curves. Fortunately, chemical connectivity often helps with such assignments.

The pronounced dependence of the atomic scattering power as a function of resolution can intuitively be understood from the fact that the diameter of the interacting electron cloud is not negligible when compared to the interplanar distance within a family of lattice planes, see below. For a scattering center of very small radius, a constant scattering length rather than a function of resolution results; this situation is encountered in neutron diffraction.

The atomic scattering factor also reflects the dynamics of a crystalline solid: the curves shown in Figure 1.4 are valid for atoms at rest, whereas *displacement* will reduce scattering exponentially as first recognized by Debye [7]. In Figure 1.5, scattering factors for a (hypothetically!) immobile and an oscillating potassium cation are compared.

Figure 1.5 shows that atomic displacement will reduce scattering, in particular for higher resolution; diffraction experiments are thus preferentially performed at low temperature. In particular for molecular crystals with low melting or decomposition points, intensity data should be collected at the lowest possible sample temperature. The most intuitive form of displacement is thermal motion; displacement parameters have therefore been addressed as *temperature parameters*. Nowadays, the term *displacement parameter* has become more common because oscillation of a scattering center about an equilibrium position and static

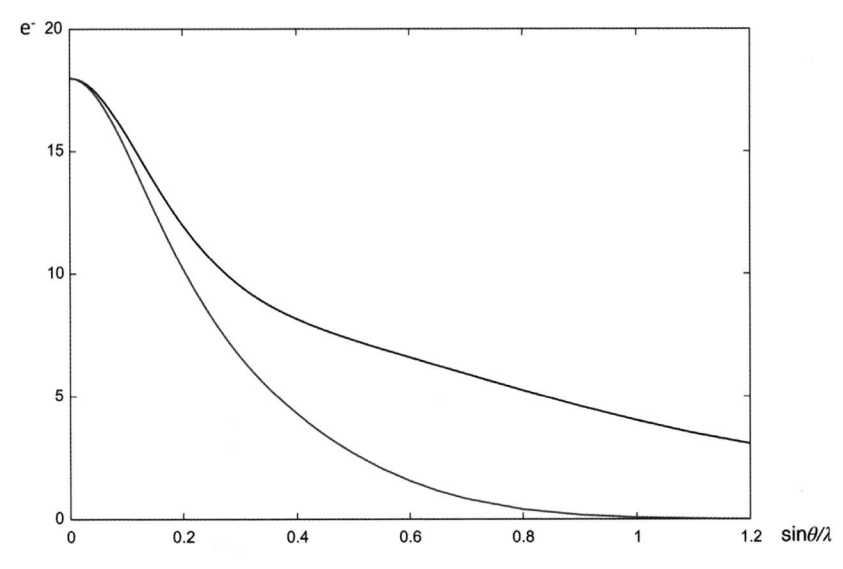

Figure 1.5 Atomic scattering factors for K$^+$ at rest (black) and K$^+$ oscillating with a mean-square displacement of 0.05 Å^2 (blue).

disorder equally reduce the scattering power and cannot be distinguished by a diffraction experiment at a single temperature. In general, atomic movement is anisotropic; within the harmonic approximation, it can be described by a *displacement ellipsoid*. This ellipsoid corresponds to a symmetric tensor associated with three principal axes and three Eulerian angles in space; six parameters are therefore refined for each atom that is assigned an *anisotropic displacement parameter*, for short adp. Displacement ellipsoid plots are the most convenient way to inspect magnitude and orientation of adps; we will come back to this aspect in Sections 1.6 and 1.7.

The discussion above refers to scattering by an individual atom – how about the many atoms in a crystal? We recall the concept of a lattice: whenever a plane passes through lattice points, a family of parallel planes with the same orientation can be drawn such that the whole family contains all lattice points. A two-dimensional projection of a lattice and two families of planes are shown in Figure 1.6 as examples. In 3D space, each family of planes can be assigned three so-called *Miller* indices *hkl*: the plane not containing the origin but closest to it divides the unit cell parameter *a* into *h*, *b* into *k*, and *c* into *l* segments.

The many atoms in a crystal are arranged regularly. Each of them may be approximated as a spherical scattering center as explained in Figures 1.4 and 1.5, and interference of the scattered rays creates the diffraction pattern of the crystal. Nonvanishing diffraction intensities, that is, constructive interference, are only encountered if the geometrical requirements given by the *Bragg equation* (Figure 1.7) are fulfilled [8].

The interplanar distance *d* from the Bragg equation is related to the Miller indices and lattice parameters.

For an orthorhombic system, the relationship $\frac{1}{d^2} = \frac{h^2}{a^2} + \frac{k^2}{b^2} + \frac{l^2}{c^2}$ holds true.

For nonorthogonal systems, the angles between the lattice parameters must be taken into account. The above equation shows that a family of lattice planes with

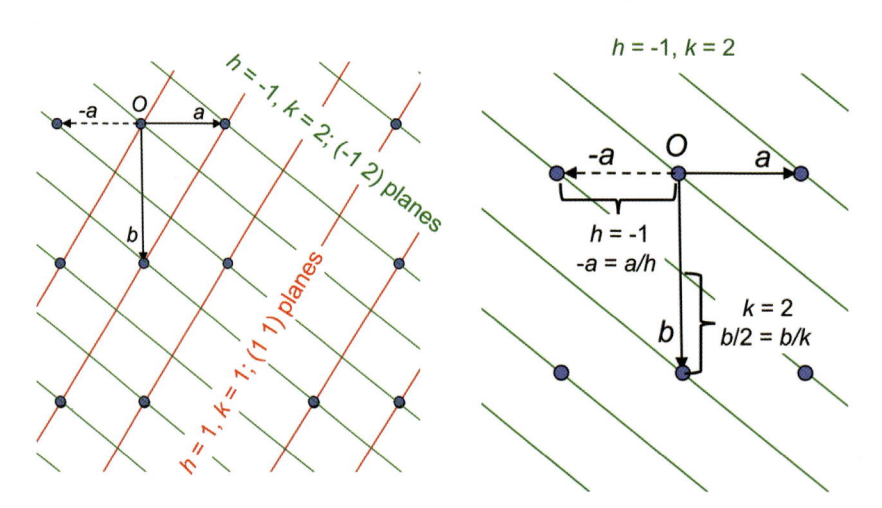

Figure 1.6 2D projection of a lattice and examples for planes and their Miller indices.

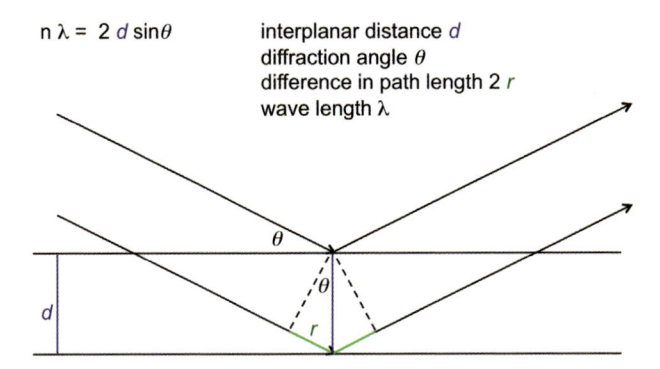

$n\lambda = 2d\sin\theta$ interplanar distance d
diffraction angle θ
difference in path length $2r$
wave length λ

Figure 1.7 Bragg equation: constructive interference ("reflection") by a family of lattice planes with interplanar distance d.

high indices h, k, l is associated with a small interplanar distance d. A convenient description of diffraction relies on reciprocal space, in which the reciprocal interplanar distance d^* is a vector parallel to d, with $|d^*| = 1/|d|$. The endpoints of all vectors d^* can be understood as the so-called reciprocal lattice, which is subtended by three reciprocal lattice parameters a^*, b^*, and c^*. In an orthogonal system, a^* is parallel to a, with $|a^*| = 1/|a|$ and so on. For the more general case of nonorthogonal coordinate systems, each reciprocal lattice vector is perpendicular to two vectors in real space, with a^* perpendicular to b and c and so on.

We can summarize the considerations from Figures 1.6 and 1.7: The translation vectors subtending the unit cell decide under which angles nondestructive interference may occur and a diffracted beam with nonzero intensity can be detected. The intensity of the reflections depends on the distribution of electron density in the unit cell. The individual scattering centers (j) in the unit cell with their atomic form factors (f_j) as explained in Figure 1.4, and their fractional coordinates (x_j, y_j, and z_j) contribute via constructive and destructive interference to the so-called structure factor F_{hkl}:

$$F_{hkl} = \sum_j f_j e^{2\pi i(hx_j + ky_j + lz_j)} \quad \text{(structure factor equation)}.$$

This may alternatively be expressed as

$$F_{hkl} = \sum_j f_j \cos(2\pi(hx_j + ky_j + lz_j)) + i \sum_j f_j \sin(2\pi(hx_j + ky_j + lz_j)).$$

An important consequence of the expression above arises for centrosymmetric structures: A center of inversion implies that for each atom with fractional coordinates x_j, y_j, z_j an atom with coordinates $-x_j$, $-y_j$, $-z_j$ must exist. As $\sin(x) = -\sin(-x)$, the imaginary contribution to the structure factor vanishes and F_{hkl} becomes a positive or negative real rather than a complex number.

The summation of the atomic scattering factors can also be visualized as vector addition in the complex plane (Figure 1.8). In such an Argand diagram, each

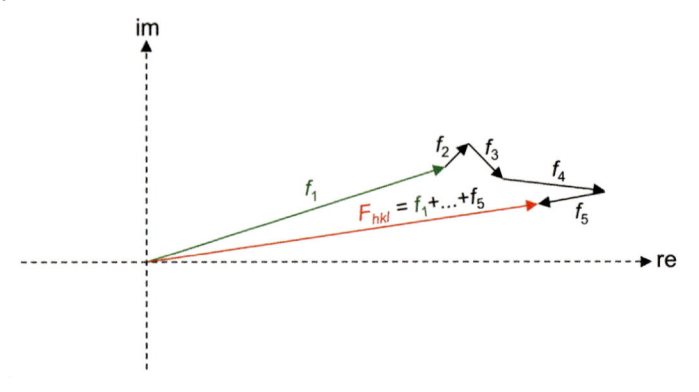

Figure 1.8 Argand diagram. Scattering by each atom is represented by a vector in the plane of complex numbers. The structure factor F_{hkl} results from vectorial addition of the individual contributions by a heavy (electron-rich) and four light atoms.

atomic contribution corresponds to a vector characterized by a modulus (f_j) and a phase angle ϕ ($\phi = 2\pi(hx_j + ky_j + lz_j)$).

The structure factor is a complex number or, in the case of centrosymmetric structures, a number associated with a sign; its squared amplitude is always positive and corresponds to the intensity that a detector will measure for the reflection:

$$I_{hkl} \sim |F_{hkl}|^2.$$

We recall the introduction of this section: X-rays interact with the electrons of the scattering centers. The structure factor equation can therefore alternatively be expressed in terms of electron density:

$$F_{hkl} = \iiint \rho_{xyz} e^{2\pi i(hx+ky+lz)} dxdydz.$$

If either the electron density in each point in space or atom type f_j and coordinates x_j, y_j, z_j for all scattering centers are known, all structure factors F_{hkl} and thus all intensities I_{hkl} can be calculated. In other words: if the structure is known or a sufficiently good model exists, its whole diffraction pattern can de deduced.

The diffraction pattern shows *at least* the symmetry of the diffracting crystal, that is, the crystallographic point group. However, even if the crystallographic point group does not contain a center of inversion, the diffraction pattern is centrosymmetric to a very good approximation. The resulting relationship

$$I_{hkl} = I_{-h-k-l}$$

is known as *Friedel's law* [9]. The symmetry of the diffraction pattern therefore corresponds exactly to one of the 32 crystallographic point groups and very closely to one of the 11 centrosymmetric groups among these, the so-called *Laue classes*.

Within the approximation of Friedel's law, a diffraction experiment cannot distinguish between left- and right-handed crystals. Our initial picture of the atomic scattering factor is, however, not yet complete. In addition to the resolution-dependent behavior shown in Figure 1.4, resonance phenomena occur and the overall scattering factor f_λ becomes a complex number:

$$f_\lambda = f_0 + f'_\lambda + f''_\lambda.$$

The usually dominant part f_0 is a real number and varies between the total number of electrons at very low scattering angles and negligibly small values at high resolution. The remaining contributions are wavelength-dependent and become relevant in the vicinity of absorption edges. f'_λ corresponds to a real number and may be positive or negative. f''_λ denotes an imaginary component and always has a positive sign, thus leading to a counter-clockwise component in a vector diagram.

Non-negligible magnitudes of f'_λ and f''_λ have two important consequences for crystallography: (i) They allow to distinguish left- and right-handed crystals; the pioneering experiment involved a strong anomalous signal and was performed by Bijvoet et al. [10]. More recent developments measure the reliability of the absolute structure assignment [11] and aim at the determination of absolute structures in the absence of strong anomalous scatterers [12–14]. (ii) Anomalous dispersion helps to solve the phase problem for structures too large for direct methods, in particular for biological structures [15].

1.4
Instrumentation

We here assume that a suitable sample is available and will not discuss crystallization and crystal growth. For the purpose of most in-house diffraction experiments, crystals with dimensions between 0.01 and 0.1 mm are required. Figure 1.9 gives a schematic overview of the relevant hardware components in such a single-crystal diffraction experiment.

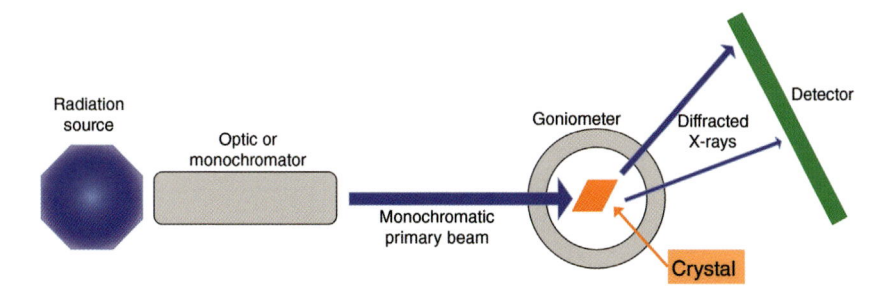

Figure 1.9 Schematic overview of hardware components involved in a single-crystal X-ray diffraction experiment.

In a laboratory, X-rays are *generated* when accelerated electrons collide with a metal anode. This classically occurs under vacuum in a sealed glass or ceramic tube; the anode is water-cooled. More power and a higher photon flux can be achieved by continually moving the target. Rotation of a solid anode is an established technique, and the corresponding generators are addressed as rotating anodes. In a recent development, the continuously regenerated anode material consists of a liquid metal or alloy. A low-power and largely maintenance-free alternative are microfocus tubes, combined with focusing multilayer optics. These microfocus tubes provide an intense primary beam of a rather small diameter and are particularly well-suited for small crystals.

In-house diffraction experiments operate with monochromatic X-rays; Mo and Cu are by far the most common anode materials. Synchrotron radiation is available in large research facilities; beamtime is usually allocated based on proposals. Synchrotron radiation offers polychromatic or tunable monochromatic wavelength. Polychromatic or "white" radiation allows to register a very large number of reflections simultaneously and in the same geometry and hence is most suitable for time-resolved studies, for example, in biocrystallography [16] or on photo-activated species [17]. With monochromatic synchrotron radiation, much smaller single crystals can be analyzed than with in-house equipment ("microdiffraction"). The wavelength can the adjusted to provide a strong anomalous signal; this is important for solving the phase problem in protein crystallography [15].

The *goniometer* allows to adjust the position of the crystal with respect to the primary beam; ideally any orientation becomes accessible. Popular technical solutions involve movement about four circles, either as a Eulerian cradle or in so-called κ geometry; in either case, three of these circles are associated with movement of the crystal and the fourth with positioning of the detector. In principle, a four circle goniometer is redundant because two rotations of the crystal are sufficient to bring lattice planes into reflecting geometry: a three-circle instrument allows to collect complete data sets, albeit under certain conditions with lower efficiency.

The most common *detectors* employed in modern instruments are image plates, CCD, CMOS, and pixel detectors. All of them are area detectors and allow to register several reflections simultaneously; data collection times are thus much shorter than in former times when point detectors were popular, and they depend much less on the size of the asymmetric unit. The physical detector signal is processed by a dedicated integration software.

The resulting data are subjected to several corrections, in particular for absorption. The relevance of this correction depends on the elements present in the sample, the wavelength, and the crystal size. In the time of the classical sealed tube, the primary beam would have a flat profile, a diameter of 0.5–0.8 mm and encase the crystal completely; absorption correction then mainly required to determine the attenuation of primary and diffracted beams by geometric considerations. Two major changes have occurred over the last decades: on the one hand, the area detectors mentioned above allow to collect much more redundant data, and on the other hand, the much smaller beam diameter and its parabolic profile require a correction for incomplete illumination of the crystal. The higher redundancy of the data is most often used for a so-called multiscan correction that concomitantly models absorption and illumination; during this process, the agreement between the intensities of symmetrically related reflections is improved. The result of the diffraction experiment is a list of intensities for each reflection (I_{hkl}) and their standard deviation, each associated with the indices of the reflecting lattice plane. We recall that the structure factor is a complex number. The phase information of this complex number is not retrieved from the diffraction experiment, thus giving rise to the so-called *phase problem* in crystallography. We will consider several strategies for its solution, sometimes addressed as *phasing*, in the next section.

1.5
Solution of the Phase Problem

We have seen in Section 1.3 that the diffraction pattern can be calculated if a structure model can be provided:

$$F_{hkl} = \sum_j f_j e^{2\pi i(hx_j + ky_j + lz_j)}.$$

Usually, crystallographers are interested in the reverse Fourier transformation and want to deduce "the crystal structure," that is, arrangement and displacement of the atoms, or, alternatively, the electron density at each point in space, from an experimentally observed diffraction pattern:

$$\rho_{xyz} = \frac{1}{V} \sum_{hkl} F_{hkl} e^{-2\pi i(hx + ky + lz)}.$$

The intensity of a diffracted beam is an observable quantity and directly related to the amplitude of the structure factor, whereas the phase is in general not experimentally accessible. Before we address approaches to tackle this phase problem, we come back to the reflection conditions mentioned in the context of Figure 1.3: Bravais centering, glide planes, and screw axes, that is, symmetry operations associated with translation, result in groups of reflections with zero intensity, so-called systematic absences. These absences may be summarized to an extinction symbol that is compatible either with a single or with only a few space groups. The *reflection conditions* or *systematic absences* therefore represent one of the most often used criteria for space group determination; alternatives will be mentioned below.

1.5.1
The Patterson Method

The *Patterson* function is the convolution of the electron density with itself. This first systematic strategy to solve the phase problem [18] exploits the fact that atoms associated with many electrons will dominate diffraction and hence be most important for a structure model. When the complex number F_{hkl} in the above equation is replaced by its squared amplitude $|F_{hkl}|^2$, the Patterson function P rather than the electron density is obtained:

$$P_{uvw} = \frac{1}{V} \sum_{hkl} |F_{hkl}|^2 e^{-2\pi i(hx+ky+lz)}.$$

This Patterson function is always centrosymmetric; its symmetry corresponds to the Bravais type followed by the point group symbol of the Laue class (rightmost item of the second line in each space group entry in the IT, Figure 1.3). The maxima of this Patterson function represent interatomic vectors, and the height of the maxima is proportional to the product of the electron count for the atoms connected by the vectors. The Patterson function is most useful when the asymmetric unit contains only one or a small number of "heavy" atoms. Its highest maxima will be associated with the vectors subtended by these atoms. The Patterson function always shows the most pronounced maximum in the origin, corresponding to the overlap of all vectors of zero length, originating and ending in the same atom; from the highest nontrivial interatomic vectors, the fractional coordinates of the dominant scattering centers may be obtained. We consider the simple example of a single many-electron atom together with several lighter atoms in the asymmetric unit of the triclinic space group P-1: The space-group symmetry implies the presence of atoms with fractional coordinates x, y, z and $-x, -y, -z$; the most relevant nontrivial interatomic vector therefore corresponds to $2x, 2y, 2z$ between the two heavy atoms in the unit cell, and division of the vector components by the factor 2 directly provides the fractional coordinates of the heavy atom. In case of symmetry operations associated with translation, certain components of the vectors between symmetry-related atoms adopt special values; such *Harker vectors* can easily be recognized and allow to assign or verify the space group.

1.5.2
Difference Fourier Synthesis

Once the dominant scattering center(s) in the asymmetric unit of a structure have been assigned fractional coordinates, they may be used as an initial structure model and their contribution to the structure factors can be calculated. We recall Figure 1.8: The contribution of a heavy atom is associated with a vector of similar phase angle as the total structure factor F_{hkl}. This relationship can be exploited for a so-called difference Fourier synthesis: The electron density is calculated with the quantity $||F_{obs}|-|F_{calc}||$, that is, the difference between the moduli of observed and calculated structure factors, and with the phases of the calculated structure factors. The maxima of such a difference Fourier synthesis correspond to "missing" sites of electron density, that is, atoms not yet assigned to the structure model. This procedure may be applied at quite different stages of a structural study: At the very beginning, the coordinates of a single or a few heavy atoms, for example, obtained from a Patterson synthesis, can be used to assign tentative phases to the observed structure factor amplitudes. At later stages, subsequent difference Fourier syntheses help to locate additional atoms; in favorable cases, even the positions of hydrogen atoms in nontrivial geometry may be recovered. After completion and refinement (see Section 1.6) of the structure model, an essentially featureless difference Fourier synthesis indicates that no major disorder has been overlooked; the residual local electron-density maxima and minima are often considered quality criteria.

1.5.3
Direct Methods

When random phase angles are assigned to the structure factors F_{hkl} in the equation

$$\rho_{xyz} = \frac{1}{V} \sum_{hkl} F_{hkl} e^{-2\pi i(hx+ky+lz)},$$

the resulting electron densities ρ_{xyz} will adopt positive or negative values. Negative electron densities do not occur in matter, and phase angles leading to such negative densities can therefore not be realistic. We will come back to a procedure connected with apparently negative electron densities in Section 1.5.4. For the structure solution strategy discussed here, the requirement for *non-negative electron density* throughout the crystal represents an important boundary condition for the *a priori* unknown phase angles of the complex structure factors [19].

The structure solution technique that relies on deducing phase angles directly from intensities is referred to as *direct methods*. Many scientists have contributed to its enormous success, which nowadays allows to solve the phase problem for many structures almost routinely, and this approach has contributed decisively to the development of (semi-)automatic expert systems. In 1985, the Nobel prize in chemistry was awarded to Herbert Hauptman and Jerome Karle for their outstanding achievements in the development of direct methods.

Strong reflections play an important role in this approach. The intensity of a reflection depends on resolution (*cf.* Figures 1.4 and 1.5), and it is therefore mandatory to establish a criterion for its relative strength. For this purpose, direct methods rely on so-called *normalized structure factors*; they are obtained according to

$$E_{hkl}^2 = \frac{F_{hkl}^2}{\langle F^2 \rangle}.$$

In this equation, $\langle F^2 \rangle$ stands for the mean-squared structure factor within a certain resolution range of the data set. Large E values range between 2 and 4. Centrosymmetric structures are associated with more large and small normalized structure factors, noncentrosymmetric structures with a higher number of intermediate E values. The overall distribution of the normalized structure factors may help to decide about (non-)centrosymmetry of the space group, and E value statistics can indicate merohedral twinning.

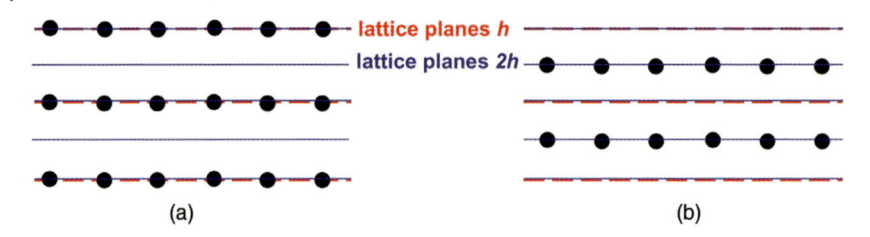

Figure 1.10 Σ1 relationship for two large normalized structure factors E_h and E_{2h}. (a) Electron density is located on the lattice planes h, E_h is large and associated with a positive phase. (b) Electron density is located in-between the lattice planes h, E_h is large and associated with a negative phase. In either case, the electron density is on the lattice planes $2h$ and E_{2h} has a positive phase.

The solution of the phase problem by direct methods is easier for centrosymmetric structures in which only a sign has to be attributed to each structure factor. An arbitrary sign is chosen for a small number of structure factors, and additional phases can be retrieved using intensity-based relationships. For two strong normalized structure factors E_h and E_{2h}, the so-called Σ1 relationship in Figure 1.10 holds true.

Figure 1.11 shows a triplet relationship between three large normalized structure factors, also known as Σ2 relationship. The phase signs of these structure factors are not independent; their product is always positive (Table 1.2).

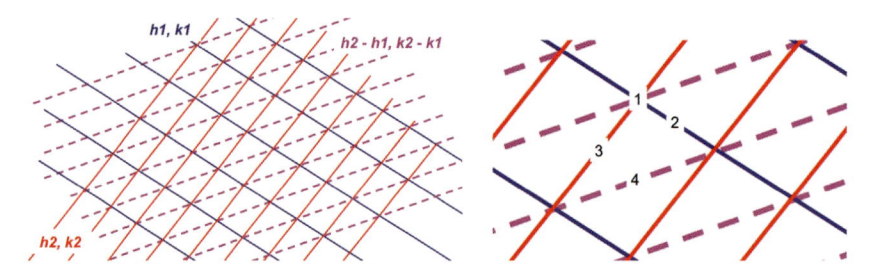

Figure 1.11 Triplett relationship for three large normalized structure factors E_1, E_2, and E_{2-1}. If electron density is located in the sites marked as 1–4, the three large E values have the phase signs explained in Table 1.2.

Table 1.2 Phase signs for the normalized structure factors E_1, E_2, and E_{2-1} shown in Figure 1.1 1.

Electron density site	E_1 planes h1, k1	E_2 planes h2, k2	E_{2-1} planes h2-h1, k2-k1
1	+	+	+
2	+	−	−
3	−	+	−
4	−	−	+

For all but the strongest E values, $\Sigma 1$, $\Sigma 2$, and similar phase relationship are probability statements rather than exact equations. Direct methods therefore follow a multisolution strategy: Different phase sets, in general some hundred, are assigned, and the most promising is selected based on a (tunable) figure of merit (FOM). Alternatively, the crystallographer may decide to test phase assignments with less favorable FOMs.

Albeit their enormous success, direct methods have their limitations: phase relationships become less reliable with increasing number of atoms in the asymmetric unit; larger biological structures therefore require different approaches. Symmetry also plays an important role: In contrast to the Patterson synthesis which only requires the correct assignment of Bravais type and Laue class, direct methods can usually not be successful unless the right space group has been assigned. Finally, phasing by direct methods requires data of atomic resolution in the range of about 1.3 Å.

1.5.4
Charge Flipping

Positive electron density was a relevant criterion for the successful solution of the phase problem by direct methods: the (many) phase combinations that lead to regions of significantly negative electron density are usually discarded. Not so in the structure solution approach discussed in this paragraph: *charge flipping* [20] owes its name to a process of density modification in real space that is schematically shown in Figure 1.12. The method is iterative: The measured structure factor amplitudes are assigned random phases. The reverse Fourier transform of the complex numbers thus obtained is a density ρ that will have positive and negative areas. Strongly negative densities are flipped to positive values, thus resulting in a modifies density ρ'. The Fourier transform of this density affords new structure factors; their phases are combined with the measured amplitudes, and the cycle restarts.

Very small phase changes in successive cycles and an abrupt drop in residuals (see Section 1.6) indicate convergence. Successful phasing by charge flipping requires diffraction data with atomic resolution. Its simplicity apart that makes it a good candidate for automatic phasing, the algorithm offers another

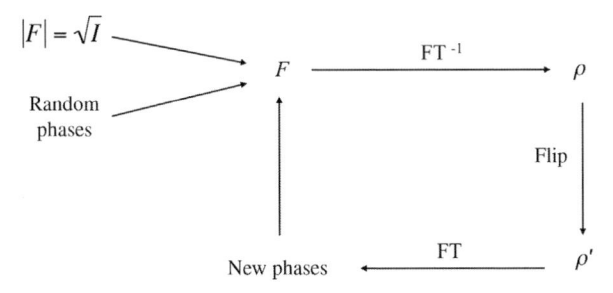

$|F| = \sqrt{I}$ — Random phases → F — FT^{-1} → ρ — Flip → ρ' — FT → New phases

Figure 1.12 Dual space strategy for solving the phase problem by charge flipping.

interesting advantage: it only requires the correct primitive cell; in 3D space, it operates in space group $P1$. The correct space group is assigned after convergence, an attractive feature for the phasing of diffraction data from modulated structures and quasicrystals [21,22].

1.5.5
Other Methods for Structure Solution

Dual Space methods allow to tackle the phase problem for data with lower resolution; [23] they are more frequently used in structural biology.

Intrinsic phasing [24] has become popular since 2014. The computer program SHELXT combines several modern crystallographic tools, from the phasing stage to a first interpretation of the electron density and its assembly into a most connected structure model. We here only mention that the space group assignment is based on the phases of the solution in $P1$ [25] rather than on an attempt to interpret the real space symmetry of the unit cell.

1.6
Refinement

A structure model afforded by one of the approaches discussed above is not the final one: the solution of the phase problem only affords approximate atomic coordinates, and only estimated isotropic displacement parameters are assigned. Based on the structure factor equation, the model allows to calculate a diffraction pattern.

1.6.1
Agreement Factors

During structure refinement, the initial model is modified and expanded, with the goal to achieve the best possible match between calculated and observed data. This match is usually expressed by agreement factors or residuals, often abbreviated as R factors. The so-called *conventional R* factor is defined as

$$R = \frac{\Sigma ||F_{obs}| - |F_{calc}||}{\Sigma |F_{obs}|},$$

and should adopt a small value; typical agreement factors amount to some percent for a final structure model. Weights can be introduced into the expressions for residuals; they take into account that not all data contribute in equally reliable way to the structure model. Nowadays, linear least-squares refinement is applied to the variables of the structure model: an overall scale factor, atomic coordinates (unless fixed by symmetry), isotropic or anisotropic displacement parameters, and sometimes additional quantities such as site occupancies or twin domain size are modified to ensure the lowest possible residual. As the relationship between refinement variables and structure factors is not really linear, refinement is an iterative process: in successive cycles, the refined parameters change less and less and finally converge. When converged, the least-squares algorithm does not only provide the optimum value for each variable but also affords its standard uncertainty, an important criterion for accuracy. Refinement based on intensities rather than structure factors was suggested quite early but only implemented with the advent of SHELXL93 [26]. The corresponding agreement factor

$$wR^2 = \sqrt{\frac{\sum w(F_{obs}^2 - F_{calc}^2)^2}{\sum w(F_{obs}^2)^2}}$$

is two or three times higher than the conventional residual. Refinement techniques other than linear least squares such as convergent gradient or nonlinear least squares can be used to minimize the target residual but are much less popular in small-molecule crystallography. We will not address the mathematical background of any minimization algorithm.

In addition to the residuals above, alternative quality criteria are relevant: After all relevant local electron density maxima have been assigned to atoms of a structure model, a subsequent difference Fourier synthesis should be flat: pronounced maxima and minima should either not occur at all or be located very close to many-electron scattering centers. The final structure model must confirm to chemical and physical standards. The former aspect requires meaningful interatomic geometries and balanced charges, and the latter is mainly reflected in magnitude, shape, and consistency of displacement parameters.

The so-called goodness of fit (GOF or S) defined as

$$S = \sqrt{\frac{\sum w(F_{obs}^2 - F_{calc}^2)^2}{n_{ref} - n_{var}}}$$

contains information about the overdetermination of a diffraction experiment. The denominator of the fraction is the difference between observed intensities (n_{ref}) and variables (n_{var}). After completion of the structure model and convergence of the final refinement, S ideally adopts a value close to 1.

1.6.2
Constraints and Restraints

A ratio of 10 between observations and variables (n_{ref}: n_{var}) is sometimes considered the minimum for a sufficiently well-determined structure; for good crystals, much higher ratios may be achieved. If the number of observations is small, for example, due to limited sample quality, or if many variables have to be refined as in the case of disordered structures, two obvious alternatives can be followed to increase the ratio n_{ref}: n_{var}:

  i) The number of refined variables can be decreased by establishing *constraints* between different variables; rigid group refinements are the most common example. Hydrogen atoms are associated with low electron density, and their contribution to the structure model is only minor. In those cases when the H-atom positions are unambiguous, for example, for an aromatic ring, they can be calculated and constrained to those of their parent atoms.

ii) The number of observations can be increased; in addition to the measured diffraction intensities, chemical or physical boundary conditions can be provided. These pseudo-observations or *restraints* may comprise assumptions about interatomic geometry in terms of distances, angles, or planarity, the shape of displacement ellipsoids (see below) or site occupancies. We will only mention two examples for the use of geometry restraints. The first occurs in the context of H-atom positions: The only electron of a hydrogen atom is involved in chemical bonding with a partner X, and the electron density maximum associated with a H is in general shifted toward X. Apparent X—H bond lengths determined by X-rays are therefore systematically shorter than the real interatomic distances. Restraints may ensure an acceptable compromise between electron density modeling and physical reality. A second frequent case refers to disordered groups. Different but similar conformers of the same object may crystallize as a solid solution, with statistical distribution of the alternative geometries. The distances between the mutually exclusive, partially occupied sites are often smaller than the resolution of the diffraction experiment and preclude independent refinement. Similarity restraints may then be used to ensure chemically reasonable geometries for either conformer. An excellent collection of refinement recipes for the popular SHELX programs is available [27].

1.6.3
Troubleshooting

Even is the phase problem can be solved and a structure model can be obtained, refinement does not always proceed smoothly.

i) Convergence problems may be encountered: in successive least-squares cycles the refined variables do not converge but rather oscillate. Such a behavior is often due to correlation, that is, an attempt to refine parameters independently that in reality are related by symmetry. Overlooked inversion symmetry is particularly popular. In such a case the space-group symmetry should be carefully checked and a search for missing symmetry in real space [28] should be performed.

ii) Unaccounted electron density not associated with any atom position may indicate disorder. It should be tested whether the remaining local electron density maxima can be assigned to an additional (minority) conformer of a molecule or ion. The alternative conformers are assigned mutually exclusive atom sites of fractional occupancy. It is advisable to analyze potential disorder before attempting a refinement with anisotropic displacement

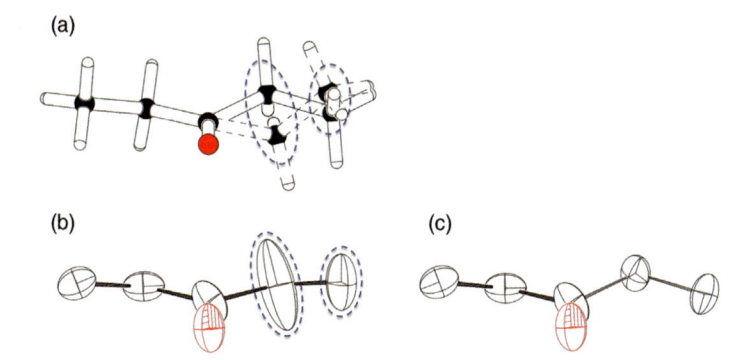

Figure 1.13 Diethyl ketone molecule in a host–guest crystal. (a) ball-and-stick plot, explaining the conformational disorder; (b) displacement ellipsoid plot without taking the disorder into account; (c) displacement ellipsoid plot for the majority conformer after the disorder has been modeled; adps in the plots at the bottom are drawn at the 50% level.

parameters as this might lead to an unrealistic model. The example in Figure 1.13 refers to diethyl ketone in a host crystal of cholic acid [29]. The upper part of the figure explains the conformational disorder in one of the ethyl arms, and the lower part shows the refinement results. Anisotropic displacements have been assigned before modeling the disorder in the left-hand part; the adps marked by dashed blue lines enclose the atom sites of both conformers. For comparison, the lower right-hand part of Figure 1.13 shows the much less anisotropic displacement parameters for the majority conformer after disorder modeling.

iii) Refinement programs provide a compilation of disagreeable reflections; if these reflections consistently show much higher observed than calculated structure factors, this is an indication for additional intensity contribution by twinning, see Section 1.6.4.

iv) Reflections collected at very low diffraction angles may have been obstructed by the beam stop. In this case, they show almost no observed intensity and should be excluded individually from the refinement.

v) If very strong reflections have much smaller observed than calculated structure factors, they may suffer from secondary extinction: lower lying lattice planes do not receive the full intensity of the primary beam because of a strong intensity loss due to diffraction in the upper lattice planes. Most refinement programs allow for empirical extinction corrections.

1.6.4

Twin Refinement

Disorder involves alternative atom positions within the coherence length of the X-ray beam; these alternative positions contribute to the structure factors by interference. The systematic intergrowth of larger domains is called twinning.

The diffraction pattern of a twinned crystal can be perceived as the sum of the individual diffraction patterns, each one due to a specific domain. If the overlap of the domains is complete, the diffraction pattern can be indexed in the same way as that of a single crystal; such twins are called (pseudo) merohedral, cf. Section 1.2. The alternative nonmerohedral case corresponds to partial overlap of the domains and can in general be detected at the stage of indexing and integration. Structure solution may be difficult for twins with two or more domains of competitive size. In contrast, if one of the domains is much larger than the others, it will dominate the joint diffraction pattern and twinning may remain undetected during structure solution. As the intensity of a reflection to which several domains contribute cannot be smaller and is in general larger than that due to a single domain, twinning can be detected *a posteriori*: if data with $F_{obs} \gg F_{calc}$ are abundant in the list of disagreeable reflections, this is an indication for twinning. Based on the indices of such obviously overlapped reflections and the unit cell parameters, the relationship between the domains, the twin law, can de deduced. In the case of merohedral and pseudomerohedral twins, the twin law associates each reflection with each domain; for nonmerohedral twins, reflections are either due to a single or to several domains. The relative domain sizes are treated as refined variables.

1.6.5
Refinement of Modulated Structures

Modulated structures can be understood as periodic in a higher dimensional superspace rather than in 3D space; their diffraction patterns can therefore not be described with three indices. The additional satellites can be described with the help of up to three modulation wavevectors. In the real space description of a modulated structure, atoms or molecules are shifted according to an extra periodicity not compatible with the simple lattice parameters of the unit cell [6]. JANA2006 is the most widespread and comfortable software package to refine modulated structures [30].

1.7
Interpretation of Diffraction Data

We shortly summarize the preceding sections: In diffraction experiments limited to standard resolution, atoms are identified as local maxima of electron density. During refinement they are treated with spherical, element-specific atomic scattering factors as discussed in Section 1.3. The direct results of a diffraction experiment are fractional coordinates and displacement parameters; unless fixed by symmetry, these quantities are associated with standard uncertainties. If an atom site is not fully occupied, its site occupancy parameter may be refined, too.

1.7.1
Geometry and Displacement Parameters

A primary interest of the crystallographer is to extract information from the refined atomic coordinates and elucidate the geometry of new compounds, either with the aim to establish qualitative aspects such as connectivity or coordination environment or in order to accurately determine interatomic distances and angles. Diffraction experiments can thus lead to structure–property relationships in many areas of science, in medicinal chemistry as well as in materials science.

Not only the fractional coordinates convey information: anisotropic displacement parameters (adps) were once regarded as the trash can for systematic errors [31], but if the diffraction experiment is competently conducted, they allow to analyze atomic movement. Adps are conveniently represented by ellipsoid plots – ORTEP [32] doubtlessly is the classical program for this purpose. Mercury [33] and PLATON [34] are available for the most common operating systems and input file formats, and many alternative visualization tools for adps exist [35–37].

Displacement parameters must at least be physically reasonable. Isotropic displacement parameters must be positive and usually adopt values smaller than $0.1\,\text{Å}^2$. Wrongly assigned electron density (wrong atom types!) will in part be compensated by unphysically small or large displacement parameters. When anisotropic displacement parameters are refined, their principal components should differ only moderately in magnitude; strongly prolate (two components much smaller than the third) or oblate ellipsoids (two components much larger than the third) are often a warning sign for problems with the data (inappropriate absorption correction, symmetry problems) or unaccounted disorder (see Figure 1.13, Section 1.6.3).

Acceptable size and shape are necessary but not sufficient conditions for anisotropic displacement parameters: Hirshfeld has pointed out that adp components along covalent bonds should be equal in magnitude [38], and this rigid-bond test has become an established quality criterion. Significant discrepancies between adp components of atoms subtending a covalent bond can be a warning sign indicating, for example, that an atom type has been wrongly assigned. Rigid building blocks in a crystal move as a whole, for example, by librating about their centers of inertia. Cruickshank pioneered the analysis of adps in terms of rigid-body motion [39] and the attempt to correct for the apparent shortening of bonds due to motion in the solid [40]. The present TLS model [41] assumes that rigid-body motion occurs as a combination of translation, liberation, and screw modes. A variety of restraints may be employed when adps adopt unexpected values during refinement; among these, rigid-bond restraints are the most substantiated.

1.7.2
Intermolecular Interactions

In addition to the geometry within the molecular or ionic components in a crystal, diffraction results provide information about intermolecular contacts.

Hydrogen bonds [42–44] count among the strongest secondary interactions; as a rule, hydrogen bonds do form when suitable donors and acceptors are available. In view of the low electron density associated with the H atom itself, donor···acceptor distances between electronegative atoms are often used to analyze hydrogen bonds. Graph set theory [45,46] has proven a suitable tool for the description of more complex hydrogen bond patterns. Halogen bonds are directional interactions between a N, O, S, or halogen atom as electron-density donor and a heavy halogen atom as electrophile [47,48] and have met increasing interest in the last decades [49–52]. The idea of arranging the constituents of a solid in a predictable way with the help of solid-state synthons [53] has lead to the flourishing field of crystal engineering [54–56]. To what extend a small number of short atom–atom pair interactions in solids are decisive for the crystal structure is under debate [57–61].

1.7.3 CIF

Dedicated software-specific file formats for data storage and exchange have a long history and were obvious requirements for databases (see below). With increasing electronic data exchange across hardware platforms running different software and operating systems, a general and reliable file format became indispensible. By now, the crystallographic information file (CIF) [62,63] represents the *de facto* standard for this purpose. As the final model of a crystal structure is obtained by refinement, refinement programs allow to generate a CIF containing symmetry information, a data collection summary, refined variables with their standard uncertainties and observed and calculated structure factors. Several structures can be concatenated into a joint CIF. The file uses ASCII text and therefore can be handled with any text editor. Additional information concerning sample properties, bibliography or even a complete manuscript for submission to a crystallographic journal may be included in the file.

The International Union of Crystallography (IUCr, www.iucr.org) maintains documentation and useful software for handling CIFs. A particularly helpful and often-used service is checkCIF, (http://journals.iucr.org/services/cif/checking/checkform.html) a web-based resource for analyzing and validating CIF files. Uploaded CIFs will be checked for syntax and then undergo a series of consistency and quality checks. The output is organized in different alert levels that intuitively adopt the color scheme of a traffic light: red level A alerts indicate more serious shortcomings than level B (orange) or C (yellow) alerts; dedicated author answers are shown in green. Submission of a CIF, often accompanied by its checkCIF output, has become mandatory for many chemical journals, and the checkCIF service is a standard tool for referees.

1.7.4 Databases

The first and highly successful computer-based approach to collect the results of diffraction experiments was due to Olga Kennard in the mid-1960s [64–66] and lead to the CSD (database c below). Since 2003, the Crystallography Open Database (COD, database f below) has rapidly gained momentum [67]. The most relevant databases address different classes of compounds rather than different diffraction methods; single-crystal diffraction results are included in all of them.

a) The CrystMet database [https://cds.dl.ac.uk/cgi-bin/news/disp?crystmet] covers metals, alloys, and intermetallics.
b) The ICSD [https://icsd.fiz-karlsruhe.de/search/index.xhtml] contains inorganic crystal structures.
c) The CSD [http://www.ccdc.cam.ac.uk/solutions/csd-system/components/csd/] contains structure analyses of carbon-containing molecules with up to one thousand atoms, peptides of up to 24 residues, and mono-, di-, and trinucleotides. Longer peptides are included in (d), higher oligonucleotides in (e).
d) The PDB [http://www.rcsb.org/pdb/home/home.do] is the repository for protein structures.
e) The NDB [ndbserver.rutgers.edu] contains information about experimentally determined nucleic acids and complex assemblies.
f) The COD [www.crystallography.net/index.php] is the open-access collection of crystal structures of organic, inorganic, metal-organic compounds and minerals; PDB-covered biopolymers are explicitly excluded.

1.7.5
Electron Density

We recall a basic statement from Section 1.3: each X-ray diffraction experiment probes the electron density of the sample. At standard resolution, local density maxima can be associated with atom sites; geometry data can be very precise, but no subtle conclusions with respect to electron density are drawn. All atoms of an element are treated as spherical scattering centers of the same type, regardless of their electronic situation and bonding environment. In principle, this limitation can be overcome: Given sufficient crystal quality and suitable equipment, diffraction experiments of high resolution can be performed. These experiments are conducted at low temperature in order to minimize thermal motion and allow to access reflections at high resolution (e.g., $\sin\theta/\lambda > 1$ in Figure 1.4) with significant intensity. Such high-order reflections owe their intensity to the innermost electron shells and essentially define the position of the core, whereas the outer valence electrons may be polarized, engage in covalent bonding or act as lone pairs. Thus, the experimental electron density in each point of the unit cell is experimentally accessible [68] and diffraction experiments can provide information concerning charge accumulation or depletion and the electron distribution along covalent bonds or in regions of short intermolecular contacts. Bader's Quantum Theory of Atoms In Molecules (QTAIM) [69] was originally intended to analyze electron densities obtained from theory and today represents a particularly popular way to interpret such experimental charge densities.

Acknowledgment

Help from Dr. Ruimin Wang is gratefully acknowledged.

References

1 Raven, W., Hermes, P., Kalf, I., Schmaljohann, J., and Englert, U. (2014) *J. Organomet. Chem.*, **766**, 34–39.
2 Hahn, T. (ed.) (2002) *International Tables for Crystallography. Vol. A*, 5th edn, Kluwer Academic Publishers, Dordrecht, (see Further Reading and Reference Books).
3 Hahn, T. (ed.) (2006) *International Tables for Crystallography Vol. A*, 1st online edn.
4 Bärnighausen, H. (1980) *MATCH*, **9**, 139–175.
5 Müller, U. (2013) *Symmetry Relationships between Crystal Structures – Applications of Crystallographic Group Theory in Crystal Chemistry*, Oxford University Press, Oxford (see Further Reading and Reference Books).
6 Wagner, T. and Schönleber, A. (2009) *Acta Crystallogr.*, **B65**, 249–268, and references cited therein.
7 Debye, P. (1914) *Ann. Phys.*, **348**, 49–92.
8 Bragg, W.L. (1912) *Proc. Cambridge Philos. Soc.*, **17**, 43–57.
9 Friedel, M.G. (1913) *Comptes Rendus Acad. Sci. Paris*, **157**, 1533–1536.
10 Bijvoet, J.M., Peerdeman, A.F., and van Bommel, A.J. (1951) *Nature*, **168**, 271–272.
11 Flack, H.D. (1983) *Acta Crystallogr.*, **A39**, 876–881.
12 Hooft, R.W.W., Straver, L.H., and Spek, A.L. (2008) *J. Appl. Cryst.*, **41**, 96–103.
13 Hooft, R.W.W., Straver, L.H., and Spek, A.L. (2010) *J. Appl. Cryst.*, **43**, 665–668.
14 Parsons, S., Flack, H.D., and Wagner, T. (2013) *Acta Crystallogr.*, **B69**, 249–259.
15 Rhodes, G. (2006) *Crystallography Made Crystal Clear*, 3rd edn, Elsevier/Academic Press, (see Further Reading and Reference Books).
16 Graber, T., Anderson, S., Brewer, H., Chen, Y.-S., Cho, H.S., Dashdorj, N., Henning, R.W., Kosheleva, I., Macha, G., Meron, M., Pahl, R., Ren, Z., Ruan, S., Schotte, F., Srajer, V., Viccaro, P.J., Westferro, F., Anfinrud, P., and Moffat, K. (2010) *J. Synchrotron Rad.*, **18**, 658–670.
17 Coppens, P., Formitchev, D.V., Carducci, M.C., and Culp, K. (1998) *J. Chem. Soc., Dalton Trans.*, 865–872.
18 Patterson, A.L. (1935) *Z. Kristallogr.*, **A90**, 517–542.
19 Cochran, W. (1952) *Acta Crystallogr.*, **5**, 65–67, and references cited therein.
20 Oslanyi, G. and Süto, A. (2004) *Acta Crystallogr.*, **A60**, 134–141.
21 Palatinus, L. (2009) *Acta Crystallogr.*, **B69**, 1–16.
22 Palatinus, L. and Chapuis, G. (2007) *J. Appl. Cryst.*, **40**, 786–790.
23 Usón, I. and Sheldrick, G.M. (1999) *Curr. Opin. Struct. Biol.*, **9**, 643–648.
24 Sheldrick, G.M. (2015) *Acta Crystallogr.*, **A71**, 3–8.
25 Burla, M.C., Carrozzini, B., Cascarano, G.L., Giacovazzo, C., and Polidori, G. (2000) *J. Appl. Cryst.*, **33**, 307–311.
26 Sheldrick, G.M. (2008) *Acta Crystallogr.*, **A64**, 112–122.
27 Müller, P., Herbst-Irmer, R., Spek, A.L., Schneider, T., and Sawaya, M. (2006) *Crystal Structure Refinement: A Crystallographer's Guide to SHELXL*, Oxford University Press, Oxford.
28 LePage, Y. (1988) *J. Appl. Cryst.*, **21**, 983–984.
29 Kalf, I. and Englert, U. (2011) *Acta Crystallogr.*, **C67**, o206–o208.
30 Petříček, V., Dušek, M., and Palatinus, L. (2014) *Z. Kristallogr.*, **229**, 345–352.
31 Schwarzenbach, D. (2012) *Z. Kristallogr.*, **227**, 52–62.
32 Johnson, C.K. (1965) *ORTEP: A FORTRAN Thermal-Ellipsoid Plot Program for Crystal Structure Illustrations*, ONRL Report #3794. Oak Ridge National Laboratory, Oak Ridge, Ten.
33 Macrae, C.F., Bruno, I.J., Chisholm, J.A., Edgington, P.R., McCabe, P., Pidcock, E.,

Rodriguez-Monge, L., Taylor, R., van de Streek, J., and Wood, P.A. (2008) *J. Appl. Cryst.*, **41**, 466–470.

34 Spek, A.L. (2009) *Acta Crystallogr.*, **D65**, 148–155.

35 Palmer, D.C. (2015) *Z. Kristallogr.*, **230**, 559–572.

36 Dowty, E. (2006) *ATOMS*, Shape Software, Kingsport, TN, USA.

37 Brandenburg, K. (1999) *DIAMOND*, Crystal Impact GbR, Bonn, Germany.

38 Hirshfeld, F.L. (1976) *Acta Crystallogr.*, **A32**, 239–244.

39 Cruickshank, D.W.J. (1956) *Acta Crystallogr.*, **9**, 754–756.

40 Cruickshank, D.W.J. (1956) *Acta Crystallogr.*, **9**, 757–758.

41 Schomaker, V. and Trueblood, K.N. (1968) *Acta Crystallogr.*, **B24**, 63–76.

42 Jeffrey, G.A. (1997) *An Introduction to Hydrogen Bonding*, Oxford University Press, Oxford.

43 Gilli, P., Bertolasi, V., Ferretti, V., and Gilli, G. (1994) *J. Am. Chem. Soc.*, **116**, 909–915.

44 Gilli, P., Bertolasi, V., Pretto, L., Ferretti, V., and Gilli, G. (2004) *J. Am. Chem. Soc.*, **126**, 3845–3855.

45 Etter, M. (1991) *J. Phys. Chem.*, **95**, 4601–4610.

46 Etter, M.C., MacDonald, J.C., and Bernstein, J. (1990) *Acta Crystallogr.*, **B46**, 256–262.

47 Metrangolo, P. and Resnati, G. (2001) *Chem. Eur. J.*, **7**, 2511–2519.

48 Wang, R., Dols, T., Lehmann, C.W., and Englert, U. (2012) *Chem Commun.*, **48**, 6830–6832.

49 Zordan, F., Purver, S.L., Adams, H., and Brammer, L. (2005) *CrystEngComm*, **7**, 350–354.

50 Zordan, F., Brammer, L., and Sherwood, P. (2005) *J. Am. Chem. Soc.*, **127**, 5979–5989.

51 Pigge, F.C., Vangala, V.R., and Swenson, D.C. (2006) *Chem. Commun.*, 2123–2125.

52 Cavallo, G., Metrangolo, P., Pilati, T., Neukirch, H., Resnati, G., Sansotera, M., and Terraneo, G. (2010) *Chem. Soc. Rev.*, **39**, 3772–3783.

53 Desiraju, G.R. (1995) *Angew. Chem.*, **107**, 2541–2558.

54 Braga, D., Grepioni, F., and Desiraju, G.R. (1998) *Chem. Rev.*, **98**, 1375–1405.

55 Braga, D., Desiraju, G.R., Miller, J.S., Orpen, A.G., and Price, S.L. (2002) *CrystEngComm*, **4**, 500–509.

56 Brammer, L. (2004) *Chem. Soc. Rev.*, **33**, 476–489.

57 Dance, I. (2003) *New J. Chem.*, **27**, 22–27.

58 Dunitz, J.D. and Gavezzotti, A. (2005) *Angew. Chem., Int. Ed.*, **44**, 1766–1787.

59 Dunitz, J.D. (2015) *IUCrJ*, **2**, 157–158.

60 Thakur, T.S., Dubey, R., and Desiraju, G.R. (2015) *IUCrJ*, **2**, 159–160.

61 Lecomte, C., Espinosa, E., and Matta, C.F. (2015) *IUCrJ*, **2**, 161–163.

62 Hall, S.R., Allen, F.H., and Brown, I.D. (1991) *Acta Crystallogr.*, **A47**, 655–685.

63 Hall, S. and McMahon, B. (2005) *International Tables for Crystallography, Volume G*, Springer, Heidelberg.

64 Kennard, O., Watson, D.G., and Town, W.G. (1972) *J. Chem. Doc.*, **12**, 14–19.

65 Allen, F.H., Kennard, O., Motherwell, W.D.S., Town, W.G., and Watson, D.G. (1973) *J. Chem. Doc.*, **13**, 119–123.

66 Allen, F.H., Kennard, O., Motherwell, W.D.S., Town, W.G., and Watson, D.G. (1974) *J. Appl. Cryst.*, **7**, 73–78.

67 Gražulis, S., Daškevič, A., Merkys, A., Chateigner, D., Lutterotti, L., Quirós, M., Serebryanaya, N.R., Moeck, P., Downs, R.T., and Bail, A.Le. (2012) *Nucl. Acids Res.*, **40**, D420–D427.

68 Coppens, P. (1997) *X-Ray Charge Densities and Chemical Bonding*, Oxford University Press, Oxford .

69 Bader, R.F.W. (1990) *Atoms in Molecules – A Quantum Theory*, Clarendon Press, Oxford.

Further Reading and Reference Books

Burns, G. and Glazer, A.M. (1990) *Space Groups for Solid State Scientists*, 2nd edn, Academic Press, San Diego.

Giacovazzo, C. (ed.) (2011) *Fundamentals of Crystallography*, 3rd edn, Oxford University Press, Oxford.

Hahn, T. (ed.) (2002) *International Tables for Crystallography. Vol. A*, 5th edn, Kluwer Academic Publishers, Dordrecht.

Luger, P. (2014) *Modern X-ray analysis on single crystals*, De Gruyter, Berlin.

Massa, W. (2011) *Crystal Structure Determination*, 2nd edn, Springer, Berlin, Heidelberg.

Müller, U. (2013) *Symmetry Relationships between Crystal Structures – Applications of Crystallographic Group Theory in Crystal Chemistry*, Oxford University Press, Oxford.

Rhodes, G. (2006) *Crystallography Made Crystal Clear*, 3rd edn, Elsevier/Academic Press.

Warren, B.E. (1990) *X-Ray Diffraction*, Dover Publications, New York.

2
Laboratory and Synchrotron Powder Diffraction
R.E. Dinnebier,[1] M. Etter,[2] and T. Runcevski[3]

[1]Röntgenographie, Max Planck Institute for Solid State Research, Heisenbergstr. 1, 70569 Stuttgart, Germany
[2]Photon Science (FS-PE), Deutsches Elektronen-Synchrotron, FS-PE, Notkestr. 85, 22607 Hamburg, Germany
[3]Dept. of Chemistry, University of California, Berkeley, CA 94720-1460, USA

2.1
Introduction

Unlike the case of an oriented single crystal, in a randomly distributed microcrystalline powder the scattering vectors of different directions but equal or similar lengths overlap systematically or accidentally, making their separation difficult or even impossible. This drawback is the main reason why powder diffraction, despite its early invention (just 4 years after the first single crystal experiment), was mainly used for phase identification and quantification for more than half a century. The amount of structural information accumulated in a powder pattern is huge, yet it is very challenging to get it revealed. A milestone was set in 1969, when Hugo Rietveld introduced a new optimization technique for crystal structure refinement from powder diffraction data, which intrinsically takes the peak overlap into account [1]. Only few years after the invention of the Rietveld method, the power of powder diffraction was recognized in *ab initio* structure determinations, using algorithms working in real space, reciprocal space, or a combination of both.

Nowadays, almost all branches of natural sciences and engineering use powder diffraction to determine crystal structures with different complexities. Within the last decade, a new generation of powder diffractometers in the laboratory and at synchrotrons was designed, providing high resolution, variable energy, and high intensity of radiation. Taking advantage of these modern setups and software, even microstructural parameters such as domain size and microstrain are routinely deduced from powder diffraction data.

Owing to the finite length of this chapter, the focus has been set on high-resolution powder diffraction, which is a prerequisite for structure determination and refinement from powder diffraction data. This chapter starts with a brief

Handbook of Solid State Chemistry, First Edition. Edited by Richard Dronskowski, Shinichi Kikkawa, and Andreas Stein.
© 2017 Wiley-VCH Verlag GmbH & Co. KGaA. Published 2017 by Wiley-VCH Verlag GmbH & Co. KGaA.

history and basics of (powder) diffraction, followed by a condensed description of the typical geometry of state-of-the-art high-resolution powder diffractometers in the laboratory and at synchrotrons. In correspondence to its importance, the Rietveld method will be explained in more detail, in particular mentioning alternative descriptions of crystal structures using symmetry modes, which are particularly suitable for the description of structural phase transitions. Among the many existing methods for structure determination from powder diffraction data, two established methods, one working in real space, the other in reciprocal space are exemplarily described. In the last part, four research projects conducted by the authors have been selected as case studies, as they allow tracing the process of powder diffraction data evaluation for typical problems in modern solid-state chemistry.

For the interested reader, more in-depth information about the method of powder diffraction can be found in two recent textbooks [2,3].

2.2
History and Basics of Powder Diffraction

Although early experiments with light were carried out by the Greek philosophers [4], it is believed that the Italian Francesco Maria Grimaldi (1618–1663) was the first who investigated the diffraction of light more rigorously. In his book *Physico mathesis de lumine, coloribus, et iride, aliisque annexis libri duo* from 1665, which was published after his death, he described the shape of a light cone, after the light has passed through a pinhole [5]. In fact, he was also the first to use the term "diffraction" to name the physical properties of his light scattering experiments.

Over the next years, the investigations and experiments with light evidenced development at a steady pace, as just the narrow spectral range of visible light was accessible. (The infrared spectral range and the ultraviolet spectra range were discovered rather late by Friedrich Wilhelm Herschel (1738–1822) in around 1800 and by Johann Wilhelm Ritter (1776–1810) in 1801.) The situation changed dramatically in 1895, when Wilhelm Conrad Röntgen (1845–1923) discovered the "X-rays," while experimenting with discharge tubes [6,7]. This new kind of rays rapidly attracted the interest of a lot of researchers, sparkling a two decades-long debate whether these new rays consist of particles or they are electromagnetic waves.

In 1912, Walter Friedrich (1883–1968), Paul Knipping (1883–1935), and Max von Laue (1879–1960) conducted their famous X-ray diffraction experiment, in which they collected a single-crystal diffractogram by irradiating a copper sulfate crystal with the (polychromatic) light from an X-ray tube [8–11]. With this experiment, they could prove two physical principles at once: that a single crystal is built by regular blocks (a required condition to allow for diffraction effects to happen) and that the X-rays are waves with a wavelength on the order of the distances between the building blocks of the crystal [12–14]. In fact, the idea for the experiment is ascribed to von Laue who also gave the first theoretical description of the observed phenomenon [9–11,13–15]. In the theoretical part of their joint publications [9–11], von Laue introduced the famous equations that later got named after him:

$$\vec{a} \cdot (\vec{s} - \vec{s_0}) = h \cdot \lambda, \qquad (2.1.1)$$

$$\vec{b} \cdot (\vec{s} - \vec{s_0}) = k \cdot \lambda, \qquad (2.1.2)$$

$$\vec{c} \cdot (\vec{s} - \vec{s_0}) = l \cdot \lambda. \qquad (2.1.3)$$

Here, $\vec{s_0}$ is the unit vector of the primary beam, $\vec{s}$ is the unit vector of the scattered beam, h, k, l are the Miller indices, $\vec{a}$, $\vec{b}$, $\vec{c}$ are the primitive lattice vectors, and λ is the wavelength. Due to the usage of unit vectors for the incident and the outgoing beams, the scalar product of a primitive lattice vector with a beam vector reduces to a projection of the primitive lattice vector (which is equal to the distance of two points) onto the beam vector. For constructive interference, the difference between the two projections must be equal to a multiple of the wavelength (see Figure 2.1a). For a crystal structure in three dimensions, diffraction occurs if all three Laue equations are fulfilled simultaneously.

In the same year of the Friedrich, Knipping, and von Laue discovery, another mathematical description of the diffraction condition was given by Sir William Lawrence Bragg (1890–1971). He published an equation [17] equivalent to the Laue equation:

$$2d_{hkl} \cdot \sin\theta = n \cdot \lambda. \qquad (2.2)$$

The geometrical interpretation is similar to the interpretation of the Laue equations. If the sine of the scattering angle θ multiplied by the doubled lattice plane spacing d_{hkl} is equal to a multiple of the wavelength λ (n is a positive integer), then diffraction occurs (see Figure 2.1b).

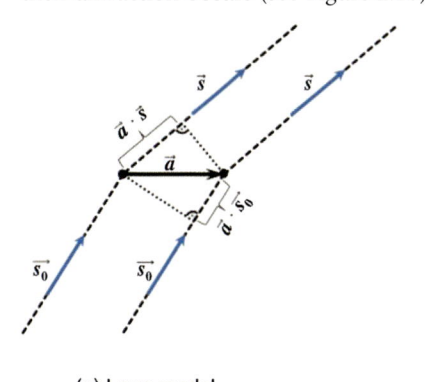

(a) Laue model

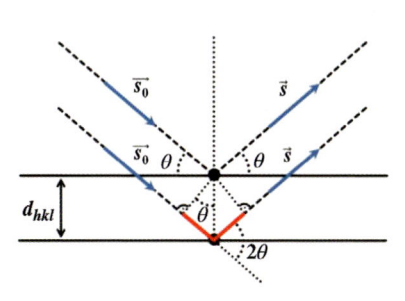

(b) Bragg model

Figure 2.1 Visualization of the (a) Laue equation and the (b) Bragg equation with two point scatters. For the Bragg equation, the optical path that must be a multiple of the wavelength is shown in red. Note that this visualization of the Bragg equation is only a simplified representation, as in reality the point scatters can lie anywhere on the lattice planes and not necessarily above each other [3,16]. (Adapted from Dinnebier 2008 [3] and Bloss 1971 [16].)

Although Bragg's and von Laue's description are equivalent, nowadays Bragg's equation is often preferred, due to its natural linkage of the lattice plane spacing d, which is a function of the lattice parameters a, b, c, α, β, γ of the unit cell and the Miller indices h, k, l with the scattering angle θ and the wavelength λ.

In addition to the Laue equations, von Laue in his milestone publication from 1912 also gave an equation for the structure factor amplitude F [9–11], which is required for the calculation of the intensity of a h, k, l-dependent Bragg reflection. The calculation of the complex structure factor F depends on an individual atomic form factor f, which itself depends on the diffraction angle, the Miller indices h, k, l, and the relative atomic coordinates x, y, z of each atom n in the unit cell:

$$F_{hkl} = \sum_n f_n \cdot e^{2 \cdot \pi \cdot i \cdot (h \cdot x_n + k \cdot y_n + l \cdot z_n)}. \tag{2.3}$$

Mathematically, the structure factor is the Fourier transform of the convolution between the real-space lattice and a motif, which is in the case of X-ray diffraction the distribution of the electron density in the unit cell and in the case of neutron diffraction the distribution of the atomic nuclei in the unit cell. Owing to the fact that the structure factor is a Fourier transform, many properties of diffraction can be predicted, for example, the invariance of the diffracted pattern if the convoluted lattice with the motif is translated. This is an important statement as it implies that there is no need to define an artificial origin of the investigated crystal. Another important property of the Fourier transform is that the measured peak width is directly correlated with the number of unit cells that build the real-space lattice. Namely, if the number of unit cells that contribute to the diffraction in three dimensions is very small, the Fourier transform will give a broader peak width, or if the number of unit cells in all three dimensions is high, the experimental peak width is no longer dominated by size effects of the crystal.

In order to calculate the intensity I of a h, k, l-dependent Bragg reflection, the square of the absolute value of the structure factor or equivalently the multiplication of the structure factor with its complex conjugate needs to be build. This product of the structure factor and its complex conjugate is proportional to the intensity:

$$I_{hkl} \propto F_{hkl} \cdot F^*_{hkl} = |F_{hkl}|^2. \tag{2.4}$$

In general, Eqs. (2.3) and (3.4) are valid for crystal structures infinitely extended in three dimensions. Nevertheless, they are sufficient approximations for finite crystals and they are even applicable to nanocrystals that can have a spatial extent ranging a few nanometers. A question arises: How can the diffracted intensity be calculated if the crystal is no longer built by regular building blocks? For example, for amorphous compounds or even for liquids, where the atoms or molecules can occupy all possible orientations? It is obvious that the deficit of regular building blocks with well-defined unit cells automatically leads to partial loss of symmetry. This issue was addressed in 1915, when Peter Debye (1884–1966) published an article about the dispersion of X-rays [18]. In his work he considered an amorphous compound where a molecule or polyatomic ions can take all possible orientations in space. He realized that the diffracted intensity of such a compound depends only on the distance r_{mn} between two individual point scatterers m and n within or between the molecules and the scattering vector k.[1] Taking into account that all intra- and intermolecular distances must be considered and that the electronic distribution of an individual point scatterer is given by the atomic form factor f, it is possible to modify the original double sum equation of Debye to obtain the commonly known Debye scattering equation:

$$I_{eu} = \sum_m \sum_n f_m \cdot f_n \cdot \frac{\sin k \cdot r_{mn}}{k \cdot r_{mn}}. \tag{2.5}$$

In this representation of the Debye scattering equation, the intensity is calculated in electronic units and it is possible to apply this equation to "gases, liquids, amorphous solids and crystalline powders" [19].

Looking at the scientific interests of Peter Debye and his particular interest into the scattering of X-rays from particles, it is not surprising that he was also involved in one of the first powder diffraction experiments, dated 1916. Although, as early as in 1913, Walter Friedrich in Germany [20] and Shoji Nishikawa (1884–1952) and S. Ono in Japan [21] carried out X-ray diffraction experiments with powders, they were not able to give the correct explanation for the diffraction rings that they observed because at that time they were not aware of the spectrum of their X-ray source [12]. As a consequence, it took another 3 years, until Peter Debye and Paul Scherrer (1890–1969) could use the knowledge of characteristic X-rays to explain the occurrence of diffraction rings, when they investigated lithium fluoride powder [22]. Almost at the same time, on the other site of the Atlantic Ocean, Albert Wallace Hull (1880–1966) conducted a powder X-ray diffraction experiment with iron powder and he gave the same explanations as those of Debye and Scherrer [23]. Unlike Debye and Scherrer, the experimental setup of Hull was much more sophisticated as he was using one of the first monochromators with a zirconium filter in order to suppress the characteristic K_β radiation and most of the unwanted Bremsstrahlung background of his molybdenum X-ray tube [23]. The diffraction patterns that Debye, Scherrer, and Hull recorded were already looking very similar to the simulated powder diffraction patterns in Figure 2.2c and d.

A descriptive interpretation of the Laue equations was given by Paul Peter Ewald (1888–1985) in 1913, when he introduced the concept of the Ewald sphere [24]. In this illustration of the reciprocal space lattice, the diffraction condition can be graphically evaluated (see Figure 2.3a). In order to find the hkl values that fulfill the diffraction condition, the incident wave vector $\vec{s_0}$ has to be drawn in a reciprocal space lattice in a way that the vector ends at the origin of the reciprocal space lattice. The direction of the incident wave vector is given by the experimental setup and the length of the vector is given by the reciprocal

---

1) The length k of the scattering vector $\vec{k}$ is defined as $k = (4\pi \cdot \sin \theta)/\lambda$ [19].

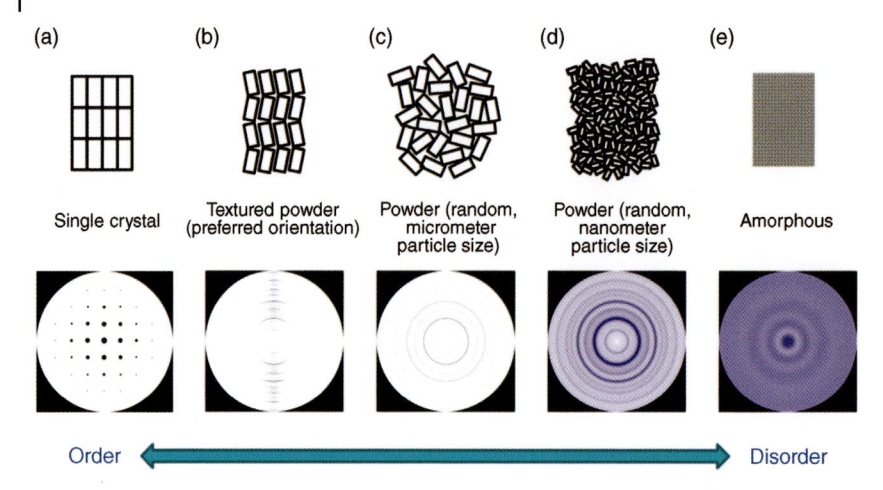

Figure 2.2 Simulated two-dimensional diffraction patterns of different types of crystalline materials. *From left to right*: The evolution of these patterns is shown, when a single crystal is crushed and the disorder is increased. (a) Diffraction pattern of a single crystal. (b) Diffraction pattern of a textured powder with preferred orientation. (c) Diffraction pattern of powder with particles in micrometer size. (d) Diffraction pattern of powder with particles in nanometer size. (e) Diffraction pattern of an amorphous crystal (also valid for gases and liquids).

value of the wavelength. Subsequently, the Ewald sphere is drawn by taking the length of the reciprocal wavelength as the radius with the starting point of the incident wave vector as the center of the sphere. The Laue conditions and therefore constructive interference are then fulfilled for all reciprocal lattice points hkl, which lie on the surface of the Ewald sphere. To such a reciprocal lattice point, it is possible to draw the outgoing wave vector $\vec{s}$ with the same length as for the incoming wave vector $\vec{s_0}$. The difference between the outgoing wave vector and the incoming wave vector gives the scattering vector $\vec{h}$:

$$\vec{h} = \vec{s} - \vec{s_0}. \tag{2.6}$$

The construction of the Ewald sphere does not solely provide information on physical values in reciprocal space, such as scattering vectors and reciprocal lattice points. If an equivalent description of the Bragg equation is taken, it is possible to give a geometrical interpretation of the angle between the incoming and the outgoing wave vectors in real space. Such a variation of the Bragg equation is given by

$$\frac{2 \cdot \sin\theta}{n \cdot \lambda} = \frac{1}{d} = |\vec{h}|. \tag{2.7}$$

Without loss of generality, it can be shown that for constructive interference to occur, the scattering vector $\vec{h}$ must be equal to a reciprocal lattice vector $\vec{d}^*_{hkl}$ and that $|\vec{d}^*_{hkl}| = 1/|\vec{d}_{hkl}|$ and $\vec{d} \parallel \vec{d}^*$. In the case of the Ewald construction, the multiple n of the wavelength can be set to 1, as the definition from above requires that

2.2 History and Basics of Powder Diffraction

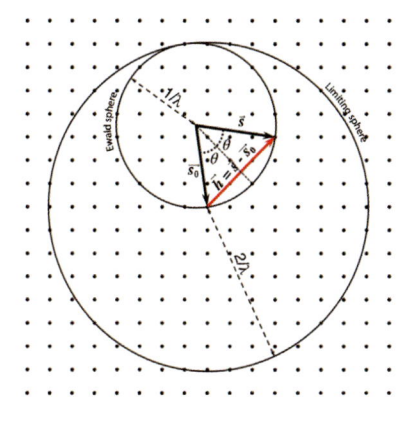

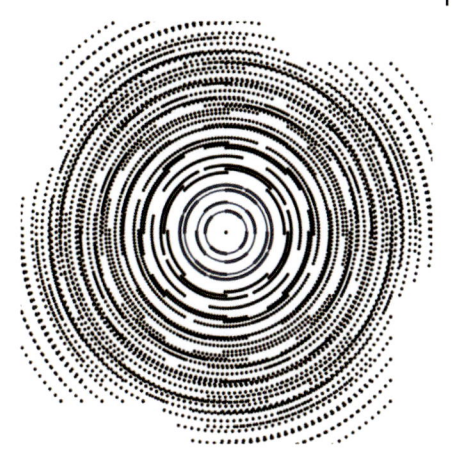

(a) Ewald construction for a single crystal

(b) Stepwise rotated reciprocal lattice

Figure 2.3 (a) Two-dimensional projection of the reciprocal space lattice with Ewald sphere and the limiting sphere (the limiting sphere determines the maximal reachable *hkl* values in a powder diffraction experiment). The radius of the limiting sphere is given by $2/\lambda$ (therefore, the minimum reachable d_{hkl} value is given by $d_{hkl} = \lambda/2$). Note that the incoming beam within the Ewald sphere does not necessarily start at a reciprocal lattice point. (b) Twenty-four single-crystal diffraction patterns each rotated by an angle of 2°. It is obvious that in a powder where ideally all possible orientations of crystal grains exist, the single spots in two dimensions will merge into a continuous diffraction ring, which becomes a continuous diffraction sphere in three dimensions.

the length of the incoming and outgoing wave vectors is $\vec{s} = \vec{s_0} = 1/\lambda$. With this rewritten Bragg equation, it is possible to derive a real-space interpretation of the angle 2θ (see Figure 2.3a). This shows the remarkable power of the Ewald construction, providing direct access to the real-space diffraction angle 2θ, which is required to position a detector in order to measure the diffracted intensity of a certain *hkl* reflection.

Although the explanation above seems to be only valid for the case of single-crystal diffraction, it can be also used to explain the occurrence of diffraction rings in a powder diffraction pattern. Figure 2.3b shows the effect when a crystal is turned during the measurement, which is the same as when a powder composed of randomly oriented crystals is measured. In a static experiment, it is only possible to measure the *hkl* reflections that lie directly on the surface of the Ewald sphere. If other *hkl* reflections shall be measured, the single crystal has to be turned until the corresponding reciprocal lattice points hits the surface of the Ewald sphere. A powder diffraction experiment is therefore comparable to a single crystal experiment in which the data integration is performed continuously, while the single crystal is rotated in three dimensions. In such a dynamical diffraction experiment, all reciprocal lattice points will hit the Ewald surface at a certain point in time. In contrast to that, the grains in a powder already have all possible orientations and therefore they contribute to the diffraction pattern at

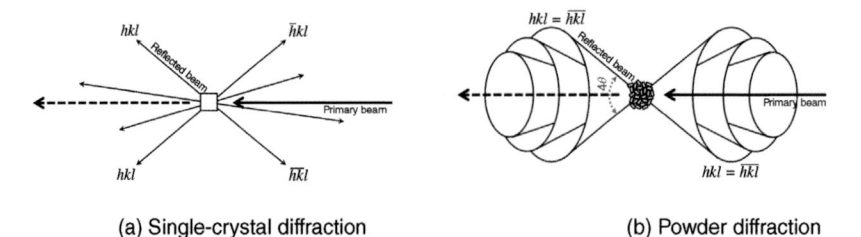

Figure 2.4 Diffracted X-rays of a (a) single-crystal specimen and diffracted X-rays in (b) Debye–Scherrer cones for a powder sample. The Debye–Scherrer rings result from a cut projection of the spheres that arise due to the smearing of reciprocal lattice points onto different spheres in reciprocal space. The cone shape is simply given by the propagation of the radiation.

the same time giving the same smearing effect of the reciprocal lattice points onto the surface of a sphere as for a dynamic single crystal experiment. Consequently, this smearing of a reciprocal lattice point onto the surface of a sphere leads not only to the reduction in dimensionality (from three dimensions to one dimension; mathematically this is a projection: $\vec{d}_{hkl}^* \to |\vec{d}_{hkl}^*|$) but also to many other effects such as systematic and accidental overlaps, which will be described later.

In Figure 2.2, the evolution of a diffraction pattern of a single crystal split into pieces is exemplarily shown. Namely, Figure 2.2a shows the diffraction pattern of a single crystal with sharp reflection spots, whereas Figure 2.2b shows a crystal consisting of different grains (=polycrystalline) with preferred orientation of these grains, resulting in smearing effect of the reciprocal lattice points. In the case of a powder (Figure 2.2c) where the grains are of micrometer size and where they take all possible orientations, a smearing of the reciprocal lattice points to a circle or diffraction ring can be observed. In general, these diffraction rings are cut projections of so-called Debye–Scherrer cones (see Figure 2.4 for a three-dimensional view of the optical path of the diffracted rays), which in turn are originating from cut projections through the above-mentioned surface of the sphere that arises due to the smearing of the reciprocal lattice points. If the powder particles are further split into nanoparticles, a severe broadening[2] of the diffraction rings can be observed (see Figure 2.2d). Finally, the material becomes completely amorphous and the diffraction rings pass over to a halo effect (see Figure 2.2e).

Expectedly, the differences in the diffraction patterns of single crystals and powders merit different data analysis approaches. In both cases, the possible hkl values have to be assigned to either the individual reciprocal lattice points or to the diffraction rings (normally the two-dimensional powder diffraction patterns

---

[2] The broadening due to the particle size can be modeled by the Scherrer equation:
FWHM(2θ) = $(K \cdot \lambda)/L \cdot \cos\theta$, where FWHM($2\theta$) is the full width at half maximum for a given diffraction angle θ, K is the Scherrer constant that depends on the particle habitus and is typically around 0.9, λ is the used wavelength, and L is the particle size [25].

are integrated along circles into a one-dimensional powder diffraction pattern as the intensity is a function of the radius, thus no information will be lost during this integration. This argument becomes also clear by taking care of the above-described reduction in dimensionality). As it can be seen in Figure 2.5, this is a challenging task in the powder case, as the smearing of the reciprocal lattice points leads to different degrees of information loss. For instance, in the single crystal case, the full three-dimensional measurement of reciprocal space provides the entire information about the kind of the reciprocal lattice and the individual intensities at the reciprocal lattice points. Accordingly, the real-space lattice and the Laue group can be deduced. (Note that because of the Fourier transform, the single-crystal diffraction pattern has always inversion symmetry and therefore two reciprocal lattice points with the same intensity exist, the

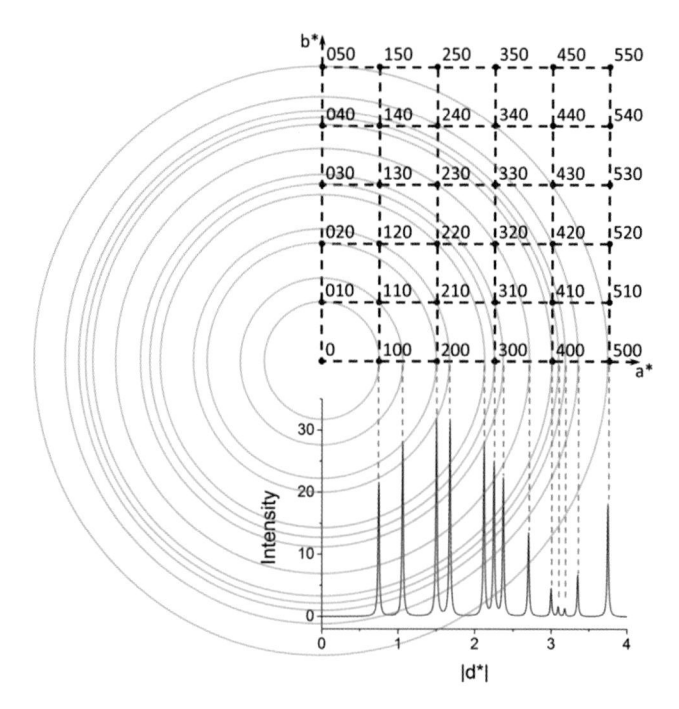

Figure 2.5 Cubic reciprocal lattice where the reciprocal lattice points are continuously smeared onto the surfaces of different spheres. If these spheres are arbitrarily cut through the center, continuous two-dimensional powder diffraction rings can be observed. Another cut projection through the center of the diffraction rings gives the one-dimensional powder diffraction pattern (in an experiment, normally the one-dimensional powder diffraction patterns are obtained by the integration of the rings along a cut that is perpendicular to the rings). Indexing of the single peaks in this powder diffraction pattern can be done by following the orange dashed lines and then by following the corresponding lines of the circle to the reciprocal lattice points. Note that, for instance, the reciprocal lattice points 100 and 010 merge into a single peak (this is the case of reflection multiplicity) as well as the reciprocal lattice points 500 and 340 merge into a single peak (this is the case of systematic overlap of reflections).

commonly known Friedel pairs). In contrast, for the powder case, the information of the Laue group and also of the real-space lattice is lost, as the intensities of radially symmetric equivalent reciprocal lattice points are merged into a single intensity and with this all orientation information is gone (e.g., reciprocal lattice points *100* and *010* in Figure 2.5 are radially symmetric equivalent). The number of radially symmetric equivalent reciprocal lattice points that merge into a single reflection depends on the Laue group and is called reflection multiplicity (if the symmetry is known, the reflection multiplicity as a systematic overlap can be easily treated, as it is just a multiple of the reflection intensity). Besides the reflection multiplicity, a systematic overlap of reflections can be observed. For example, if different independent hkl values lead to exactly the same d_{hkl} value, a systematic overlap occurs, as it can be seen, for example, for the *(500)* reflection and the *(340)* reflection for a cubic crystal structure in Figure 2.5. The last possible overlap of reflections is occasional or accidental. For instance, if two reflections are very close in reciprocal space, broad peaks can merge into a single peak and cannot be any longer distinguished. This phenomenon can be often observed for high values of the measured 2θ range, where different reflections often come very close.

In addition to the reflection's multiplicity and the systematic overlap, peak broadening and estimated standard deviations in each measurement of the angle 2θ are a severe problem, which often makes it difficult to find the right indexing for a powder diffraction pattern. For instance, the error (=estimated standard deviation) that arises in the determination of the correct d_{hkl} due to a shift of the 2θ value can be estimated by the curves in Figure 2.6.

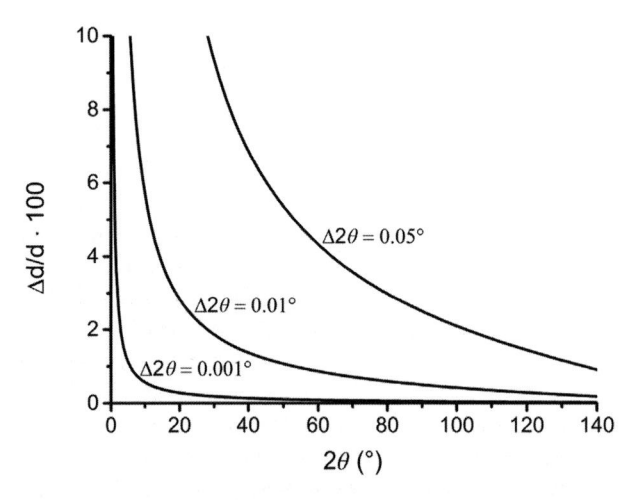

Figure 2.6 The deviation of the scattering angle by a constant angular misalignment leads to different percentage errors for the obtained $\Delta d/d$ values as can be seen by the different curves. For instance, a misalignment of $\Delta 2\theta = 0.01°$ for a measured peak at $2\theta = 20°$ and at a wavelength of $\lambda = 1.54059$ Å leads to an error in d of $\Delta d = 0.057$. The curves can be calculated through the equation $\left|\frac{dd_{hkl}}{d_{hkl}}\right| \approx \frac{d\theta}{\tan\theta}$, which can be obtained by calculating the total differential of the Bragg equation [3].

Table 2.1 Equations for the calculation of the d_{hkl} values dependent on the real-space unit cell parameters for the different crystal systems.

System	$1/d_{hkl}^2$
Cubic	$\dfrac{h^2 + k^2 + l^2}{a^2}$ (1)
Tetragonal	$\dfrac{h^2 + k^2}{a^2} + \dfrac{l^2}{c^2}$ (2)
Orthorhombic	$\dfrac{h^2}{a^2} + \dfrac{k^2}{b^2} + \dfrac{l^2}{c^2}$ (3)
Hexagonal and trigonal (P)	$\dfrac{4}{3 \cdot a^2} \cdot (h^2 + k^2 + h \cdot k) + \dfrac{l^2}{c^2}$ (4)
Trigonal (R)	$\dfrac{1}{a^2} \cdot \left(\dfrac{(h^2 + k^2 + l^2) \cdot \sin^2\alpha + 2(h \cdot k + h \cdot l + k \cdot l) \cdot (\cos^2\alpha - \cos\alpha)}{1 + 2\cos^3\alpha - 3\cos^2\alpha} \right)$ (5)
Monoclinic	$\dfrac{h^2}{a^2 \cdot \sin^2\beta} + \dfrac{k^2}{b^2} + \dfrac{l^2}{c^2 \cdot \sin^2\beta} - \dfrac{2 \cdot h \cdot l \cdot \cos\beta}{a \cdot c \cdot \sin^2\beta}$ (6)
Triclinic	$\left[\dfrac{1}{1 - \cos^2\alpha - \cos^2\beta - \cos^2\gamma + 2\cos\alpha \cdot \cos\beta \cdot \cos\gamma} \cdot \left(\dfrac{h^2}{a^2} \cdot \sin^2\alpha + \dfrac{k^2}{b^2} \cdot \sin^2\beta + \dfrac{l^2}{c^2} \cdot \sin^2\gamma + \dfrac{2 \cdot h \cdot k}{a \cdot b} \cdot (\cos\alpha \cdot \cos\beta - \cos\gamma) + \dfrac{2 \cdot h \cdot l}{a \cdot c} \cdot (\cos\alpha \cdot \cos\gamma - \cos\beta) + \dfrac{2 \cdot k \cdot l}{b \cdot c} \cdot (\cos\beta \cdot \cos\gamma - \cos\alpha) \right) \right]$ (7)

Source: Adapted from Altomare 2008 [27] and Bail 2004 [42].

In order to find the correct indexing values for each Bragg reflection, the following indexing equation must be solved:

$$\frac{1}{d_{hkl}^2} = d_{hkl}^{*2} = h^2 \cdot a^{*2} + k^2 \cdot b^{*2} + l^2 \cdot c^{*2} + 2 \cdot h \cdot k \cdot a^* \cdot b^* \cdot \cos\gamma^* + 2 \cdot h \cdot l \cdot a^* \cdot c^* \cdot \cos\beta^* + 2 \cdot k \cdot l \cdot b^* \cdot c^* \cdot \cos\alpha^*. \tag{2.8}$$

This equation is the general form for the triclinic case and it simplifies in the case of higher symmetries. The equation follows from equation 7 in Table 2.1, if the relationships between direct and reciprocal lattice parameters from equations 1–6 in Table 2.2 are used. In theory, the general triclinic case needs at least six independent observed d_{hkl} values in order to index the diffraction pattern and to find the correct reciprocal lattice parameters a^*, b^*, c^*, α^*, β^*, γ^*. However, if there is a mentionable uncertainty of each of this six measured d_{hkl} values, it is almost impossible to find the correct reciprocal lattice parameters. Due to this reason, reliable indexing results of most of the computer algorithms used today can be obtained by providing more d_{hkl} values (usually between 20 and 30 values at all) than mathematically required. In the following text, a short overview of different historic and modern algorithms is given.

Table 2.2 Relations between the real/direct space lattice parameters and the reciprocal space lattice parameters.

$$a^* = \frac{1}{V} \cdot |\vec{b} \times \vec{c}| = \frac{b \cdot c \cdot \sin \gamma}{V} \quad (8)$$

$$b^* = \frac{1}{V} \cdot |\vec{a} \times \vec{c}| = \frac{a \cdot c \cdot \sin \beta}{V} \quad (9)$$

$$c^* = \frac{1}{V} \cdot |\vec{a} \times \vec{b}| = \frac{a \cdot b \cdot \sin \alpha}{V} \quad (10)$$

$$\cos \alpha^* = \frac{\cos \beta \cdot \cos \gamma - \cos \alpha}{\sin \beta \cdot \sin \gamma} \quad (11)$$

$$\cos \beta^* = \frac{\cos \alpha \cdot \cos \gamma - \cos \beta}{\sin \alpha \cdot \sin \gamma} \quad (12)$$

$$\cos \gamma^* = \frac{\cos \alpha \cdot \cos \beta - \cos \gamma}{\sin \alpha \cdot \sin \beta} \quad (13)$$

$$V = a \cdot b \cdot c \sqrt{1 - \cos^2 \alpha - \cos^2 \beta - \cos^2 \gamma + 2 \cos \alpha \cdot \cos \beta \cdot \cos \gamma} \quad (14)$$

Source: Further relations can be found in Ref. [26].

In 1917, Runge proposed the first applicable approach for the systematic indexing of powder diffraction patterns [28]. The works of Ito in 1949 [29] and of de Wolff in 1957 [30] used similar concept. In 1969, Visser published a computer program (nowadays known as "ITO" programs) based on these concepts [31]. In general, the Runge-Ito-de Wolff method implemented by Visser is a zone[3] indexing algorithm that is very powerful for the indexing of powder diffraction patterns with lower symmetries [27,32]. Another algorithm of that time is the one proposed by Werner in 1964 [33]. This technique is a semiexhaustive trial-and-error method where the Miller indices for the observed diffraction lines are permuted [27,34]. The corresponding computer program is named "TREOR." Later in 1972, another computer algorithm, based on the successive dichotomy method, was developed by Louër and Louër [35]. The computer program was named "DICVOL" and was expanded to include monoclinic [36] and triclinic symmetries [37]. This method varies the cell parameters in direct space and tries to reduce the possible solution space [27,34]. In addition to these algorithms, meaningful figure of merits (FOM) were given by de Wolff in 1968 with the M_{20}-FOM [38] and by Smith and Snyder in 1979 with the F_N-FOM [39], which allowed the judgment of the quality of the calculated unit cell parameters.

Several other indexing programs were developed in recent time. Examples of indexing algorithms that use direct space approaches are, for instance, approaches with genetic algorithms [40] or Monte Carlo approaches such as the indexing by singular value decomposition [41] or the McMaille approach [42].

Additionally, some of the indexing programs used nowadays also suggest probable space groups by checking the pattern for systematically absent

---

3) A crystallographic zone is a family of planes that have parallel cutting edges. The direction of the cutting edges is known as zone axis.

reflections (there is no possibility to determine all space groups unambiguously as some of them obey the same extinction rules).

2.3
Modern Powder Diffractometers in the Laboratory and at the Synchrotron

The geometry of a common laboratory powder diffractometer is either of Bragg–Brentano (Figure 2.7) or Debye–Scherrer type (Figure 2.8). If high resolution is needed, monochromatic radiation is mandatory, which is commonly generated by a curved primary beam monochromator in Johann or Johansson geometry, typically reflecting pure $K_{\alpha 1}$ radiation of the anode target. The reflecting planes of most crystals used as primary beam monochromators are either (111) or (220) for silicon and germanium or (101) for quartz. Usual target materials of the X-ray tubes are copper ($\lambda = 1.5406$ Å), molybdenum ($\lambda = 0.70930$ Å), or silver ($\lambda = 0.559410$ Å). It should be noted that in contrast to parallel beam synchrotron radiation, for divergent beam geometry, the absolute resolution decreases with decreasing wavelength (increasing energy).

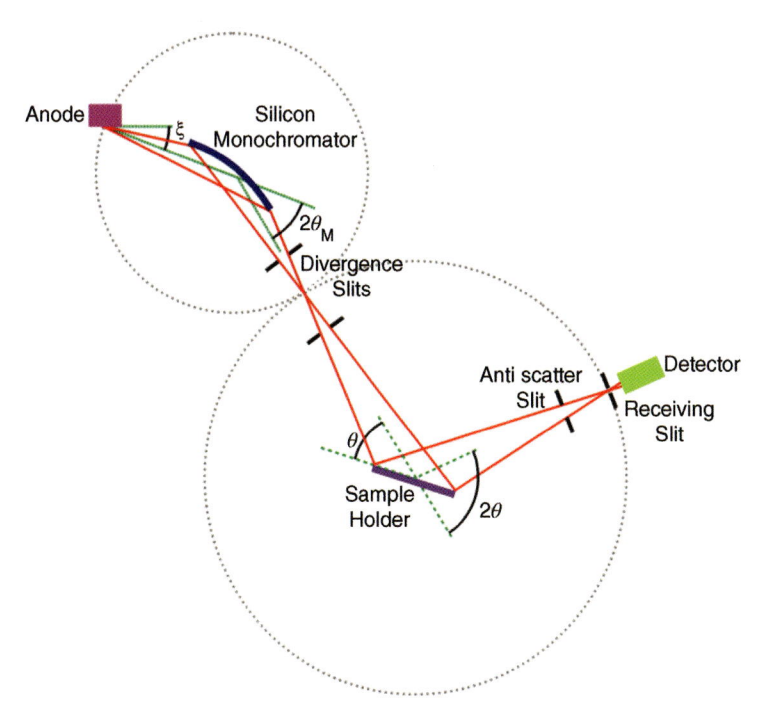

Figure 2.7 Sketch of Bragg–Brentano geometry of a laboratory powder diffractometer with a primary (before sample) beam curved silicon monochromator for separation of $K_{\alpha 1}$ and $K_{\alpha 2}$ radiation. (From Ref. [3].)

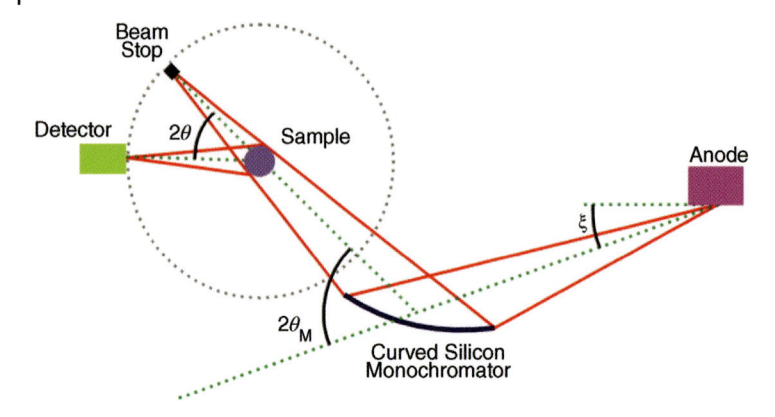

Figure 2.8 Parafocusing Debye–Scherrer geometry of a laboratory powder diffractometer with a primary (before sample) beam curved silicon monochromator for separation of $K_{\alpha 1}$ and $K_{\alpha 2}$ radiation. The focus is on the detector. (From Ref. [3].)

A typical powder pattern of a LaB_6 line profile standard recorded on a laboratory powder diffractometer in Debye–Scherrer geometry using $MoK_{\alpha 1}$ radiation from a curved Ge(220) primary beam monochromator and a silicon strip detector with an opening of 3.5° 2θ is shown in Figure 2.9. The entire reflection profile can be described by only four fundamental parameters related to the geometry of the diffractometer.

In general, instead of slow point detectors such as scintillation or proportional counters, modern one-dimensional position sensitive silicon strip detectors

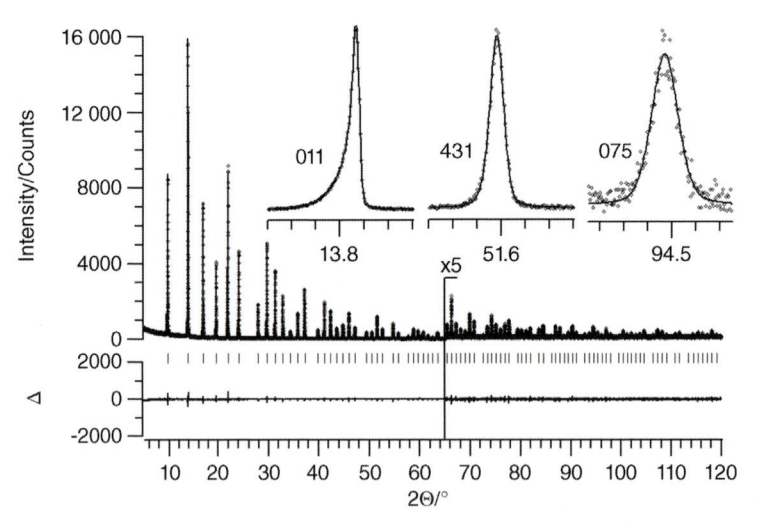

Figure 2.9 Whole powder pattern fit of the LaB_6 standard measured with a wavelength of $\lambda = 0.70\,930$ Å in the laboratory. The high angle part starting at 65° 2θ is magnified by a factor of 5 for clarity. Three reflections from different parts of the powder pattern are magnified in the *inset*. (Reproduced with permission from Ref. [43]. Copyright 2013, John Wiley & Sons.)

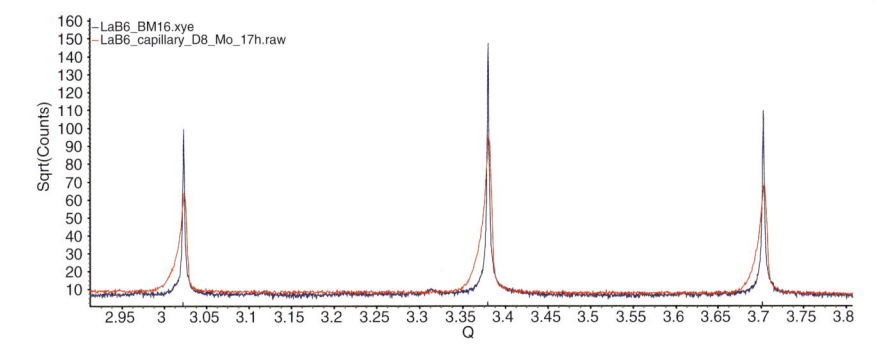

Figure 2.10 Comparison of two high-resolution powder patterns of the LaB$_6$ standard measured in Debye–Scherrer geometry with a wavelength of $\lambda = 0.70\,930$ Å in the laboratory (orange) and with a wavelength of $\lambda = 0.85$ Å at the synchrotron (BM16, ESRF) (blue). The x-axis is therefore plotted in $Q = 2\pi/d$ for better comparison.

usually covering 3–12° 2θ are widely used in the laboratory, leading to an intensity gain of several orders of magnitude while keeping the high resolution. The width of the individual stripes can be considered a kind of receiving slit width, thus being a main contributor to the instrumental resolution. The spatial resolution of a modern high-resolution powder diffractometer is on the order of $\Delta d/d \approx 3 \times 10^{-3}$ for copper radiation (about 8 keV). If better resolution and higher intensity is needed, synchrotron powder diffraction is the method of choice.

The story of synchrotron X-ray powder diffraction (SXPD) began already in the 1970s, when, for example, in 1976 B. Buras, J. Staun Olsen, and L. Gerward reported their first powder diffraction results that they obtained using an energy-dispersive powder diffractometer set up at the Deutsches Elektronen-Synchrotron (DESY) in Hamburg, Germany [44]. Nowadays the fourth generation of synchrotron sources is available at spatial resolution on the order of $\Delta d/d \approx 1.5 \times 10^{-4}$ at 10 keV (for example, at the ID22 (former ID31) beamline at ESRF/Grenoble or at the I11 beamline at Diamond/Oxfordshire) (Figure 2.10).

A schematic view of a modern synchrotron is given in Figure 2.11a. To create synchrotron radiation, an electron gun produces high-energy electrons (typically 90 keV), which are then accelerated to approximately 100 MeV by a linear accelerator (LINAC) (1) before they are injected into the booster ring (2) where they are further accelerated to 2–6 GeV and conditioned. Bunches of electrons are then injected to the storage ring. The electrons travel now almost at the speed of light and show relativistic behavior. X-ray photons are then mainly produced by bending magnets, superconducting wigglers, or undulators (Figure 2.11b).

Highly collimated X-rays are sent to optical (6) and experimental (7) hutches to conduct experiments. While the beam current and therefore the number of emitted photons of earlier synchrotrons were slowly decaying with time, more modern synchrotron sources keep the beam current constant by top up filling of

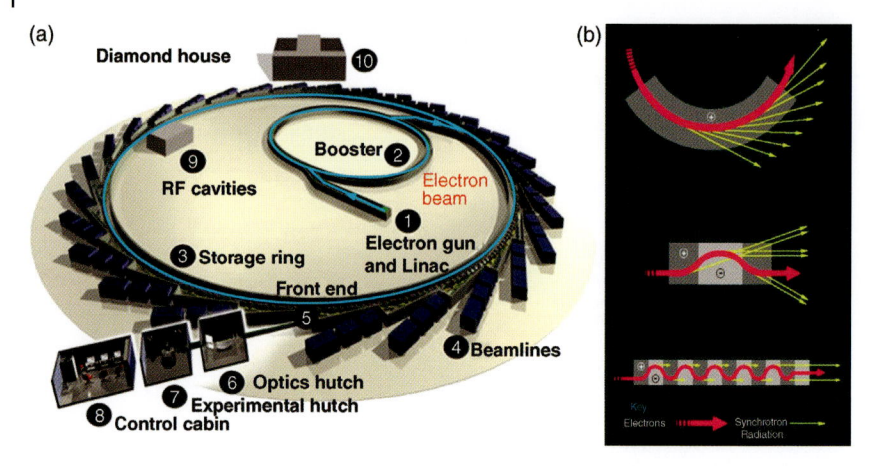

Figure 2.11 (a) Schematic view of a modern synchrotron (Diamond, RAL). (b) Photon production by a bending magnet (*top*), superconducting wiggler (*middle*), and a MPW or undulator (*bottom*). (Courtesy of Chiu Tang, I11, Diamond.)

the storage ring. Typical characteristic numbers for an undulator beam are vertical divergence ~25 mrad (~0.001°), horizontal divergence ~100 mrad (~0.006°), photon beam at 48 m from source ~0.5 × 2.5 mm² ($E = 15$ keV, $\lambda = 0.827$ Å).

Among the many advantages of synchrotron radiation are the high brilliance, high intensity, high collimation of the X-ray beam, and the tunability of the wavelength, making it extremely useful for powder diffraction experiments. The high intensity and resolution allow for high-quality data being collected in a short amount of time with a high spatial resolution and good statistics, while the tunability of the wavelength can either be used to avoid absorption edges or to do resonant scattering experiments at absorption edges. Nonambient sample environments, for example, cryogenic systems and furnaces, can easily be accommodated. Most SXPD experiments are conducted in Debye–Scherrer geometry (Figure 2.12). The reasons are manifold. With a spinning capillary, preferred orientation is greatly reduced, and diffracted X-rays can be detected with zero-, one-, or two-dimensional detectors, depending on the task.

In case of a zero-dimensional point detector, the use of an analyzer crystal (Figure 2.12a) or even more effective, a multianalyzer stage (Figure 2.12b) leads to narrow peaks (sample limited) and accurate peak positions and stringently defines a true 2θ angle rather than infers 2θ from the position of a slit (Figure 2.13). Therefore, peak positions are insensitive to misalignment, transparency, and effects due to specimen size or shape. In addition, the peak widths are independent of any $\theta/2\theta$ parafocusing condition and non-Bragg scattering, such as fluorescence, Compton scattering, and so on, is suppressed. Radiation damage can be reduced by moving the capillary perpendicular to the beam during measurement. It should be noted that even for multianalyzer stages, scanning is relatively slow (typically on the order of minutes for a full powder pattern) and large crystallites cause intensity errors.

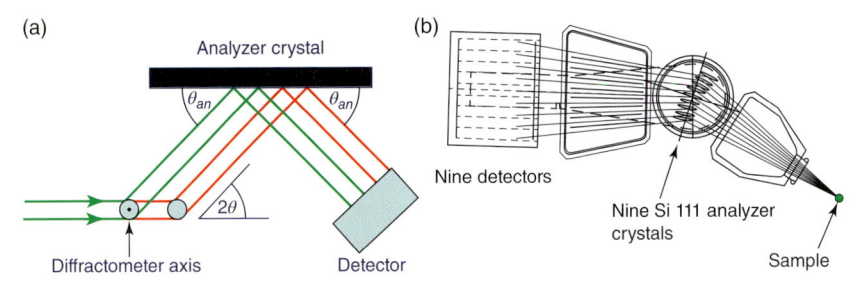

2.3 Modern Powder Diffractometers in the Laboratory and at the Synchrotron

Figure 2.12 (a) Schematic view of an analyzer crystal in Bragg–Brentano geometry. (b) Multianalyzer stage with nine silicon (111) analyzer crystals for Debye–Scherrer geometry. (Courtesy of Andy Fitch, ID22, ESRF.)

If faster measurements of full powder patterns in the second regime are required, curved one-dimensional Si-strip position-sensitive detectors (PSDs), which are available for soft and intermediate energies, are a good choice (e.g., Mythen 1 K with 1280 channels, 50 μm step, 4.83° per unit, readout time of 250 μs).

Alternatively, to obtain a full Q range in one shot (e.g., for fast pair distribution function (PDF) analysis) or to follow fast chemical reactions, large flat-panel digital 2D PSD that can fast record entire Debye–Scherrer rings or arc sections of those are common (Figure 2.14). In particular, 2D detectors based on amorphous silicon show a high quantum yield for high energies ≥60 keV. For the latter, fast powder diffraction scans down to a millisecond and below are possible. Since analyzer crystals cannot be used for this setup, the best resolution with a 1D or 2D PSD is obtained by focusing the beam onto the detector. Traditionally, this is achieved via a curved metal-coated mirror (set at grazing incidence) with an undulator, refractive lenses can be employed because the high horizontal collimation directs much of the beam into the lens's limited aperture.

Capillaries are susceptible to 2θ-dependent absorption μ. In order to keep absorption to a manageable level, the capillary diameter ($2r$) should be adjusted so that $\mu r < 1.5$. The effect is usually negligible at energies higher than 40 keV.

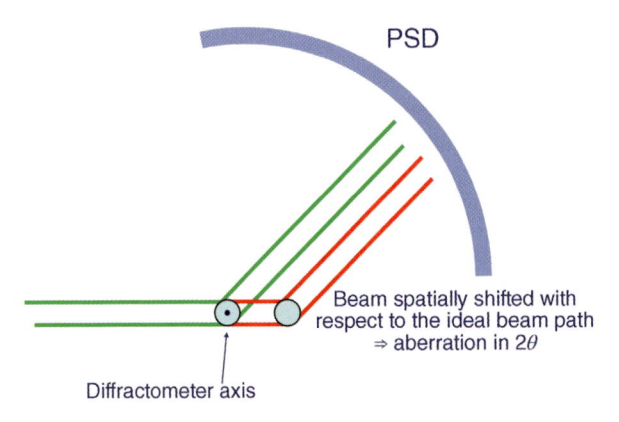

Figure 2.13 Angular aberration for Debye–Scherrer geometry without the use of an analyzer crystal. (Courtesy of Andy Fitch, ID22, ESRF.)

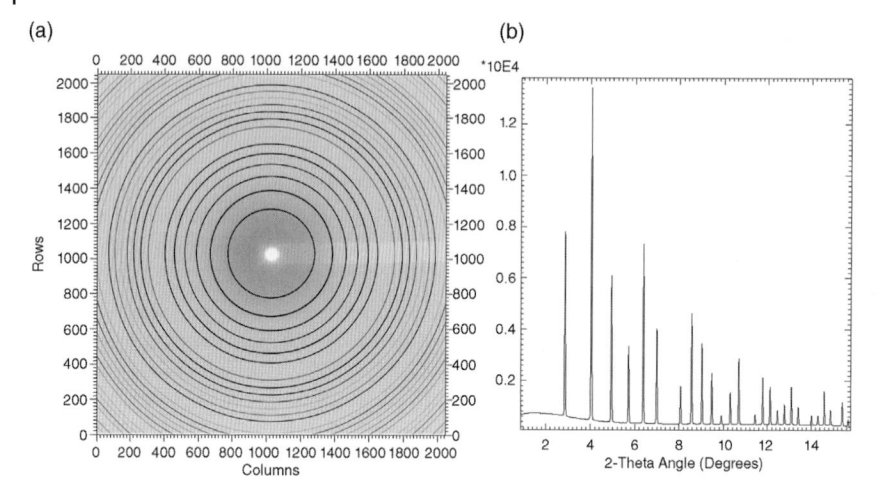

Figure 2.14 LaB$_6$ at 60 keV measured with a flat-panel digital amorphous silicon 2D detector (PerkinElmer XRD1621). (a) Raw data showing the Debye–Scherrer rings. (b) Integrated powder diffraction pattern.

Today, the investigation of powders by synchrotron radiation with powder diffraction methods has become a standard procedure used by hundreds of scientists every day all over the world. The data acquisition is often highly automated, only interrupted by changing environmental parameters or by changing of samples (some synchrotron beamlines provide automatic sample changing by robots). In a few hours it is possible to collect dozens or even hundreds of powder patterns. This is especially true in the case of the most modern synchrotrons where one can collect a powder pattern almost every second, which produces a huge number of data sets, requiring sophisticated software for data analysis.

2.4
The Rietveld Method

The method of choice in order to fit an entire one-dimensional powder diffraction pattern by the refinement of crystal structures is a structure-based whole powder pattern fitting (WPPF) method developed by Hugo Rietveld in the late 1960s. This method, nowadays known as the Rietveld refinement method or just Rietveld method, uses a least-squares algorithm in order to refine a calculated powder diffraction pattern against an experimentally observed powder diffraction pattern.

Rietveld made the first successful attempt to use a least-squares algorithm to refine directly the background-corrected integrated intensities of X-ray and neutron powder patterns [45]. Although this first attempt was not well recognized by the crystallographic community, he continued his work and 1 year later he published the first WPPF analysis of tungsten trioxide, where he could show that even a severe peak overlap of reflections can be treated by his refinement algorithm [46]. Finally, in 1969, he published one of the most cited scientific articles in crystallography, where he demonstrated that with a powerful computer machine, a WPPF can be performed with almost all important parameters that determine a crystal structure such as lattice parameters, atomic positions, components of a magnetic vector, and so on [1] (for a full list, see also Refs [2,3]). However, within the first 8 years, the Rietveld method was solely applied to neutron powder diffraction due to the simplicity of the obtained peak shapes (neutron powder diffraction peak shapes can be satisfactorily modeled by the assumption of a simple Gaussian peak shape). After this period, in 1977, three different groups published nearly simultaneously Rietveld refined X-ray powder diffraction patterns by the application of more sophisticated peak shapes [47–49].

Although the basis of the Rietveld method was laid almost 50 years ago, the mathematical description remains unchanged. In general, a refinement of a crystal structure by a least-squares minimization between the calculated and observed powder diffraction patterns can be done by the following equation [1]:

$$\text{Min} = \sum_{2\theta_i} (Y_{\text{obs}}(2\theta_i) - Y_{\text{calc}}(2\theta_i))^2, \qquad (2.9)$$

where Min is the desired global minimum of the refinement, Y_{obs} is the experimentally observed powder diffraction pattern, and Y_{calc} is the calculated diffraction pattern for which a detailed description is given below. Normally, the 2θ space is a continuum, but due to the data collection procedure, the 2θ space is discretized, which is denoted by the running index i whose integer value reflects the current data point. Usually Eq. (2.9) is modified with a weighting factor w in order to guarantee that peaks with a high intensity are not overestimated [1]:

$$\text{Min} = \sum_{2\theta_i} w(2\theta_i)(Y_{\text{obs}}(2\theta_i) - Y_{\text{calc}}(2\theta_i))^2, \qquad (2.10)$$

where the weighting w is, for example, given by $w(2\theta_i) = 1/(Y_{\text{obs}}(2\theta_i))$, assuming Poisson statistics using scintillation or proportional counters [2]. Besides the weighting with the inverse observed intensity, other weighting schemes can be chosen. For instance, in the TOPAS software [50] and in some textbooks [43], a weighting factor w with the inverse of the squared variance of the observed intensity is chosen (while it is assumed that all covariance between different observed intensities are zero [43]):

$$w(2\theta_i) = \frac{1}{(\sigma(Y_{\text{obs}}(2\theta_i)))^2}, \qquad (2.11)$$

where $\sigma(Y_{\text{obs}}(2\theta_i))$ denotes the variance (which is often simply the square root of the estimated standard deviation/error of the measurement) of the experimentally observed intensity. Please note that Rietveld already stated in his publication from 1969 that if the variance of the observed intensity from counting statistics is equal to the observed intensity, then the weighting scheme becomes equal to the weighting with the inverse observed intensity (if not, the squared variance is used as above in Eq. (2.11) [1]).

The above-described calculated intensity is formally given by the following equation:

$$Y_{\text{obs}}(2\theta_i) = \sum_p S_p \sum_{\{h,k,l\}_p} \left(\left| F_{\text{calc}}\left(\{h,k,l\}_p\right) \right|^2 \cdot \Phi_{\{h,k,l\}_p}\left(2\theta_i - 2\theta_{\{h,k,l\}_p}\right) \cdot \text{Corr}_{\{h,k,l\}_p}(2\theta_i) \right) + \text{Bkg}(2\theta_i), \tag{2.12}$$

where S_p is a phase p dependent scale factor, $\{h,k,l\}_p$ denotes a tupel of three Miller indices that depend on the phase p, $F_{\text{calc}}(\{h,k,l\}_p)$ is the structure factor of a certain phase-dependent Bragg reflection, $\Phi_{\{h,k,l\}_p}(2\theta_i - 2\theta_{\{h,k,l\}_p})$ is the normalized peak profile at the peak position $2\theta_{\{h,k,l\}_p}$, $\text{Corr}_{\{h,k,l\}_p}(2\theta_i)$ is a product of different correction functions, which depend on the Bragg reflection and/or the discrete $2\theta_i$ value, and finally $\text{Bkg}(2\theta_i)$ is the background, which is normally fitted by point interpolation or polynomials.

In addition to the minimization Eqs. (2.10) and (2.11), the least-squares algorithm requires residual values (R values) in order to judge the quality of the refinement. These R values, also known as agreement factors, are often defined differently:

$$R_p = \frac{\sum_{2\theta_i} |Y_{\text{obs}}(2\theta_i) - Y_{\text{calc}}(2\theta_i)|}{\sum_{2\theta_i} Y_{\text{obs}}(2\theta_i)}, \tag{2.13}$$

$$R_{\text{wp}} = \sqrt{\frac{\sum_{2\theta_i} w(2\theta_i)(Y_{\text{obs}}(2\theta_i) - Y_{\text{calc}}(2\theta_i))^2}{\sum_{2\theta_i} w(2\theta_i)(Y_{\text{obs}}(2\theta_i))^2}}, \tag{2.14}$$

$$R_{\text{exp}} = \sqrt{\frac{M - P}{\sum_{2\theta_i} w(2\theta_i)(Y_{\text{obs}}(2\theta_i))^2}}, \tag{2.15}$$

$$\text{GOF} = \chi^2 = \frac{R_{\text{wp}}}{R_{\text{exp}}} = \sqrt{\frac{\sum_{2\theta_i} w(2\theta_i)(Y_{\text{obs}}(2\theta_i) - Y_{\text{calc}}(2\theta_i))^2}{M - P}}, \tag{2.16}$$

$$R_B = \frac{\sum_{\{h,k,l\}_p} \left| I_{\text{obs}}\left(\{h,k,l\}_p\right) - I_{\text{calc}}\left(\{h,k,l\}_p\right) \right|}{\sum_{\{h,k,l\}_p} I_{\text{obs}}\left(\{h,k,l\}_p\right)}. \tag{2.17}$$

Here R_p is the profile residual, R_{wp} is the weighted profile residual, R_{exp} is the expected residual with M as the number of data points and P as the number of the refined parameters, GOF is the goodness of fit, and R_B is the Bragg residual. Note that the GOF given here is the one that is defined in the TOPAS software [50]. In textbooks [2,51] and in the GSAS software [52], the GOF is defined as follows:

$$\text{GOF} = \chi^2 = \left(\frac{R_{\text{wp}}}{R_{\text{exp}}}\right)^2 = \frac{\sum_{2\theta_i} w(2\theta_i)(Y_{\text{obs}}(2\theta_i) - Y_{\text{calc}}(2\theta_i))^2}{M - P}. \tag{2.18}$$

In addition to the above given definitions for Eqs. (2.13)–(2.15), the R_p, R_{wp}, and R_{exp} can be defined in a background-corrected version if a low peak-to-background ratio makes such a definition necessary [50].

2.5
Sequential and Parametric Rietveld Refinements

Recent progress in instrumentation allows for the collection of huge amounts of one- or two-dimensional powder patterns, especially when *in situ* powder diffraction experiments are carried out. In such cases the data sets depend on external variables such as temperature, pressure, time, and so on. Subsequent treatment of these large numbers of powder patterns by WPPF methods is often very time-consuming, but in some cases the data handling can be simplified.

For instance, if the step width of the external variable is small, the refinement results of one diffraction pattern can be used as starting values for the consecutive diffraction pattern, provided that no dramatic changes in the refined values will occur (which can happen if a phase transition sets in). This method, using the refinement results of a powder pattern for the consecutive one, is called sequential refinement or in the case of using the Rietveld method, sequential Rietveld refinement.

In 2007, Stinton and Evans published the first successful attempt of a parametric Rietveld refinement (also called surface refinement) [53], where different powder diffraction patterns, which depend on common variables, are treated simultaneously. In this approach, one or several parameters of different powder diffraction patterns are constrained by one or more equations where the independent variable is the external variable of the *in situ* measurement. It is obvious that any type of equation can be chosen, as long as the equation is physically or empirically connected with the information stored in a single or over a group of powder diffraction patterns. In numerous publications, it could be shown that parametric Rietveld refinements have the potential to reduce the correlation between parameters, to reduce the final standard uncertainties, avoid false minima in individual powder diffraction pattern refinements, and most importantly allow a direct modeling of parameters that are normally not part of the refinement as they are first introduced by the applied equations [53]. Another welcome effect is the reduction of the total number of refined parameters, which is believed to give a further stabilization of the refinement in a least-squares minimization process.

2.6
Whole Powder Pattern Decomposition

Derived from the Rietveld method are the whole powder pattern decomposition (WPPD) methods according to Pawley [54] and Le Bail *et al.* [55], which do not require the knowledge of a crystal structure. Instead of using the full information of the crystal structure as it is used in the Rietveld WPPF, the WPPD methods require only the approximate knowledge of the lattice parameters and the space group. Other parameters such as the peak shape parameters or the zero-shift can be obtained by the refinement as they are independent from the crystal structure. The squared structure factors (reflection intensities) are not calculated from atomic positions but freely refined.

In the Pawley WPPD method, the peak intensities are individually refined, which allows in general, two closely overlapping peaks: one peak can become positive, whereas the other peak can become negative.[4] In such a case, the correlation matrix can be used as a measure for the accuracy of individual reflection intensities and only the sum of the intensities of the two peaks would be subjected to, for example, the structure determination process.

The Le Bail WPPD method uses a different approach with respect to the Pawley method to obtain intensities. Instead of performing a direct least-squares refinement of the intensities as the Pawley method does, the Le Bail method uses an iterative process to keep the intensities positive. After each least-squares refinement cycle, the obtained intensities from the Rietveld formula are used as squared structure factor amplitudes for the next refinement cycle. This process is continued iteratively, until the refinement converges. In general, this is an adequate method to keep the intensities in most cases positive [2,43].

2.7
Concept of Symmetry Modes

The concept of symmetry modes, also known as distortion modes, is an intriguingly natural concept in order to describe structural, occupational, or magnetic changes in a crystal structure, as it connects on a mathematical basis the higher symmetry (HS) of an undistorted crystal structure with the lower symmetry (LS) of the distorted version, as long as the space group of the distorted crystal structure with the lower symmetry is a subgroup of the parental space group of the undistorted crystal structure.

In crystallography, a group–subgroup relation between space groups always implies that a phase transition of a particular crystal structure between two of these groups can be in principle regarded as a quasi-continuous transformation, independent of the fact whether the real-phase transition is of first or second order. Due to this fact, such a quasi-continuous transformation can be described by a set of distortion vectors (also known as polarization vectors) and

---

4) Nowadays, most of the refinement software used disallow negative peak intensities in a Pawley as well as in a Le Bail refinement.

2.7 Concept of Symmetry Modes

corresponding amplitudes, which are responsible for the distortion of the entire crystal structure.[5] Although the description of a crystal structure change by distortion vectors sounds quite easy, the calculation of a specific distortion vector requires a profound knowledge of group theory and representation theory, therefore only a superficial explanation will be given here.

From diffraction experiments, it is known that each quasi-continuous phase transformation of a crystal structure to a lower symmetry and therefore to a subgroup will generate additional Bragg reflections, which are sometimes also called superstructure reflections. These superstructure reflections will appear at special k-points in the first Brillouin zone[6] of the parent crystal structure and are therefore connected with one or more propagation vectors $\vec{k}$, which point from the gamma point of the Brillouin zone to these k-points[7] [56]. For each propagation vector $\vec{k}$, it is possible to find a set of symmetry operations from the parental space group, for which the rotational part of these symmetry operators leave the propagation vector invariant. This set of symmetry operators is then called the group of the wave vector or the propagation vector group, or simply the "little group" [57]. By mapping, these symmetry operations of the little group can be linked to a finite number of irreducible representations [58]. Interestingly, mathematically it can be shown that each irreducible representation stands for a set of parental symmetry operations that can be broken[8] [58]. Furthermore, if a phase transition breaks only symmetries of a specific irreducible representation, then this irreducible representation is linked to an order parameter[9],[10] [58]. Besides the connection with order parameters, each irreducible representation is also associated with basis vectors (mathematical term for the terms polarization vectors, distortion vectors, distortion modes, or symmetry modes). If now a certain parameter is changed in the crystal structure upon crossing the phase transition, then this can be expressed as the sum of different basis or distortion vectors [62]:

$$r_{LS} = r_{HS} + \sum_m A_m \cdot \varepsilon_m, \tag{2.19}$$

where r_{LS} is the parameter value in the LS phase, r_{HS} is the parameter value in the HS phase, and A_m is the amplitude of a certain polarization vector ε_m.

---

5) This also explains why phase transitions of first order can be equally described with this calculus, as the amplitude of a certain distortion, which shows first-order behavior, will be simply discontinuous and makes a jump at the critical phase transition point, whereas the distortion vector will not be affected. This is also the reason why the author is calling this transformation as quasi-continuous, as a first-order phase transition of a crystal structure that has group–subgroup relationships can still be described with the same calculus.
6) The first Brillouin zone in physics is defined as the primitive Wigner–Seitz cell in reciprocal space.
7) The simplest case is that the superstructure reflections appear only at one k-point and are therefore connected only with a single propagation vector $\vec{k}$.
8) In crystallographic terms, this means that each irreducible representation has the ability to lead to at least one or more different subgroups of the parental space group.
9) This implies that a phase transition that breaks multiple irreducible representations can have several order parameters.
10) The term "order parameter" refers to the order parameter from Landau theory [59–61].

Besides the simply explained concept of the symmetry modes, there are a lot of implications that arise from the mathematical basis; for instance, for a displacive–structural phase transition, the number of distortion vectors is equal to the number of variable atomic coordinate parameters [58]. Apart from that example, other effects as well as a rigorous calculus can be found in the literature given, for instance, by Miller and Love [63], Stokes et al. [64], Stokes and Hatch [65], Dove [61], Hatch and Stokes [66], Campbell et al. [56,58], Orobengoa et al. [67], and Perez-Mato et al. [62].

Nowadays, for the exploration of the different possible crystal structures of a given parent structure, two very powerful tools, which are available online, can be used: ISODISTORT [56] and AMPLIMODES [67]. In ISODISTORT, different modes can be used if the crystallographic information of a crystal structure is provided. This includes, for instance, a search of possible subgroups by a given propagation vector $\vec{k}$ or the decomposition of a given crystal structure of the subgroup into symmetry modes. In contrast to ISODISTORT, AMPLIMODES provides only the possibility to perform the decomposition of given crystal structures of a group–subgroup pair into symmetry modes.

2.8
Structure Determination from Powders

All algorithms for crystal structure determination from powder diffraction data use single peak intensities, the entire powder pattern, or a combination of both. The conventional single-crystal reciprocal space techniques often fail due to an unfavorable ratio between available observations and structural parameters. This situation changed dramatically with the introduction of the "charge flipping" technique (Figure 2.15) [68,69] and the development of global optimization methods in direct space, of which the simulated annealing technique (Figure 2.16) [70,71] is the most prominent representative. In particular, the introduction of chemical knowledge into the structure determination process using rigid bodies [72] or the known connectivity of molecular compounds concerning bond lengths and angles strongly reduces the number of necessary parameters. In other words, instead of three positional parameters for every single atoms, only the external (and few internal) degrees of freedom of groups of atoms need to be determined. It is this reduction of structural complexity that makes the powder method a real alternative to single-crystal analysis.

The basic idea of charge flipping, which is a four-stage cyclic process, is outlined in Figure 2.15. At the beginning, random phases are assigned to the structure factor amplitudes derived from the observed intensities, allowing the plot of a random charge density map ρ. In the first stage, the signs of all electron densities in the map below a small threshold value $\delta>0$ are reversed, which results in a modified electron density map g. In the second stage, a set G of temporary structure factors is calculated by a fast Fourier transformation (FFT). In the third stage, the phases of these Gs are combined with the moduli of the original

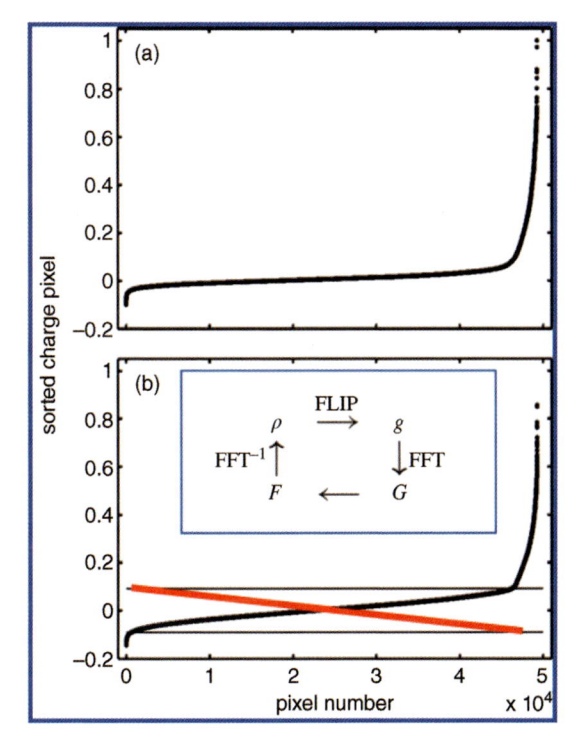

Figure 2.15 Charge flipping is an iterative structure solution method based on the alternating modification of an electron density in direct space and structure factors in reciprocal space.

structure factors, leading to a new set of structure factors F, whose phases are no longer random. In the fourth stage, a new charge density map ρ is calculated from the new Fs using the method of inverse FFT. Now the iteration restarts with stage one until convergence is reached. Although the charge flipping method can solve crystal structures without knowledge about symmetry and cell content, this type of information speeds up the process, in particular in case of powder diffraction. A detailed explanation of the method is given in Ref. [69].

There is no reason other than computational efficiency why the minimization algorithm used in the Rietveld method could not be a more robust global optimizer and this capability is now starting to be implemented in modern Rietveld codes. The most common and most easily implemented global optimizer, although one of the least efficient, is the Metropolis, or *simulated annealing* (SA) algorithm [73]. The most common implementation is actually as a "Regional" optimizer where the updates to parameters such as atomic position are constrained not to be too far from the previous values in such a way that the algorithm makes a random walk through the parameter space. This algorithm can get out of a local minimum by walking uphill since changes to the parameters that produce a worse agreement may be accepted according to a Boltzmann criterion, $\exp(-(\Delta R/kT))$. The temperature in this expression is a fictitious

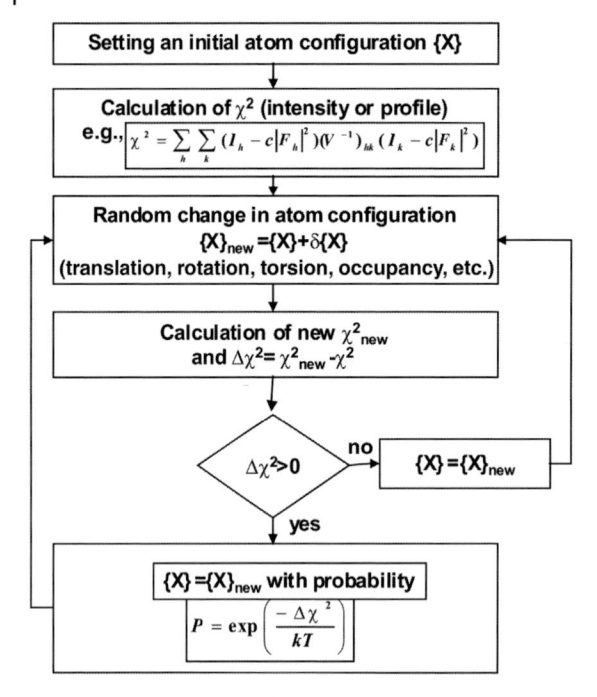

```
                    ┌─────────────────────────────────────┐
                    │ Setting an initial atom configuration {X} │
                    └─────────────────────────────────────┘
                                       ↓
         ┌──────────────────────────────────────────────────────┐
         │   Calculation of χ² (intensity or profile)           │
         │ e.g., $\chi^2 = \sum_h \sum_k (I_h - c|F_h|^2)(V^{-1})_{hk}(I_k - c|F_k|^2)$ │
         └──────────────────────────────────────────────────────┘
                                       ↓
         ┌──────────────────────────────────────────────────────┐
         │       Random change in atom configuration            │
    →    │              {X}_new = {X} + δ{X}                    │  ←
         │  (translation, rotation, torsion, occupancy, etc.)   │
         └──────────────────────────────────────────────────────┘
                                       ↓
                    ┌─────────────────────────────────────┐
                    │   Calculation of new χ²_new         │
                    │   and Δχ² = χ²_new − χ²             │
                    └─────────────────────────────────────┘
                                       ↓
                              ╱ Δχ² > 0 ╲   no
                              ╲         ╱ ─────→  {X} = {X}_new
                                  ↓ yes
                    ┌─────────────────────────────────────┐
                    │  {X} = {X}_new with probability     │
                    │  $P = \exp\left(\dfrac{-\Delta\chi^2}{kT}\right)$ │
                    └─────────────────────────────────────┘
```

Figure 2.16 Flow diagram of a simulated annealing procedure used for structure determination from powder diffraction data. (Reproduced with permission from Ref. [43]. Copyright 2013, John Wiley & Sons.)

temperature (i.e., it does not refer to any real temperature) and ΔR is the change in the agreement produced by the trial update. The temperature plays the role of tuning the probability of accepting a bad move. It is initially chosen to have a high value, giving a high probability of escaping a minimum and allowing the algorithm to explore more of the parameter space. Later during the run the temperature is lowered, trapping the solution into successively finer valleys in the parameter space until it settles into (hopefully) the global minimum (Figure 2.17). The calculation of R can be based on the entire profile, or on integrated intensities. For the latter, the correlation between partially or fully overlapping reflections must be taken into account (as outlined in Figure 2.16).

A flow diagram of a typical SA algorithm as used for structure determination from powder diffraction data is shown in Figure 2.16. Parameters that can be varied during the SA runs include internal and external degrees of freedom such as translations (fractional coordinates or rigid body locations), rotations (Cartesian angles, Eulerian angles, or quaternions, describing the orientation of molecular entities), torsion angles, fractional occupancies, temperature factors, and so on. Figure 2.17 shows a typical simulated annealing run in which the integrated $R = \chi^2$ value falls dramatically in the first few thousand moves, indicating that the scattering is dominated by the positioning of heavier atoms or globular molecules. Several million trial structures are usually generated before a minimum

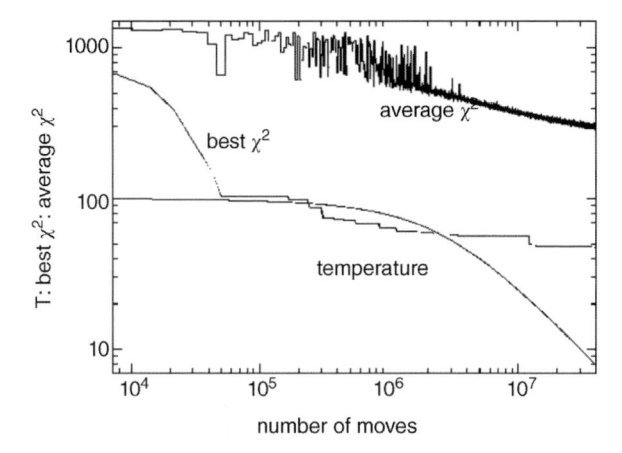

Figure 2.17 (χ^2 (cost function) and "temperature" in dependence on the number of moves during simulated annealing run. (Reproduced with permission from Ref. [43]. Copyright 2013, John Wiley & Sons.)

can be reached. At the end of the simulated annealing run, Rietveld refinement is used to find more rapidly the bottom of the global minimum valley.

Usually no special algorithms are employed to prevent close contact of atoms or molecules during the global optimization procedure. In general, these have not been found necessary, as the fit to the structure factors alone quickly moves the molecules to regions of the unit cell where they do not grossly overlap with neighboring molecules. A consecutive Rietveld refinement in which only the scale and overall temperature factors are refined will immediately show if further refinement of bond lengths and bond angles is necessary or not. Since unconstrained refinement often results in severe distortions from the ideal molecular geometry, either rigid bodies or soft constraints on bond lengths, planarity of flat groups, and bond angles can be used to stabilize the refinement. Another advantage of the simulated annealing technique is that hydrogen atoms can often be included at calculated positions from the beginning if their relative position with respect to other atoms is fixed, as it is often the case for molecular structures.

In particular, for inorganic crystal structures, the identification of special positions or the merging of defined rigid bodies is useful during the final stages of structure solution. This can be accomplished by a so-called occupancy merge procedure [74]. Here, the site occupancy of the sites is rewritten in terms of their fractional atomic coordinates. The sites are thought of as spheres with a radius r. In this way any number of sites can be merged when their distances are less than $2r$. As an example, the crystal structure solution of minium (Pb_3O_4) is shown in Figure 2.18. In this example, special positions are identified when two oxygen or lead atoms approach within a distance less than the sum of their respective merging radii, which is estimated as 0.7 Å. The occupancies of the sites then become $1/(1 + \text{intersection fractional volumes})$.

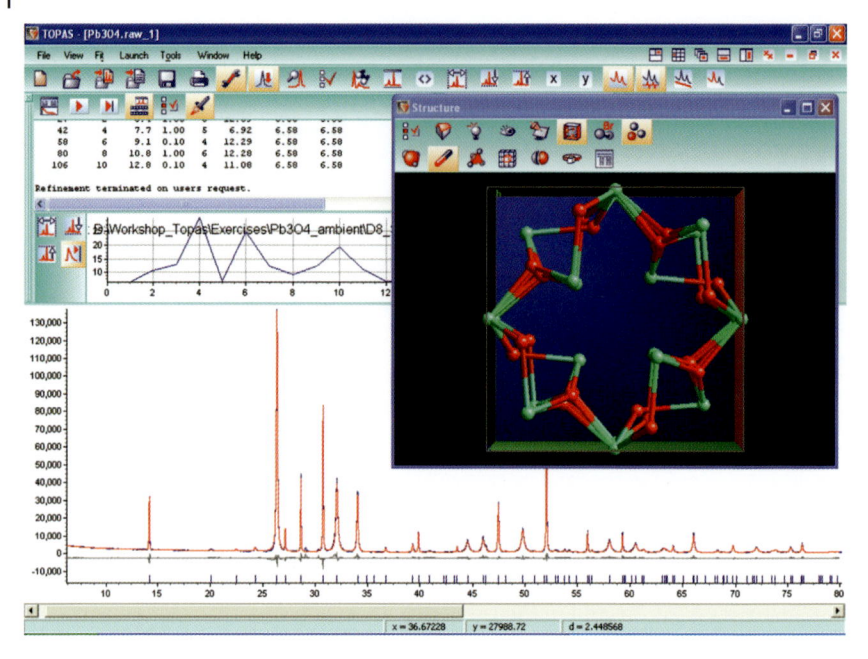

Figure 2.18 Screenshot (using the program TOPAS 4.1 (2007)) of a simulated annealing run on Pb_3O_4 measured with a laboratory diffractometer in Bragg–Brentano geometry. (Reproduced with permission from Ref. [43]. Copyright 2013, John Wiley & Sons.)

2.9
Case Studies

2.9.1
Case Study 1: Crystal Structure Determination of Codeine Phosphate Sesquihydrate by a Direct Space Global Optimization Method of Simulated Annealing Using Rigid Bodies [75]

Codeine phosphate sesquihydrate (COP-S), an important hydrate salt of codeine, is a commercialized, active pharmaceutical ingredient (API), extensively used as a narcotic, analgesic, antitussive, and antidiarrheal medicine. Although this form was known for a long time, its crystal structure was only recently reported [75]. The reason for the scarce structural knowledge can be traced to the crystallization behavior of COP-S. Namely, this compound crystallizes as a polycrystalline bulk with micrometer-sized crystallites. In the following text, the process of crystal structure determination, using laboratory XRPD data, is comprehensively explained, with the hope to attract the attention of, *inter alia*, organic solid-state chemists and pharmacists interested in the solid-state properties of APIs.

A sample of COP-S was purchased by Alkaloid AD Skopje. As seen from scanning electron microscopy (SEM) images (Figure 2.19a), COP-S usually crystallizes in the form of plate- shaped crystallites, several micrometers in length and

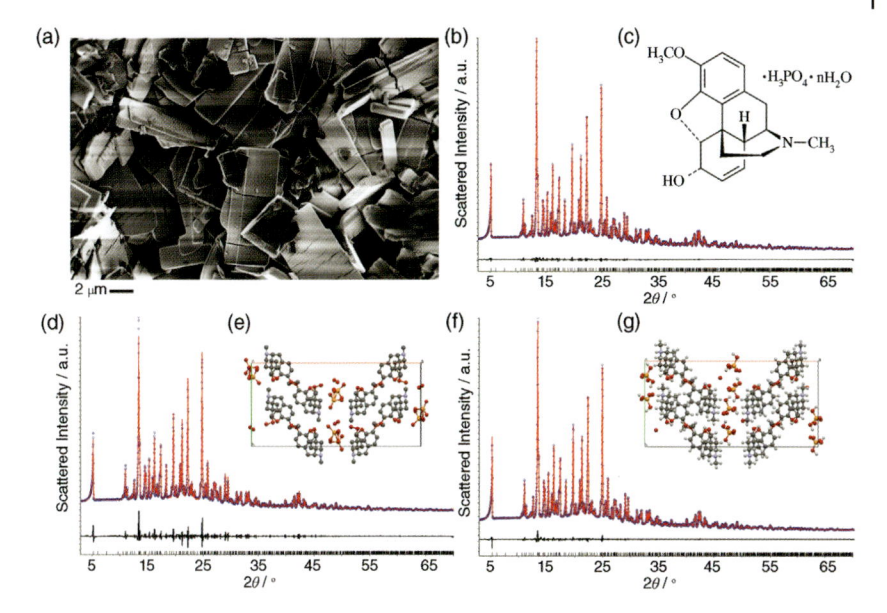

Figure 2.19 SEM image of the COP-S powder sample (a). Pawley fit of the diffraction pattern (b). Molecular structure of codeine phosphate hydrate (c). Measured and simulated patterns (d) of the crystal structure model (e) obtained by the SA method. Rietveld plot (f) of the COP-S crystal structure (g). The measured pattern is presented with blue circles, the calculated patterns with red lines, and the difference curves with black lines.

width, with thickness of less than a micrometer. Prior to data collection, the sample was carefully grinded in a pestle in order to reduce the preferred orientation of the plate-like particles. The Debye–Scherrer geometry was used to further reduce the preferred orientation, with the sample placed in a borosilicate capillary (0.5 mm diameter, Hilgenberg glass No. 50). For better particle statistics, the capillary was spun during data collection.

High-resolution XRPD pattern was collected on a laboratory powder diffractometer *Stoe Stadi-P* with $CuK_{\alpha 1}$ radiation from primary Ge(111) primary beam monochromator. Dectris-MYTHEN 1 K strip, position-sensitive detector (PSD), with an opening angle of 12° in 2θ was used to record the scattered intensity. Diffraction data were collected during 24 h in the 4°–70° 2θ range.

The powder data analysis (pattern indexing, profile fitting, crystal structure solution, and refinement) were performed using the program TOPAS 4.2 [50]. For determination of the unit cell parameters and symmetry, the positions of the first 28 reflections were carefully estimated. Indexing was performed by iterative use of singular value decomposition (or latent semantic indexing, LSI) [41], leading to a huge number of possible solutions. The solutions were predominately clustered around two volume values: 1933 and 3867 Å^3. The unit cells with smaller volume were described in the monoclinic crystal system (with the most probable space group being $P2_1$) and with unit cell parameters that relate to the crystal structure of codeine phosphate hemihydrate, recently solved by single-crystal diffraction [76]. However, few low-angle diffraction peaks were not accounted using any of these solutions. The unit cells with larger volume, described in the orthorhombic crystal system, accounted for all observed diffraction peaks in the pattern. The systematic absences of reflections pointed at $P2_12_12_1$ as the most probable space group. Precise lattice parameters were determined by a structureless Pawley fitting [54]. During the full profile decomposition, the lattice parameters, strain, and crystal size contributions were refined. In addition, Chebyshev polynomials were used to model the background. The unit cell parameters were refined to $a = 33.4761(7)$ Å, $b = 16.0612(3)$ Å, $c = 7.1921(2)$ Å, and $V = 3866.9(2)$ Å^3. The refinement converged quickly, reaching R_{wp} value of 2.5% (Figure 2.19b).

Once the unit cell parameters and symmetry were established (with confidence due to a high indexing figure of merit and satisfactory Pawley fit), structure determination was initiated. Solving crystal structures of organic molecules from powder diffraction is challenging due to the low scattering abilities of the atoms in the structure. Reciprocal space methods or dual-space approaches (e.g., charge flipping) are seldom applicable. The global optimization methods in real space are better suited for this purpose. It could be argued that the method of simulated annealing (SA) stands as a favorite choice in the powder crystallographic community [71]. As mentioned previously, during the SA runs, atoms are moved within the unit cell following an algorithm that minimizes a cost function (usually the difference between the observed and calculated patterns).

From the volume increments, it was found that the asymmetric unit of COP-S is composed of two codeine molecules, two phosphate groups, and three water molecules. Each codeine base (Figure 2.19c) is made of 22 nonhydrogen atoms. Therefore, with a classical global optimization approach, the positions of as much as 57 atoms need to be freely varied. The huge number of degrees of freedom directly translates in incredibly long computational time. It is not exaggerated to say that the crystal structure of COP-S cannot be solved by fully unconstrained SA using a personal computer during a reasonable length of time. Moreover, in many cases, free variation of all atoms leads to wrong solutions with lower R_{wp} values compared to the true structure. Luckily, the degrees of freedom can be significantly reduced if the rightful connectivity of atoms is considered. The solid-state molecular conformation of the codeine base was established by crystal structure determinations performed on several codeine salts [76]. Therefore, instead of varying the x, y, and z fractional coordinates of all 22 atoms per codeine base, a model of the molecule can be build and used in the SA runs. A convenient way of introducing the molecular geometry as a rigid body is by using z-matrix notation. Figure 2.20 presents the codeine molecule explained in a z-matrix notation (written for a TOPAS input file), as observed in the structure of codeine phosphate hemihydrate [76]. Note that prior to the structure solution, the possibility of isomerization around the chiral centers was examined by other chemical methods. During the SA process, the unit cell parameters were taken from the Pawley fitting and were kept fixed to further reduce the degrees of freedom. The same applied to the fitted coefficients for

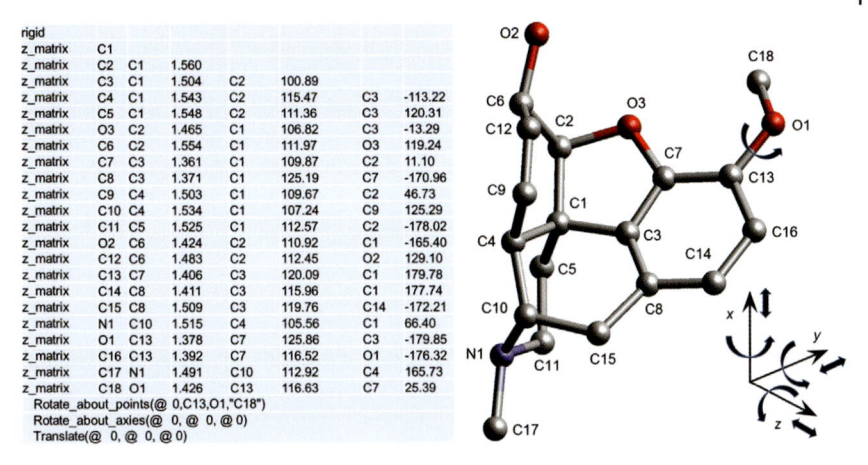

```
rigid
z_matrix    C1
z_matrix    C2   C1    1.560
z_matrix    C3   C1    1.504   C2    100.89
z_matrix    C4   C1    1.543   C2    115.47   C3   -113.22
z_matrix    C5   C1    1.548   C2    111.36   C3    120.31
z_matrix    O3   C2    1.465   C1    106.82   C3   -13.29
z_matrix    C6   C2    1.554   C1    111.97   O3    119.24
z_matrix    C7   C3    1.361   C1    109.87   C2    11.10
z_matrix    C8   C3    1.371   C1    125.19   C7   -170.96
z_matrix    C9   C4    1.503   C1    109.67   C2    46.73
z_matrix    C10  C4    1.534   C1    107.24   C9    125.29
z_matrix    C11  C5    1.525   C1    112.57   C2   -178.02
z_matrix    O2   C6    1.424   C2    110.92   C1   -165.40
z_matrix    C12  C6    1.483   C2    112.45   O2    129.10
z_matrix    C13  C7    1.406   C3    120.09   C1    179.78
z_matrix    C14  C8    1.411   C3    115.96   C1    177.74
z_matrix    C15  C8    1.509   C3    119.76   C14  -172.21
z_matrix    N1   C10   1.515   C4    105.56   C1    66.40
z_matrix    O1   C13   1.378   C7    125.86   C3   -179.85
z_matrix    C16  C13   1.392   C7    116.52   O1   -176.32
z_matrix    C17  N1    1.491   C10   112.92   C4    165.73
z_matrix    C18  O1    1.426   C13   116.63   C7    25.39
Rotate_about_points(@ 0,C13,O1,"C18")
Rotate_about_axies(@ 0, @ 0, @ 0)
Translate(@ 0, @ 0, @ 0)
```

Figure 2.20 Rigid body of the codeine molecule explained using the z-matrix notation, prepared for a TOPAS input file.

the zero-error shift, simple axial model, peak profile, and background. Only the scale factor was varied. In the unit cell, two codeine molecules were independently translated and rotated along the axes. In each of them, one torsion angle was varied, to account for the methoxide rotational flexibility. In addition, two phosphate groups (defined also as rigid bodies) were translated and rotated, together with free variation of three oxygen atoms representing water molecules. An overall displacement parameter for each atom type was included in the process. The number of degrees of freedom accounting for translations and rotations of the rigid bodies and oxygen atoms equaled 35 (plus the scale and displacement parameters). Note that if the rigid body notation were not used, the number of degrees of freedom would have been 171 (plus the scale and displacement parameters).

SA was performed on a personal computer and in about an hour a reasonable model was obtained. Figure 2.19d shows the measured powder pattern together with a simulated pattern of the crystal structure model shown in Figure 2.19e. The difference in curve and the obtained R_{wp} value of 7.5% indicated that the found solution is close to the real crystal structure. Moreover, the visual inspection of the crystal structure revealed a reasonable crystal packing of the molecules.

Once a global minimum was found, the model was subjected to Rietveld refinement [1]. All profile and lattice parameters, as well as the zero-shift, simple axial model, and background coefficients, were freely refined. The anisotropy of width and asymmetry of the Bragg reflections was successfully modeled by applying symmetry-adapted spherical harmonics of low order to Gaussian, Lorentzian, and exponential distributions, which were then convoluted with geometrical and instrumental contributions to the final peak profile. The bond distances and angles of the rigid bodies were also refined. Unconstrained and unrestrained refinement lead to slightly longer P—O bonds than expected, without significant influence on the overall hydrogen bonding scheme. Despite the use of capillaries in Debye–Scherrer geometry, a small amount of preferred orientation originating from the plate-like crystals was detected and was adequately described by the use of symmetry-adapted spherical harmonics (improving the R_{wp} value for less than 0.5%). At the last stage, hydrogen atoms were added on calculated positions using the software Mercury [77]. The final R_{wp} value reached 0.0348%. The final Rietveld plot is given in Figure 2.19f and a plot of the crystal structure is presented in Figure 2.19g.

2.9.2
Case Study 2: Application of *In Situ* X-Ray Powder Diffraction in Elucidating of Physical Phenomena: The Thermosalient (Jumping Crystal) Effect [78]

The thermosalient (TS) phenomenon, or the "jumping crystals" effect, is a rare property of some crystals to be ferociously self-actuated by an active ballistic event when taken over a phase transition [78]. These crystals are able to travel over distances that can reach up to thousands times their size, in timescales of seconds or shorter. Despite that this interesting phenomenon provides a conceptually new platform for direct conversion of heat into mechanical work, it remains mechanistically poorly understood. Drawing its mechanistic picture is challenged by the degradation of single crystals upon the phase transition. Very often, the jumping single crystal cannot withstand the generated stress, and thus disintegrates. In such cases, single-crystal diffraction is applicable only for solving the crystal structure before the phase transition happens, but not after the process. Moreover, the TS phase transitions sometimes proceed via metastable phases, present in a very narrow temperature region. The crystal structures of these phases are crucial for understanding and explaining the jumping crystal effect. XRPD is a very suitable method for studying TS active systems due to several reasons: The disintegration of single crystals or severe increase of mosaicity does not hinder structure solution or refinement, the method can be applied *in situ*, time-resolved and/or temperature-resolved, and the time of data collection can be made considerably short, making the technique suited to detect metastable species. The following text explains how XRPD can contribute to elucidate and decipher the mechanism of this remarkable phenomenon with the case study of (phenylazophenyl)palladium(II) hexafluoroacetylacetonate (PHA), the first published example of an TS active system.

In the early 1980s, it was reported that single crystals of PHA when heated on one side "literally fly off the hot stage" [79]. The authors solved the crystal structures of two polymorphs from single-crystal diffraction data at room temperature (RT). However, the processes behind the jumping effect have remained unexplained for over three decades. Among the reasons for the lack of knowledge on this remarkable system was the need for high-resolution, time- and temperature-resolved *in situ* studies. Thermoanalytical and spectroscopic methods can provide valuable information and deserve special attention. However, for visualizing the structural changes, temperature-resolved crystallographic experiments are needed.

Recently, powder diffraction data of PHA were collected upon several cooling and heating cycles. One single crystal of PHA was powdered by gentle grinding in a pestle and loaded in a borosilicate capillary (0.5 mm diameter, Hilgenberg glass No. 50). For better particle statistics, the capillary was spun during data collection. XRPD patterns were collected in Debye–Scherrer geometry on a laboratory powder diffractometer *Stoe Stadi-P* with $CuK_{\alpha 1}$ radiation from a primary Ge(111)-Johann-type monochromator. A Dectris-MYTHEN 1 K strip, position-sensitive detector, with an opening angle of 12° in 2θ was used for recording the scattered intensity. Hot and cold air blowers were used for heating and cooling. Scattered intensity was collected in a relatively short 2θ range of 5°–30°, for 10 min, at every 5 K. The following temperature control program was used: cooling from 293 to 98 K, heating from 98 to 388 K, cooling from 388 to 98 K, heating from 98 to 420 K, and cooling from 420 to 293 K. Besides the short time of data collection and the short 2θ range, *in situ* collected patterns revealed clear changes of the scattered intensity, reflecting the extraordinary rich crystal chemistry of PHA.

Figure 2.21a presents a section of the 2D presentation of scattered intensity plotted as a function of diffraction angle and temperature. Observing the changes in the powder patterns, the following phase transitions are detected: At room temperature, the form α is stable (solved by single-crystal diffraction) [79], by cooling it transforms into a new polymorph (form δ) via a second-order phase transition. By further cooling, a sharp change of the scattered intensity is evidenced, indicative of a first-order phase transition and form δ is transformed into a new form ε. All of these phase transformations are reversible and mirrored upon heating, transforming PHA again to the α polymorph at RT. Heating of form α at moderately high temperatures gives the TS phenomenon, where crystals jump for large distances. In the 2D presentation of *in situ* collected patterns, it is evident that this TS-active transition proceeds via a metastable γ phase (the α to γ phase transition is responsible for the jumping) to a thermodynamically stable phase β (solved by single-crystal diffraction after recrystallization) [79]. Heating and cooling XRPD experiments performed on form β did not reveal any phase transformations (note that low-temperature single-crystal diffraction indicated that the unit cell of form β is doubled) [78]. Figure 2.21b summarizes the thermo-induced phase transitions of PHA. Using *in situ* XRPD, three new polymorphs of PHA were detected, including one metastable form directly responsible for the TS effect. Single-crystal diffraction is not applicable for solving any of the new crystal structure: Single crystals disintegrate when taken over $\alpha \to \delta$ and $\alpha \to \gamma$ phase transformations. The $\alpha \to \delta$ resembles a second-order phase transformation, thus the crystal packing of both forms is expected to be similar. Moreover, this transition is not TS active. Therefore, our attention was primarily focused on the sharp, first-order phase transitions occurring at low temperature $(\alpha \to) \delta \to \varepsilon$ (TS inactive) and at high temperature $\alpha \to \gamma \to \beta$ (TS active).

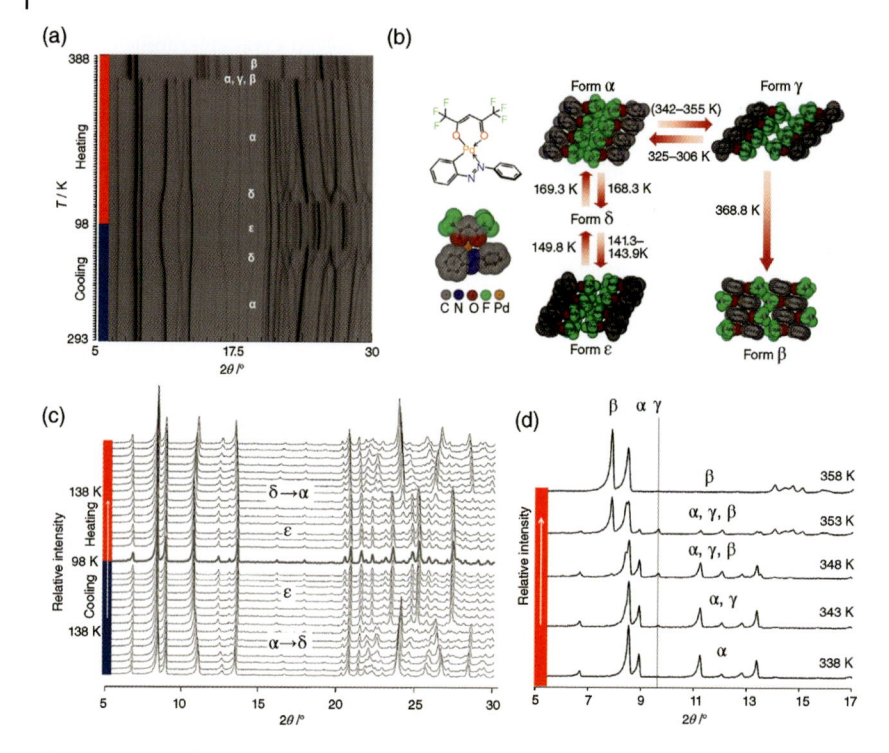

Figure 2.21 Two-dimensional projection of the scattered intensity of PHA collected *in situ* upon cooling from 293 to 98 K and subsequent heating to 388 K. Region where given polymorphic modifications are stable are noted with their symbols (a). Molecular structure of the PHA together with schematic representation of the phase transitions upon change of temperature (b). Selected diffraction patterns of PHA at low (c) and high (d) temperatures, revealing the phase transitions.

To get structural information on the ε polymorph, the low-temperature region was examined in great detail (Figure 2.21c). It was found that this PHA polymorph is stable and exists as a single phase at low temperatures. Even though the powder diffraction pattern was collected for a short period of time and only up to 30° in 2θ, structure solution was attempted. For the structure determination, the same procedure as for the codeine phosphate sesquihydrate was followed (more information can be found in the respective case study): indexing from first principles (indicating a triclinic space group, $P1$ or $P\bar{1}$), full-pattern decomposition and Pawley fitting (to refine the lattice and profile parameters), structure solution by global optimization method of simulated annealing (using rigid bodies to decrease the degrees of freedom), and finally Rietveld refinement of the crystal structure (Figure 2.22a). The crystal structure was successfully refined to an R_{wp} value of 10.02%. Expectedly, the uncertainties of atomic positions, thus bond lengths and angles, were high (on the second decimal place for the fractional coordinates). The thermal displacement factors were also high, especially for the rotationally free $-CF_3$ groups. Therefore, data were collected at

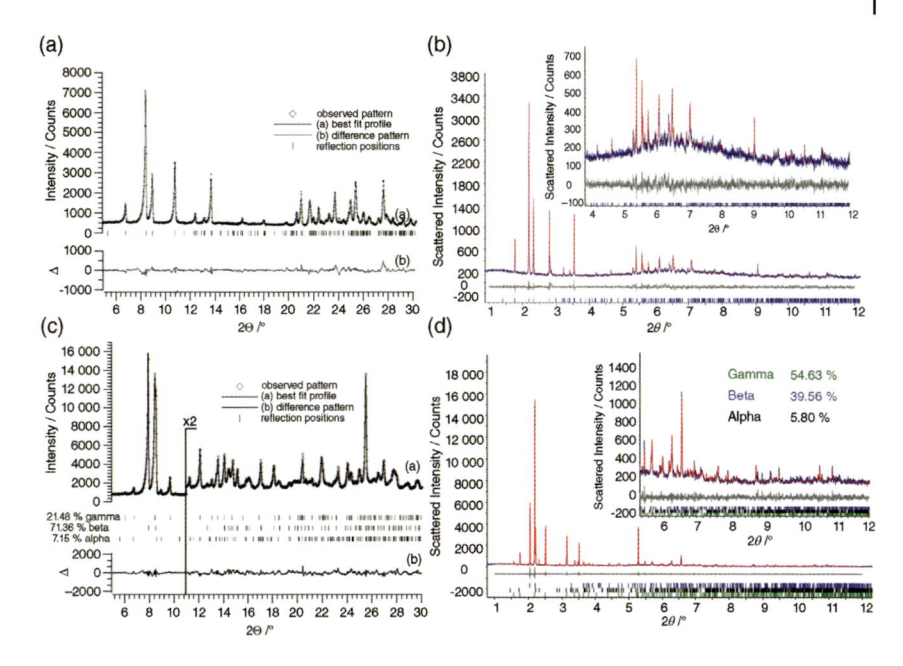

Figure 2.22 Rietveld plots of the refined crystal structure of ε form of PHA performed on (a) laboratory and (b) synchrotron XRPD data. Rietveld plots of the refined crystal structure of γ form of PHA performed on (c) laboratory and (d) synchrotron XRPD data.

the European Synchrotron Radiation Facility (ESRF) at the ID31 High-Resolution Powder Diffraction Beamline with a wavelength of 0.40 Å, using crystal analyzer detector at 90 K. Rietveld refinement of the crystal structure of ε-PHA performed using the high-resolution synchrotron data (Figure 2.22b) improved the model. The refined crystal structure was significantly more accurate, with R_{wp} of 9.87%, uncertainties of the fractional coordinates on the third decimal place, and overall lower thermal displacement factors. The crystal structures refined from laboratory and synchrotron data exhibited the same crystal packing, proving the usefulness and applicability of laboratory XRPD in studying relatively complicated systems.

To elucidate the TS phenomenon exhibited by PHA, attention was paid on the processes occurring in the high-temperature region (Figure 2.21d). It was found that the jumping effect of crystals is triggered by a phase transition where the α-phase transforms into a new metastable phase γ, which readily transforms into a thermodynamically more stable form β. Several samples of α-PHA were heated to high temperatures under different conditions (modifying the heating rate, duration of grinding, amount of sample used, pretreatments, etc.). Unfortunately, in all of the cases, a triple mixture (with different relative amount) of α, β, and γ was obtained. The crystal structures of both α and β phases were previously solved by single-crystal diffraction [79]. Therefore, crystal structure determination of the γ polymorph from the triple mixture was possible, although challenging. Surprisingly, following the previously outlined procedure for structure determination gave a reasonable model of the crystal structure besides the low-quality diffraction data (Figure 2.22c). Expectedly, the accuracy of the model was very low, as evidenced by high estimated standard deviations and thermal displacement factors. Similarly, as for the low-temperature processes, high-resolution diffraction data were collected *in situ* at the ESRF powder diffraction beamline. The model of the structure was refined against the synchrotron diffraction data, yielding significantly improved crystal structure description.

The occurrence of both TS and non-TS phase transitions of PHA and the determined crystal structures of the polymorphs provides basis to establish the mechanism of the TS effect. It is not surprising that the phase transitions at low temperatures are not thermosalient, given the similarity of the crystal structures before and after the phase transitions. Even after the first-order phase transition, the layers of face-to-face oriented and head-to-head stacked PHA molecules in α form remain the same as in form ε (and probably in form δ). The similarity of the room and low-temperature polymorphs warrants reversibility of the process. On the contrary, on heating the form α, its unit cell expands mainly along the a-axis and the molecules get separated. Simultaneously, a rotation of the phenyl ring causes continuous changes of the packing and contraction of the c-axis. The shear strain caused by these distortions is accumulated to the point where it overweighs the cohesive interactions, and the structure switches to form γ, which is characterized by a high increase of the unit cell volume per molecule. The molecules have retained their head-to-head disposition within the layers, but given the increase of free space in the unit cell, the columns are separated with respect to each other to a great extent. The system releases the acquired shear stress by forceful jumps. The crystal structure of form β is significantly different (exhibiting head-to-tail disposition) (Figure 2.21b) and thermodynamically more stable [78].

In summary, crystals of PHA exhibit a multitude of extraordinary phenomena that were successfully revealed by time- and temperature-resolved, *in situ* XRPD experiments. This case study also shows that laboratory XRPD can be very useful for *in situ* studies, making the solution of crystal structures of even metastable species possible. In addition, the diffraction data were used for refining the unit cell parameters and calculating the thermal expansion characteristics; for further details, the reader is referred to Ref. [78], where the axial thermal expansion coefficients and discussion are given.

2.9.3
Case Study 3: Crystal Structure Determination of Corrosion Product by a Dual-Space Charge Flipping Algorithm, Aided by Inspection of Difference Fourier Maps [80,81]

Many artifacts and historic objects suffer degradation by formation of secondary phases frequently found on the surface of the material that is prone to react with chemicals present in the environment, such as air pollutants, volatile organic materials, vapors, to name but a few. Knowledge on the composition and structure of these products is of ultimate importance to address the degradation processes. Moreover, quantitative-phase Rietveld (QPR) analysis (a routine, straightforward, and nondestructive method) requires crystallographic information. Often, these secondary phases crystallize as microparticles forming mixtures with the reactants and single crystals cannot be grown. Therefore, powder diffraction is a remaining alternative.

One example of a degradation phenomenon is the so-called glass-induced metal corrosion, occurring at the interface of metal alloys and glass (Figure 2.23a). Sodium copper formate hydroxide oxide hydrate, $Cu_4Na_4O(HCOO)_8(H_2O)_4(OH)_2$ (Corr1, Figure 2.23b), is among the most frequently observed phases. Sample of Corr1 was scratched from the surface of an artifact and carefully grinded in a pestle to reduce the preferred orientation of the particles. The Debye–Scherrer geometry was used for diffraction data collection, with the sample placed in a borosilicate capillary (0.5 mm diameter, Hilgenberg glass No. 50). For better particle statistics, the capillary was spun.

The XRPD pattern (Figure 2.23c) was collected at room temperature on a high-resolution laboratory powder diffractometer *Stoe Stadi-P* with $CuK_{\alpha 1}$

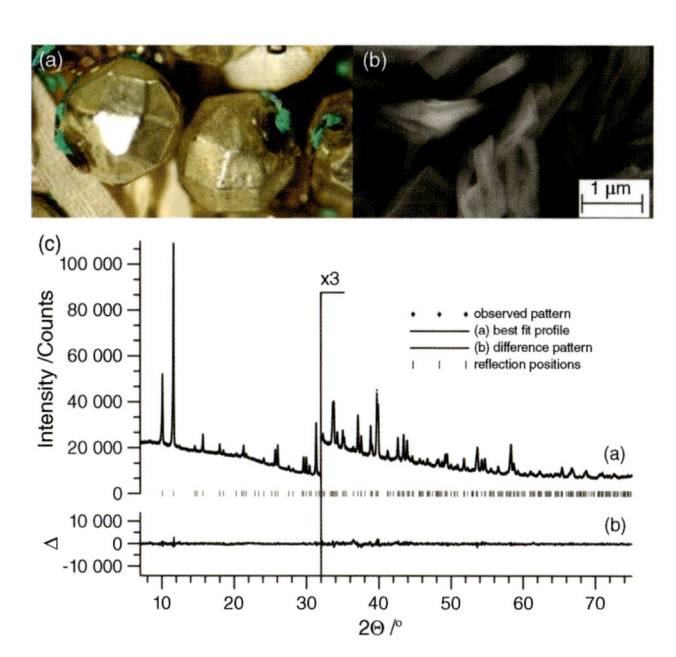

Figure 2.23 (a) Glass beads on a traditional Black Forest headdress from the Franziskanermuseum Villingen Schwenningen (Germany). (b) Field emission scanning electron microscopy image of Corr1. (c) Scattered X-ray intensities of Corr1 at ambient condition, as a function of diffraction angle. The observed pattern (diamonds) measured in Debye–Scherrer geometry, the best Rietveld fit profiles (line), and the difference curve between the observed and the calculated profiles (below) are shown. The high angle part is enlarged for clarity.

radiation from a primary Ge(111)-Johann-type monochromator. A Dectris-MYTHEN 1 K strip, position-sensitive detector, with an opening angle of 12° in 2θ, was used to record the scattered intensity. Diffraction data were collected during 20 h in the 5°–75° 2θ range.

Data analysis (pattern indexing, profile fitting, crystal structure solution, and refinement) were performed using the program TOPAS 4.2 [50]. The unit cell parameters and symmetry were determined using the positions of the first 24 observed reflections. Indexing was performed by iterative use of singular value decomposition (LSI). Considering the systematic absences of reflections in the pattern, the tetragonal space group $P4_2/n$ was selected as the most probable space group. Precise lattice parameters were determined by Pawley fitting of the pattern [54], using the FP approach. During the full profile decomposition, the lattice parameters, strain, and crystal size contributions were freely refined. Chebyshev polynomials were added to model the background. The final unit cell parameters were determined to be $a = 8.4251(1)$ Å, $c = 17.4796(3)$ Å, and $V = 1240.75(4)$ Å^3. The fitting converged quickly, reaching R_{wp} value of 1.22%.

Once the unit cell parameters were precisely determined, the structure solution process was initiated. As it can be seen in Figure 2.23c, the powder pattern contains relatively well-resolved reflections even at very high angles. Therefore, the charge flipping approach was adopted [68,69]. Figure 2.24 shows the results

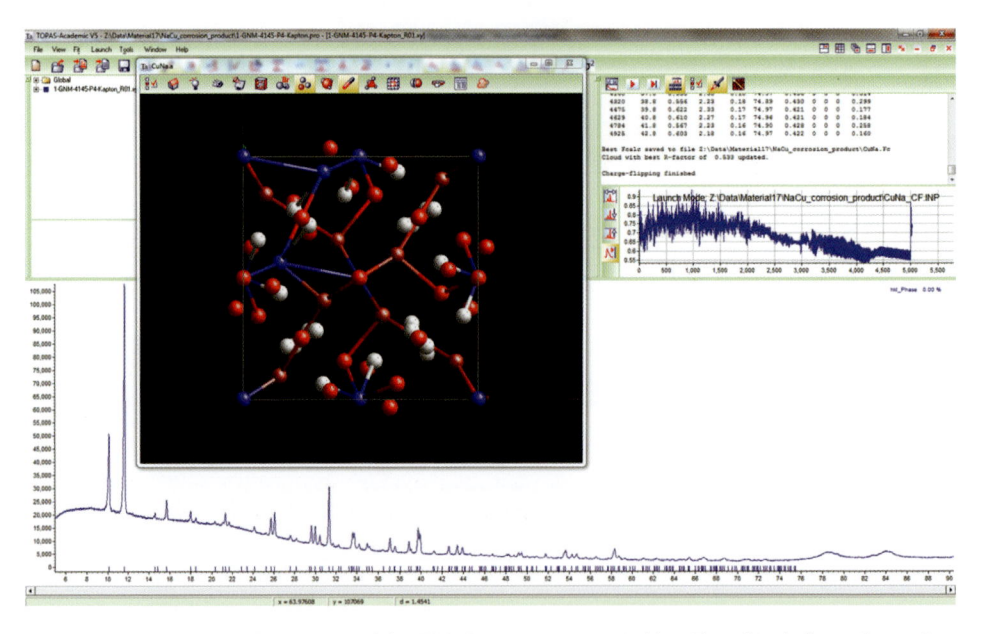

Figure 2.24 Print screen of the TOPAS program performing the charge flipping structure solution of Corr1. As it can be seen, the positions of the heavier atoms are determined, revealing the crystal packing characteristics. Cu atoms in dark red, O atoms in red, and Na atoms in blue. The white balls are "atoms" placed by the program in areas with excess of electron density. Note that these are not real and are omitted during the later stages of structure solution.

obtained after relatively short computing time. The positions of the Cu atoms as well as most of the O and Na atoms are readily obtained by this process. The model obtained by charge flipping was incomplete and one oxygen atom (from a water molecule) was missing. In order to find the position of that atom, the difference Fourier map using the model derived by the charge flipping algorithm was carefully inspected. The model was completed by adding hydrogen atoms at calculated positions using the program Mercury [77]. Finally, the positions of the atoms as well as the zero-error, profile and unit cell parameters were refined by the Rietveld method, and the final plot is given in Figure 2.23c. The refinement converged giving the following values for the unit cell: $a = 8.425\ 109(97)$ Å, $c = 17.47\ 962(29)$ Å, and $V = 1240.747(35)$ Å^3. The figures of merit read $R_{exp} = 1.042\%$, $R_p = 1.259\%$, $R_{wp} = 1.662\%$, and $R_{Bragg} = 0.549\%$.

2.9.4
Case Study 4: Utilizing Symmetry Modes for the Investigation of High-Pressure Phase Transitions in the Rare Earth Orthoferrite LaFeO$_3$ [82,83]

The investigation of phase transitions in solid-state science is an important task, as in many cases a phase transition can lead to a complete change of the inherent physical and chemical properties of a material, which affects instantaneously its applicability in technology. Therefore, tremendous research is mandatory in order to explore the stability of a given phase and to find new interesting phases, which are of technological relevance (for instance, sometimes metastable phases can be quenched from high/low temperatures or high pressures to ambient conditions and then used for technological purposes). Among the technologically relevant material classes, the perovskites gained a lot of attention during the last few decades, as they exhibit many different phenomena such as multiferrocity, superconductivity, and so on. However, not only due to the different occurring effects, but also due to the variety of chemical compositions and doping capabilities, this material class often acts nowadays as a physical and chemical model system. Taking the rare earth orthoferrite lanthanum ferrite (LaFeO$_3$) as such a perovskite model system, the crystallographic applicability of symmetry modes to high-pressure powder diffraction data can be investigated.

Polycrystalline samples of LaFeO$_3$ were synthesized using a solid-state reaction of a stoichiometric mixture of La$_2$O$_3$ and Fe$_2$O$_3$ powders as described by Peterlin-Neumaier and Steichele [84] and Selbach et al. [85]. The obtained perovskite powders were loaded into different membrane-driven diamond-anvil cells (DAC) together with different pressure media (N$_2$, Ar, methanol–ethanol 4: 1) in order to perform high-pressure synchrotron X-ray powder diffraction measurements at beamline ID09 at the European Synchrotron Radiation Facility (ESRF). In all experiments, the pressure was gradually increased along the room temperature isotherm. The value of the current pressure was determined by utilizing the ruby luminescence method according to Mao et al. [86]. Diffracted X-rays were collected on image plates that were read out by an offline molecular dynamics image plate scanner. The so obtained two-dimensional data sets were subsequently reduced to one-dimensional powder diffraction patterns, using integration parameters from a silicon reference sample and the software FIT2D [87]. Due to geometric restrictions of common high-pressure experiments using diamond anvil cells, reasonable data analysis could solely be performed in a 2θ range of $5°–20.8°$.

From a capillary measurement at ambient conditions at the same beamline, it could be proven that the sample is phase pure and that the LaFeO$_3$ perovskite crystallizes in an orthorhombic crystal structure with space group *Pbnm* [82,83,88,89], which is typical for many perovskite systems. Carrying out a Rietveld refinement using the TOPAS 4.2 software, the following lattice parameters could be determined: $a = 5.5549(1)$ Å, $b = 5.5663(1)$ Å, and $c = 7.8549(2)$ Å, with a resulting volume of $V = 242.876(9)$ Å^3 (Figure 2.24).

Starting with the determined values from ambient condition, the powder diffraction patterns at higher pressures could be consecutively refined. From these refinements, two observations below 30 GPa can be made. One observation is that the lattice parameters and the volume show a nonlinear shrinkage of their values with increasing pressure (Figure 2.26a). This is a typical behavior for most solids under pressure and can be described by using equations of state [90–92]. The other observation is that the pressure-induced continuous movement of the *y*-coordinate of the lanthanum cation stops at a fixed position at a pressure of approximately 21 GPa (Figure 2.26b). Such behavior can indicate a pressure-dependent continuous-phase transition as no obvious jump in the lattice parameters or the volume can be observed (see Figure 2.26a). Although the data quality usually decreases with increasing pressure (see Figure 2.25b and c), it was possible to determine the higher symmetric space group *Ibmm* above 21 GPa, which allows an adequate description of the high-pressure crystal structure of LaFeO$_3$ (see Figure 2.25b).

Interestingly, between both employed space groups, a group–subgroup relationship can be established; therefore, the crystal structure of the LaFeO$_3$ perovskite at lower pressure can also be seen as a distorted version of the higher symmetric crystal structure. This observation leads directly to the symmetry mode or distortion mode approach, where on a mathematical basis all possible shifts of atomic coordinates are naturally calculated out of distortion vectors. Although the calculation of atomic positions out of refined amplitudes of symmetry modes seems to be on the first sight more complicated than a direct refinement, it has profound benefits. One of the most important benefits is that in many cases, the amplitudes of some distortion vectors are so small that they can simply be neglected. In principle, this leads to a greater stabilization in the Rietveld refinement as the parameter space of the refinement is reduced (for more details on symmetry modes and other benefits, see Refs [56,58,61–67,93–95]).

Although the space group *Ibmm* can be used as parental crystal structure for the LaFeO$_3$ orthoferrite, it is more common for perovskites to use the highest possible space group $Pm\bar{3}m$, even if such a phase is experimentally not (yet) observed. In the case of the LaFeO$_3$ perovskite, such a symmetry mode-based

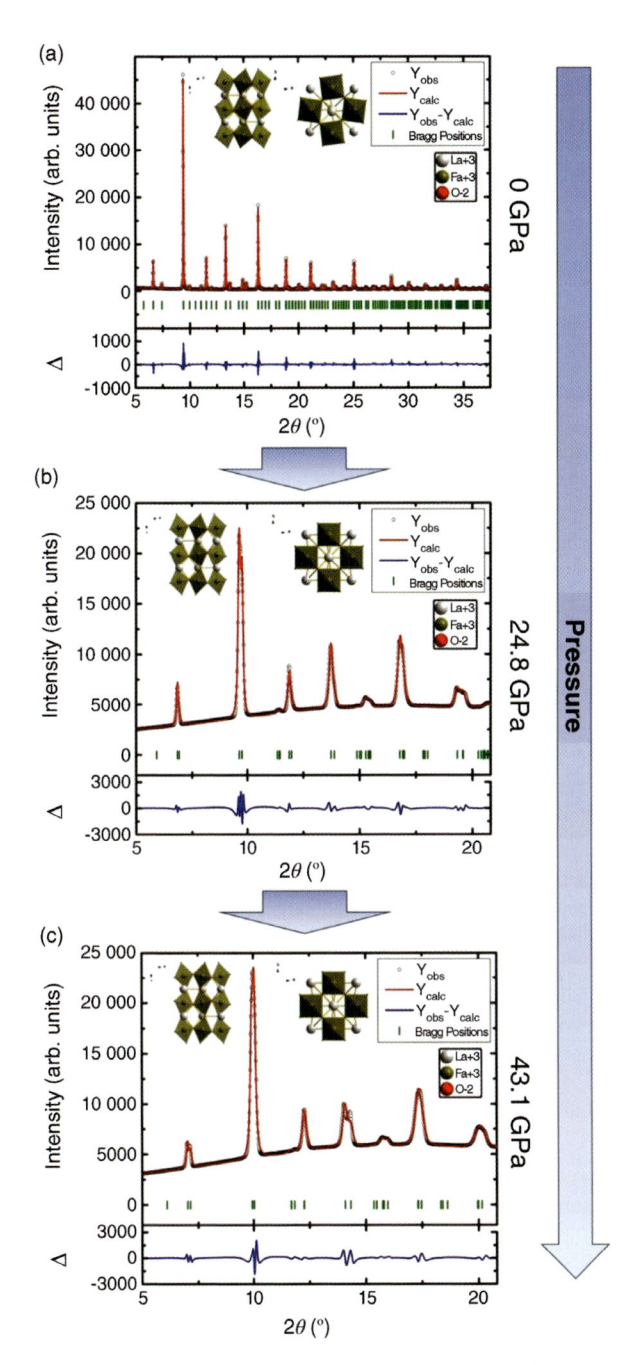

Figure 2.25 Rietveld plots of LaFeO$_3$ at different pressures along the room-temperature isotherm. (a) 0 GPa measured in a capillary. (b and c) 24.8 and 43.1 GPa, respectively, both measured in a diamond anvil cell (DAC). With increasing pressure, it is obvious that the data quality decreases.

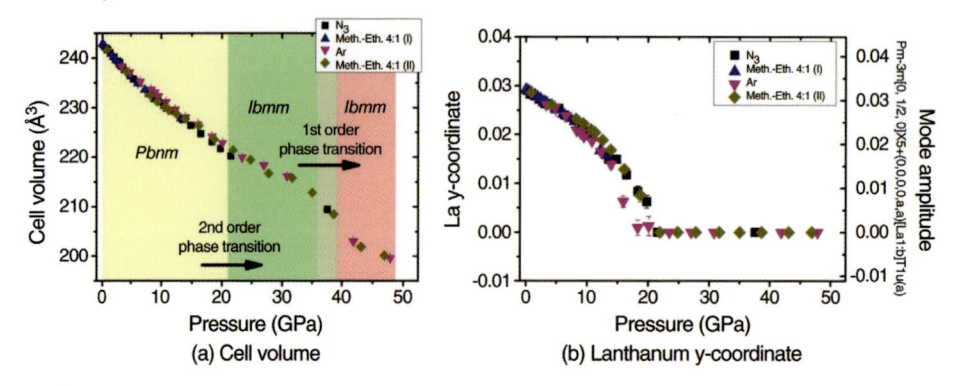

Figure 2.26 (a) Cell volume of LaFeO$_3$ over pressure. A continuous second-order phase transition to a higher symmetry occurs approximately at 21 GPa. Around 38 GPa, a transition region can be found where a discontinuous first-order isosymmetric phase transition, due to a high-spin ($S = 5/2$) to low-spin ($S = 1/2$) transition of the Fe^{3+} cation, takes place. (b) Observation of the y-coordinate of the lanthanum cation over pressure. The right scale shows the amplitude of the symmetry mode, which can also be used for the description of the atomic movement.

Rietveld refinement, with a cubic parental symmetry, resulted in the same quality at lower pressures and with a tiny improvement in quality at higher pressures than for a traditional atomic coordinate refinement [83].

Apart from the continuous-phase transition of second order at approximately 21 GPa, another phase transition can be observed. At around 38 GPa, the volume drops approximately by 3%, which indicates a discontinuous first-order phase transition in the LaFeO$_3$ perovskite (Figure 2.26a). Rietveld refinements with different crystal structure models showed that this sluggish phase transition can be ascribed to an isosymmetric phase transition, which is caused by a high-spin ($S = 5/2$) to low-spin ($S = 1/2$) transition of the Fe^{3+} cation [96–100] (Figure 2.25c). Despite the fact that this isosymmetric phase transition in space group *Ibmm* is discontinuous, the symmetry modes can still be used for the modeling of the atomic coordinates, as they will perform a jump in their amplitude values, just as the atomic coordinate would jump if they were directly applied.

References

1 Rietveld, H.M. (1969) A profile refinement method for nuclear and magnetic structures. *J. Appl. Crystallogr.*, **2**, 65.

2 Pecharsky, V.K. and Zavalij, P.Y. (2009) *Fundamentals of Powder Diffraction and Structural Characterization of Materials*, 2nd edn, Springer Science+Business Media, LLC.

3 Dinnebier, R.E. and Billinge, S.J.L. (eds) (2008) *Powder Diffraction: Theory and Practice*, The Royal Society of Chemistry.

4 Hecht, E. (2009) *Optik*. Oldenbourg Wissenschaftsverlag.

5 Grimaldi, F.M. (1665) *Physico mathesis de lumine, coloribus, et iride, aliisque annexis libri duo*, Bernia.
6 Röntgen, W.C. (1895) Über eine neue art von strahlen. Aus den Sitzungsberichten der Würzburger Physik.-medic. Gesellschaft, p. 3.
7 Röntgen, W.C. (1898) Über eine neue art von strahlen. *Ann. Phys (Berlin)*, **300**, 1.
8 Friedrich, W., Knipping, P., and von Laue, M. (1912) *Interferenzerscheinungen bei Röntgenstrahlen*, Bayerische Akademie der Wissenschaften/ Mathematisch,-Physikalische Klasse, Sitzungsberichte, p. 303.
9 Friedrich, W., Knipping, P., and von Laue, M. (1913) Interferenzerscheinungen bei Röntgenstrahlen. *Ann. Phys (Berlin)*, **346**, 971.
10 von Laue, M. (1952) Interferenzerscheinungen bei Röntgenstrahlen: Theoretischer Teil. *Naturwissenschaften*, **39**, 361.
11 Friedrich, W. and Knipping, P. (1952) Interferenzerscheinungen bei Röntgenstrahlen: Experimenteller Teil. *Naturwissenschaften*, **39**, 364.
12 Etter, M. and Dinnebier, R. (2014) A century of powder diffraction: a brief history. *Z. Anorg. Allg. Chem.*, **640**, 3015.
13 Ewald, P.P. (ed.) (1962) *Fifty Years of X-Ray Diffraction*, International Union of Crystallography.
14 Authier, A. (2013) *Early Days of X-Ray Crystallography*, International Union of Crystallography (Oxford University Press).
15 von Laue, M. (1913) Eine quantitative prüfung der theorie für die interferenzerscheinungen bei röntgenstrahlen. *Ann. Phys (Berlin)*, **346**, 989.
16 Bloss, F.D. (1971) *Crystallography and Crystal Chemistry*, Holt, Rinehart and Winston, Inc.
17 Bragg, W.L. (1912) The diffraction of short electromagnetic waves by a crystal. *Proc. Camb. Philos. Soc.*, **17**, 43.
18 Debye, P. (1915) Zerstreuung von Röntgenstrahlen. *Ann. Phys (Berlin)*, **351**, 809.
19 Warren, B.E. (1990) *X-Ray Diffraction*, Dover Publications Inc.
20 Friedrich, W. (1913) Eine interferenzerscheinung bei röntgenstrahlen. *Ann. Phys.*, **14**, 317.
21 Nishikawa, S. and Ono, S. (1913) Transmission of X-rays through fibrous, lamellar and granular substances. *Proc. Tokyo Math. Phys. Soc.*, **7**, 131.
22 Debye, P. and Scherrer, P. (1916) Interferenzen an regellos orientierten Teilchen im Röntgenlicht: I. *Nachrichten Gesellschaft Wissenschaften Göttingen, Math. Phys. Klasse*, **1916**, 1.
23 Hull, A.W. (1917) A new method of X-ray crystal analysis. *Phys. Rev.*, **10**, 661.
24 Ewald, P.P. (1913) Zur theorie der interferenzen der röntgenstrahlen in kristallen. *Phys. Z.*, **11**, 465.
25 Scherrer, P. (1918) Bestimmung der Größe und der inneren Struktur von Kolloidteilchen mittels Röntgenstrahlen. *Nachrichten Gesellschaft Wissenschaften Göttingen, Math. Phys. Klasse*, 98.
26 Giacovazzo, C. (2002) Crystallographic computing, in *Fundamental of Crystallography*, 2nd edn, Oxford University Press, pp. 67–151.
27 Altomare, A., Giacovazzo, C., and Moliterni, A. (2008) Chap. Indexing and space group determination, *Powder Diffraction: Theory and Practice*, The Royal Society of Chemistry, pp. 206–226.
28 Runge, C. (1917) Die Bestimmung eines kristallsystems durch röntgenstrahlen. *Phys. Z.*, **18**, 509.
29 Ito, T. (1949) A general powder X-ray photography. *Nature*, **164**, 755.
30 de Wolff, P. (1957) On the determination of unit-cell dimensions from powder diffraction patterns. *Acta Crystallogr.*, **10**, 590.
31 Visser, J.W. (1969) A fully automatic program for finding the unit cell from powder data. *J. Appl. Crystallogr.*, **2**, 89.
32 Shirley, R. (2003) Overview of powder-indexing program algorithms: history and strengths and weaknesses. International Union of Crystallography – Commission on Crystallographic Computing, Newsletter, **2**, 48.
33 Werner, P.E. (1964) Trial-and-error computer methods for the indexing of

unknown powder patterns. *Z. Kristallogr.*, **120**, 375.
34 Louër, D. (1991) Indexing of powder diffraction patterns. *Mater. Sci. Forum*, **79–82**, 17.
35 Louër, D. and Louër, M. (1972) Méthode d'essais et erreurs pour l'indexation automatique des diagrammes de poudre. *J. Appl. Crystallogr.*, **5**, 271.
36 Louër, D. and Vargas, R. (1982) Indexation automatique des diagrammes de poudre par dichotomies successives. *J. Appl. Crystallogr.*, **15**, 542.
37 Boultif, A. and Louër, D. (1991) Indexing of powder diffraction patterns for low-symmetry lattices by the successive dichotomy method. *J. Appl. Crystallogr.*, **24**, 987.
38 de Wolff, P.M. (1968) A simplified criterion for the reliability of a powder pattern indexing. *J. Appl. Crystallogr.*, **1**, 108.
39 Smith, G.S. and Snyder, R.L. (1979) FN: a criterion for rating powder diffraction patterns and evaluating the reliability of powder-pattern indexing. *J. Appl. Crystallogr.*, **12**, 60.
40 Kariuki, B.M., Belmonte, S.A., McMahon, M.I., Johnston, R.L., Harris, K.D.M., and Nelmes, R.J. (1999) A new approach for indexing powder diffraction data based on whole-profile fitting and global optimization using a genetic algorithm. *J. Synchrotron Radiat.*, **6**, 87.
41 Coelho, A.A. (2003) Indexing of powder diffraction patterns by iterative use of singular value decomposition. *J. Appl. Crystallogr.*, **36**, 86.
42 Le Bail, A. (2004) Monte Carlo indexing with McMaille. *Powder Diffr.*, **19**, 249.
43 Dinnebier, R. and Müller, M. (2013) Modern Rietveld refinement, a practical guide, in *Modern Diffraction Methods*, Wiley-VCH Verlag GmbH, Germany, pp. 27–60.
44 Buras, B., Olsen, J.Staun., and Gerward, L. (1976) X-ray energy-dispersive powder diffractometry using synchrotron radiation. *Nucl. Instrum. Meth.*, **135**, 193–195.
45 Rietveld, H.M. (1966) The crystal structure of some alkaline earth metal uranates of the type M_3UO_6. *Acta Crystallogr.*, **20**, 508.
46 Rietveld, H.M. (1967) Line profiles of neutron powder-diffraction peaks for structure refinement. *Acta Crystallogr.*, **22**, 151.
47 Malmros, G. and Thomas, J.O. (1977) Least-squares structure refinement based on profile analysis of powder film intensity data measured on an automatic microdensitometer. *J. Appl. Crystallogr.*, **10**, 7.
48 Khattak, C.P. and Cox, D.E. (1977) Profile analysis of X-ray powder diffractometer data: structural refinement of $La_{0.75}Sr_{0.25}CrO_3$. *J. Appl. Crystallogr.*, **10**, 405.
49 Young, R.A., Mackie, P.E., and von Dreele, R.B. (1977) Application of the pattern-fitting structure-refinement method to X-ray powder diffractometer patterns. *J. Appl. Crystallogr.*, **10**, 262.
50 Bruker AXS (2008) TOPAS V4.2: General Profile and Structure Analysis Software for Powder Diffraction Data. User's Manual, Bruker AXS, Karlsruhe, Germany.
51 R. B., von Dreele. (2008) Rietveld refinement, in *Powder Diffraction: Theory and Practice*, The Royal Society of Chemistry, pp. 266–281.
52 Larson, A.C. and Dreele, R.B.V. (2004) General structure analysis system (GSAS), Los Alamos National Laboratory Report LAUR, pp. 86–748.
53 Stinton, G.W. and Evans, J.S.O. (2007) Parametric Rietveld refinement. *J. Appl. Crystallogr.*, **40**, 87.
54 Pawley, G.S. (1981) Unit-cell refinement from powder diffraction scans. *J. Appl. Crystallogr.*, **14**, 357.
55 Le Bail, A., Duroy, H., and Fourquet, J.L. (1988) *Ab-initio* structure determination of $LiSbWO_6$ by X-ray powder diffraction. *Mater. Res. Bull.*, **23**, 447.
56 Campbell, B.J., Stokes, H.T., Tanner, D.E., and Hatch, D.M. (2006) ISODISPLACE: a web-based tool for exploring structural distortions. *J. Appl. Crystallogr.*, **39**, 607.
57 Rodríguez-Carvajal, J. and Bourée, F. (2012) Symmetry and magnetic structures. *EPJ Web Conferences*, **22**, 1.

58 Campbell, B.J., Evans, J.S.O., Perselli, F., and Stokes, H.T. (2007) Rietveld refinement of structural distortion-mode amplitudes. International Union of Crystallography – Commission on Crystallographic Computing, Newsletter, **8**, 81.

59 Landau, L.D. and Lifshitz, E.F. (1980) *Statistical Physics – Part 1, Course of Theoretical Physics 5*, 3rd edn, Pergamon Press.

60 Salje, E.K.H. (1993) *Phase Transitions in Ferroelastic and Co-Elastic Crystals*, Student edn, Cambridge Topics in Mineral Physics and Chemistry, Cambridge University Press.

61 Dove, M.T. (1997) Theory of displacive phase transitions in minerals. *Am. Miner.*, **85**, 212.

62 Perez-Mato, J.M., Orobengoa, D., and Aroyo, M.I. (2010) Mode crystallography of distorted structures. *Acta Crystallogr. A*, **66**, 558.

63 Miller, S.C. and Love, W.F. (1967) *Tables of Irreducible Representations of Space Groups and Co-Representations of Magnetic Space Groups*, Pruett Press, Boulder, CO.

64 Stokes, H.T., Hatch, D.M., and Wells, J.D. (1991) Group-theoretical methods for obtaining distortions in crystals: application to vibrational modes and phase transitions. *Phys. Rev. B*, **43**, 11010.

65 Stokes, H.T. and Hatch, D.M. (1988) *Isotropy Subgroups of the 230 Crystallographic Space Groups*, World Scientific Publishing Co. Pte. Ltd.

66 Hatch, D.M. and Stokes, H.T. (2001) Complete listing of order parameters for a crystalline phase transition: a solution to the generalized inverse Landau problem. *Phys. Rev. B*, **65**, 014113.

67 Orobengoa, D., Capillas, C., Aroyo, M.I., and Perez-Mato, J.M. (2009) AMPLIMODES: symmetry-mode analysis on the Bilbao Crystallographic Server. *J. Appl. Crystallogr.*, **42**, 820.

68 Oszlanyi, G. and Suto, A. (2004) *Ab initio* structure solution by charge flipping. *Acta Crystallogr. A*, **60** (2), 134–141.

69 Palatinus, L. (2013) The charge-flipping algorithm in crystallography. *Acta Crystallogr. B*, **69** (1), 1–16.

70 Newsam, J.M., Deem, J.M., and Freeman, C.M. (1992) *NIST Special Publication 864: Accuracy in Powder Diffraction II: Proceedings of the International Conference May 26–29, 1992* (eds E. Prince and J.K. Stalick), NIST, U.S. Department of Commerce, Gaithersburg, MD.

71 Coelho, A.A. (2000) Whole-profile structure solution from powder diffraction data using simulated annealing. *J. Appl. Crystallogr.*, **33** (3), 899–908.

72 Dinnebier, R.E. (1999) Rigid bodies in powder diffraction: a practical guide. *Powder Diffr.*, **14** (2), 84–92.

73 Metropolis, N., Rosenbluth, A., Rosenbluth, M., Teller und, A., and Teller, E. (1953) Equation of state calculations by fast computing machines. *J. Chem. Phys.*, **21**, 1087–1092.

74 Favre-Nicolin, V. and Černý, R. (2004) FOX, modular approach to crystal structure determination from powder diffraction. *Mater. Sci. Forum*, **443–444**, 35–38.

75 Runčevski, T., Petrusevski, G., Makreski, P., Ugarkovic, S., and Dinnebier, R.E. (2014) On the hydrates of codeine phosphate: the remarkable influence of hydrogen bonding on the crystal size. *Chem. Commun.*, **50**, 6970.

76 Langes, C., Gelbrich, T., Griesser, U.J., and Kahlenberg, V. (2009) Codeine dihydrogen phosphate hemihydrate. *Acta Crystallogr. C*, **56**, 419.

77 Macrae, C.F., Edgington, P.R., McCabe, P., Pidcock, E., Shields, G.P., Taylor, R., Towler, M., and van de Streek, J. (2006) Mercury: visualization and analysis of crystal structures. *J. Appl. Crystallogr.*, **39**, 453.

78 Panda, M.K., Runčevski, T., Sahoo, S.C., Belik, A.A., Nath, N.K., Dinnebier, R.E., and Naumov, P. (2014) Colossal positive and negative thermal expansion and thermosalient effect in a pentamorphic organometallic martensite. *Nat. Commun.*, **5**, 4811.

79 Etter, M.C. and Siedle, A.R. (1983) Solid-state rearrangement of (phenylazopheny1) palladium

hexafluoroacetylacetonate. *J. Am. Chem. Soc.*, **105**, 641.

80 Dinnebier, R.E., Runčevski, T., Fischer, A., and Eggert, G. (2015) Solid-state structure of a degradation product frequently observed on historic metal objects. *Inorg. Chem.*, **54**, 2638.

81 Wahlberg, N., Runčevski, T., Dinnebier, R.E., Fischer, A., Eggert, G., and Iversen, B.B. (2015) Crystal structure of thecotrichite, an efflorescent salt on calcareous objects stored in wooden cabinets. *Cryst. Growth Des.*, **15**, 2795.

82 Etter, M., Müller, M., Hanfland, M., and Dinnebier, R. (2014) High-pressure phase transitions in the rare-earth orthoferrite $LaFeO_3$. *Acta Crystallogr. B*, **70**, 452.

83 Etter, M., Müller, M., Hanfland, M., and Dinnebier, R. (2014) Possibilities and limitations of parametric Rietveld refinement on high pressure data: the case study of $LaFeO_3$. *Z. Kristallogr.*, **229**, 246.

84 Peterlin-Neumaier, T. and Steichele, E. (1986) Antiferromagnetic structure of $LaFeO_3$ from high resolution TOF neutron diffraction. *J. Magn. Magn. Mater.*, **59**, 351.

85 Selbach, S.M., Tolchard, J.R., Fossdal, A., and Grande, T. (2012) Non-linear thermal evolution of the crystal structure and phase transitions of $LaFeO_3$ investigated by high temperature X-ray diffraction. *J. Solid State Chem.*, **196**, 249.

86 Mao, H.K., Xu, J., and Bell, P.M. (1986) Calibration of the ruby pressure gauge to 800 kbar under quasi-hydrostatic conditions. *J. Geophys. Res.*, **91**, 4673.

87 Hammersley, A.P., Svensson, S.O., Hanfland, M., Fitch, A.N., and Hausermann, D. (1996) Two-dimensional detector software: from real detector to idealised image of two-theta scan. *High Pressure Res.*, **14**, 235.

88 Geller, S. and Wood, E.A. (1956) Crystallographic studies of perovskite-like compounds: I. Rare earth orthoferrites and $YFeO_3$, $YCrO_3$, $YAlO_3$. *Acta Crystallogr.*, **9**, 563.

89 Marezio, M. and Dernier, P.D. (1971) The bond lengths in $LaFeO_3$. *Mater. Res. Bull.*, **6**, 23.

90 Angel, R.J. (2000) Chapter 4, in *Transformation Processes in Minerals*, (eds S. Redfern, M. Carpenter, and P. Ribbe), Reviews in Mineralogy & Geochemistry Volume 39, Mineralogical Society of America, Washington DC., pp. 85–104.

91 Angel, R.J. (2000) Chapter 2, in *High-Temperature and High-Pressure Crystal Chemistry*, (eds R. Hazen, R. Downs, and P. Ribbe), Reviews in Mineralogy & Geochemistry Volume 41, Mineralogical Society of America, Geochemical Society, Washington DC., pp. 35–59.

92 Etter, M. and Dinnebier, R.E. (2014) Direct parameterization of the pressure-dependent volume by using an inverted approximate Vinet equation of state. *J. Appl. Crystallogr.*, **47**, 384.

93 Aroyo, M.I., Perez-Mato, J.M., Orobengoa, D., Tasci, E., de la Flor, G., and Kirov, A. (2011) Crystallography online: Bilbao Crystallographic Server. *Bulg. Chem. Commun.*, **43**, 183.

94 Aroyo, M.I., Perez-Mato, J.M., Capillas, C., Kroumova, E., Ivantchev, S., Madariaga, G., Kirov, A., and Wondratschek, H. (2006) Bilbao Crystallographic Server I: databases and crystallographic computing programs. *Z. Kristallogr.*, **221**, 15.

95 Aroyo, M.I., Kirov, A., Capillas, C., Perez-Mato, J.M., and Wondratschek, H. (2006) Bilbao Crystallographic Server II: representations of crystallographic point groups and space groups. *Acta Crystallogr. E*, **62**, 115.

96 Hearne, G.R., Pasternak, M.P., Taylor, R.D., and Lacorre, P. (1995) Electronic structure and magnetic properties of $LaFeO_3$ at high pressure. *Phys. Rev. B*, **51**, 11495.

97 Xu, W.M., Naaman, O., Rozenberg, G.K., Pasternak, M.P., and Taylor, R.D. (2001) Pressure-induced breakdown of a correlated system: the progressive collapse of the Mott–Hubbard state in $RFeO_3$. *Phys. Rev. B*, **64**, 094411.

98 Javaid, S., Akhtar, M.J., Ahmad, I., Younas, M., Shah, S.H., and Ahmad, I. (2013) Pressure driven spin crossover and isostructural phase transition in $LaFeO_3$. *J. Appl. Phys.*, **114**, 43712.

99 Javaid, S. and Akhtar, M.J. (2014) Pressure-induced magnetic, structural, and electronic phase transitions in LaFeO$_3$: a density functional theory (generalized gradient approximation)+U study. *J. Appl. Phys.*, **116**, 023704.

100 Rozenberg, G.K., Pasternak, M.P., Xu, W.M., Dubrovinsky, L.S., Carlson, S., and Taylor, R.D. (2005) Consequence of pressure-instigated spin crossover in RFeO$_3$ perovskites: a volume collapse with no symmetry modification. *Europhys. Lett.*, **71**, 228.

3
Neutron Diffraction

Martin Meven[1,2] and Georg Roth[1]

[1]*RWTH Aachen University, Institute of Crystallography, Jägerstraße 17-19, 52056 Aachen, Germany*
[2]*Forschungszentrum Jülich GmbH, Jülich Centre for Neutron Science at Heinz Maier Leibnitz Zentrum, Lichtenbergstraße 1, 85747 Garching, Germany*

3.1
Introduction

Many mechanical, thermal, optical, electrical, and magnetic properties of solid matter depend decisively on its atomic structure. Therefore, any fundamental understanding of the physical properties requires knowledge of the spatial arrangement of atoms, ions, molecules, and also magnetic moments in the solid. If the material is available in crystalline form (either single- or polycrystalline), diffraction is the most powerful tool to answer questions about the atomic and/or magnetic structure. In a physicist's description, diffraction is equivalent to coherent, elastic scattering, where the term "coherent" stands for "able to interfere" and "elastic" means "scattering without change of energy." Diffraction experiments may be performed with a number of different types of radiation: X-rays, electrons, and neutrons. Modern electron microscopy (HRTEM) can achieve atomic resolution in "imaging mode," but more detailed and quantitative information on the 3D atomic arrangement in crystals and on 3D magnetic structures and magnetization densities requires the application of diffraction rather than imaging methods.

The basic theory of diffraction used for structural analysis, the so-called kinematical theory, is similar for X-rays (see Section 3.3.1) and neutrons but usually cannot be applied to electron diffraction. Due to the different properties of X-rays, neutrons and electrons, and their specific interaction with matter, complementary information is obtained from experiments with different types of radiation. In this chapter, we focus on neutron diffraction exclusively. X-ray diffraction (see the corresponding section in this volume) will only be discussed to demonstrate this complementarity. Inelastic or incoherent neutron scattering

Handbook of Solid State Chemistry, First Edition. Edited by Richard Dronskowski, Shinichi Kikkawa, and Andreas Stein.
© 2017 Wiley-VCH Verlag GmbH & Co. KGaA. Published 2017 by Wiley-VCH Verlag GmbH & Co. KGaA.

and the information content derived from such experiments are discussed elsewhere [1].

We will start by recollecting a number of unique properties of the neutron (Section 3.2), followed by a brief description of instrumentation for neutron diffraction (Section 3.3). Section 3.4 will discuss a number of examples chosen to represent typical applications of neutron diffraction in solid-state chemistry.

3.2
Neutrons as Structural Probes

3.2.1
Characteristic Properties of Neutrons

For a typical diffraction experiment, the wavelength of the radiation is chosen such that it matches the size of the object to be resolved. As the experiment is expected to resolve atom positions and interatomic distances, such experiments use wavelengths in the range ($0.5\,\text{Å} < \lambda < 2.5\,\text{Å}$). Neutrons in this range are classified according to their energy as "hot" ($0.3\,\text{Å} < \lambda < 1.0\,\text{Å}$), thermal ($1.0\,\text{Å} < \lambda < 2.0\,\text{Å}$), and "cold" ($\lambda > 2.0\,\text{Å}$). Figure 3.1a shows the energy–wavelength relation for X-rays, thermal neutrons, and electrons.

A wavelength of $1\,\text{Å}$ (horizontal bar in Figure 3.1a corresponds, in the X-ray case, to a photon energy of ~10 keV, neutrons with such a wavelength have energies on the order of 20 meV. The scattering process is also very much different: While X-rays are scattered by the atom's electron shell, neutrons are scattered by

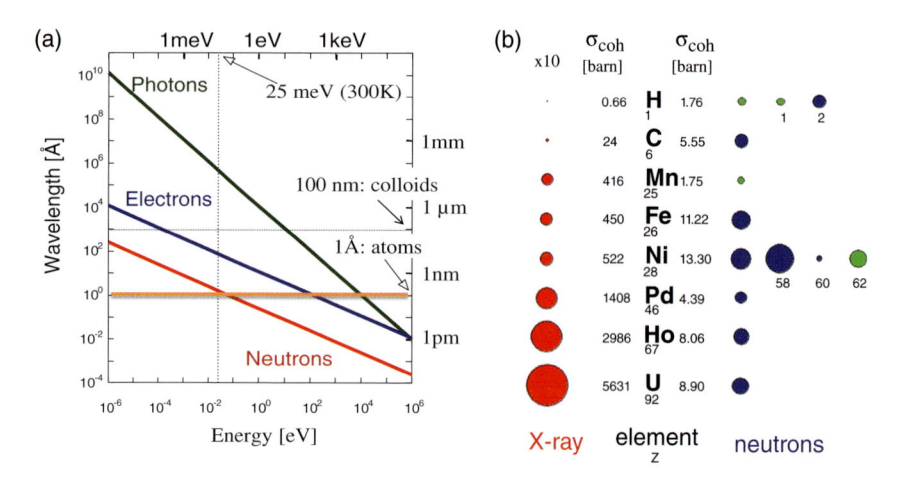

Figure 3.1 (a) Energy–wavelength diagram for photons (X-rays), neutrons, and electrons [2]. (b) Scattering power (coherent cross section, 1 barn = 10^{-24} cm^2) of selected atoms for X-rays (red) and neutrons with wavelengths around $1\,\text{Å}$ (blue = positive values, green = negative values) [2].

the atom's nucleus. This leads to the following characteristic differences between X-ray and neutron diffraction: While the scattering cross section for X-rays increases with the square of the element number (number of shell electrons), the variation of the corresponding neutron scattering cross section is less systematic and has a much smaller range of values (Figure 3.1b). On the other hand, even different isotopes of the same element may have, due to their different nuclear structure, quite divers neutron scattering amplitudes. Figures 4.1b and 4.2a show this for hydrogen and deuterium as well as for a number of different Ni-isotopes.

Various properties of elements/isotopes, including coherent scattering cross sections are available in Ref. [3]. An important fact in neutron scattering is also that hydrogen and particularly deuterium is a strong scatterer. This gives neutron diffraction the unique opportunity to determine hydrogen positions in crystals and also possible hydrogen disorder with high precision. The reason is not only the high scattering power of hydrogen, but also the fact that neutrons give the true position of the hydrogen nucleus, while X-rays would at best detect the maximum of the bonding electron density that is systematically closer to the bonding partner than the true position.

The numbers in Figure 3.1b also show that the scattering cross sections for neutrons are generally orders of magnitude smaller than those for X-rays, resulting in much weaker intensities of neutron reflections. The weaker interaction, combined with the intrinsically smaller neutron flux that can be obtained from the source (compared to laboratory X-ray sources and even more so to synchrotron sources), leads to much weaker intensities in the neutron case and consequently to higher demands on the crystal or sample size (typical values: X-ray: crystal 100 μm, powder: 100 mg, neutrons: crystal >1 mm, powder: >1 g). At the same time the large neutron penetration depths for metallic components make neutron tomography and neutron stress analysis very attractive for certain industrial applications.

Also obvious from Figure 3.2a is the fact that elements that are close to each other in the PSE can have quite different scattering amplitudes. It is therefore,

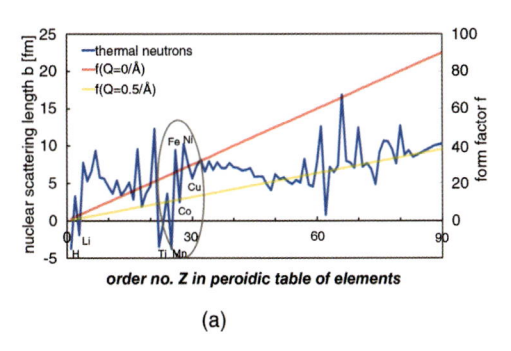

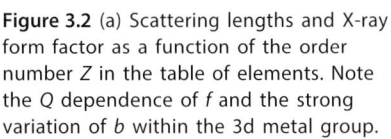

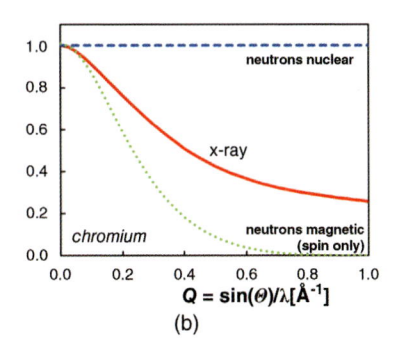

Figure 3.2 (a) Scattering lengths and X-ray form factor as a function of the order number Z in the table of elements. Note the Q dependence of f and the strong variation of b within the 3d metal group. (b) Normalized form factors of chromium (absolute value of the scattering amplitude) as a function of the scattering angle for X-ray and nuclear and magnetic neutron scattering.

much easier to distinguish, for instance, neighboring 3d or 4d transition metal elements and characterize their occupation probabilities with neutrons than with X-rays.

In the X-ray case, the *atomic form factor* (see discussion in Section 3.3.1) falls off with increasing scattering angle (Figure 3.2b). This is the result of destructive interference of X-rays scattered from different parts of the electron cloud and occurs because the scattering object (electron shell) is comparable in size to the wavelength of the scattered radiation. In the neutron case, the scattering object (the nucleus) is much smaller than the wavelength ($10^{-14} \ldots 10^{-15}$ m versus 10^{-10} m), and hence destructive interference does not occur and there is no "form factor fall-off." As a consequence, neutron diffraction yields useful data even at very large diffraction angles, where X-ray diffraction intensities have already fallen off to almost zero. This leads to a much higher precision of the atomic coordinates and atomic displacement parameters (see Section 3.3.1) if derived from a neutron diffraction experiment.

In quite a number of cases, additional information can be drawn from the complementarity of X-ray and neutron diffraction: Neutrons allow us to very precisely determine atom positions (position of the nuclei) and anisotropic displacement parameters while X-rays probe the total electron density distribution in the crystal. The combination of both sources of information allows the experimental determination of the bonding electron density in the crystal, via the so-called X-N-synthesis [4], and the result can then be compared to the results of quantum chemical calculations.

Another peculiarity of the neutron is its *magnetic moment* that interacts with magnetic moments in the sample and leads to "magnetic neutron scattering." The scattering objects in this case are unpaired electrons in the atom's electron shell and consequently, magnetic neutron scattering suffers from a form factor fall-off just as X-ray scattering does (Figure 3.2b). Magnetic and nuclear neutron diffraction intensities are typically of comparable magnitude. Magnetic structures are therefore, quite easy to measure experimentally. Neutrons can also be *polarized* and, if so, the nuclear and magnetic scattering contribution can interact coherently, partly alleviating the so-called "phase problem of crystallography." The combination of polarized incident neutrons and polarization analysis on the detector side allows determining very complex, noncollinear, and possibly incommensurately modulated magnetic structures, evaluate magnetic domains, and also determine the magnetic chirality of structures [5]. Magnetic X-ray diffraction (particularly resonant X-ray scattering), on the other hand, has been demonstrated successfully at synchrotron sources [6] but the signals are usually very weak and the intensities hard to interpret quantitatively.

Absorption in matter is described by the Lambert–Beer law:

$$I = I_0 \exp(-\mu x), \; \mu(\text{cm}^{-1}) = \text{linear absorption coefficient}, \; x \,(\text{cm})$$
$$= \text{mean path through sample}.$$

The linear absorption coefficient is a macroscopic isotropic property depending on the wavelength and kind of radiation. X-ray penetration depths range only to a few millimeters or below (e.g., for silicon with $\mu_{MoK\alpha} = 1.546\,\text{mm}^{-1}$ and $\mu_{CuK\beta} = 14.84\,\text{mm}^{-1}$ penetration depths are 3 mm and 0.3 mm, respectively). For experiments with sample environments like cryostats or furnaces, this can become a severe limitation. In contrast to that thermal neutrons have for most elements a penetration depth of several centimeters. As this holds true for many construction materials, such as steel or aluminum, it becomes very easy to use bulky sample environments (cryostats, furnaces, magnets, pressure cells, etc.) on neutron diffractometers without severely obstructing the incoming and diffracted beams. Only few isotopes are very strong absorbers, for example, ^{10}B, ^{113}Cd, and especially some in the group of the lanthanides such as Gd, Dy, and Sm. If possible, one should strictly avoid them by using less absorbing isotopes if available. As an alternative applicable also for X-rays, one can reduce absorption using smaller wavelengths (gamma radiation and hot neutrons). The latter has been demonstrated successfully for several compounds at the diffractometer HEiDi (e.g., GdMnO$_3$ [7], EuFe$_2$As$_2$, [8,9]). If absorption cannot be neglected, its bias to intensities due to different scattering paths through the sample for different reflections needs to be corrected. Even for a spherical sample the mean path lengths depend on 2θ. In addition, the absorption of the sample environment might need to be considered.

3.2.2
Generation of Neutrons and Neutron Sources

Neutrons unite particle and wave specific aspects: They have a mass $m_n = 1.675 \times 10^{-27}$ kg, no electric charge $Q = 0$ C, a spin ½ and a magnetic dipole moment $\mu_n = -9.6623647(23) \times 10^{-27}$ J/T. The following relationship exists between its kinetic energy E, velocity v, wavelength λ, wave vector k, and temperature (k_B Boltzmann constant, $\hbar = h/2\pi$ reduced Planck constant):

$$E = m_n v^2/2 = k_B T = (\hbar k)^2/2m_n, \quad k = 2\pi/\lambda = m_n v/\hbar$$

Thermal neutrons have an energy of approximately 25 meV, a velocity of about 2500 m/s, and a wavelength of 1.8 Å. Unbound neutrons have a mean lifetime ($\tau \approx 880$ s) with $n \rightarrow p + e^- + \nu_e + 0.78$ MeV. On the Earth there are no natural sources of noticeable neutron flux. In fact the discovery of neutrons took place in the 1930s when the irradiation of beryllium atoms with alpha particles generated carbon atoms and neutrons. Nowadays, this is known as the radium beryllium source. To generate high neutron flux, the following are the two major possibilities:

i) *Reactor sources:* Here, a thermalized neutron hits a ^{235}U nucleus. The resulting ^{236}U nucleus is unstable and splits into two separate nuclei and 2.4 fast neutrons on average. Within the reactor vessel these neutrons are moderated to thermal energies. A part of them is used to keep the fission process running while another part is used for experiments. An example of such a reactor source is the FRM II at the Heinz Maier-Leibnitz Zentrum

(MLZ) in Garching with a thermal power of 20 MW. It generates about 5×10^{14} neutrons/cm^2/s near the fuel element. These are moderated in a D$_2$O-tank in the reactor vessel to a broadband of thermal energies. Via channels in the reactor vessel, neutrons reach experiments in the experimental and neutron guide halls. The source generates a continuous flux of neutrons until the ^{235}U fuel has been burned up. Therefore, this kind of source is called a *continuous source*. Inside Europe, the BER II reactor of the Helmholtz Zentrum Berlin (HZB) in Berlin, the ORPHÉE reactor of the Laboratoire Léon Brillouin (LLB) at Saclay (F), and the high-flux reactor of the Institute Laue-Langevin (ILL) at Grenoble (F) should also be noted as continuous neutron sources for research.

ii) *Spallation sources:* These sources require strong proton accelerators to strip neutrons from heavy nuclei like tungsten in a target [10]. Each impact of a proton into the target generates a very short but intense "shower" of neutrons with fluxes several orders of magnitude higher than for a continuous source. As the accelerators produce only short pulses of protons, this kind of source is called a *pulsed source*. Inside Europe, the ISIS in Oxfordshire (GB) operates since 1984. The most recent spallation source is the upcoming European Spallation Source (ESS) near Lund[1] in Sweden. The proton accelerator will have an average beam power of 5 MW with a peak power of 125 MW. The target will be a helium cooled tungsten wheel with a neutron production rate of 10^{18} neutrons/s.

3.3
Neutron Diffraction

3.3.1
Diffracted Intensities and Corrections

X-ray and neutron diffraction methods share the same principle but also show noteworthy differences. In order to keep things simple in the following description, we will focus on the so-called angular dispersive techniques, also known as constant wavelength techniques, for neutrons and ignore so-called energy dispersive techniques, for example, time-of-flight (TOF) and Laue diffraction. As mentioned in the section about single-crystal X-ray diffraction, there are two sources of information that yield insight into crystalline structures. The first one is the Bragg equation:

$$2d_{hkl} \cdot \sin \theta_{hkl} = \lambda.$$

It establishes a relationship between a selected lattice plane $(h\ k\ l)$ with lattice spacing d_{hkl}, radiation wavelength λ, and the diffraction angle θ_{hkl} that allows us to *determine the unit cell* for a specific crystal.

---

1) europeanspallationsource.se/.

The second source of information is the measured intensities I_{hkl} of the Bragg reflections that contain the corresponding structure factors F_{hkl}:

$$I_{hkl} \sim |F_{hkl}|^2, \text{ with } F_{hkl} = \Sigma_{i=1}^{n} b_i(\mathbf{Q}) \exp(2\pi \iota (hx_i + ky_i + lz_i)),$$

with $n =$ total number of atoms in the unit cell and i representing the ith atom.

For neutrons, the scattering length b is replacing the atomic scattering factor f that is used to describe the scattering power of atoms in the case of X-rays. The neutron diffraction experiment senses the nuclear (and magnetic) density distribution within the unit cell. Measuring a set of Bragg reflections allows *determining atomic positions* or more strictly the positions of the nuclei.

Neutrons scatter isotope specific and one needs to consider that the scattering length b is a statistical average over all scattering lengths b_j of N isotopes, $1 \leq j \leq N$, of one or more elements including vacancies ☐ on a site $<b> = \Sigma_{j=1}^{N} p_j b_j + p_0^* ☐$ with $1 = \Sigma_{j=0}^{N} p_j \cdot p_j$ are the probabilities of the occurrence of the j isotope or vacancy ($j = 0$). One can easily extend this description to *study vacancies and site disorder* (= different elements on same site). In contrast to X-rays, some nuclei have negative scattering lengths (an additional phase shift between incoming and outgoing radiation) for thermal neutrons. This unique feature can be used to tailor mixed compounds that show no Bragg scattering ($<b> = 0$), for example, for sample environment.

Thermal movements of atoms around their average positions are treated exactly as in the X-ray case yielding a *mean square displacement* (MSD) B_i (see Section 3.3.4). The treatment can be extended to anisotropic (ADP) and even to anharmonic displacements. The impact of thermal movement on the structure factor is an additional angular dependent factor to be multiplied with the scattering length, the so-called Debye–Waller factor:

$$b_i(\mathbf{Q}) \rightarrow b_i(\mathbf{Q})^* \exp(-B_i (\sin \theta_Q / \lambda)^2).$$

With increasing temperature the Debye–Waller factor attenuates the Bragg reflection intensities. Furthermore, it depends on Q and weakens reflections at large scattering angles stronger than those at small angles. An important point is the fact that neutron diffraction data are much better suited for obtaining accurate displacement parameters because the missing form factor fall-off allows to measure high angle reflections more reliably. It should also be noted that the Q-independent scattering lengths of neutrons are advantageous for studies on ADP in two ways: X-ray form factors show a similar Q dependence as ADP causing a severe intensity reduction at higher Q values plus potential correlation problems during refinement while neutrons do not.

The structure factor equation is based on the so-called kinematical approximation, stating that the scattered beams are always weak compared to the primary beam and the interaction between neutrons and matter can be described as a single scattering event. This assumption is often not strictly valid for neutrons and the resulting *multiple scattering* events give rise to artifacts like extinction and "Umweg-Anregung" (the Renninger-effect) (see Figure 3.3a and b).

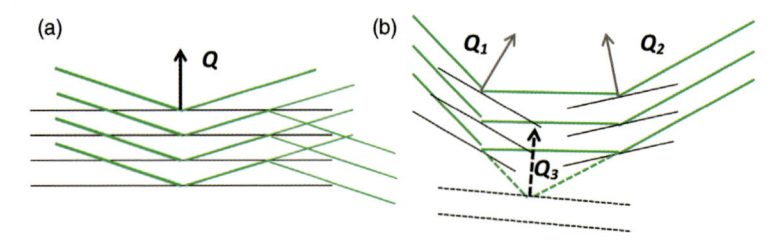

Figure 3.3 (a) Extinction effect. (b) Umweg-Anregung.

Multiple scattering effects emerge if one takes into account that the incident beam interacts not only once with the set of lattice planes in a crystal but that every diffracted beam can act as a new incident beam scattering at the same or different layers. Multiple scattering affects Bragg intensities in two possible ways:

The *extinction effect* describes the weakening effect on Bragg intensities caused by multiple scattering on the same set of layers as the diffracted beam becomes partially backscattered toward the direction of the very first incident beam. High perfectness of a crystal amplifies this effect. Especially, for very strong reflections it can reduce intensities dramatically (up to 50% and more). Theoretical models that include a quantitative description of the extinction effect were developed by Zachariasen and Becker and Coppens [11–13]. These models are based on an ideal spherical mosaic crystal with a very perfect single crystal (primary extinction) or different mosaic blocks with almost perfect alignment (secondary extinction) to describe the strength of the extinction effect. It is also possible to take into account anisotropic extinction effects if the crystal quality is also anisotropic. Nowadays, most refinement programs [14–17] include extinction correction. In general, extinction is a problem of sample quality and size and therefore, more commonly a problem for neutron diffraction and not so often for X-ray diffraction with much smaller samples and stronger absorption. Sometimes short wavelengths, where extinction effects become weaker, can be used as a solution.

For multiple scattering from more than one set of lattice planes in the crystal, the diffracted beam of the first lattice plane $(h_1\ k_1\ l_1)$ acts as incident beam for a second not equivalent lattice plane $(h_2\ k_2\ l_2)$ that by accident also fulfils Bragg's law. This is called *Umweg-Anregung* or "Renninger" effect. Umweg-Anregung can severely affect measured intensities of Bragg-reflections. Even more important, the effect can generate intensities at reflection positions forbidden by space group symmetry. This is quite often the origin of wrong assumptions about the crystallographic space group symmetry. Multiple scattering depends strongly on the specific properties of a sample and on the instrumental setup. Although there are tools to deal with this effect [18,19], its correction can become quite cumbersome.

3.3.2
Concepts and Components of Instrumentation

The following scheme (Figure 3.4) represents a diffraction experiment using the "constant wavelength" (angle dispersive) method. The beam from a "white"

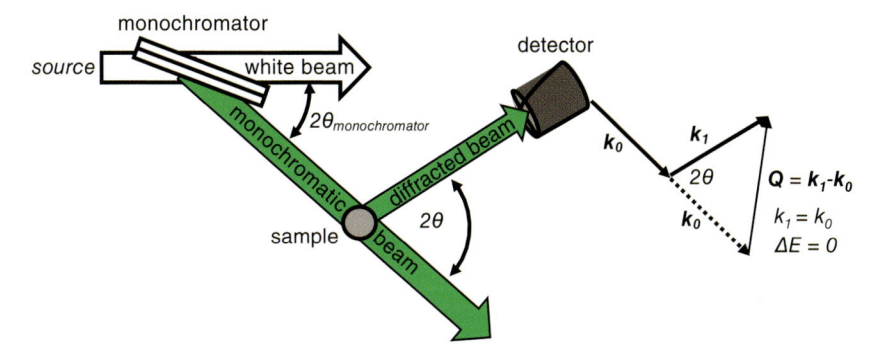

Figure 3.4 Schematic representation of a constant wavelength diffractometer.

source is diffracted by a monochromator with given d spacing and a fixed Bragg angle $2\theta_M$ to select a specific wavelength λ. Collimators can be used to limit the beam divergence to α_1 and α_2 of the beam before the sample and to α_3 for the diffracted beam (diffraction angle 2θ) after the sample, respectively.

X-ray tubes in laboratory sources generate "out of the box" strong quasi monochromatic radiation by exciting characteristic lines K_α and K_β of a metallic target, for example, Cu or Mo. Neutron (and also synchrotron) sources deliver a much broader band of energies and wavelengths, a "white" beam. Therefore, one has to equip the diffractometer with a crystal monochromator as shown in Figure 3.4. The monochromator selects a band of wavelengths ($\lambda \pm \Delta\lambda/\lambda$) from the beam. The corresponding *bandwidth* $\Delta\lambda/\lambda$ can be influenced by the mosaic spread of the monochromator crystal itself but also by collimation before and after the monochromator. A small bandwidth improves the angular resolution but also reduces the total flux toward the sample. Diffraction with a monochromatic beam typically uses a narrow band of energies/wavelengths on the order of $\Delta\lambda/\lambda < 10^{-2}$–$10^{-3}$ or even smaller. For such a small bandwidth, the flux of neutrons is several orders of magnitude smaller than the flux of X-rays of a corresponding synchrotron source or even an X-ray tube in the laboratory. This and the small scattering power of matter for thermal neutrons have to be compensated by a larger sample size of several millimeters to centimeters.

A closer look reveals that in all cases already discussed, the monochromatic beam becomes contaminated by additional bands of wavelengths and this introduces an ambiguity between the d spacing of a crystal lattice and its set of Bragg angles. Monochromator crystals diffract for a given angle not only a specific wavelength λ but also its higher orders $\lambda/2$, $\lambda/3$, and so on corresponding to the higher order reflections ($2h$, $2k$, $2l$), ($3h$, $3k$, $3l$), and so on. For neutrons, this so-called $\lambda/2$ *contamination* can be suppressed by using neutron filters (e.g., Er foil with a strong absorption resonance around 0.4 Å). Another way to attenuate short wavelengths is to use the scattering from materials like beryllium or graphite. These filters use the fact that there is no Bragg diffraction for $\lambda > 2d_{max}$, where d_{max} is the largest interplanar spacing of the unit cell. For such long wavelengths, the reflection condition (Bragg equation) cannot be fulfilled. Below this critical wavelength, the neutron beam is attenuated by diffraction and this can be used to suppress higher order reflections very effectively. Frequently used materials are polycrystalline beryllium and graphite. Due to their unit cell dimensions, they are particularly suitable for experiments with long wavelength (cold) neutrons because they block wavelengths smaller than about 3.5 Å and 6 Å, respectively.

The angular resolution of any diffractometer is an important characteristic. Figure 3.9 shows the resolution function (reflection full width at half maximum (FWHM) as a function of scattering angle) for the four-circle single-crystal neutron diffractometer HEiDi at FRM II shown on the left. The resolution depends on a number of factors, among them being the collimation, the monochromator type and quality, the 2θ and (hkl) of the reflection used for monochromatization and so on. One should also note that for diffraction usually there is no analysis of the energy of the scattered beam behind the sample. Therefore, the dominant elastic scattering is accompanied by quasielastic and inelastic scattering contributions that also hit the detector.

Again in close similarity to X-rays, two further instrumental effects influence the measured intensities and need to be corrected for. One is the *Lorentz factor L*, a geometrical factor that takes into account the 2θ dependence of the *effective* time a set of lattice planes rotates through the Bragg position (details can be found in textbook discussions about construction of Ewald sphere). The second effect is the *polarization* of the beam. It should be noted that the physical background of this correction is fundamentally different for X-ray and neutron diffraction. In the X-ray case, it is due to the (partial) polarization that occurs in any diffraction event of an electromagnetic wave. For neutrons, this effect is absent and the term polarization relates to the magnetic dipole moment of the neutron [1].

Monochromatic powder diffraction experiments yield $I/2\theta$ diagrams while single-crystal diffraction experiments yield a list of corresponding intensities and their standard deviations for a set of Bragg reflections that includes full spatial information. In both cases, a structure refinement has to be performed (see Section 3.3.1) that optimizes the agreement between experimental data and data calculated from structural models. The concepts and definitions of the various numbers measuring the quality of the agreement (crystallographic reliability factors: internal R-value R_{int}, *unweighted R-value*: R_u, the weighted R-value R_w, Goodness of Fit S, etc.) remain the same as for the X-ray case (see corresponding section). Please note, that for powder diffraction, other residual values are also important such as R_{Bragg} and R_p.

3.3.3
Powder Neutron Diffraction

Crystalline powder consists of a large number of randomly oriented microscopic single crystals. A powder sample in a small capillary or on a flat sample holder generates cones of diffracted radiation, the so-called Debye Scherrer cones, if irradiated by a directed monochromatic beam. Following Bragg's law, the discrete angles at which these intensities occur can be measured by a single detector or an area detector. From these results one can derive the crystallographic unit cell and the crystal system (see Section 3.3.1). The resulting powder diffractogram represents the intensity distribution versus 2θ or Q. The relatively short time for data acquisition is one main advantage of powder diffraction over single crystal diffraction. At high flux sources, a complete diffractogram can be collected within a few minutes. Thus, one can not only perform data analysis routinely and quickly for many samples but also do *in situ* experiments as a function of temperature, pressure, and time to follow chemical reactions with the possibility to even observe intermediate states. This holds true especially for usage of area detectors with large angular range. Another advantage is that the principle of $\theta/2\theta$-scans on powders reveals the full intensity distribution within a given 2θ range yielding a complete overview (compared to a single crystal experiment with point detector). One final advantage of powder samples is their availability as it is much easier to get a single-phase powder sample of a particular compound than to grow a large single crystal.

Although powder diffraction provides quickly results, its main disadvantage is that three-dimensional diffraction data are projected to one-dimension ($I/2\theta$) and the information content is severely reduced by reflection overlap or coincidence. The instrumental (angular) resolution of a powder diffractometer is therefore, of prime importance. Also, a nonrandom orientation of crystallites within a powder sample, the so-called preferred orientation (or texture), can affect the intensity distribution within the Debye Scherrer cones and generate erroneous diffraction intensities (Figure 3.5).

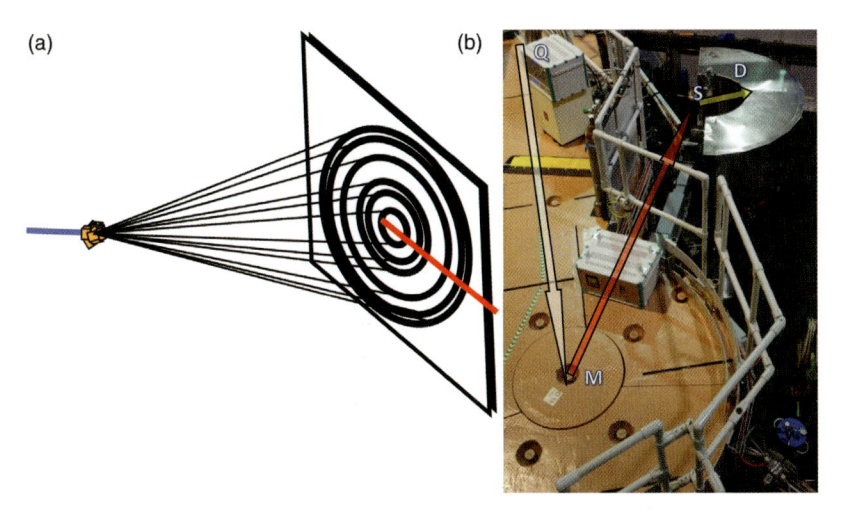

Figure 3.5 (a) Sketch of a powder diffraction experiment, diffraction cones are recorded on a 2D or 1D detector (red line). (Reproduced from Braggs world, talk by Th. Proffen, ebookbrows.com/proffen-talk-bragg-pdf-d59740269. (retrieved: May 05, 2015).) (b) Neutron powder diffractometer SPODI at MLZ (Q: source, M: monochromator, S: sample, D: detector).

Instrumentation: Depending on the type of the neutron source (continuous versus pulsed), different instrumental concepts have been developed to take advantage of the specific properties of the primary beam. In this contribution, we only discuss constant wavelength powder diffractometry as opposed to time of flight methods [1,20]. An example for a thermal neutron powder diffractometer with fixed wavelengths of 1.11 Å, 1.55 Å, or 2.54 Å is the instrument SPODI at the MLZ/Garching. The broad spectrum of thermal neutrons has to be monochromatized by a monochromator crystal first (e.g., Ge-331, Ge-551, Ge-711) before it hits the sample. For fast data collection, the detector consists of a set of ^{3}He detectors covering an angular range of about 160° in 2θ that count simultaneously. Due to the high penetration depth of neutrons, a variety of sample environments are applicable to cover a temperature range from 0.5 K up to more than 2000 K and magnetic fields up to several Tesla and tensile rig to study tensile stress or compression stress or torsion with $F_{max} = 50$ kN, $M_{max} = 100$ Nm [21].

Neutron Rietveld analysis: As already described, the conversion from 3D- to 1D-intensity data caused by the averaging over all crystallite orientations in a powder sample restricts the informative value of powder diffraction experiments. Even with optimized resolution, the overlap of reflections on the 2θ-axis often prohibits the extraction of reliable integrated intensities for individual reflections from the experiment. Instead, the Rietveld method, also referred to as *full pattern refinement*, is used to refine a given structural model against powder diffraction data. The method, which is now also widely used in powder X-ray-diffraction, was invented by Hugo Rietveld in 1966 for the structural analysis from powder neutron data [22,23]. Full pattern refinement means that along with the structural parameters (atomic coordinates, thermal displacements, site occupations) that are also optimized in a single-crystal structure refinement, additional parameters such as the shape and width of the reflection profiles and their 2θ-dependence, background parameters, lattice parameters, and so on can be refined. Even more, in the case of multiphase samples this method allows to split up overlapping intensity contributions according to the different structural models.

Powder neutron diffraction experiments profit from the particular properties of neutrons as structural probes, among them being a vastly different contrast between atom types being close to each other in the PSE, large scattering power of weak X-ray scatterers such as hydrogen, and the sensitivity of neutrons to (ordered) magnetic moment in the sample (magnetic neutron diffraction). The latter point is exemplified in Figure 3.6 that shows the low angle part of the neutron powder diffractogram of $CoGeO_3$ above (large panel) and below (inset) the magnetic ordering phase transition.

3.3.4
Single-Crystal Diffraction

A single-crystal experiment allows access to the full 3D diffraction data. The scheme in Figure 3.7a shows a single crystal in the center of an Eulerian cradle

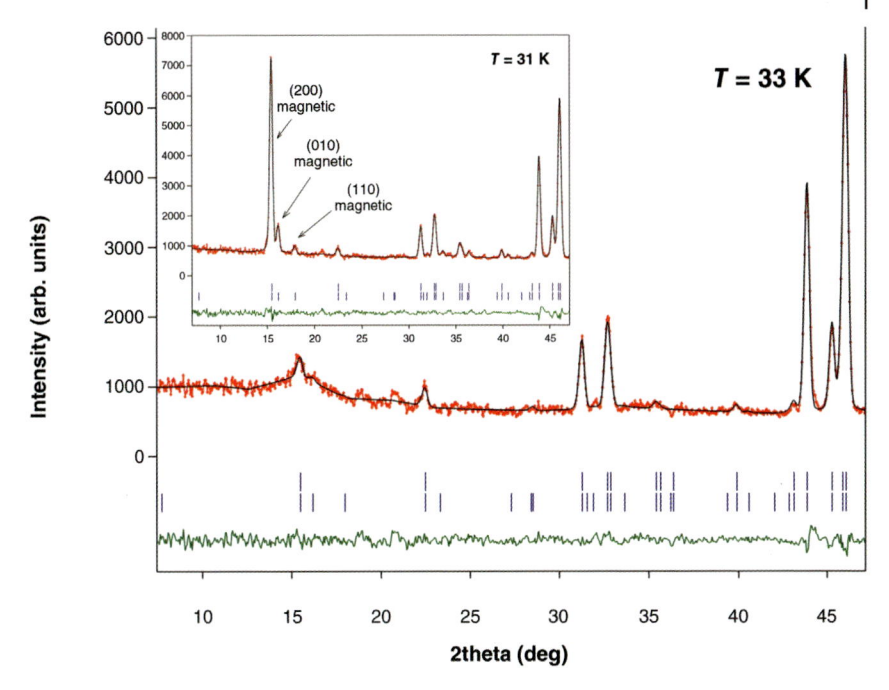

Figure 3.6 Rietveld refinement at the magnetic phase transition of CoGeO$_3$ [24], red: measured intensity I_{obs}, black: model calculation I_{calc}, green: difference, blue: tick-marks at allowed reflection positions. Figure shows low-angle part of two diffractograms measured at SPODI at 33 and 31 K. Note the strong magnetic reflection appearing below magnetic ordering transition (inset).

with its three axes ω, χ, φ. These angles allow the crystallographic coordinate system of the sample to be oriented within the coordinate system of the diffractometer toward any direction. Virtually any lattice plane can be oriented to make its diffracted beam hit the detector in the diffraction plane defined by the 2θ axis. As each lattice plane can be measured separately, one overcomes the problem of

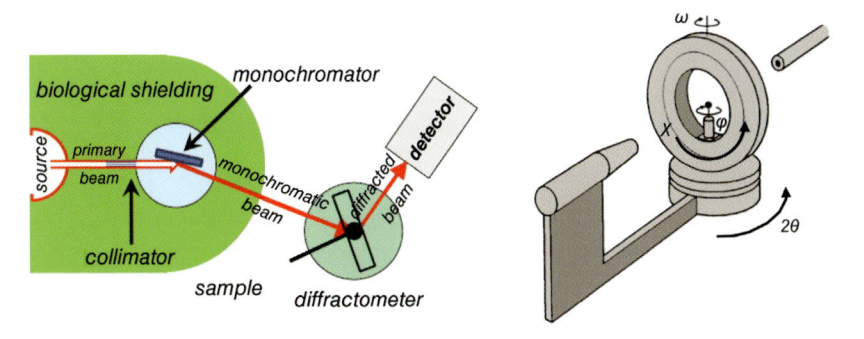

Figure 3.7 Scheme of a traditional single crystal diffractometer with single detector and Eulerian cradle.

reflection overlap that is inherent to powder diffraction. Furthermore, one can study certain anisotropic effects hidden in powder diffractometry like orientation dependent extinction. In spite of these two significant advantages, one has to keep in mind that the sequential measurement of individual Bragg intensities can become very time consuming, especially in the case of single-crystal neutron diffraction for two reasons: First, the low flux at neutron sources compared to X-ray sources defines a lower limit for the required data collection times that ranges in minutes per reflection and not in seconds. Second, the weak interaction of neutrons with matter requires single-crystal samples in the millimeter range instead of the 10–100 μm range for X-rays. Not all compounds can be grown to this size. These problems can only be partly overcome by utilizing 2D position sensitive neutron detectors.

In large-scale neutron facilities, several single-crystal diffractometers are installed to cover the need of different applications. For instance, at the MLZ there are four single-crystal diffractometers: BioDIFF for studies on large/biological structures using cold neutrons, RESI for standard studies on crystal structures with thermal neutrons, HEiDi for detailed studies on small unit cells focusing on detailed studies of ADP and highly absorbing compounds using hot neutrons, and POLI designed for studies on magnetic compounds using hot polarized neutrons. Other neutron large-scale facilities like the ones mentioned in Section 2.2 offer similar suites of instrumentation.

In what follows, the single-crystal neutron diffractometer HEiDi will be discussed as an example in more detail. Again, it should be noted that there are other instrumental concepts like Laue or TOF (time of flight), which are for example, optimized for pulsed neutron sources [25].

3.3.4.1 Instrumentation

An example for a classical neutron single-crystal diffractometer is HEiDi, operated jointly by RWTH Aachen and JCNS at the hot source of FRM II at MLZ. The instrument offers discrete wavelengths between 1.2 Å and 0.4 Å and a large Q range using a rotating stage to choose between different monochromator crystals and angles (Figure 3.8a and b).

The resolution function of the instrument and two tables concerning the available wavelengths and maximum Q ranges are listed in Figure 3.9. Data collection is performed collecting integral intensities for a list of *hkl* reflections. This is done by rotating the crystal for each reflection to a position where the diffracted beam hits the detector and rotating the sample afterward while the detector stays on a fixed position (Rocking scan/ω-scan) or with synchronously moving detector ($\omega/2\theta$ scan). Special software transforms the collected Bragg intensities into a list of intensities of all reflections, including a list of the corresponding standard deviations.

3.3.4.2 Sample Environment

The standard sample environment on HEiDi covers a temperature range between 2.5 K and up to more than 1300 K and the possibility of external

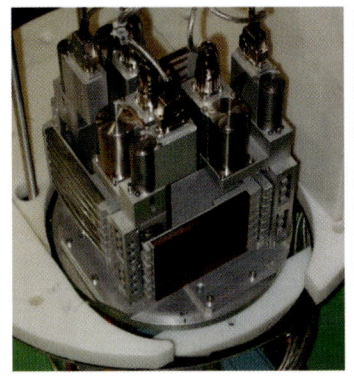

Monochromator HEiDi				
crystal	Ge(311)	Cu(220)	Ge(422)	Cu(420)
2θ$_{Mono.}$	Wavelengths λ			
20°	0.593	0.443	0.408	0.280
40°	1.168	0.870	0.793	0.552
50°	1.443	1.079	0.993	0.680
	Q range			
20°	1.46	1.95	2.12	3.09
40°	0.74	0.99	1.09	1.57
50°	0.60	0.80	0.87	1.27

Figure 3.8 (a) Monochromator stage of HEiDi (in front: Cu(220), to the left: Ge(311)). (b) Wavelengths and Q ranges for different monochromator crystals and monochromator angles 2θ$_{Mono.}$.

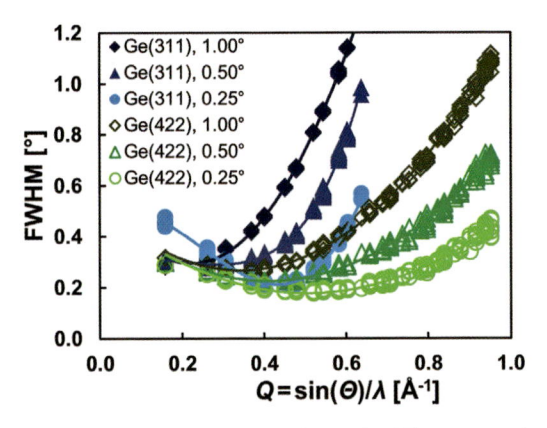

Figure 3.9 Resolution function of HEiDi for different monochromators; 2θ$_{Mono.}$ = 40°, collimations (1°, 0.5°, 0.25°).

electrical fields (Figure 3.10). Other instruments such as POLI are specialized on studies of magnetic compounds in zero-field or applied magnetic fields up to several Tesla or can use special sample environments for studies down to temperatures in the milliKelvin range such as RESI.

3.4 Examples

In this chapter, we briefly describe a few typical examples for the application of neutron diffraction in solid-state chemistry. They were selected in order to

Figure 3.10 HEiDi (a) Closed cycle cryostat ($T_{min} = 2.5$ K), (b) Optical furnace ($T_{max} > 1000$ °C).

demonstrate some of the distinguishing properties of neutrons in solid-state structural research.

3.4.1
The Hydrogen Challenge

The determination of the structural parameters (coordinates, displacement parameters) of hydrogen atoms in crystals is a special problem that illustrates impressively the different properties of X-rays and neutrons. It is obvious that H or D atoms with $Z = 1$ give only a small contribution to the electron density and, therefore, they are hardly visible in X-ray structure analysis, particularly if heavy atoms are also present in the structure. However, there is an even more fundamental problem: The single electron of H or D is engaged in the chemical bonding and is by no means localized at the proton/deuteron position. Therefore, bond distances from X-ray diffraction involving hydrogen are notoriously wrong and any comparison with quantum mechanical calculations is quite hard to perform. This lack of sound experimental information is in sharp contrast to the importance of hydrogen bonding in solids, particularly in biological molecules like proteins, where hydrogen bonds govern to a large extent structures and functionalities of these "biocatalysts". A combination with neutron diffraction experiments is important to determine the structural parameters of the H/D atoms properly. More generally, the structure analysis by neutron diffraction yields separately and independently from the X-ray data the structure parameters of all atoms, including the mean square displacements due to static and dynamic (even anharmonic) effects.

3.4.1.1 H/D Ordering in Ferroelectric RbH$_2$PO$_4$ (RDP)

The hydrogen problem in crystal structure analysis is of special importance for structural phase transitions driven by proton ordering. KH$_2$PO$_4$ (KDP) is the

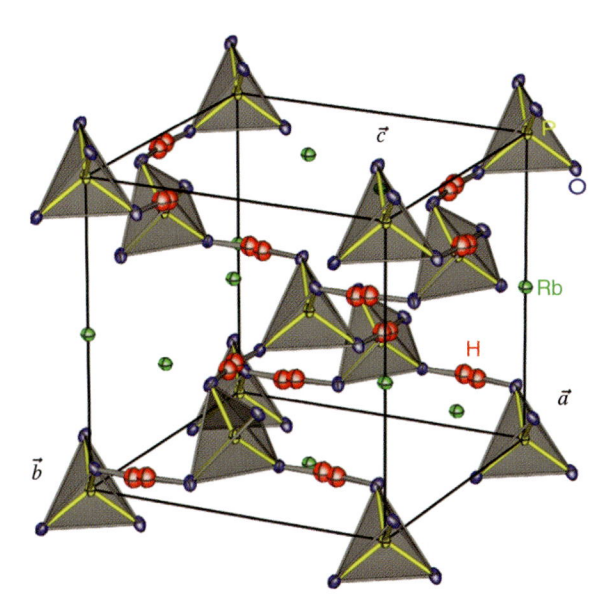

Figure 3.11 Crystal structure of the paraelectric phase of RDP (RbH$_2$PO$_4$) with a split-model representation of the hydrogen disorder [26].

most well-known representative of hydrogen-bonded ferroelectrics. Here, we discuss the isotypic RbH$_2$PO$_4$ (RDP). The crystal structure consists of a three-dimensional network of PO$_4$ groups linked by strong hydrogen bonds (Figure 3.11).

In the paraelectric phase at room temperature, KDP as well as RDP crystallize in the tetragonal space group *I*42*d*, where the H-atoms are dynamically disordered in symmetric O$\cdots$H$\cdots$O bonds, which are almost linear with short O—O distances, typically in the range of 2.5 Å. The disordered H-distribution may be interpreted as corresponding to a double-well potential [27].

Figure 3.12a and b show the corresponding results for RDP, obtained from single crystal neutron diffraction [26].

The two very close hydrogen positions with 50% occupation probability (Figure 3.13) are, of course, an artifact of the time-space averaging that is inherent to diffraction. In this case, the hydrogen disorder is assumed to be a dynamic hopping process between the two energetically degenerate sites.

At $T_c = 147$ K, RDP transforms to a ferroelectric phase of orthorhombic symmetry (space group *Fdd*2) in which the protons order in short asymmetric O—H$\cdots$O bonds (Figure 3.12b). The PO$_4$-tetrahedra show a characteristic deformation with two shorter and two longer P-O distances due to a transfer of electron density to the covalent O—H bonds. The electrical dipole moments are oriented $||z$ that give rise to a polarization along the *c*-direction.

The phase transition temperatures of KDP-type compounds change drastically when H is substituted by D. For instance, for K(H,D)$_2$PO$_4$ the para- to

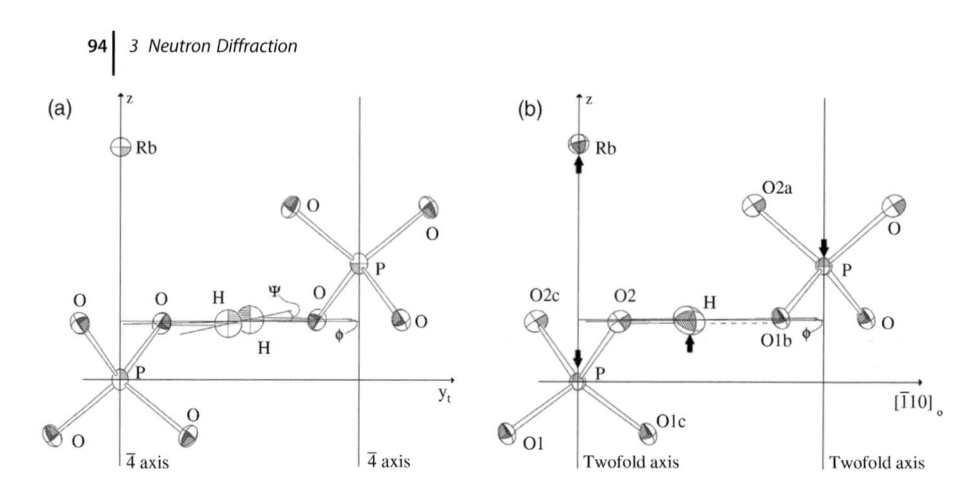

Figure 3.12 (a) Local configuration of two PO$_4$-tetrahedra in the paraelectric phase of RDP at $T_c + 4$ K linked by a strong, disordered hydrogen bond. (b) Ferroelectric, hydrogen-ordered structure of RDP close to the phase transition at $T_c - 1$ K (major changes indicated by arrows) [26].

ferroelectric T_C changes from 122 K in the protonated to 229 K in the deuterated compound. This huge H/D-isotope effect proves that hydrogen ordering and dynamics is the major factor controlling this phase transition. Clearly, the use of neutron diffraction is detrimental to a better understanding of these compounds and their interesting physical properties.

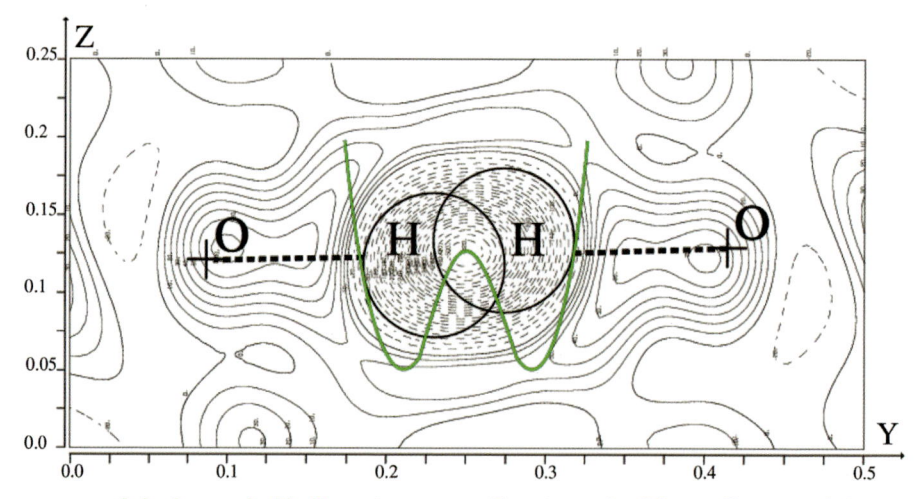

model: dynamic H-disorder according to a double-well potential

Figure 3.13 Difference–Fourier plot of the negative proton density in the hydrogen bond of paraelectric RDP indicated by the broken contour lines [26]. The double-well potential model used to describe this density is inscribed in green.

3.4.2
Diffraction Contrast

3.4.2.1 Crystal Structure and Site Occupation of $(Mn_{1-x}Cr_x)_{1+\delta}Sb$

As an example that exploits the neutron diffraction contrast of atoms with similar atomic numbers, the combination of X-ray and neutron diffraction information is demonstrated for the intermetallic compounds $(Mn_{1-x}Cr_x)_{1+\delta}Sb$ (with $0 \leq x \leq 1$) [28]. This solid solution system is interesting for its magnetic properties: One end member of the solid solution series $(Mn_{1+\delta}Sb)$ shows ferromagnetic behavior while the other one $(Cr_{1+\delta}Sb)$ is a uniaxial antiferromagnet. Intermediate compositions are characterized by competing magnetic interactions leading to a complex magnetic phase diagram. The crystal structure is closely related to the hexagonal NiAs-type structure (space group $P6_3/mmc$) with some additional partial occupation (≤ 0.14) of the interstitial site 2(d) (see Figure 3.14).

Conventional X-ray diffraction can hardly differentiate between chromium ($Z_{Cr} = 24$) and manganese ($Z_{Mn} = 25$) but still yields information on the overall occupation probabilities by (Mn, Cr) for site 2(a) (denoted as a) and site 2(d) (denoted as d). The Sb position is assumed to be fully occupied, thus serving as an internal standard for the scattering power. The compound formula can now be reformulated site-specifically as

$$\underbrace{(Mn_{1-y}Cr_y)_a}_{\text{site 2(a)}} \underbrace{(Mn_{1-z}Cr_z)_d}_{\text{site 2(d)}} Sb,$$

corresponding to a chemical composition of $Mn_{[(1-y)a + (1-z)d]} Cr_{[ya + zd]} Sb$.

In contrast to the X-ray case, the nuclear scattering lengths of Cr and Mn for neutron diffraction are extremely different with $b_{Cr} = +3.52$ fm and $b_{Mn} = -3.73$ fm [3].

In the structure analysis of the neutron data, site-specific effective scattering lengths b_{eff} (2a) and b_{eff} (2d) are refined, which in turn are expressed as

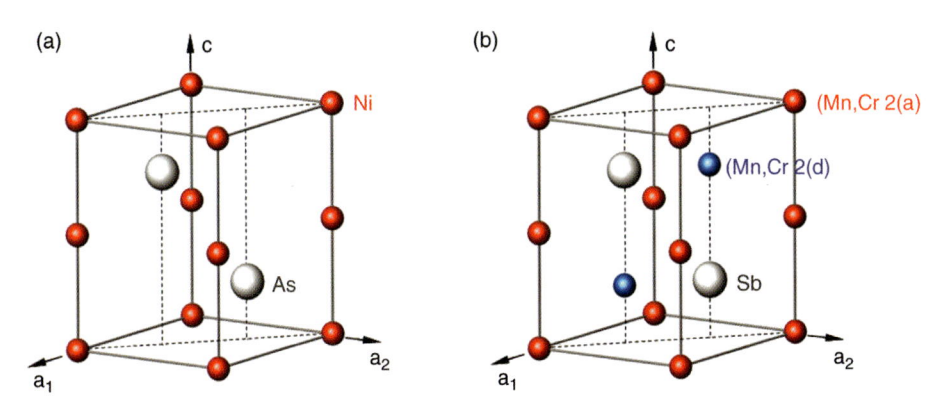

Figure 3.14 (a) NiAs structure. (b) $(Mn_{1-x}Cr_x)_{1+\delta}Sb$ structure.

the following:

$$b_{\text{eff}}(2a) = a \cdot [(1-y) \cdot b_{\text{Mn}} + y \cdot b_{\text{Cr}}] \quad \text{and} \quad b_{\text{eff}}(2d) = d \cdot [(1-z) \cdot b_{\text{Mn}} + z \cdot b_{\text{Cr}}],$$

solving for the unknown parameters y and z gives

$$y = [b_{\text{eff}}(2a)/a - b_{\text{Mn}}]/[b_{\text{Cr}} - b_{\text{Mn}}] \quad \text{and} \quad z = [b_{\text{eff}}(2d)/d - b_{\text{Mn}}]/[b_{\text{Cr}} - b_{\text{Mn}}].$$

The combination of the overall occupation probabilities a and d – from conventional X-ray studies – with the effective scattering lengths $b_{\text{eff}}(2a)$ and $b_{\text{eff}}(2d)$ determined in a neutron diffraction experiment allows the evaluation of the Cr and Mn concentrations on the different sites 2(a) and 2(d).

It is evident that the individual (Cr, Mn) distributions on the two crystallographically different sites 2(a) and 2(d) are not accessible merely by a chemical analysis. For most of the samples studied, the site 2(a) was found to be fully occupied: $a \approx 1.0$ (not necessarily $d \approx \delta$ though). But the formula $(Mn_{1-x}Cr_x)_{1+\delta}Sb$ used normally is only correct for the special case of equal Cr:Mn ratios on both sites:

$$x = y = z \quad \text{and} \quad 1 + d = a + d.$$

Note that, in general, a statistical occupation of one crystallographic site by *three* kinds of scatterers – for example, Mn, Cr, and "vacancies" – requires at least *two* independent diffraction experiments with sufficiently different relative scattering power of the atoms involved to determine the fractional occupancies.

The detailed information on the (Cr, Mn) distribution is needed to explain the magnetic properties of these intermetallic compounds, but we will not further elaborate on this.

3.4.3
Static Disorder and Displacements in HT Superconductors

$La_{2-x}Sr_xCuO_4$ is one of the cuprate high-T_c superconductors with K_2NiF_4 structure (layered perovskite) (Noble prize in physics 1988, Bednorz and Müller [29]). Pure La_2CuO_4 is an insulator while doping with earth alkali metals (Ca^{2+}, Sr^{2+}, Ba^{2+}) on the La^{3+} lattice position generates, depending on the degree of doping, superconductivity with a maximum T_c of 38 K for Sr doping of $x = 0.15$ [30].

Pure La_2CuO_4 undergoes at $T_{t-o} = 530$ K a structural phase transition from the tetragonal high temperature phase (HTT) $F4/mmm$ ($a = b = 5.384$ Å, $c = 13.204$ Å, $\alpha = \beta = \gamma = 90°$ at $T = 540$ K) to the orthorhombic low temperature phase (LTO) $Abma$ ($a = 5.409$ Å, $b = 5.357$ Å, $c = 13.144$ Å, $\alpha = \beta = \gamma = 90°$ at room temperature). For $La_{2-x}Sr_xCuO_4$ the phase transition temperature T_{t-o} drops with increasing doping and disappears above $x = 0.2$.

The following peculiarities of structure analysis with neutrons can be seen from this example:

Light elements: The phase transition is driven by a displacement of the oxygen atoms (see Figure 3.15). Together with the Cu atoms the oxygens form CuO_6 octahedra with four oxygens in the ab-plane (O_{inplane}) and 2 oxygens above/below

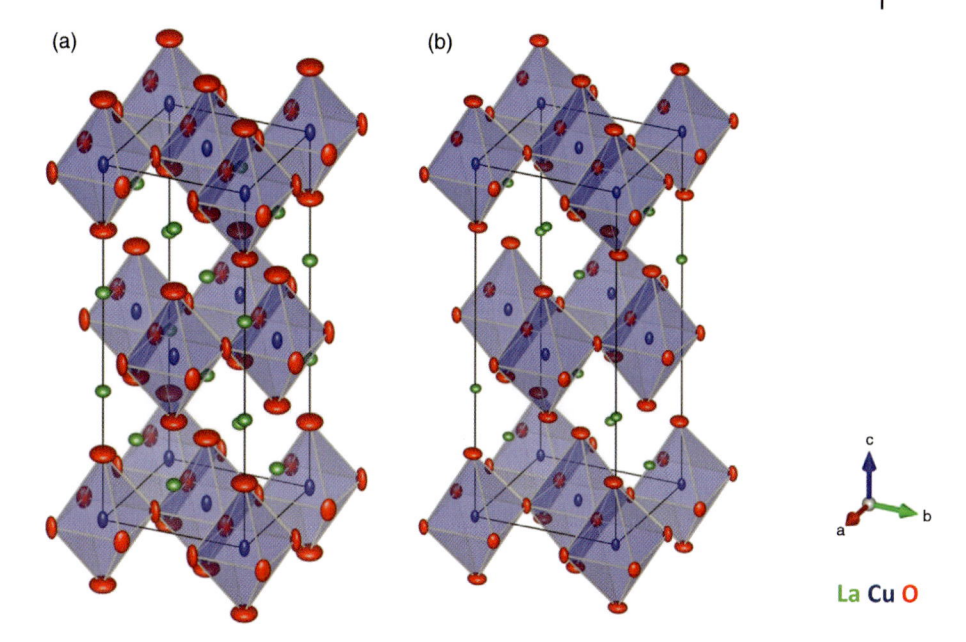

Figure 3.15 Tetragonal HTT phase (a) and orthorhombic LTO phase of La_2CuO_4 (b).

(O_{apical}) each Cu atom. As the oxygen atoms are much lighter than any other element in this compound, the accurate observation of these displacements depends strongly on the chosen type of radiation. As the atomic positions of Cu and La do not change significantly between the HTT and LTO phase, the structure factor for the superstructure reflections can be written as

$$F(hkl) \sim \Sigma_i \, s_i \cdot \exp(-2\pi i(hx_i + ky_i + lz_i)) = F(hkl)_{\text{apex oxygen}} + F(hkl)_{\text{in plane oxygen}} + F(hkl)_{\text{remains}}$$
$$\to F(hkl)_{\text{apex oxygen}} + F(hkl)_{\text{in plane oxygen}}.$$

As the apex oxygen moves away from the $x=0$ position to $(x0z)$, the corresponding superstructure reflection for odd h has the structure factor

$$F(hkl)_{\text{apex oxygen}} = \sin(2\pi hx) \cdot \cos(2\pi lz).$$

In the case of X-rays, the weak scattering power of the oxygen ($Z=8$) compared to Cu ($Z=29$) and La ($Z=57$) atoms makes this intensity contribution almost invisible (« 1% of main reflections). In the case of neutrons, the scattering lengths of all atoms are of the same order of magnitude ($b_O = 5.803$ fm, $b_{Cu} = 7.718$ fm, $b_{La} = 8.24$ fm) and, therefore, also the superstructure reflections yield easily measurable intensities significantly larger than 1% of the strongest main structure reflections.

Mean square displacements: Pure La_2CuO_4 shows a linear change of the mean square displacements with temperature. Deviations from this harmonic behavior of the Debye–Waller factors could be an indication of a close by order–disorder phase transition. As the compound $La_{1.85}Sr_{0.15}CuO_4$ shows the highest $T_c = 38$ K,

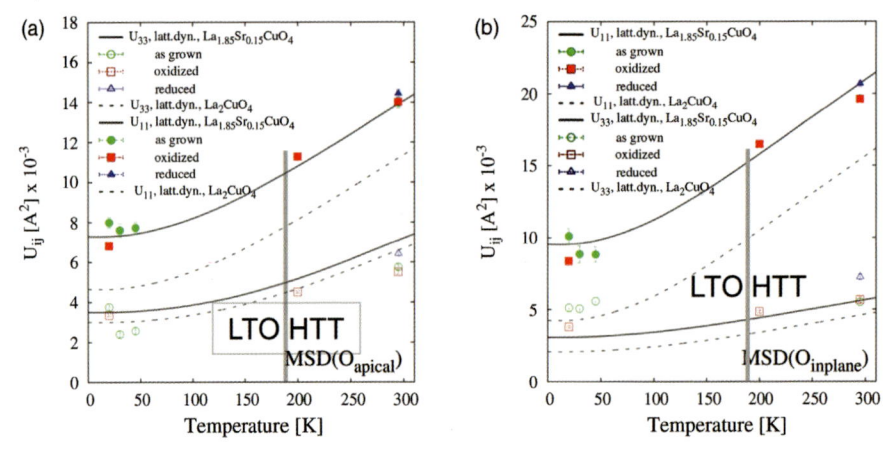

Figure 3.16 MSD of O_{apical} (a) and $O_{inplane}$ (b) for $La_{1.85}Sr_{0.15}CuO_4$ [31]. The dotted lines in the middle of the diagrams for O_{apical} and $O_{inplane}$ are the corresponding calculated values of the undoped parent compound La_2CuO_4.

it was discussed whether an order/disorder phase transition could be related to its superconductivity. Bragg data sets were taken with neutron single-crystal diffraction at three temperatures above and below the structural phase transition ($T = 186$ K) and the superconducting state ($T = 38$ K) for pure ($x = 0$) and doped ($x = 0.15$) $La_{2-x}Sr_xCuO_4$ samples on the single-crystal diffractometer 5C2 at the LLB/Saclay and the single-crystal diffractometer D9 at the ILL/Grenoble. The displacement curves show no anomalies for all atoms, including the two oxygens O_{apical} and $O_{inplane}$ (Figure 3.16). The only noticeable difference between the pure and doped compound is constantly increased values of $U_{11}(O_{apical})$ and $U_{33}(O_{inplane})$ for *all* temperatures for the doped sample. Harmonic lattice dynamical calculations from experimentally determined phonon dispersion curves taking into account the Sr doping are in good agreement with this observation. Thus, the random distribution of Sr atoms on La sites introduces static disorder into the structure [31].

3.4.4
Crystal Structure and Magnetic Order in Co_2SiO_4

As already discussed, neutron diffraction is very useful for obtaining precise atomic coordinates and displacement parameters. The improved accuracy (compared to X-rays) stems mainly from the absence of the form factor fall-off. We will use measurements on Cobalt–olivine, Co_2SiO_4, (crystal size $3 \times 2 \times 2$ mm) taken at the four-circle diffractometer HEiDi at the MLZ/Garching with hot neutrons ($\lambda = 0.552$ Å) to demonstrate this advantage for the x-coordinate of one of the Co-atoms.

The olivine structure (Figure 3.17) consists of chains of two types of edge-sharing CoO_6 octahedra connected by SiO_4 tetrahedra. A large data set with

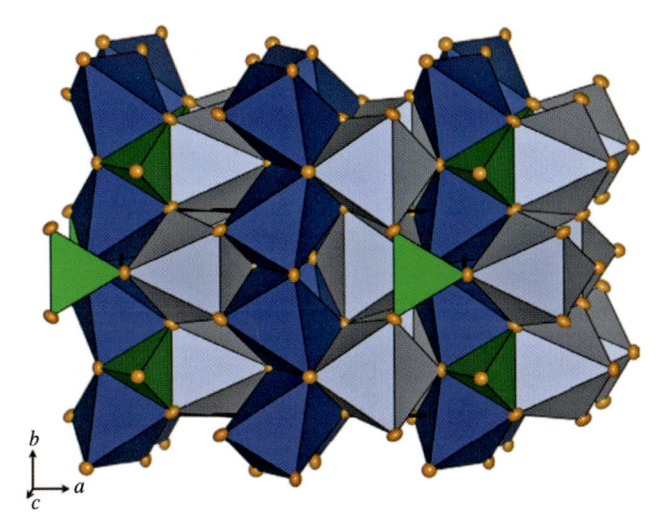

Figure 3.17 Structure of Co_2SiO_4 olivine at room temperature, projected along c. Green: SiO_4 tetrahedra, dark blue: CoO_6 octahedra at the Co1 site, and light blue: CoO_6 octahedra at the Co2 site. Displacement ellipsoids are plotted at the 95% probability level. (From Ref. [32].)

1624 independent reflections up to $\sin\theta/\lambda = 1.05$ Å^{-1} was measured. Then, the data were successively cut off in shells of $\sin\theta/\lambda$ and the resulting partial data sets were used to analyze the resolution dependent variation of structural parameters.

Figure 3.18 shows two interesting observations: First of all, the precision improves significantly with increasing $(\sin\theta/\lambda)_{max}$, as is evident from the decreasing size of the error bars. In the X-ray case, high angle reflections are usually very weak and their measurement does not often lead to improved

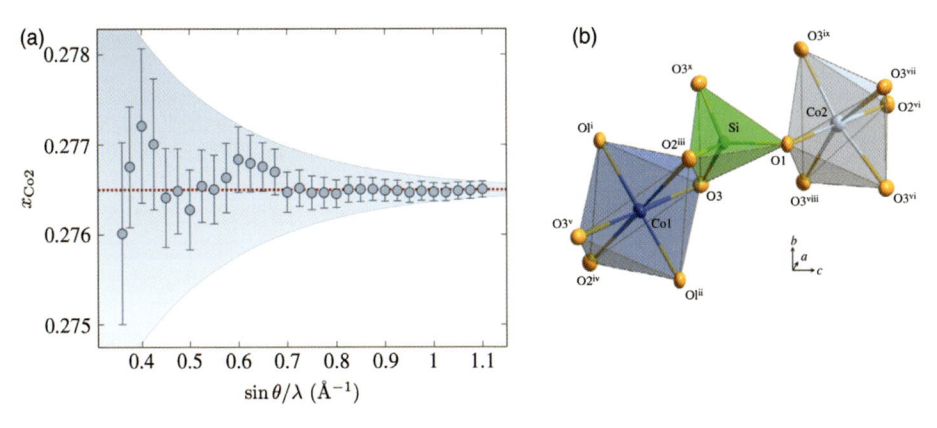

Figure 3.18 (a) Values and error bars of the x coordinate of Co_2 in Co_2SiO_4 as a function of the sin $(\theta)/\lambda$ range from single-crystal neutron diffraction data at room temperature [32]. (b) Clinographic view of the CoO_6 and SiO_4 polyhedra in Co_2SiO_4 at room temperature [32].

precision. Second, there are apparent oscillations of the value of the coordinate that only diminish for very high data resolution. The high d_{hkl} value resolution that can be obtained from neutron diffraction is also useful to derive precise temperature-dependent displacement parameters.

Just as in the case of high quality single crystal X-ray diffraction data, anisotropic displacement parameters can be determined as well. In addition to that, the quality of single crystal neutron data also often allows refining anharmonic displacement parameters. Anharmonic oscillations of atoms in crystals occur if the atoms are vibrating in a nonparabolic potential well. In such cases, the harmonic approximation, which is the basis of the description of thermal displacements by the Debye–Waller factor, fails. An analysis of the anharmonic displacements allows reconstructing the nonparabolic potential at the site of the vibrating atom.

3.4.5
Magnetic Structures from Neutron Diffraction

Cobalt–Olivine, Co_2SiO_4, orders magnetically below about 50 K. The magnetic moments of the Co^{2+} ions turn from a paramagnetic phase with no long-range order of the magnetic moments into an antiferromagnetically ordered arrangement. We use Co_2SiO_4 again to briefly demonstrate the application of neutron diffraction to the analysis of magnetic structures. This time, a powder neutron diffraction experiment was performed at the diffractometer D20 (ILL, France) in its high-resolution mode, at temperatures between 70 and 5 K with a neutron wavelength of $\lambda = 1.87$ Å and approximately 2 g of powdered Co_2SiO_4 [32].

At about 50 K and below, new magnetic reflections (001), (100), (110), (300), and so on appear (Figures 3.19 and 3.20). The nuclear reflections do not change much at the magnetic phase transition. The new reflections can be indexed within the same unit cell as the nuclear reflections, but they are forbidden in the paramagnetic phase with space group *Pnma*. Obviously, the symmetry has changed at the magnetic ordering transition. The task is now – just as in "ordinary" structure determination – to find a structural model (i.e., magnetic moments on the magnetic ions, here Co^{2+} and their orientation) that fits the observed positions and intensities of the magnetic Bragg peaks. Magnetic structure determination is outside the scope of this chapter, but assuming such a model has been constructed, it can be refined – in the case of powder data by the Rietveld method (Figure 3.20).

The lower trace (blue) is the difference $I_{obs} - I_{calc}$ on the same scale (I_{obs} the experimentally determined data, $I_{calc} \sim |F_{calc}|^2$ are the corresponding values calculated from the model). The upper row of the green marks shows Bragg reflections corresponding to the nuclear phase and the lower row represents the allowed positions of the magnetic peaks. Some of the Bragg peaks are indexed. "N" and "M" denote the nuclear and magnetic contributions, respectively [32]. Note that the magnetic Bragg peaks are only visible at low diffraction angles due to the form factor fall-off of magnetic neutron diffraction (see Figure 3.2b).

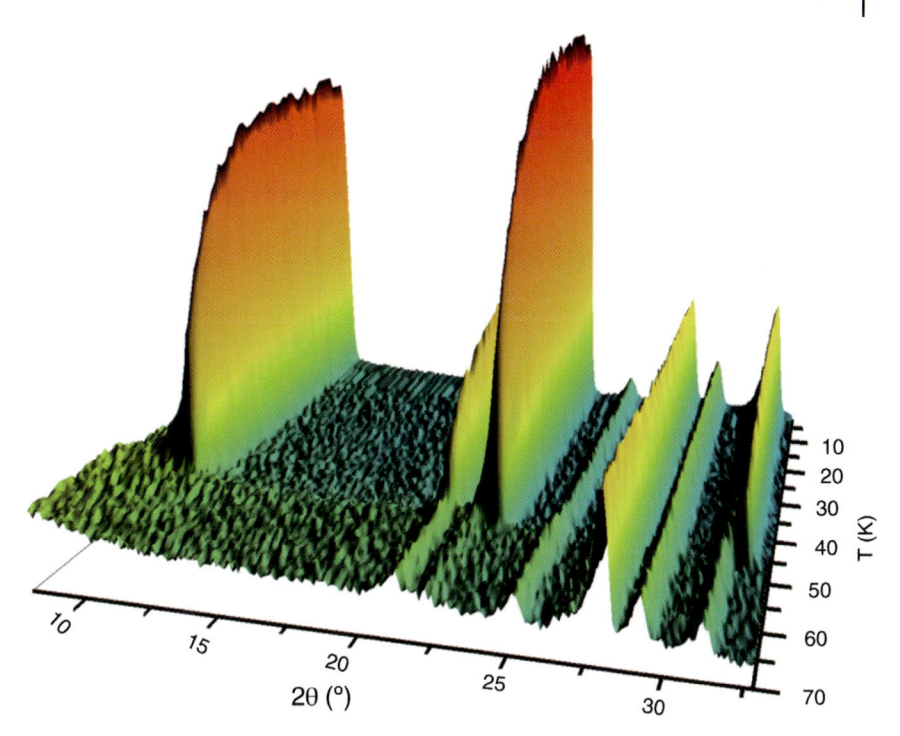

Figure 3.19 Thermal evolution of the neutron powder diffraction pattern (low angle part) of Co$_2$SiO$_4$ [32].

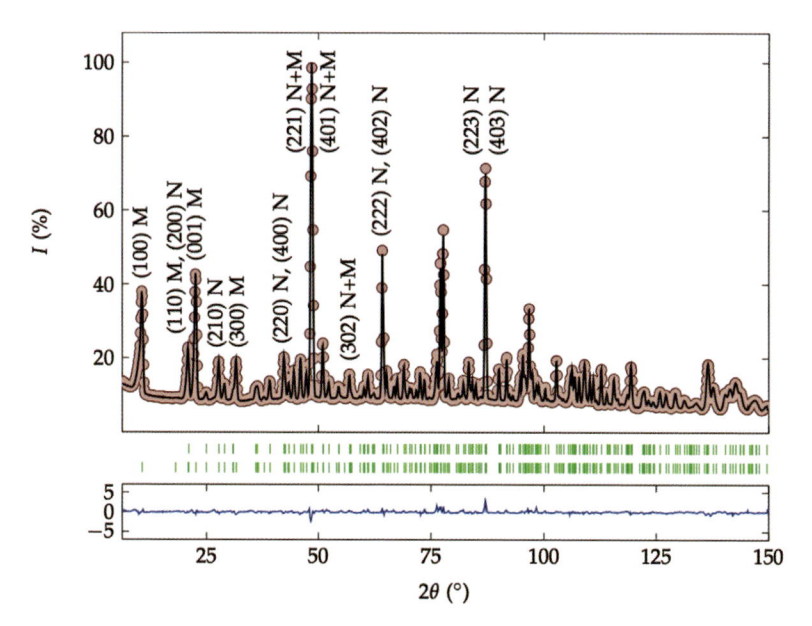

Figure 3.20 Neutron powder diffraction pattern (dots), Rietveld fit (black line), difference (blue), and allowed Bragg reflections (green marks) at 5 K of Co$_2$SiO$_4$ [32].

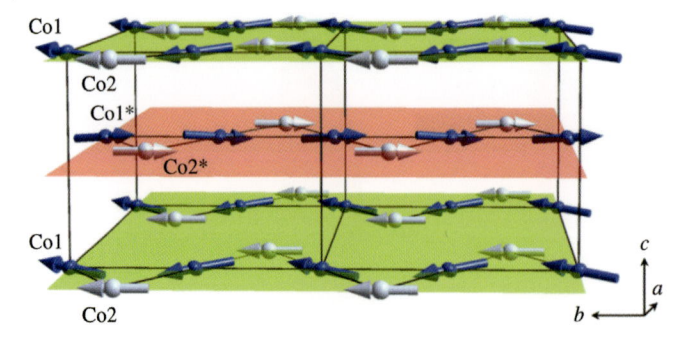

Figure 3.21 Graphical representation of the magnetic structure of Co_2SiO_4 below 50 K. The nonmagnetic atoms (Si and O) are excluded for simplicity. The figure shows the magnetic moments arranged in zigzag chains of Co1 and Co2 in layers perpendicular to the c-axis [32].

From the Rietveld refinements, one can derive the exact spin orientation (Figure 3.21) as well as parameters describing quantitatively the magnetic moments on the two symmetrically nonequivalent Co^{2+} sites (Table 3.1). However, magnetic neutron diffraction from single crystals often gives additional and more accurate information. Table 3.1 shows the components of the magnetic moments at the Co1 and Co2 sites according to the single-crystal neutron diffraction data at 2.5 K. The directions of the magnetic moments for other cobalt ions in the unit cell can be obtained by applying the symmetry operations of the magnetic space group (Schubnikov group) *Pnma*.

3.4.6
Electron Densities from X-Rays and Neutrons

Another advanced application of neutron diffraction in structural analysis is the determination of 3-dimensional high-resolution maps of the electron density in the unit cell to study, for instance, details of the chemical bonding. The most involved method of electron density studies (called x-N-synthesis) uses a

Table 3.1 Cartesian (M_x, M_y, and M_z) and spherical (M, φ, and θ) components of the Co1 and Co2 magnetic moments at 2.5 K.

	Co1(0,0,0,)	Co2(x,1/4,z)
M_x (μ_B)	1.18 ± 0.05	–
M_y (μ_B)	3.61 ± 0.04	3.37 ± 0.04
M_z (μ_B)	0.66 ± 0.18	–
M (μ_B)	3.86 ± 0.05	3.37 ± 0.04
φ (°)	71.9 ± 0.7	90
θ (°)	80.2 ± 2.7	90

$\chi^2 = 2.23$, $R[F^2 > 2\sigma(F^2)] = 0.033$, $\omega R(F^2) = 0.044$.

combination of high quality single-crystal neutron and X-ray diffraction experiments. In the present case, a single crystal of Co_2SiO_4 with dimensions $3 \times 2 \times 2$ mm was measured on the four-circle diffractometer HEiDi at MLZ/Garching at $\lambda = 0.552$ Å, the single-crystal X-ray (synchrotron) experiment was performed on the diffractometer D3 at the synchrotron facility HASYLAB/DESY in Hamburg with a Co_2SiO_4-sphere, diameter 150 μm as the sample and an X-ray wavelength of $\lambda = 0.5$ Å.

The first step of data evaluation is to take the X-ray data and to do a Fourier-transform (Fourier synthesis) to obtain the electron density map:

$$\rho(r) = 1/V \cdot \Sigma_\tau F(\tau) \cdot \exp[2\pi i (\tau \cdot r)], \text{ with } F(\tau) = |F(\tau)| \cdot \exp[i\varphi(\tau)],$$

r and τ being 3D vectors in real and reciprocal space, respectively, and V the unit cell volume.

The phases of the structure factors $\varphi(\tau)$ are calculated from the atomic model (structure factor equation), the moduli $|F(\tau)|$ are taken from the measured X-ray intensities. The result is a 3D map of the total electron density $\rho(r)$ within the unit cell (Figure 3.22a).

In favorable cases, such a map already shows interesting features of the (anisotropic) bonding electron density, however, the information content of the map can be very significantly improved by taking the coordinates and displacement parameters from the more accurate neutron diffraction experiment (see above for the reasons) and calculate, in a second step, the so-called deformation density. This is done by subtracting from the total electron density $\rho(r)$ the density $\rho(r)_{spherical}$ corresponding to a superposition of spherical atoms at the nuclear positions. More specifically, atomic positions x_j, y_j, z_j and thermal displacements T_j of atoms j derived from the neutron experiment are "decorated" with the calculated spherical single atom electron densities.

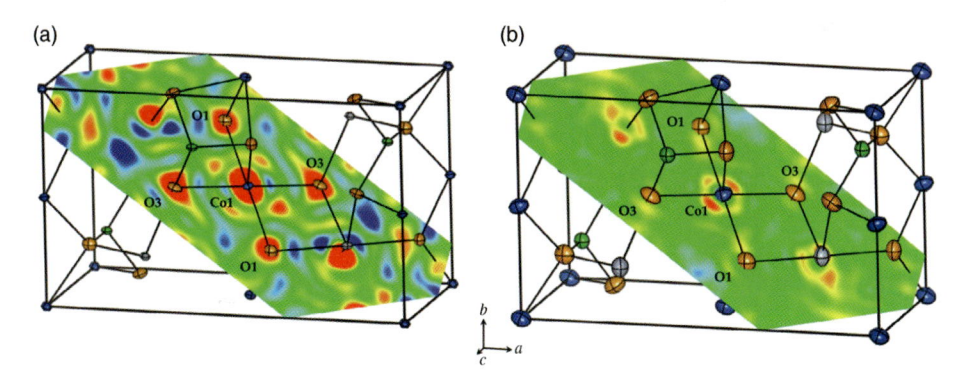

Figure 3.22 (a) Electron density distribution $\rho(r)$ of Co_2SiO_4 at 12 K from Fourier synthesis of X-ray data. Contours range from -8 e/Å^3 (blue) to 10 e/Å^3 (red). A plane that intersects the CoO_6 octahedron containing the Co1, O1, and O3 atoms is shown together with a sketch of the crystal structure. (b) Deformation density from the x-N-difference Fourier map of Co_2SiO_4 at 300 K: Section through the O1–Co1–O3 plane. The difference density varies from -1.25 e/Å^3 (blue) to 1.15 e/Å^3 (red) [32].

$\rho(r)_{\text{deform}} = \rho(r) - \sum \rho(r)_{\text{spherical}}$, where the sum runs over all atoms in the unit cell.

$\rho(r)_{\text{spherical}}$ corresponds to the expectation value of the electron density within the unit cell without any effects that are due to chemical bonding. The deformation density then represents the deformation of the charge distribution as a result of the formation of chemical bonds. Figure 3.22b shows such a deformation density map for Co_2SiO_4. In favorable cases, the electron density in the hybridized bonding orbitals (in this case of Co3d and O2p character) can be observed directly.

3.4.7
Magnetization Densities from Neutron Diffraction

As another example for the application of neutron diffraction in solid-state chemistry, we briefly sketch how a three-dimensional map of the magnetization density, that is, the density of magnetic moments (spin as well as orbital moments) within the unit cell can be determined. These maps are sometimes lucidly called "spin density maps," but in systems with nonvanishing orbital moments, the term "magnetization density" is really the correct one.

The experiment is performed by polarized neutron diffraction on a single crystal using the flipping ratio method, which is not discussed here for brevity (see, for instance, Ref. [5]). The flipping ratio method allows separating nuclear and

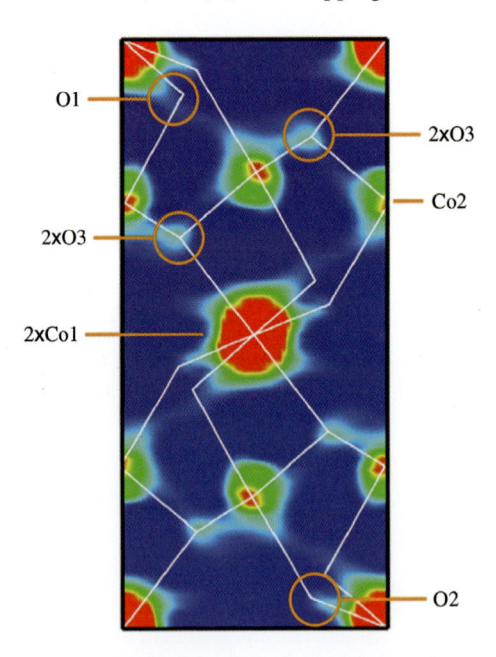

Figure 3.23 Reconstruction of the magnetic density (projected along the b-axis) corresponding to the observed magnetization distribution of Co_2SiO_4 at 70 K with contours ranging from $0\,\mu_B/\text{Å}^3$ (blue) to $2\,\mu_B/\text{Å}^3$ (red) [32].

magnetic contributions to the diffracted intensities. It is performed *above* the magnetic phase transition in the paramagnetic state (in the case of Co_2SiO_4 above $T_N = 50$ K) and the sample is in a strong external magnetic field (here: 7 T). 207 Bragg reflection flipping ratios were measured at the diffractometer 5C1 at the LLB/Saclay for Co_2SiO_4 at 70 K up to $\sin \theta/\lambda \approx 0.62$ Å^{-1} at a neutron wavelength of $\lambda = 0.845$ Å. Given the flipping ratios and the nuclear structure factors, the magnetic structure factors can be calculated that are then Fourier transformed to give the spatially resolved magnetization density shown in Figure 3.23 in a section through the unit cell of Co_2SiO_4.

One of the most interesting features of this map is the observation of magnetization density on the, nominally nonmagnetic, oxygen atoms coordinating the Co^{2+} ions. These "transferred moments" are a direct experimental evidence for the hybridization of the oxygen 2p- with Co-3d-orbitals that is responsible not only for the covalent bonding but also for the magnetic exchange interaction along the Co—O—Co bond network.

3.4.8
Spatially Resolved In Situ Neutron Diffraction – ZEBRA® cell

The development of sodium metal halide batteries began in the late 1970s. They are known as ZEBRA® cells as the first company dealing with this type of cell was the South African company Zebra Power Systems and Beta R&D Ltd, where the abbreviation stands for *Zero Emission Battery Research Activities* (Figure 3.24 and 4.25a). This power cell type offers two advantages that make it attractive for commercial use especially in the mobile sector: First, the ingredients for the cell are rather common and cheap elements, see cell setup in Figure 3.25. Second, they operate with a voltage of 2.58 V and a high specific power of 120 Wh/kg (four times higher than lead accumulators). A disadvantage is the operating temperature of about 300 °C to keep the Na liquid that takes up to 20% from the total power that on the other hand, makes the battery independent from the surrounding temperature.

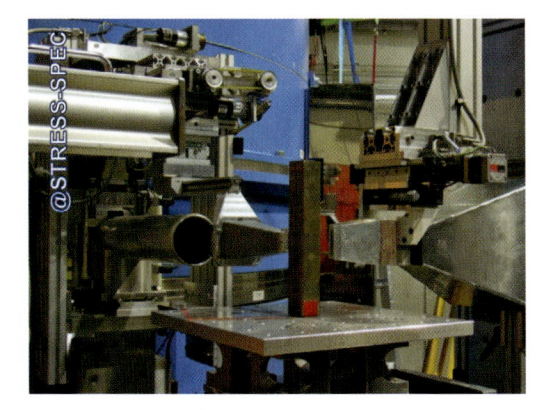

Figure 3.24 ZEBRA® power cell. (cell on STRESS-SPEC) [33].

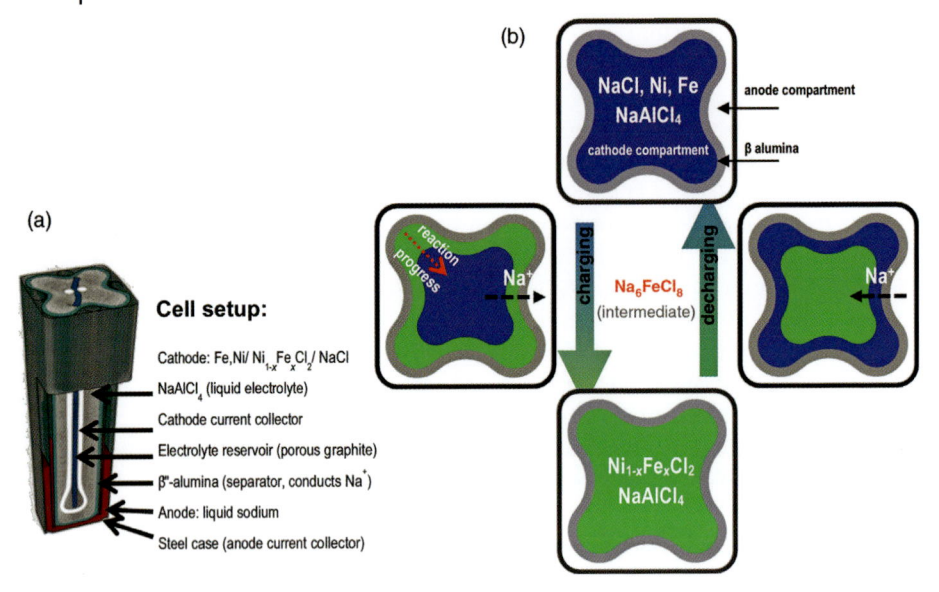

Figure 3.25 ZEBRA® power cell. (a) Cell design (35 × 35 × 240 mm³) [33]. (b) Reaction process [34].

The charge process can be described rather simply with $2\,NaCl + Fe \rightarrow FeCl_2 + 2\,Na$.

By using the high penetration depth of neutrons one can analyze *in situ* the reaction front inside the battery during operation, for example, during the charging process. Taking a fixed position inside the battery several diffractograms

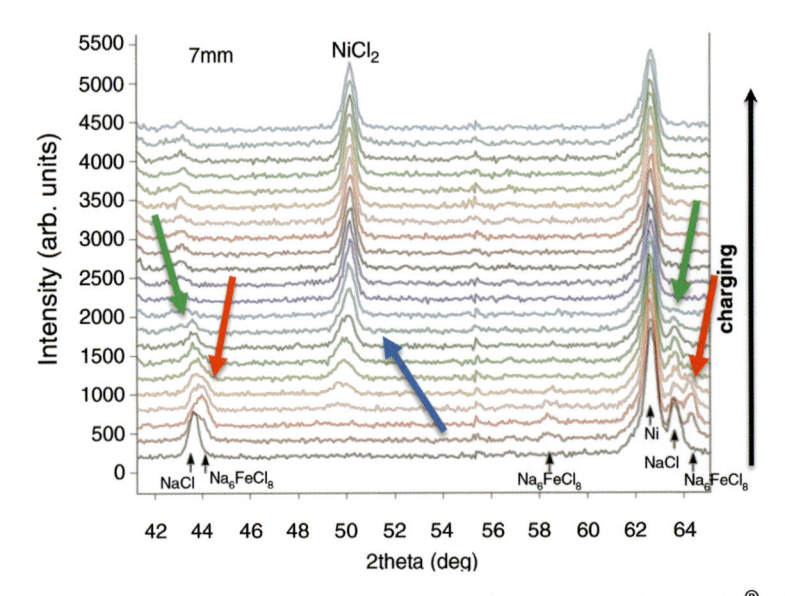

Figure 3.26 Evolution of various phases during the charge process within a ZEBRA® cell [34].

were taken (6 min/diffractogram, every 26 min, diffractometer STRESS-SPEC at MLZ/Garching [33]), it was shown that the cell reaction proceeds via intermediate phases, as first the NaCl reacts to Na_6FeCl_8, then both disappear, and formation of $Ni_{1-x}Fe_xCl_2$ takes place. This data was then used to develop a microscopic model of the processes within the cell during charge or discharge (Figure 3.26) [34].

3.5 Outlook

Powder and single-crystal neutron diffractometry are the most powerful tools for detailed studies of chemical and magnetic structures. Due to the different interactions of X-rays and neutrons with matter, neutron and X-ray (synchrotron) diffraction techniques are often complementary to each other allowing these techniques to contribute essential information to solid-state chemistry, physics, biology, and material sciences. With the new, brighter neutron sources becoming available in the near future (like the European Spallation Source ESS), neutron diffraction and more generally neutron scattering will further develop toward *in situ*, time resolved studies of the structure and properties of solids.

Acknowledgments

We would like to thank all colleagues involved in the preparation of this chapter, namely, the instrument scientists of the instruments mentioned within this contribution and those who supported us with suitable examples.

References

1 Furrer, A., Mesot, J., and Strässle, Th. (2009) *Neutron Scattering in Condensed Matter Physics*, World Scientific, Singapore.

2 Brückel, T. (2010) Neutron Primer, in *Neutron Scattering*, Lectures of the JCNS Laboratory Course (eds T. Brückel, G. Heger, D. Richter, G. Roth, and R. Zorn), Forschungszentrums Jülich, Jülich, pp. 2:1–32.

3 NIST Center for Neutron Research (2013) Neutron scattering lengths and cross sections. Available at http://www.ncnr.nist.gov/resources/n-lengths. (accessed July 7, 2016)

4 Coppens, P. (1978) *Neutron Diffraction: Volume 6 of Topics in current physics*, (ed. H. Dachs), Springer, Berlin, pp. 71–11.

5 Roessli, B. and Böni, P. (2000) Polarized neutron scattering, doi: arXiv:cond-mat/0012180.

6 Paolasini, L. (2014) Resonant and magnetic X-ray diffraction, *Synchrotron Radiation*, Springer, pp. 361–387.

7 Möchel, A. (2008) Magnetische Struktur und Anregungen von SEMnO$_3$ multiferroika (SE = Gd,Tb). Diploma thesis, Rheinische Friedrich-Wilhelms-Universität Bonn.

8 Jin, W.T., Li, Wei, Su, Y., Nandi, S., Xiao, Y., Jiao, W.H., Meven, M., Sazonov, A.P., Feng, E., Chen, Yan, Ting, C.S., Cao, G.H., and Brückel, Th. (2015) *Phys. Rev. B*, **91**, 064506.
9 Xiao, Y., Su, Y., Meven, M., Mittal, R., Kumar, C.M.N., Chatterji, T., Price, S., Persson, J., Kumar, N., Dhar, S.K., Thamizhavel, A., and Brueckel, Th. (2009) *Phys. Rev.*, **B 80**, 174424.
10 Russell, G.J. (1990) Spallation physics – an overview, XI International Collaboration on Advanced Neutron Sources KEK, Tsukuba, Oct. 20–26, 1990.
11 Becker, P.J. and Coppens, P. (1974) *Acta Crystallogr.*, **A 30**, 129–147; Becker, P.J. and Coppens, P. (1974) *Acta Crystallogr.*, **A 30**, 148–153.
12 Coppens, P. and Hamilton, W.C. (1970) *Acta Crystallogr.*, **A 26**, 71–83.
13 Zachariasen, W.H. (1965) *Acta Crystallogr.*, **18**, 703; Zachariasen, W.H. (1965) *Acta Crystallogr.*, **18**, 705.
14 Petříček, V., Dušek, M., and Palatinus, L. (2014) *Z. Kristallogr.*, **229**, 345–352.
15 Rodríguez-Carvajal, J. (1993) *Physica B*, **192**, 55–69.
16 Sheldrake, G.M. (2008) *Acta Crystallogr. A*, **64**, 112–122.
17 Zucker, U.H., Perenthaler, E., Kuhs, W.F., Bachmann, R., and Schulz, H. (1983) *J. Appl. Crystallogr.*, **16**, 358.
18 Rosmanith, E. (2006) *Acta Crystallogr.*, **A62**, 174–177; (2007) *Acta Crystallogr.*, **A 63**, 251–256.
19 Sazonov, A. et al. (2016) *J. Appl. Crystallogr.*, **49**, 556.
20 Houben, A., Schweika, W., Brückel, Th., and Dronskowski, R. (2012) *Nucl. Instrum. Methods A*, **680**, 124–133.
21 Hoelzel, M., Senyshyn, A., Juenke, N., Boysen, H., Schmahl, W., and Fuess., H. (2012) *Nucl. Instr.*, **A 667**, 32–37.
22 Rietveld, H.M. (1967) *Acta Crystallogr.*, **22**, 151.
23 Rietveld, H.M. (1969) *J. Appl. Crystallogr.*, **2**, 65.
24 Redhammer, G. et al. (2010) *Phys. Chem. Miner.*, **37**, 311–332.
25 Ohhara, T. et al. (2016) *J. Appl. Crystallogr.*, **49**, 120–127.
26 Mattauch, S., Heger, G., and Michel, K.H. (2004) *Cryst. Res. Technol.*, **39**, 1027.
27 Nelmes, R.J., Kuhs, W.F., Howard, C.J., Tibballs, J.E., and Ryan, T.W. (1985) *J. Phys. C: Solid State Phys.*, **18**, L711.
28 Reimers, W., Hellner, E., Treutmann, W., and Heger, G. (1982) *J. Phys. C: Solid State Phys.*, **15**, 3597.
29 Bednorz, J. and Müller, K. (1986) *Z. Phys.*, **B 64**, 189.
30 Birgeneau, R.J. and Shirane, G. (1989) *Physical Properties of High Temperature Superconductors I* (ed. D.M. Ginsberg), World Scientific, Singapore.
31 Braden, M., Meven, M., Reichardt, W., Pintschovius, L., Fernandez-Diaz, M.T., Heger, G., Nakamura, F., and Fujita, T. (2001) *Phys. Rev.*, **B 63**, 140510.
32 Sazonov, A. (2009) Pecularities of crystal and magnetic structures of synthetic Co-olivine, Co_2SiO_4. Ph.D. thesis, RWTH Aachen; Sazonov, A. et al. (2009) *Acta Crystallogr.*, **B 65**, 664–675.
33 Hofmann, M., Gilles, R., Gao, Y., Rijssenbeek, J.T., and Mühlbauer, M.J. (2012) *J. Electrochem. Soc.*, **159**, A1833.
34 Zinth, V. et al. (2015) *J. Electrochem. Soc.*, **162** (3), A384–A391.

4
Modulated Crystal Structures

Sander van Smaalen

University of Bayreuth, Laboratory of Crystallography, Universitatsstrasse 30, 95440 Bayreuth, Germany

4.1
Order Without Translational Symmetry

The faceting morphology of minerals has attracted attention for the past many centuries. The modern scientific approach to these crystals started with the work of Steno (1669) [1], who developed an essentially modern theory of the growth of crystals, which explained that the angles between facets are always the same set of angles for a single substance. Toward the end of the eighteenth century, Haüy showed that a faceted crystal can be constructed as a regular stacking of a single parallelepiped building block (the unit cell in modern language) [2]. The shape of the parallelepiped of a particular crystalline substance is directly related to the observed interfacial angles between facets of this crystal. A major development of the work of Haüy was that various crystal shapes could be obtained from this model through so-called stepped surfaces (Figure 4.1). They appear as flat facets, because of the tiny nature of the building blocks.

Haüy's model [2] firmly established translation symmetry into the theory of the crystalline state. The next step was the development of the mathematical theory of these periodic crystals. Hessel (1830) [4,5] derived the 32 crystallographic point groups; Bravais (1850) [6] established the 14 space lattices, named Bravais lattices; and Fedorov (1891) [7] and Schoenflies (1891) [8] obtained the complete list of 230 space groups. This theory could explain all physical properties of crystals by a translational symmetric structure. Finally, the publication of the first X-ray diffraction pattern by Friedrich *et al.* in 1912 [9] provided the ultimate experimental proof of the lattice structure of crystals. It should therefore not be surprising that since about 1900, translational symmetry has been considered a key property of crystals [10].

Nevertheless, the crystalline state has been defined in the nineteenth century as a homogeneous and anisotropic solid possessing three-dimensional long-range order. Indeed, 3D translational symmetry implies long-range order, and it

Handbook of Solid State Chemistry, First Edition. Edited by Richard Dronskowski, Shinichi Kikkawa, and Andreas Stein.
© 2017 Wiley-VCH Verlag GmbH & Co. KGaA. Published 2017 by Wiley-VCH Verlag GmbH & Co. KGaA.

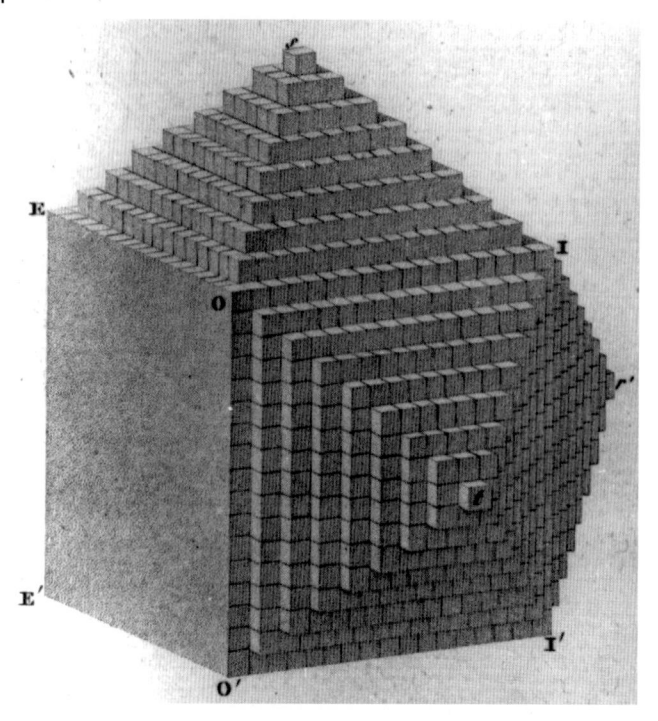

Figure 4.1 Model of a crystal according to Haüy, demonstrating the formation of stepped facets from a decrement of unit cells. (Reprinted from Ref. [3].)

has been assumed for a considerable time that the properties of translational symmetry and long-range order could be used interchangeably [11]. We now know that this is not true. Long-range order of matter can be realized in different ways than by a periodic arrangement of the atoms: These are the aperiodic crystals, discussed in the present chapter.

Before 1912 diffraction by lattices and by distorted lattices was known from the theory of optical gratings. Dehlinger (1927) [12] was the first to consider X-ray diffraction by a crystal, in which the atoms were displaced out of periodic positions according to a wave of a wavelength that does not match any of the periodicities of the periodic basic structure. He showed that the diffraction observed for mechanically distorted metals could be explained by a periodic distortion of a lattice structure, thus giving rise to additional Bragg reflections, the so-called "Gittergeister," or satellite reflections as they are now called. Subsequently, metastable modulated structures were observed in alloys [13,14]. These observations, rare but documented in widely read journals, did not incite the need for modifying theories of growth and constitution of crystals, because they could be attributed to metastable states. This situation changed around 1950, when several compounds were discovered with a modulated structure of a thermodynamically stable state. They include the incommensurate atomic order in Cu_3Pd (1957) [15], spiral magnetic order in certain rare earth elements (1961) [16], and the incommensurate displacive modulated structure of NaNO$_2$ (1961) [17]. The thermodynamic stability of these phases can be inferred from the fact that they are obtained through reversible, temperature-dependent phase transitions.

The same is true for sodium carbonate, for which de Wolff and coworkers discovered satellite reflections at incommensurate positions in the X-ray powder diffraction of the γ-phase, stable for temperatures T between 170 and 605 K (1964) [18]. These observations were the basis for the development by de Wolff of the superspace theory for the description of incommensurately modulated structures (1974) [19,20]. Together with the independent discovery of superspace symmetry by Janner and Janssen (1977) [21,22], and the use of this theory for the determination of the modulated crystal structure of γ-Na$_2$CO$_3$ (1976) [23], these developments led to the idea that crystals with incommensurately modulated crystals constitute a state of matter essentially different from periodic crystals.

Support for aperiodic ground states also became available from theory. The Frenkel–Kontorova model describes a one-dimensional periodic chain of particles in an external potential that has a different period. Frank and van der Merwe (1949) [24] showed that the state of lowest energy of this model could be periodic or incommensurately modulated, depending on the ratio of these periods and the strength of the potential. Around 1980 other models have been developed, which describe systems with a fully ordered but aperiodic ground state (see the discussion in Ref. [25]). These works provided clear evidence that an aperiodically ordered arrangement of atoms can be the ground state of matter.

Next to modulated structures, loosely defined above, two more states of aperiodic order have been found in matter. Composite crystals comprise two or more subsystems with different modulated structures that are intertwined into a single compound [26]. Quasicrystals possess point symmetry according to a noncrystallographic point group, for example, involving 5-fold or 10-fold rotation axes (1984) [27]. Again, exact knowledge gained in the nineteenth century – translational symmetry is not compatible with fivefold rotational symmetry – has been translated in the twentieth century into the loose statement that crystals cannot possess fivefold rotational symmetry. Indeed, noncrystallographic rotations as symmetries of quasicrystals imply that the latter do not possess translational symmetry, a feature that is experimentally confirmed. Nevertheless, quasicrystals are perfectly ordered and exhibit all the physical properties commonly assigned to the crystalline state.

The crystallography of modulated materials has been firmly established through the work of Yamamoto [28], who determined and refined the modulated crystal structures of several compounds in the early 1980s. By now, aperiodic order has been found as the ground state of compounds in all classes of crystalline matter. Aperiodic crystals are found among minerals [29], inorganic solids [30], molecular crystals [31,32], alloys [15,33], proteins [34], and the chemical elements at high pressures or low temperatures [35]. They include ferroelectric materials [30], high-T_C superconductors [36], charge density wave (CDW) compounds [37], and magnetically ordered systems [38].

This chapter gives a brief introduction to incommensurately modulated structures. Comprehensive treatments of structural analysis and symmetry can be found in textbooks by Janssen et al. [25] and van Smaalen [26], and for quasicrystals by Steurer and Deloudi [39]. Superspace symmetry, including a list of all 775 $(3+1)$-dimensional superspace groups for modulated structures with a single modulation wave, is presented in Ref. [40]. A complete list of $(3+d)$D superspace groups for $d = 1, 2, 3$ is given in Ref. [41]. An overview of compounds with 2D and 3D modulations $(d = 2, 3)$ can be found in Ref. [42]. The history of crystals and crystallography has been discussed by Burke [43]. A brief summary, including a discussion of aperiodicity, has been given in Ref. [44].

4.2
Displacive Modulation of the Basic Structure

X-ray diffraction patterns of incommensurately modulated structures exhibit sharp Bragg reflections and possess true point symmetry (Figure 4.2). Sharp Bragg reflections are the result of long-range order of the crystal structure in 3D space. The superspace theory of de Wolff et al. [45] translates this hidden order into translational symmetry of a structure model in a space of dimension $(3+d)$: the superspace. The integer d counts the number of independent modulation waves. Most modulated compounds possess a one-dimensional modulation $(d=1)$. Two-dimensional modulated compounds $(d=2)$ are regularly found, while $d=3$ occurs in rare cases [42]. One of the major achievements of the superspace theory is that it translates the observed rotational symmetry of the diffraction into rotational symmetry of the $(3+d)$D structure model of the

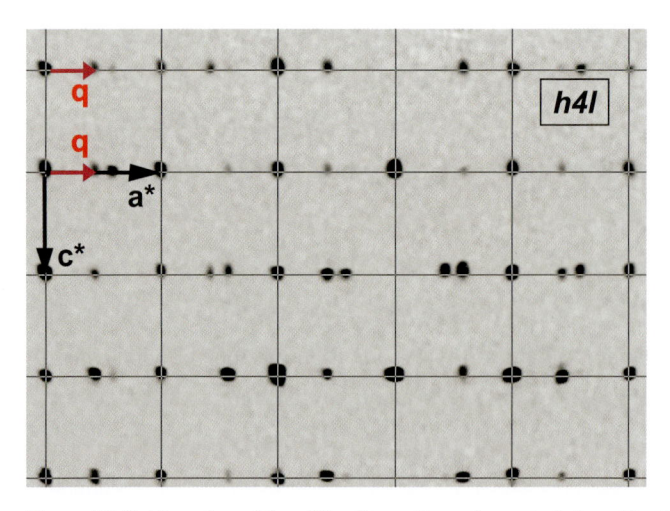

Figure 4.2 ($h\,4\,l$) section of the diffraction pattern of morpholinium–BF_4, (reprinted from Ref. [46].)

incommensurate crystal, although the crystal structure in 3D lacks both translational and rotational symmetries.

The description of a modulated crystal structure starts with a basic structure possessing translational symmetry in 3D space. The description of such a periodic structure can be found in any standard text on crystallography, where, however, the three basis vectors of the lattice are usually described by the symbols **a**, **b**, and **c**. For practical purposes, these three basis vectors of the periodic basic structure are presently numbered by subscripts:

$$\{\mathbf{a}_1, \mathbf{a}_2, \mathbf{a}_3\}. \tag{4.1}$$

The position of atom μ in the unit cell of the periodic basic structure is usually given by relative coordinates (x_μ, y_μ, z_μ) with respect to the basis vectors in Eq. (4.1). Here we denote these coordinates by $\left(x_1^0(\mu), x_2^0(\mu), x_3^0(\mu)\right)$. The position of atom μ in the unit cell of the periodic basic structure then is

$$\mathbf{x}^0 = x_1^0(\mu)\mathbf{a}_1 + x_2^0(\mu)\mathbf{a}_2 + x_3^0(\mu)\mathbf{a}_3. \tag{4.2}$$

X-ray structural analysis of periodic structures aims at the determination of these coordinates for the $\mu = 1, \ldots, N$ atoms in the unit cell. The structure model is completed by isotropic or anisotropic atomic displacement parameters (ADPs) for each atom. Furthermore, it may include parameters for fractional occupancies of atomic sites or for directions and magnitudes of atomic magnetic moments. Additional parameters may be introduced through the multipole model describing chemical bonding [47], or through the Gram–Charlier parameters describing anharmonic thermal motion (anharmonic ADPs) [48].

The lattice of the basic structure (Eq. (4.1)) has a reciprocal lattice defined by the reciprocal basis vectors $\{\mathbf{a}_1^*, \mathbf{a}_2^*, \mathbf{a}_3^*\}$. The basic structure is 3D periodic and therefore produces Bragg reflections at the nodes of this reciprocal lattice:

$$\mathbf{H}_{\text{basic}} = h\,\mathbf{a}_1^* + k\,\mathbf{a}_2^* + l\,\mathbf{a}_3^*. \tag{4.3}$$

The indices h, k, and l are integers and each triplet $(h\,k\,l)$ describes a so-called main reflection. As is demonstrated in Figure 4.2, additional reflections due to the modulation appear at distances $\pm\mathbf{q}$ from each main reflection, where $\mathbf{q}$ is the modulation wave vector. Accordingly, Bragg reflections of a modulated crystal can be indexed with four integers $(h\,k\,l\,m)$ according to

$$\mathbf{H} = h\,\mathbf{a}_1^* + k\,\mathbf{a}_2^* + l\mathbf{a}_3^* + m\mathbf{q}. \tag{4.4}$$

$m = 0$ represents the main reflections $(h\,k\,l\,0)$; $m = \pm 1$ are first-order satellite reflections $(h\,k\,l \pm 1)$; and satellites of order $|m|$ have the fourth index equal to $\pm|m|$. In case of more than one modulation wave (more than one independent vector **q**), the diffraction pattern will be indexed by five or six or more integers. Reciprocal superspace of a modulated crystal with a 1D modulation is obtained by considering the Bragg reflections $(h\,k\,l\,m)$ to be projections of a reciprocal lattice in four-dimensional space. This reciprocal lattice defines $(3+1)$D superspace.

Incommensurately modulated structures can be understood without reference to superspace. To this end, we need the coordinates of all atoms in the crystal beyond the coordinates of the atoms in the first unit cell given in Eq. (4.2). For the periodic basic structure, they are

$$\bar{x}_i = l_i + x_i^0(\mu), \quad \text{for} \quad i = 1, 2, 3, \tag{4.5}$$

where l_i are integers defining the translational symmetry, and $0 \leq x_i^0(\mu) < 1$ for atoms in the first unit cell. For an incommensurately modulated crystal with displacive modulation, the atomic positions are described by the basic structure position (Eq. (4.5))

$$\bar{\mathbf{x}} = \bar{x}_1 \mathbf{a}_1 + \bar{x}_2 \mathbf{a}_2 + \bar{x}_3 \mathbf{a}_3, \tag{4.6}$$

together with a usually small deviation from this position. Long-range order exists, because the deviation follows a strict recipe, although this recipe results in different values of the atomic shifts in different unit cells, thus destroying 3D translational symmetry. Atomic displacements are described by modulation wave functions (Figure 4.3). In analogy with phonons, the modulation is based on a modulation wave vector describing the direction and wavelength of the wave. The modulation wave vector has coordinates $(\sigma_1, \sigma_2, \sigma_3)$ with respect to the reciprocal lattice vectors $\{\mathbf{a}_1^*, \mathbf{a}_2^*, \mathbf{a}_3^*\}$ of the basic structure:

$$\mathbf{q} = \sigma_1 \mathbf{a}_1^* + \sigma_2 \mathbf{a}_2^* + \sigma_3 \mathbf{a}_3^*. \tag{4.7}$$

At least one of the components σ_i needs to be an irrational number for the crystal to be incommensurately modulated. By definition, a modulation wave function is any periodic function of $\mathbf{q} \cdot \bar{\mathbf{x}}$, where $\cdot$ indicates the scalar product between vectors. Allowing for an initial phase t, the modulation function $\mathbf{u}^\mu(\bar{x}_{s,4})$ of atom μ is defined to be a function of (Eqs 1.5–1.7)

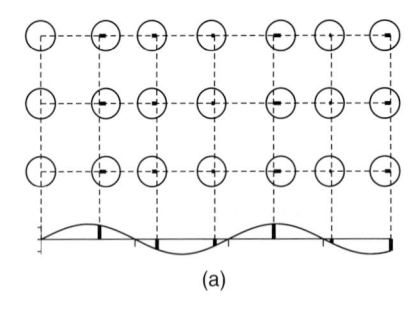

$$\begin{aligned}\bar{x}_{s,4} &= t + \mathbf{q} \cdot \bar{\mathbf{x}} \\ &= t + \sigma_1 \bar{x}_1 + \sigma_2 \bar{x}_2 + \sigma_3 \bar{x}_3.\end{aligned} \tag{4.8}$$

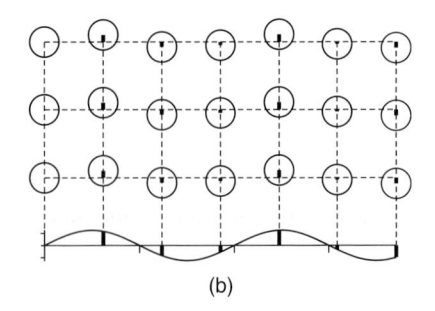

(a)

(b)

Figure 4.3 Crystal structures with displacement modulations. (a) Longitudinal modulation and (b) transverse modulation. Dashed lines indicate the lattice of the basic structure. Atomic displacements follow from the values of the sinusoidal modulation functions (shown at the bottom) at the corresponding basic structure positions. (Adapted from Ref. [26].)

4.2 Displacive Modulation of the Basic Structure

The modulation function is periodic with period one:

$$\mathbf{u}^{\mu}(\bar{x}_{s,4} + 1) = \mathbf{u}^{\mu}(\bar{x}_{s,4}). \tag{4.9}$$

This notation anticipates the superspace approach, where $\bar{x}_{s,4}$ will be identified with the relative coordinate along the fourth axis of $(3+1)$D superspace. Positions of the atoms in 3D space are described by the following coordinates with respect to the basic structure basis vectors (Eq. (4.1)):

$$x_i = \bar{x}_i + u_i^{\mu}(\bar{x}_{s,4}), \quad \text{for} \quad i = 1, 2, 3. \tag{4.10}$$

Structural analysis of modulated crystals requires the determination of the modulation functions in addition to the $3N$ atomic coordinates (Eq. (4.2)) and the $6N$ anisotropic ADPs of the basic structure. Since *a priori* knowledge is not available for the shapes of the modulation functions, an infinite number of parameters should be determined, for example, as given by the infinite number of parameters in its Fourier series expansion. In practice, only one or a few low-order harmonic parameters of the Fourier series can be determined on the basis of the diffraction experiment. Each of the N atoms in the basic structure unit cell attains $6N_{\text{harm}}$ additional parameters, where N_{harm} is the number of harmonic waves ($i = 1, 2, 3$):

$$u_i^{\mu}(\bar{x}_{s,4}) = \sum_{n=1}^{N_{\text{harm}}} A_i^n(\mu)\sin\left[2\pi n \bar{x}_{s,4}\right] + B_i^n(\mu)\cos\left[2\pi n \bar{x}_{s,4}\right]. \tag{4.11}$$

The incommensurately modulated structure thus is completely specified by the parameters $x_i^0(\mu)$, $A_i^n(\mu)$, and $B_i^n(\mu)$ for $i = 1, 2, 3$ for each of the N atoms. Furthermore, ADPs as well as possibly modulation functions for ADPs, site occupancy factors, or other aspects may be added to the structure model. The phase parameter t does not appear in this list. The reason is the incommensurability between the basic structure periods (Eq. (4.1)) and the periodicity of the modulation functions (irrational components of $\mathbf{q}$ – Eq. (4.7)). Because of this irrational ratio, each value of t leads to the same modulated structure, although shifted in space by a vector dependent on the exact value of t. This property is used in t-plots, where complete information on the structure is provided by considering structural parameters for t between 0 and 1 only (Section 4.4).

The superspace theory is indispensable for structural analysis of aperiodic crystals, because it provides symmetry restrictions on the structural parameters, like $x_i^0(\mu)$, $A_i^n(\mu)$, and $B_i^n(\mu)$. A complete presentation of this method, including superspace symmetry, can be found elsewhere [25,26]. For an understanding of the structural chemistry of modulated crystals, it appears sufficient to notice that the argument of the modulation functions is to be identified with the fourth coordinate along an axis perpendicular to the 3D space. The $(3+1)$D structure model is periodic along the fourth axis, because of the periodicity of the modulation functions (Eq. (4.9)). The loss of translational symmetry in 3D space upon the transition from basic to modulated structure is codified by replacing each of the three basis vectors $\mathbf{a}_i$ by a basis vector $\mathbf{a}_{s,i}$ of translational symmetry in

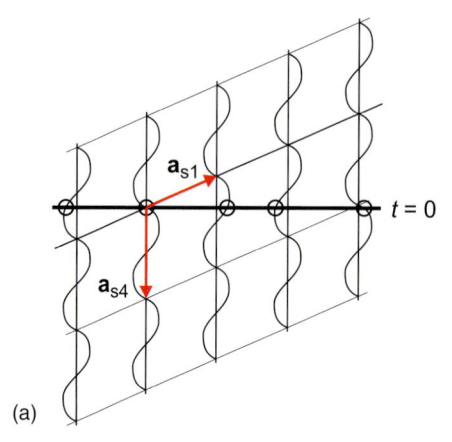

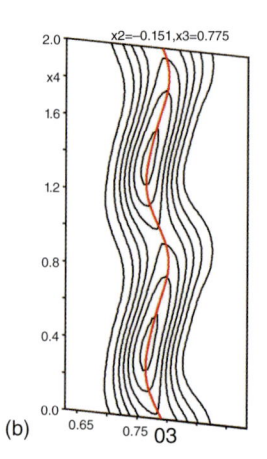

Figure 4.4 The (3 + 1)D structure model in superspace. (a) $(x_{s,1}, x_{s,4})$ section of (3 + 1)D superspace providing 4 × 3 superspace unit cells with one modulated atom at the origin. $\mathbf{a}_{s1}$ and $\mathbf{a}_{s4}$ are basis vectors of the superspace lattice. Circles represent the atoms in 3D space marked $t = 0$; their arrangement obviously lacks periodicity. (b) $(x_{s,1}, x_{s,4})$ section of the Fourier map of $Zn_2P_2O_7$ at the position of atom O3 $[(x_1^0(O3), x_2^0(O3), x_3^0(O3)) = (0.782, -0.151, 0.775)]$. Contour lines of equal electron density are given at different densities with the highest density at the center. (Reprinted from Ref. [49].)

superspace. $\mathbf{a}_{s,i}$ differs from $\mathbf{a}_i$ by a component perpendicular to 3D space, that is, along the fourth axis (Figure 4.4a).

Incommensurately modulated crystals are described by a periodic structure model in (3 + 1)D superspace (Figure 4.4a). Complete information on the crystal structure is contained in one unit cell of superspace. Instead of point-like atoms in the 3D periodic structure of a periodic crystal, the (3 + 1)D structure model involves strings (wavy lines), on the average parallel to the fourth coordinate axis. A (3 + 1)D Fourier map contains string-like electron density features. The trace of its maximum along the string exactly follows the modulation functions of the atoms. As an example, Figure 4.4b shows a 2D section of the (3 + 1)D Fourier map of $Zn_2P_2O_7$ at the positions $x_2^0 = 0.151$ and $x_3^0 = 0.775$ of the oxygen atom O3 [49]. The $(x_{s,1}, x_{s,4})$ section shows two periods along $x_{s,4}$ of the modulated electron density. The trace of the maximum of this density closely follows the modulation function given by the red line.

4.3
Modulation Functions and Fourier Maps

Understanding and interpretation of modulated crystal structures may start with a visual inspection of the (3 + 1)D superspace structure model. It is, of course, not possible to make a drawing of something existing in (3 + 1) dimensions. The maximum one could do is to produce a perspective view involving three out of the four coordinate axes in a so-called 3D plot or in a stereo view, as is often used in protein crystallography for 3D periodic structures. With four, five, or six dimensions, the most common representations of modulated crystals are 2D sections and projections. Atoms of a modulated crystal show up in the superspace structure model as wavy lines, on the average parallel to the fourth axis ($x_{s,4}$ coordinate) (Figure 4.4a). Alternatively, modulated atoms are responsible for local maxima in the Fourier map of diffraction data. This maximum follows a trace that represents the modulated position of the atom (Figure 4.4b). Any map known from 3D crystallography can be generalized to superspace. Fourier maps and difference Fourier maps are particularly helpful in establishing the structure model and solving for the modulation functions.

As an example, consider the 1D modulated crystal structure of γ-Na_2CO_3, for which Dusek et al. [50] have obtained an extensive X-ray diffraction data set, including satellite reflections up to order $|m| = 4$. The basic structure contains three crystallographically independent sodium atoms, two oxygen atoms, and one carbon atom (Figure 4.5a). The Fourier map of observed structure factors with phases from the structure model clearly shows an excellent match between electron density and the modulated atomic positions, thus confirming the model [50]. For example, the $(x_{s,2}, x_{s,4})$ section of the $(3+1)$D Fourier map at the position of Na3 reveals that the displacement modulation of Na3 in the direction of $x_{s,2}$ is different from a simple sinusoidal wave (Figure 4.6a). Instead, the shape of this function would be better approximated by a zigzag function. In general, Fourier maps may provide information about nonsinusoidal shapes of modulation functions. Special functions that have been regularly encountered as modulation functions include the block wave or crenel function, the saw-tooth function, and the zigzag function [51].

Employing a model with four harmonic waves for the displacement modulation functions ($N_{harm} = 4$ in Eq. (4.11)), Dusek et al. [50] noticed that the

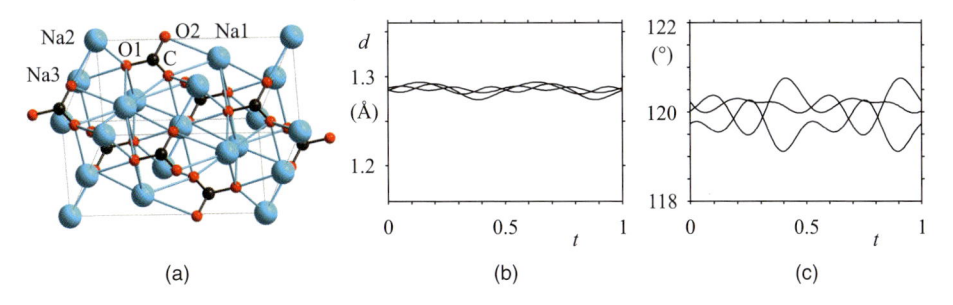

Figure 4.5 (a) Basic structure of γ-Na_2CO_3. CO_3^{2-} groups are composed of one O1 and two O2 atoms. C-centered monoclinic lattice with $a = 8.934$, $b = 5.268$, $c = 6.057$ Å, and $\beta = 101.55°$. Modulation with **q** = (0.1776, 0, 0.3252) and superspace group $C2/m(\sigma_1\ 0\ \sigma_3)0s$. (Graphics made with Diamond software employing atomic coordinates from Ref. [50].) (b) t-Plot of C—O bond distances within one carbonate anion. (c) t-Plot of O—C—O bond angles within one carbonate anion. (t-Plots made with JANA2006 employing atomic coordinates from Ref. [50].)

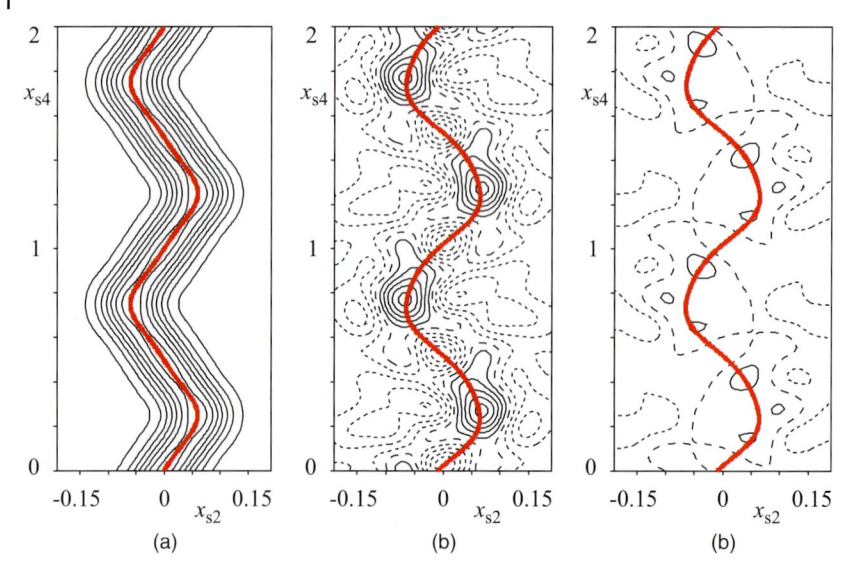

Figure 4.6 Modulated structure of Na$_2$CO$_3$ in (3 + 1)D superspace [50]. Each panel gives a $(x_{s,2}, x_{s,4})$ section of superspace showing a width of 2 Å along $x_{s,2}$ and two periods along $x_{s,4}$. (a) Fourier map at atom Na3. (b) Difference Fourier map at Na1 for a model comprising displacive modulations only. (c) Difference Fourier map at Na1 for a model including modulation functions for the ADPs. The thick red curve represents the refined atomic position. Positive contours of equal electron density are given by full lines; long-dashed lines are zero contours; dashed lines represent negative values. Densities are integrated for ±1 Å along the $x_{s,1}$ and $x_{s,3}$ axes. (Computed with Jana2006 according to the Jana cookbook.) (Reproduced with permission from Ref. [52]. Copyright 2014, De Gruyter.)

difference Fourier map – computed on the basis of the difference between the observed and calculated structure factors – features a very specific pattern of local maxima and local minima (negative peaks). Local maxima appear where the modulation function of atom Na1 describes maximum displacement either in the positive or in the negative direction (Figure 4.6b). Local minima appear at the position of Na1, where the modulation function is approximately zero. This pattern indicates a modulated ADP of Na1. At maximum displacement, the ADP is smaller than average, indicating an amplitude of atomic vibrations that is smaller than average. This is explained by the tighter environment of Na1 for these values of $x_{s,4}$. At zero displacement, the ADP is larger than average, corresponding to a less tight environment. Indeed, incorporation of modulation functions for ADPs with up to second-order harmonics results in a better fit to the diffraction data (lower R values) as well as the disappearance of features in the difference Fourier map (Figure 4.6c).

While excellent for uncovering the last details about modulation functions, Fourier maps are less suited for analyzing the correlations between the modulations of neighboring atoms. The latter is most favorably done with aid of so-called t-plots (see Section 4.4).

4.4
Crystal Chemical Analysis: t-Plots

The superspace approach has been highly successful in providing a complete structural analysis of modulated crystals, including a description of their symmetries and structure refinements against diffraction data. Nevertheless, matter exist in our space of three dimension. The fourth superspace coordinate is a fictitious construct, facilitating the analysis, but being of an essentially different character than the three dimensions of our physical space. For example, a surface structure with two independent modulations requires $(2+2)$D superspace, while a bulk crystal with one modulation wave is described in $(3+1)$D superspace. Although both descriptions refer to a 4D space, properties of $(2+2)$D superspace and $(3+1)$D superspace are different in all essential aspects. A consequence of the prominent role of the three physical dimensions is that modulation functions can only possess meaningful amplitudes in three physical space directions (usually taken along the three basis vectors of the basic structure unit cell). A further consequence is that thermal vibrations of the atoms have components in physical space only, thus having prevented phonon theory to be generalized toward modulated crystals and superspace [25,26].

Given the superspace structure model, the atomic positions in physical space are easily constructed by taking the intersection of the wavy lines representing atoms with physical space. The latter is composed of the three dimensions perpendicular to the fourth superspace axis, illustrated in Figure 4.7a by the horizontal line along $\mathbf{a}_2$ and marked $t=0$ (Eq. (4.8)). Atoms in physical space are represented by circles on the line $t=0$. They obviously are not arranged periodically, thus illustrating the aperiodic character of modulated crystals. The simple crystal structure of Figure 4.7 has one atom at the origin of the unit cell of the basic structure. The atom in the first unit cell is given the number 0 (this atom being at $(0,0,0)$). Atoms in consecutive unit cells along $\mathbf{a}_2$ are numbered $1, 2, \ldots$. Interatomic distances between an atom and its neighbor in the direction of $\mathbf{a}_2$ then are denoted by d_{01}, d_{12}, and so on (Figure 4.7a). The basic structure has but one value for this distance (the value $a_2 = 3$ Å), but many different values are found in the modulated structure. They differ from the basic structure value by an amount depending on the values of the modulation function. In particular, they are all different because of the incommensurability of the modulation. Figure 4.7a displays four such different distances.

By virtue of the translational symmetry in superspace, any atom pair (each of the distances d_{pq}) has a counterpart within the first unit cell of superspace (Figure 4.7a). Mapping a distance toward the first unit cell modifies the value of t as indicated. Any pair of atoms in physical space maps onto a different value of t within the interval $0 \leq t < 1$, since one period along $\mathbf{a}_{s,4}$ encompasses values of t from 0 to 1. It is now possible to replace the table of infinitely many different distances d_{pq} by the continuous dependence of the distance d_{01} on t, restricted to one period. All possible distances between an atom and its neighbors thus are obtained by the values of the function $d_{01}(t)$ for $0 \leq t < 1$ (Figure 4.7b). The

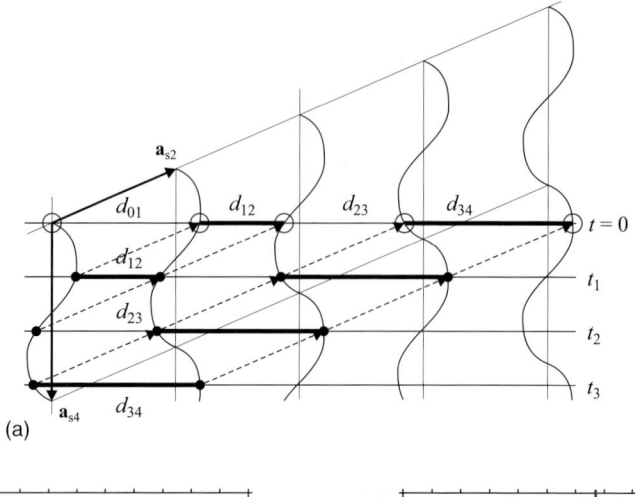

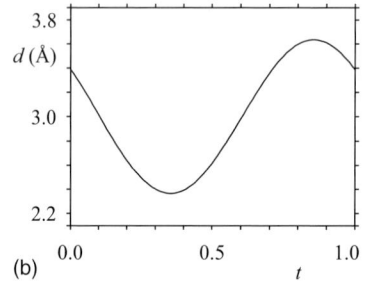

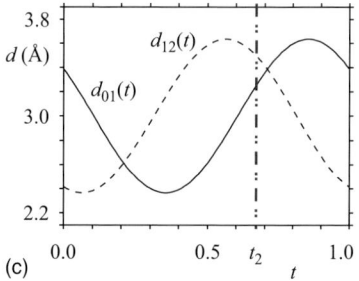

Figure 4.7 Two-dimensional section of (3 + 1)D superspace for a simple modulated structure, demonstrating the construction of t-plots of interatomic distances. (a) Structure with one atom at the origin of the unit cell; its modulated position is described by a wavy line, on average parallel to $\mathbf{a}_{s,4}$. (b) The distance d_{01} between atoms 0 and 1 as a function of t. (c) Distances between atom 1 and its neighbor toward the left ($d_{01}(t)$; solid curve) and its neighbor toward the right ($d_{12}(t)$; dashed curve). The red dashed–dotted line marks the t-value t_2.

other way around, any distance within the first unit cell of superspace represents a distance somewhere in physical space (somewhere on the line $t = 0$). As a result of the incommensurability, slightly different values of t give slightly different distances, which, however, correspond to atom pairs far apart in physical space. Except for the plotted distance (Figure 4.7b), information on relations between atoms occurring within the same region of physical space are lost in the t-plot (but see below). One important property of t-plots is that of proportionality: All distances that occur for a certain fraction of t values (e.g., for $0.4 < t < 0.5$ representing 10% of the t values) exactly represent the same fraction of distances in physical space (10% of the distances are approximately 2.4 Å in the structure of Figure 4.7b).

The most important property of t-plots is that each value of t gives a representation of physical space, which are all equal and only differ by a translation. This allows correlations between the modulations to be analyzed for any set of atoms. In the example of Figure 4.7a, we can consider the two neighbors of an atom. Taking atom 1 as reference, the distance d_{01} gives the distance toward the left, while the distance d_{12} gives the distance toward the right neighbor. The t-plot of both functions then provides a picture of the environment of an atom, as one steps through the incommensurate structure (Figure 4.7c). For example, at $t = 0$, Figure 4.7c gives $d_{01}(0) \approx 3.4$ Å and $d_{12}(0)$ much smaller at ~ 2.4 Å. These values appear to be in agreement with the values of d_{01} and d_{12} in the section $t = 0$ of Figure 4.7a. Agreement can also be checked for $t = t_2$ (red dashed–dotted line in Figure 4.7c), where $d_{01}(t_2)$ is smaller than $d_{12}(t_2)$ in agreement with $d_{23} < d_{34}$ (Figure 4.7a).

These properties of t-plots can be used to establish the structural chemistry of modulated crystals. γ-Na$_2$CO$_3$ contains carbonate anions, for which it is known that the C—O bond length can assume values within a narrow range. For the example of γ-Na$_2$CO$_3$, the t-plot of the three C—O distances within a single CO$_3^{2-}$ ion indeed shows variations of less than ± 0.01 Å (Figure 4.5b), despite individual modulation amplitudes exceeding 0.4 Å. The t-plot of the O—C—O bond angles shows a maximum deviation of $\pm 1°$ from the ideal value of $120°$ (Figure 4.5c). These small variations demonstrate the rigidity of the carbonate group. Modulations of the carbon and oxygen atoms are correlated, such that the C—O distances and O—C—O angles remain approximately equal to their optimal values. These observations illustrate a general property: Modulations respect the chemistry known from periodic crystals. The bond lengths of fully developed chemical bonds are never modulated, because shortening of these bonds would be too costly in energy.

Again, for γ-Na$_2$CO$_3$, one can consider the t-plots of the Na—O distances. Like for the carbonate ion, variations are small for the individual Na—O distances around the Na2 and Na3 atoms (Figure 4.8b and c). However, the t-plots

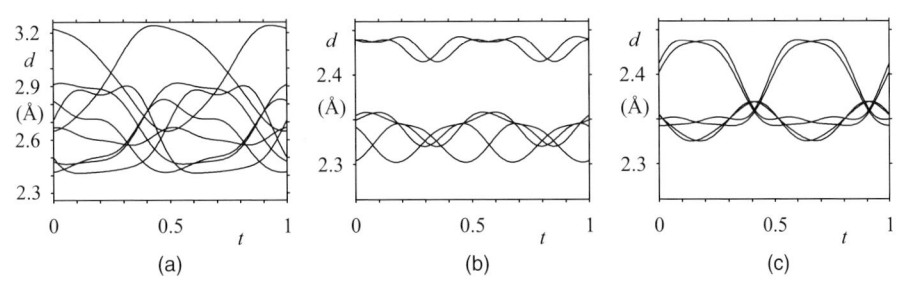

Figure 4.8 Interatomic distances in γ-Na$_2$CO$_3$. (a) Na1—O. (b) Na2—O. (c) Na3—O. Notice the different distance scales. Other than the six Na—O contacts shown, Na—O distances shorter than 3.26 Å do not occur for Na2 and Na3 atoms. (t-Plots made with JANA2006 employing atomic coordinates from Ref. [50].)

also show the correlations between the modulations of different atoms. The six curves in Figure 4.8b represent the distances from atom Na2 to the six oxygen atoms in its first coordination shell. The six Na—O distances obtained at a single value of t describe the coordination polyhedron of Na2 somewhere in the modulated crystal structure; the six distances at another value of t describe the coordination polyhedron elsewhere. Accordingly, these t-plots demonstrate that the crystallographically independent atoms Na2 and Na3 are in sixfold coordination everywhere in the modulated crystal (Figure 4.8b and c). Variations of individual Na—O bonds are small. Nevertheless, the modulations of the six Na—O bonds exhibit a clear correlation: If one Na—O distance has its minimum value, other Na—O distances are at their maximum values. This compensating effect of modulations of neighboring atoms has been observed in many modulated crystals.

The situation is quite different for Na1. Up to a distance of 3.26 Å, nine Na—O contacts can be found in the structure (Figure 4.8a). Individual Na—O distances may show large variations, but rules known from the chemistry of periodic crystals are respected: None of the Na—O contacts fall below the value of 2.3 Å found for Na2 and Na3. Actually, the shortest Na—O distance is about 0.1 Å longer for Na1 compared to Na2 and Na3. The higher coordination number thus is compensated by longer distances for the individual Na—O contacts, such that Na1 even appears to be slightly underbonded compared to Na2 and Na3. The exceptional coordination of Na1 along with the large variations of its Na—O bond lengths have led to the interpretation that the less than optimal coordination of Na1 is at the origin of the complex phase diagram of Na_2CO_3, featuring incommensurate and commensurate superstructures [50].

4.5
Modulated Molecular Crystals

The origin of incommensurability of crystal structures is found in the principle of competing interactions. The state of minimum energy according to one type of interaction would be a periodic arrangement of the atoms, while another type of interaction then is not optimally satisfied or, taken by itself, would have led to a structure with a different period. Many compounds form simple crystal structures possessing translational symmetries, despite competing interactions. In other compounds, nature finds some intermediate solution to the frustration, and an incommensurately modulated crystal is formed. As a function of temperature, phase transitions are quite often found between a periodic structure at high temperatures, an ordered superstructure at low temperatures, and an incommensurately modulated structure at intermediate temperatures.

Frustration of interactions is most easily recognized in molecular compounds. It is well established that molecular conformations in crystals can be different from those in the gas phase or in solution. Apparently, intermolecular interactions favor a molecular conformation that is different from the conformation optimizing intramolecular interactions. The change of conformation of the molecule upon entering the solid state is facilitated if the cost of energy is marginal for the conformational change. This situation is achieved if conformational changes occur through modifications of torsion angles. Several molecular crystals are known, for which the competition between these two interactions have resulted in an incommensurate variation of the intramolecular torsion throughout the crystal [31,32,53].

A simple example of this principle is provided by 2-phenylbenzimidazole ($C_{13}H_{10}N_2$) [54]. The molecule comprises a phenyl ring and a benzimidazole moiety, which are connected through a C—C bond. Both moieties are planar and rigid. Torsion about the bond between them is described by the dihedral angle between the planes of the two moieties, with coplanarity defined by $\phi = 180°$ (Figure 4.9a). The optimal dihedral angle of a single molecule is $\phi = 160°$ [32]. Crystal packing effects push this angle toward 180°. Instead of a single conformation, frustration is solved by an incommensurate modulation with different modulation functions for the torsion angles of the two crystallographically independent molecules. This feature is demonstrated by the t-plot of the torsion angles, which reveal a variation of ϕ by $\pm 5°$ for molecule A and by $\pm 3°$ for molecule B, while the average torsion angle of molecule B deviates from coplanarity by approximately 5° [54].

A different mechanism of incommensurability is found in phenazine–chloranilic acid (Phz–H_2ca). The crystal structure of Phz–H_2ca is made of infinite chains of alternating Phz and H_2ca molecules. The H_2ca molecule is donor for an O—H⋯N hydrogen bond with each of its two Phz neighbors. The crystal is centrosymmetric monoclinic at room temperature with H_2ca located at the inversion center, making its two opposing O—H⋯N hydrogen bonds equivalent by symmetry [56]. Below room temperature, the loss of inversion center allows for incomplete proton transfer in one of the two hydrogen bonds of each molecule, resulting in a ferroelectric moment of the crystal [56]. Further cooling leads to an incommensurate ferroelectric phase and a twofold superstructure

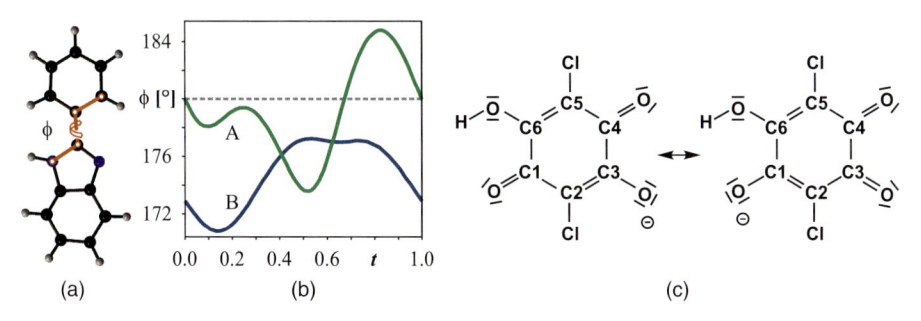

Figure 4.9 (a) Molecular structure of 2-phenylbenzimidazole. (b) t-Plot of the torsion angle ϕ for molecule A (green) with an amplitude of $\pm 5°$, and molecule B (blue) with an amplitude of $\pm 3°$. The dashed gray line indicates the value $\phi = 180°$ corresponding to a flat molecule. (Reprinted from Ref. [31].) (c) Resonance on the chain O4—C1—C2—C3—O1 of the chloranilic acid ion Hca⁻. (Reprinted from Ref. [55].)

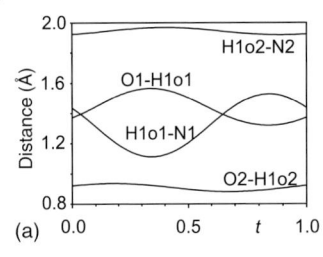

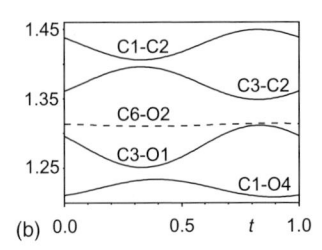

Figure 4.10 t-Plots of selected interatomic distances in Phz–H_2ca. (a) O—H and N—H distances within the two intermolecular hydrogen bonds. (b) C—C and C—O distances of the resonance system O4—C1—C2—C3—O1 of Hca$^-$, as well as the C6—O2 distance not involved in resonance. (Reprinted from Ref. [55].)

below 137 K. The latter ferroelectric phase features proton transfer in half of the chains [56].

The Hca$^-$ ion is stabilized by resonance involving half of the molecule (Figure 4.9c). The formally double bonds C1—O4 and C3—C2 and single bonds C1—C2 and C3—O1 should be different in H_2ca and Hca$^-$ ions because of this resonance. This feature is demonstrated by the crystal structure of the incommensurate phase. The O2—H1o2···N2 hydrogen bond is not modulated (Figure 4.10a), while the position of the hydrogen atom within the O1—H1o1···N1 hydrogen bond depends on t, effectively modulating between neutral (H_2ca) and ionic (Hca$^-$) states [55]. The t-plot of the distances reveals that the modulations of the carbon and oxygen atoms are correlated to those of O1—H1o1···N1. Within the region of neutral molecules ($t \approx 0.82$ – distance O1—H1o1 is at a minimum (Figure 4.10a)), the distances C1—O4 and C3—C2 are at a minimum too, indicating fully developed double bonds (compare with Figure 4.9c). At the same value of t, C1—C2 and C3—O1 are at maximum separation, in agreement with single bonds. Within regions where the proton is transferred toward nitrogen ($t \approx 0.32$–distance O1—H1o1 is at a maximum), the distances C1—O4 and C3—C2 are at a maximum, indicating the admixture of single-bond character into these formally double bonds. Similarly, C1—C2 and C3—O1 are at minimum separation for $t \approx 0.32$, indicating the admixture of double-bond character into formally single bonds. Through careful analysis of t-plots, the incommensurately modulated structure of Phz–H_2ca demonstrates the validity of the concept of resonance-stabilized proton transfer within hydrogen bonds formed by H_2ca [55]. At the same time, the observed features strongly support the model that proton transfer is responsible for the ferroelectric properties [56].

4.6
Charge Order and Related Phenomena

Instead of atomic displacements as primary modulation wave, nonstoichiometric compounds may develop an incommensurate ordering of the vacancies. An example is Pyrrhotite, $Fe_{1-x}S$, which may exist for a range of compositions with $x < 0.2$. The compound $x = 0.09$ is incommensurately modulated. The superspace model revealed ordering of the vacancies on the iron site, accompanied by a displacement modulation of both Fe and S [57]. The latter can be understood as being induced by the vacancy ordering because vacant and occupied sites require a different environment. A related phenomenon is substitutional modulation of two chemical elements alternately occupying the same site of a basic structure. An example is the ionic conductor $(Bi_2O_3)_{0.76}(Nb_2O_5)_{0.24}$ [58]. The complicated 3D modulation describes the formation of a network of octahedral chains NbO_6. The channels are filled with oxygen-deficient Bi_2O_5, which is responsible for the high ionic conductivity [58].

Yet another class of materials with potentially aperiodic crystal structures are mixed-valence compounds. Recently, a series of iron oxides have been discovered at high pressures, which differ in their Fe/O ratios and therefore in the relative amounts of Fe^{2+} and Fe^{3+} [59]. The compound Fe_4O_5 develops an incommensurate modulation below $T \approx 150$ K, which affects the distribution of Fe^{2+} and Fe^{3+} over the iron sites. The periodic structure at room temperature contains three crystallographically independent, collinear chains of iron atoms, containing Fe1, Fe2, and Fe3 atoms, respectively (Figure 4.11a). The periodic structure at room temperature is preserved as basic structure in the modulated phase. Application of the bond valence method [60] results in average values of the computed bond valence sums (BVS) of 2.1 for Fe1, 2.69 for Fe2, and 2.55 for Fe3. These values are close to the formal valences $Fe1^{2+}$, $Fe2^{2.67+}$, and $Fe3^{2.67+}$, which represent ideal values for Fe_4O_5 (equal amounts of Fe^{2+} and Fe^{3+}), if the mixed-valence chains comprise 1/3 Fe^{2+} and 2/3 Fe^{3+}. The Fe1 chain does not participate in the modulation, in agreement with the valence 2+ and with the different coordination of Fe1 (trigonal prismatic coordination by oxygen) than the octahedral coordination of Fe2 and Fe3. Largest amplitudes of the incommensurate, displacive modulation are the displacements along **a**, that is, along the chains, of the Fe2 and Fe3 atoms. t-Plots show that Fe–Fe distances along these chains exhibit a variation of up to 0.44 Å (Figure 4.11b and c).

Further analysis requires consideration of the modulation wave vector $\mathbf{q} = (1/3, 0.729, 0)$. The commensurate component along $\mathbf{a}^*$ leads to the peculiar property that each chain forms a threefold superstructure, while chains shifted in the direction of **b** possess different structures due to the incommensurate component along $\mathbf{b}^*$. The t-plots show the correlated variations of interatomic distances between five consecutive atoms, Fe^{iii}–Fe^{ii}–Fe–Fe^{i}–Fe^{iv}, along a chain (compare with Figure 4.7). Defining distances shorter than the Fe1–Fe1 distance of 2.861 Å as "bonds," the t-plot can be divided into six regions of equal width: three regions refer to the formation of trimers, while the other three regions refer to the formation of dimers (Figure 4.11b and c). For a single chain, modulation functions are sampled at exactly three values of their phases: t_0, $t_0 + 1/3$, and $t_0 + 2/3$; different chains have different t_0. The t-plots demonstrate that the modulation of a single chain samples regions of t with exclusively trimers or exclusively dimers. The modulated crystal structure thus

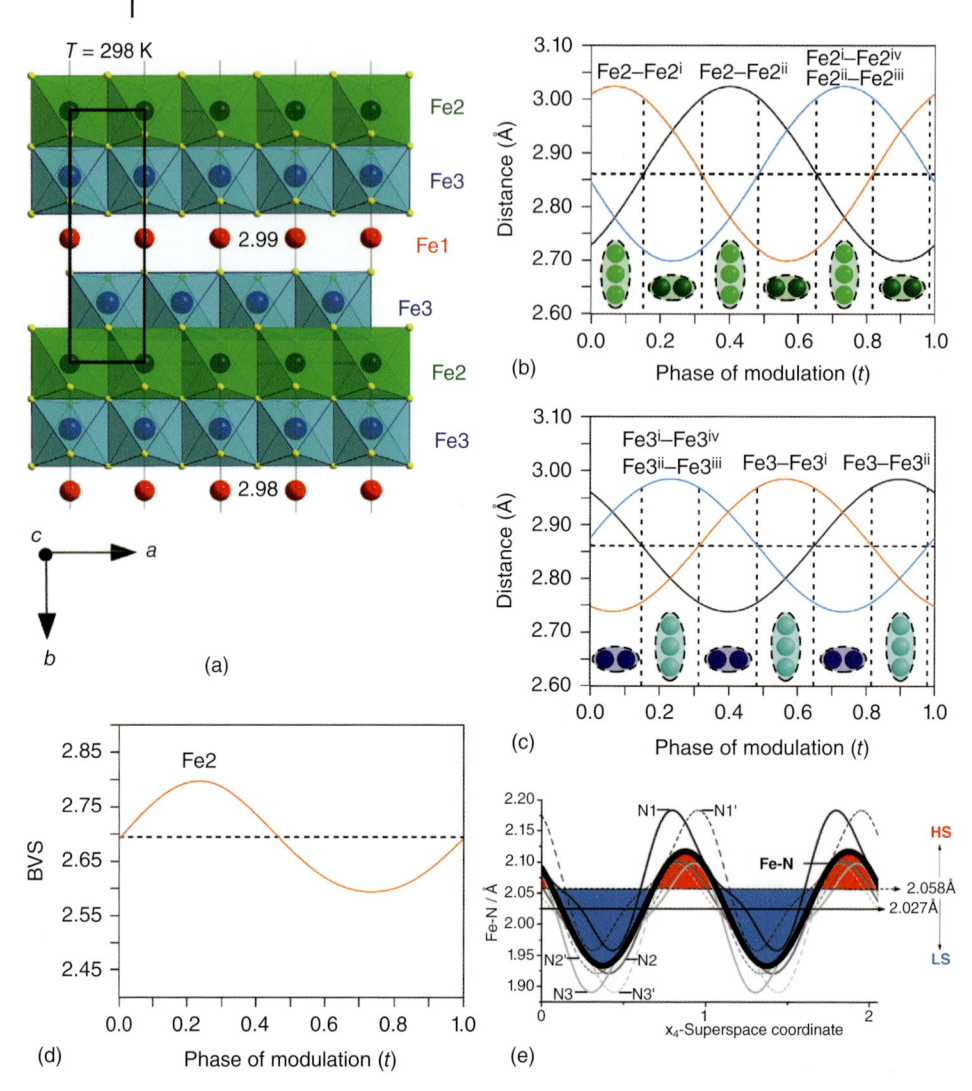

Figure 4.11 (a) Periodic structure of Fe_4O_5 at 295 K with Fe1 in trigonal prismatic and Fe2 and Fe3 in octahedral environments; $a = 2.891$ Å, as indicated. (b) t-Plot of Fe–Fe distances along the Fe2 chains for the modulated crystal structure at $T = 4$ K. (c) t-Plot of Fe–Fe distances along the Fe3 chains. t-Plots involve five consecutive atoms along each chain according to Fe^{iii}–Fe^{ii}–Fe–Fe^i–Fe^{iv}. (d) t-Plot of the BVS of Fe2. (Reprinted from Ref. [61].) (e) t-Plot of Fe–N and average (thick line) Fe–N distances in modulated $Fe[(Hg(SCN)_3)_2][4,4'-bipy]_2$. (Reproduced with permission from Ref. [62]. Copyright 2016, John Wiley & Sons.)

contains trimer chains and dimer chains. Dimers can be interpreted as a Fe^{2+}–Fe^{3+} pair. Trimers formally are Fe^{3+}–Fe^{2+}–Fe^{3+}. They are reminiscent of the trimerons proposed to be the key structural element of the complex superstructure of magnetite Fe_3O_4 [63]. The exact location of the additional electron (of Fe^{2+}) could not be unequivocally determined. The BVS is not of much help here, as it reveals only a minor variation of the valence with t (Figure 4.11d), precluding any interpretation toward the real nature of charge order on single chains [61].

A completely different but related problem concerns the magnetic state of iron ions. A weak crystal field (longer Fe–ligand distances) imposes a high-spin (HS) state with 3d orbitals singly occupied by electrons with parallel spins. A strong crystal field (shorter Fe–ligand distances) brings iron into the low-spin (LS) state, with low-lying 3d orbitals doubly occupied. So-called spin-crossover compounds may exhibit a phase transition between the HS and LS states. $Fe[(Hg(SCN)_3)_2][4,4'\text{-bipy}]_2$ (bipy = bipyridine) is a spin-crossover compound with Fe in sixfold coordination by nitrogen. It develops an incommensurate modulation upon cooling toward ~102 K [62]. A t-plot of the Fe–N distance averaged over the distances toward the six ligands (thick line in Figure 4.11(e)) reveals a clear dependence of the environment on t, resulting in 34% of Fe being in the HS state. The modulation leads to stripes of HS and LS states [62].

4.7
Outlook

Careful analysis of modulated structures through t-plots will uncover the origin of incommensurability in each crystal. It may contribute to unraveling the origin of phase transitions and discovering the mechanisms lying at the roots of various properties, including superconductivity, magnetic order, and crystallization. t-Plots have shown that in all modulated compounds, the local chemistry (bond lengths, coordination polyhedra, atomic valences) is the same as in periodic crystals, thus demonstrating that incommensurability is just another normal state of matter.

References

1 Steno, N. (1669) De solido intra solidum naturaliter contento dissertationis prodromus, Florence.
2 Haüy, A.R.J. (1784) Essai d'une théorie sur la structure des cristaux, Paris.
3 Haüy, A.R.J. (1801) Traité de minéralogie, Paris.
4 Hessel, J.F.C. (1830) Krystall, in *Gehler's Physikalisches Wörterbuch Bd. 5* (eds H.W. Brandes, L. Gmelin, J. Horner, G.W. Muncke, and C.H. Pfaff), E. B. Schwickert, Leipzig, Germany, pp. 1023–1340.
5 Hessel, J.F.C. (1897) Krystallometrie oder krystallonomie und krystallographie, in *Ostwalds's Klassiker der Exakten Wissenschaften*, vols. **88–89** (ed. E. Hess), Wilhelm Engelmann, Leipzig, Germany.
6 Bravais, A. (1850) Les systèmes formés par des points distribués regulièrement sur un plan ou dans l'espace. *J. De l'Ecole Polytechnique*, **XIX**, 1–128.
7 Fedorov, E.S. (1891) Simmetriia pracil'nykh sistem figur. *Zap. Min. Obshch.*, **28**, 1–146.
8 Schoenflies, A. (1891) *Kristallsysteme und Kristallstruktur*, B. G. Teubner, Leipzig, Germany.

9 Friedrich, W., Knipping, P., and von Laue, M. (1912) Interferenzerscheinungen bei röntgenstrahlen. eine quantitative prüfung der theorie für die interferenzerscheinungen bei röntgenstrahlen. Bayerische Akademie der Wissenschaften, Mathematisch-Physikalische Klasse, Sitzungsberichte 1912, pp. 303–322.

10 Groth, P. (1921) *Elemente der Physikalischen und Chemischen Krystallographie*, R. Oldenbourg, Munich, Germany.

11 Tutton, A.E.H. (1924) *The Natural History of Crystals*, Kegan Paul, Trench, Trubner & Co., London.

12 Dehlinger, U. (1927) Über die verbreiterung der debyelinien bei kaltbearbeiteten metallen. *Z. Kristallogr.*, **65**, 615–631.

13 Preston, G.D. (1938) The diffraction of X-rays by age-hardening aluminium copper alloys. *Proc. R. Soc. Lond. USA*, **167**, 526–538.

14 Daniel, V. and Lipson, H. (1943) An X-ray study of the dissociation of an alloy of copper, iron and nickel. *Proc. R. Soc. Lond. A*, **181**, 368–378.

15 Fujiwara, K. (1957) On the period of out-of-step of ordered alloys with anti-phase domain structure. *J. Phys. Soc. Japan*, **12** (1), 7–13.

16 Wilkinson, M.K., Koehler, W.C., Wollan, E.O., and Cable, J.W. (1961) Neutron diffraction investigation of magnetic ordering in dysprosium. *J. Appl. Phys. Suppl.*, **32** (3), 48S–49S.

17 Tanisaki, S. (1961) Microdomain structure in paraelectric phase of $NaNO_2$. *J. Phys. Soc. Japan*, **16**, 579–1579.

18 Brouns, E., Visser, J.W., and de Wolff, P.M. (1964) An anomaly in the crystal structure of Na_2CO_3. *Acta Crystallogr.*, **17**, 614–614.

19 de Wolff, P.M. (1974) The pseudo-symmetry of modulated crystal structures. *Acta Crystallogr. A*, **30**, 777–785.

20 de Wolff, P.M. (1977) Symmetry operations for displacively modulated structures. *Acta Crystallogr. A*, **33**, 493–497.

21 Janner, A. and Janssen, T. (1977) Symmetry of periodically distorted crystals. *Phys. Rev. B*, **15**, 643–658.

22 Janner, A. and Janssen, T. (1979) Superspace groups. *Physica A*, **99**, 47–76.

23 van Aalst, W., den Hollander, J., Peterse, W.J.A.M., and de Wolff, P.M. (1976) The modulated structure of $\gamma\text{-}Na_2CO_3$ in a harmonic approximation. *Acta Crystallogr. B*, **32**, 47–58.

24 Frank, F.C. and van der Merwe, J.H. (1949) One-dimensional dislocations: I. Static theory. *Proc. R. Soc. Lond. A*, **198**, 205–216.

25 Janssen, T., Chapuis, G., and de Boissieu, M. (2007) *Aperiodic Crystals: From Modulated Phases to Quasicrystals*, Oxford University Press, Oxford.

26 van Smaalen, S. (2012) *Incommensurate Crystallography*, Oxford University Press, Oxford.

27 Shechtman, D., Blech, I., Gratias, D., and Cahn, J.W. (1984) Metallic phase with long-range orientational order and no translational symmetry. *Phys. Rev. Lett.*, **53**, 1951–1953.

28 Yamamoto, A. (1982) Structure factor of modulated crystal structures. *Acta Crystallogr. A*, **38**, 87–92.

29 Dam, B., Janner, A., and Donnay, J.D.H. (1985) Incommensurate morphology of calaverite ($AuTe_2$) crystals. *Phys. Rev. Lett.*, **55**, 2301–2304.

30 Cummins, H.Z. (1990) Experimental studies of structurally incommensurate crystal phases. *Phys. Rep.*, **185**, 211–409.

31 Schönleber, A. (2011) Organic molecular compounds with modulated crystal structures. *Z. Kristallogr.*, **226**, 499–517.

32 Pinheiro, C.B. and Abakumov, A.M. (2015) Superspace crystallography: a key to the chemistry and properties. *IUCrJ*, **2** (1), 137–154.

33 Steurer, W. and Dshemuchadse, J. (2016) *Intermetallics: Structures, Properties, and Statistics*, Oxford University Press, Oxford.

34 Lovelace, J.J., Simone, P.D., Petricek, V., and Borgstahl, G.E.O. (2013) Simulation of modulated protein crystal structure and diffraction data in a supercell and in superspace. *Acta Crystallogr. D*, **69** (6), 1062–1072.

35 McMahon, M.I. and Nelmes, R.J. (2006) High-pressure structures and phase transformations in elemental metals. *Chem. Soc. Rev.*, **35**, 943–963.

36 Petricek, V., Gao, Y., Lee, P., and Coppens, P. (1990) X-ray analysis of the incommensurate modulation in the 2: 2: 1: 2 Bi–Sr–Ca–Cu–O superconductor including the oxygen atoms. *Phys. Rev. B*, **42**, 387–392.

37 van Smaalen, S. (2005) The Peierls transition in low-dimensional electronic crystals. *Acta Crystallogr. A*, **61**, 51–61.

38 Perez-Mato, M., Ribeiro, J.L., Petricek, V., and Aroyo, M.I. (2012) Magnetic superspace groups and symmetry constraints in incommensurate magnetic phases. *J. Phys. Condens. Matter*, **24**, 163–201.

39 Steurer, W. and Deloudi, S. (2009) *Crystallography of Quasicrystals: Concepts, Methods and Structures*, Springer Series in Materials Science, Springer, Berlin.

40 Janssen, T., Janner, A., Looijenga-Vos, A., and de Wolff, P.M. (2006) Incommensurate and commensurate modulated structures, in *International Tables for Crystallography*, vol. **C** (ed. E. Prince), Kluwer Academic Publishers, Dordrecht, The Netherlands, pp. 907–955.

41 Stokes, H.T., Campbell, B.J., and van Smaalen, S. (2011) Generation of (3+d)-dimensional superspace groups for describing the symmetry of modulated crystalline structures. *Acta Crystallogr. A*, **67**, 45–55.

42 van Smaalen, S., Campbell, B.J., and Stokes, H.T. (2013) Equivalence of superspace groups. *Acta Crystallogr. A*, **69**, 75–90.

43 Burke, J.G. (1966) *Origins of the Science of Crystals*, University of California Press, Berkeley.

44 van Smaalen, S. (1995) Incommensurate crystal structures. *Crystallogr. Rev.*, **4**, 79–202.

45 de Wolff, P.M., Janssen, T., and Janner, A. (1981) The superspace groups for incommensurate crystal structures with a one-dimensional modulation. *Acta Crystallogr. A*, **37**, 625–636.

46 Noohinejad, L. (2016) *Aperiodic Molecular Ferroelectric Crystals*, PhD thesis, University of Bayreuth, Bayreuth, Germany.

47 Gatti, C. and Macchi, P. (2012) *Modern Charge-Density Analysis*, Springer, Dordrecht, The Netherlands.

48 Kuhs, W.F. (1992) Generalized atomic displacements in crystallographic structure analysis. *Acta Crystallogr. A*, **48**, 80–98.

49 Stoger, B., Weil, M., and Dusek, M. (2014) The alpha ↔ beta phase transitions of $Zn_2P_2O_7$ revisited: existence of an additional intermediate phase with an incommensurately modulated structure. *Acta Crystallogr. B*, **70**, 539–554.

50 Dusek, M., Chapuis, G., Meyer, M., and Petricek, V. (2003) Sodium carbonate revisited. *Acta Crystallogr. B*, **59**, 337–352.

51 Petricek, V., Eigner, V., Dusek, M., and Cejchan, A. (2016) Discontinuous modulation functions and their application for analysis of modulated structures with the computing system JANA2006. *Z. Kristallogr.*, **231**, 301–312.

52 Petricek, V., Dusek, M., and Palatinus, L. (2014) Crystallographic computing system JANA2006: general features. *Z. Kristallogr.*, **229**, 345–352.

53 Wagner, T. and Schonleber, A. (2009) A non-mathematical introduction to the superspace description of modulated structures. *Acta Crystallogr. B*, **65**, 249–268.

54 Zuniga, F.J., Palatinus, L., Cabildo, P., Claramunt, R.M., and Elguero, J. (2006) The molecular structure of 2-phenylbenzimidazole: a new example of incommensurate modulated intramolecular torsion. *Z. Kristallogr.*, **221**, 281–287.

55 Noohinejad, L., Mondal, S., Ali, S.I., Dey, S., van Smaalen, S., and Schonleber, A. (2015) Resonance-stabilized partial proton transfer in hydrogen bonds of incommensurate phenazine-chloranilic acid. *Acta Crystallogr. B*, **71**, 228–234.

56 Horiuchi, S., Kumai, R., and Tokura, Y. (2009) Proton-displacive ferroelectricity in neutral cocrystals of anilic acids with phenazine. *J. Mater. Chem.*, **19**, 4421–4434.

57 Yamamoto, A. and Nakazawa, H. (1982) Modulated structure of the NC-type (N=5.5) pyrrhotite, $Fe_{1-x}S$. *Acta Crystallogr. A*, **38**, 79–86.

58 Ling, C.D., Schmid, S., Blanchard, P.E.R., Petricek, V., McIntyre, G.J., Sharma, N., Maljuk, A., Yaremchenko, A.A., Kharton,

V.V., Gutmann, M., and Withers, R.L. (2013) A (3+3)-dimensional "hypercubic" oxide-ionic conductor: type II Bi_2O_3-Nb_2O_5. *J. Am. Chem. Soc.*, **135** (17), 6477–6484.

59 Sinmyo, R., Bykova, E., Ovsyannikov, S.V., McCammon, C., Kupenko, I., Ismailova, L., and Dubrovinsky, L. (2016) Discovery of Fe_7O_9: a new iron oxide with a complex monoclinic structure. *Sci. Rep.*, **6**, 32852.

60 Brown, I.D. (2009) Recent developments in the methods and applications of the bond valence model. *Chem. Rev.*, **109**, 6858–6919.

61 Ovsyannikov, S.V., Bykov, M., Bykova, E., Kozlenko, D.P., Tsirlin, A.A., Karkin, A.E., Shchennikov, V.V., Kichanov, S.E., Gou, H., Abakumov, A.M., Egoavil, R., Verbeeck, J., McCammon, C., Dyadkin, V., Chernyshov, D., van Smaalen, S., and Dubrovinsky, L. (2016) Charge-ordering transition in iron oxide Fe_4O_5 involving competing dimer and trimer formation. *Nat. Chem.*, **8**, 501–508.

62 Trzop, E., Zhang, D., PiCeiro-Lopez, L., Valverde-MuCoz, F.J., MuCoz, M.C., Palatinus, L., Guerin, L., Cailleau, H., Real, J.A., and Collet, E. (2016) First step towards a devil's staircase in spin-crossover materials. *Angew. Chem., Int. Ed.*, **55**, 8675–8679.

63 Senn, M.S., Wright, J.P., and Attfield, J.P. (2012) Charge order and three-site distortions in the Verwey structure of magnetite. *Nature*, **481** (7380), 173–176.

5
Characterization of Quasicrystals

Walter Steurer

ETH Zurich, Department of Materials, Leopold-Ruzicka-Weg 4, 8093 Zurich, Switzerland

5.1
Introduction

Quasicrystals (QCs) are binary or ternary intermetallic phases with quasiperiodic crystal structures (Figure 5.1). They can be characterized on all length scales by the same experimental techniques and methods as periodic intermetallics. Quantitative QC structure analysis, however, is still a challenge and requires a specific approach that is in the focus of this chapter. For an introductory review on the topic of QCs in general, see Ref. [1], for instance, and for a comprehensive book on QCs see Ref. [2].

QCs belong to the class of "aperiodic crystals," which can be best defined in Fourier space. The reciprocal space is also called Fourier space, because direct space (crystal structure) and reciprocal space (diffraction pattern) are related by Fourier transform.

Ideal d-dimensional (dD) crystal structures, be they periodic or aperiodic, are characterized by their pure-point Fourier spectra, which are supported on a $\mathbb{Z}$-module (vector module) of rank n. With other words, one needs a set of n reciprocal basis vectors, $\mathbf{a}_i^*$, for indexing the diffraction pattern of a dD periodic crystal with integers, h_i. If $n = d$, the crystal structure is periodic, for $n > d$ it is aperiodic. Apart from QCs, there are two more classes of aperiodic crystals known: those with incommensurately modulated structures (IMSs), and those with composite (host/guest) structures (CSs).

By definition, QCs are the only aperiodic crystals that can but do not need to feature a rotational ("noncrystallographic") symmetry that is incompatible with 3D lattice periodicity. The thermodynamically stable intermetallic QCs known so far, all show either 5-, 10-, or 12-fold rotation axes (Figure 5.2). The structures of QCs with 3D-lattice compatible rotational ("crystallographic") symmetries could be also described as IMSs. The QC description would be more appropriate in the cases where the diffraction data could not be clearly grouped

---

Handbook of Solid State Chemistry, First Edition. Edited by Richard Dronskowski, Shinichi Kikkawa, and Andreas Stein.
© 2017 Wiley-VCH Verlag GmbH & Co. KGaA. Published 2017 by Wiley-VCH Verlag GmbH & Co. KGaA.

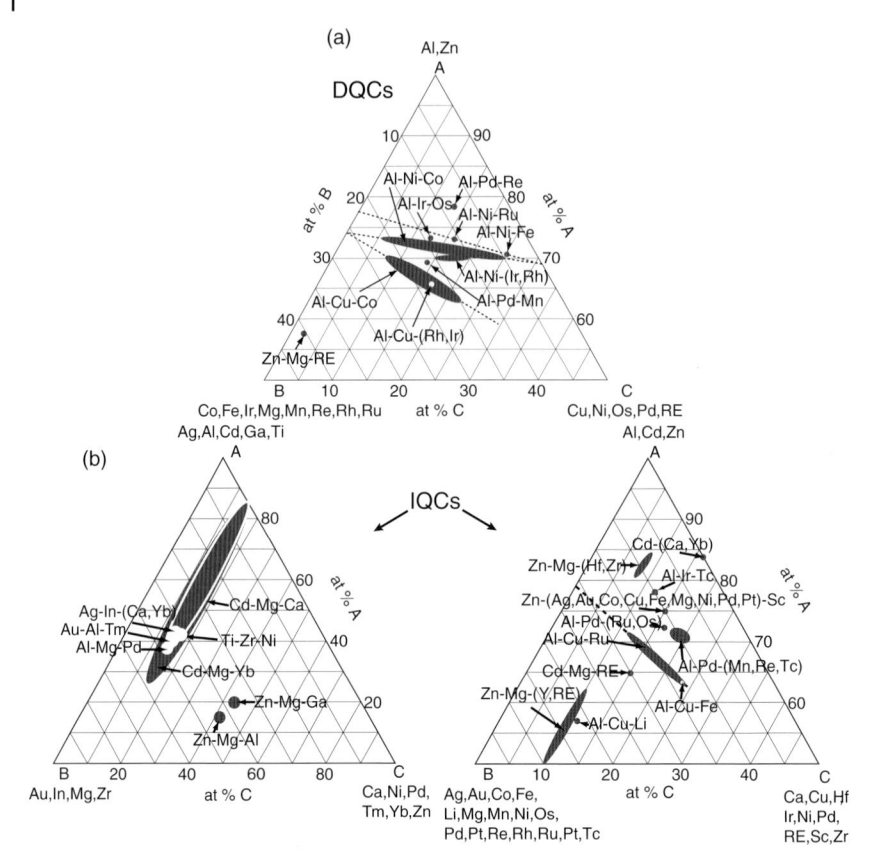

Figure 5.1 Compositions of decagonal QCs (DQCs (a)) and icosahedral QCs (IQCs (b)). Most ternary QCs show extended stability fields, indicating intrinsic substitutional disorder. The elements that can be substituted for A, B, and C in the general formula A–B–C are listed above or below these letters at the corners of the concentration triangles. An abbreviation such as Cd-(Ca,Yb), for example,, does not mean a mixed (Ca,Yb) site, it is just the shorthand notation for the two IQCs in the systems Cd–Ca and Cd–Yb, respectively.

into sets of main and satellite reflections, respectively, and if the structure could be well described based on tilings. In contrast, the IMS description should be chosen in all cases where the structures could be described as substitutional and/or displacive modulations of underlying periodic basic structures, leading to clearly separable sets of main and satellite reflections. QCs can show periodic and/or quasiperiodic superstructures.

Quasiperiodic structures can be described either based on the "higher dimensional (nD) approach" [3,4] or on the 3D "tiling-decoration method" [5]. The symmetry analysis can be performed either within the nD description leading to the "nD space groups" or, fully equivalently, in 3D reciprocal space leading to the 3D "quasicrystallographic space groups" [6,7]. The structure solution and

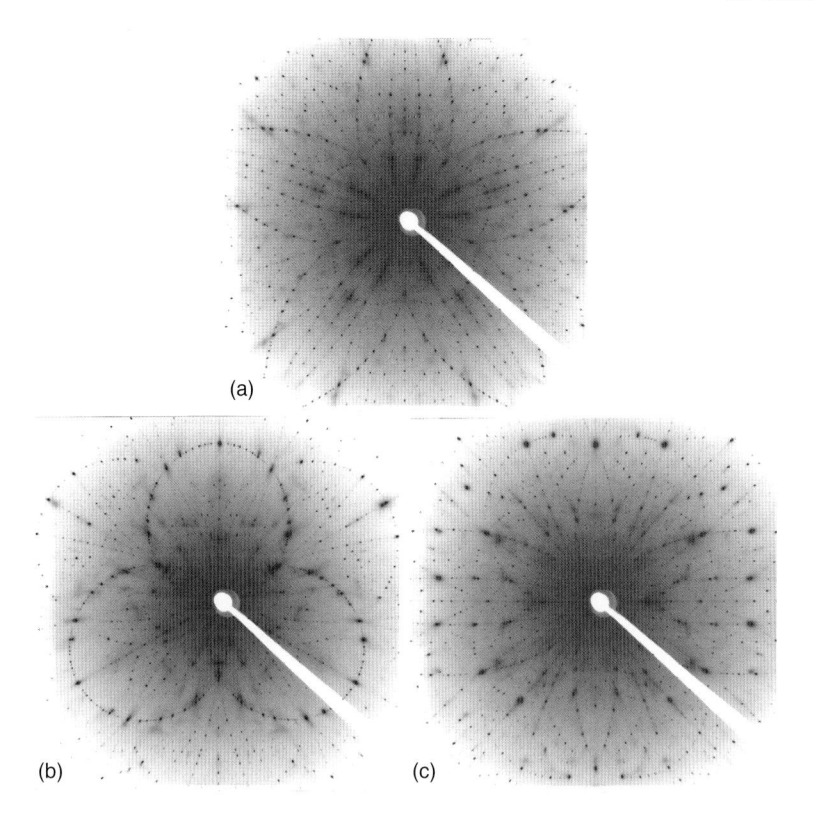

Figure 5.2 In-house X-ray Laue photographs of icosahedral Al–Mn–Pd taken along (a) the fivefold axis, (b) the threefold axis, and (c) the twofold axis. Some diffuse scattering, caused by phasonic and other kinds of structural disorder, is visible.

refinement, however, is best done using the nD approach. This allows a closed, unit-cell-based description of the infinite quasiperiodic structure, and makes the comparison of different QC structures easier.

In intermetallic systems featuring QCs, frequently approximants are observed at slightly different chemical compositions. Approximants are crystals with periodic structures that are built from the same structural subunits (usually termed "clusters") as that of the related QCs. If their structures can be described by the nD approach in the same way as QCs, then they are called "rational approximants." "Rational," because the irrational algebraic number defining the kind of quasiperiodicity is replaced by one of its rational approximations. For instance, the golden mean $\tau = \left(1 + \sqrt{5}\right)/2 = 2\cos\pi/5 = 1.618\ldots$, which is relevant for the description of QCs with fivefold symmetry, is replaced by one of its approximations $1/1, 2/1, 3/2, 5/3, 8/5\ldots$. Knowing the structure of rational approximants helps in setting up starting models for QC-structure analysis, since they carry information about the structure of the fundamental clusters and the possible ways to arrange them.

Until 2004, the term "quasicrystal" was exclusively used for intermetallic phases with quasiperiodic structures. This has changed with the discovery of mesoscopic QCs, that is, self-assembled colloidal quasiperiodic systems [8]. Furthermore, it has been demonstrated by computer simulations that quasiperiodic order is more universal, that it can take place on any scale provided that at least two different interaction length scales with a specific ratio exist in the structure (see, e.g., Refs. [9,10]).

5.2
Structure Analysis of Quasicrystals

We are interested in the crystal structure of a chemical compound because its knowledge is crucial for understanding its stability and the origin of its physical properties. We can gain a deeper insight into the kind of atomic interactions (chemical bonding) present in the crystal structure if it is analyzed as a function of composition, temperature, pressure, and, in particular cases, also of other external fields.

A crystal structure is defined by its symmetry (space group), metrics (lattice parameters), occupancy of the atomic sites (Wyckoff positions) in the unit cell, atomic displacement parameters (ADPs), and the kind of chemical and structural disorder, if any. The electron density distribution as well as the dynamics of the structure (phonon dispersion spectrum, vibrational density of states) can elucidate the kind of chemical bonding present. Further insight can be gained by quantum mechanical calculations on the structure model. The main goal of this effort is to extend the data basis of structure/property relationships for improving our ability to predict and tailor the composition of any material according to its required properties.

In the case of QC structure analysis, all the parameters listed above have to be described in terms of the nD approach. Not only the positions of the nD hyperatoms in the nD unit cell have to be determined but also the partition of their "atomic surfaces" (see below) into subdomains, and the chemical composition, occupancy, phasonic ADP, and physical space shift of each subdomain. The atoms generated by each subdomain have a specific coordination by the neighboring atoms.

In the case of 3D periodic structures, the periodic long-range order (lro) can be described by one of the 14 Bravais lattices, which can be derived by qualitative inspection of the diffraction pattern. In contrast, the knowledge of the nD Bravais lattice is not sufficient for describing the quasiperiodic lro. That is coded in the shapes and positions of the atomic surfaces, which result from structure solution and refinement only. This is the problem in the case that the structure should be analyzed based on the tiling decoration method, because one has to choose between an infinite number of different 3D tilings ("quasilattices"), which are not reliably known *a priori*. Electron microscopy can help in this case.

Unfortunately, due to the lack of periodicity, it is not possible yet to perform quantum mechanical calculations for quasiperiodic structures. These can be carried out only on structure models of rational approximants so far. If done for a series from lower to higher approximants, the information on the chemical bonding and electronic structure can be extrapolated to the case of QCs. But, due to limited computer power, we can still not fully understand why a particular structure is quasiperiodic, and whether the quasiperiodic ordering state is a ground state (stable at 0 K) or entropy stabilized.

In general, diffraction methods are the methods of choice for quantitative crystal structure analysis, while electron microscopy can give additional information on local ordering phenomena and the defect structure. This is also true for QCs. The determination of the symmetry and metrics of a QC is a straightforward task, because the nD lattice parameters and the nD space group can be easily derived from high-quality diffraction experiments. As for periodic crystals, convergent-beam electron diffraction (CBED) can be a valuable help for distinguishing between centrosymmetric and noncentrosymmetric nD space groups.

A full structural characterization also includes the determination of structural (displacive and substitutional) disorder. For QCs, one distinguishes between structural disorder of phasonic origin, which is specific for QCs, and one of other origin. One has to keep in mind that the existence of an extended compositional stability field (see Figure 5.1) already implies intrinsic chemical disorder accompanied by structural relaxations, leading to diffuse scattering in the diffraction patterns. For a general review, on diffuse scattering phenomena see, for instance, Ref. [11].

Spectroscopic methods, such as extended X-ray absorption fine structure (EXAFS) spectroscopy, for instance, can be applied to QCs in the same way as to periodic intermetallics. One has to keep in mind, however, that the local information obtained in this way is always averaged over the whole sample.

5.3
Transmission Electron Microscopy for Quasicrystal Structure Imaging

The local structure of intermetallic phases can be explored with either phase-contrast electron microscopy (high-resolution transmission electron microscopy, HRTEM) or scanning transmission electron microscopy (STEM). In the latter case, we distinguish between high-angle annular dark field (HAADF)-STEM and annular bright field (ABF)-STEM. HRTEM uses a coherent electron beam transmitting the whole sample at a time, while the STEM methods scan the sample with a focused beam either incoherently (HAADF-STEM, Figure 5.3) or coherently (ABF-STEM). The focus diameter is with ≈ 2 Å smaller than an atomic diameter. In the case of spherical-aberration (Cs) corrected instruments, a spatial resolution < 1 Å is possible. One has to keep in mind, however, that the structural information is averaged over the whole specimen thickness (50 – 100 Å, that is, 20–40 atomic layers).

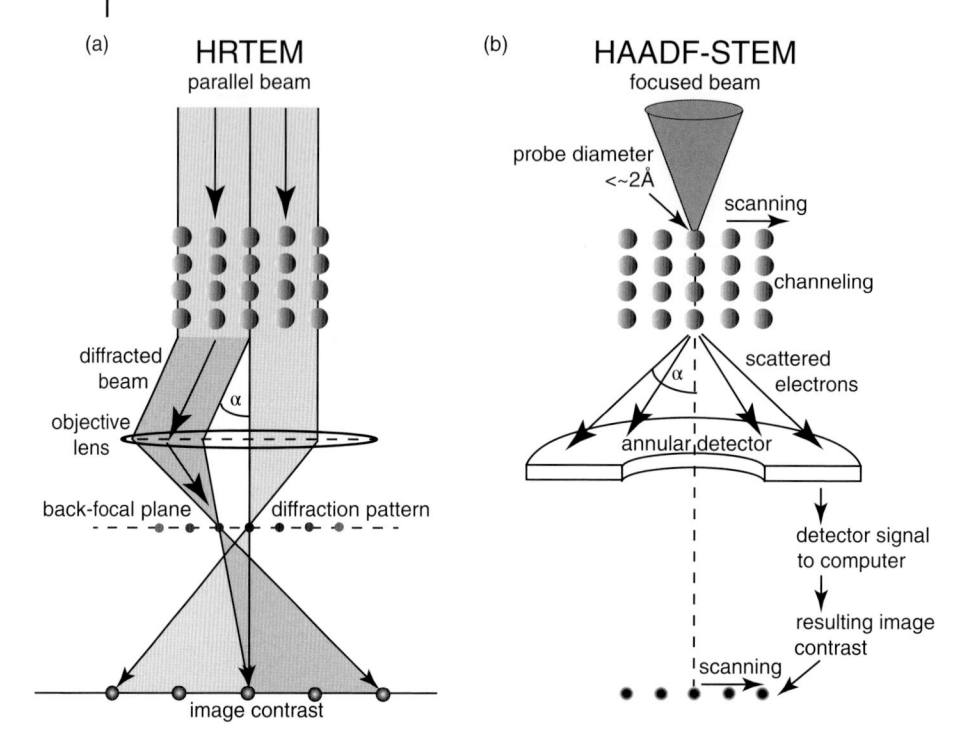

Figure 5.3 Principles of HRTEM (a) and HAADF-STEM (b) imaging. In the case of HRTEM, a parallel coherent electron beam is used and, depending on the setting of the focus of the first intermediate lens, either the selected-area electron diffraction (SAED) pattern or the image can be viewed. In the case of HAADF-STEM (Z contrast) a focused electron beam is moved over the specimen, and the inelastically scattered electrons are detected by an annular detector. Subsequently, the image is pixelwise reconstructed in the computer. (Reproduced from Ref. [12] with permission of JEOL Ltd).

Electron microscopy works for QCs in the same way as for periodic crystals since only the local structure is imaged. If most atoms are arranged in columns and/or on flat or only slightly corrugated layers, then the electron micrographs, which reflect the projected structure, will show contrasts that can be related to well-defined structural subunits such as clusters. Fortunatley, the atoms in the structures of both DQCs and IQCs are arranged on each other interpenetrating flat or slightly puckered layers, so that most atoms form columns along the projection directions. This is illustrated on the example of the structure of the DQC in the system Al–Cu–Co (Figure 5.4) [13].

The goal of electron-microscopic studies of QCs is to identify the "true" symmetry of clusters, and the ways the clusters are arranged and overlap, in order to confirm or exclude particular classes of tilings. For instance, whether the tiling underlying the structure of a DQC could be described by a pentagon/rhomb tiling rather than by a hexagon/boat/star tiling or just a rhomb tiling or even a

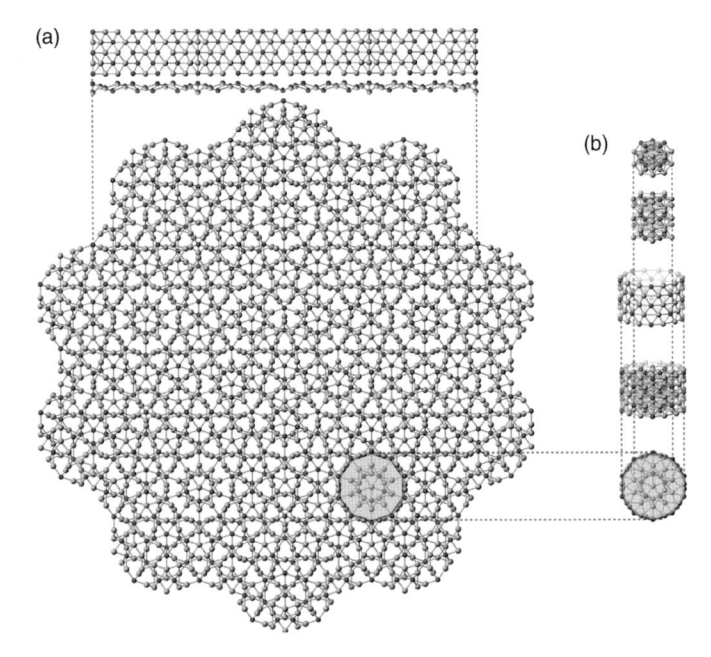

Figure 5.4 Along the periodic tenfold axis projected two-layer structure of the supercluster constituting decagonal $Al_{65}Cu_{14.6}Co_{20.4}$ (for a detailed discussion see Refs. [13,14]). (a) The subcluster structure can be clearly seen as well as the only slightly puckered layers perpendicular to the projection plane, of which one is shown on top. (b) The structure of such a columnar subcluster with ≈ 14 Å diameter is illustrated in an exploded view.

random tiling. This information cannot always be reliably obtained from diffraction data alone due to the globally averaging effect of this technique.

5.3.1
High-Resolution Transmission Electron Microscopy (HRTEM)

In the case of HRTEM (Figure 5.3a), coherent elastically scattered electrons are used for imaging. The sample is illuminated by the electron beam as a whole; the electron waves that are diffracted in the same direction are focused by the objective lens into points on the back-focal plane, leading to the electron diffraction pattern of the sample. The coherent diffracted electron beams, each carrying information from all atoms of the sample, propagate further to the image plane, creating there the image by interference. The image contrasts cannot be directly related to projected atomic columns due to dynamical scattering effects, which can cause drastic image contrast changes as a function of sample thickness and defocus value. The image contrasts have to be interpreted by image simulations based on a structure model. One of the advantages of HRTEM is that the sample is illuminated as a whole what leads to faster imaging, which allows to study dynamic changes (phason flips) in the structure as a function of temperature, for instance [15].

Selected-area electron diffraction (SAED) (see, e.g., Figure 5.5a) allows to identify orientation, metrics, and symmetry of a selected sample region such as individual clusters, for instance. The evaluation of the SAED reflection

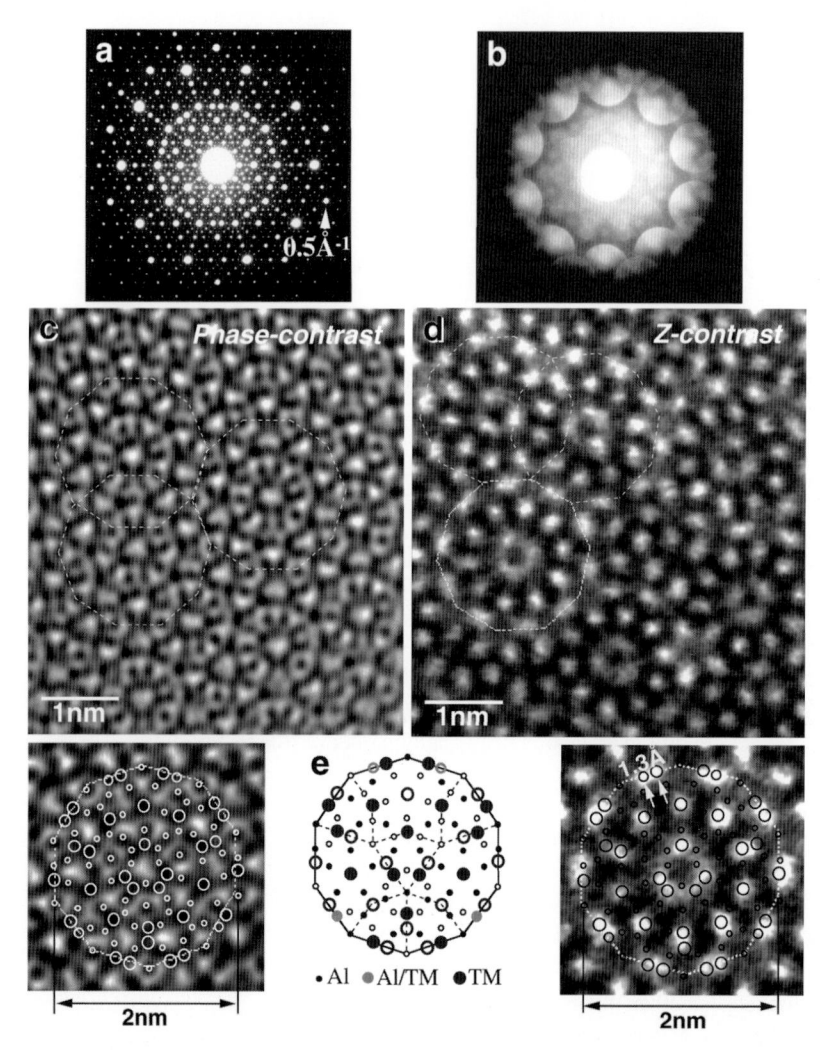

Figure 5.5 (a) Selected area (SAED) and (b) convergent beam electron diffraction (CBED) patterns taken along the tenfold symmetry axis of decagonal $Al_{72}Co_8Ni_{20}$. The CBED pattern allows deriving the point group symmetry, which is 10mm in this projection. The structure images were taken by (non-Cs corrected) (c) HRTEM and (d) HAADF-STEM from a 50 Å and a 100 Å thick sample region, respectively. The encircled regions in the images (c) and (d) are slightly enlarged reproduced below these images, with the structure model of a single cluster (e) overlaid. Solid and open circles correspond to atoms on different layers along the tenfold axis. (Reproduced from Ref. [12] with permission of JEOL Ltd.)

intensities for structure analysis is only possible to some extent due to multiple scattering. If only crystals were available that are too small for a single-crystal X-ray structure analysis, electron crystallography together with powder X-ray diffraction can be the method of choice, however (for an overview see Ref. [16], for instance).

5.3.2
High-Angle Annular Dark-Field and Annular Bright-Field Scanning Transmission Electron Microscopy (HAADF- and ABF-STEM)

In the case of HAADF-STEM (Figure 5.3b), an incoherent focused electron beam (focus diameter < 2 Å, convergence angle ≈ 10 mrad) is scanning over the specimen. The electrons channel through the sample, being mainly inelastically and incoherently scattered by the thermally vibrating atoms. The electrons leaving the sample within a certain angular range (> 50 mrad at 200 kV) are counted by an annular detector. The resulting signal is proportional to the scattering cross sections of the atoms contributing to the scattering, that is, the intensity is proportional to the square of the number of electrons of the atoms (Z^2). Therefore, this technique is also called Z-contrast electron microscopy. Dynamical scattering effects, phase relationships, and defocus do not play a role for imaging. Consequently, atoms always appear as bright contrasts making the interpretation of the images almost straightforward. A CBED pattern and an HAADF-STEM image are shown in Figure 5.5b and d, respectively.

The diffraction patterns in Figure 5.5a and b clearly reflect the tenfold symmetry of the DQC. A comparison of its HRTEM and HAADF-STEM images shows that one can identify structural subunits (clusters with diameters of ≈ 20 Å) and their arrangements in both cases. However, in the left image one might intuitively choose a τ-times smaller cluster (≈ 12 Å) than in the right one, and one could not assign all the atoms properly without image simulations. In contrast, the right micrograph quite clearly depicts the TM atoms, without being able distinguishing between Co and Ni atoms with atomic numbers, Z, differing by just one ($Z_{Co} = 27$, $Z_{Ni} = 28$); Al atoms ($Z_{Al} = 13$) are hardly visible. In the CBED patterns, the Bragg reflections appear as disks (Figure 5.5b). Their intensity distribution allows the derivation of the actual point group symmetry, not only of the Laue groups (centrosymmetric point groups). Due to dynamical effects, Friedel's law, $I(\mathbf{H}) = I(-\mathbf{H})$, does not apply.

A focused electron beam is also used by ABF-STEM, which is a scanning method using phase-contrast imaging. The convergence angle α is smaller (11–22 mrad) than in the case of HAADF-STEM, so that the direct beam can interfere with the electrons that are elastically scattered under small angle. Consequently, the annular detector is also smaller than that used for the HAADF technique. This method shows much less Z-contrast. Therefore, a combination of HAADF-STEM and ABF-STEM micrographs can be very powerful in differentiating and imaging light atoms next to heavier ones in order to get a complete structural picture [17,18].

5.3.3
Surface Structure Analysis of Quasicrystals

The surface structures of QCs can be studied in the same way as those of periodic crystals, that is, by atomic force microscopy (AFM), scanning tunneling microscopy (STM), low-energy electron diffraction (LEED), and other surface-sensitive diffraction techniques (for reviews see, e.g., Refs. [19–21]). The surfaces of the QCs investigated so far do not reconstruct. Consequently, if terraced crystal surfaces are studied, one can get information not only about the most stable terminating atomic layer but also about the layers beneath, which are accessible from neighboring terraces. In this way, even information about the bulk structure can be reconstructed and structure models obtained with other techniques can be verified. With LEED, one can get quantitative 3D information about the structures and distances of the atomic layers close to the surface, if one succeeds in calculating the dynamic structure factors of the quasiperiodic layers.

5.4
Diffraction Methods for the Analysis of Quasicrystal Structures

For the determination of the kind of long-range order present in a QC, the single-crystal diffraction analysis by X-rays, neutrons, and/or electrons, yielding a space- and time-averaged structure, is the method of choice. Space averaging means that the structure information from the irradiated area of the sample is taken modulo the assigned nD unit cell, and time averaging takes place, because each reflection intensity is integrated over its measurement time.

The first step in single-crystal X-ray diffraction structure analysis is the collection of a good data set, best with synchrotron radiation, and a single-photon noise-free counting pixel-detector (see, e.g., Refs. [22,23]). A high dynamic range ($> 10^{10}$) is of importance for measuring the required data set with a sufficient number of weak and very weak reflections as well as diffuse-scattering intensities. This high dynamic range can only be achieved if the detector features energy discrimination in order to suppress fluorescence radiation (see, e.g., Figure 5.2 of Ref. [23]).

In the case of QCs, the Bragg reflections can be integer-indexed based on a set of n reciprocal basis vectors, with $n = 5$ (Figure 5.6) in the case of DQCs, and $n = 6$ in the case of IQCs. If $n > d = 3$, with d the dimension of the physical par(allel)-space, then the reflections occupy countably dense the 3D reciprocal space. Consequently, in the limited reciprocal-space region experimentally accessible, there will be an infinite number of reflections, which can be seen as a proper projection of an nD reciprocal lattice. While the spatial resolution in the dD par-space is limited by the maximum experimentally achievable diffraction angle θ, the resolution in the $(n-d)$D complimentary perp(endicular)-space depends on the number of reflections with large perp-space diffraction vector components, $|\mathbf{H}^\perp|$, measured within the given reciprocal space section.

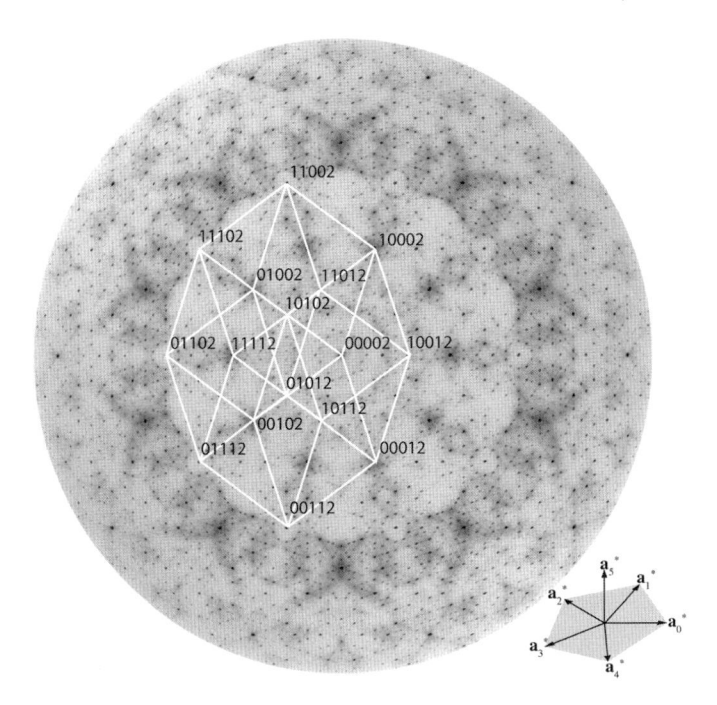

Figure 5.6 $h_1h_2h_3h_42$ reciprocal space layer of decagonal $Al_{71}Co_{13}Ni_{16}$ reconstructed from image-plate data taken at SNBL/ESRF. One projected 4D hyperrhombohedral reciprocal subunit cell is marked with the reflections at its 16 vertices indexed. The actual unit cell is 5D, the 5th dimension runs along $\mathbf{a}_5^*$, that is, the periodic [00001] direction. The set of basis vectors is shown at bottom right $(\mathbf{a}_0^* = -(\mathbf{a}_1^* + \mathbf{a}_2^* + \mathbf{a}_3^* + \mathbf{a}_4^*))$.

For periodic structures, it is possible to collect a "full data set" within a given range of the diffraction angle θ, for instance, $3° \leq \theta \leq 30°$ for MoK$_\alpha$ radiation. A full data set contains both observed and unobserved reflections, both carry structural information. In contrast, in the case of QCs one can collect only a "relatively full data set," consisting of intensities above a given threshold value that are experimentally accessible. The threshold value is defined by the dynamic range of the detector, the synchrotron beam time available for the data collection, and, of course, by the perfection of the QC. One has to keep in mind that intermetallic QCs are grown from the melt, annealed at high temperatures for homogenization and thermal equilibration, and finally quenched to ambient temperature. Consequently, the samples represent high-temperature states, and will have a significant amount of phasonic and other kinds of structural disorder. This reduces the number of observable reflections with large $|\mathbf{H}^\perp|$, and therewith also the accessible perp-space resolution (see, e.g., Figure 5.1 of Ref. [23]).

It should be kept in mind that once a specific set of reciprocal basis vectors has been chosen, the underlying fundamental structure model has been accepted automatically. With other words, if the reflections were indexed based on the nD approach, the quasiperiodicity of the structure to be determined was already taken for granted. This means, one could not check anymore whether the structure of the studied sample be actually quasiperiodic or not, because one could not get beyond quasiperiodic solutions. Furthermore, an erroneously chosen quasiperiodic model structure could even nicely fit the experimental data of an actual nonquasiperiodic structure such as an orientationally twinned approximant, for instance (see, e.g., Ref. [24]).

As already mentioned above, QC-structure solution based on diffraction data is best done within the nD approach, which allows to represent an infinite quasiperiodic structure by the content of a single nD unit cell. The Fourier spectrum $M_F^* = \{F(\mathbf{H})\}$ of a QC consists of δ-peaks, weighted with the structure factors $F(\mathbf{H})$, on a Z-module (an additive Abelian group)

$$M^* = \left\{ \mathbf{H} = \sum_{i=1}^{n} h_i \mathbf{a}_i^* | h_i \in Z \right\}, \tag{5.1}$$

of rank $n > d$, with basis vectors $\mathbf{a}_i^*$, $i = 1, \ldots, n$, and d the dimension of the QC structure, which in our case always equals the dimension of the 3D physical space. The nD embedding space V can be separated into two orthogonal subspaces (Figure 5.7) both preserving the point group symmetry according to the nD space group

$$V = V^{\|} \oplus V^{\perp}, \tag{5.2}$$

with the physical par-space $V^{\|} = \mathrm{span}(\mathbf{v}_1, \mathbf{v}_2, \mathbf{v}_3)$ and the complementary perpspace $V^{\perp} = \mathrm{span}(\mathbf{v}_4, \ldots, \mathbf{v}_n)$. The vectors $\mathbf{v}_i$ refer to a Cartesian coordinate system. The structure factor, the Fourier transform of the nD unit cell, is defined as

$$F(\mathbf{H}) = \sum_{k=1}^{m} T_k\left(\mathbf{H}^{\|}, \mathbf{H}^{\perp}\right) f_k\left(|\mathbf{H}^{\|}|\right) g_k\left(\mathbf{H}^{\perp}\right) e^{2\pi i \mathbf{H} \mathbf{r}_k}. \tag{5.3}$$

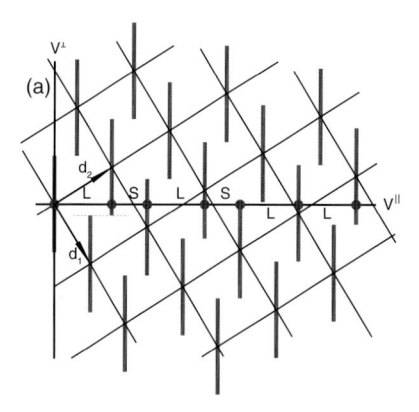

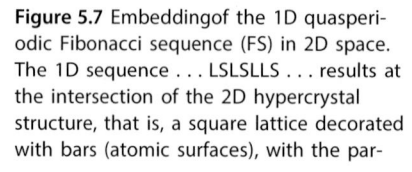

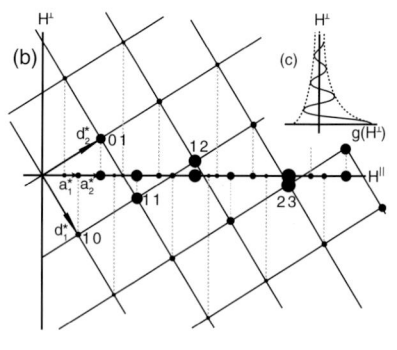

Figure 5.7 Embedding of the 1D quasperiodic Fibonacci sequence (FS) in 2D space. The 1D sequence ... LSLSLLS ... results at the intersection of the 2D hypercrystal structure, that is, a square lattice decorated with bars (atomic surfaces), with the par-space $V^{\|}$. (b) The 1D diffraction pattern of the FS results from the projection of the 2D reciprocal space onto the reciprocal par-space. (c) Fourier transform of one atomic surface illustrating the rapid decay of the perp-space atomic form factor $g(\mathbf{H}^{\perp})$.

In par-space, one gets the conventional atomic scattering factor $f_k\left(\left|\mathbf{H}^{\|}\right|\right)$ and the ADP (temperature factor) $T_k\left(\mathbf{H}^{\|}\right)$. In perp-space, the Fourier transform of the atomic surfaces, the geometrical form factor $g_k\left(\mathbf{H}^{\perp}\right)$, results to

$$g_k\left(\mathbf{H}^{\perp}\right) = \frac{1}{A_{UC}^{\perp}} \int_{A_k} e^{2\pi i \mathbf{H}^{\perp} \mathbf{r}^{\perp}} d\mathbf{r}^{\perp}, \tag{5.4}$$

with $A_{UC}^{\perp}$ the volume of the nD unit cell projected onto $V^{\perp}$, and A_k the volume of the k-th atomic surface. The perp-space component $T_k\left(\mathbf{H}^{\perp}\right)$ of the ADP, the phason ADP, describes the effect of phason fluctuations along the perp-space. These fluctuations, originate either from phason modes or from random phason flips.

The n-star of rationally independent vectors defining the Z-module M^* can be considered as appropriate projection, $\mathbf{a}_i^* = \pi^{\|}(\mathbf{d}_i^*)$ ($i = 1, \ldots, n$), of the basis vectors $\mathbf{d}_i^*$ of an nD reciprocal lattice Σ^*,

$$M^* = \pi^{\|}(\Sigma^*). \tag{5.5}$$

In direct space, the QC structure results at the intersection of the periodic nD hypercrystal structure with the nD par-space, $V^{\|}$ [3]. The nD hypercrystal structure can be described as an nD lattice Σ decorated with nD hyperatoms. This is shown in Figure 5.7 on the example of the 1D Fibonacci sequence (FS). The $(n-d)$D perp-space component of an nD hyperatom is called atomic surface (occupation domain), the dD par-space component corresponds to the actual atoms. The basis vectors of Σ are obtained via the orthogonality condition of direct and reciprocal space

$$\mathbf{d}_i \mathbf{d}_j^* = \delta_{ij}. \tag{5.6}$$

The atomic positions in par-space thus depend on the kind of nD lattice, as well as on size and shape of the atomic surfaces. Cutting a hypercrystal structure with par-space at different perp-space positions will result in different par-space structures. This is a consequence of the irrational slope of the par-space section with respect to the n-dimensional lattice. All sections with different perp-space components belong to the same local isomorphism class (that is, they are homometric) and will show identical diffraction patterns. Consequently, only QCs belonging to different local isomorphism classes can be distinguished by diffraction experiments. Only if the par-space intersection runs through the origin of the nD hyperlattice, the quasiperiodic structure will have a singular point reflecting the lattice symmetry.

A diffraction data set contains the observed intensities, $I(\mathbf{H})$, which have to be properly corrected ("data reduction") in the usual way. The reduced intensities are proportional to the squared structure amplitudes, $|F(\mathbf{H})|^2$. The electron density distribution can be calculated by Fourier transforming the structure factors, that is, not only the structure amplitudes but also their phases, which cannot be directly obtained from standard diffraction experiments ("phase problem of crystallography"). The most convenient method for the solution (phase determination) of both periodic and aperiodic crystal structures, is presently "charge-flipping" (CF) in combination with "low-density elimination" (LDE) [25], algorithms coded in the powerful program, SUPERFLIP [26].

5.4.1
Charge Flipping (CF) and Low-Density Elimination (LDE)

The method of choice for structure solution in any dimension is presently the combined application of the iterative direct-space methods CF [25,27] and LDE [28]. These methods do not require atomicity as the standard "direct methods" do. The only input data needed are the nD unit cell parameters and the structure amplitudes. The obtained electron diffraction maps can be improved, however, if the data are merged based on the symmetry information.

5.4.1.1 Charge Flipping (CF)

First, a starting set of structure factors $F^{(n=0)}(\mathbf{H})$ is created by assigning random phases to the experimentally obtained structure amplitudes $|F_{obs}(\mathbf{H})|$. Then each iteration involves four steps in the following way [26]:

1) A trial electron density $\rho^{(n)}$, sampled on voxels with values $\rho_i, i = 1, \ldots, N_p$, is obtained by inverse Fourier transform of the structure factors $F^{(n)}(\mathbf{H})$:

$$\rho^{(n)} = \mathrm{FT}^{-1}\left[F^{(n)}(\mathbf{H})\right]. \tag{5.7}$$

2) A modified density $\sigma_i^{(n)}$ is obtained from $\rho_i^{(n)}$ by reversing the sign (charge flipping) of all pixels i with density below a certain positive threshold δ:

$$\sigma_i^{(n)} = \begin{cases} +\rho_i^{(n)}, & \text{if } \rho_i^{(n)} > \delta, \\ -\rho_i^{(n)}, & \text{if } \rho_i^{(n)} \leq \delta. \end{cases} \tag{5.8}$$

3) The structure factors $G^{(n)}(\mathbf{H})$ of this modified density are obtained by Fourier transform of $\sigma^{(n)}$

$$G^{(n)}(\mathbf{H}) = \mathrm{FT}\left[\sigma^{(n)}\right]. \tag{5.9}$$

4) The structure factors $F^{(n+1)}(\mathbf{H})$ are obtained from $F_{obs}(\mathbf{H})$ and $G^{(n)}(\mathbf{H}) = |G^{(n)}(\mathbf{H})|\exp\left[2\pi i \phi_G(\mathbf{H})\right]$ according to the following scheme:

$$F^{(n+1)}(\mathbf{H}) = |F_{obs}(\mathbf{H})|\exp\left[2\pi i \phi_G(\mathbf{H})\right], \tag{5.10}$$

for $F(\mathbf{H})$ observed and strong,

$$F^{(n+1)}(\mathbf{H}) = |G^{(n)}(\mathbf{H})|\exp\left\{2\pi i \left[\phi_G(\mathbf{H}) + 1/4\right]\right\}, \tag{5.11}$$

for $F(\mathbf{H})$ observed and weak,

$$F^{(n+1)}(\mathbf{H}) = 0, \tag{5.12}$$

for F(H) unobserved, and

$$F^{(n+1)}(\mathbf{H}) = G^{(n)}(\mathbf{H}), \tag{5.13}$$

for $\mathbf{H} = 0$.

The iteration cycles are repeated until convergence. The threshold value δ, which determines how fast the iterations converge, can be determined by trial and error in an automated way. Another crucial parameter is the number of reflections considered weak in the fourth iteration step. Flipping the phases of weak reflections can significantly improve the performance of the algorithm in cases of more complex structures [27]. The algorithm seeks a Fourier map that is stable against repeated flipping of all density regions below. Obviously, a large number of missing reflections, which cause termination ripples, will make the algorithm less efficient. A method for a better performance of CF for incomplete data sets has been developed by Ref. [26].

5.4.1.2 Low-density Elimination (LDE)

The low-density elimination method [28] was modified for nD structure analysis of QC by Refs. [29,30]. The principle behind this iterative approach is that all (electron) density values below a given threshold δ are set to zero.

First, a starting set of structure factors $F^{(n=0)}(\mathbf{H}) = |F_{obs}(\mathbf{H})|\exp(2\pi i\phi_{rand})$ is created by assigning random phases, ϕ_{rand}, to the experimentally derived structure amplitudes $|F_{obs}(\mathbf{H})|$ and a trial electron density $\rho^{(0)}$ is obtained by inverse Fourier transform of the weighted structure factors $F^{(0)}$:

$$\rho^{(0)} = \mathrm{FT}^{-1}\left[w(\mathbf{H})F^{(0)}(\mathbf{H})\right]. \tag{5.14}$$

Then each iteration cycle n involves the following steps:

1) The density $\rho_i^{(n)}$ in the ith voxel is modified to $\sigma_i^{(n)}$ according to

$$\sigma_i^{(n)} = \begin{cases} \rho_i^{(n)}\left\{1 - \exp\left[-\frac{1}{2}\left(\frac{\rho_i^{(n)}}{0.2\rho_c}\right)^2\right]\right\}, & \text{if } \rho_i^{(n)} > \delta, \\ 0, & \text{if } \rho_i^{(n)} \leq \delta. \end{cases} \tag{5.15}$$

ρ_c is the expected average peak height in the unit cell. It can be estimated by determining the average of the maximum peak height $\rho_{max}^{(j)}$ in each of the M sections:

$$\rho_c = \frac{1}{M}\left(\sum_{j+1}^{M}\rho_{max}^{(j)}\right). \tag{5.16}$$

2) The structure factors $G^{(n)}(\mathbf{H})$ of this modified density result from the Fourier transform of $\sigma^{(n)}$

$$G^{(n)}(\mathbf{H}) = \mathrm{FT}\left[\sigma^{(n)}\right]. \tag{5.17}$$

3) The structure factors $F^{(n+1)}(\mathbf{H})$ are obtained from $F_{\text{obs}}(\mathbf{H})$ and $G^{(n)}(\mathbf{H}) = |G^{(n)}(\mathbf{H})|\exp[2\pi i\phi_G(\mathbf{H})]$ as

$$F^{(n+1)}(\mathbf{H}) = |F_{\text{obs}}(\mathbf{H})|\exp[2\pi i\phi_G(\mathbf{H})]. \quad (5.18)$$

4) The new electron density $\rho^{(n+1)}$ results from the inverse Fourier transform of the weighted structure factors $w(\mathbf{H})F^{(n+1)}$:

$$\rho^{(n+1)} = \text{FT}^{-1}\left[w(\mathbf{H})F^{(n+1)}(\mathbf{H})\right] \quad (5.19)$$

with

$$w(\mathbf{H}) = \tanh\left[\frac{|G^{(n+1)}(\mathbf{H})F^{n+1}(\mathbf{H})|}{\langle G^{(n+1)}(\mathbf{H})\rangle\langle F^{n+1}(\mathbf{H})\rangle}\right]. \quad (5.20)$$

Then the iteration cycles are repeated until convergence, which can be defined in a way that phase changes in each cycle are smaller than 0.5°, for instance. Subsequently, the weight is set to one and several cycles more are calculated to obtain the final electron density maps. An example for the performance of CF calculations on a simulated high-quality reflection data set of a decagonal QC model structure is shown in Figure 5.8 [31].

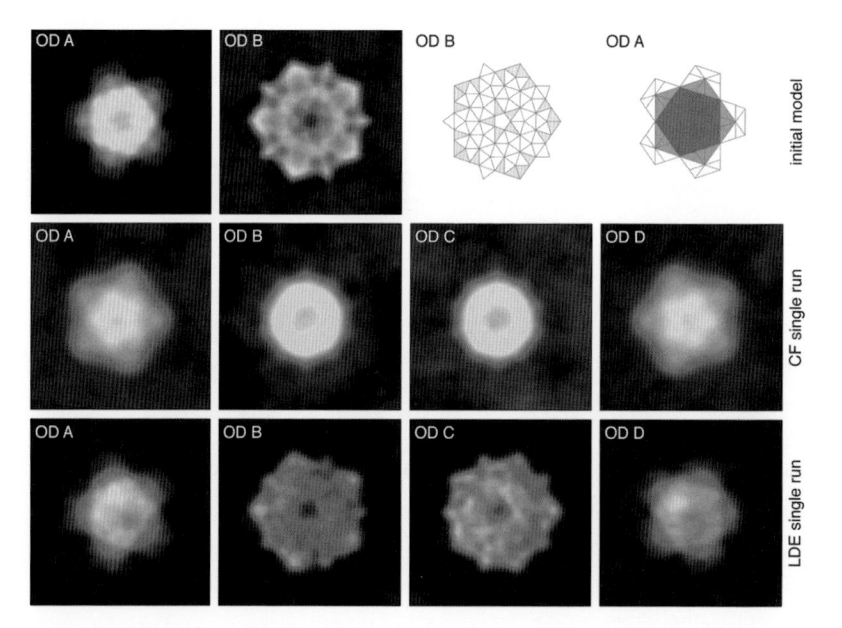

Figure 5.8 Reconstructed atomic surfaces of the structure model for decagonal $Al_{72}Co_8Ni_{20}$ [30] based on single CF and LDE runs phasing 164 304 reflections in $P1$ (≈ 4000 unique reflections). The pentagonal-shaped OD A, B and C, D related by an inversion operation. No symmetry averaging was performed. (For details see Ref. [31].)

5.4.2
nD and 3D Model Structure Refinement, Respectively

The solution of a structure is the first step within the course of a structure determination. The rough structure model obtained by CF and LDE, for instance, has to be parameterized for the refinement process, where it will be iteratively modified until the differences between calculated and observed diffraction intensities will have converged to a minimum. Figures of merit are the reliability factor R and the goodness of fit (GoF) in the same way as for conventional structure analysis. Additionally, $\log|F_{obs}(\mathbf{H})|/\log|F_{clc}(\mathbf{H})|$-plots should be calculated to check for systematic deviations from the expected distribution.

The crucial step in setting up a good starting model for the refinements is its parametrization, since the atomic surfaces define both the local and the global structures, that is, the kind of tiling underlying the structure and its decoration. In the case of IQCs, the Ammann tiling, or one that is directly derivable from it, is usually taken as the underlying quasilattice, which is decorated by the basic clusters ("cluster-embedding approach"). Since these clusters constitute both QCs and their rational approximant(s), and can be easily determined from the latter, one can already build in this way a starting structure model for the refinements.

The situation is much more complex in the case of DQCs, where an infinite number of possible underlying tilings exists. Examples are the generalized Penrose tilings [32] and the Masakova tilings [33–35]. This means that in this case one has to refine at the same time both the structure of the clusters and their long-range order, that is, the kind of tiling underlying the structure. While the "cluster-embedding approach" relies on the validity of a preselected tiling, the "atomic-surfaces-modeling method" can yield a less-biased structure model [36].

If a specific tiling can be reliably chosen for the quasilattice of a structure, and one has a good starting model for the cluster structure, then the quasiperiodic model structure can be directly refined in par-space as well. This has some advantages because the structure model can be optimized in a similar way as in the case of periodic structures. An example is shown in the work of Ref. [5].

Although QC structure refinements can converge to R factors comparable to those obtained in standard structure refinements of complex intermetallics, there remains a quality problem. In the low-intensity region of the $\log|F_{obs}(\mathbf{H})|/\log|F_{clc}(\mathbf{H})|$ – plots, the values of the calculated structure amplitudes are frequently systematically underestimated relative to those of the observed ones (see, e.g., Refs. [5,37]). One reason for this effect can be a too high value for the phason ADP used in the refinements, which weakens mainly the intensities of reflections with large $|\mathbf{H}^\perp|$. With a large phason ADP, the weaknesses of a model structure, which most strongly influence the large-$|\mathbf{H}^\perp|$ reflections, can be compensated for in the refinements to some extent.

Another contribution to this effect can originate from multiple scattering, which is virtually omnipresent in the case of QCs. Multiple diffraction means that at least two Bragg reflections, with intensities $I(\mathbf{H})$ and $I(\mathbf{G})$, are simultaneously excited by the primary X-ray beam with wave vector $\mathbf{k}_0$ (Figure 5.9). Then, the coupling reflection $I(\mathbf{H} - \mathbf{G})$ is excited as well, with the reflected beam $\mathbf{k}_\mathbf{G}$ acting now as the (much weaker) primary beam. The reflected beams $\mathbf{k}_\mathbf{H}$ and $\mathbf{k}_{\mathbf{H}-\mathbf{G}}$ point into the same direction, and the resulting interference wave with intensity $I = |F(\mathbf{H}) + F(\mathbf{H} - \mathbf{G})|^2$ will be detected instead of $I(\mathbf{H})$ alone.

Fortunately, multiple diffraction only plays a role if $I(\mathbf{G})$ and $I(\mathbf{H} - \mathbf{G}) > I(\mathbf{H})$. Strong reflections must have rather small values for the perp-space component of the diffraction vectors (see Figure 5.7c). If $|\mathbf{G}^\perp|$ and $|\mathbf{H}^\perp - \mathbf{G}^\perp|$ are both small, then $|\mathbf{H}^\perp|$ is small as well and $I(\mathbf{H}^\perp)$ could be strong, consequently. Only if $I(\mathbf{H}^\perp)$ is weak and the coupling reflection $I(\mathbf{H} - \mathbf{G}) \gg I(\mathbf{H})$ strong, then the resulting interference wave will strongly bias the $I(\mathbf{H}^\perp)$. However, the majority of very weak unobservable reflections, that is, those with large values of $\mathbf{H}^\perp$, will rarely be enhanced sufficiently by multiple diffraction to become observable.

Generally, the situation for QCs is comparable to that of complex intermetallic phases with large unit cells where multiple diffraction is usually no problem at all for structure analysis. Significant multiple diffraction in QC diffraction experiments will mainly take place for special diffraction geometries such as rotation around particular diffraction vectors (see, e.g., Ref. [38]).

A program, QUASI07_08, used for most QC structure refinements so far, can be downloaded from Ref. [39]. The program is described in Ref. [40], examples for its application can be found in Refs. [41,42], for instance.

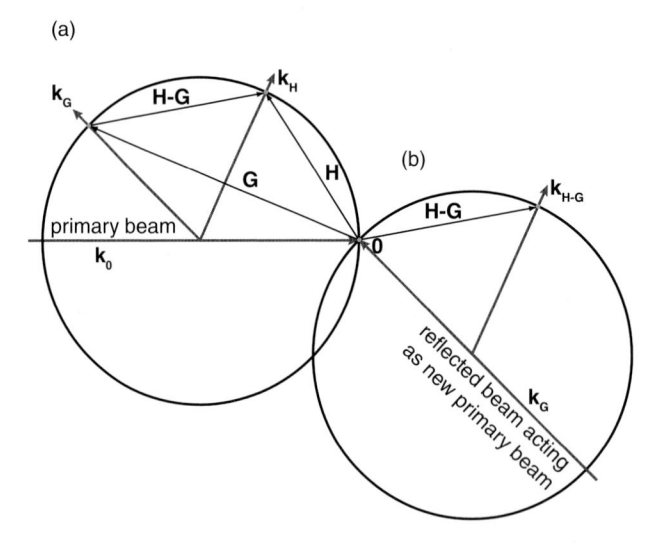

Figure 5.9 Multiple diffraction illustrated by the Ewald construction. The primary X-ray beam $\mathbf{k}_0$ creates the two reflected beams $\mathbf{k}_\mathbf{H}$ and $\mathbf{k}_\mathbf{G}$, since both reflections are located on the (a) Ewald sphere at the same time. (b) Ewald construction shows how the reflected beam $\mathbf{k}_\mathbf{G}$ acts now as primary beam, and that the beam is reflected now into the direction $\mathbf{k}_{(\mathbf{H}-\mathbf{G})}$. The wave vector $\mathbf{k}$ has the modulus $1/\lambda$, with λ the wavelength of the X-ray beam.

5.5
Spectroscopic Methods

There exists a large number of different spectroscopic methods and their variants that can be employed for the study of QCs. Since they do not give any information related to the long-range order of QC structures, their application and interpretation does not principally differ from that for intermetallic phases in general. In other words, there is nothing to consider what is specific for quasiperiodic structures. Consequently, we will briefly discuss only some of the more frequently used methods, and refer to some examples in literature.

For chemical analysis of QCs on the microscale, electron microprobe analysis (EMPA) is the most accurate technique; element concentrations in a sample can be measured at levels as low as 10 ppm in the best cases. Less accurate but more common, however, is energy- or wavelength-dispersive spectroscopy (EDS or WDS), at least one of the two being part of standard scanning electron microscopes (SEMs). The energy resolution of EDS is with $\approx 1\%$ lower than the corresponding wavelength resolution of WDS systems. The element detection sensitivity is of the order of 1%. High-resolution EDS on atomic scale can be performed in combination with spherical aberration-corrected HAADF-STEM, for instance [43].

Analytical electron microscopy can be a powerful method for identifying the chemical composition of local structural subunits. However, one has to take into account that the respective methods, energy-filtered transmission electron microscopy (EFTEM) and electron-energy-loss spectroscopy (EELS) in combination with HRTEM and HAADF-STEM, respectively, average the chemical information over the sample thickness of ≈ 10 nm. Consequently, this method will give much clearer results for decagonal QCs than for icosahedral QCs, if imaged along the periodic direction. Another analytical method that is carried out in a TEM is the atom-location channeling enhanced microanalysis (ALCHEMI). It allows to determine the site-occupation of atoms by evaluating their characteristic X-ray emission rates [44].

Extended X-ray absorption fine structure (EXAFS) and X-ray absorption near-edge structure (XANES) are powerful methods to probe the local atomic and electronic structure of crystalline, quasicrystalline, and amorphous materials, or even of supercooled liquid alloys [45]. For a general introductory review see, for instance, Ref. [46]. This method is particularly useful for the comparison of the local atomic structure of alloys in different ordering states. It has been shown, for example, that the ordering around Ni atoms is different in amorphous, crystalline, and quasicrystalline Ti–Zr–Ni alloys [47]. Another example can be found in the study by Menushenkov and Rakshun (2007) [48], where a combination of both EXAFS and XANES is employed for the study of icosahedral $Al_{65}Cu_{22}Fe_{13}$ and the tetragonal phase Al_7Cu_2Fe. Based on the spectra above the Cu and Fe K-absorption edges, they found a rearrangement of Al atoms caused by a rotation of the square Cu structural units for the phase transformation from the crystalline to the quasicrystalline phase.

EXAFS is also well suitable for *in-situ* studies of phase transformations as a function of pressure [49]. Investigating single crystalline samples by polarized EXAFS allows to identify the anisotropic features of the local surrounding. This is particularly useful studying anisotropic quasicrystals such as decagonal phases, which are quasiperiodic in two dimensions and periodic perpendicular to it [50]. Solid-state nuclear magnetic resonance (NMR) experiments on single (quasi) crystals can reveal the local structure and its anisotropy [51] as well as information on magnetic properties [52].

Spectroscopic techniques such as EXAFS, XANES, and NMR applied to intermetallics allow to study the local environment of specific target atoms. One has to keep in mind, however, that the information is obtained from all atoms of the same kind as the target atom at the same time; consequently, the information about the local environments is averaged globally. Techniques such as X-ray photoelectron spectroscopy (XPS) and soft-X-ray emission spectroscopy (SXES) can give information about the valence band density of states [53,54].

Finally, it should be mentioned that the characterization of physical properties of QCs in general can be performed in exactly the same way as for periodic intermetallics. There is nothing special for experiments on QCs, just the evaluation and modeling may be more challenging. Examples for the determination of the magnetic, thermal, thermoelectric, and transport properties of QCs can be found in Refs. [55,56], for instance.

5.6
Summary

The characterization of QCs differs from that of periodic crystals only if the atomic long-range order has to be taken into account, this means mainly for crystal structure analysis. The combination of electron microscopic techniques with diffraction methods is the high road for the determination of QC structures. Even if diffraction patterns with sharp Bragg peaks are indicative of well-defined quasiperiodic average structures with large correlation lengths, the local deviations from it, particularly phasonic disorder can be quite significant. This can be easily observed on electron microscopic images. Furthermore, besides the local cluster structure, electron microscopic images also allow to identify the ways the clusters can overlap, and how they are arranged on a scale of tenth of nanometers. This helps in the selection of the tilings that can be used for the description of a quasiperiodic structure. This problem does not exist in the case of periodic structures, where the Bravais lattices can be easily identified just based on diffraction patterns.

References

1 Steurer, W. and Deloudi, S. (2008) Fascinating quasicrystals. *Acta Crystallogr. A*, **64**, 1–11.

2 Steurer, W. and Deloudi, S. (2009) *Crystallography of Quasicrystals: Concepts,*

Methods and Structures, Springer Series in Materials Science, Springer, Berlin.

3 Janssen, T. (1988) Aperiodic crystals: a contradictio in terminis? *Phys. Rep.*, **168**, 57–113.

4 Janssen, T. and Janner, A. (2014) Aperiodic crystals and superspace concepts. *Acta Crystallogr. B*, **70**, 617–651.

5 Kuczera, P., Wolny, J., and Steurer, W. (2012) Comparative structural study of decagonal quasicrystals in the systems Al-Cu-Me (Me = Co, Rh, Ir). *Acta Crystallogr. B*, **68**, 578–589.

6 Rabson, D.A., Mermin, N.D., Rokhsar, D.S., and Wright, D.C. (1991) The space-groups of axial crystals and quasi-crystals. *Rev. Mod. Phys.*, **63**, 699–733.

7 Mermin, N.D. (1992) The space groups of icosahedral quasicrystals and cubic, orthorhombic, monoclinic, and triclinic crystals. *Rev. Mod. Phys.*, **64**, 3–49.

8 Zeng, X.B., Ungar, G., Liu, Y.S., Percec, V., Dulcey, S.E., and Hobbs, J.K. (2004) Supramolecular dendritic liquid quasicrystals. *Nature*, **428** (6979), 157–160.

9 Haji-Akbari, A., Engel, M., and Glotzer, S.C. (2011) Phase diagram of hard tetrahedra. *J. Chem. Phys.*, **135**, 94101.

10 Barkan, K., Engel, M., and Lifshitz, R. (2014) Controlled self-assembly of periodic and aperiodic cluster crystals. *Phys. Rev. Lett.*, **113**, 8304.

11 Welberry, T.R. and Weber, T. (2016) One hundred years of diffuse scattering. *Crystallogr. Rev.*, **22**, 2–78.

12 Abe, E. and Tsai, A.P. (2001) Structure of quasicrystals studied by atomic-resolution electron microscopy. *JEOL News*, **36 E**, 18–21.

13 Deloudi, S., Fleischer, F., and Steurer, W. (2011) Unifying cluster-based structure models of decagonal Al-Co-Ni, Al-Co-Cu and Al-Fe-Ni. *Acta Crystallogr. B*, **67**, 1–17.

14 Steurer, W. and Deloudi, S. (2014) Decagonal quasicrystals – what has been achieved? *C. R. Phys.*, **15**, 40–47.

15 Edagawa, K., Suzuki, K., and Takeuchi, S. (2000) High resolution transmission electron microscopy observation of thermally fluctuating phasons in decagonal Al-Cu-Co. *Phys. Rev. Lett.*, **85**, 1674–1677.

16 McCusker, L.B. and Baerlocher, C. (2013) Electron crystallography as a complement to X-ray powder diffraction techniques. *Z. Kristallogr.*, **228**, 1–10.

17 Hiraga, K. and Yasuhara, A. (2013) Arrangements of transition-metal atoms in three types of Al-Co-Ni decagonal quasicrystal studied by Cs-corrected HAADF-STEM. *Mater. Trans.*, **54**, 493–497.

18 Hiraga, K. and Yasuhara, A. (2013) The structure of an AlCoNi decagonal quasicrystal in an $Al_{72}Co_8Ni_{20}$ alloy studied by Cs-corrected scanning transmission electron microscopy. *Mater. Trans.*, **54**, 720–724.

19 Ledieu, J. and Fournee, V. (2014) Surfaces of quasicrystals. *C. R. Phys.*, **15**, 48–57.

20 McGrath, R., Smerdon, J.A., Sharma, H.R., Theis, W., and Ledieu, J. (2010) The surface science of quasicrystals. *J. Phys. Condens. Matter*, **22**, 84022.

21 Thiel, P.A. (2008) Quasicrystal surfaces. *Ann. Rev. Phys. Chem.*, **59**, 129–152.

22 Johnson, I., Bergamaschi, A., Billich, H., Cartier, S., Dinapoli, R., Greiffenberg, D., Guizar-Sicairos, M., Henrich, B., Jungmann, J., Mezza, D., Mozzanica, A., Schmitt, B., Shi, X., and Tinti, G. (2015) Eiger: a single-photon counting X-ray detector. *J. Instr.*, **9**, C05032.

23 Weber, T., Deloudi, S., Kobas, M., Yokoyama, Y., Inoue, A., and Steurer, W. (2008) Reciprocal-space imaging of a real quasicrystal. A feasibility study with PILATUS 6 M. *J. Appl. Crystallogr.*, **41**, 669–674.

24 Estermann, M.A., Haibach, T., and Steurer, W. (1994) Quasicrystal versus twinned approximant: a quantitative analysis with decagonal $Al_{70}Co_{15}Ni_{15}$. *Philos. Mag. Lett.*, **70**, 379–384.

25 Oszlányi, G. and Sütő, A. (2004) *Ab initio* structure solution by charge flipping. *Acta Crystallogr. A*, **60**, 131–141.

26 Palatinus, L., Steurer, W., and Chapuis, G. (2007) Extending the charge-flipping method towards structure solution from incomplete data sets. *J. Appl. Crystallogr.*, **40**, 456–462.

27 Oszlányi, G. and Sütő, A. (2005) *Ab initio* structure solution by charge flipping. II.

Use of weak reflections. *Acta Crystallogr. A*, **61**, 147–152.

28 Shiono, M. and Woolfson, M.M. (1992) Direct-space methods in phase extensions and phase determination. I. Low-density elimination. *Acta Crystallogr. A*, **48**, 451–456.

29 Takakura, H., Shiono, M., Sato, T.J., Yamamoto, A., and Tsai, A.P. (2001) *Ab initio* structure determination of icosahedral Zn-Mg-Ho quasicrystals by density modification method. *Phys. Rev. Lett.*, **86**, 236–239.

30 Takakura, H., Yamamoto, A., Shiono, M., Sato, T.J., and Tsai, A.P. (2002) *Ab initio* structure determination of quasicrystals by density modification method. *J. Alloys Compd.*, **342**, 72–76.

31 Fleischer, F., Weber, T., Deloudi, S., Palatinus, L., and Steurer, W. (2010) *Ab initio* structure solution by iterative phase retrieval methods: performance tests on charge flipping and low-density elimination. *J. Appl. Crystallogr.*, **43**, 89–100.

32 Pavlovitch, A. and Kleman, M. (1987) Generalised 2D Penrose tilings: structural properties. *J. Phys. A Math. Gen.*, **20**, 687–702.

33 Masakova, Z., Patera, J., and Zich, J. (2003) Classification of Voronoi and Delone tiles in quasicrystals: I. General method. *J. Phys. A Math. Gen.*, **36**, 1869–1894.

34 Masakova, Z., Patera, J., and Zich, J. (2003) Classification of Voronoi and Delone tiles in quasicrystals: II. Circular acceptance window of arbitrary size. *J. Phys. A: Math. Gen.*, **36**, 1895–1912.

35 Masakova, Z., Patera, J., and Zich, J. (2005) Classification of Voronoi and Delone tiles in quasicrystals: III. Decagonal acceptance window of any size. *J. Phys. A Math. Gen.*, **38**, 1947–1960.

36 Cervellino, A., Haibach, T., and Steurer, W. (2002) Structure solution of the basic decagonal Al-Co-Ni phase by the atomic surfaces modelling method. *Acta Crystallogr. B*, **58**, 8–33.

37 Takakura, H., Gomez, C.P., Yamamoto, A., De Boissieu, M., and Tsai, A.P. (2007) Atomic structure of the binary icosahedral Yb-Cd quasicrystal. *Nat. Mater.*, **6**, 58–63.

38 Lee, H., Colella, R., and Shen, Q. (1996) Multiple Bragg diffraction in quasicrystals: the issue of centrosymmetry in Al-Pd-Mn. *Phys. Rev. B*, **54**, 214–221.

39 Yamamoto, A. (2008) Quasi07_08. Available at http://wcp-ap.eng.hokudai.ac.jp/yamamoto/ (November 28, 2016).

40 Yamamoto, A. (2008) Software package for structure analysis of quasicrystals. *Sci. Tech. Adv. Mater.*, **9**, 13001.

41 Strutz, A., Yamamoto, A., and Steurer, W. (2009) Basic Co-rich decagonal Al-Co-Ni: average structure. *Phys. Rev. B*, **80**, 4102.

42 Strutz, A., Yamamoto, A., and Steurer, W. (2010) Basic Co-rich decagonal Al-Co-Ni: superstructure. *Phys. Rev. B*, **82**, 064107.

43 Botton, G. (2007) Analytical electron microscopy, in *Science of Microscopy* (eds. P.W. Hawkes and J.C.H. Spence), vol. **I**, Springer, pp. 273–405.

44 Saitoh, K. and Tsai, A.P. (2007) ALCHEMI study of chemical order in Al–Cu–Co decagonal quasicrystals. *Philos. Mag.*, **87**, 2741–2746.

45 Holland-Moritz, D., Jacobs, G., and Egry, I. (2000) Investigations of the short-range order in melts of quasicrystal-forming Al–Cu–Co alloys by exafs. *Mater. Sci. Eng.*, **294**, 369–372.

46 Li, Z.R., Dervishi, E., Saini, V., Zheng, L.Q., Yan, W.S., Wei, S.Q., Xu, Y., and Biris, A.S. (2010) X-ray absorption fine structure techniques. *Part. Sci. Technol.*, **28**, 95–131.

47 Sadoc, A., Huett, V.T., and Kelton, K.F. (2007) Icosahedral ordering in Ti–Hf–Ni alloys? *J. Non. Cryst. Solids*, **353**, 3689–3692.

48 Menushenkov, A.P. and Rakshun, Y.V. (2007) Exafs spectroscopy of quasicrystals. *Crystallogr. Rep.*, **52**, 1006–1013.

49 Sadoc, A., Itie, J.P., Polian, A., and Lefebvre, S. (1997) Pressure-induced phase transition in icosahedral Al-Li-Cu quasicrystals. *Philos. Mag. A*, **74**, 629–639.

50 Zaharko, O., Meneghini, C., Cervellino, A., and Fischer, E. (2001) Local structure of Co and Ni in decagonal AlNiCo investigated by polarized EXAFS. *Eur. Phys. J. B.*, **19**, 207–213.

51 Jeglic, P., Vrtnik, S., Bobnar, M., Klanjsek, M., Bauer, B., Gille, P., Grin, Y., Haarmann, F., and Dolinsek, J. (2010) M-Al-M groups trapped in cages of $Al_{13}M_4$

(M = Co, Fe, Ni, Ru) complex intermetallic phases as seen via NMR. *Phys. Rev. B*, **82**, 4201.

52 Rau, D., Gavilano, J.L., Beeli, C., Hinderer, J., Felder, E., Wigger, G.A., and Ott, H.R. (2005) Symmetry-dependent Mn-magnetism in $Al_{69.8}Pd_{12.1}Mn_{18.1}$. *Eur. Phys. J. B*, **46**, 281–287.

53 Zurkirch, M., Erbudak, M., and Kortan, A.R. (1998) Electronic-structure analysis of decagonal $Al_{70}Co_{15}Ni_{15}$ by XPS and EELS. *J. Electron Spectros. Relat. Phenomena*, **94**, 211–215.

54 Rogalev, V.A., Groning, O., Widmer, R., Dil, J.H., Bisti, F., Lev, L.L., Schmitt, T., and Strocov, V.N. (2015) Fermi states and anisotropy of Brillouin zone scattering in the decagonal Al-Ni-Co quasicrystal. *Nat. Commun.*, **6**, 8607.

55 Nakayama, M., Tanaka, K., Matsukawa, S., Deguchi, K., Imura, K., Ishimasa, T., and Sato, N.K. (2015) Localized electron magnetism in the icosahedral Au-Al-Tm quasicrystal and crystalline approximant. *J. Phys. Soc. Jpn.*, **84**, 24721.

56 Takagiwa, Y. and Kimura, K. (2014) Metallic-covalent bonding conversion and thermoelectric properties of Al-based icosahedral quasicrystals and approximants. *Sci. Tech. Adv. Mater.*, **15**, 44802.

6
Transmission Electron Microscopy

Krumeich Frank

ETH Zürich, Laboratory of Inorganic Chemistry, Vladimir-Prelog-Weg 1, 8093 Zürich, Switzerland

6.1
Introduction

Electron microscopy methods are of high value for solid-state chemistry as they provide indispensable information in the course of a comprehensive characterization of materials that often cannot be gained with other analytical tools. Modern electron microscopy comprises a wide range of different advanced methods that use the various signals resulting from the interactions of the electron beam with the sample to characterize the structure, morphology, and composition with high lateral resolution. The success story of transmission electron microscopy (TEM) started in 1931 when E. Ruska and M. Knoll constructed the first transmission electron microscope [1]. This practical utilization of electromagnetic lenses to focus an electron beam verified earlier calculations of H. Busch. In spite of its quite small magnification of just 17 times, this first TEM provided the experimental proof that images can be formed with electrons using electromagnetic lenses. This breakthrough was honored by awarding the Nobel Prize to Ruska – although quite late – in 1986. Various technical improvements – in particular of the electron optics for TEM – led to a quick increase of the achieved resolution so that the resolution of the light microscope was surpassed by the TEM only a few years after its invention. Around 1970, the best microscopes had a resolution of about 3.5 Å, which was sufficient to resolve columns of metal atoms in many oxide structures with octahedral frameworks like those of niobium oxides [2]. These studies can be regarded as the beginning of high-resolution TEM (HRTEM) as we know it today. A more recent breakthrough was the construction and application of aberration correctors for the image forming lenses so that advanced TEMs routinely reach sub-Å resolution today [3,4]. In addition, new insights into many materials properties are gained by the unprecedented energy and spatial resolution provided by up-to-date analytical techniques [5,6].

The aim of this introduction is to explain the basics of the most important electron microscopy methods in a qualitative way. After describing the

Handbook of Solid State Chemistry, First Edition. Edited by Richard Dronskowski, Shinichi Kikkawa, and Andreas Stein.
© 2017 Wiley-VCH Verlag GmbH & Co. KGaA. Published 2017 by Wiley-VCH Verlag GmbH & Co. KGaA.

experimental set-ups and operating principles of the different types of electron microscopes, the electron–matter interactions and their applications will be treated. Typical examples for research applying the different electron microscopy methods will demonstrate their power and versatility. For a more comprehensive understanding of TEM theory and practice, however, the study of textbooks is recommended [7–10].

6.2
Operation Principle

6.2.1
Properties of the Electron

An electron e is an elementary particle with a negative charge that when moving exhibits also wave characteristics. According to de Broglie, the wavelength λ is related to the momentum p:

$$\lambda = h/p = h/mv,$$

where h is Planck constant, m is the mass, and v is the velocity.

Taking into account that an electron accelerated in an electric field U reaches an energy $E = eU$, which corresponds to a kinetic energy $E_{kin} = mv^2/2$, an equation for the relation of wavelength and acceleration voltage U is derived from the de Broglie equation:

$$\lambda = h / \sqrt{2m_0 eU},$$

where U is the acceleration voltage, e is the electron charge $= -1.602176487 \times 10^{-19}$ C, and m_0 is the rest mass of the electron.

If the energy of an electron exceeds 100 keV, a relativistic correction must be used for calculating the electron wavelength:

$$\lambda = h / \sqrt{2m_0 eU (1 + eU/2m_0 c^2)}.$$

Values for wavelengths, relative weight, and velocity of the electrons for different acceleration voltages are given in Table 6.1.

Table 6.1 Properties of electrons depending on the acceleration voltage U (rest mass of an electron: $m_0 = 9.109 \times 10^{-31}$ kg; speed of light in vacuum: $c = 2.998 \times 10^8$ m/s).

U (kV)	λ (pm)	$m \times m_0$	$v/10^8$ m/s
100	3.70	1.20	1.64
200	2.51	1.39	2.09
300	1.97	1.59	2.33
400	1.64	1.78	2.48
1000	0.87	2.96	2.82

6.2.2
The Electron Lens

A electron lens is a magnetic lens that consists of a coil of copper wires inside a cylinder made of an iron-based alloy (Figure 6.1a). An electric current through the coils creates a magnetic field where an electron experiences the Lorentz force F:

$$F = -e(E + v \times B),$$

with $|F| = evB \sin(v, B)$ for a magnetic lens.

Here E is the strength of the electric field, B is the strength of the magnetic field, and e/v is the charge/velocity of electrons.

The magnetic field is rotationally symmetric but radially inhomogeneous: It is weak in the center of the lens and becomes stronger close to the bore. Electrons moving close to the lens center are thus less strongly deflected than those passing the lens far from there. The overall effect is that a beam of parallel electrons is focused into a spot (the so-called *cross-over*).

The focusing effect of a magnetic lens increases with the magnetic field B that is controlled by the current running through the coils. As the force acting on an electron by the magnetic field is a vector product of the electron velocity v and the field strength B, the force F is perpendicular to both, v and B. This results in rather complicated helical trajectories of the electrons so that an image is rotated in respect of the object (magnetic rotation). Apart from this effect, magnetic lenses influence electrons in a similar way as convex glass lenses do with light.

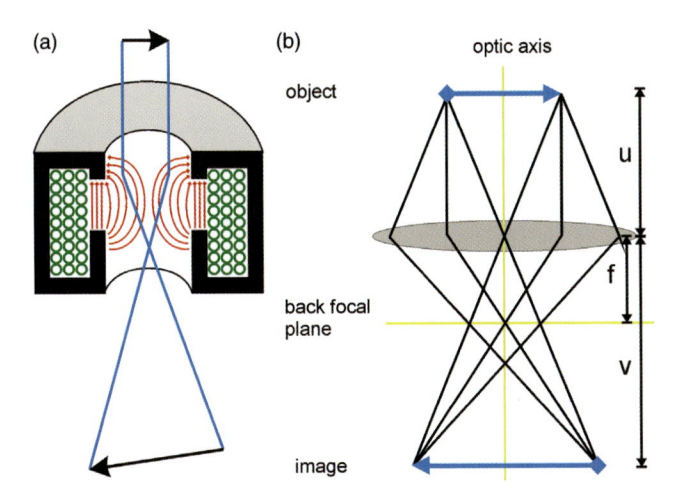

Figure 6.1 Schematic cross section of a magnetic lens (a). Copper wire (green) is coiled inside an iron cylinder (black). The rotationally symmetric magnetic field (red) has a focusing effect on the electron beam (blue). Like in the light optical analogue with a convex glass lens (b), a magnified image of an object appears in the image plane. The lens equation $1/u + 1/v = 1/f$ is valid as well in electron optics (f: focal length; u: object distance; v: image distance).

Thus, similar diagrams can be drawn to describe the respective ray paths (Figure 6.1b). The vertical line through the centers of the lenses is called *optic axis* in electron and light optics.

In electron microscopes, magnetic lenses perform two different tasks: (i) They generate an electron beam with certain characteristics. This is the purpose of the lenses in the condenser system that forms either a parallel beam as required for TEM or a focused beam for scanning techniques. (ii) They form an image and magnify it (objective, diffraction, intermediate, and projective lenses in TEM).

Similar to glass lenses in light optics, magnetic lenses have several imperfections: (i) Spherical aberration C_s: Off-axis electrons are deflected stronger than those traveling close to the optic axis. As a result, the focal point of the beam is spread. (ii) Chromatic aberration C_c: Electrons of different wavelengths are deflected differently, leading to a focus spread as well. In light microscopy, the spherical aberration of a convex lens can be compensated by a concave lens. Such lenses are not possible for electrons. Instead, a carefully designed corrector system, the so-called aberration corrector, usually consisting of two hexapol elements and two transfer lenses, can indeed achieve a compensating dispersion effect analogous to a concave glass lens [3].

Moreover, the iron alloy pole pieces are not perfectly circular, which causes a distortion of the magnetic field leading to a deviation from rotational symmetry. The resulting astigmatism of the objective lens can distort the image seriously and it thus must be corrected. This can fortunately be achieved by using quadrupole elements, so-called stigmators, which generate an additional electric field compensating the inhomogeneity of the magnetic field.

6.2.3
Transmission Electron Microscope

In the electron gun located at the top of a TEM, the electrons are generated and then accelerated to a user-selected high energy, typically in the range 100–300 keV. To produce an electron beam, materials with a low work function for the electron emission are utilized as the source. There are different working principles: In a thermoionic electron gun, the filament (W fiber or LaB$_6$ crystal) is heated by a current that provides the energy for the electron emission. The result is a broad, intense electron beam. The alternative is the **f**ield **e**mission **g**un (FEG), in which electrons are extracted from a sharply pointed tip of a tungsten needle by applying an electric field. The generated electron beam is highly coherent and can be focused into a small spot.

The electrons then propagate down the column and are focused into a particular beam state by a condenser system that consists of different magnetic lenses and apertures. This makes it possible to set the illumination conditions generating either a parallel beam for TEM or a convergent beam with selected convergence angles for **s**canning **t**ransmission **e**lectron **m**icroscopy (STEM).

The thereby formed beam then interacts with the specimen. Most modern microscopes have side-entry goniometers in which the sample holders are inserted into the high vacuum of the column. Several types of holders are available, including the standard single- and double-tilt holders. It is often important to study a specimen along a certain direction, for example, crystals along a crystallographic zone axis. To achieve this, the often arbitrarily oriented sample is tilted in two perpendicular directions using a double-tilt holder. Of course, it is an advantage to have large tilt angles, but the trade-off is the resolution that is better for small polepiece gaps. Other sample holders enable *in situ* heating, cooling, and straining experiments, just to mention a few.

The objective lens is positioned below the sample and represents the most important lens in the microscope since it generates the first intermediate image as well as the first diffraction pattern (Figure 6.2). The quality of

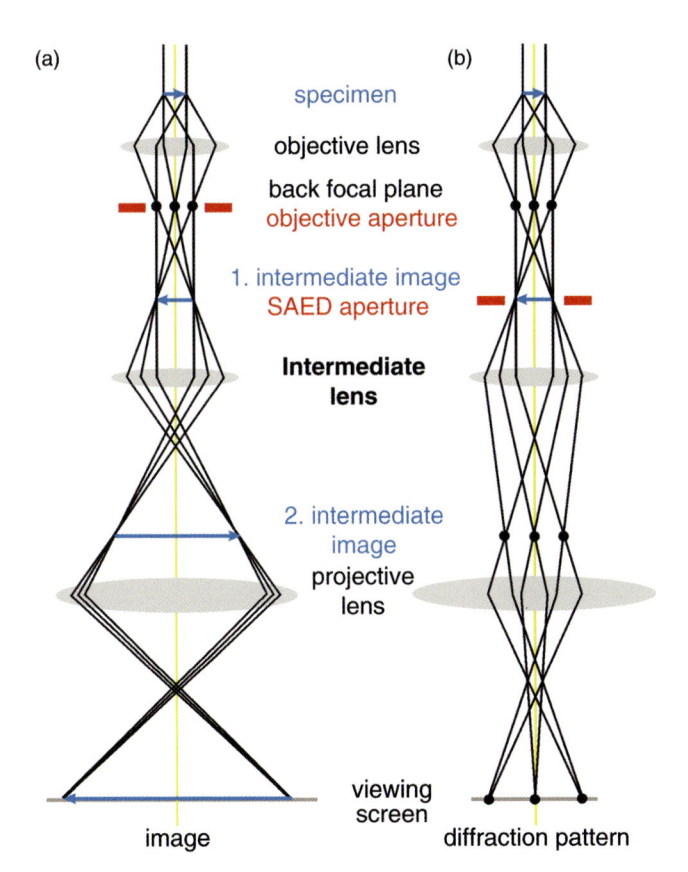

Figure 6.2 Operating modes of a transmission electron microscope (TEM): simplified ray diagram for (a) imaging mode and (b) diffraction mode. The first lens (intermediate or diffraction lens) below the objective lens is either focused onto the first intermediate image (imaging mode) or onto the back focal plane (diffraction mode). Consequently, an image or a diffraction pattern appears in the plane of the second intermediate image and further magnified on the viewing screen.

the image formed by the objective lens indeed determines the resolution of the whole microscope. The intermediate lens system after the objective lens can be focused either on the image or on the diffraction pattern and thereby determines which of them is further magnified by the projective lenses. This diffraction or intermediate lens allows one to switch between imaging and diffraction mode just by pressing a knob. Working in real space and in reciprocal space at the same area of a specimen is a unique feature of the TEM.

In diffraction mode, moreover, an objective aperture can be inserted in the back focal plane of the objective lens to select one or more beams, which then form an image with particular information. An aperture is a metal disk with a hole in its center. If only the direct beam is passing through the hole, a **b**right-field TEM image (BF-TEM) results. The aperture can be shifted to select one or more diffracted beams forming a **d**ark-field (DF) TEM image). To ensure a high image quality, however, this aperture has to stay on the optic axis of the microscope, and for this purpose the electron beam is tilted before the sample in such a way that the selected diffracted beam passes through the aperture. If a large aperture is selected, the direct beam as well as several diffracted beams are collected so that they interfere with each other in the imaging plane. This is the setup used for **h**igh-resolution TEM (HRTEM). Another aperture, the selected area aperture, placed in the plane of the first intermediate image enables to choose the specimen region of which a **s**elected **a**rea **e**lectron **d**iffraction (SAED) pattern is obtained (Figure 6.2).

The final image can directly be observed on the viewing screen in the projection chamber or on a TV or CCD camera mounted below the microscope column. Nowadays, images are recorded on slow-scan CCD or CMOS cameras.

It is important that because of the strong interaction of electrons with any matter, the amount of gas particles in the microscope column must be minimized to avoid unwanted scattering, which diminishes the instrument performance. The required high vacuum is maintained by a vacuum system typically comprising a rotary pump (prevacuum pump), a diffusion pump, and one or more ion getter pumps.

6.2.4
Scanning Transmission Electron Microscope

In contrast to TEM working with a parallel illumination, STEM as well as scanning electron microscopy (SEM) uses a tiny, convergent electron beam that is scanned spot-by-spot across a defined area of the sample. At each spot, the generated signal is recorded by selected detectors, building up an image on a screen pixel by pixel. As the resolution of a STEM is linked to the diameter of the focused electron beam, the correction of the spherical aberration of the probe-forming lens leads to substantial improvements of the imaging properties. Forming an electron beam with a diameter of less than 0.1 nm gives access to the sub-ångström regime as it is achieved in aberration-corrected TEMs as well.

Three main types of detectors are used to obtain STEM images (Figure 6.3):

1) *BF detector*: The bright field (BF) detector is placed on the optic axis of the microscope, that is, at the same site as the objective aperture in *BF-TEM*, and detects the intensity in the direct beam after passing a point on the specimen.
2) *ADF detector*: The annular dark field (ADF) detector is a semiconductor disk with a hole in its center (collection angle ~10–50 mrad) where the BF detector is installed. The ADF detector uses electrons scattered or diffracted into

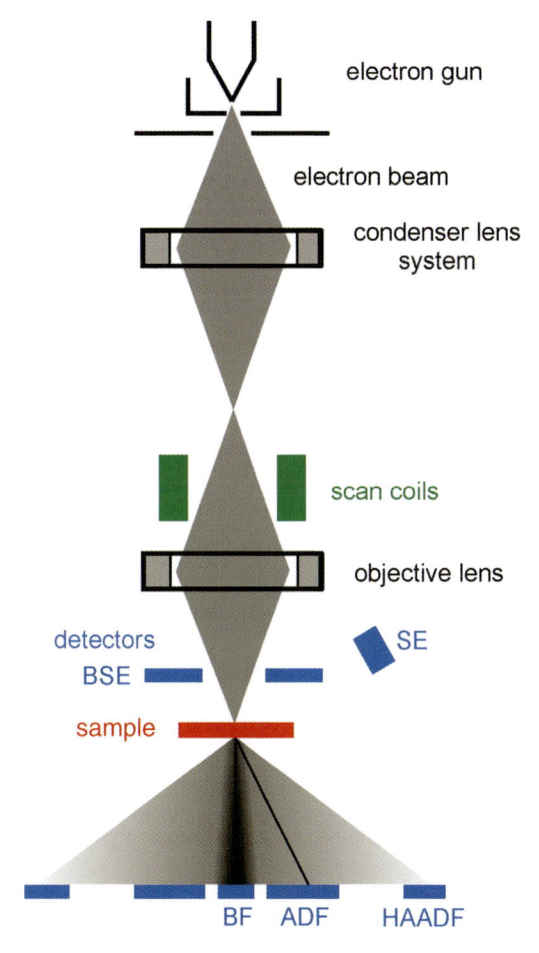

Figure 6.3 Schematic of a scanning transmission electron microscope (STEM) with the detector arrangement below the thin sample (BF: bright field; (HA)ADF: (high angle) annular dark field). Most electrons pass a thin sample without any interaction or are scattered into small angles only. The intensity distribution is indicated by the different gray values. Diffracted beams as exemplified by a black line hit the ADF detector. In a conventional scanning electron microscope (SEM), the secondary electrons (SE) and backscattered electrons (BSE) are collected for image formation on the upper side of the usually thick sample side.

small angles for image formation, similar to the *DF*-TEM mode. The measured contrast mainly results from electrons diffracted in crystalline areas but is superimposed by incoherent scattering (see below).
3) *HAADF detector*: The high-angle annular dark field (HAADF) detector is also a disk with a hole, but the inner disk diameter (>50 mrad) is much larger than in the ADF detector. Thus, it detects electrons that are scattered to high angles and almost only incoherent scattering contributes to the image. Thus, atomic number (*Z*) *contrast* is achieved (see below).

In addition, there is the option to install a detector for secondary electron (SE) and backscattered electrons (BSE) above the sample like in a SEM microscope and to obtain additional morphological and topological information. In fact, the boundaries between SEM and STEM are rather open in up-to-date instruments: Besides SE and BSE imaging in the STEM, electron transparent samples can be investigated in a SEM by employing special holders providing BF- and ADF-STEM images at low voltage ($U_{acc} \leq 30\,\text{kV}$) in addition to the conventional SE and BSE images.

6.2.5
Sample Preparation

As the interaction between the electron beam and any material is strong, the sample has to be thin to allow electrons to pass through and to give interpretable signals. The thickness depends on the material and on the technique applied. For HRTEM, the samples must be thinner than 10 nm for optimum results, while for other TEM investigations like defect analyses the thickness might be in the range of several 100 nm. In any case, compact samples must be prepared in such a way that the area of interest is thin and that the whole specimen fits into the holder of the TEM; often disks of 3 mm in diameter are required. Various methods such as mechanical grinding, ion milling, chemical etching, electropolishing, and combinations thereof are applied for specific bulk materials. These abrasive thinning methods might affect the structure and chemistry of the sample so that the danger of generating artefacts exists and the experimental results thus have to be interpreted with care. A rather recent development is the focused ion beam (FIB), a versatile method that is applicable for sample preparation in various fields. In many cases, it is sufficient to grind the material, prepare a suspension, put some droplets on a thin carbon film supported on a metal grid, and let them dry. Nanoparticles that are intrinsically thin are often directly deposited on a TEM grid.

Considering that with the aberration-corrected microscopes perfect imaging tools are available today, the preparation of a perfect and clean sample has become increasingly important for getting interpretable results, in particular when reaching the resolution limits. The most common problem is the contamination of the sample by a steadily growing carbon layer during the TEM and especially the STEM study. This deposition of a carbon layer on the specimen surface is caused by the decomposition of residual organic molecules attached to the surface of any TEM specimen. The sample can be purified by treating it in a vacuum chamber with a plasma. An alternative way to reduce contamination inside a TEM, a procedure called "beam shower," is the irradiation of a large area of interest with an intense electron beam for several minutes.

6.3
Electron–Matter Interactions and Their Applications

Electron microscopy comprises different methods that offer unique possibilities to gain insights into structure, topology, morphology, and composition of a material. The wealth of very different information that is obtainable is caused by the multitude of signals that arise when electrons interact with a specimen (Figure 6.4). Gaining a basic understanding of these interactions is an essential prerequisite to understand the results obtained by any electron

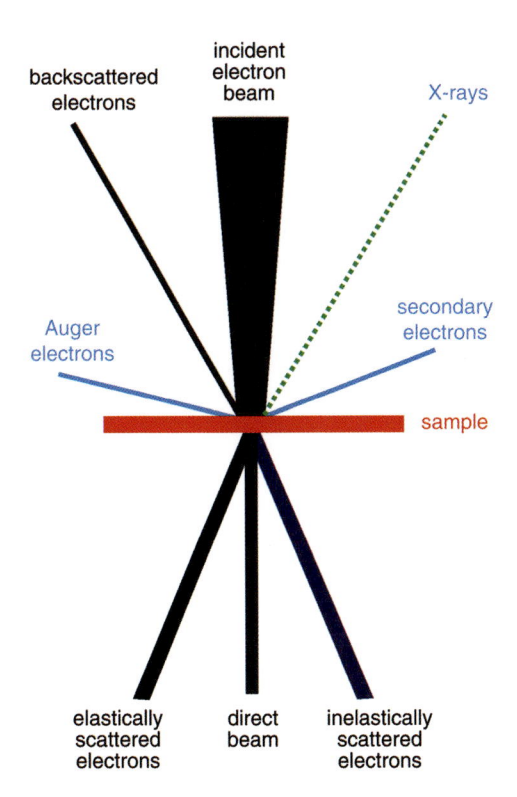

Figure 6.4 Overview of the most important electron–matter interactions arising from the impact of an electron beam onto a specimen. Note that a signal below the sample is only observable if the thickness is small enough to allow at least a part of the electrons to pass through.

microscopy method. For the discussion of the electron–matter interactions, it is helpful to classify them into two types, namely, elastic and inelastic interactions. Some basic physics and the applications of the different interactions will be treated in the following section.

6.3.1
Elastic Interactions

During an elastic interaction, no energy is transferred from the energy-rich electron to the sample and thus the electron leaving the sample still has its original energy E_0: $E_{el} = E_0$. This is of course also the case if the electron passes the sample without any interaction at all. Such electrons contribute to the direct beam that contains the electrons that pass the sample in direction of the incident beam (Figure 6.4). Furthermore, elastic scattering happens if the electron is deflected from its path by Coulomb interaction with the nucleus. By this process, the primary electron loses no energy or – to be accurate – a negligible amount of energy.

6.3.1.1 Incoherent Scattering

For the description of the elastic scattering of an electron by an atom, it is sufficient to regard it as a negatively charged particle and neglect its wave properties.

An electron penetrating into the electron cloud of an atom is attracted by the positive potential of the nucleus (electrostatic or Coulombic interaction), and its path is deflected toward the core as a result (Figure 6.5a). The Coulombic force F responsible for such an interaction is defined as

$$F = Q_1 Q_2 / 4\pi \varepsilon_0 r^2,$$

where r is the distance between the charges Q_1 and Q_2, and ε_0 is the dielectric constant.

The closer the electron comes to the nucleus, that is, the smaller the r, the larger the F and consequently the larger the scattering angle. In rare cases, electrons can even be turned backward, generating the so-called backscattered electrons. These electrostatic electron–matter interactions can be treated as elastic. Note that for an accurate description of this process, however, phonon scattering and **thermal diffuse scattering** (TDS) must be considered, which are in fact inelastic effects but result in minute energy losses of a few millielectronvolts only that cannot be measured.

Because of its dependence on the charge, the force F, with which an electron is attracted by the nucleus, is stronger if there are more positives charges, that is, more protons. Thus, the Coulomb force increases with increasing atomic number Z. If electrons scattering into high angles or even backward are collected for image formation, the brightness increases with Z (Z contrast), giving rise to chemical information. This effect is utilized in HAADF–STEM and SEM with BSEs. An example is Au nanoparticles (Au NPs) on TiO_2 appearing with bright contrast in the HAADF–STEM image (Figure 6.5c).

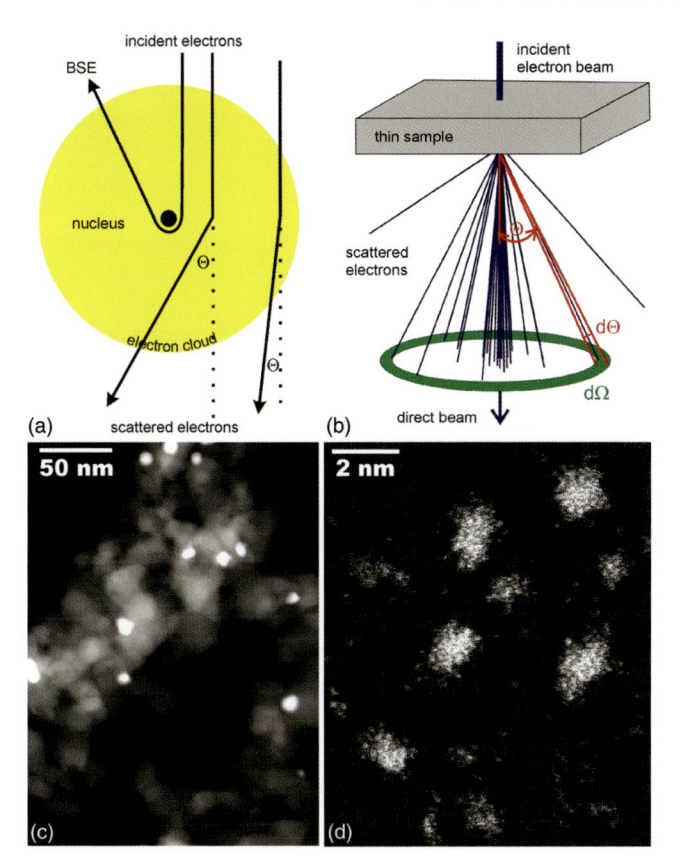

Figure 6.5 Schematic of (a) the scattering of an electron inside the electron cloud of an atom and (b) the scattering in a thin sample. HAADF–STEM images of Au NPs on TiO_2 (c) and of Pt clusters and single atoms on carbon (d).

In general, the Coulombic interaction of electrons with matter is strong and multiple scattering of electrons passing a sample happens frequently. When an electron passes through a specimen, it may be scattered not at all, once (single scattering), several times (plural scattering), or even many times (multiple scattering). Although electron scattering occurs most likely in the forward direction, there is a small probability for backscattering (Figure 6.5a). The statistical predictions of the outcome of the electron–matter interaction require the probability laws of quantum mechanics. The probability of an interaction is expressed using the cross section concept, which is based on the simple model of an effective interaction area: If an electron passes within this area, an interaction will certainly occur. Dividing the cross section of an atom by its actual area gives a probability for an interaction event. Consequently, the likelihood for a certain interaction increases with increasing cross section. It should be mentioned here that each type of possible interaction of electrons with a material has a characteristic cross section that depends on the electron beam energy.

The angular distribution of electrons scattered by an atom is described by the differential cross section $d\sigma/d\Omega$. Electrons are scattered by an angle Θ and collected within a solid angle Ω (Figure 6.5b). If the scattering angle Θ increases, the cross section σ decreases so that scattering into high angles is rather unlikely. The differential cross section is important since it is the measured quantity in annular dark field STEM. Owing to the high Z contrast, small clusters and even single atoms of a heavy metal on a light support can be observed by HAADF–STEM (Figure 6.5c and d).

6.3.1.2 Mass Thickness Contrast

In the bright field modes of the TEM and the STEM, only the direct beam is used for image formation. In the TEM, a small objective aperture is inserted into the back focal plane of the objective lens in such a way that exclusively the direct beam is allowed to pass its central hole and to build up the image (Figures 6.2 and 6.6a). Scattered and diffracted electrons are efficiently blocked by the aperture. In a STEM, a bright field detector is placed on the optic axis to measure the intensity of the direct beam (Figure 6.3).

It is essentially the reduction of the direct beam's intensity that is detected in BF images. A main component of this weakening is due to the mass thickness effect. The effect of large differences in mass dominates the contrast in a BF-TEM image of gold particles on TiO_2 (Figure 6.6b). The particles with a size of several 10 nm appear black since Au is by far the heaviest element in this sample and therefore scatters strongest. Furthermore, the Au particles are crystalline

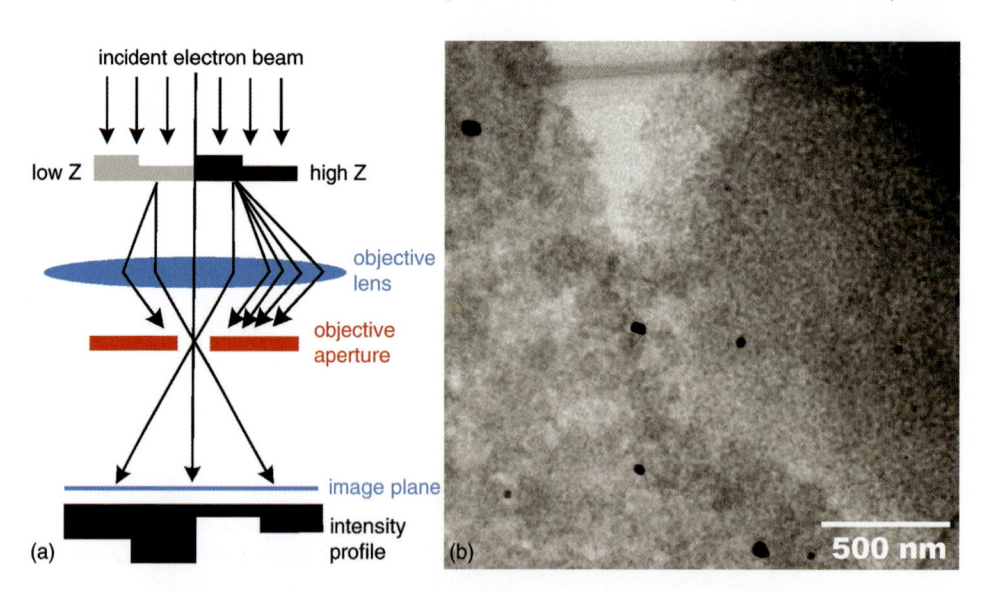

Figure 6.6 Schematic representation of the contrast generation in BF-TEM mode (a). In BF-TEM, only the direct beam passes through the central hole and contributes to the final image, whereas scattered and diffracted electrons are caught by the objective aperture. (b) BF-TEM image of Au NPs on TiO_2.

and thus Bragg contrast contributes to the dark contrast as well (see below). The TiO_2 support appears as almost uniformly gray material. However, the area in the upper right corner of the image is thicker than the center as indicated by the darker contrast there (thickness contrast). Note that in the HAADF–STEM image of a similar sample, the contrast is inverted (cf. Figure 6.5c).

6.3.1.3 Coherent Scattering

Mostly incoherent scattering of the largely coherent electrons takes place if the scattering centers are arranged in an irregular way like in amorphous compounds. If, on the other hand, a crystalline specimen is transmitted by electrons, then coherent scattering takes place. All atoms act as scattering centers that deflect a part of the incoming electrons from their direct paths due to the electrostatic interaction of the negatively charged electrons with the nucleus (Figure 6.7a). Since the spacing between the scattering centers is regular, constructive interference of the scattered electron waves occurs in certain directions and thereby diffracted beams are generated (Bragg diffraction).

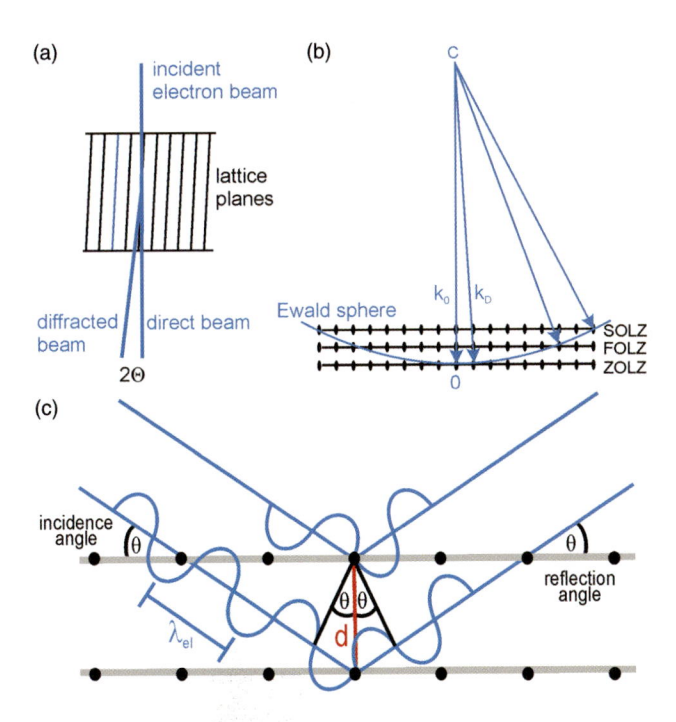

Figure 6.7 (a) Diffraction of an electron beam by a crystal that is schematically represented by a set of parallel equidistant lattice planes. (b) Ewald sphere construction (point 0: origin of reciprocal lattice, k_0: wave vector of the incident wave, k_D: wave vector of a diffracted wave, ZOLZ (FOLZ, SOLZ): zero (first, second)-order Laue zone). (c) Reflection of two parallel electron beams by a set of two parallel lattice planes resulting in constructive interference.

The Bragg equation describes the relation between diffraction angle, electron wavelength, and interplanar distance (Figure 6.7c). In this simplified model, the diffraction of an incoming electron wave from a set of equidistant lattice planes is treated as a reflection of the beam by the lattice planes. This description leads to a general equation that is valid not only for diffraction of electrons but also for that of X-rays and neutrons as well. Two incident electron waves that are in phase with each other (left side in Figure 6.7c) are reflected by the lattice planes, after which they have to be in phase again for constructive interference (right side). For this, the distance that the wave with the longer path follows needs to be an integer multiple of the electron wavelength. This path depends on the incident angle Θ, the distance between the lattice planes d and the electron wavelength λ_{el}. The path length is two times $d \cdot \sin\Theta$. Since this value must be a multiple of λ_{el}, it follows (Figure 6.7c):

$$2d \sin \Theta = n\lambda.$$

From this equation, an idea about the value of the diffraction angles can be obtained. For $U = 300$ kV, the wavelength is $\lambda = 0.00197$ nm. For a d value of 0.2 nm, an angle $\Theta = 0.28°$ results. As a rule, diffraction angles in electron diffraction are quite small as a consequence of the small wavelength and typically in the range of $0° < \Theta < 1°$.

Diffraction can be described in reciprocal space by the Ewald sphere construction (Figure 6.7b) in which a sphere with radius $1/\lambda$ is drawn so that it contains the origin of the reciprocal lattice. Then, for each reciprocal lattice point that is located on the Ewald sphere of reflection, the Bragg condition is satisfied and a diffraction spot arises. Due to the small wavelength of electrons (e.g., $\lambda = 1.97$ pm for 300 keV electrons), the radius of the Ewald sphere ($1/\lambda$) is quite large. Furthermore, the lattice points in the reciprocal lattice of thin samples are elongated so that the Ewald sphere intersects several of these rods even if the Bragg condition is not exactly fulfilled and many reflections appear simultaneously. Thus, electron diffraction patterns correspond to 2D cuts of the reciprocal lattice. If the interplanar distance in direction of observation is small in reciprocal space, higher order Laue zones (HOLZ) may be observed as well (Figure 6.7b).

6.3.1.4 Electron Diffraction

Different types of SAED patterns are observed depending on the size of the investigated crystallites. If a diffraction pattern of a single crystal is recorded, then the reflections appear on well-defined points of reciprocal space that are characteristic for the crystal structure and its lattice parameters (Figure 6.8a). The example shown here corresponds to $YbSi_{1.4}$ observed in orientation along the c-axis, that is, in the zone axis [001]. The distances of the reflections correspond to d values characteristic for the crystal structure; missing reflections such as (100) and (010) provide valuable information about the crystal symmetry [11].

If more than one crystal of a phase contributes to the diffraction pattern, as is the case for polycrystalline samples, then the diffraction patterns of many

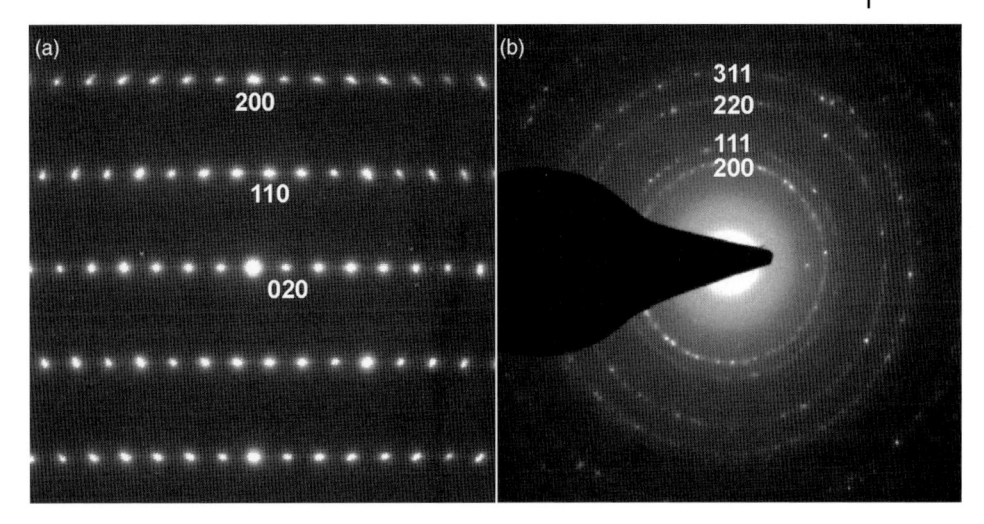

Figure 6.8 (a) Selected area electron diffraction (SAED) patterns of (a) a single crystal of YbSi$_{1.4}$ and (b) polycrystalline platinum. Some indices are assigned to the corresponding spots and circles, respectively.

micro- or nanocrystals are superimposed. Since the d values and thus the distances in reciprocal space are the same for all of them, they have the same distance from the center, that is, the origin of reciprocal space, and thus the spots are then located on rings (Figure 6.8b). Some large, strongly diffracting crystallites present in the sample investigated here produce bright spots that are located on the diffraction rings. The ring diameters correspond to d values that can be attributed to certain lattice planes and assigned with indices, here with respect to the face-centered cubic structure of platinum.

The quantitative evaluation of the reflection intensities as done in X-ray diffraction is impeded in electron diffraction by frequent multiple diffraction. This effect can be minimized by tilting away from the zone axes as done when using precession methods, and data sets for structure determination such as by single-crystal X-ray diffraction are obtainable [12]. Besides, SAED obtained with a parallel electron beam, a convergent beam can be applied to get additional information, for example, about high-order Laue zones (convergent beam electron diffraction CBED).

6.3.1.5 Diffraction Contrast

If a sample is crystalline, then another type of contrast appears in BF- and DF-TEM and STEM images, namely, diffraction or Bragg contrast. If a crystal is oriented close to a zone axis, many electrons are strongly scattered to contribute to the intensities in the diffraction pattern. Therefore, these areas then appear dark in the BF image. On the other hand, such areas may appear bright in the DF-TEM image if they diffract into the area of reciprocal space that is selected by the objective aperture.

ZrO$_2$ microcrystals appear with different contrast in the BF-TEM image (Figure 6.9a). Thickness contrast appears as the specimen is wedge shaped and thin close to the hole at the lower right side. Of course, the brightness of the hole is maximal as all electrons pass there and contribute to the BF-TEM image. Remarkably, some crystallites show up with higher darkness than those in their neighborhood, although the thickness is similar. These dark crystals are oriented, by chance, close to a zone axis where much more electrons are diffracted than in an arbitrary orientation. Thus, the intensity of the direct beam that solely contributes to the image contrast in the BF-TEM mode is reduced there, and such crystals appear relatively dark as a result. In the DF-TEM (Figure 6.9b), a single or several diffracted beams are allowed to pass through the objective aperture and build up the image. This means that crystallites diffracting into that particular area of reciprocal space appear bright, whereas others remain dark. Note that the hole in the sample appears black as no scattering occurs there (Figure 6.9b).

One should be aware that coherent and incoherent mechanisms of contrast generation, namely, mass thickness and diffraction contrast, occur simultaneously in specimens that are at least partly crystalline. This renders the interpretation of BF- and DF-TEM and STEM images complex and quite difficult in many cases.

6.3.1.6 High-Resolution Transmission Electron Microscopy (HRTEM)

To obtain lattice images, a large objective aperture has to be selected that allows many beams including the direct one to pass (Figure 6.2a). The image is formed then by the interference of diffracted beams with each other and with the direct beam leading to phase contrast. If the point resolution of the microscope is sufficiently high and a suitable crystalline sample is oriented along a zone axis, then

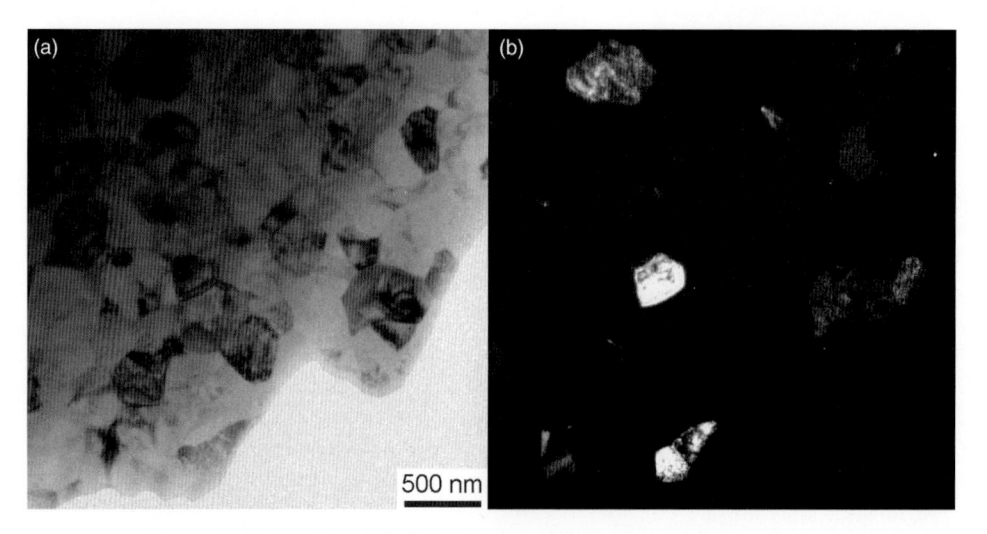

Figure 6.9 (a) BF-TEM and (b) DF-TEM image of ZrO$_2$ microcrystals.

HRTEM images are obtained. The diameter of the objective aperture is usually selected in such way that reflections with high frequencies k up to the point resolution of the microscope are collected.

The incident parallel electron beam, ideally a plane wave, interacts elastically with the crystal potential while passing through the specimen, and the resulting modulations of its phase and amplitude are present in the electron wave leaving the specimen. This wave, the object exit wave $o(r)$, thus contains information about the object's structure. The objective lens is not ideal but has aberrations (astigmatism, spherical C_s, and chromatic C_c aberration) that reduce the image quality so that a point in the sample becomes a disk in the image. In thin samples (so-called weak-phase objects), the intensity distribution of the exit wave function is described by the phase-contrast transfer function (PCTF) $\sin(-\chi(k))$, which gives the phase shifts of diffracted beams with respect to the direct beam:

$$\chi(k) = \pi \lambda \Delta f k^2 + 1/2\pi C_s \lambda^3 k^4.$$

The complicated curve $\sin\chi(k)$ strongly depends on C_s (spherical aberration coefficient of the image-forming objective lens), on λ (electron wavelength defined by accelerating voltage), on the defocus, and on the spatial frequency k. While this function is zero at the origin, it becomes negative for intermediate values of k (Figure 6.10). In this region of k, all information is transferred with negative phase contrast, that is, atom positions appear dark in the HRTEM image, and consequently the image is directly interpretable. The point resolution of a TEM corresponds to the first crossing of the PCTF with the k axis when the area of negative information transfer is maximally extended toward high frequencies. This particular defocus is called Scherzer defocus. Both the Scherzer

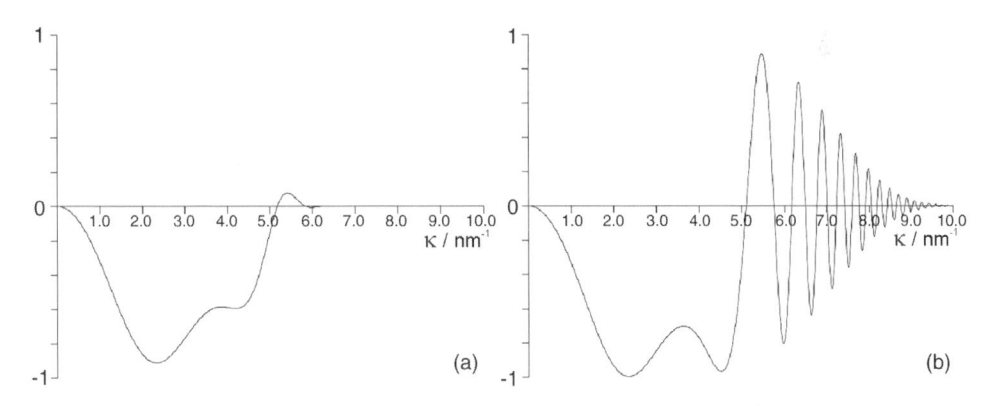

Figure 6.10 PCTFs of two 300 kV TEMs close to Scherzer defocus ($C_s = 1.15$ mm, $C_c = 1.5$ mm). (a) LaB$_6$ cathode; PCTF crosses the κ axis at about 5.1 nm^{-1}, which corresponds to a resolution of about 1.9 Å. (b) Field-emission gun; in the coherent electron beam of the FEG, the effect of damping is reduced compared to the LaB$_6$ cathode. As a result, there is additional information beyond the point resolution in the oscillating PCTF till the information limit at about 1.0 Å.

defocus Δf_s and the point resolution Δx are functions of C_s and λ:

$$\Delta f_s = -\sqrt{\frac{4}{3} C_s \lambda},$$

$$\Delta X = 0.66 \sqrt[4]{C_s \lambda^3}.$$

Depending on the defocus, the CTF may oscillate strongly (Figure 6.10). At larger k vectors, it is strongly damped mainly due to the effects of chromatic aberration, focus spread, and energy instabilities. Note that damping is strong for electrons coming from a LaB_6 cathode but much less for the more coherent electron beam produced by a FEG (Figure 6.10). Since the defocus is variable and can be adjusted at the microscope, an adequate value can be chosen to optimize the imaging conditions. As HRTEM images are formed by phase contrast, image simulations must be performed to understand their contrast, which depends not only on the crystal structure and the microscope's properties but also on the defocus and sample thickness.

6.3.1.7 Examples of HRTEM Studies

$Nb_8W_9O_{47}$ crystallizes with a threefold superstructure of the tetragonal tungsten bronze (TTB) type: space group $P2_12_12$ with $a = 1.2239$, $b = 3.6577$, $c = 0.39403$ nm [13,14]. Such quasi-two-dimensional structures are suitable to be observed by HRTEM in projection as the polyhedra are stacked upon each other linked by corner-sharing along a short crystallographic axis. The TTB unit consists of four five-rings of MO_6 octahedra (M = Nb,W) arranged around a central square. The threefold superstructure is achieved by a systematic occupation of 1/3 of the five-rings with metal–oxygen strings resulting in a pentagonal bipyramid (yellow in Figure 6.11b). The SAED pattern (Figure 6.11a) demonstrates that the investigated crystal is oriented along the crystallographic c-axis and that a threefold TTB superstructure along b^* is present. There was much interest in such structures in the 1970s, and the structure of $Nb_8W_9O_{47}$ was among the first that had been characterized by HRTEM images by observing all metal positions in projection [15]. Close to Scherzer defocus, the positions of all metals, which are by far the strongest scattering atoms here and thus have the highest contribution to the imaged projected crystal potential, are observed as dark patches in the HRTEM image (Figure 6.11c). That this direct interpretation of the HRTEM image is valid has to be confirmed by image simulations. Indeed, the experimental and the image simulated for a thickness of 3.9 nm and a defocus of 30 nm (*inset* in Figure 6.11c) match well (microscope: CM30 with $U = 300$ kV and $C_s = 1.2$ mm).

The particular strength of the HRTEM method is to detect and characterize defects and less ordered structures. This is demonstrated in Figure 6.11d, which shows a large area of a $Nb_8W_9O_{47}$ crystal. On the first view, this area seems to be perfectly ordered, but, when looking at the occupied pentagonal rings of MO_6 octahedra, the presence of a grain boundary between perfectly ordered areas becomes obvious. In the boundary, the pairs of occupied pentagonal units are

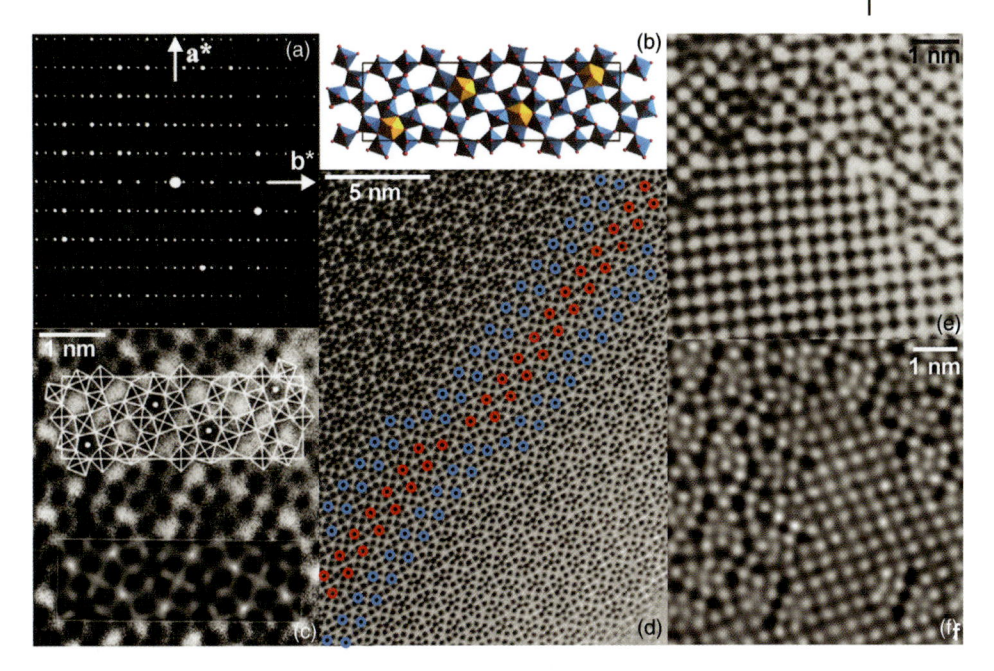

Figure 6.11 Nb$_8$W$_9$O$_{47}$ with a threefold superstructure of the tetragonal tungsten bronze (TTB) observed along [001]. (a) SAED pattern. (b) Representation of the structure. (c) HRTEM image with the structure representation and a simulation fitted in. (d) HRTEM image of a large area showing a dislocation of the occupied pentagonal units. (e) HRTEM and (f) HAADF–STEM images of an oxidation product of Nb$_4$W$_{13}$O$_{47}$.

shifted by [1 1/3 0]. Note that the basic TTB substructure is not affected by this defect.

Another example is shown in Figure 6.11e: The oxidation product of Nb$_4$W$_{13}$O$_{47}$ represents an intergrowth of the TTB-type structure and segregations of WO$_3$ [16]. As in the case of Nb$_8$W$_9$O$_{47}$, all metal positions are recognizable in the HRTEM image recorded along the short crystallographic axis. WO$_3$ crystallizes in a monoclinic distorted ReO$_3$-type structure and appears in the image as a square array of dark spots. With the structure chemical knowledge gained in the course of the investigation of Nb$_8$W$_9$O$_{47}$, all dark patched in the TTB-type structure can be assigned to either an MO$_6$ octahedron or an MO$_7$ pentagonal bipyramid and thereby a structural model can be derived from the HRTEM image. Remarkably, the HAADF–STEM image of a similar area (Figure 6.11f) provides even more information: Besides the positions of the metal atoms imaged now as bright spots, their brightness depends on the ratio W:Nb of the respective atomic column (Z contrast). As W has a higher Z than Nb ($Z_W = 74 > Z_{Nb} = 41$), W-rich atom positions appear brighter than Nb-rich ones (intensity increases with $\sim Z^2$) [17,18]. In addition to visualizing the metal positions, the equatorial oxygen positions are recognizable in the phase and modulus of the specimen exit plane wave obtained by focus series reconstruction, and therefore this method provides valuable information about distortions of the MO_6 octahedra in bronze-type oxides [19,20].

The studies presented here have been performed on conventional microscopes, and many other remarkable results combining HRTEM and electron diffraction investigations could have been mentioned such as the recent discovery of a germanium clathrate structure [21], ordering variants of the Zintl phase $YbSi_{1.4}$ [22], and the characterization of mesoporous structures [23]. It should be stressed that aberration-corrected TEM and STEM microscopes are now available worldwide and frequently applied in solid-state chemistry for the precise characterization of defects as well as of modulated and disordered structures with highest possible resolution and with minimized information delocalization [24,25].

6.3.2
Inelastic Interactions

If energy is transferred from the incident electrons to the sample, then the energy of the electron after its inelastic interaction with the sample is consequently reduced: $E_{el} < E_0$. The energy transferred to the specimen can cause different signals that depend on the material and are exploited by the various methods of analytical electron microscopy.

6.3.2.1 Overview
If a part of the energy that an electron carries is transferred to the specimen, several processes can take part generating the following signals:

1) *Inner shell ionization:* The incident electron transfers a part of its energy to an electron localized in an inner electron shell of an atom. Depending on the amount of energy transferred, the electron is promoted into an empty electronic state or ejected into the vacuum. Of course, the resulting electronic state is energetically unstable as an inner shell with a low energy has an electron vacancy, whereas levels of higher energy are occupied. This triggers an electron to drop down from a higher level and fill the vacancy. The excess energy of the electron dropping to a lower state can be either emitted as a characteristic X-ray (see below) or neutralized by the generation of an Auger electron.
2) *Braking radiation (Bremsstrahlung):* The deceleration of electrons by the Coulomb force of the nucleus generates X-rays that can carry any amount of energy up to that of the incident beam: $E \leq E_0$.
3) *Secondary electrons (SE):* The work function of electrons located in the valence or conduction band is small and thus only the transfer of a small amount of energy is necessary to eject them into the vacuum. Typically such SEs carry energies below 50 eV and are utilized in SEM for observing morphology and surface topography.
4) *Phonons:* By the uptake of energy from the electron beam, collective oscillations of the atoms in a crystal lattice can be initiated, which is equivalent to heating up the specimen. This may lead to beam damage.

5) *Plasmons:* If the electron beam passes through an assembly of free electrons, like in the conduction band of metals, the transfer of energy can induce collective oscillations of the electrons, the plasmons.
6) *Cathodoluminescence:* Electron–hole pairs are generated by promoting an electron from the valence band into the conduction band. This excited state is energetically instable, and it can relax by filling the electron hole by the electron dropping down from the conduction band. This recombination process leads to the emission of a photon carrying the difference energy $E = h\nu$, which can be used to determine the bandgap in semiconductors.

6.3.2.2 X-Ray Spectroscopy

After the ionization of an atom leading to a vacancy in an inner electron shell, this electron hole is filled up by an electron from a higher energy level (Figure 6.12a). The energy difference might be emitted as high-energetic electromagnetic radiation in the form of an X-ray quantum. This energy differences between the electronic states are typical for a certain element and thus the energies of the emitted X-rays (characteristic peaks) can be used for a qualitative analysis (Figure 6.12b). Note the presence of an uncharacteristic background due to braking radiation. Furthermore, the intensities of the signals can be determined and be used for a quantitative analysis. Mostly, energy-dispersive X-ray spectrometers (EDXS) are used currently as they can easily be attached to any electron microscope and their parallel measuring mode provides a quick overview over the whole energy range. A better energy resolution can be achieved with wavelength dispersive spectrometers (WDS) but with the drawback of long measuring times.

6.3.2.3 Electron Energy Loss Spectroscopy (EELS)

All inelastic electron–matter interactions described earlier need energy that is taken from an electron in the incoming beam. Consequently, this electron suffers a loss of energy, which can be measured by EELS. An EEL spectrum essentially comprises three different signals and energy ranges (Figure 6.12c) [5]:

1) *Zero loss (ZL) peak:* This peak contains all electrons that have passed the specimen without any interaction at all or with an elastic interaction only. If the sample is thin, the ZL peak is by far the most intense signal.
2) *Low loss region:* This region includes the energy losses between the ZL peak and about 100 eV. Here, the plasmon peaks are the predominant feature. From the intensity of these peaks, information about the sample thickness can be derived: intense plasmon peaks point to a thick sample area.
3) *High loss region:* Starting at an energy loss of about 100 eV, the signal intensity drops rapidly. The continuous background results from energy losses causing unspecific signals, most importantly the braking radiation. As in an X-ray spectrum, there are additional peaks above the background in the EELS. These ionization edges appear at electron energy losses that are again typical for a specific element and thus a qualitative analysis of a material is possible by EELS as well. The onset of such an ionization edge corresponds

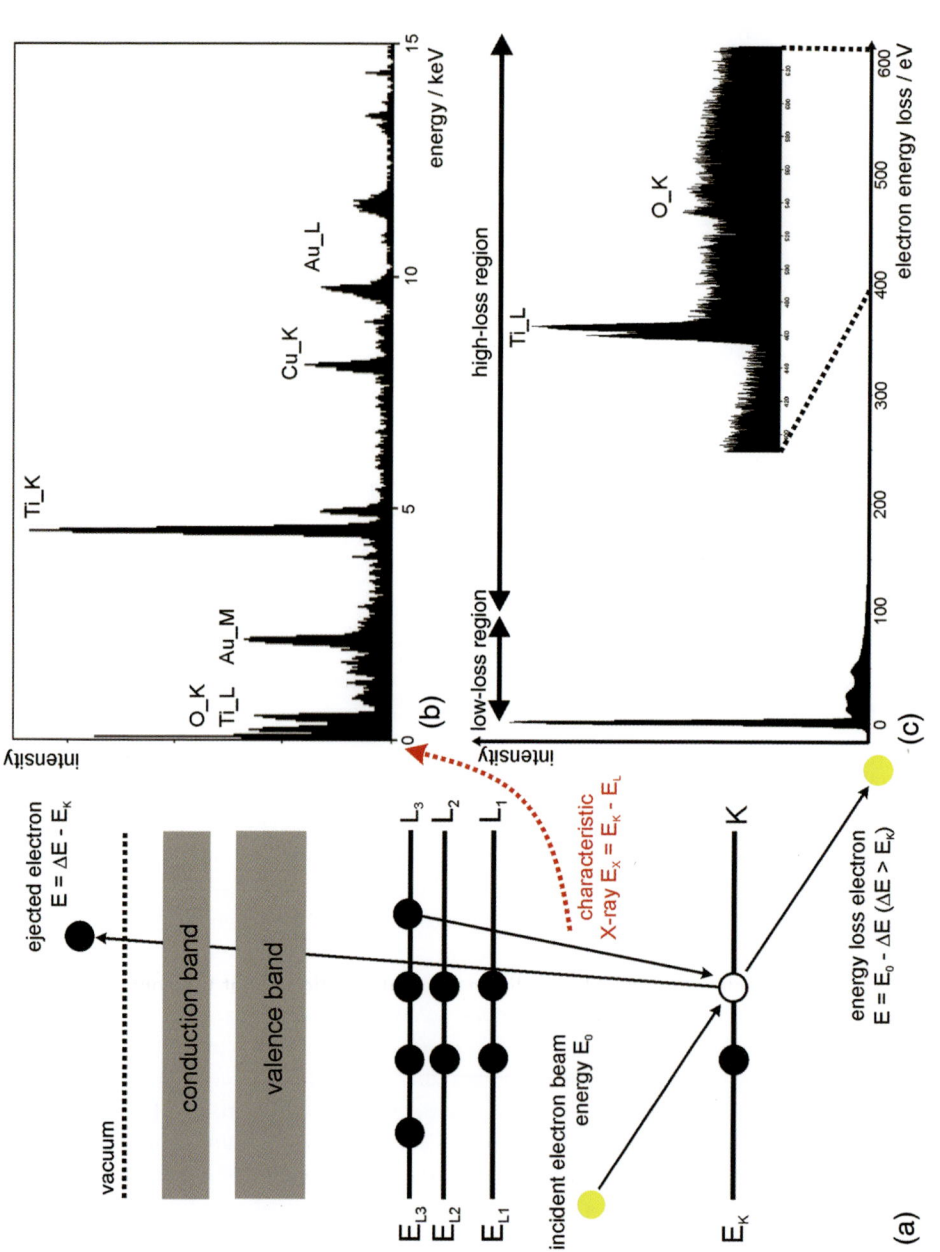

Figure 6.12 (a) Scheme of the generation of a characteristic X-ray. (b) Energy-dispersive X-ray spectrum (EDXS) of an Au/TiO$_2$ composite. Cu peaks arise from the TEM grid. (c) Electron energy loss spectrum (EELS) of TiO$_2$.

to the threshold energy that is necessary to promote an inner shell electron from its energetically favored ground state to the lowest unoccupied level. This energy is specific for a certain shell and for a certain element. Above this threshold energy, all energy losses are possible since an electron transferred to the vacuum might carry any amount of kinetic energy. If the material has a well-structured density of states (DOS) around the Fermi level, not all transitions are equally likely. This gives rise to a fine structure in the energy range close to the ionization edge that reflects the DOS. The method utilizing this information is called electron energy loss near edge structure (ELNES) and can be used to probe the oxidation state, for example, of iron and manganese [26]. Even valence mapping is feasible, as demonstrated for the distribution of $Ce^{3+/4+}$ in cerium oxide [27]. From a careful evaluation of the fine structure farther away from the edge, information about coordination and interatomic distances is obtainable (extended energy loss fine structure, EXELFS).

EEL and X-ray spectroscopy can be regarded as complimentary analytical methods. Both utilize the ionization of atoms by an electron beam and the signals can be used for compositional analyses. EELS works best in an energy loss region below about 1000 eV because at even higher energy losses the intensity decreases drastically. In this region, the K edges of the light elements occur that are less efficiently detected by X-ray spectroscopy. The energy resolution in EELS (well below 1 eV) is much better than that in X-ray spectroscopy (~100 eV) and enables one to observe fine structures of the ionization edges. The ionization edges of O_K and Ti_L, for instance, are well separated in EELS, while the corresponding peaks overlap in EDXS (Figure 6.12b and c). To get all details present in the EEL spectra provides valuable information so that this method profits most from the improved energy resolution provided by a cold field emitter ($\Delta E \approx 0.35$ eV) or a monochromator ($\Delta E \approx 0.15$ eV) compared to a standard FEG ($\Delta E \approx 0.5$–1 eV).

6.3.2.4 Analytical Electron Microscopy

The highly localized signal from the specimen that is obtained in STEM can be exploited in analytical electron microscopy by the combination with spectroscopic methods such as EDXS and EELS. Performing spectroscopy to get information about the chemistry at a spot, along a line or in an area has become a main application in practical work. The maximum of obtainable information is represented by the so-called data cube: For each pixel in the STEM image, the full spectral information is available [5]. From this, information about the composition at any spot or area in the image is retrievable. This STEM spectrum imaging method is used to generate elemental maps that visualize the elemental distribution in selected specimen areas.

An alternative method to STEM-coupled spectroscopy is energy-filtered TEM (EFTEM): An imaging filter is either mounted below the column of a TEM (post-column filter) or integrated in the microscope column (in-column filter) [5]. It is used to record images at certain energy loss ranges. The imaging filter consists of two main parts: the magnetic prism in which electrons with different energies are dispersed in a curved magnetic field and the optical column, which forms either images with electrons of a selected energy range (energy-spectroscopic imaging, ESI) or measures the EEL spectrum with a CCD camera. By recording images at subsequent energy ranges, the data cube can be filled by this method as well but with a reduced energy resolution compared to STEM-coupled EELS.

The so-called three-window method is mostly applied for elemental mappings using EFTEM: an image is taken after a suitable ionization edge of the corresponding element (postedge image) and two additional images (preedges 1 and 2) are recorded at energy losses smaller than the ionization edge. Only the electrons passing through the selected energy slit contribute to these images. The preedge images are used for an approximate determination of the unspecific background, which is then subtracted from the postedge image leading to an elemental map with enhanced contrast. Alternatively, the postedge image can be divided pixel by pixel by a preedge image (ratio method) [28].

Results obtained for a Ba–Ce oxide are shown in Figure 6.13. The BF-TEM image shows inhomogeneous particles (Figure 6.13a). The colored elemental

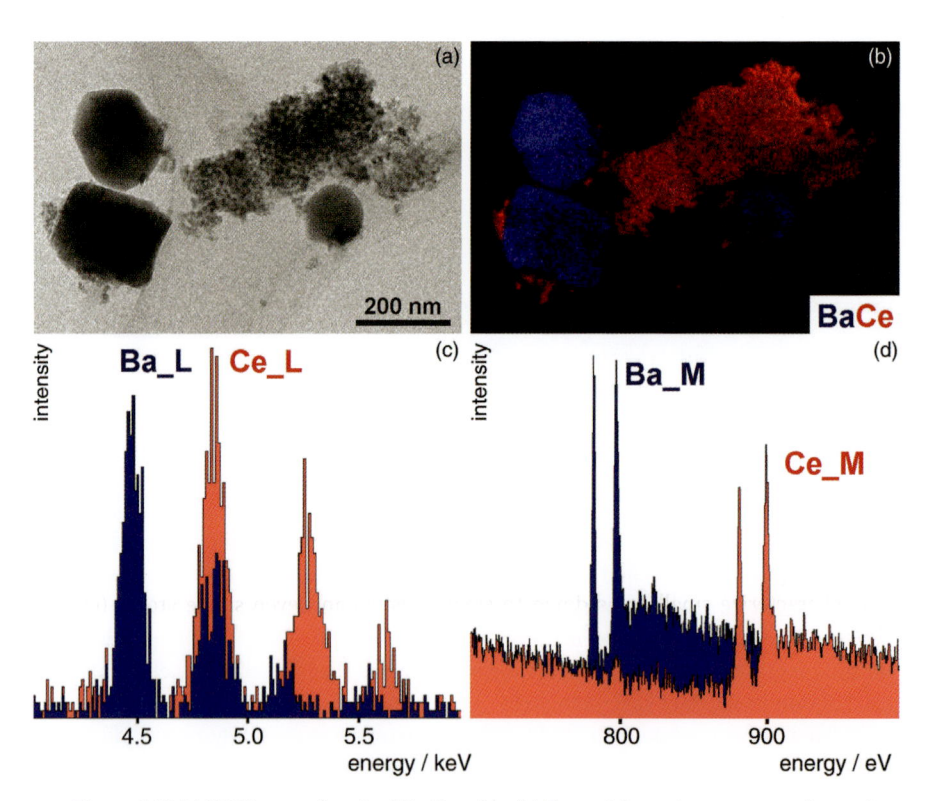

Figure 6.13 (a) TEM image of a mixed Ba–Ce oxide. (b) Elemental map by energy-spectroscopic imaging (ESI) reveals the inhomogeneous distribution of Ba and Ce. Relevant cuttings of the (c) EDX and (d) EEL spectra of areas consisting of BaO (blue) and CeO_2 (red), respectively.

distribution map, which was constructed from the individual elemental maps obtained by the three-window method, reveals that the large crystals contain Ba while the smaller particles contain Ce (Figure 6.13b). Sections of the EDX spectra of the two different areas (Figure 6.13c) reveal the presence of Ba and Ce there as indicated by the corresponding L emission peaks. Note that a part of these peaks overlap, a common problem that in some cases impedes an unambiguous identification of elements. In Figure 6.13d, the relevant part of the EELS is shown with the M ionization edges of the two elements as prominent features. The M_4 and M_5 edges of Ba (781 and 796 eV) and Ce (883 and 901 eV) are clearly separated, demonstrating the superior energy resolution of EELS.

6.4
The Impact of Aberration Correction

6.4.1
High Resolution

The correction of the spherical aberration (C_s) of the objective lens was achieved around 2000 based on a corrector design proposed by Rose [3]. With the correction of the C_s aberration of the objective lens, resolutions down to about 0.5 Å have been reached both in TEM and STEM microscopes. The availability of microscopes with such high resolution capability has opened new fascinating perspectives to characterize materials with unprecedented precession. In addition, it is of utmost importance that corrected TEMs allow one to investigate structures without the disturbing effect of the delocalization of the information as caused by the spherical aberration. Especially, the study of interface structures has become more straightforward.

Another important application of C_s correction is TEM investigation at lower voltages than usual ($U_{acc} \geq 100$ kV). For example, the knock-on damage of graphene happens at about 90 kV. Today many carbon structures are therefore observed at low voltages ($U_{acc} \leq 80$ kV) to avoid such unwanted effects [29,30].

To have a precisely focused electron beam in probe-corrected STEM improves not only the resolution of imaging but also that of analytical studies: HAADF–STEM coupled with EELS and EDXS is an outstanding tool to image and chemically characterize small areas down to small clusters and even single atoms (cf. Figure 6.5d). Technique and applications of aberration correction are the topics of excellent textbooks [31–33].

6.4.2
In Situ Studies

Usually chemical reactions are carried out *ex situ* under ambient conditions, and the resulting changes of the structure and composition are characterized by investigating the starting material and the reaction products separately in the electron microscope. Many insights into structural modifications caused, for example, by heating and straining have been gained from experiments performed *in situ* in a TEM using specially designed specimen holders. However, it is always the question how the results obtained in vacuum relate to those obtained under gas atmosphere. An old idea to overcome this problem and to observe reactions *in situ* is to let a limited amount of gas into the specimen area in the TEM. While the sample should be fully immersed in gas, the thickness of this gas layer has to be as thin as possible to minimize unspecific scattering of the electron beam, which will decrease the resolution of the microscope. Today, two different concepts to achieve this goal are utilized experimentally:

1) *Differential pumping:* Gas is let into the specimen area and additional apertures separate this area from the rest of the microscope column. An additional pump removes the gas close to the specimen area and thereby helps to maintain the high vacuum in other sections of the TEM. Microscope equipped with differential pumping stages are designated as environmental TEMs (ETEMS) [34,35].
2) *Closed cells:* Special holders are used on which the sample is located between two electron-transparent windows consisting of a thin film of light element materials (e.g., SiC, SiN, or graphene). The holder is constructed in such a way that one or more gases can be let into the space between the windows that furthermore can be heated up by an electric current. Thus, it is possible to perform and observe solid–gas reactions inside the TEM [36].

The resolution will be reduced in both cases by unwanted scattering processes. On microscopes equipped with a C_s corrector, it is possible to achieve high resolution even under such adverse conditions. By this combination of environmental options and aberration correction, observations of solid–gas reactions and even reactions in a liquid have been become feasible in TEM – certainly a step forward to a real chemical lab inside the microscope [37,38].

6.5
Conclusions

Electron microscopy, as it is understood today, is not just a single technique but a diverse collection of different ones that offer unique possibilities to gain insights into structure, topology, morphology, and composition of materials. Various imaging and spectroscopic methods utilizing the multitude of electron–matter interactions represent indispensable tools for the characterization of all kinds of structures on smaller and smaller size scales with the ultimate limit of a single atom. In this contribution, only the most important and versatile TEM-based techniques applied in solid-state chemistry have been treated. It should be mentioned that other, more specialized methods have been omitted here, such as Lorenz microscopy (for the study of magnetic structures), electron holography (for imaging electric and magnetic fields), and electron tomography (to get a full three-dimensional picture of materials and biological structures). Because the specimens that can be observed include inorganic and organic materials, biological objects, as well as composite structures, the impact of electron microscopy on various fields of solid-state science can hardly be overestimated[1].

Acknowledgments

Dr. A. Sologubenko, Dr. S. Gerstl, and the reviewers are gratefully acknowledged for valuable comments on this chapter.

References

1 Ruska, E. (1987) The development of the electron microscope and of electron microscopy. *Angew. Chem., Int. Ed.*, **26**, 595–605.
2 Gruehn, R. and Mertin, W. (1980) High-resolution transmission electron microscopy re-examined as a tool in solid state chemistry. *Angew. Chem., Int. Ed.*, **19**, 505–520.
3 Rose, H. (2009) Historical aspects of aberration correction. *J. Electron Microsc.*, **58**, 77–85.
4 Urban, K. (2009) Is science prepared for atomic resolution electron microscopy? *Nat. Mater.*, **8**, 260–262.
5 Sigle, W. (2005) Analytical transmission electron microscopy. *Annu. Rev. Mater. Res.*, **35**, 239–314.
6 Muller, D.A. (2009) Structure and bonding at the atomic scale by scanning transmission electron microscopy. *Nat. Mater.*, **8**, 263–270.
7 Williams, D.B. and Carter, C.B. (1996) *Transmission Electron Microscopy*, Plenum Press, New York.
8 Thomas, J. and Gemming, T. (2014) *Analytical Transmission Electron Microscopy: An Introduction for Operators*, Springer, Berlin.
9 Reimer, L. and Kohl, H. (2008) *Transmission Electron Microscopy: Physics of Image Formation*, 5th edn, Springer, Berlin.
10 Spence, J.C.H. (2003) *High-Resolution Electron Microscopy*, 3rd edn, University Press, Oxford.
11 Morniroli, J.P. and Steeds, J.W. (1992) Microdiffraction as a tool for crystal structure identification and determination. *Ultramicroscopy*, **45**, 219–239.
12 Mugnaioli, E., Gorelik, T., and Kolb, U. (2009) "Ab initio" structure solution from electron diffraction data obtained by a combination of automated diffraction tomography and precession technique. *Ultramicroscopy*, **109**, 758–765.
13 Sleight, A.W. (1966) The crystal structure of $Nb_{16}W_{18}O_{94}$, a member of a $(MeO)_xMeO_3$ family of compounds. *Acta Chem. Scand.*, **20**, 1102–1112.
14 Krumeich, F., Hussain, A., Bartsch, C., and Gruehn, R. (1995) Zur Präparation und Struktur gemischtvalenter Niob-Wolframoxide der Zusammensetzung $(Nb, W)_{17}O_{47}$. *Z. Anorg. Allg. Chem.*, **621**, 799–806.
15 Iijima, S. and Allpress, J.G. (1974) Structural studies by high-resolution electron microscopy: tetragonal tungsten bronze-type structures in the system Nb_2O_5-WO_3. *Acta Crystallogr. A*, **30**, 22–29.
16 Krumeich, F., Bartsch, C., and Gruehn, R. (1995) Oxidation products of $Nb_4W_{13}O_{47}$: a transmission electron microscopy study. *J. Solid State Chem.*, **120**, 268–274.

---

1) Several figures have first been published on the author's Web site: www.microscopy.ethz.ch.

17 Krumeich, F. and Nesper, R. (2006) Oxidation products of the niobium tungsten oxide $Nb_4W_{13}O_{47}$: a high-resolution scanning transmission electron microscopy study. *J. Solid State Chem.*, **179**, 1857–1863.

18 Kirkland, A.I. and Saxton, W.O. (2002) Cation segregation in $Nb_{16}W_{18}O_{94}$ using high angle annular dark field scanning transmission electron microscopy and image processing. *J. Microsc.*, **206**, 1–6.

19 Kirkland, A.I. and Meyer, R.R. (2004) "Indirect" high-resolution transmission electron microscopy: aberration measurement and wavefunction reconstruction. *Microsc. Microanal.*, **10**, 401–413.

20 Kirkland, A.I., Sloan, J., and Haigh, S. (2007) Ultrahigh resolution imaging of local structural distortions in intergrowth tungsten bronzes. *Ultramicroscopy*, **107**, 501–506.

21 Guloy, A.M., Ramlau, R., Tang, Z., Schnelle, W., Baitinger, M., and Grin, Y. (2006) A guest-free germanium clathrate. *Nature*, **443**, 320–323.

22 Kubata, C., Krumeich, F., Wörle, M., and Nesper, R. (2005) The real structure of $YbSi_{1.4}$: commensurately and incommensurately modulated silicon substructures. *Z. Anorg. Allg. Chem.*, **631**, 546–555.

23 Willhammar, T., Yun, Y., and Zou, X. (2014) Structural determination of ordered porous solids by electron crystallography. *Adv. Funct. Mater.*, **24**, 182–199.

24 Van Tendeloo, G., Hadermann, J., Abakumovab, A.M., and Antipov, E.V. (2009) Advanced electron microscopy and its possibilities to solve complex structures: application to transition metal oxides. *J. Mater. Chem.*, **19**, 2660–2670.

25 Urban, K., Jia, C.-L., Houben, L., Lentzen, M., Mi, S.-B., and Tillmann, K. (2009) Negative spherical aberration ultrahigh-resolution imaging in corrected transmission electron microscopy. *Phil. Trans. R. Soc. A*, **367**, 3735–3753.

26 Schmid, H.K. and Mader, W. (2006) Oxidation states of Mn and Fe in various compound oxide systems. *Micron*, **37**, 426–432.

27 Goris, B., Turner, S., Bals, S., and Van Tendeloo, G. (2014) Three-dimensional valency mapping in ceria nanocrystals. *ACS Nano*, **8**, 10878–10884.

28 Hofer, F., Warbichler, P., and Grogger, W. (1995) Imaging of nanometer-sized precipitates in solids by electron spectroscopic imaging. *Ultramicroscopy*, **59**, 15–31.

29 Krivanek, O.L., Dellby, N., Murfitt, M.F., Chisholm, M.F., Pennycook, T.J., Suenaga, K., and Nicolosi, V. (2010) Gentle STEM: ADF imaging and EELS at low primary energies. *Ultramicroscopy*, **110**, 935–945.

30 Kurasch, S., Kotakoski, J., Lehtinen, O., Skakalova, V., Smet, J.H., Krill, C., Krasheninnikov, A.V., and Kaiser, U. (2012) Atom-by-atom observation of grain boundary migration in graphene. *Nano Lett.*, **12**, 3168–3173.

31 Brydson, R. (ed.) (2011) *Aberration-Corrected Analytical Transmission Electron Microscopy*, John Wiley & Sons, Ltd., Chichester.

32 Erni, R. (2015) *Aberration-Corrected Imaging in Transmission Electron Microscopy*, 2nd edn, Imperial College Press, London.

33 Pennycook, S.J. and Nellist, P.D. (eds) (2011) *Scanning Transmission Electron Microscopy*, Springer, New York,

34 Boyes, E.D. and Gai, P.L. (1997) Environmental high resolution electron microscopy and applications to chemical science. *Ultramicroscopy*, **67**, 219–232.

35 Jinschek, J.R. (2014) Advances in the environmental transmission electron microscope (ETEM) for nanoscale *in situ* studies of gas–solid interactions. *Chem. Commun.*, **50**, 2696–2706.

36 Creemer, J.F., Helveg, S., Hoveling, G.H., Ullmann, S., Molenbroek, A.M., Sarro, P.M., and Zandbergen, H.W. (2008) Atomic-scale electron microscopy at ambient pressure. *Ultramicroscopy*, **108**, 993–998.

37 Ramachandramoorthy, R., Bernal, R., and Espinosa, H.D. (2015) Pushing the envelope of *in situ* transmission electron microscopy. *ACS Nano*, **9**, 4675–4685.

38 Wu, J., Shan, H., Chen, W., Gu, X., Tao, P., Song, C., Shang, W., and Deng, T. (2016) *In situ* environmental TEM in imaging gas and liquid phase chemical reactions for materials research. *Adv. Mater.*, **28**, 9686–9712.

7
Scanning Probe Microscopy

Marek Nowicki[1] and Klaus Wandelt[1,2]

[1]*University of Wroclaw, Institute of Experimental Physics, pl. M. Borna 9, 50-204 Wroclaw, Poland*
[2]*University of Bonn, Institute of Physical and Theoretical Chemistry, Wegelerstr. 12, 53115 Bonn, Germany*

7.1
Introduction

Surfaces play a crucial role in solid-state chemistry; they are the interface between phases, and as such the location of reactions.

The physicist Wolfgang Pauli (1945) is often cited with the statement "God created the bulk, but the surface was made by the devil," which accentuates the diabolic "broken symmetry" at the surface of a solid (or liquid). A consequence of this broken symmetry are gradients, which are a driving force for processes at surfaces such as phase transitions and chemical reactions. Even a solid surface exposed to "nothing," that is, vacuum, is affected by the existence of this gradient: The properties at the very surface of a solid *must* be different from those in the bulk, due to the different environment of the outermost surface atoms [1–8]. This is even more true if a surface is in contact with atoms from an adjacent phase. Thus, the understanding and control of chemical reactions at and via surfaces and interfaces call for a detailed investigation of their evolution in space and time. Moreover, since these reactions are often an important part in modern materials design, which tends toward smaller and smaller dimensions (ultrathin films, nanowires, clusters) and even atomic manipulation, these investigations desirably require the application of experimental methods of highest, if possible, atomic resolution.

The first attempt to investigate a solid surface with atomic resolution was made by Erwin Müller, who developed field emission microscopy (FEM) in 1936 [9]. FEM enabled to image the surface of a tungsten tip emitter by applying a strong electric field to it. Later in 1951 he published the principle of field ion microscopy (FIM) [10]. In FIM the imaging process involves helium as a working gas. Both microscopies enabled to image individual atoms adsorbed on facets and along step edges on the tip surface.

Handbook of Solid State Chemistry, First Edition. Edited by Richard Dronskowski, Shinichi Kikkawa, and Andreas Stein.
© 2017 Wiley-VCH Verlag GmbH & Co. KGaA. Published 2017 by Wiley-VCH Verlag GmbH & Co. KGaA.

In 1981, Gerd Binnig and Heinrich Rohrer from the IBM Laboratory, Zürich, developed the scanning tunneling microscope [11–14], which was a revolutionary breakthrough in surface science. With this device, the first images of the reconstructed Si(111)-(7×7) [14] and Au(110)-(1×n), n = 2, 3, . . . [15], were recorded with atomic resolution. In contrast to FEM and FIM, in scanning tunneling microscopy (STM) the test samples no longer were required to be sharp tips, but could rather be macroscopically flat surfaces of any *conducting* material. In 1986, Binnig and Rohrer were honored with the Nobel Prize for this achievement. Figure 7.1 illustrates the rapid progress of STM within the first few years. While in the line scan representation of Figure 7.1a just the unit cell of the (7×7)-reconstruction of the Si(111) surface (marked by white dots) is visible, Figure 7.1b shows its atomic structure including irregularities (arrow) sharply resolved [14,16].

Since then a variety of further scanning probe techniques (SPM), as they are called in general, as well as different detection modes of the various techniques have been developed. Most important, in 1985, Binnig, Quate, and Gerber [17] invented the atomic force microscope, which opened up the possibility to investigate also *nonconducting* samples. Common to all scanning probe techniques is a fine, preferably monoatomically sharp tip that approaches the surface in question to a distance of less than a nanometer. At this proximity, the various techniques are tuned to probe different interactions between the tip and the surface, such as electronic conductivity (STM), mechanical (AFM) and magnetic (MFM) force, heat conductivity, and so on [18–24]. Depending on the detection mode of the respective signal, the measurements provide microscopic or spectroscopic information about surfaces with (even sub-) atomic spatial resolution.

While early SPM studies were performed in ultrahigh vacuum (UHV), nowadays these measurements can also be carried out under ambient conditions, that is, in a gas atmosphere [25–32] or at solid–liquid interfaces [33–35], including ionic liquids [36–40]. Moreover, specific designs of the instruments enable very high detection rates, and thus the production of movies of dynamic surface processes on the atomic level [41,42].

In the References section, we recommend several specialized books and chapters that provide a broad and in-depth understanding of the general principles and wide applications of scanning probe techniques [21–24].

In this chapter, we first elucidate the basic physical detection mechanisms of scanning tunneling microscopy and atomic force microscopy (AFM) and then describe the basic instrumental aspects for the realization of these microscopes and their operation. Subsequently, we present and discuss several representative examples that illustrate the incredible capability of these techniques. These examples include the identification of surface and adsorbate structures on the atomic scale, the growth of ultrathin films and nanostructures, the direct observation of surface atom motion, the controlled manipulation of individual atoms and molecules, the local spectroscopy of electronic and vibrational properties, and the visualization of the individual steps of a surface-confined Ullmann reaction.

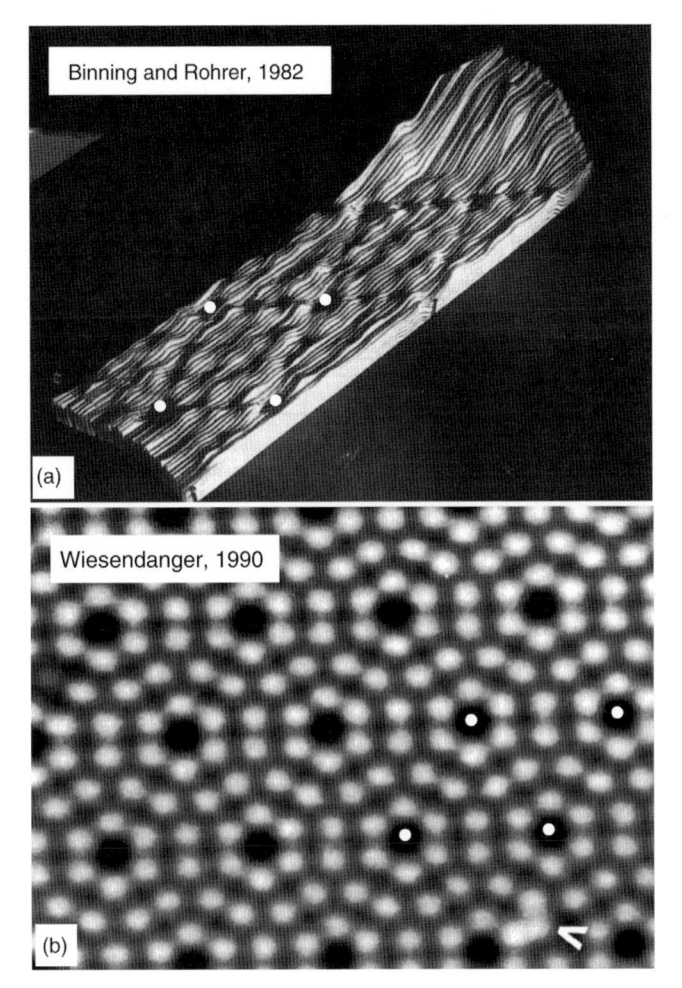

Figure 7.1 (a) The first STM image taken in 1982 from a reconstructed Si(111)(7×7) surface showing two complete 7×7 units cells and represented in relief form. (From Ref. [14].) (b) An STM image of the same type of surface 8 years later showing every single Si atom and, thus, also atomic defects (arrow). (From Ref. [16].) The bright dots in both panels mark the corner of a rhombic units cell whose diagonals have the crystallographic values of 46.56 and 26.88 Å, respectively.

7.2
Scanning Tunneling Microscopy

7.2.1
Physical Background

Figure 7.2 shows two energy diagrams with an energy barrier (II) of height Φ and width z. According to classical physics, an electron of energy $E_1 < \Phi$ (Figure 7.2a)

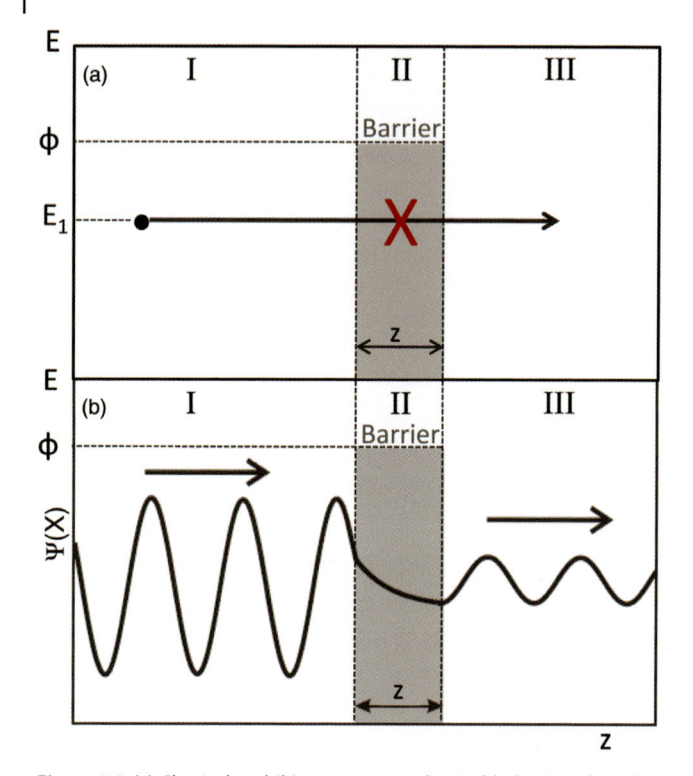

Figure 7.2 (a) Classical and (b) quantum mechanical behavior of an electron facing an energy barrier of height φ and width d. Regions I and III represent metals, and region II represents a vacuum gap.

is unable to ever overcome this energy barrier from left (I) to right (III). In turn, in quantum mechanics, the state of an electron is described by a wave function, which is a solution of the Schödinger equation and which can permeate the energy barrier (II) with the same frequency (energy) but decreased amplitude from left (I) to right (III) in Figure 7.2b; it is said that the electron "tunnels" elastically through the energy barrier with some probability. If I and III represent two metals and II a vacuum gap between them, wave functions from either side leak into the vacuum gap and may overlap. As a consequence, electrons can "tunnel" from one metal to the other even though the two metals have no contact. If one of the two metals is designed as a very fine tip, ideally monoatomic, above an extended surface of the other metal, the tunneling process between both metals is spatially confined to the very position of the tip with respect to the surface. This is the basic principle of a tunneling microscope. The only further aspect to consider is the Pauli principle: The electron can only tunnel through the barrier if there are empty "quantum states" on the other side. This is illustrated in Figure 7.3. As long as the Fermi levels of both metals are aligned, that is, $U_B = 0$, all electronic states below E_F are filled on both sides. In this case, an electron can only tunnel from one metal to the other if simultaneously

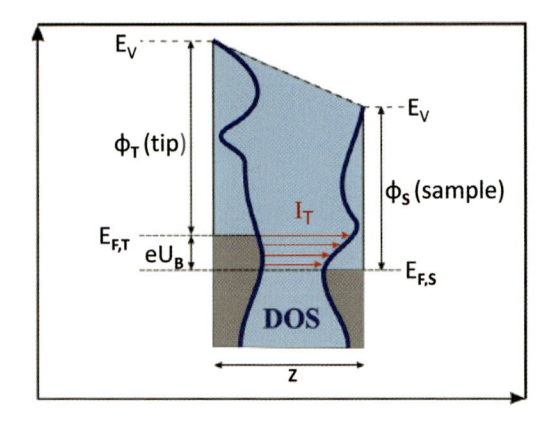

Figure 7.3 Energy diagram of a tunneling junction between tip (T) and sample (S). E_V = vacuum level, E_F = Fermi level, Φ = work function, z = tunneling gap, U_B = applied bias voltage between the tip and the sample, I_T = tunneling current, DOS = density of states.

another electron tunnels in the opposite direction. Thus, a measurable electron current will only occur if there is a net excess flow of electrons in one direction. This can be achieved by applying a small so-called bias voltage $U_B \neq 0$ between both metals, which causes a relative shift of both Fermi levels of tip (T) and sample (S), $E_{F,T}$ and $E_{F,S}$, as illustrated in Figure 7.3. Now, electrons from occupied states within the interval $\Delta E_F = eU_B$ can tunnel from the left into empty states above $E_{F,S}$ on the right, causing a measurable tunneling current [18–24]:

$$I_T \sim U_B \cdot e^{(-kz\sqrt{\Phi})} \rho(E_{F,S}); \quad (eU_B \ll \Phi), \tag{7.1}$$

which depends linearly on the bias voltage U_B, but exponentially on the barrier height Φ and barrier with z. $\rho(E_{F,S})$ is the local density of states near the Fermi level of the sample. By far most of the electrons tunnel elastically, that is, into a state of same energy. z is the distance between the tip and the sample, and the barrier height is given by $\Phi = 1/2(\Phi_T + \Phi_S)$.

According to a quantum mechanical treatment [21–24], the tunneling current is given by

$$I_T = 2\frac{\pi}{\hbar} e \sum_{\mu,\nu} [f(E_\nu) - f(E_\mu)] \cdot |M_{\mu\nu}|^2 \delta(E_\nu + eU_B - E_\mu), \tag{7.2}$$

where $f(E)$ is the Fermi function, U_B is the applied bias voltage, and $M_{\mu\nu}$ is the tunneling matrix element between the electron state ψ_μ and ψ_ν of sample and tip. E_μ is the energy eigenvalue of ψ_ν relative to the Fermi level $E_{F,S}$ of the sample, and E_ν is the energy eigenvalue of ψ_ν relative to the Fermi level $E_{F,T}$ of the tip. Replacing the Fermi functions by their zero Kelvin values, that is, by step functions and assuming a small bias voltage, Eq. (7.2) simplifies to

$$I_T = 2\frac{\pi}{\hbar} e^2 U_B \sum_{\mu,\nu} |M_{\mu\nu}|^2 \delta(E_\mu - E_F) \delta(E_\nu - E_F). \tag{7.3}$$

Bardeen [43] showed that the tunneling matrix element $M_{\mu\nu}$ may be given by

$$M_{\mu\nu} = \frac{\hbar^2}{2}m \int dS(\psi_\mu^* \nabla \psi_\nu - \psi_\nu \nabla \psi_\mu^*), \quad (7.4)$$

where $*$ stands for complex conjugated, and integration is performed over the whole sample surface S. If the tip as source of the tunneling electrons is idealized as a mathematical point r_S, $M_{\mu\nu}$ is proportional to the amplitude of ψ_μ, and the tunneling current becomes

$$I_T \propto \sum_\mu |M_\mu(r_S)|^2 \delta(E_\mu - E_\nu) \equiv \rho_S(r_S, E_F), \quad (7.5)$$

where $\rho_S(r_S, E_F)$ is the local density of states (LDOS) at the Fermi level of the sample surface.

Since the representation of the tip by a mathematical point, of course, is unrealistic, Tersoff and Hamann [44,45] introduced the so-called s-wave model, in which the tip apex is modeled by a hemisphere of radius R. Under these conditions, Eq. (7.5) remains valid; however, the tunneling current becomes dependent on R, because $|\psi_\mu(r_S)|^2$ is proportional to $e^{-2k(R+z)}$, where $k = \hbar^{-1}(2m\Phi)^{1/2}$ is the so-called decay length and z again is the separation between tip and sample surface and $\Phi = 1/2(\Phi_T + \Phi_S)$. In particular, Lang [46] could show that also for larger bias voltages and a tip apex of realistic extension, the tunneling current is described by

$$I_T \propto \int_{E_F}^{E_F + eU_B} dE \rho_S(r_S, E) \rho_T(E - eU_B), \quad (7.6)$$

where $\rho_S(r_S, E) = \rho_S(E) \cdot T$. Here

$$T \approx \exp\left\{-z\sqrt{4\frac{m}{\hbar^2}(\Phi_s + \Phi_T + eU_B - 2E)}\right\} \quad (7.7)$$

is the "transmission probability." The integration includes all states of tip and sample, which at given bias voltage contribute to the tunneling current and yields an approximately linear relationship between tunneling current and bias voltage.

It is the exponential dependence on the barrier width z (tunneling gap) that distinguishes a "tunneling contact" from an "ohmic contact" and that, most importantly, causes a strong spatial confinement of the tunneling process to the *shortest* distance between the surface and the tip, the basis of the high spatial resolution of STM. Depending on the polarity of U_B, the tunneling current can either flow from the tip into empty electron states above E_F of the sample or, vice versa, from the sample into empty states above E_F of the tip, which can be exploited to probe the density of electron states at the surface of the metal (scanning tunneling spectroscopy (STS)) [47] or of adsorbed molecules.

The derivative of Eq. (7.6) with respect to U_B yields

$$\frac{dI_T}{dU_B} = A\rho_S\rho_T \cdot T + B\frac{\rho_S}{U_B} + C\frac{dT}{dU_B}, \qquad (7.8)$$

where A, B, and C are proportionality factors. Assuming $\rho_T =$ constant and T only weakly dependent on U_B gives

$$\frac{dI_T}{dU_B} \propto \rho_S. \qquad (7.9)$$

The differential conductivity is, to first approximation, proportional to the local density of states (LDOS) at the given surface position.

Speaking about absorbed molecules within the tunneling gap, one has to consider that these molecules also possess, besides electronic, mechanical (translation, rotation, vibration) degrees of freedom, which may be excited by the electrons tunneling between the metal substrate and the tip. Accordingly, the tunneling electrons will lose characteristic quanta by exciting these degrees of freedom, which is shown in Figure 7.4. By varying U_B accordingly, electrons from level E_F may now tunnel via *two* channels from left to right, (i) by elastic tunneling and (ii) by inelastic tunneling, by loosing, for example, $\Delta E = eU_B = h\nu$

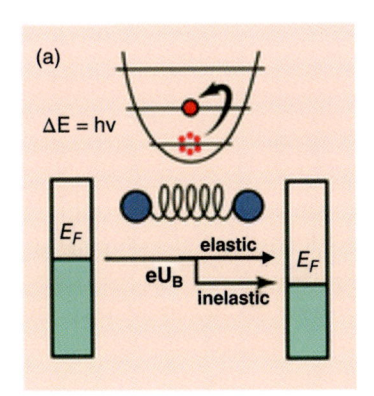

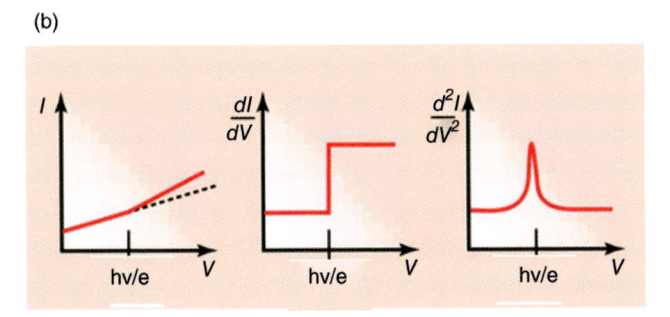

Figure 7.4 (a) Elastic and inelastic tunneling of electrons. (b) The resulting changes of the tunneling current and its first and second derivatives.

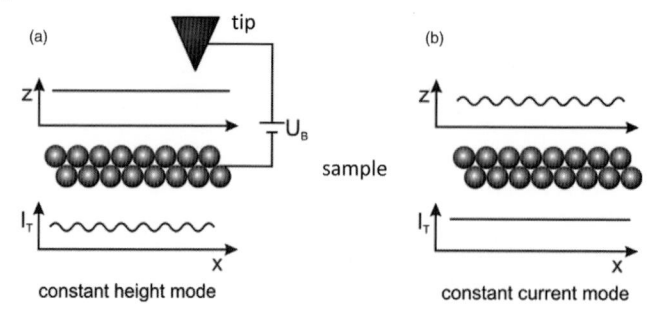

Figure 7.5 Illustration of (a) the constant height and (b) constant current mode of STM imaging.

to excite a characteristic vibrational mode of an adsorbed molecule. The opening of the extra "inelastic channel" manifests itself by a distinct change of the measured tunneling current at U_B, which may be accentuated by taking the first or second derivative (Figure 7.4b). This enables vibrational spectroscopy of individual adsorbed molecules, or even of only parts of an adsorbed molecule (see Ref. [48]) and thereby provides some chemically specific information.

7.2.2
Instrumental Aspects

7.2.2.1 Control of Tip Position

In order to take advantage of the behavior of the tunneling current as shown in Figures 7.3 and 7.4, the tip apex must be brought to a distance where the wave functions of tip and surface start to overlap (Figure 7.3) and tunneling occurs (≤ 1 nm). The tip is then moved (scanned) in x- and y-directions across the surface either by keeping the tip apex at constant z-position or by keeping the tunneling current I_T constant. In the first case (constant height mode), any surface roughness changes the width of the tunneling gap and thereby the measurable tunneling current (Figure 7.5a). This mode should only be applied for flat surfaces in order to avoid tip crash with high surface protrusions. In the second case (constant current mode), the z- position of the tip apex must be changed by means of a feedback loop in order to keep the current constant. Since in this case the tip follows the contours of the surface, tip crashes can be avoided, but due to the continuous height adjustment this detection mode is slower. Some typical values of tunneling currents in the nanoampere range and bias voltages in the millivolt range are given in the figure captions.

The instrumental realization of these detection modes obviously has to obey the following most important conditions:

i) The tip must be brought from a macroscopic distance (mm) to the tunneling distance (nm).

ii) The position of the tip at any point above the surface must be controllable with subatomic (≈ 0.01–0.001 nm) precision.
iii) The tip should be mechanically stable and as sharp as possible, preferably monoatomic.

While all three conditions need to be obeyed for all types of instruments independent of their use in UHV, gas atmosphere, or liquids, the preparation of tips to be used in liquids requires special attention as described further below.

Conditions (i) and (ii) can be realized by the use of piezoelectric actuators and the suppression of any other external disturbing influence on the tunneling gap. The most common technical realization of an STM is the beetle design [49] – an example is illustrated in Figure 7.6a. The plate R is carried by three piezotubes $P_1 - P_3$ (tripod) and stands on a basis G with three ramped 120° sectors. In the center of plate R is the fourth piezotube, the actual scanner P_4, which holds the tip, pointing downward to the sample that is mounted, face up, in the center between P_1 and P_3. Likewise, any other combination of arrangements of the ramped 120° sectors (on R or G), the three piezolegs $P_1 - P_3$ (in R or G) as well as the scanner piezo P_4, and sample facing each other is possible. Obviously, in the case when the sample is submersed in a liquid, the scanner must be pointing down (Figure 7.9a). The function of the piezotubes $P_1 - P_3$ and the piezoscanner P_4 is shown in detail in Figure 7.7. These piezotubes are coated with separate metal electrodes, four outside and one inside. In 1880, Jacque and Pierre Curie discovered that some crystalline or ceramic materials (quartz, lithium niobate,

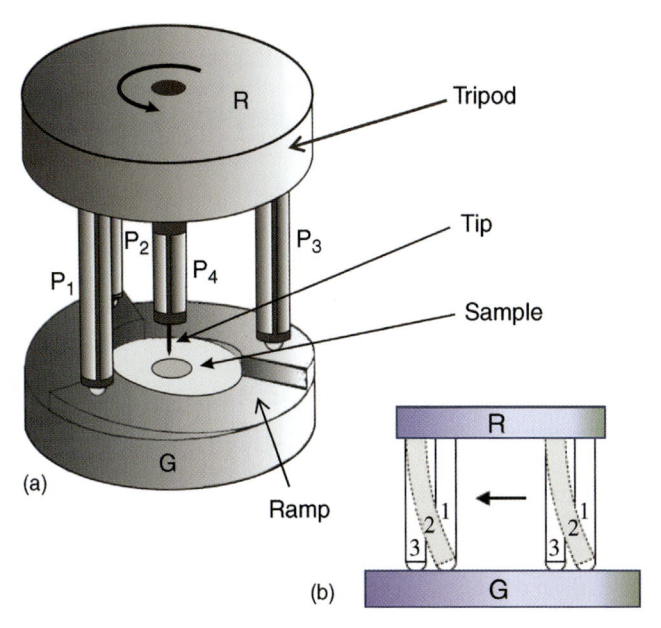

Figure 7.6 (a) Beetle-type setup of an STM. (b) Illustration of the stick and slip mechanism for tip approach (see text).

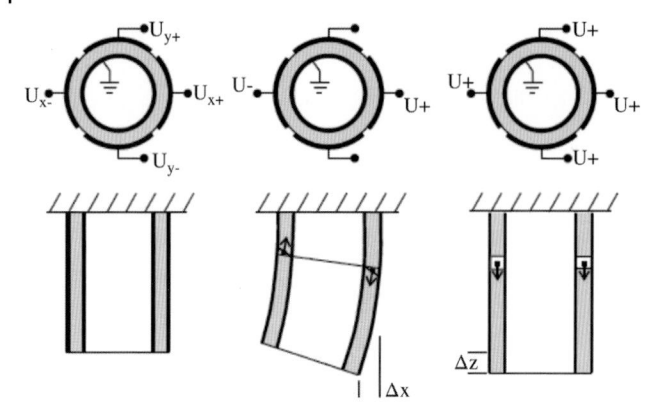

Figure 7.7 Operation of piezotube actuators for use in STM.

lead zirconate titanate, etc.) show the so-called piezoeffect, the induction of electronic charges on their surfaces and, thus, an internal electric field, under the influence of an external mechanical stress. Inversely, under the influence of an external electric field, these materials exhibit deformation (inverse piezoeffect). The latter effect is used in the design of an STM: By applying appropriate voltages U_+, U_- between any of the outer and the inner electrodes, respectively, the piezotube can be made to bend in $-x-$, $+x-$, $-y-$, $+y-$ or stretch/contract in z-direction roughly on the order of 1 nm per applied volt. In this way the tip can be positioned or scanned in a controlled way above the surface taking advantage of the so-called stick and slip mechanism shown in Figure 7.6b: Applying slowly a voltage to the piezo, originally in position 1, makes the piezo bend into position 2 and moves the plate R. If the voltage is abruptly reduced, the piezo jumps into position 3, while the heavy (inert) plate R remains in the new position, and so on. Through this stick and slip mechanism, the tripod walks up or down the ramps of plate G in Figure 7.6a and thereby enables the coarse approach between the tip and the sample.

Obviously, particular care has to be exercised so that no external mechanical (sound, building vibrations, thermal drift, etc.) and electromagnetic perturbations have any significant effect on the geometry of the tunneling gap and, thus, on the tunneling current. These perturbations can be routinely suppressed nowadays by effective vibration damping systems, the choice of suitable materials to avoid or compensate for thermal expansion and drift, and electromagnetic shielding (for details, see specialized literature [18–24]). A further improvement of the stability, and thus the resolving power, of an STM can be achieved by cooling the surrounding of the tunneling gap to very low temperatures (using liquid nitrogen or liquid helium), which, of course, is only possible in UHV. This freezes in all motional degrees of freedom, including those of the atoms/molecules in the tunneling gap.

7.2.2.2 STM Setup in UHV

A typical design of a combined UHV STM/AFM is shown in Figure 7.8. The whole instrument is mounted on a flange. The base-plate G is suspended on springs housed within the four tubes $S_1 - S_4$ for vibration damping. The interlocked toothed copper (Cu) and stainless steel (SS) rings serve for further eddy-current damping. The actual STM as shown in Figure 7.6 or AFM shown in Figure 7.13 is placed in the center (arrow). Leads, wires, and extra tubes pointing toward the center of the installation serve as electric connections and for the control of the sample temperature in a wide temperature range (variable (VT) and low temperature (LT) STM).

As an alternative to the spring suspension system inside the UHV chamber, a small apparatus could externally suspended as a whole by a pneumatic system or by springs as shown in Figure 7.9c.

7.2.2.3 Electrochemical STM Setup

Figure 7.9 shows a small home-built STM system developed at the University of Bonn [50–53] and dedicated to measurements at solid–liquid, specifically at metal–electrolyte, interfaces. This electrochemical scanning tunneling microscope (EC-STM) consists of a single-tube scanner in an assembly, as shown in Figures 7.6 and 7.9a. The electrochemical cell can be mounted onto the STM basis. The cell holder allows a variable connection of electrodes. Apart from the

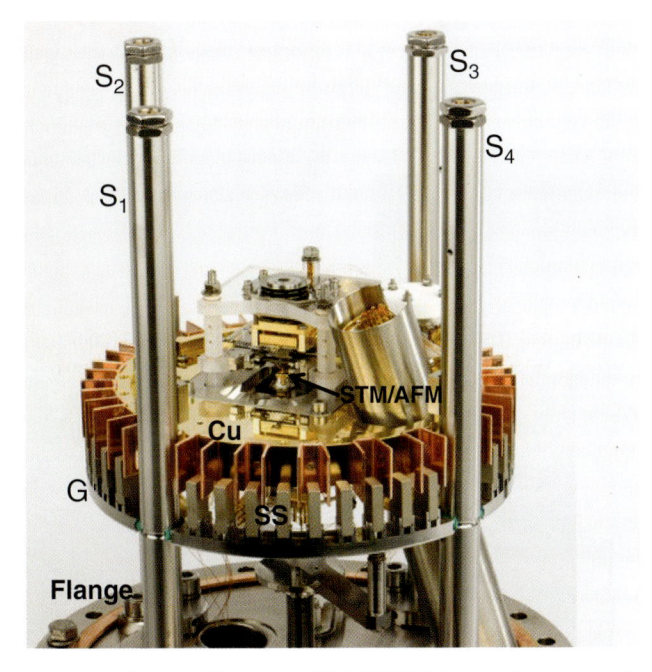

Figure 7.8 Commercial Omicron UHV STM/AFM (see text), www.scientaomicron.com.

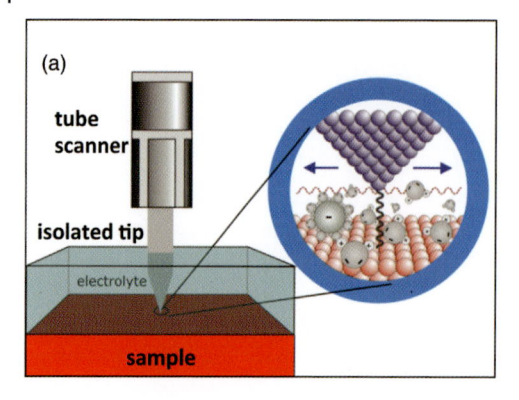

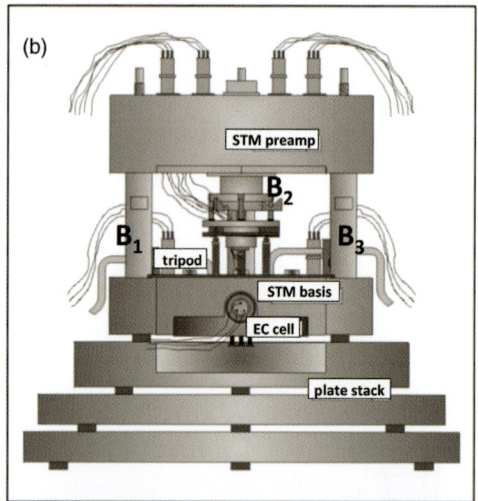

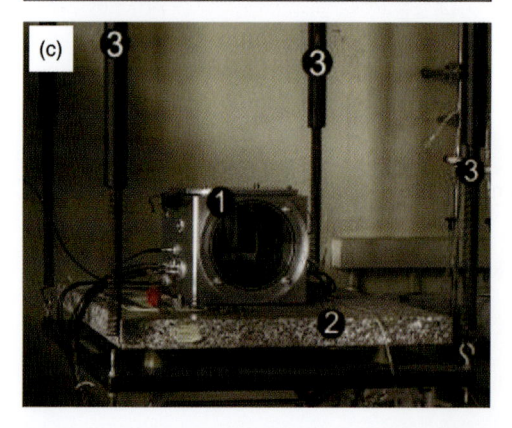

Figure 7.9 Electrochemical scanning tunneling microscope (EC-STM). (a) Principal scheme. (b) Construction of the "Bonn EC-STM" [35]. (c) Overview of the aluminum housing (1) shielding the EC-STM inside the external vibration damping system (2,3).

sample (working electrode), which is connected to virtual ground potential, a counter electrode, an internal reference electrode, and a generator electrode for *in situ* dissolution of metals into the electrolyte can be connected [50–53]. The base carries also the electrolyte flow system and the three piezolegs ($P_1 - P_3$ in Figure 7.6) for the coarse approach. An external reference electrode, for example, a reversible hydrogen electrode (RHE), can be connected via a Luggin capillary. The STM head (Figure 7.9b) rests on three spacer bolts ($B_1 - B_3$) and contains the scanner and the preamplifier, very close to the tunneling tip. Both are further shielded from electromagnetic interferences by aluminum in order to minimize the noise on the tunneling current signal. The whole system is mounted on a stack of three brass plates separated by rubber spacers in an aluminum housing (1 in Figure 7.9c). The aluminum housing shields the STM from acoustic, thermal, and electromagnetic disturbances. This complete setup rests on a rubber mat on a heavy granite slab (2), which is suspended from the ceiling by springs (3). This suspension, the granite slab and the stack of metal plates, inside the aluminum housing serve as damping for low- and high-frequency vibrations. This design offers some key advantages:

  i) Despite the compact setup, the electrochemical cell contains a volume of about 2.5 ml, which is sufficiently large to allow a Nernst diffusion layer to build and to avoid concentration changes due to evaporation. Therefore, electrochemical studies, such as cyclic voltammetry, can be carried out in the same cell as the STM measurements. The sample is fixed against a cutout in the cell bottom so that only a defined area of the single-crystal surface of the sample is in contact with the electrolyte.
  ii) The electrolyte feed and drain system allows changing of the electrolyte without removing the scanner or even opening the aluminum housing, greatly increasing experimental flexibility and limiting contamination. The housing also allows air-free operation under inert gas atmosphere.
  iii) A bipotentiostat independently controls the potential of the tip and the sample versus the reference electrode, enabling potentiodynamic measurements in which the working electrode potential is changed during an ongoing STM scan. Another option is to change the bias potential independent of the sample potential, which allows tunneling spectroscopy measurements.

A detailed description of the system, including the electronic setup of the STM controller, has been published in Refs [50,51].

7.2.2.4 Tip Preparation

Various protocols have been developed to fulfill condition (iii) in Section 7.2.2.1, namely, to produce stiff tips as sharp as possible, preferably monoatomic. The stiffness is particularly important if high scan rates are to be employed (Video-STM), in order to avoid bending [41,42]. Therefore, most tips are prepared from tungsten, but also from Pt–Ir alloy wires. Sharpness is actually easy to achieve,

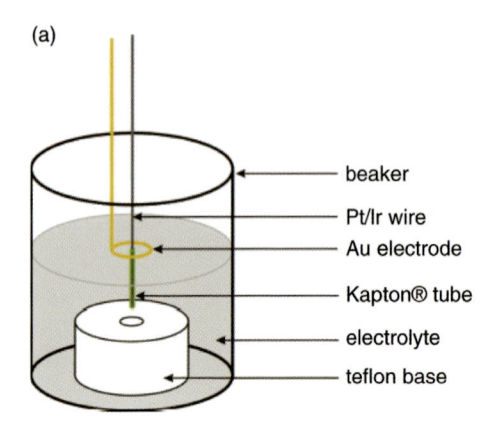

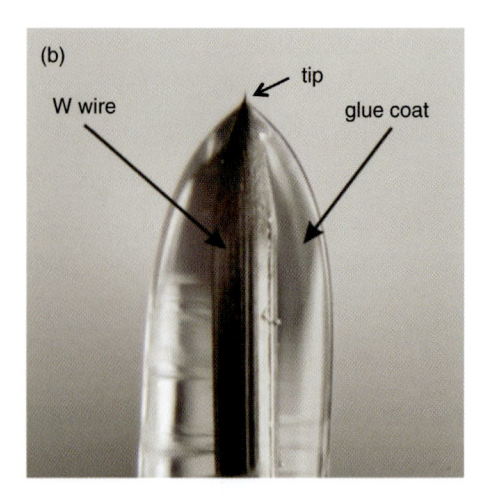

Figure 7.10 (a) Setup for tip etching, and (b) glue insulation of a W-tip for electrochemical measurements.

no matter how the wire is disrupted. The wire could simply be cut by pliers or torn apart; it is not unlikely that after separation one atom sticks out furthest on either end of the rupture. Most commonly, however, STM tips are prepared by electrochemical etching (Figure 7.10). A gold wire loop as a working electrode is placed in the surface plane of a solution (2 M KOH for tungsten tips; 2 M KOH/ 4 M KSCN for Pt–Ir tips). The wire to be etched is partly protected by a Kapton tube and immersed inside the gold loop into the electrolyte such that only a very small unprotected area of its upper part is exposed to the electrolyte (Figure 7.10a). This part is etched by applying a DC voltage (e.g., 2 V) superimposed by a rectangular AC voltage of 10 V and a frequency of 1 kHz until the lower part falls off under its own weight into a hole of the teflon base (Figure 7.10a). After carefully removing the Kapton tube, the tip is rinsed with high-purity water and dried. Alternatively, only the gold wire loop could be repeatedly filled with a lamella of electrolyte through which the tip is pierced (without further Kapton shielding).

If the tip is to be used for measurements at metal–electrolyte interfaces, a further very important preparation step must be undertaken. Since the biased metallic tip inside the electrolyte solution may act as an electrode itself, possible Faradaic currents through the tip would interfere with the tunneling current. Therefore, tips to be used in electrolytes need to be largely insulated. This can be achieved by coating the whole tip shaft with hot glue (Figure 7.10b). Due to the very high curvature, the hot glue will retreat from the very tip apex, leaving, after drying, only the very apex of just a few tens of nanometers uncovered so that any Faradaic current through this open part is two or more orders smaller than the actually desired measurand, the tunneling current.

7.2.3
Interpretation of STM Images

As a first example, Figure 7.11 shows an atomically resolved STM image of the (111) surface of a Ni_3Al alloy. This bulk alloy is chemically ordered: According to the unit cell shown in panel (a), Al atoms occupy the eight corner positions, while the Ni atoms sit at the face centers. As a consequence, in a (111) plane of this material (highlighted in panel (a)), every second site is occupied by an Al atom with an Al–Al separation of $a_0 = 0.507$ nm as displayed in panel (b). This is exactly the distance between bright dots in measured STM image (Figure 7.11c). Thus, the image does not show every atom but only every second, and one may intuitively be tempted to interpret the bright dots as the Al atoms. One good reason for this interpretation may be the fact that the Al atoms are bigger than Ni atoms and therefore stick out of the surface plane further than the Ni atoms, as has indeed been concluded from electron diffraction experiments [54]. A line profile along a row of bright dots in Figure 7.11c suggests a corrugation of about 15 pm.

However, caution should be exercised. As mentioned earlier, STM images are determined not only by surface topography but also by the electron density (see Section 7.2.1). The electron density, however, is lower on Al sites than on Ni

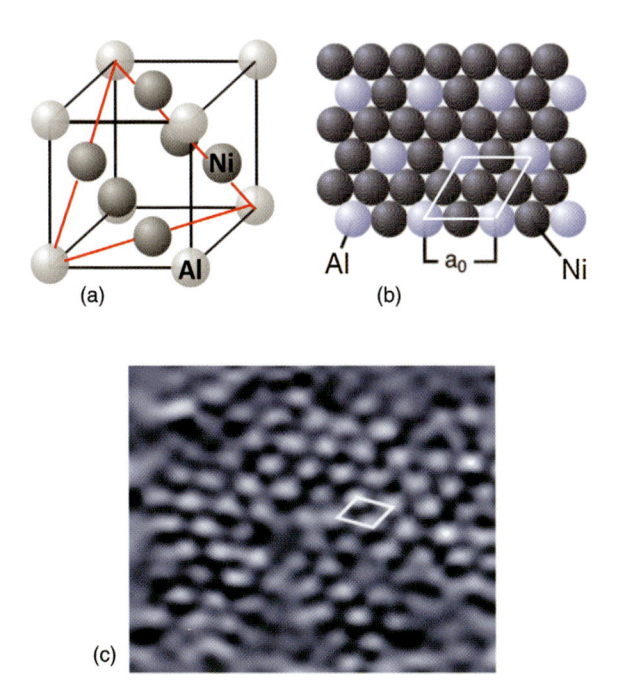

Figure 7.11 (a) Bulk unit cell of an ordered Ni_3Al alloy. (b) Hard sphere model of the (111) surface of the Ni_3Al alloy with unit cell. (c) Atomically resolved STM image (6.2 nm × 7.7 nm) of a $Ni_3Al(111)$ surface with unit cell.

sites. Moreover, in metals, the electron density is particularly high *between* the atoms (metal bond). Hence, in principle, it could very well be that the STM image in Figure 7.11c does not preferentially show the (protruding) Al atoms but the electron density-rich threefold Ni hollow sites between the Al atoms, whose separation, of course, is also 0.507 nm. In cases like this and even more so in the case of adsorbed molecules (see, for example, Figure 7.32), a definitive decision can only be made on the basis of theoretical calculations of the respective electronic structure and a simulated STM image thereof. Density functional theory (DFT) [55–58], for instance, enables to calculate the ground-state properties of the atomic and molecular systems in question and has become a powerful tool in material science to calculate and predict the electronic and structural properties of solids and surfaces. Simulated STM images thereof are compared with experimental ones in order to arrive, in combination with STS measurements, at correct interpretations of structures seen in STM images like in Figure 7.11c. An alternative approach based on a formalism developed by Keldysh [59] and adopted in the STM theory [60,61] was used by Jurczyszyn *et al.* [62] to interpret Figure 7.11c. The best fit between these experimental results and the theoretical simulations clearly indicate that the tunneling process is dominated by s- and p_z-contributions connected with surface Al atoms. Thus, the theoretical calculations support the assignment of the bright dots in Figure 7.11c to Al atoms.

The situation is, of course, even more complex with STM images taken *in situ* from solid–liquid interfaces due to the presence of solution species in the tunneling gap. Experimentally measured tunneling barriers are significantly lower than in UHV, which can be explained by the adsorption of the polar water molecules on the electrode surface; water commonly decreases the work function of metal surfaces. Moreover, the local barrier height should be modulated by a two-dimensional network of fluctuating hydrogen and oxygen atoms in the molecular water layer between the tip and the surface [63]. This effect should, in principle, be detectable because the time an electron needs to tunnel through a tunneling gap of ~1 nm is at least two orders of magnitude faster than the movement of the water molecules [64]. However, since a large number of individual tunneling events are necessary to register a sizable tunneling current, a too slow sampling rate of the STM controller electronics limits the speed of data processing. Therefore, the variation in the barrier height is averaged over a number of tunneling events at different orientations of the water molecules so that the recorded tunneling current is the result of an effective (averaged) barrier height.

There is one more important difference between tunneling through an empty or a water-filled gap: In UHV the tunneling current depends strictly exponentially on the distance between the tip and the surface (see Section 7.2.1). In an electrochemical environment, however, several groups have detected a modulation of the tunneling current as a function of the tip–surface distance [65,66]. These oscillations are ascribed to individual layers of water molecules parallel to the surface.

7.3
Atomic Force Microscopy

7.3.1
Physical Background

Atomic force microscopy makes use of interaction *forces* between a tip and a sample in order to probe the atomic structure and material-specific properties of the surface, and was introduced by Binnig et al. [17] shortly after the invention of the scanning tunneling microscope. Since in this case neither the tip nor the sample need to be conducting, AFM is an ideal supplement to STM in order to study *insulators* [21–24,67].

In principle, there are several different types of forces acting between the tip and the sample surface, whose relative strength depends on the tip–sample distance. It is, therefore, useful to distinguish between short- and long-range forces. The long-range forces act on the whole tip and partly even the cantilever (see Figure 7.13) that holds the tip. The short-range forces involve only the atoms at the very apex of the tip and those atoms at the surface nearest to the tip. The long-range forces include electrostatic and magnetic forces that, of course, can be attractive or repulsive, and van der Waals (vdW) forces. van der Waals forces are attractive and act between all atoms and molecules, and result from interactions between permanent dipoles (Keesom forces), between permanent and induced dipoles, the latter being induced by the former (Debye forces), as well as between fluctuating dipoles of nonpolar atoms (dispersion forces). Classically, the motion of electrons in an atom leads to momentary distortions of the spherical charge distribution and, thus, to short-lived fluctuating dipoles that interact with induced dipoles in neighboring atoms (which by the way explains also the attraction and, thus, condensation of rare gases). In total, van der Waals forces vary with r^{-7}, and thus the van der Waals potential with r^{-6} as a function of distance r between two interacting atoms. Near the surface, the tip atoms interact with all sample atoms so that integration over a certain sample volume ($\sim r^3$) gives r^{-4} and r^{-3} for the force and the potential, respectively.

At short distances, the short-range forces dominate and are ultimately decisive for the atomic resolution of surface structures. These short-range forces, also termed "chemical forces" result from the same interactions that eventually lead to ionic, covalent, and metallic bonds or in real systems to mixtures between them, whose description requires involved quantum mechanical calculations. At very short distance, the orbitals of two atoms start to overlap and the electrons have to obey the Pauli principle, according to which the states of two involved electrons must differ in at least one quantum number. The fulfillment of this condition may require the excitation of electrons to higher (unoccupied) energy levels that together with the Coulomb repulsion between the atom cores causes the so-called very short-ranged Pauli repulsion ($\sim r^{-12}$).

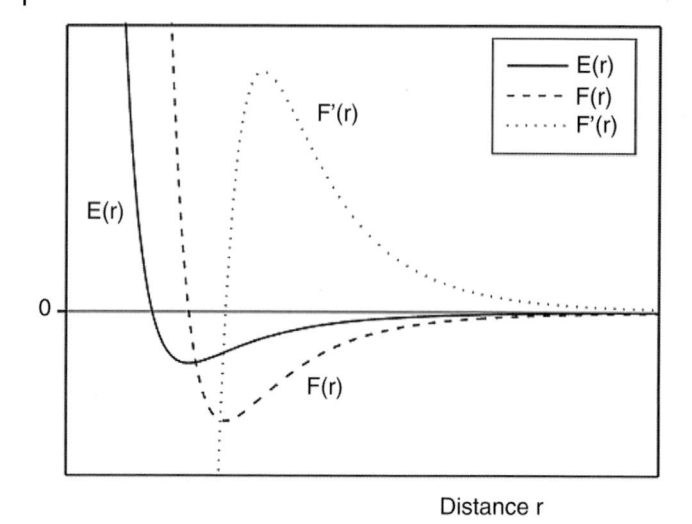

Figure 7.12 The Lennard-Jones potential E(r), the resultant distance dependence of the interaction force F(r), and the force gradient F'(r).

The most commonly used description of the superposition of all long- and short-ranged interactions is the Lennard-Jones (L-J) potential:

$$E_{L-J}(r) = -\frac{A}{r^6} + \frac{C}{r^{12}}, \tag{7.10}$$

with A and C being material-specific constants and r is the distance between two atoms. At large distances, the first attractive van der Waals term $-A/r^6$ prevails, while at short distances the repulsive $+C/r^{12}$ term dominates. Figure 7.12 shows the Lennard-Jones potential $E(r)$, the distance dependence of the interaction force $F(r)$, and the gradient $F'(r)$. Note that upon approach from large distances r, both the potential energy $E(r)$ and the interaction force $F(r)$ decrease only slowly, while, conversely, both increase very steeply at very short distances. The minimum of $E(r)$ occurs at the equilibrium or bond distance, respectively; while the $F(r)$ minimum corresponds to the maximum adhesion force. Both regimes of the force curve, $F(r)<0$ or $F(r)>0$, are exploited to perform AFM measurements.

7.3.1.1 AFM Contact Mode

The contact mode exploits the very strong distance dependence of $F(r)$ in the very steep repulsive regime, $F(r)>0$, that is, the tip is in mechanical contact with the surface. Figure 7.13 displays a setup that enables to measure the force. The tip is located at the end of a cantilever (see *inset*), both mostly prepared from single-crystal silicon. The backside of the cantilever reflects an incident infrared light beam shining from an LED and a mirror 1 via mirror 2 onto the center of a

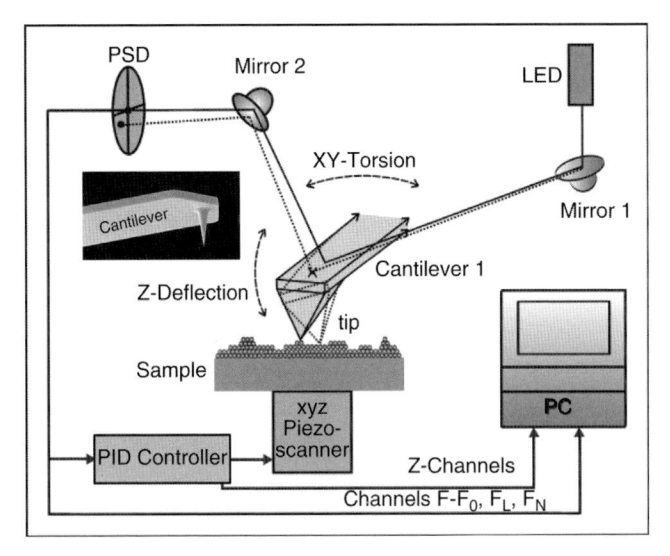

Figure 7.13 Experimental setup of atomic force microscopy (see text). *Inset:* Cantilever with AFM tip.

position-sensitive photodetector (PSD) with four quadrants. A force between tip and surface causes a bending or torsion of the cantilever and thereby a deflection of the beam on the PSD. A purely vertical (normal) force between the tip and the surface causes a shift of the reflected beam toward the lower or upper half of the PSD. Measurement of the intensity *difference* registered between the upper and lower PSD quadrants enables to determine the deflection of the cantilever, which according to Hookes law $F_N = \kappa z$ is proportional to the acting normal force F_N as long as the deflection z (normal to the surface) is small; κ is the force constant of the cantilever. Likewise, lateral friction forces in x- and y-directions lead to torsion of the cantilever and can be measured by a resultant shift of the reflected beam toward the left or right half of the PSD.

Scanning the surface is achieved by moving the sample by means of piezoactuators in x- and y-directions relative to the tip (Figure 7.13), and like in STM two different detection modes are possible. In the "constant force" mode (comparable to the constant current mode in STM), an electronic feedback loop adjusts the z-distance between the tip and the surface such that the normal force F_N remains constant while the sample is moved in x- and y-directions. Since in this case the deflection of the cantilever is constant, this mode is also called "static force microscopy." The voltage that is applied to the z-scanner in order to compensate for changes of F_N between the tip and the sample serves to generate a topographic image of the surface. Such image can not only be generated from the forward and backward scan direction but also from the laterally varying friction forces (F_L) and from $\Delta F = F - F_0$, with F being the actually measured and F_0 the chosen set force (load).

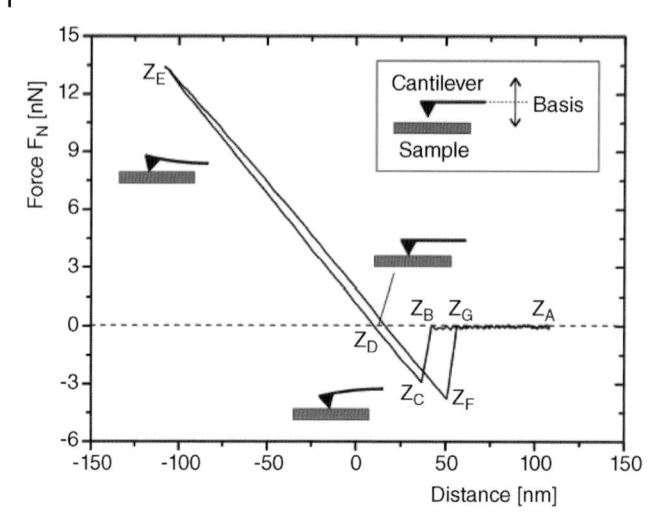

Figure 7.14 Force–distance curve (for details see text).

In the "constant height" mode (comparable to the constant height mode in STM), the z-piezo remains inactive. As a consequence, any height modulation of the surface when scanning in x- and y-directions leads to related changes of the measured normal and lateral force, that is, to deflections of the light beam on the PSD, which serve to construct a topographic image of the surface.

If for *fixed* coordinates x and y the normal force F_N is measured as a function of distance z, one speaks about "force spectroscopy." In this case, the tip is positioned above a specific point on the surface, and the tip–surface separation is systematically varied. The concomitant bending of the cantilever of force constant κ in z-direction yields the normal force F_N. Figure 7.14 shows such an experimental force–distance curve. At very large tip–surface separation (z_A), changes of z have no influence on the measured force, which serves as reference, $F_N = 0$. Once the tip–surface distance reaches z_B, the attractive vdW forces exceed the restoring force of the cantilever and the tip is pulled into contact with the surface (z_C), the cantilever bends in negative direction (toward the surface). Upon further approach, the steeply rising repulsive forces dominate and the measured force increases until at z_D the cantilever is again unbent, repulsive and attractive forces are in equilibrium. Even further approach leads to positive forces and bending of the cantilever in positive direction up to a set value F_N at z_E. The linear increase serves to calibrate the normal force F_N from a known distance $\Delta z = z_D - z_E$ and the force constant κ. In the reverse scan direction, the "pull-off-contact" occurs at the distance $z_F \neq z_C$ and the cantilever restores its force-free equilibrium state at z_G (like at z_A). $F_N(z_C)$ and $F_N(z_F)$ represent the maximum attraction force and the maximum adhesion force, respectively. The hysteresis in this force–distance curve (z_B, z_C, z_F, z_G) may be an indication of inelastic effects in the tip–surface contact (or "creep" of the piezoactuator).

Deviations between force–distance curves measured at different x,y-positions on the same surface may be indicative of chemical heterogeneity of the surface.

An important aspect of AFM in contact mode are deformations of tip and surface as a function of the applied normal force. The deformations obviously depend on the acting force (load) and may be elastic or plastic. In both cases, the contact area may increase and, as a consequence, the microscopic resolution decrease. Moreover, plastic deformations leave irreversible destruction behind, as illustrated in Figure 7.15a and b [68]. A $Co_{89}Hf_{11}$ surface was first scanned within the dark region of panel (a) with $F_N = 3$ nN, and then once again over the full image size with $F_N = 1$ nN. As shown by the height profile in panel (b), the first imaging conditions created a depression of ~1 nm. This effect, however, can also be used advantageously to free a surface area from contaminants and deposits. As an example, Figure 7.15c and d shows this "cleaning" effect for a graphene surface. While in panel (c) the contaminants are still agglomerated along graphene ripples (bright lines) in the central part, continuous scanning moves these contaminants to the border of the scanned quadratic area, as seen in panel (d) [67]. This effect can also be used to remove adsorbed molecules from a part of a surface, which afterward enables to compare the lattices of the substrate surface and the adsorbate layer in one and the same image and to determine thereby the exact adsorption site of the adsorbed molecules.

7.3.1.2 AFM Noncontact Mode

The potential risk of contact-mode AFM to cause mechanical destruction of the surface can be avoided by applying the so-called noncontact mode. In this case, the cantilever with the tip is excited to vibrate normal to the surface and bend periodically (dynamic force microscopy), as illustrated in Figure 7.16. As long as the vibrating tip is (and remains) far away from the surface, the cantilever can vibrate without interacting with the surface and behave like a harmonic oscillator with a parabolic potential $E_p(z) = \kappa z^2$, as shown in Figure 7.17 by the dotted line. Its motion is described by a sinusoidal oscillation:

$$z(t) = A_0 \sin(2\pi \nu_0 t), \tag{7.11}$$

with an amplitude A_0 and the resonance frequency

$$\nu_0 = \frac{1}{2\pi}\sqrt{\frac{\kappa}{m}} \tag{7.12}$$

with κ = force constant and m = vibrating mass. If the sample is approached toward the cantilever such that at $-A_0$ the tip "feels" an interaction with the surface as described by the dashed Lennard-Jones potential in Figure 7.17, both potentials $E_p(z)$ and $E_{L-J}(z)$ act simultaneously on the tip/cantilever with the consequence that the oscillation of the cantilever is no longer driven by a harmonic but by an effective (anharmonic) potential $E_{eff}(z)$. As a consequence, its vibration

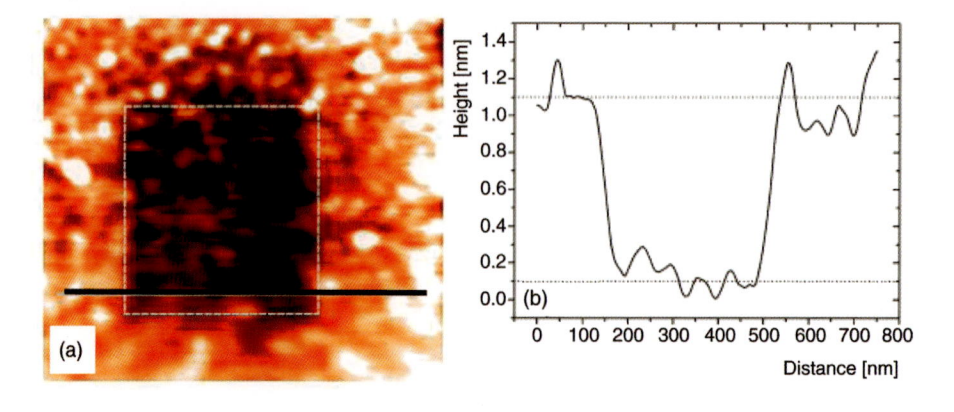

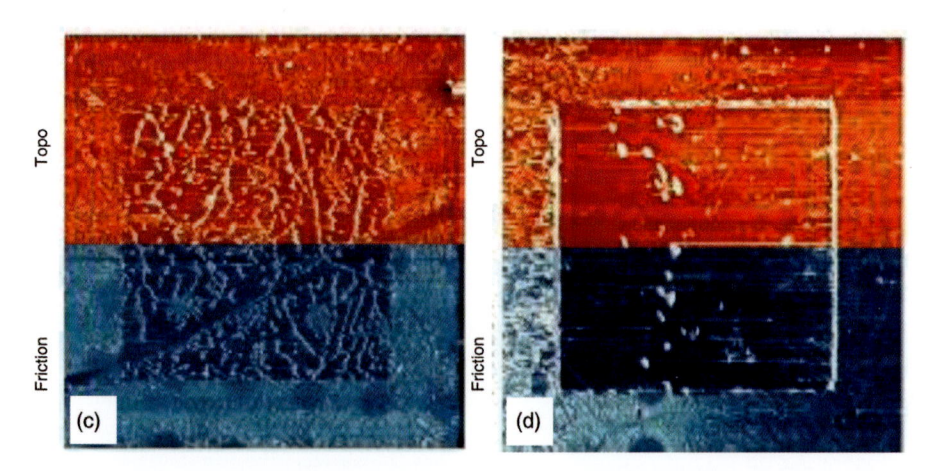

Figure 7.15 (a and b) Deformation of a Co$_{89}$Hf$_{11}$ alloy surface caused by an AFM probe operating in contact mode. (a) c-AFM image (800 nm × 700 nm) registered with F_N = 1 nN, the dark area within the marked square was previously scanned with F_N = 3nN. (b) Height profile along black line in part (a). (From Ref. [68].) (c and d) Pushing aside of surface contaminants: Contaminants at ripples of a graphene surface (white lines) in panel (c) are moved to the borders of the quadratic scan area shown in panel (d). (From Ref. [67].) Topo = topographic mode, that is, registration of F_N, friction = measurement of the lateral friction force F_L (see Figure 7.13).

frequency also changes:

$$\nu = \frac{1}{2\pi}\sqrt{\frac{\kappa_{\text{eff}}}{m}}. \tag{7.13}$$

The more the distortion of the potential, the closer the tip vibrates toward the surface. If the interaction between the tip and the surface (at the surface-near turning point) is attractive, the tip stays a little longer near the surface, it is slowed down, so that $\nu<\nu_0$. Conversely, if the tip–surface interaction is repulsive,

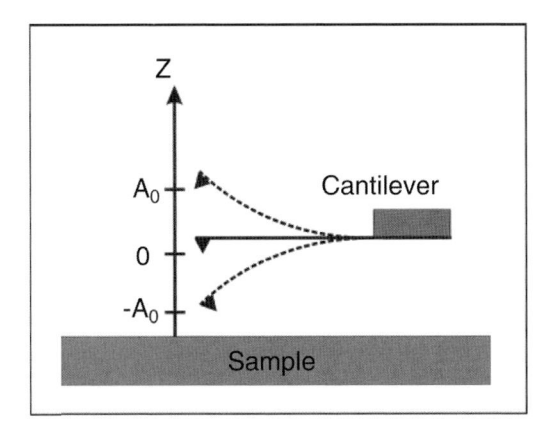

Figure 7.16 Vibration of an AFM cantilever working in the noncontact mode.

the tip is pushed faster away from the surface, so that $\nu > \nu_0$. It is the basis of noncontact AFM to measure $\Delta\nu = \nu - \nu_0$ (Figure 7.18), while the tip never "touches" the surface.

Nowadays new types of sensors, as described in the following sections, enable to perform STM and noncontact atomic force microscopy (nc-AFM) with one single sensor and to benefit from both complementary information. For details of the frequency measurement and further dynamic modes, for example, the so-called intermittent or tapping mode, we refer the reader to the specialized literature [15,21–24].

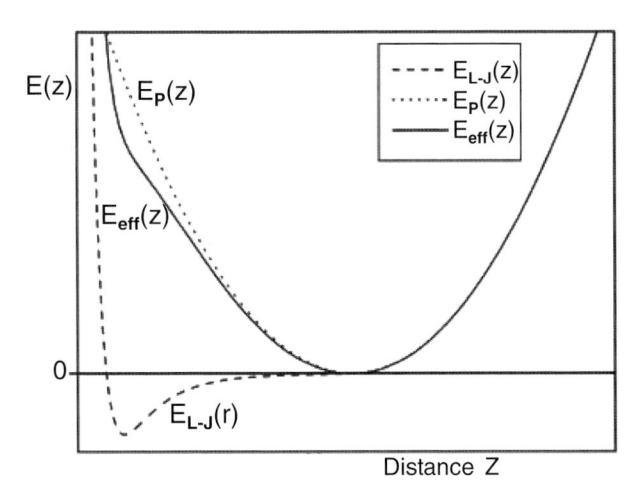

Figure 7.17 Parabolic potential $E_p(z)$ of a harmonically vibrating undisturbed AFM cantilever; superposition with a Lennard-Jones potential $E_{L-J}(z)$ leads to the effective (anharmonic) potential $E_{eff}(z)$.

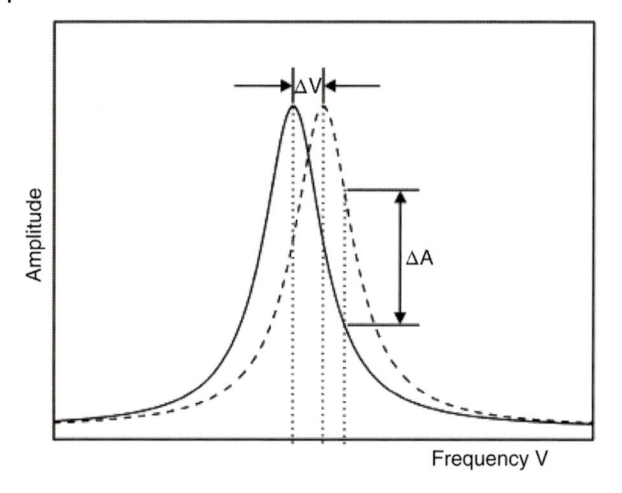

Figure 7.18 Frequency shift in noncontact AFM of the cantilever vibrating freely (dashed) and interacting with the surface (solid). The maxima correspond to resonance conditions.

7.3.1.3 Tuning Fork Sensor

A tuning fork sensor (TFS) [69] is a quartz oscillator with a very sharp resonance frequency, which is widely used in clocks and watches. This tuning fork is glued with one prong onto an insulating plate, which is carried by a piezoactuator, as shown in Figure 7.19a. The oscillation frequency of the (piezoelectric) tuning fork is monitored via two contacts K_1 and K_2. A metal tip (e.g., Pt–Ir) is attached to the end of the free prong such that it is electrically insulated from the tuning fork electrodes, but connected to a thin wire to register a tunneling current.

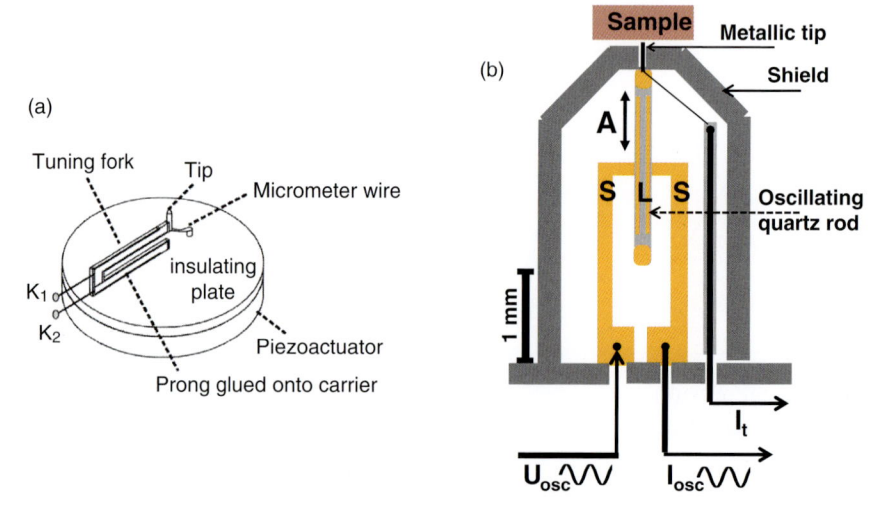

Figure 7.19 (a) Tuning fork sensor and (b) length extension resonator for STM and noncontact AFM measurements (see text).

Similar to Figure 7.13, the sample is approached to the tip in z-direction and scanned relative to the tip in x- and y-directions by a piezoactuator. During scanning, three different imaging modes can be employed: (i) constant current I_T, (ii) constant $\Delta\nu$, and (iii) constant z, as already described in Sections 7.2.2.1 and 7.3.1.2.

7.3.1.4 Length-Extension Resonator

A disadvantage of a vibrating cantilever (Figure 7.16) is its relatively large oscillation amplitude of several nanometers with the tip being in the tunneling and short-range force interaction regime only at the near-surface turning point of each oscillation cycle. An alternative is a so-called length-extension resonator (LER). This is a stiff sensor that consists of a LER quartz and enables combined STM/nc-AFM measurements with fast scanning speeds and operation at sub-nanometer oscillation amplitudes [70–73].

A schematic drawing of a "KolibriSensor™" [70] (The name KolibriSensor is a registered trademark of SPECS GmbH, Berlin) is shown in Figure 7.19b. This device is based on a rectangular piezoelectric quartz LER (L) whose sidewalls are gold coated. At its center, it is supported by two arms S to a base plate, ensuring a highly symmetric geometry of the oscillating rod L. Oscillations of this quartz rod are excited by applying a sinusoidal voltage (U_{osc}) to one of its sidewall electrodes. The oscillating electric field across the rod causes an extension or contraction of the quartz rod across its width due to the piezoelectric effect. As a consequence of this transversal extension/contraction of the quartz rod, its length changes also periodically in the longitudinal direction with an amplitude A (see Figure 7.19b). This purely electric operation of the resonator ensures an excitation of only the desired resonance mode perpendicular to the surface, for example, at a relatively high resonance frequency of about 1 MHz that enables faster data acquisition and faster scanning speeds. The oscillation amplitude A is as low as 1 pm. The use of a metallic tip that is separately contacted by a thin wire (see Figure 7.19b) enables simultaneous nc-AFM- ($\Delta\nu$) and STM measurements (I_T). The stiffness of the resonators makes it also particularly useful for measurements under environmental conditions, that is, in gases and liquids [74].

7.4
Other Types of Scanning Probes

For the sake of completeness, we briefly mention here other types of scanning probes that provide specific complementary information about surfaces.

Magnetic force microscopy (MFM) exploits magnetic forces between a magnetic surface and a magnetized tip and provides information about the strength and orientation of magnetic moments at surfaces [75–78].

Electrostatic force microscopy (EFM) is sensitive to electrostatic Coulomb forces [24,67,79]. Depending on the charge of the tip apex (positive, negative), the interaction of the tip with anions or cations at the surface will be attractive or repulsive, which can be measured.

Kelvin force microscopy (KFM) [80,81] is a version of nc-AFM with a conducting tip. The metallic tip (T), a metal surface (S), and the gap between them form a very localized capacitor. If tip and sample are different metals, their different work functions result in a contact potential $\Delta\Phi = \Phi_T - \Phi_S$ between them. Vibrations of the tip normal to the surface cause a periodic distance change and thereby capacity change dC with a concomitant alternating current,

$$I_K(t) = \frac{dQ}{dt} = \Delta\Phi \frac{dC}{dt}, \qquad (7.14)$$

between the tip and the sample that can be measured. If an external voltage $U_0 = -\Delta\Phi$ is applied to the tip–surface gap, this current I_K vanishes and U_0 is a measure of the work function *difference* between the tip and the sample. This technique can be used to monitor local work function *changes*, for instance, due to adsorption at differing surface sites such as chemically and structurally deviating patches, defects, clusters, and so on during the measurement (provided the adsorbant does not affect the work function Φ_T of the tip).

7.5
Exemplary Results

In this section, we present experimental results that illustrate the broad range of information that can be obtained from STM, STS, and AFM measurements. The examples are chosen to focus on *chemistry*-related phenomena such as adsorption, dissociation, bond formation, reaction, and so on, and since STM and AFM are primarily microscopic techniques, the results are best presented by images, "a picture paints a thousand words."

7.5.1
Investigations in UHV

Elucidation of chemical processes at solid surface, of course, starts with a precise knowledge about the properties of the surface itself. This is seemingly guaranteed if the study is done according to the so-called idealized "surface science approach," namely, the use of well-defined single-crystal surfaces that, under ultrahigh vacuum (UHV) conditions, are exposed to a well-controlled pressure or beam of atoms or molecules [2–8]. But as mentioned in Section 7.1, the "broken symmetry" at the surface, that is, the different environment of the surface atoms compared to that of atoms in the bulk, unavoidably results in new surface-specific properties. For simple thermodynamic reasons, the atomic structure, electronic and vibrational properties, and the chemical composition at surface *must* be different than in the bulk. This needs to be known and considered for the understanding and description of any process at solid (or liquid) surfaces. In the following, we start with a presentation of specific properties of bare surfaces (in UHV) as obtained by STM/STS and AFM measurements, and later move on to measurements in real gas pressures as well as in liquids.

7.5.1.1 Surface Structure: Reconstruction

The formation of a surface by cutting through the bulk is associated with breaking bonds. The broken (unsaturated) bonds at the surface cause an increase of the total energy of the solid. This excess surface energy can be reduced by decreasing the number or the energy of broken bonds, which can be achieved by changing the geometric structure and/or the chemical composition of the surface; the first leads to "surface reconstruction" and the second to "surface segregation" [1–8,82,83].

Surface reconstruction results in an arrangement of the surface atoms different from that in a parallel crystallographic plane in the bulk. Figure 7.20 displays two examples. Panels (a) and (b) show ball models of a Si(111)- and a Pt(100) bulk plane, respectively. The (111) plane corresponds to a regular hexagonal arrangement of atoms, while that of the (100) plane exhibits a quadratic lattice. The STM images in panels (c) and (d) reveal totally different structures. The famous Si(111)-(7×7) reconstruction was already predicted from electron microscopic measurements [84] and interpreted in terms of the so-called DAS-model, but conclusively proven by STM imaging (see also Figure 7.1) [11]. Likewise, the arrangement of atoms at the Pt(100) surface is not quadratic, but nearly hexagonal with a ∼20% higher atom density [82,85,86]. The mismatch between the normal quadratic second Pt layer and the hexagonal first layer (which in addition is rotated by 0.7° with respect to parallel bulk planes) leads to the

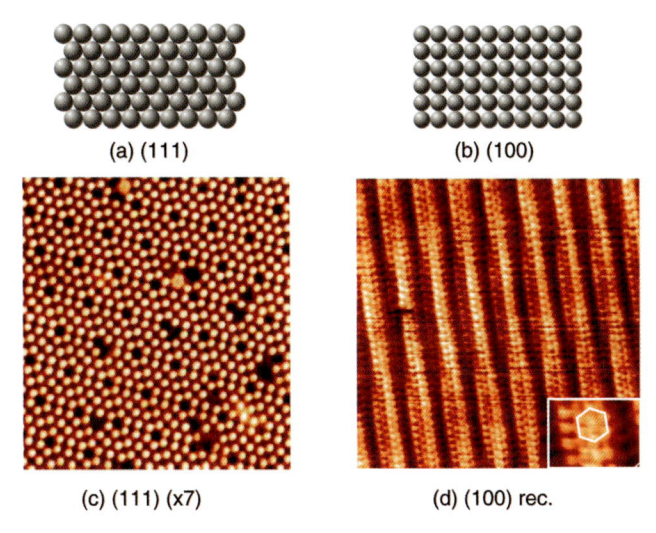

Figure 7.20 (a and b) Hard sphere models of the unreconstructed Si(111) and Pt(100) surface. (c and d) STM image of the respective reconstructed surface. Image size: Si: 178 Å × 178 Å; Pt: 110 Å × 110 Å.

appealing "cable stitch" pattern at the surface. The driving force for both the Si (111) and Pt(100) reconstruction is a lowering of the surface free energy; for instance, the number of broken (unsaturated) bonds of a Pt atom in the unreconstructed Pt(100) surface is four, while that in a hexagonally close-packed (111)-like surface is only three.

Reconstruction is a general phenomenon of semiconductor surface. Further bare metal surfaces known to reconstruct are, for example, Ir(100), Au(100), Au (111), and Pt(111) [82].The rearrangement and densification of the atomic arrangement at solid surfaces, obviously, requires atoms to diffuse and, thus, thermal activation.

7.5.1.2 Surface Composition: Alloys

The catalytic and reactive properties of alloy surfaces depend, obviously, on the surface composition, but *also* on the distribution of the constituent species within the surface. As mentioned in Section 7.5.1.1, nonelementary solids exhibit the phenomenon of surface segregation, that is, the enrichment of one component (or more) at the very surface compared to the bulk composition [2–5,83]. Driving force for this process is the lowering of the total energy of the system. Roughly, too voluminous atoms within the bulk lattice segregate to the surface that leads to relaxation of lattice stress. Enrichment of the component of lower sublimation energy reduces the number of higher energy unsaturated bonds at the surface [87]. Typical surface spectroscopies like Auger electron spectroscopy (AES), X-ray photoelectron spectroscopy (XPS), and ion scattering spectroscopy (ISS) are well suited to determine the "true" surface composition [4]. However, in order to arrive at a full understanding of surface reactivity and catalytic activity/selectivity of surfaces, it is also necessary to know the distribution of the surface constituents, that is, the size and composition of local atomic "ensembles," as well as the electronic interactions between the atoms in these ensembles. It appears intuitively clear that within an hexagonal ensemble, the electronic properties of, for example, a Pt atom surrounded by six Cu atoms (ligands) are different compared to a pure Pt_7 ensemble [88]. These subtle but important local "ensemble-" and "ligand-effects" can only be assessed by scanning probe techniques.

Figure 7.21 shows an STM image of a well-annealed, that is, equilibrated, (100) surface of a binary PtRh alloy, whose bulk composition is 50% Pt and 50% Rh. The STM shows clear chemical contrast between Rh (bright) and Pt (dark) atoms, and simple counting (averaged over many images) reveals that there are about 31% Rh and 69% Pt atoms at the surface. The black spots are caused by residual carbon. Both the Pt and Rh atoms tend to cluster in small groups (ensembles) of the same species [89,90].

This grouping appears less pronounced in the case of a PtCo surface. Figure 7.22a shows an STM image of a PtCo alloy surface after exposure to carbon monoxide at room temperature. The brightest spots correspond to adsorbed CO molecules. The distribution of Co, Pt atoms, and CO molecules within the selected area is identified in panel (b). A close inspection of this distribution

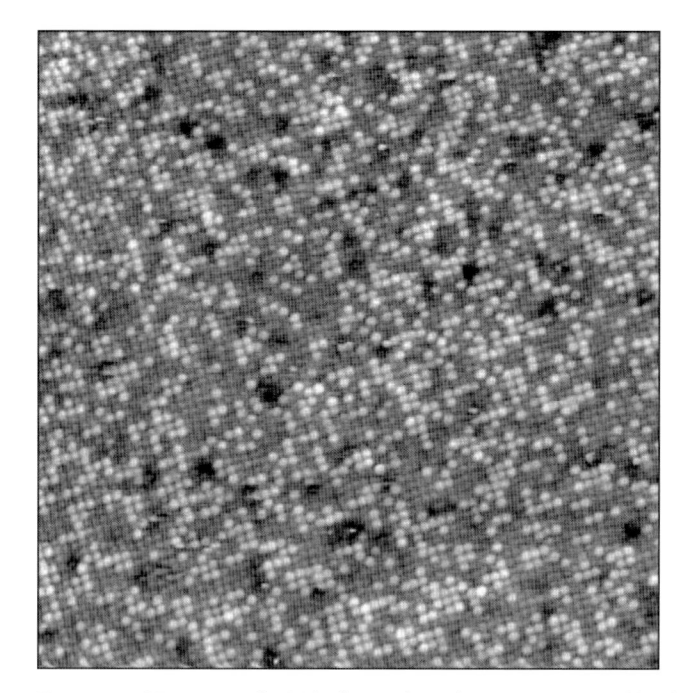

Figure 7.21 STM image of a PtRh alloy surface. Rh atoms are imaged brighter than Pt atoms. The black spots are caused by residual carbon. Image size 200 Å × 200 Å. (From Ref. [90].)

shows that the CO molecules preferably adsorb on top of Pt atoms; the higher the number of their nearest-neighbor Co atoms the more likely the CO adsorption (panel (c)) [91]. This finding represents a clear manifestation of the so-called ligand effect [92]. This effect describes the modification of the adsorption properties of a given surface site due to an electronic influence on the part of the neighboring atoms (ligands). In the specific case of the PtCo alloy, the "dilution" of the Pt(111) surface layer by Co atoms leads to an upward shift of the Fermi level and thereby to a strengthening of the Pt—CO bond due to an enhancement of electron back-donation into the molecular $CO(2\pi^*)$ level [93–95].

7.5.1.3 Surface Oxidation

An important reaction is surface oxidation. The formation of surface oxide films is the result of undesired corrosion processes, of desired surface passivation, or of deliberate growth of thin insulating oxide films in the production of, for example, electronic devices. Oxygen adsorption and surface oxide formation have therefore been a subject of intense surface research since many decades. In particular, the controlled growth of ultrathin insulating oxide films of uniform thickness is an important requirement for, for example, the ongoing miniaturization of electronic circuitry. Ultrathin single-crystalline, that is, epitaxial, oxide films are also grown as a support for metal clusters in order to model

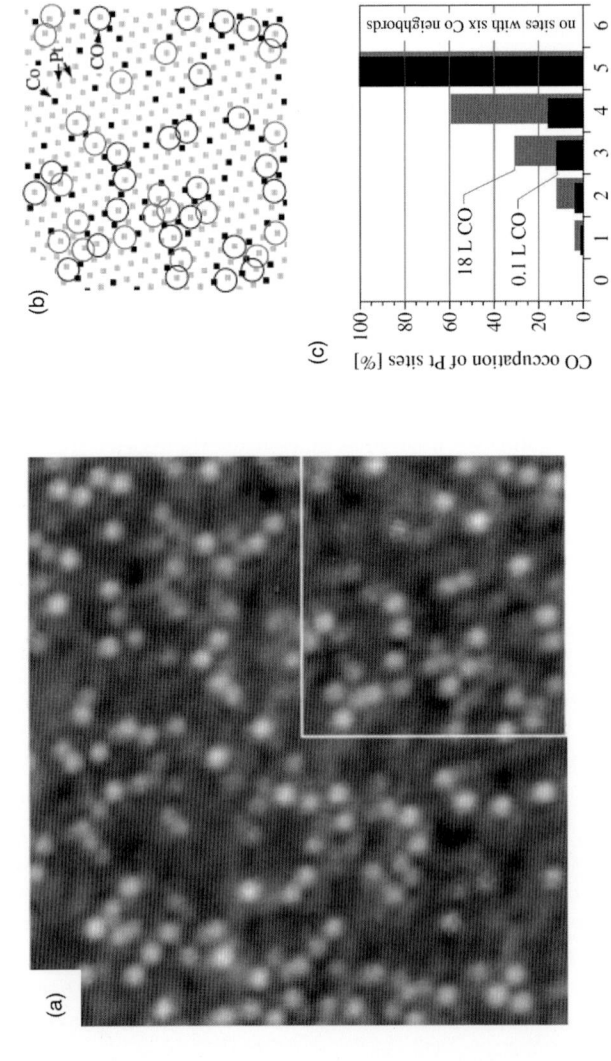

Figure 7.22 (a) STM image (100 Å × 100 Å) of a PtCo alloy surface covered with CO molecules (bright spots). (b) Identification of the CO adsorption sites with respect to the surface atom positions. (c) Histogram of the relative CO population of Pt on-top sites with 0, 1, 2, 3, 4, 5, or 6 nearest-neighbor Co atoms illustrating the so-called ligand effect. $U_B = -0.31$ V, $I_T = 0.68$ nA. (From Ref. [91].)

heterogeneous catalysts. A particularly interesting example is the growth of bilayer Al oxide films on a Ni$_3$Al(111) substrate, which was already discussed in Section 7.2.3 and shown in Figure 7.11. The formation of these Al oxide films and their characterization have extensively been studied by STM and AFM.

Adsorption on the highly reactive Ni$_3$Al(111) surface at 800 K results in the formation of triangular islands of same orientation and similar size, as seen in Figure 7.23a [96,97]. Zooming into these islands (panels (b) and (c)) discloses a structure that is similar to that found after oxygen adsorption on a bare Al(111) surface at 440 K. This suggests that, as a first step, adsorbing oxygen "assembles" the aluminum atoms within the Ni$_3$Al(111) surface to form an O–Al bilayer on the alloy substrate (panel (d)), because the interaction of oxygen with aluminum is much stronger than that with nickel as suggested by the heats of formation of their oxides (ΔH_f(NiO) = 240 kJ/mol; ΔH_f(Al$_2$O$_3$) = 1675 kJ/mol). The necessary convergence of the Al atoms in the *alloy* surface explains the higher activation temperature of 800 K compared to 440 K on the bare Al(111) surface in order to produce the similar structure.

Exposing the Ni$_3$Al(111) surface at the higher temperature of 1000 K to oxygen enables even segregation of aluminum atoms from the subsurface region and the formation of a closed oxygen-terminated aluminum O–Al–O–Al double-bilayer film on the alloy substrate as proven by angular resolved XPS

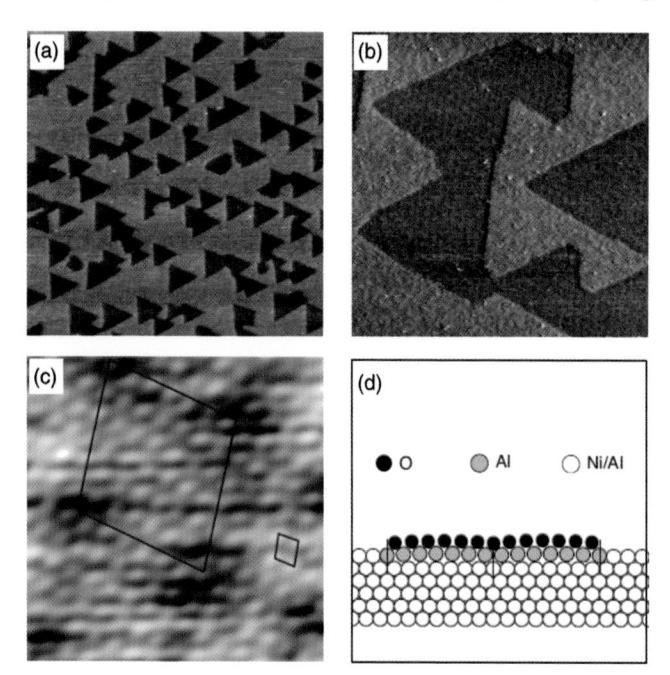

Figure 7.23 (a–c) Successively zooming-in STM images of a Ni$_3$Al(111) surface oxidized with a low dose of oxygen at 800 K. Image size: (a) 290 nm × 290 nm, U_B = 500 mV, I_T = 0.15 nA. (b) 60 nm × 60 nm, U_B = 545 mV, I_T = 0.15 nA. (c) 4 nm × 4 nm. (d) Model illustrating the oxygen-induced assembly of Al *surface* atoms. (From Ref. [96].)

measurements (Rosenhahn, A. and Fadley, C.S. unpublished results). This aluminum oxide film has extremely interesting structural and electronic properties that have exhaustively been studied with a large number of surface techniques, including STM and AFM [96–101]. Briefly, the aluminum oxide film possesses a Moiré-type superstructure (Figure 7.23c) that arises from the mismatch between the hexagonal lattices of the underlying (111) structure of the substrate and of the oxide film. On the long range, this superstructure exhibits two hexagonal sublattices as revealed by STM, the so-called net structure and the dot structure of 2.4 and 4.1 nm periodicity length, respectively, as shown in Figure 7.24. While the first (upper half) arises from a topographic modulation, the second (lower half) results from a respective modulation of the electronic properties at every $\sqrt{3}$ site of the net structure as verified by simply changing the bias voltage between the surface and STM tip as seen in the upper and lower half in Figure 7.24. Local STS measurements within the unit cell of the net structure (Figure 7.25a) reveal different electronic properties at all "dot"-positions compared to the rest of the oxide film. The STS-spectra in Figure 7.25b are taken at the corresponding points in panel (a). The peak at 3 eV above the Fermi level and just below the top of the bandgap of the oxide film is found exclusively at the "dot" positions and replaces the peak at 2.4 eV detected everywhere else [102,103]. DFT simulations suggested that the "dot positions" may possibly be just a tiny hole in the oxide film down to the metallic substrate [101]. It shall further be mentioned that both superstructures of these Al oxide films, the net and the dot structures, have proven to be excellent templates for the growth of ordered arrays of metal clusters [101,104].

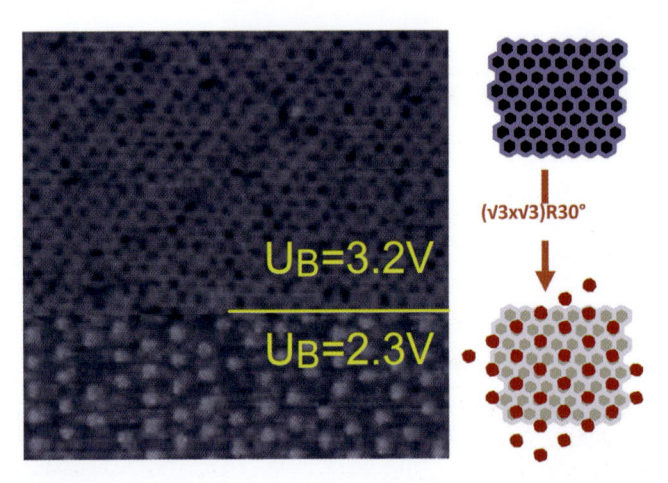

Figure 7.24 Topographic "net-" (upper half) and electronic "dot" (lower half) – structure within the Moirè-type superstructure of an Al oxide double-layer grown at 1000 K on a Ni$_3$Al (111) alloy surface; both structures can be imaged at the same surface area by just changing the bias voltage from 3.2 to 2.3 V. The "dots" appear at $\sqrt{3}$-positions of the "net"-structure. Image size: 55 nm × 55 nm. (From Ref. [96].)

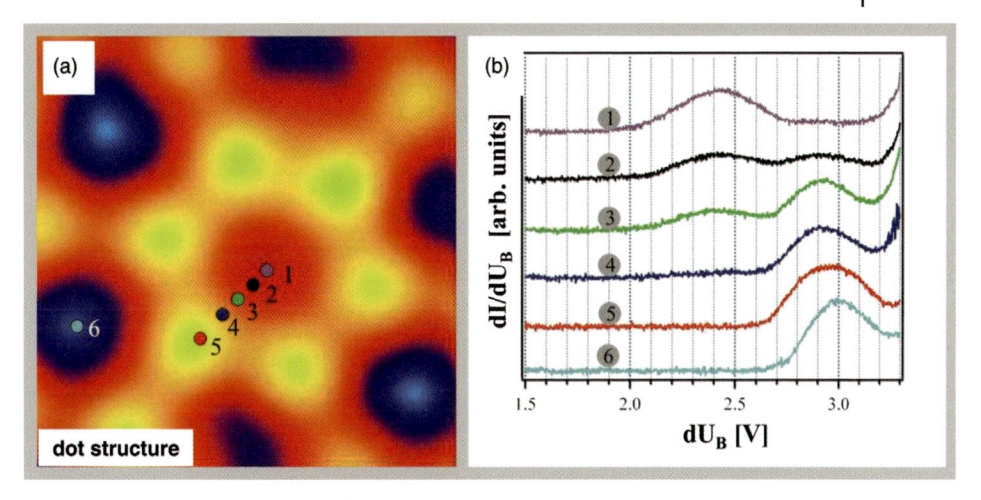

Figure 7.25 STS measurements at different points near a dot position (a) and the corresponding spectra (b). (From Ref. [103].)

Noncontact AFM studies have disclosed even further structural details within the Moiré coincidence mesh of these Al oxide films. As depicted in Figure 7.26a, very detailed analysis of the raw nc-AFM results shown in panel (a) reveals the presence of ensembles of hexagonal building blocks within the unit cell of the dot structure (panel (b)) [98].

7.5.1.4 Amorphous Structures

Even though glasses belong to the oldest man-made materials, by far not as much is known about their detailed structure as about crystalline solids. For a long time, their "random network" structure as proposed by Zachariasen [105] could only be described in terms of "pair–distance distributions" or "pair correlation functions" that provide information only about the probability of relative distances between their atomic or molecular building blocks, but not about directions. STM (applicable only in case of sufficient conductivity) and AFM, as real space imaging techniques, were, for a long time, *per se* considered unsuitable to resolve irregular atomic structures, until first c-AFM results about the random network structure of SiO_4 units in the surface of bulk quartz and barium–silicate surfaces were published by Raberg *et al.* [106–108]. In the meantime, the much improved nc-AFM technique has led to spectacular insights into the real space structure of the random network structure of thin vitreous silica films grown on a metal substrate. Figure 7.27a shows the schematic 2D random network according to Zachariasen [105] in which the SiO_4 units of a vitreous silica layer are connected via Si–O–Si bridges forming different angles and mesh sizes. Figure 7.27b and c shows an atomically resolved constant height nc-AFM($\Delta\nu$) and an STM(I_T) image, respectively, registered simultaneously on the same area of a thin vitreous silica film grown on a Ru(0001) substrate [109]. Both images are atomically resolved, but whereas the $\Delta\nu$ representation (b) shows the Si

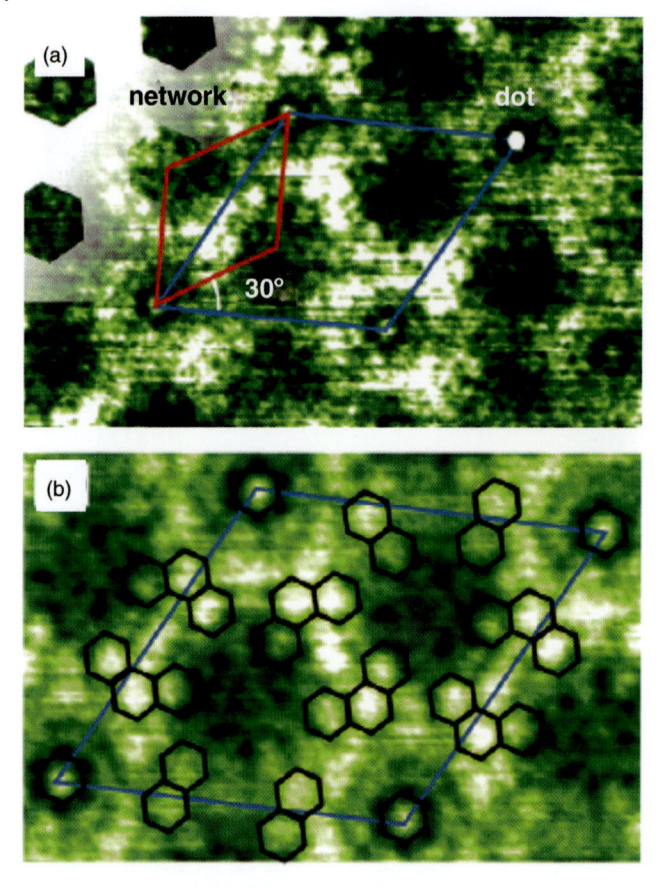

Figure 7.26 Noncontact AFM images of the Moiré structure of an Al oxide film on Ni$_3$Al(111). Image size: (a) 74 Å × 48 Å, (b) 48 Å × 31 Å. (From Ref. [98].)

positions, the I_T image (c) reveals the O positions of the topmost silica layer. This assignment results from a close inspection of the bright protrusions. In panel (b) or (c), four or three bright protrusions respectively form one "threefold star," and the length of the "rays" of these stars is longer or (shorter) in panel (b) or (c), respectively. This supports the notion that the protrusions in panel (b) correspond to Si atoms, while the bright protrusions in panel (c) represent O atoms. A corresponding atomic model of the imaged topmost layer of this silica film is superimposed onto the lower right corner of both images [109].

7.5.1.5 Adsorption

Dissociative Oxygen Adsorption
Any reaction of a solid surface starts with the adsorption of the reactant. Staying with the oxidation of surfaces, the first step is adsorption of molecular oxygen.

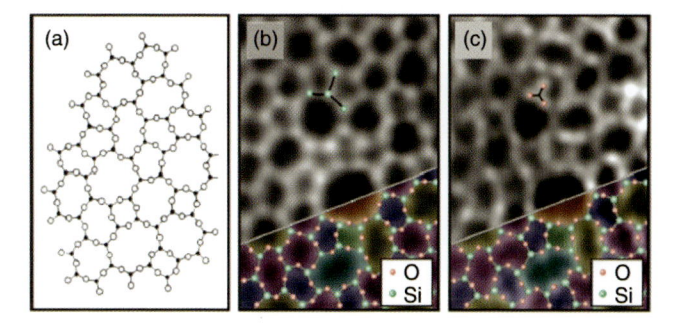

Figure 7.27 (a) Zachariasen's scheme of a random network (black dots: cations, white circles: anions). (b and c) Scanning probe images representing a single atomically resolved constant height measurement, where panel (b) shows the nc-AFM- and panel (c) shows the STM channel. Image size: 2.7 nm × 3.9 nm. An atomic model of the topmost layer of the silica film is superimposed onto the lower right corner of the images in panels (b) and (c) (green balls: Si atoms, red balls: O atoms).

Numerous experimental and theoretical studies have been made, for instance, of the initial oxidation of Si(111) surfaces that, as shown in Figures 7.1 and 7.28, are reconstructed to form a (7×7) superstructure. While the final stable oxidation product involves insertion of O *atoms* into Si—Si bonds (see previous section), a stable precursor state of *molecularly* adsorbed O_2 species has also been detected by its vibrational and photoelectron spectra [110–112]. Figure 7.28a shows an STM image of a Si(111)-(7×7) surface recorded after the exposure of a very small dose of 0.05 L O_2 (1 Langmuir (L) = 1×10^{-6} Torr·s) at room temperature [113]. The very bright dot (arrow) in the image is attributed to an inserted O atom (B-configuration in panel (d)). A further exposure to 0.03 L O_2 leads to a change at the same surface site (arrow) in panel (b), which the authors identify with an additionally bound long-lived *molecular* O_2 species (D-configuration in panel (d)), which however in a next step under the bombardment with electrons of 7 eV energy (due to a +7 V bias voltage at the sample; see also Section 7.5.1.7) disappears again and restores the B-configuration (arrow in panel (c)).

On most metal surfaces at room temperature, the lifetime of a molecular species is too short to show up in STM images. Impinging O_2 molecules dissociate spontaneously. On a Pt(111) surface, for instance, this happens already at 160 K. Figure 7.29 shows pairs of dark dots that correspond to pairs of oxygen *atoms* deriving from one parent O_2 molecule. Each oxygen atom sits, not surprisingly, within a threefold hollow site on this hexagonally close-packed surface as seen in panel (a). Interesting, however, are the distinctly different atomic distances within the pairs (panel (b)). These distances correspond to $1a_{Pt}$, $\sqrt{3}a_{Pt}$, $2a_{Pt}$, $\sqrt{7}a_{Pt}$, and $3a_{Pt}$, a_{Pt} being the Pt–Pt nearest-neighbor distance in the Pt(111) surface, as illustrated in panel (c). From a very careful analysis of the frequency of these distances over a large number of atom pairs, it is concluded that at these low temperatures the two separated atoms reach their final positions by nonthermal transient motion. The thermal equilibration with a separation of $2a_{Pt}$

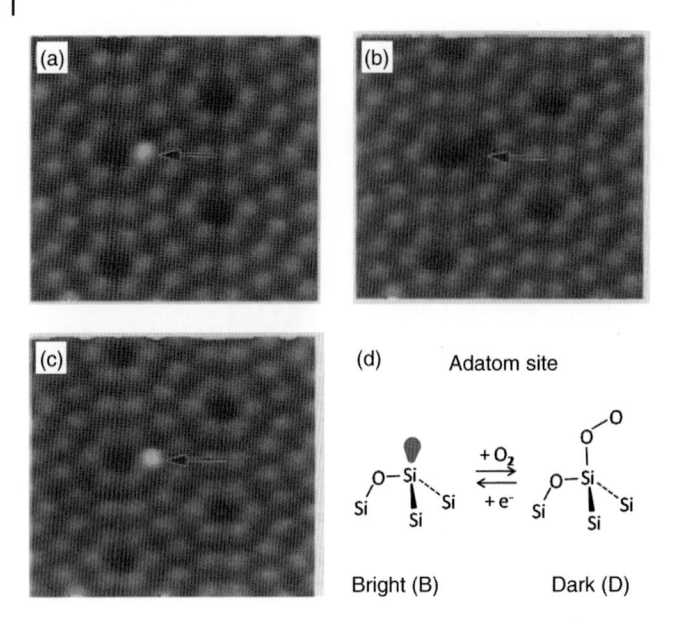

Figure 7.28 (a–c) Very initial steps of Si(111) oxidation (see text). (d) Models representing the chemical structure in panels (a–c), that is, the bright dot in panels (a) and (c), and the dark vacancy (arrow) in panel (b). Image size: 61 Å × 68.5 Å.

sets in around 200 K [114]. The resultant O atoms may (and eventually will) not only stay *on* the surface but may be activated to penetrate below the surface.

Subsurface Atoms/Impurities

Figure 7.30 shows a patchy Pd(111) surface. Each bright patch in the image, overlaid by the hexagonal mesh of the surface Pd atoms (small dots), is attributed to a subsurface carbon atom (below the first layer). Its existence affects the

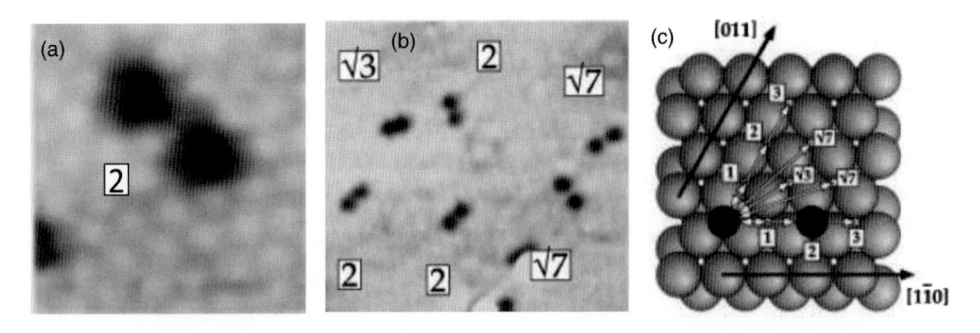

Figure 7.29 (a and b) STM images of oxygen atoms (black dots) on a Pt(111) surface after dissociation of impinging O_2 molecules at 160 K. Image size: (a) 1.9 nm × 1.9 nm. (b) 11.0 nm × 9.2 nm. The numbers in panels (a) and (b) give O–O distances in terms of Pt–Pt nearest-neighbor distances as illustrated in panel (c).

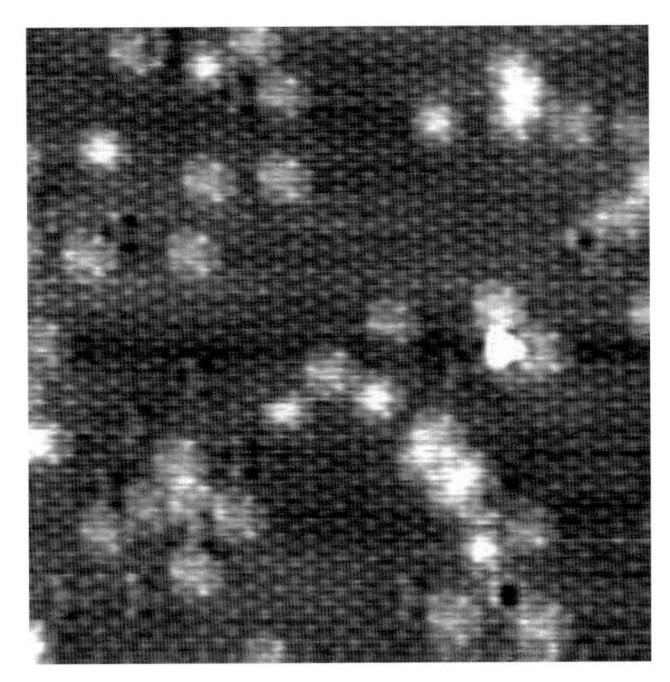

Figure 7.30 STM image of subsurface carbon atoms at a Pd(111) surface. Image size: 100 Å × 100 Å. (Courtesy from M. Salmeron.)

electronic structure of a number of outer Pd atoms that manifests itself in the "halo" around each carbon atom. The image in Figure 7.30 is actually taken from a movie of STM images that shows even the hopping of the C atoms between interstitial subsurface sites [115].

Molecular Adsorbates

Adsorption of organic molecules is a fundamental process in, for example, catalysis and modern materials design. Heterogeneous catalysis facilitates specific reactions and enables to steer their selectivity. Modern materials design relies more and more on low-dimensional structures of self-organized organic molecules on solid surfaces. In both cases, a detailed picture of the interactions among the adsorbed molecules as well as the adsorbates and the substrate is the basis for an understanding of their induced reactivity or structural arrangement. Here STM and AFM provide unique insights even on the submolecular level.

As an example, Figure 7.31a shows local inelastic tunneling spectra (IETS) of normal (C_2H_2) and deuterated (C_2D_2) acetylene molecules (see also Figure 7.4) adsorbed on a Cu(100) surface [116]. The tunneling electrons suffer energy losses of 358 and 266 mV due to the excitation of C–H or C–D stretching vibrations, respectively (spectra 1 and 2). These values are in close agreement with results obtained with other vibrational spectroscopies (HREELS, IR) and their difference reflects the mass ratio of 2 of the two hydrogen isotopes on the

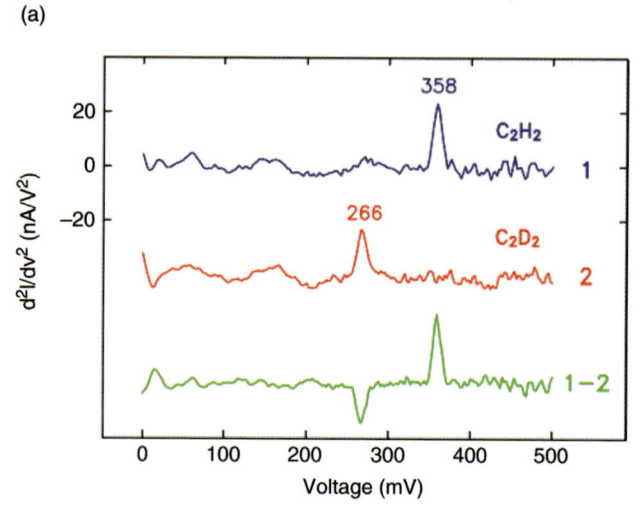

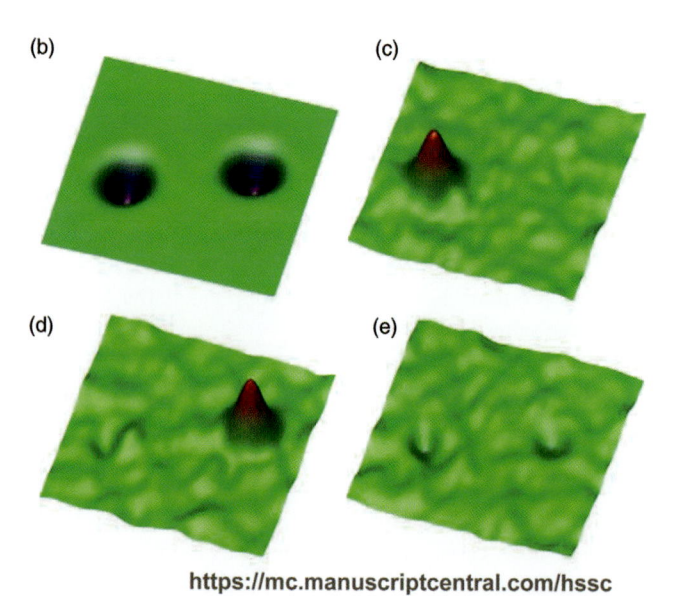

Figure 7.31 (a) Background subtracted d^2I/dV^2 tunneling spectra of C_2H_2 (1) and C_2D_2 (2) adsorbed on Cu(100) and difference spectrum (1–2) showing energy loss peaks at 358 and 266 mV, respectively. (b) Regular constant current STM image of a C_2H_2 (*left*) and a C_2D_2 molecule (*right*) (48 × 48 Å^2); d^2I/dV^2 images of the same area recorded at (c) 358 mV, (d) 266 mV, and (e) 311 mV. (From Ref. [116].)

resonance frequency $\nu = 1/2\pi(\kappa/m)^{1/2}$. This ability to discriminate spectroscopically between molecules with STM opens the possibility of chemically sensitive microscopy. Figure 7.31b first shows a regular constant current STM image of a

C$_2$H$_2$ (*left*) and a C$_2$D$_2$ molecule (*right*) and, as expected, shows no contrast between both acetylene species. However, during scanning at each surface point, the feedback loop is turned off and a bias voltage modulation turned on in order to record dI/dV and d^2I/dV^2, the inelastic contributions to the tunneling current can be accentuated, as illustrated in Figure 7.4. In this way panel (c) in Figure 7.31 was obtained at a bias voltage of 358 mV showing only the contribution of the C$_2$H$_2$ molecule, while panel (d) was generated by setting a bias voltage of 266 mV showing only C$_2$D$_2$-related intensity. Obviously, this single-molecule vibrational spectroscopy adds enormous power to STM. The round appearance of the molecular images is attributed to the rotation of the molecules on their site between two equivalent orientations on this quadratic Cu(100) surface during the measurement.

The bonding of a molecule to a metal surface, of course, involves coupling of the molecular orbitals with electronic states of the substrate. The image in panel (e) of Figure 7.31 was registered at a bias voltage of 311 mV, that is, between the resonance energies of C–H and C–D vibrational excitation. The shallow dips in the image of panel (e) are taken as indication for this adsorbate-induced modification of the *substrate* electronic density of states at the sites of the two molecules.

However, not only the local electronic properties of the substrate are affected, but also the orbitals of the adsorbed molecule are modified, by both direct coupling to the electronic states of the substrate and the mutual electronic coupling of different *molecular* states *through the metal surface*. As a result, experimental images of the adsorbate orbitals may show hardly any resemblance with the genuine orbitals of the free molecule. A sensible assignment of measured "orbital images" is then only possible on the basis of involved calculations of the full adsorbate complex molecule + surface, or by comparing experimental images of the adsorbate with and without the aforementioned molecule–substrate interactions. The latter can, in fact, be achieved by decoupling the adsorbate from the metal states using an insulating spacer layer. The lower set of six panels in Figure 7.32 shows *experimental* STM images of a pentacene molecule (panel (a)) adsorbed on a bilayer NaCl film on Cu(111) registered with a metal (W) tip that is clean (b–d) or modified by a pentacene molecule (e–g). While the STM images (c and f) taken with bias voltages in the HOMO–LUMO gap are relatively featureless, the images taken with bias voltages above the HOMO (b and e) or LUMO (d and g) show pronounced structures that resemble very much the electron densities of the HOMO and LUMO orbitals calculated with DFT for the free molecule [117].

Although very impressive, the images in Figure 7.32 do not yet permit to conclude on the constitutional formula of the molecule. This, however, is possible with nc-AFM. Using a tuning fork AFM, Gross *et al.* [118] obtained the image of a pentacene molecule adsorbed directly on a Cu(111) surface shown in Figure 7.33. The five fused benzene rings, the positions of the carbon atoms, and the outward pointing hydrogen atoms are clearly seen. This high resolution was obtained with a metal tip "decorated" with a CO molecule. This CO molecule is

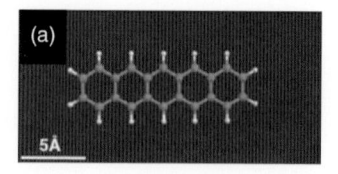

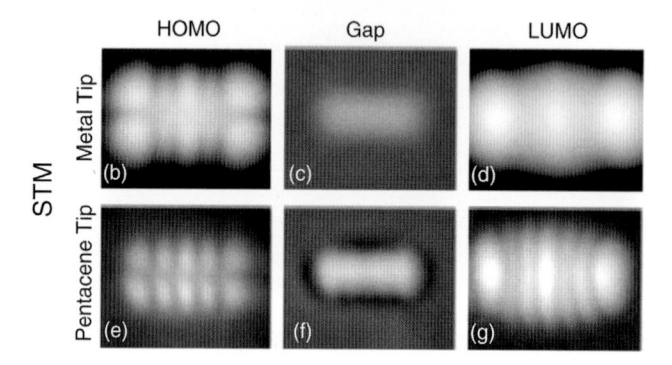

Figure 7.32 (a) Structure model of a pentacene molecule. (b–g) STM images of a pentacene molecule adsorbed on a bilayer NaCl film on Cu(111) registered with a metal (W) tip that is clean (b–d) or modified by a pentacene molecule (e–g) taken with bias voltages in the HOMO–LUMO gap (c and f), above the HOMO (b and e) or LUMO (d and g). (From Ref. [117].)

bound via the C atom on the tip apex. The outward pointing O atom is smaller, is repelled at shorter distances during tip approach, and therefore penetrates deeper into the C-rings of the adsorbed pentacene molecule than a bare metal apex atom.

With STM, it is also possible to follow the dynamics of individual molecules. The round shape of the images of C_2H_2 and C_2D_2 in Figure 7.31 was attributed to rotations that these molecules execute while staying at their adsorption site. On the other hand, Figure 7.34 shows STM images of a copper phthalocyanine (CuPc) molecule on a Ag(100) surface that during imaging performs site

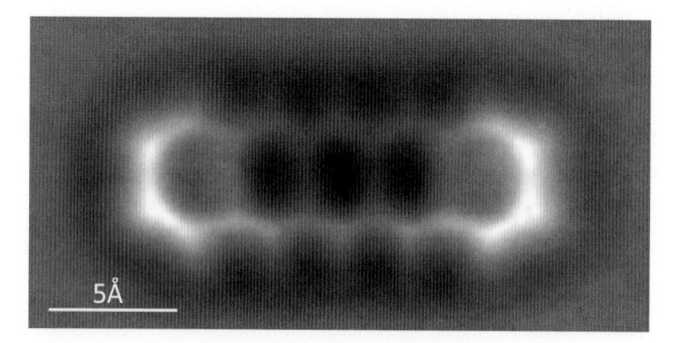

Figure 7.33 Constant height AFM image of a pentacene molecule adsorbed on a bilayer NaCl film on Cu(111) acquired with a CO-modified tip. (From Ref. [118].)

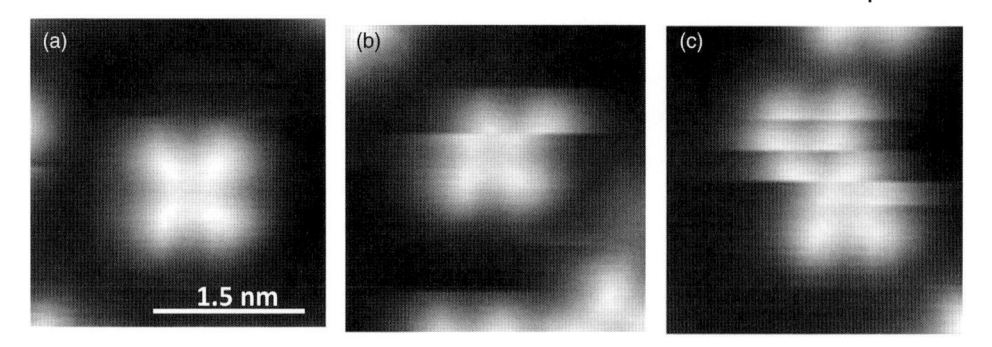

Figure 7.34 (a–c) STM images of a copper phthalocyanine (CuPc) molecule on a Ag(100) surface that during imaging performs site exchanges. (From Ref. [119].)

exchanges [119]. Panel (a) exhibits the stationary molecule at low temperatures. At increased temperatures, the molecule is seen to make one (b) or four jumps (c) during the scan time. The distribution of such molecular displacements enables to retrieve quantitative information about the migration process such as the activation energy for site exchange.

7.5.1.6 Atomic Manipulation

An STM tip can also be used as a tool to manipulate single atoms or molecules on surfaces, for example, by moving them across a surface or by breaking and reestablishing bonds. A well-known example is the "quantum corral" of Fe atoms on a Cu(111) surface, whose stepwise construction is illustrated in Figure 7.35 [120]. The starting point is the Cu(111) surface covered with a low concentration of Fe atoms at liquid helium temperature in order to suppress spontaneous diffusion. Individual atoms are then picked up and transferred by the STM tip, and redeposited at the desired surface site. The final configuration in Figure 7.35d shows the completed ring of Fe atoms, as well as a characteristic pattern of standing electron waves, a direct visualization of the quantum nature of a two-dimensional electron gas. The Fe atoms act as scattering centers for the electron wave functions within the Cu(111) surface. The interference pattern depends on the size and the shape of the enclosed region, and the mere fact that it exists only inside the corral indicates different (size-dependent) properties of the Cu surface inside and outside the ring. Adsorbed molecules like CO may also be pushed (pulled) across the surface by means of repulsive (attractive) interactions between the tip and the molecule, without actually leaving the surface. The involved activation energies to lift, push, or pull a particle may also be measured [121].

7.5.1.7 Tip-Induced Surface Reaction

The ability to use the STM as tool to manipulate atoms and molecules at surfaces (see previous section) has also been used to deliberately induce chemical reactions between individual adsorbates. One example, namely, an Ullmann

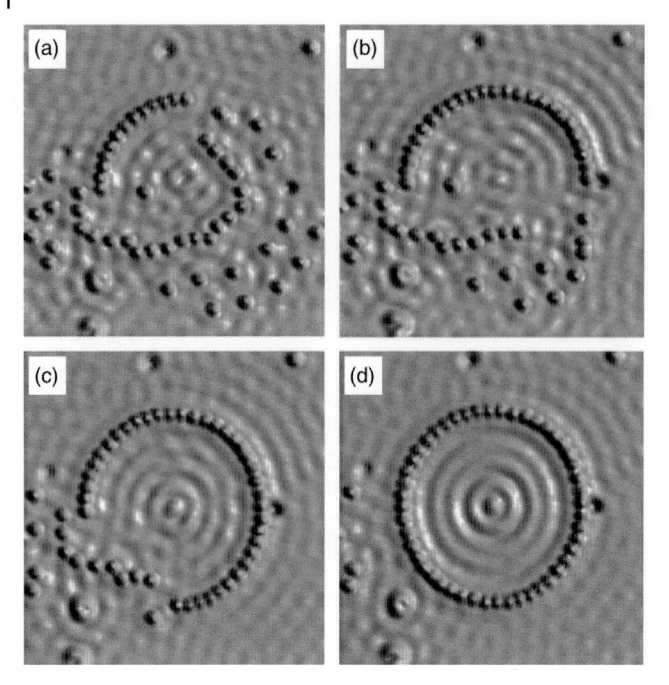

Figure 7.35 Formation of a "quantum corral" by atomic manipulation of Fe atoms on a Cu(111) surface. The final ring of 48 Fe atoms in panel (d) has a diameter of 71.3 Å. (From Ref. [120].)

reaction between two iodobenzene molecules (C_6H_5I)

$$2\,\text{\Large\textcircled{}}\text{-I} + 2\,\text{Cu} = \text{\Large\textcircled{}}\text{-\Large\textcircled{}} + 2\text{CuI} \tag{7.15}$$

is displayed in Figure 7.36 [122]. Two C_6H_5I molecules are adsorbed at a step edge of a Cu(111) surface (panel (a)). By positioning the STM tip successively above these two molecules and applying shortly a bias voltage of 1.5 V leads to electron-induced dissociation of these molecules into I atoms and phenyl radicals (C_6H_5) (panels (b) and (c)). Removing the I atoms and pushing the two C_6H_5 species together leads to coupling between them and formation of a biphenyl (panels (d–f)). The stable bond formation between the two radicals was shown by *pulling* the newly created species along the step edge, which proved their firm coupling (panels (g) and (h)).

7.5.2
Scanning Tunneling Microscopy at High Gas Pressure

The experimental results on gas–solid interactions in UHV, albeit their high scientific value, are always questioned concerning their transferability to real and industrial conditions ("pressure- " and "materials gap"). An important trend in

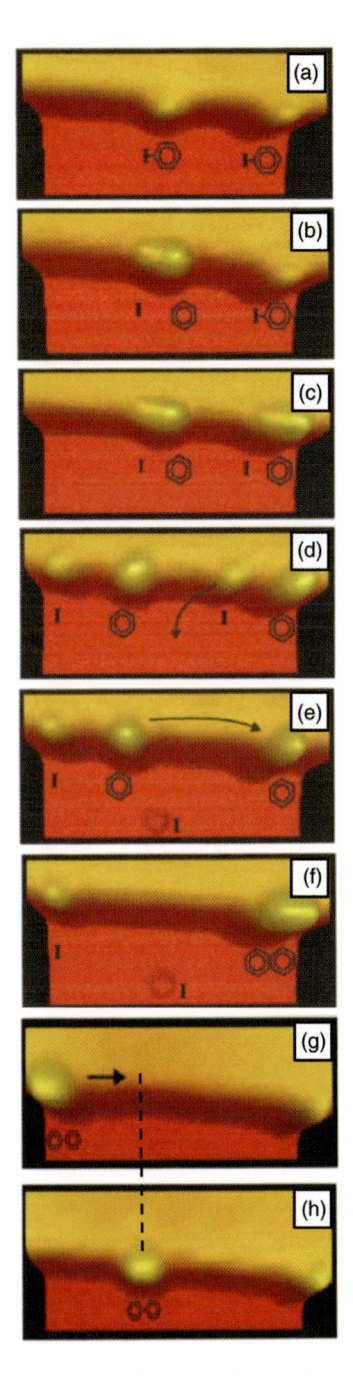

Figure 7.36 STM images showing (a) two adsorbed iodobenzene molecules at a step of a Cu(111) surface. (b and c) The tip-induced dissociation of both molecules. (d) The removal of the iodine atom from both molecules. (e) The tip-induced approach of the two phenyl radicals. (f) The association of the two phenyl radicals. (g and h) The stable bond between the two coupled phenyl radicals. Image size: 7.0 nm × 3.0 nm. (From Ref. [123].)

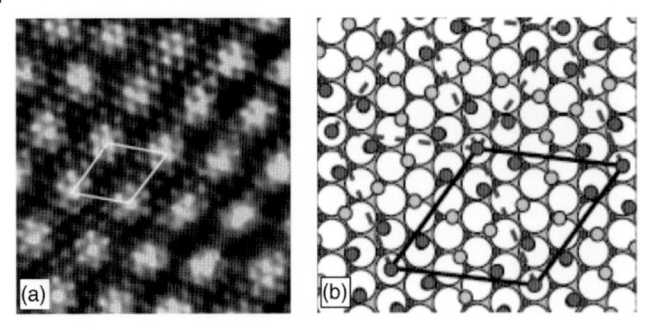

Figure 7.37 (a) High-resolution STM image (55 Å × 51 Å) of a CO overlayer on Pt(111) obtained at 1 bar CO and room temperature. (b) Structure model with unit cell: Molecules near on-top (near hollow) positions are imaged brighter (darker) in panel (a). (From Ref. [122].)

surface and interface science therefore is the development of methods that enable to "bridge these gaps." It came as a kind of surprise that soon after its realization STM was shown to work not only in UHV but also under real ambient conditions like gas pressures [27–35] or liquids [36–40]. Here we present one example of high-pressure STM and then concentrate on the more extreme case of STM in liquids in the following sections.

Figure 7.37a shows an STM image of a Pt(111) surface in the presence of 1 bar CO at room temperature [124]. The image exhibits clearly the individual CO molecules (small bright dots) and a Moiré-type superstructure. A model of the layer of adsorbed CO molecules is shown in panel (b). Molecules adsorbed at and near Pt on-top positions are imaged brighter than those in and near hollow sites. Both the structure and the coverage of 0.68 ML (monolayer) CO are identical to the structure obtained for CO adsorption at low temperature (170 K) in UHV. Thus, raising the pressure at room temperature leads to the same result as low-temperature adsorption in UHV, as long as thermodynamic equilibrium is maintained [124].

7.5.3
Scanning Tunneling Microscopy in Solution

STM in liquids, in particular also in conducting electrolytes, can nowadays be done with nearly the same precision as in UHV. Atomic scale investigations of metal–electrolyte interfaces are to provide insight into basically the same properties and processes as in UHV, namely, surface morphology and structure, adsorption and reaction of anions (corrosion), electrodeposition (plating), molecular self-assembly, electrocatalysis, and so on. In particular, the *in situ* combination of cyclic voltammetry (CV) and STM has proven to be a powerful approach to combine thermodynamic and kinetic studies on interfacial charge transfer processes with detailed structural information from the electrode surfaces.

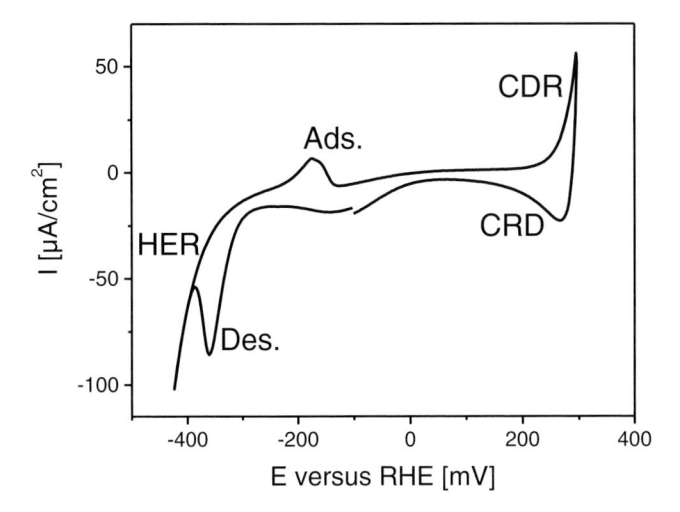

Figure 7.38 Cyclic voltammogram of Cu(111) in 10 mM HCl solution.

7.5.3.1 Surface Structure

The components of an aqueous electrolyte are water, anions, and cations. Water interacts weakly with (noble or coinage) metals, and anions and cations can be prevented from interaction with the electrode surface by applying appropriate electric potentials to the (working) electrode. These potentials can be determined by cyclic voltammetry. Figure 7.38 shows a cyclic voltammogram of a Cu(111) electrode surface in dilute HCl solution [125]. The cyclic voltammogram is limited by the hydrogen evolution reaction (HER) at negative (cathodic) potentials (decomposition of the solvent) and the copper dissolution reaction (CDR) at positive (anodic) potentials (electrode dissolution), both accompanied by a corresponding cathodic and anodic Faraday current. The anodic peak at $-177\,\text{mV}$ versus RHE and the cathodic peak at $-360\,\text{mV}$ versus RHE arise from Cl anion adsorption and desorption, respectively. This indicates that at negative potentials below the desorption peak, the surface is free of adsorbate. In this potential range, *in situ* STM images, indeed, reveal clearly the hexagonal, quadratic, and rectangular surface lattice symmetry of the bare and unreconstructed Cu(111), Cu(100), and Cu(110) surface with atomic resolution (Figure 7.39). In turn, Au and Pt single-crystal surfaces in aqueous solution show the same reconstructed superstructures as they are well known from UHV (Figure 7.40). As discussed in Section 7.5.1.1, the reconstruction is accompanied by an increase in surface atom density that upon lifting causes the appearance of the small islands seen in Figure 7.40c [126]. Unlike in UHV, however, in solution the structural transition between reconstructed and unreconstructed phase is, potential-driven, fully reversible, a feature unique to the electrode–electrolyte interface.

| Cu(111) | Cu(100) | Cu(110) |

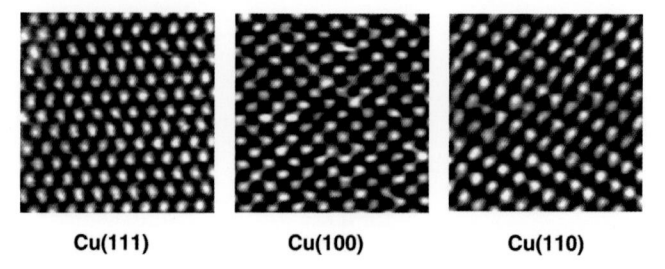

Figure 7.39 STM images of an *adsorbate-free* Cu(111)-, Cu(100)-, and Cu(110) electrode surface acquired in solution. Image size: Cu(111) 26.4 Å × 26.4 Å; Cu(100) 30 Å × 30 Å; Cu(110) 30 Å × 30 Å; typical tunneling parameters: $U_B = 5$ mV, $I_T = 1$ nA.

7.5.3.2 Electrochemical Double-Layer

A basic concept of the metal–electrolyte interface is the model of the electrochemical double-layer [127]. Ions adsorb on the electrode surface, either strongly in direct contact with the surface losing at least part of their hydration sphere (called "specific adsorption") and thereby forming the "inner Helmholtz plane"

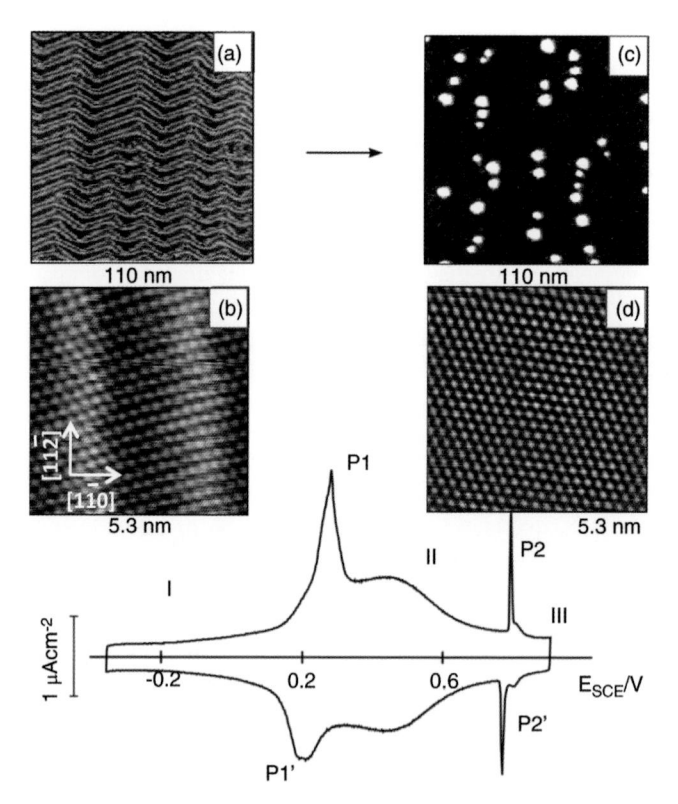

Figure 7.40 Cyclic voltammogram of a Au(111) electrode surface in sulfuric acid solution and *in situ* STM images of the reconstructed (a and b) and unreconstructed (c and d) surface at negative and positive potentials, respectively. (From Ref. [125].)

or weakly with intact hydration sphere forming the "outer Helmholtz plane." In particular, anions that are larger than cations tend to give off their hydration sphere more easily (their hydration energy is lower) and adsorb specifically. It is therefore important to gain insight into the influence of these strongly binding anions on the processes of interest. EC-STM provides insight into the structure of the adsorbed anion layer, for example, as a function of electrode potential.

7.5.3.3 Anion Adsorption on Cu(111)

According to Figure 7.38, chloride is adsorbed on the Cu(111) electrode at potentials more positive than $-150\,\mathrm{mV}$. *In situ* STM reveals a highly ordered ($\sqrt{3} \times \sqrt{3}$) R30°-Cl superstructure parallel to the surface, as depicted in Figure 7.41 [125]. In order to gain insight into the structural relation between adsorbate and substrate, that is, to determine the absolute adsorption site of the Cl anions, one might think of taking an image of the adsorbate-covered surface, change the potential to a regime negative of the desorption peak, and then take another image of the adsorbate-free surface and overlay the two images to correlate both structures. The problem with this approach, however, is that while the periodicity of both lattices can be compared with satisfactory accuracy, the exact relative atom positions can deviate due to thermal drift between the two images. This problem can be circumvented by making use of the spectroscopic capability of the STM. If both the adsorbate and the substrate have a density of states near the Fermi level, they can both contribute to the total STM signal. Thus, by varying the bias voltage, it is possible to obtain an STM image that is either dominated by substrate states or adsorbate states (Figure 7.41a). If, however, the bias voltage is chosen such that *both* layers contribute to a *composite* image, it is possible to separate their different lattice periodicity by Fourier transformation (see Figure 7.41a). Since the separated images originate from the same scan with exactly the same tip position during the scan, their superposition unambiguously shows the relative atom positions of adsorbate and substrate. In the case of Cl/Cu(111), this approach confirms the expectation that the chloride ions are adsorbed in threefold hollow sites on the Cu (111) substrate (Figure 7.41b), like in UHV [128–130].

Iodide also forms an ordered monolayer on the Cu(111) surface. But this layer is commensurate only at very negative potentials near the onset of the hydrogen evolution reaction where it forms also a ($\sqrt{3} \times \sqrt{3}$)R30°-I overlayer (Figure 7.42a). Upon increase of the potential, the iodide layer starts to be compressed in one direction and forms a layer that is incommensurate in the compressed direction but still in registry in the orthogonal direction (*electrocompressibility*). As a consequence, the STM images start to show a one-dimensional long-range height modulation with a wavelength l that decreases as a function of the applied electrode potential (Figure 7.42b and c) [131]. The reduced symmetry of the compressed adlayer is, of course, accompanied by the appearance of three rotational domains on the Cu(111) surface that are indeed observed in the STM images at rotation angles of 120°. At even more positive potentials, the growth of an epitaxial CuI compound films has also been followed with atomic resolution [132].

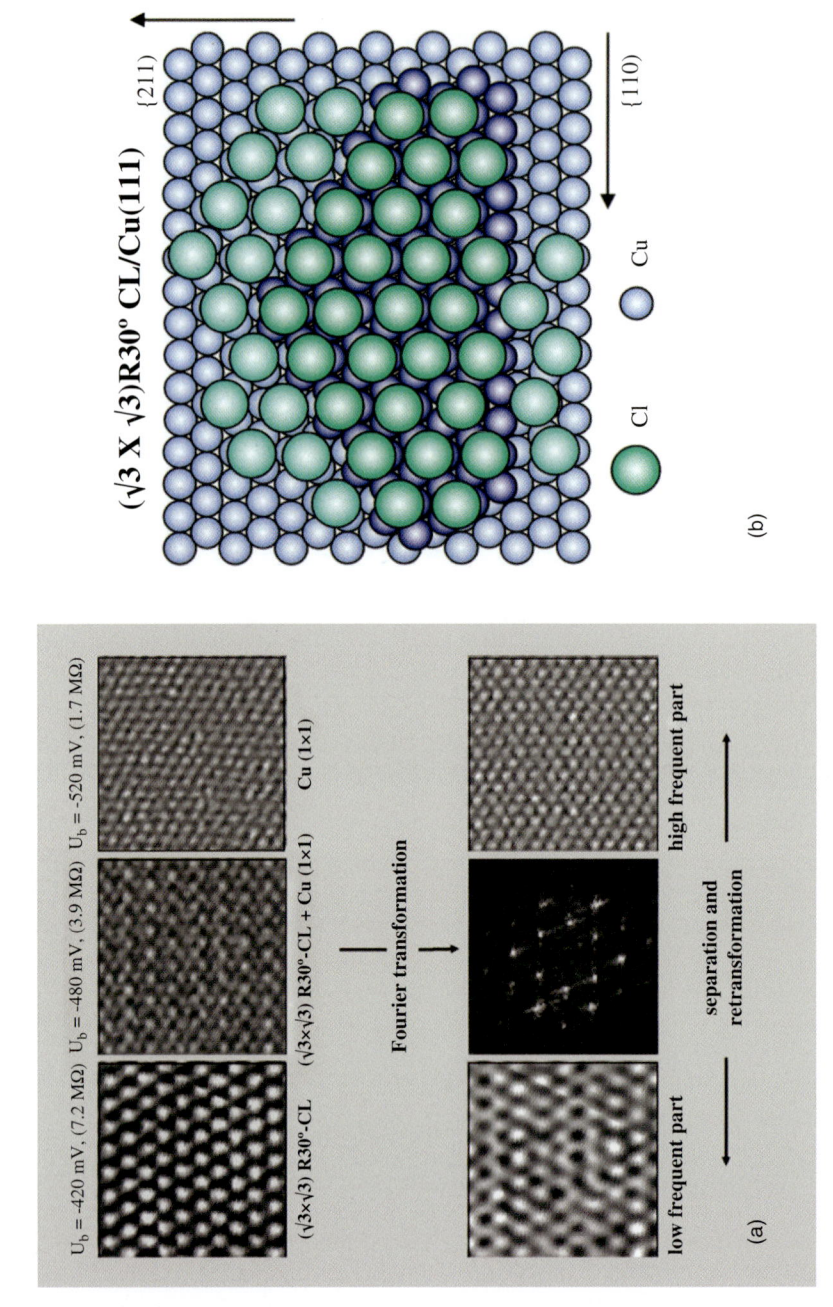

Figure 7.41 (a) STM images of a Cu(111) electrode surface in 10 mM HCl solution taken at different bias voltage U_b. (b) Structure model obtained from a Fourier transformation of the images (see text). Scanned area 38 Å × 38 Å. (From Ref. [124].)

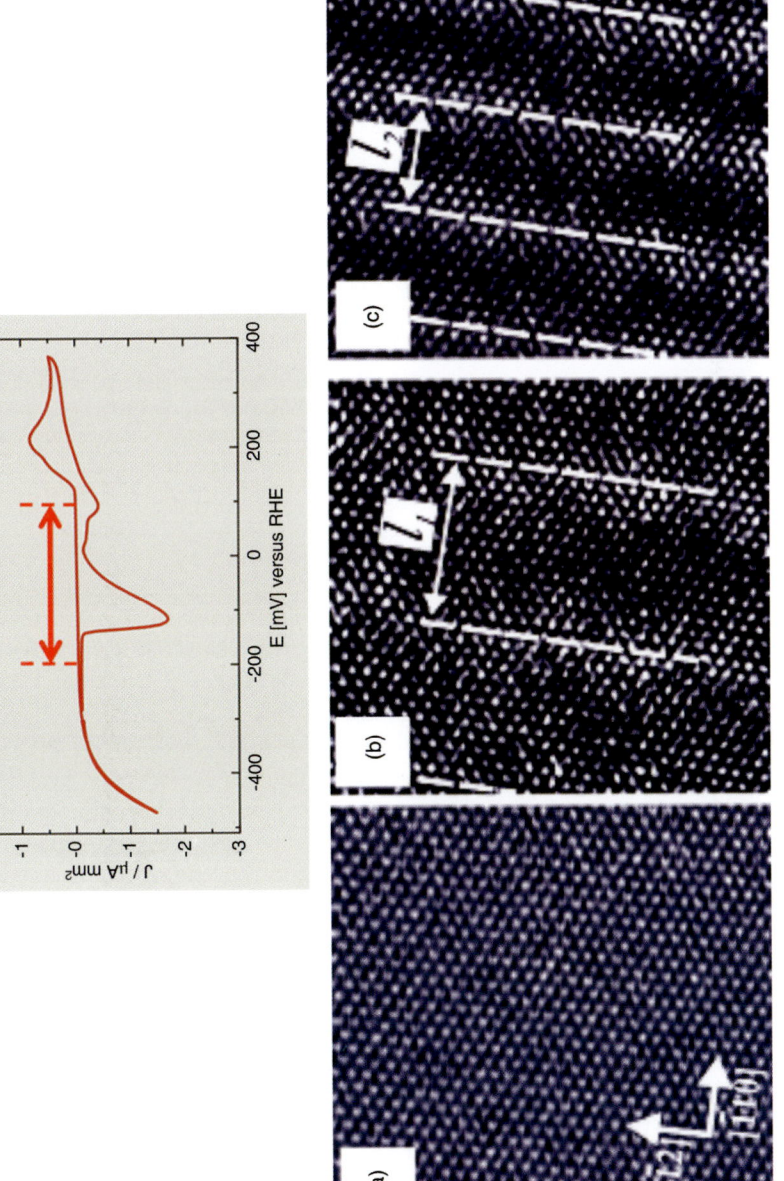

Figure 7.42 Cyclic voltammogram of Cu(111) in an iodide-containing H_2SO_4 solution and *in situ* STM images showing the potential-dependent electrocompression of the adsorbed iodide layer. Image size of all three panels: 13.6 nm × 13.6 nm.

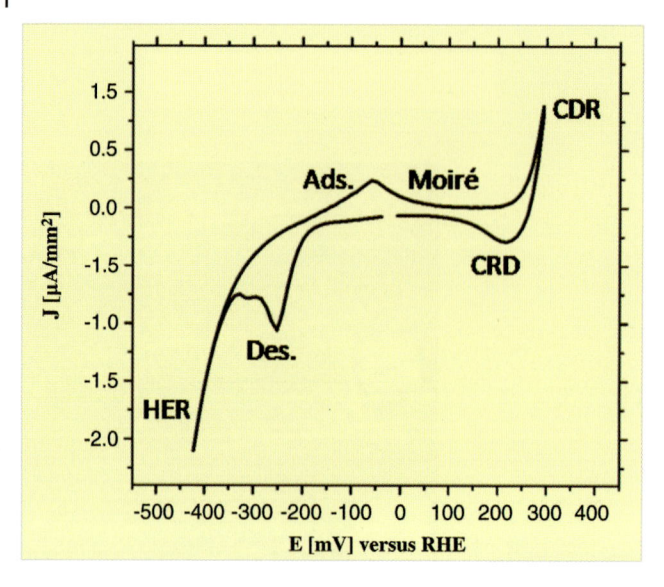

Figure 7.43 Cyclic voltammogram of Cu(111) in 10 mM sulfuric acid solution. HER = hydrogen evolution reaction; CDR = copper dissolution reaction; CRD = copper redeposition; Des./Ads. = desorption and adsorption of SO_4^{2-} anions; Moirè = potential regime of SO_4^{2-} induced Moiré-type superstructure formation.

The Cu(111) surface in contact with dilute sulfuric acid represents a particular interesting case. The cyclic voltammogram for this system (Figure 7.43) exhibits a particularly large hysteresis between the SO_4^{2-} adsorption (at −55 mV) and SO_4^{2-} desorption peak (at −250 mV versus RHE) hinting to a strong kinetic hindrance of either one or both of these processes. Structural investigations by *in situ* STM provide clear explanation for this hysteresis. Figure 7.44 shows a series of STM images taken very near to the SO_4^{2-} adsorption peak [42,133]. Starting from a step edge, the formation of a hexagonal superstructure is visible, which progressively expands onto the upper terrace (see arrows 1–3 in Figure 7.44a and b). This structure grows relatively slowly and is accompanied by the appearance of small islands that are covered by the same structure (Figure 7.44d, arrow 4). In panel (d) only small disordered, but SO_4^{2+} covered regions are visible (arrows 5 and 6). The periodicity length of the (nearly) hexagonal superstructure is about 3 nm, and an atomic scale close-up is shown in Figure 7.45 [133]. This latter high-resolution image reveals three structural elements: (i) the big bright dots correspond to a $(\sqrt{3} \times \sqrt{7})R30°$-like structure of the adsorbed SO_4^{2-} anions whose mutual distances, however, are slightly larger than $\sqrt{3}$-times and $\sqrt{7}$-times the Cu–Cu distance of the unreconstructed Cu (111) surface; (ii) the zigzag rows of smaller spots between the rows of SO_4^{2-} anions that are assigned to H_2O or H_3O^+ species forming a stabilizing 2D analog of the SO_4^{2-} hydration sphere, and (iii) the hexagonal arrangement of dark

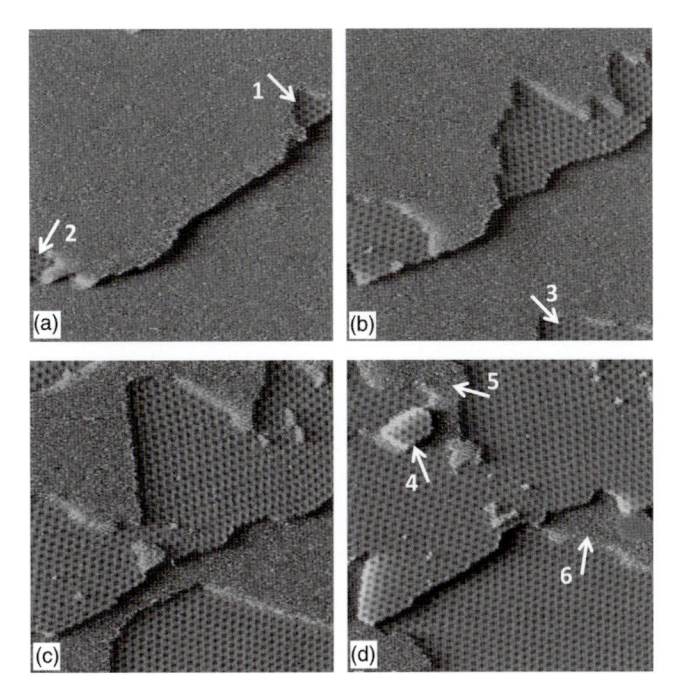

Figure 7.44 Growth of the sulfate-induced Moiré structure on Cu(111) in sulfuric acid solution (see text). Image size: 101 nm × 101 nm. (From Ref. [50].)

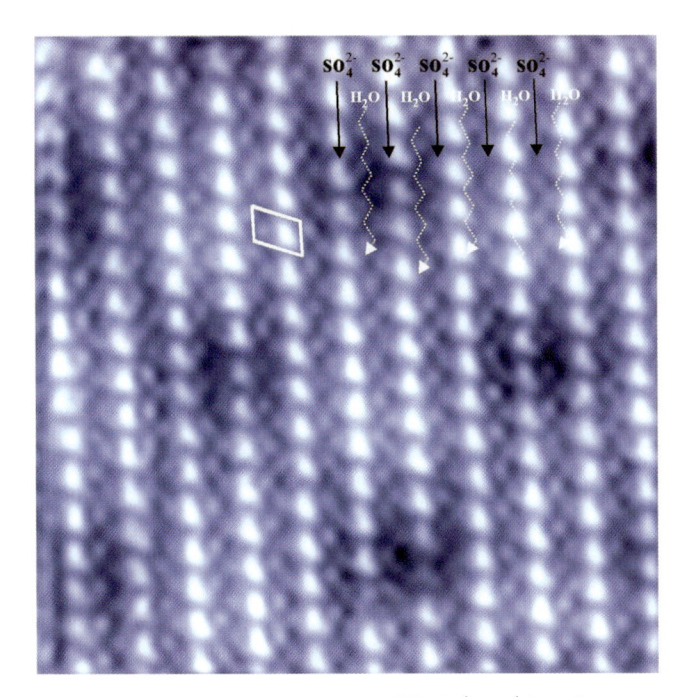

Figure 7.45 High-resolution STM image of the ($\sqrt{3} \times \sqrt{7}$) coadsorption structure of SO_4 anions (big bright dots) and water molecules (zigzag rows of small bright dots), superimposed by the Moiré structure (darker regions). Image size: 7.6 nm × 7.6 nm.

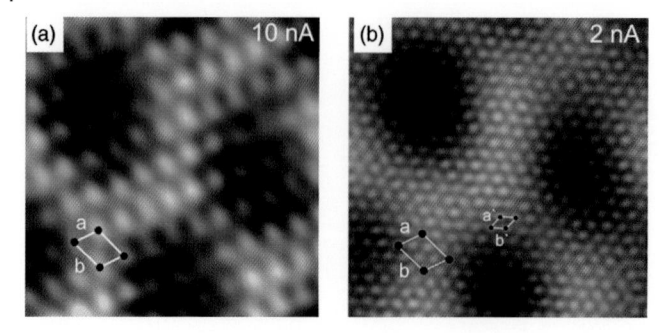

Figure 7.46 State-selective *in situ* STM images of the sulfate-covered Cu(111) electrode surface acquired at different tunneling currents I_T. (a) Dominant contribution from the sulfate overlayer ($U_B = 2$ mV, $I_T = 10$ nA); (b) dominant contribution from Cu substrate states ($U_B = 2$ mV, $I_T = 2$ nA).

regions that cause the 3 nm superstructure seen to grow in Figure 7.44. Applying again the spectroscopic STM mode explained in Figure 7.41, it can unambiguously be shown that the first Cu atom layer below the SO_4^{2-} anion layer is expanded by ≈6% compared to the bare Cu(111) surface (Figure 7.46). Thanks to the very high resolution of the images in Figures 7.44–7.46 [133], all these observations can consistently be explained by a SO_4^{2-} adsorption-induced reconstruction process in which the Cu(111) surface layer is expanded by ≈6%. The mismatch between the expanded first and the normal second layer of the Cu substrate causes the long-range 3 nm Moiré-type superstructure. The expansion is also accompanied by the displacement of Cu atoms that assemble into the extra islands on top (arrows in Figure 7.44d) again covered by the Moiré structure, and explains the strong hysteresis in the CV (Figure 7.43) as a consequence of the transport of Cu atoms out of the first layer and the formation of the extra islands again agglomerated into full Moiré units. The slow reconstruction kinetics becomes also apparent when changing the potential scan rate. After a potential jump from −350 to +100 mV versus RHE, the image shows a rather irregular structure with only a few ordered Moiré patches [133].

7.5.3.4 Underpotential Metal Deposition

A classical subject of interfacial electrochemistry is galvanization, that is, potential-driven reduction of metal cations and deposition of metal layers. In the form of electroplating, this process has become the basis of modern electronic chip production. The dissolution (oxidation) and deposition (reduction) equilibrium potential of a given metal is described by the Nernst equation – as long as dissociation and redeposition of the metal from/onto itself is considered. If, however, the metal in question is electrochemically deposited (desorbed) on (from) an unlike metal surface, an extra chemical interaction comes into play and leads to deposition and dissolution respectively *before and after* the Nernst potential is reached [127].

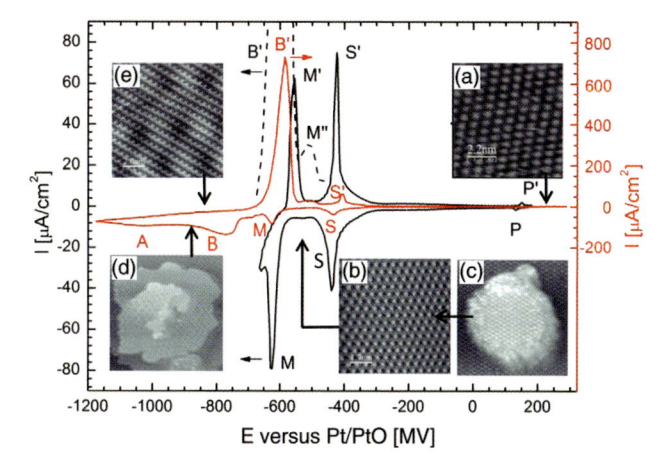

Figure 7.47 Cyclic voltammograms of Au(111) in CuSO$_4$-containing sulfuric acid solution recorded in different potential ranges. Black trace: regime of underpotential Cu deposition; red trace: regime of overpotential Cu deposition. (a–e) In situ STM images of the different surface structures at the indicated potentials (see text). Image size: (a) 8.5 nm × 8.5 nm; (b) 8.5 nm × 8.5 nm; (c) 15 nm × 15 nm; (d) 90 nm × 90 nm; (e) 10 nm × 10 nm.

The "underpotential deposition (UPD)" (before reaching the Nernst potential) of copper on a Au(111) electrode surface from a 0.1 mM CuSO$_4$ containing 0.1 m H$_2$SO$_4$ solution is shown in Figure 7.47. The black voltammogram was recorded in the region where UPD of Cu from the solution takes place. At very positive potentials, the gold surface is still uncovered, the Cu^{2+} ions are not yet attracted and reduced. The small peaks P'/P correspond to an order–disorder phase transition of sulfate on the bare Au(111) – (1 × 1) substrate; at potentials more positive than P', the SO$_4^{2+}$ anions form a ($\sqrt{3} \times \sqrt{7}$)R30° structure (panel (a)). During the potential sweep in cathodic direction, the peaks S and M reflect the adsorption of 2/3 ML and then 1/3 ML of Cu together with the coadsorption of SO$_4^{2-}$, respectively. The potential sweep in the reverse (anodic) direction results first in the desorption of 1/3 ML of Cu, reflected by the maximum M', and then desorption of 2/3 ML of copper, indicated by the peak S' [134].

The red voltammogram recorded in the broad potential range reflects similar peaks in the UPD range but also the maxima B'/B and the shoulder A at more negative potentials in the overpotential deposition (OPD) region (more negative than the Nernst potential of Cu reduction). The maxima B'/B are associated with the adsorption/desorption of multilayer copper deposit [134]. The large current (peak B') associated with the Cu multilayer desorption hides the maximum M' in the red voltammogram. The various deposition steps can now be complemented by structural information from in situ STM measurements. The high-resolution STM image recorded at $E = -500$ mV (Figure 7.47b) reflects the $\left(\sqrt{3} \times \sqrt{3}\right)$ structure of SO$_4^{2-}$ on the 2/3 ML of Cu. This sulfate structure is also observed on two-dimensional Au islands covered by the 2/3 ML Cu layer, as seen in

Figure 7.47c, with the Au islands originating from the lifting of the Au(111) reconstruction.

The STM image (Figure 7.47d) recorded at $E = -880$ mV after the multilayer copper deposition reveals a similar but not identical sulfate structure (panel (e)) as on bulk Cu(111), as shown in Figure 7.44. The surface is characterized by screw dislocations (Figure 7.47d) penetrating the Cu overlayer, most likely originating from the first stages of Cu growth at step edges and the two-dimensional Au islands (Figure 7.47c).

7.5.3.5 Molecular Self-Assembly at Metal–Electrolyte Interfaces

A major trend in modern materials design is the modification of surfaces by organic molecules with specific functionalities. The reproducibility of such layers, of course, is correlated with that of their microscopic structure. Therefore, there is a great interest in finding the right conditions for self-assembly of organic layers for applications in heterogeneous catalysis, sensorics, molecular electronics, and so on. The vapor deposition of organic films, however, is only possible as long as the molecular building blocks are intact volatile. More complex organic species can only be deposited from solution, either by spray deposition or directly from solution. The following two examples describe prototypical systems whose electrochemical or electroless deposition and self-assembly at solid–liquid interfaces have extensively been studied with *in situ* STM.

As the first example, Figure 7.48 shows *in situ* STM images of a monolayer of (*N*-methyl-4-pyridyl) porphyrine cations (H_2TMPyP^{4+}; see *inset*) adsorbed on an iodide precovered Cu(111) electrode (see Figure 7.42) in 5 mV H_2SO_4 solution [135]. The molecules are nicely ordered in rows, and individual molecules are rotated by $-17°$ (A) or $+17°$ (B) with respect to the direction of the respective row. Depending on whether molecules in adjacent rows are rotated in the same or in the opposite direction, the molecular layer can be described by a unit cell $(\bar{a} \times \bar{b})_{AA} = (\bar{a} \times \bar{b})_{BB}$ or $(\bar{a} \times \bar{b})_{AB}$, the latter being twice as large. It is interesting to note that these are rectangular unit cells even though the substrate surface exhibits threefold symmetry. This is a consequence of the iodide interlayer and, in particular, the inherent fourfold symmetry of the molecules.

The second example refers to the electrochemical adsorption and self-assembly of dibenzylpyridinium or dibenzylviologen (DBV) dications (see *inset* in Figure 7.49) on a Cu(100) surface from 10 mM HCl solution [123,136,137]. In this solution, the Cu(100) surface is precovered with a $(\sqrt{2} \times \sqrt{2})R45°$-Cl (= c(2×2)-Cl) layer that has a quadratic symmetry (like the substrate). The DBV cations are redox active and can undergo two one-electron transfer processes:

$$DBV^0 \underset{-e^-}{\overset{+e^-}{\rightleftarrows}} DBV^{+\bullet} \underset{-e^-}{\overset{+e^-}{\rightleftarrows}} DBV^{2+} \tag{7.16}$$

of which only the reduction from the dication to the monocation radical is fully reversible. At more positive electrode potentials, the lengthy dications in solution adsorb as dications and form, rather surprisingly, quadratic ensembles (Figure 7.49). Each of these quadratic ensembles consists of four dications, which

7.5 Exemplary Results | 237

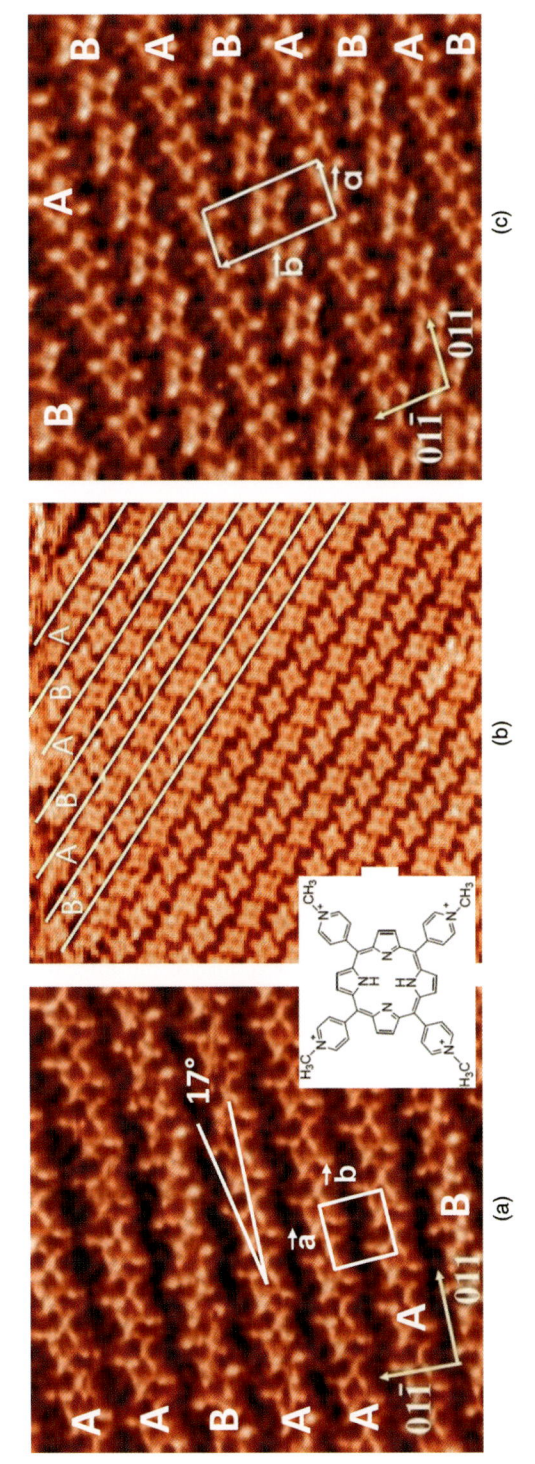

Figure 7.48 *In situ* STM images of self-assembled layers of porphyrin molecules (*see inset*) on a Cu(111) electrode surface. Arrows mark crystallographic directions of the substrate as well as the unit cells of the organic overlayer ($\vec{a}$, $\vec{b}$); A and B indicate rows of differently rotated molecules (see text). Image size: (a) 10.9 nm × 10.9 nm; (b) 25 nm × 25 nm; (c) 10.5 nm × 10.5 nm.

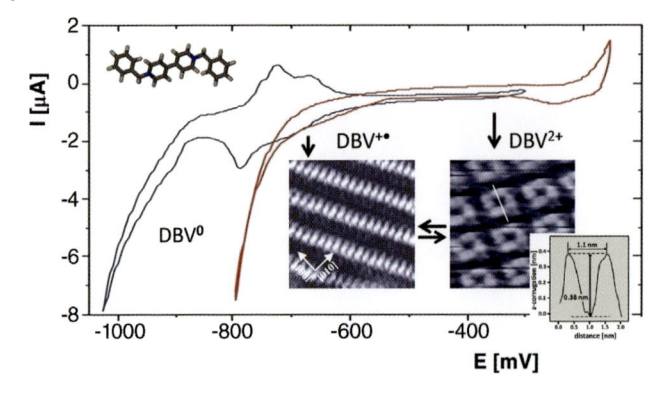

Figure 7.49 Cyclic voltammograms of a Cu (100) electrode in pure HCl solution (red) and dibenzylviologen (DBV, *inset*) containing HCl solution (black). The STM images show the structures formed by the organic molecules as dication (DBV^{2+}) and monocation radical (DBV$^{+\bullet}$) at the indicated potentials, respectively. *Inset:* Height profile along white line in the quadratic DBV^{2+} structure. Image size: *left:* 5.8 nm × 5.8 nm; *right:* 7.16 nm × 7.16 nm.

can actually occur in two chiral versions [123,136,137]. The monolayer as a whole is actually a racemate of equally abundant domains of both enantiomers. Upon lowering the potential, this highly ordered quadratic structure decays and undergoes a drastic transformation into a structure of parallel rows (Figure 7.49). Independent *ex situ* XPS measurements, after emersion of the respective structure, prove that the quadratic structure consists of dications, while the row structure is assembled from monocation radicals [137]. The more positively charged dications adopt the quadratic symmetry of the underlying grid of negatively charged chloride ions. The assembly of the less positively charged monocation radicals, instead, is dominated by π–π stacking and electron pairing of the neighboring aromatic radicals. This really deep insight into the structural details of these viologen layers is nowadays possible by *in situ* STM.

7.6
Summary

The realization of scanning probe microscopes has revolutionized the research of properties and processes at surface and interfaces. This new way of mapping properties with atomic resolution has opened previously unattainable possibilities for basic research and various applications. Different versions of these instruments allow carrying out investigations in different environment, and are therefore nowadays used by scientists in many disciplines such as physics, chemistry, engineering, biology, and medicine.

In this chapter, we have concentrated on STM, including its spectroscopic capabilities (STS, IETS), and atomic force microscopy. The physical principles of both detection techniques are described, and various examples are presented and discussed that demonstrate the wealth of information obtainable on the atomic scale. The selected examples specifically focus on fundamental problems of chemistry at solid surfaces under ultrahigh vacuum conditions, at elevated gas pressures, and in electrolytes, and include the determination of surface structures of metals, alloys, semiconductors, and insulators, the elucidation of elementary surface-confined reactions such as adsorption, dissociation, reaction, and self-assembly, and controlled atomic manipulation.

References

1. Chiarotti, G. (ed.) (1993) *Physics of Solid Surfaces*, III/24a, Springer, Berlin.
2. Somorjai, G.A. (1994) *Introduction to Surface Chemistry and Catalysis*, John Wiley & Sons, Inc., New York.
3. Ibach, H. (2006) *Physics of Surfaces and Interfaces*, Springer, Berlin.
4. Wandelt, K. (ed.) (2015) *Surface and Interface Science*, vol. **1–6**, Wiley-VCH Verlag GmbH, Weinheim.
5. Gross, A. (2009) *Theoretical Surface Science: A Microscopic Perspective*, Springer, Berlin.
6. Ertl, G. (2009) *Reactions at Solid Surface*, John Wiley & Sons, Inc., Hoboken.
7. King, D.A. and Woodruff, D.P. (eds) (1983) *The Chemical Physics of Solid Surfaces and Heterogeneous Catalysis*, vol. **1–3**, Elsevier, Amsterdam.
8. Rhodin, T.N. and Ertl, G. (eds) (1979) *The Nature of the Surface Chemical Bond*, North Holland, Amsterdam.
9. Müller, E.W. (1937) *Z. Phys.*, **106**, 541.
10. Müller, E.W. (1951) *Z. Phys.*, **131**, 136.
11. Binnig, G., Rohrer, H., Gerber, Ch., and Weibel, E. (1982) *Appl. Phys. Lett.*, **40**, 178.
12. Binnig, G., Rohrer, H., Gerber, Ch., and Weibel, E. (1082) *Phys. Rev. Lett.*, **49**, 57.
13. Binnig, G., Rohrer, H., Gerber, Ch., and Weibel, E. (1982) *Physica B*, **109/110**, 2075.
14. Binnig, G., Rohrer, H., Gerber, Ch., and Weibel, E. (1083) *Phys. Rev. Lett.*, **50**, 120.
15. Binnig, G., Rohrer, H., Gerber, Ch., and Weibel, E. (1983) *Surf. Sci.*, **131**, L379.
16. Wiesendanger, R., Tarrach, G., Scandella, L., and Güntherodt, H.-J. (1990) *Ultramicroscopy*, **32**, 291.
17. Binnig, G., Quate, C.F., and Gerber, Ch. (1986) *Phys. Rev. Lett.*, **56**, 930.
18. Chen, C.J. (2007) *Introduction to Scanning Tunneling Microscopy*, 2nd edn, Oxford University Press, ISBN 13: 9780199211500.
19. Bonnell, D. (2001) *Scanning Probe Microscopy and Spectroscopy: Theory, Techniques and Applications*, 2nd edn, Wiley-VCH Verlag GmbH, Weinheim.
20. Voigtländer, B. (2015) *Scanning Probe Microscopy*, Springer, Berlin.
21. Wiesendanger, R. (1994) *Scanning Probe Microscopy and Spectroscopy: Methods and Applications*, Cambridge University Press, ISBN: 0521428475.
22. Güntherodt, H.-J. and Wiesendanger, R. (eds) (1994) *Scanning Probe Microscopy I, II, III*, Springer Series in Surface Science, vols. **20, 28, 29**, Springer, Berlin (1994, 1995, 1996: ISBN: 978-3-642-9255-7; ISBN: 978-3-642-79366-0; ISBN: 978-3-642-80118-1, respectively.
23. Meyer, E., Hug, H.J., and Bennewitz, R. (2004) *Scanning Probe Microscopy: The Lab on a Tip*, Springer, Berlin, ISBN: 3-540-43180-2.
24. Kühnle, A. and Reichling, M. (eds) (2012) Scanning probe techniques, in *Surface and Interface Science*, vol. **1** (ed. K. Wandelt,), Wiley-VCH Verlag GmbH, Weinheim, ISBN: 978-3-527-41156-6.
25. Kuipers, L., Loos, R.W.M., Neerings, H., ter Horst, J., Ruwiel, G.J., de, Jongh, A.P., and Frenken, J.W.M. (1995) *Rev. Sci. Instrum.*, **66**, 4557.
26. Laegsgaard, E., Österlund, L., Thostrup, P., Rasmussen, P.B., Stensgaard, I., and

Besenbacher, F. (2001) *Rev. Sci. Instrum*, **72**, 3537.
27 Jensen, J.A., Rider, K.B., Chen, Y., Salmeron, M., and Somorjai, G.A. (1999) *J. Vac. Sci. Technol.*, **B17**, 1080.
28 Jensen, J.A., Rider, K.B., Salmeron, M., and Somorjai, G.A. (1998) *Phys. Rev. Lett.*, **80**, 1228.
29 Excudero, C. and Salmeron, M. (2013) *Surf. Sci.*, **607**, 2.
30 Herbschleb, C.T., van der Tuijn, P.C., Roobol, S.B., Navarro, V., Bakker, J.W., Liu, Q., Stoltz, D., Canas-Ventura, M.E., Verdoes, G., Spronsen, M.A.van., Bergman, M., Crama, L., Taminiau, I., Ofitserov, A., Baarle, G.J.C.van., and Frenken, J.W.M. (2014) *Rev. Sci. Instrum.*, **85**, 083703.
31 Tao, F., Nguyen, L., and Zhang, S. (2013) *Rev. Sci. Instrum.*, **84**, 034101.
32 Tao, F., Tang, D., Salmeron, M., and Somorjai, G.A. (2008) *Rev. Sci. Instrum.*, **79**, 084101.
33 Lustenberger, P., Rohrer, H., Christoph, R., and Siegenthaler, H. (1988) *J. Electroanal. Chem. Interfacial Electrochem.*, **243**, 225.
34 Itaya, K. and Tomita, E. (1988) *Surf. Sci.*, **201**, L507.
35 Wilms, M., Kruft, M., Bermes, G., and Wandelt, K. (1999) *Rev. Sci. Instrum.*, **70**, 3641.
36 Wen, R., Rahn, B., and Magnussen, O.M. (2015) *Angew. Chem., Int. Ed.*, **54**, 6062.
37 Wen, R., Rahn, B., and Magnussen, O.M. (2016) *J. Phys. Chem. C*, **120**, 15765.
38 Endres, F., Borisenko, N., ElAbedin, S.Z., Hayes, R., and Atkin, R. (2012) *Faraday Dicuss.*, **154**, 221.
39 Hu, X., Chen, C., Tang, S., Wang, W., Yan, J., and Mao, B. (2015) *Sci. Bull.*, **60**, 877.
40 Endres, F., MacFarlane, D., and Abbott, A. (eds) (2008) *Electrodeposition from Ionic Liquids*, John Wiley & Sons, Inc., ISBN: 978-3-527-31565-9.
41 Besenbacher, F., Laegsgaard, E., and Stensgaard, I. (2005) *Mater. Today*, **8**, 26.
42 Zitzler, L., Gleich, B., Magnussen, O.M., and Behm, R.J. (2000) *Localized In Situ Methods for Investigating Electrochemical Interfaces: Proceedings of the International Symposium* (eds A.C. Hillier, M. Seo, and S.R. Taylor), Electrochemical Society, pp. 29–38.
43 Bardeen, J. (1961) *Phys. Rev. Lett.*, **6**, 57.
44 Tersoff, J. and Hamann, D.R. (1983) *Phys. Rev. Lett.*, **50**, 1998.
45 Tersoff, J. and Hamann, D.R. (1985) *Phys. Rev.*, **B31**, 805.
46 Lang, N.D. (1986) *Phys. Rev.*, **B34**, 5947.
47 Feenstra, R.M. (1994) *Surf. Sci.*, **299–300**, 965.
48 Lauhon, L.J. and Ho, W. (1999) *Phys. Rev.*, **B69**, R8525.
49 Besocke, K. (1987) *Surf. Sci.*, **181**, 145.
50 Wilms, M. (1999) Potentiodynamische Rastertunnelmikroskopie an Fest/Flüssig Grenzflächen: Apparative Entwicklung und Untersuchungen zur Sulfat-Adsorption auf Cu(111). Ph.D. thesis, University of Bonn, Germany.
51 Wilms, M., Kruft, M., Bermes, G., and Wandelt, K. (1999) *Rev. Sci. Instrum.*, **70**, 3641.
52 Gentz, K. and Wandelt, K. (2012) *Chimia*, **66**, 1.
53 Gentz, K. and Wandelt, K. (2013) Electrochemical scanning tunneling microscopy, in *Fundamentals of Picoscience* (ed. K. Sattler), Taylor & Francis, p. 209.
54 Sondericker, D., Jona, F., and Marcus, P.M. (1986) *Phys. Rev.*, **B34**, 6770.
55 Parr, R.G. and Yang, W. (1989) *Density Functional Theory of Atoms and Molecules*, Oxford University Press, New York.
56 Michaelides, A. and Scheffler, M. (2012) An introduction to the theory of crystalline elemental solids and their surfaces, in *Surface and Interface Science*, vol. 1 (ed. K. Wandelt), Wiley-VCH Verlag GmbH, Weinheim.
57 Jurczyszyn, L., Rosenhahn, A., Schneider, J., Becker, C., and Wandelt, K. (2003) *Phys. Rev.*, **B68**, 115425.
58 Gonzales, C., Benito, I., Ortega, J., Jurczyszyn, L., Blanco, J.M., Perez, R., Flores, F., Kampen, T.U., Zahn, D.R.T., and Braun, W. (2004) *J. Phys. Condens. Mater.*, **16**, 2187.
59 Keldysh, L.V. (1964) *Zh. Eksp. Teor. Phys.*, **47**, 1515 (*JETP*, **20** (1965), 1018).
60 Martin-Rodero, A., Flores, F., and March, N.H. (1988) *Phys. Rev.*, **B38**, 10047.

61 Ferrrer, J., Martin-Rodero, A., and Flores, F. (1988) *Phys. Rev.*, **B38**, 10113.
62 Jurczyszyn, L., Rosenhahn, A., Schneider, J., Becker, C., and Wandelt, K. (2003) *Phys. Rev.*, **B68**, 115425.
63 Schmickler, W. (1996) *Chem. Rev.*, **96** (8), 3177.
64 Sebastian, K.L. and Doyen, G. (1993) *Surf. Sci. Lett.*, **290**, L703.
65 Schindler, W. and Hugelmann, M. (2003) *Surf. Sci.*, **541**, L643.
66 Wandlowski, T. and Nagy, G. (2003) *Langmuir*, **19**, 10271.
67 Fessler, G., Eren, B., Gysin, U., Glatzel, T., and Meyer, E. (2014) *Appl. Phys. Lett.*, **104**, 041910.
68 Kayser, T. (2004) Rastersondenmikroskopische Untersuchungen an amorphen Systemen und deren Rekristallisationsverhalten. Ph. D. thesis, University of Bonn, Germany.
69 Giessibl, F.J. (1998) *Appl. Phys. Lett.*, **73**, 3956.
70 Torbrügge, S., Schaff, O., and Rychen, J. (2010) *J. Vac. Sci. Technol.*, **B28**, C4E12.
71 An, T., Eguchi, T., Akiyama, K., and Hasegawa, Y. (2005) *Appl. Phys. Lett.*, **87**, 133114.
72 Heike, S. and Hasizume, T. (2003) *Appl. Phys. Lett.*, **83**, 3620.
73 Giessibl, F.J., Pielmeier, F., Eguchi, T., An, T., and Hasegawa, Y. (2011) *Phys. Rev.*, **B84**, 125409.
74 Froning, J.P., Xia, D., Zhang, S., Lægsgaard, E., Besenbacher, F., and Dong, M. (2015) *J. Vac. Sci. Technol.*, **B33**, 021801–1.
75 Hug, H.J., Moser, A., Jung, Th., Fritz, O., Wadas, A., Parashikov, I., and Güntherrodt, H.-J. (1993) *Rev. Sci. Instrum.*, **64**, 2920.
76 Kubetzka, A., Bode, M., Pietzsch, O., and Wiesendanger, R. (2002) *Phys. Rev. Lett.*, **88**, 057201.
77 Smith, A.R. (2006) *J. Scanning Probe Microsc.*, **1**, 3.
78 Wulfhekel, W. and Kirschner, J. (2007) *Annu. Rev. Mater. Res.*, **37**, 69.
79 Hirth, S., Ostendorf, F., and Reichling, M. (2006) *Nanotechnology*, **17**, 5148.
80 Nonnenmacher, M., O'Boyle, M.P., and Wickramasinghe, H.K. (1991) *Appl. Phys. Lett.*, **58**, 2921.
81 Jacobs, H.O., Leuchtmann, P., Homan, O.J., and Stemmer, S. (1998) *J. Appl. Phys.*, **84**, 1168.
82 Heinz, K. and Starke, U. (2012) Surface crystallogarphy, in *Surface and Interface Science*, vol. **2** (ed. K. Wandelt), Wiley-VCH Verlag GmbH, Weinheim.
83 Kerscher, T.C. and Müller, S. (2014) Surface properties of alloys, in *Surface and Interface Science*, vol. **3** (ed. K. Wandelt), Wiley-VCH Verlag GmbH, Weinheim.
84 Takayanagi, K., Tanishiro, I., Takahashi, S., and Takahashi, M. (1985) *Surf. Sci.*, **164**, 367.
85 Heinz, K., Lang, E., Strauss, K., and Müller, K. (1982) *Surf. Sci.*, **120**, L401.
86 Ritz, G., Schmid, M., Varga, P., Borg, A., and Rønning, M. (1997) *Phys. Rev.*, **B56**, 10518.
87 Abraham, F.F. (1981) *Phys. Rev. Lett.*, **46**, 546.
88 Schneider, U., Castro, G.R., and Wandelt, K. (1993) *Surf. Sci.*, **287/288**, 146.
89 Wouda, P.T., Nieuwenhuys, B.E., Schmid, M., and Varga, P. (1996) *Surf. Sci.*, **359**, 17.
90 Hebenstreit, E.L.D., Hebenstreit, H., Schmid, M., and Varga, P. (1999) *Surf. Sci.*, **441**, 441.
91 Gauthier, Y., Schmid, M., Padovani, S., Lundgren, E., Bus, V., Kresse, G., Redinger, J., and Varga, P. (2001) *Phys. Rev. Lett.*, **87**, 036103-1.
92 Sachtler, W.M.H. (1973) *Le Vide*, **28**, 67.
93 Blyholder, G. (1964) *J. Phys. Chem.*, **68**, 2772.
94 Bagus, P. and Pacchioni, G. (1992) *Surf. Sci.*, **278**, 427.
95 Hammer, B., Morikawa, Y., and Nørkov, J.K. (1996) *Phys. Rev. Lett.*, **76**, 2141.
96 Rosenhahn, A. (2000) Ultradünne Aluminiumoxidfilme auf $Ni_3Al(111)$: Template für nanostrukturiertes Metallwachstum. Ph.D. thesis, University of Bonn, Germany.
97 Rosenhahn, A., Schneider, J., Becker, C., and Wandelt, K. (2000) *J. Vac. Sci. Technol.*, **A18**, 1923.
98 Gritschneder, S., Degen, S., Becker, C., Wandelt, K., and Reichling, M. (2007) *Phys. Rev.*, **B76**, 014123.
99 Hamm, G., Barth, C., Becker, C., Wandelt, K., and Henry, C.R. (2006) *Phys. Rev. Lett*, **97**, 126106.

100 Degen, S., Krupski, A., Kralj, M., Langner, A., Sokolowski, M., and Wandelt, K. (2005) *Surf. Sci.*, **576**, L57.

101 Schmid, M., Kresse, G., Buchsbaum, A., Napetschnig, E., Gritschneder, S., Reichling, M., and Varga, P. (2007) *Phys. Rev. Lett.*, **99**, 196104.

102 Maroutian, T., Degen, S., Becker, C., Wandelt, K., and Berndt, R. (2003) *Phys. Rev.*, **B68**, 155414.

103 Degen, S. (2005) Aufbau eines Tieftemperaturrastertunnelmikroskops und Messungen an $Ni_3Al(111)$. Ph.D. thesis, University of Bonn, Germany.

104 Becker, C. and Wandelt, K. (2010) The template route to nanostructured model catalysts, in *Scanning Tunneling Microscopy in Surface Science* (eds M. Bowker and P.R. Davies), Wiley-VCH Verlag GmbH, Weinheim.

105 Zachariasen, W.J. (1932) *J. Am. Chem. Soc.*, **54**, 3841.

106 Raberg, W., Lansmann, V., Jansen, M., and Wandelt, K. (1997) *Angew. Chem., Int. Ed. Engl.*, **109**, 2760.

107 Raberg, W. and Wandelt, K. (1998) *Appl. Phys.*, **66**, S1143.

108 Raberg, W., Ostadrahimi, A.H., Kayser, T., and Wandelt, K. (2005) *J. Non-Cryst. Solids*, **351**, 1089.

109 Lichtenstein, L., Heyde, M., and Freund, H.-J. (2012) *J. Phys. Chem.*, **C116**, 20426.

110 Hollinger, G. and Himpsel, F.J. (1983) *Phys. Rev.*, **B28**, 3651.

111 Ibach, H., Bruchmann, H.D., and Wagner, H. (1982) *Appl. Phys.*, **A29**, 113.

112 Edamoto, K., Kubota, Y., Kobayashi, H., Onchi, M., and Nishijima, M. (1985) *J. Chem. Phys.*, **83**, 428.

113 Martel, R., Avouris, Ph., and Lyo, I.-W. (1996) *Science*, **272**, 385.

114 Wintterlin, J., Schuster, R., and Ertl, G. (1996) *Phys. Rev. Lett.*, **77**, 123.

115 Rose, M.K., Borg, A., Mitsui, T., Ogletree, D.F., and Salmeron, M. (2001) *J. Chem. Phys.*, **115**, 10927.

116 Stipe, B.C., Rezaei, M., and Ho, W. (1998) *Science*, **280**, 1732.

117 Repp, J. and Meyer, G. (2005) *Phys. Rev. Lett.*, **94**, 026803-1.

118 Gross, L., Mohn, F., Moll, N., Liljerodt, P., and Meyer, G. (2009) *Science*, **325**, 1110.

119 Trembulowicz, A. and Antczak, A. (2015) Surface diffusion in atomic scale, in *Reference Module in Chemistry, Molecular Science and Chemical Engineering* (eds J. Reedijk *et al.*), Elsevier.

120 Crommie, M.F., Lutz, C.P., Eigler, D.M., and Heller, E.J. (1995) *Surf. Rev. Lett.*, **2**, 127.

121 Morgenstern, K., Lorente, N., and Rieder, K.-H. (2015) Controlled manipulation of single atoms and small molecules using scanning tunneling microscopy, in *Surface and Interface Science*, vol. **3** (ed. K. Wandelt), Wiley-VCH Verlag GmbH, Weinheim.

122 Hla, S.-W., Bartels, L., Meyer, G., and Rieder, K.-H. (2000) *Phys. Rev. Lett.*, **85**, 2777.

123 Safarowsky, C., Wandelt, K., and Broekmann, P. (2004) *Langmuir*, **20**, 8269.

124 Vestergaard, E.Kruse., Thostrup, P., An, T., Laegsgaard, E., Stensgaard, I., Hammer, B., and Besenbacher, F. (2002) *Phys. Rev. Lett.*, **88**, 259601-1.

125 Stuhlmann, C., Wohlmann, B., Park, Z., Kruft, M., Broekmann, P., and Wandelt, K. (2003) Chloride adsorption on Cu (111) electrodes: electrochemical behavior and UHV transfer experiments, in *Solid-Liquid Interfaces: Macroscopic Phenomena – Microscopic Understanding* (eds K. Wandelt and S. Thurgate), Springer, Heidelberg, p. 199.

126 Dretschkow, T. and Wandlowski, T. (2003) Structural transitions in organic adlayers: a molecular view, in *Solid-Liquid Interfaces: Macroscopic Phenomena – Microscopic Understanding* (eds K. Wandelt and S. Thurgate), Springer, Heidelberg, p. 259.

127 Schmickler, W. and Santos, E. (2010) *Interfacial Electrochemistry*, 2nd edn, Springer, Heidelberg.

128 Kadodwala, M.F., Davis, A.A., Scragg, G., Cowie, B.C.C., Kerkar, M., Woodruff, D.P., and Jones, G.R. (1995) *Surf. Sci.*, **324**, 122.

129 Shard, A.G., Ton-That, C., Campbell, P.A., and Dhanak, V.R. (2004) *Phys. Rev.*, **B70**, 144409.

130 Peljhan, S. and Kokalj, A. (2009) *J. Phys. Chem.*, **C113**, 14363.

131 Obliers, B., Broekmann, P., and Wandelt, K. (2003) *J. Electroanal. Chem.*, **554–555**, 183.

132 Hai, N.T.M., Huemann, S., Hunger, R., Jaegermann, W., Broekmann, P., and Wandelt, K. (2007) *J. Phys. Chem.*, **C111**, 14768.

133 Broekmann, P., Wilms, M., Arenz, M., Spänig, A., and Wandelt, K. (2003) Atomic structure of Cu(111) surfaces in dilute sulfuric acid solution, in *Solid-Liquid Interfaces: Macroscopic Phenomena – Microscopic Understanding* (eds K. Wandelt and S. Thurgate), Springer, Heidelberg, p. 141.

134 Madry, B., Wandelt, K., and Nowicki, M. (2015) *Surf. Sci.*, **637–638**, 77.

135 Hai, N.T.M., Gasparovic, B., Wandelt, K., and Broekmann, P. (2007) *Surf. Sci.*, **601**, 2597.

136 Pham, D.-T-., Gentz, K., Zörlein, C., Hai, N.T.M., Tsay, S.-L., Kirchner, B., Kossmann, S., Wandelt, K., and Broekmann, P. (2006) *New. J. Chem.*, **30**, 1439.

137 Breuer, S., Pham, D.T., Huemann, S., Gentz, K., Zörlein, C., Hunger, R., Wandelt, K., and Broekmann, P. (2008) *New J. Phys.*, **10**, 125033.

8
Solid-State NMR Spectroscopy: Introduction for Solid-State Chemists

Christoph S. Zehe, Renée Siegel, and Jürgen Senker

Inorganic Chemistry III, University of Bayreuth, Universitätsstrasse 30, 95447 Bayreuth, Germany

8.1
Introduction

Since the earliest days of NMR spectroscopy, this technique has been used to determine materials properties in solids with one of the first examples being the determination of the proton–proton distance for water in gypsum $CaSO_4 \cdot 2\,H_2O$ by Pake in 1948 [1]. In spite of its early successes, the lack in resolution and sensitivity prevented its broad application in chemistry, material sciences, geology, and biology. This was changed fundamentally by introducing *magic-angle spinning* (MAS) [2], *Fourier transform* NMR spectroscopy [3,4], and *multiple-pulse* [5,6] methods. While MAS improves the spectral resolution by imprinting a "liquid-like" motion on the sample based on spatial averaging [2], the latter two techniques made NMR spectroscopy much more flexible and versatile. Indeed, the combination of MAS with *multiple-pulse* methods such as proton decoupling [7] to gain even higher resolution, recoupling techniques for selectively reintroducing desired nuclear spin interactions [8,9], and signal enhancing techniques such as cross-polarization [7] matured solid-state NMR spectroscopy (ssNMR) into one of the most expressive techniques to solve structural problems and derive information about dynamical processes.

Meanwhile, ultrafast MAS, with spinning speeds up to more than 100 kHz, allows for almost cancelling even the strongest dipole couplings, consequently approaching liquid-like resolution. A large battery of individual NMR techniques is now at the disposal of experimentalists enabling them to separate and correlate nuclear spin interactions such as the chemical shift and dipole as well as quadrupolar couplings, in order to answer questions concerning both structure and dynamics [10]. In particular, the anisotropic part of these interactions, directly measurable only in solids, provides an additional wealth of information that can be exploited for deriving connectivities, and distances on local and intermediate length scales, and for studying motional mechanisms [6,11].

Handbook of Solid State Chemistry, First Edition. Edited by Richard Dronskowski, Shinichi Kikkawa, and Andreas Stein.
© 2017 Wiley-VCH Verlag GmbH & Co. KGaA. Published 2017 by Wiley-VCH Verlag GmbH & Co. KGaA.

ssNMR has proven to be at its best for systems for which conventional analytic techniques such as single-crystal diffraction alone fail. An increasing number of *ab initio* structure solutions was reported for microcrystalline powders based on the combination of ssNMR, powder diffraction, and computational modelling – often referred to as NMR crystallography [12]. Since ssNMR does not depend on long-range order, disordered materials such as nanoscopic photocatalysts [13] and porous frameworks [14], but also amorphous systems such as glasses [15] and polymers [6], and even inhomogeneous materials such as bone tissue can be probed [10].

Nevertheless, ssNMR is still a complex technique with many individual facets. Therefore, this chapter intends to provide the necessary background and a brief overview of the most important techniques for chemists and material scientists who wish to use ssNMR to unravel structural properties and dynamical processes.

8.2
Theory and Methods

8.2.1
Important Interactions for NMR Spectroscopy

The nuclear spin interactions ($\hat{\mathcal{H}}_{int}$) include the Zeeman ($\hat{\mathcal{H}}_0$), the chemical shift ($\hat{\mathcal{H}}_{CS}$), the quadrupolar ($\hat{\mathcal{H}}_Q$), and the spin rotation ($\hat{\mathcal{H}}_{SR}$) interactions as well as the direct dipole–dipole ($\hat{\mathcal{H}}_{DD}$), the J ($\hat{\mathcal{H}}_J$), and for paramagnetic materials also the hyperfine ($\hat{\mathcal{H}}_{HF}$) couplings [16]. Additionally, external radio frequency (rf) fields ($\hat{\mathcal{H}}_{RF}$) allow for manipulating spin states. $\hat{\mathcal{H}}_0$ describes the interaction between nuclear spins and static external magnetic fields $\vec{B}_0$, which is nowadays on the order of several tesla, and is often the largest interaction for NMR spectroscopy. Therefore, all other interactions are treated as perturbations of $\hat{\mathcal{H}}_0$. $\hat{\mathcal{H}}_{CS}$ and $\hat{\mathcal{H}}_{HF}$ both originate from couplings of nuclei with magnetic fields generated by surrounding paired and unpaired electrons, respectively. Dipole–dipole interactions between spins may take place directly (through-space, $\hat{\mathcal{H}}_{DD}$) or indirectly via electrons in covalent bonds (through-bond, $\hat{\mathcal{H}}_J$). $\hat{\mathcal{H}}_Q$ describes the interaction of nuclear spins with the electric field gradient of the chemical environment, while $\hat{\mathcal{H}}_{SR}$ refers to nuclear spins coupled to angular momenta of molecular rotations. Compared to the other interactions, $\hat{\mathcal{H}}_J$ and $\hat{\mathcal{H}}_{SR}$ both are relatively small and thus are less frequently exploited in ssNMR. Therefore, we will not discuss the latter two in this chapter (for further information, see Refs [11,17]).

8.2.1.1 Tensors and Rotations

$\hat{\mathcal{H}}_{int}$ is usually anisotropic and may be described via a second-rank tensor V, which decodes its orientation-dependent magnitude. $\hat{\mathcal{H}}_{int}$ is then expressed in the following form [5]:

$$\hat{\mathcal{H}}_{int} = \vec{I} \cdot V \cdot \vec{J}, \qquad (8.1)$$

where V couples two vectors $\vec{I}$ and $\vec{J}$. $\vec{J}$ may be a magnetic field vector or a spin vector, whereas $\vec{I}$ is always a spin vector. The interaction tensor V may be decomposed into a symmetric $V_{\text{sym}} = (V + V^\dagger)/2$ and an antisymmetric part $V_{\text{asym}} = (V - V^\dagger)/2$ so that $V = V_{\text{sym}} + V_{\text{asym}}$, with $V_{\text{sym}} = V_{\text{sym}}^\dagger$ and $V_{\text{asym}} = -V_{\text{asym}}^\dagger$. Since only V_{sym} influences the shape of NMR spectra in first-order perturbation [10], we neglect V_{asym} for the following discussion. For simplicity, we thus identify V with V_{sym}. It should be noted, however, that V_{asym} plays an important role in the quantitative treatment of relaxation processes [18,19]. Under these circumstances, a coordinate system may be chosen for which the matrix representation of V is diagonal. This coordinate system is called the principal axis system (PAS) and the diagonal elements V_{XX}, V_{YY}, and V_{ZZ} are called the principal axis values. Tensors are often specified [10] using the isotropic value σ_{iso}, the anisotropy σ_{aniso}, and the asymmetry η:

$$\sigma_{\text{iso}} = \frac{V_{XX} + V_{YY} + V_{ZZ}}{3}, \quad \sigma_{\text{aniso}} = V_{ZZ} - \sigma_{\text{iso}}, \quad \eta = \frac{V_{YY} - V_{XX}}{\sigma_{\text{aniso}}}.$$

(8.2)

Hereby, the convention $|V_{ZZ} - \sigma_{\text{iso}}| \geq |V_{XX} - \sigma_{\text{iso}}| \geq |V_{YY} - \sigma_{\text{iso}}|$ is used, which leads to $0 \leq \eta \leq 1$. σ_{aniso} reflects the maximal anisotropic interaction, whereas η is determined by the deviation of the interaction strength in the two remaining directions.

The free induction decay (FID) is detected in the laboratory axis system (LAS), where commonly $\vec{B}_0$ is chosen to point along the z-axis. In contrast, the interaction tensor V depends on the chemical structure that defines the PAS and its orientation with respect to the LAS. Since both coordinate systems are orthonormal, they are linked by the so-called Euler transformations [11]. The graphical interpretation of V is hereby often given by drawing V as an ellipsoid in its PAS (Figure 8.1a) and the orientation of $\vec{B}_0$ with respect to the PAS may then be described by the two polar angles θ and ϕ [6].

However, if several reference frames are needed, a more general approach [6,20] is advantageous: the relative orientation of two coordinate frames A (e.g., PAS) and B (e.g., LAS) may be specified by three Euler angles α, β, and γ (Figure 8.1b). These angles determine an overall Euler transformation $\hat{\mathcal{R}}(\alpha, \beta, \gamma)$ of the initial frame into the final one by three subsequent rotations (Figure 8.1c). If the Euler transformation of frame A into frame B is given by $\hat{\mathcal{R}}(\alpha, \beta, \gamma)$, any tensor V^A defined in frame A might be expressed by V^B in frame B through the similarity transformation $V^B = \hat{\mathcal{R}}(\alpha, \beta, \gamma) \cdot V^A \cdot \hat{\mathcal{R}}(\alpha, \beta, \gamma)^\dagger$. In this notation, the orientation of $\vec{B}_0$ (frame B with $z_B = z_{\text{LAS}}$) in the PAS (frame A) is given by $\alpha = \phi$, $\beta = \theta$, and γ is arbitrary, since the x- and y-axes of the LAS may be freely chosen. Note that for axially symmetric tensors ($\eta = 0$), $\alpha = \phi$ becomes arbitrary. The actual calculation of rotations might be performed expressing $\hat{\mathcal{H}}$ in either Cartesian or, which is more commonly used, in spherical coordinates [6]. The latter simplifies the outcome of rotations since they allow decomposing the tensors in parts with different rotational rank, that is, with distinct rotational behaviors (Section 8.3.3).

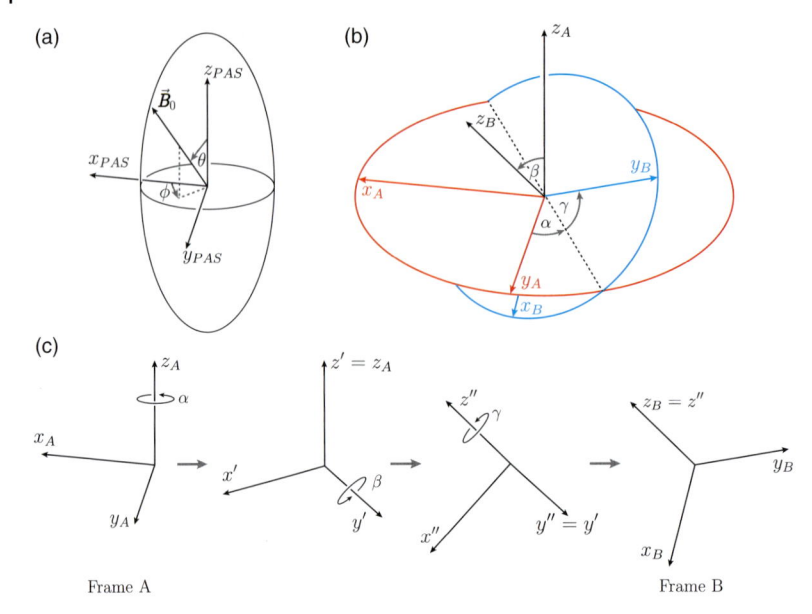

Figure 8.1 (a) The tensor **V** is represented graphically by an ellipsoid (flattened into a disk-like shape if $\eta \neq 0$), with a PAS {x_{PAS}, y_{PAS}, z_{PAS}}. The orientation of an external field $\vec{B}_0$ in the PAS is specified by the two polar angles θ and ϕ [6]. (b) The relative orientation of two coordinate frames {x_A, y_A, z_A} and {x_B, y_B, z_B} may be described by the Euler angles α, β and γ [6]. α appears graphically between the y_A axis and the node line (dashed line; intersection of the red $x_A y_A$ plane with the blue $x_B y_B$ plane); a similar consideration applies for γ. $\beta = \theta$ is the angle between z_A and z_B. (c) The Euler angles may also be regarded as the angles for three subsequent rotations about (i) the z_A axis of the initial frame A, (ii) the y' axis, and (iii) the z'' axis of the intermediate frames, yielding an overall transformation $\hat{\mathcal{R}}(\alpha, \beta, \gamma)$ of frame A into frame B [20].

The Euler transformation for consecutive frame transformations is then simply given by the product of the rotation matrices [20]. If, for example, the transformation of the PAS with respect to a common reference frame for all spins within the same crystallite (the crystal axis system (CAS)) is given by $\hat{\mathcal{R}}(\alpha_{PC}, \beta_{PC}, \gamma_{PC})$ and the orientation of the CAS with respect to the LAS is represented by $\hat{\mathcal{R}}(\alpha_{CL}, \beta_{CL}, \gamma_{CL})$, the overall Euler transformation from the PAS into LAS is determined by $\hat{\mathcal{R}}(\alpha_{PL}, \beta_{PL}, \gamma_{PL}) = \hat{\mathcal{R}}(\alpha_{PC}, \beta_{PC}, \gamma_{PC}) \cdot \hat{\mathcal{R}}(\alpha_{CL}, \beta_{CL}, \gamma_{CL})$. This is extensively used for numerical simulations [21], where not only the orientational average of a powdered sample but also the quantum mechanical time evolution of spin systems require such rotations.

8.2.1.2 Zeeman Interaction

The quantum mechanical treatment of angular momentum leads to energy quantization [16]. For spins with a quantum number I, $2I + 1$ energetically degenerate wave functions $|\Psi_m\rangle$ with magnetic quantum numbers $m = -I, -I + 1, \ldots, I - 1, I$ exist. Within a static magnetic field $\vec{B}_0 = (0, 0, B_0)$, this degeneracy is broken [18] according to

$$\hat{\mathcal{H}}_0 = -\gamma \vec{I} \cdot \vec{B}_0 = -\gamma B_0 \hat{I}_z, \tag{8.3}$$

with $\vec{I} = (\hat{I}_x, \hat{I}_y, \hat{I}_z)$ (please note that all $\hat{\mathcal{H}}_{int}$ are given in units of $\hbar$). $\hat{\mathcal{H}}_0$ is called *Zeeman interaction* and is, as already mentioned, usually significantly larger than other nuclear spin interactions. To simulate NMR spectra, the high-field (or secular) approximation [6] is thus applied to discard all terms of $\hat{\mathcal{H}}_{int}$, which do not commute with $\hat{\mathcal{H}}_0$. Moreover, a transformation of $\hat{\mathcal{H}}_{int}$ into a frame rotating with ω_0 around the z-axis, where $\hat{\mathcal{H}}_0$ disappears [22], is usually applied for calculating the time evolution of spin systems (Section 8.2.2).

8.2.1.3 Chemical Shift

The local magnetic field $\vec{B}$ at the site of a nucleus is composed of two terms: the external field $\vec{B}_0$ and an induced magnetic field $\vec{B}_{ind}$ shielding the nucleus [5]. The latter is caused by movements of surrounding electrons in response to $\vec{B}_0$, resulting in $\vec{B} = \vec{B}_0 - \vec{B}_{ind}$. Based on Eq. (8.1), the chemical shift interaction $\hat{\mathcal{H}}_{CS}$ is given by

$$\hat{\mathcal{H}}_{CS} = \gamma \vec{I} \cdot \vec{B}_{ind} = \gamma \vec{I} \sigma \vec{B}_0, \tag{8.4}$$

where σ is the chemical shielding tensor [18], which describes strength and orientation of $\vec{B}_{ind}$ with respect to $\vec{B}_0$. σ is determined by the chemical environment and may be extracted from quantum mechanical calculations [23,24]. It should be noted that the description of the hyperfine interaction $\hat{\mathcal{H}}_{HF}$ for paramagnetic materials is analogous to Eq. (8.4), with σ representing the hyperfine coupling tensor [18]. Another common convention is the use of the chemical shift tensor [22] $\delta = -\sigma$, reverting the sign of Eq. (8.4); this is more common for practical applications and will hence be used in the following. The high-field approximated chemical shift interaction [22],

$$\hat{\mathcal{H}}_{CS} = \gamma \sigma_{zz} B_0 \hat{I}_z = -\gamma \delta_{zz} B_0 \hat{I}_z = \omega_{cs} \hat{I}_z, \tag{8.5}$$

when combined with Eq. (8.3) results in [6]

$$\hat{\mathcal{H}}_0 + \hat{\mathcal{H}}_{CS} = (\omega_0 + \omega_{cs}) \hat{I}_z = \omega_0 (1 + \delta_{zz}) \hat{I}_z, \tag{8.6}$$

where $\omega_0 = -\gamma B_0$ represents the Larmor frequency of an unshielded nucleus and $\omega_{cs} = -\omega_0 \sigma_{zz} = \omega_0 \delta_{zz}$ the chemical shift frequency. The latter may be expressed [6] in the LAS as

$$\omega_{cs} = -\gamma B_0 \delta_{iso} - \frac{1}{2} \gamma B_0 \delta_{aniso} \left(3\cos^2\theta - 1 - \eta \sin^2\theta \cos 2\phi \right), \tag{8.7}$$

where θ and ϕ are given in Figure 8.1 and the three parameters $\delta_{iso} = -\sigma_{iso}$, $\delta_{aniso} = -\sigma_{aniso}$, and η are defined by the Haeberlen–Mehring–Spiess convention [5] according to Eq. (8.2).

8.2.1.4 Dipolar Interactions

Nuclear spins are perturbed not only by external magnetic fields but also by each other through magnetic dipole–dipole interactions [18]. This coupling may take place directly between spins in spatial proximity, referred to as direct dipole–dipole interaction $\hat{\mathcal{H}}_{DD}$ [17]. $\hat{\mathcal{H}}_{DD}$ may be derived from classic magnetic dipole–dipole interactions using the canonical quantization leading to [5,10]

$$\hat{\mathcal{H}}_{DD} = -\left(\frac{\mu_0 \gamma_I \gamma_S \hbar}{4\pi r^3}\right)\left(\vec{I}\cdot\vec{S} - 3\frac{(\vec{I}\cdot\vec{r})(\vec{S}\cdot\vec{r})}{r^2}\right) = -2\vec{I}\cdot\mathbf{D}\cdot\vec{S}, \qquad (8.8)$$

where μ_0 is the magnetic field constant, γ_I and γ_S are the gyromagnetic ratios of the coupled spins I and S with an internuclear vector $\vec{r}$ having the modulus r representing the distance between I and S and the dipole coupling tensor $\mathbf{D}$. The latter is often given using the dipolar coupling constant $d = (\mu_0 \gamma_I \gamma_S \hbar)/(4\pi r^3)$, with $D_{XX} = -(d/2)$, $D_{YY} = -(d/2)$, and $D_{ZZ} = d$ in its principal axis representation [10]. Thus, $\mathbf{D}$ is traceless ($\sigma_{iso} = 0$) and axially symmetric ($\eta = 0$) according to Eq. (8.2). The high-field approximation leads to

$$\hat{\mathcal{H}}_{DD}^{homo} = -d\left\{\underbrace{\hat{I}_z\hat{S}_z(3\cos^2\theta - 1)}_{A} - \underbrace{\frac{1}{4}\left[\hat{I}^+\hat{S}^- + \hat{I}^-\hat{S}^+\right](3\cos^2\theta - 1)}_{B}\right\} \qquad (8.9)$$

for the homonuclear dipole–dipole interaction [6], where the polar angle θ is the angle between $\vec{B}_0$ and $\vec{r}$ (i.e., $\vec{r}$ points along z_{PAS} in Figure 8.1) and $\hat{I}^\pm = \hat{I}_x \pm i\hat{I}_y$. The first term is commonly called the A term and the second one the B term [10]. For the heteronuclear interaction, only the A term remains, due to the large difference between the resonance frequencies of the I and S spins [6] resulting in

$$\hat{\mathcal{H}}_{DD}^{hetero} = -d\hat{I}_z\hat{S}_z(3\cos^2\theta - 1). \qquad (8.10)$$

8.2.1.5 Quadrupolar Interaction

Nuclei with $I > 1/2$ exhibit electric multipole moments due to nonspherical charge distributions within the nuclei with the quadrupolar moment Q being by far the largest one. Q couples to the electric field gradient (EFG) arising from the nuclear environment [20] leading to $\hat{\mathcal{H}}_Q$ given by [5,18]

$$\hat{\mathcal{H}}_Q = \frac{eQ}{2I(2I-1)\hbar}\vec{I}\cdot\mathbf{V}\cdot\vec{I}, \qquad (8.11)$$

where e is the elementary charge and $\mathbf{V}$ is the traceless EFG tensor. $\mathbf{V}$ exhibits the three principal axis values V_{XX}, V_{YY}, and V_{ZZ} and is most commonly characterized using the quadrupolar coupling constant C_Q and the biaxiality parameter η_Q defined [6] as

$$C_Q = \frac{eV_{ZZ}Q}{h}, \qquad \eta_Q = \frac{V_{YY} - V_{XX}}{V_{ZZ}}. \qquad (8.12)$$

Since the quadrupolar coupling strength is often quite strong, treating $\hat{\mathcal{H}}_Q$ as perturbation of $\hat{\mathcal{H}}_0$ usually requires two correction terms [16,25] added to the Zeeman splitting: the first-order $\hat{\mathcal{H}}_Q^{(1)}$ and the second-order $\hat{\mathcal{H}}_Q^{(2)}$ quadrupolar coupling:

$$\hat{\mathcal{H}}_Q^{(1)} = -\frac{2\pi C_Q}{8I(2I-1)} \left(\hat{I}^2 - 3\hat{I}_z^2\right) \left[3\cos^2\theta - 1 - \eta_Q \cos 2\phi \sin^2\theta\right]$$

$$\hat{\mathcal{H}}_Q^{(2)} = \frac{1}{2\omega_0} \left[\frac{2\pi C_Q}{2I(2I-1)}\right]^2 \hat{I}_z \Bigg\{ -\frac{1}{5}\left(\hat{I}^2 - 3\hat{I}_z^2\right)\left(3 + \eta_Q^2\right)$$

$$+ \frac{1}{28}\left(8\hat{I}^2 - 12\hat{I}_z^2 - 3\right)\left[\left(\eta_Q^2 - 3\right)(3\cos^2\theta - 1) - 6\eta_Q \sin^2\theta \cos 2\phi\right]$$

$$+ \frac{1}{8}\left(18\hat{I}^2 - 34\hat{I}_z^2 - 5\right)\left[\frac{1}{140}\left(18 + \eta_Q^2\right)(35\cos^4\theta - 30\cos^2\theta + 3)\right.$$

$$\left. -\frac{3}{7}\eta_Q \sin^2\theta(7\cos^2\theta - 1)\cos 2\phi + \frac{1}{4}\eta_Q^2 \sin^4\theta \cos 4\phi\right]\Bigg\},$$

(8.13)

where θ and ϕ denote the polar angles as given in Figure 8.1.

8.2.1.6 Radio Frequency Pulses

In modern Fourier transform NMR, additional pulsed magnetic rf fields are used to manipulate the spin states. Here, the nuclei are subjected to an oscillating magnetic field [18,20]:

$$\vec{B}_1(t) = B_1(\vec{e}_z \cos\theta_{rf} + \vec{e}_x \sin\theta_{rf})\cos\left(\omega t + \phi_{rf}\right) \quad (8.14)$$

where ϕ_{rf} is the phase of the pulse, ω is the frequency, and θ_{rf} is the angle created between $\vec{B}_1$ and $\vec{B}_0$ by the tilt of the rf coil. Within the secular approximation, the corresponding Hamiltonian $\hat{\mathcal{H}}_{rf}$ reads for an on-resonance pulse (i.e., $\omega = \omega_0 + \omega_{CS}$) in the rotating frame [20]:

$$\hat{\mathcal{H}}_{rf} = -\omega_{nut}\left(\hat{I}_x \cos\phi_{rf} + \hat{I}_y \sin\phi_{rf}\right), \quad (8.15)$$

where $\omega_{nut} = |(1/2)\gamma B_1 \sin(\theta_{rf})|$. In contrast to the secular parts of all other nuclear spin interactions we encountered so far (with the exception of homonuclear dipolar couplings), $\hat{\mathcal{H}}_{rf}$ is dominated by $\hat{I}_{x/y}$ and hence $\hat{I}_\pm$ operators, which allow mixing different spin states and creating superpositions thereof. Thus, pulses are crucial building blocks for every modern NMR experiment.

8.2.2
Time Evolution of Spin Systems

To observe and to separate the nuclear interactions, sophisticated experiments are performed in ssNMR [4]. The interpretation of those often requires to simulate the time evolution of spin systems under specific interactions, which is derived by the time-dependent Schrödinger equation [16]:

$$\frac{d}{dt}|\psi\rangle(t) = -i\hat{\mathcal{H}}(t)|\psi\rangle(t), \tag{8.16}$$

where both the wave function $|\psi\rangle(t)$ and $\hat{\mathcal{H}}(t)$ are in general time dependent. The time dependence of $|\psi\rangle$ is caused by rf pulses and the quantum mechanical evolution of non-eigenstates and is described using the spin density formalism. Time dependence in the Hamiltonian is mostly imposed by a physical sample rotation, the magic-angle spinning, which alters the orientation of the interaction tensor V (Eq. (8.1)) periodically (Section 8.2.3).

8.2.2.1 Spin Density Formalism

Since a macroscopic sample always contains a large number of spins, a quantum statistical density formalism [18] is used to describe interactions and time evolution. The spin density matrix [20] $\hat{\rho}$ is defined as the ensemble average of the ket-bra products of the wavefunctions $|\psi_i\rangle = (c_1^i, c_2^i, \ldots, c_n^i)$ of the N individual spin systems:

$$\hat{\rho} = \frac{1}{N}\sum_{i=1}^{N}|\psi_i\rangle \otimes \langle\psi_i| = \overline{|\psi_i\rangle \otimes \langle\psi_i|} \tag{8.17}$$

and consists of N^2 elements $c_i c_j^*$ with $i,j \in \{1,\ldots,N\}$. The diagonal elements $\rho_{jj} = \overline{c_j c_j^*}$ are the probabilities of finding an individual spin system in the state $|\psi_j\rangle$ and are therefore commonly called populations [20]. The off-diagonal elements $\rho_{ij} = \overline{c_i c_j^*}$ ($i \neq j$) are called $\pm\Delta m$ coherences [20] (e.g., ± 1, ± 2, ...), according to the difference $\Delta m = m_j - m_k$. Since the existence of coherence implies that different spins systems are coherently in a superposition state, they cannot exist in thermal equilibrium due to the maximization of entropy [20].

In a static external magnetic field, the Zeeman interaction leads to $\hat{\rho} \sim \hat{I}_z$ in thermal equilibrium, that is, only populations exist weighted by the Boltzmann distribution. The time evolution of $\hat{\rho}$, subjected to a Hamiltonian $\hat{\mathcal{H}}$, is described by the von Neumann equation [4]:

$$\frac{\partial}{\partial t}\hat{\rho} = [\hat{\rho}, \hat{\mathcal{H}}]. \tag{8.18}$$

The density matrix $\hat{\rho}(t)$ at a time t is then derived from the density matrix at a time t_0 by [4]

$$\hat{\rho}(t) = \hat{\mathcal{U}}(t_0 \to t)\hat{\rho}(t_0)\hat{\mathcal{U}}^\dagger(t_0 \to t), \tag{8.19}$$

with a propagator $\hat{\mathcal{U}}(t_0 \to t)$:

$$\hat{\mathcal{U}}(t_0 \to t) = e^{-i\int_{t_0}^{t}\hat{\mathcal{H}}(t')dt'}. \tag{8.20}$$

The result of a measurement using quadrature detection may be derived [4] by

$$s(t) \propto \mathrm{Tr}\left\{\hat{\mathcal{I}}^+ \hat{\rho}(t)\right\}. \tag{8.21}$$

In addition to this quantum mechanical time evolution, relaxation also leads to a time dependence of the nuclear spin states [4]. Hereby, two types of relaxation are macroscopically differentiated. First, the longitudinal or spin–lattice relaxation, characterized by a time constant T_1, transfers magnetization $M_z(t)$ parallel to the external static B_0 field into the thermal equilibrium state according to $M_z(t) \propto 1 - e^{-t/T_1}$. This requires a change of the populations of the spin density matrix and, therefore, energy exchange with the environment. Second, the transversal or spin–spin relaxation, characterized by a constant T_2, relaxes magnetization perpendicular to B_0 by $M_{\mathrm{trans}}(t) \propto e^{-t/T_2}$. This type of relaxation implies a decay of coherence, which is driven purely by the entropy maximization and does not lead to energy exchange.

8.2.2.2 Simulation of NMR Experiments

If the Hamiltonian can be regarded as time independent between t_0 and t, Eq. (8.20) simplifies to

$$\hat{\mathcal{U}}(t_0 \to t) = e^{-i\hat{\mathcal{H}}\Delta t}, \tag{8.22}$$

with $\Delta t = t - t_0$. Dividing the course of a pulse sequence in such Δt intervals – usually on the order of microseconds or less – allows applying Eq. (8.22) iteratively to simulate numerically the evolution of a complex spin system with selected interactions and pulses [21].

The large amount of uniformly distributed crystallite orientations in a powdered sample is taken into account by various averaging schemes described in the literature [26]. The most common ones are the Lebedev [27], the Zaremba–Conroy–Wolfsberg [28–30], and the REPULSION [26] methods. The orientations provided by these methods are given as sets of weighted Euler angles, where each set describes one orientation of the CAS with respect to the LAS, and simulations are carried out for each set. Various simulation packages exist, such as SIMPSON [21], SPINEVOLUTION [31], GAMMA [32] to name only the most common ones, and the progress in computing possibilities has greatly enhanced their capabilities during recent years [33]. However, the size of the spin density matrix grows exponentially with the number of spins (Eq. (8.17)), which still limits the tractable size of spin systems.

8.2.3
Lineshapes in Solid-State NMR Spectroscopy

8.2.3.1 Magic-Angle Spinning

All nuclear spin interactions $\hat{\mathcal{H}}_{\mathrm{int}}$ are anisotropic in the solid state and thus depend on the orientation of the interaction tensor V (Eq. (8.1)) with respect to $\vec{B}_0$. Since V is defined by the chemical structure of a given material, mechanical rotation of the sample alters the interaction strength and thus the resonance frequency. If the rotation is sufficiently fast and the rotation axis is inclined by the so-called magic angle ($\theta_{MAS} = 54.74°$) with respect to $\vec{B}_0$, the time average of $\hat{\mathcal{H}}_{int}$ over a full rotor period reduces to the isotropic value σ_{iso} (Eq. (8.2)) within first-order perturbation. σ_{iso} is nonzero in case of the chemical shift and the J-coupling and zero for the first-order quadrupolar and the direct dipole–dipole interactions, which allows deriving the isotropic chemical shift directly from the high-resolution solid-state NMR spectra. This setup is called magic-angle spinning [2] and is extensively used in modern NMR spectroscopy.

The effect of the uniaxial rotation is easier to understand by invoking the rotor axis system (RAS), in which the macroscopic sample rotates around the magic-angle spinning axis. The interaction tensors V of the individual nuclei are then rotated from their PAS first into a joint crystal axis system (CAS) and from there into the RAS. The latter is then rotated into the LAS, whereby this rotation becomes periodically time dependent due to MAS [5,21]. These successive transformations introduce a dependence on the second Legendre polynomial $P_2(\cos \theta) = (1/2) \cdot (3\cos^2\theta - 1)$ for first-order perturbations of $\hat{\mathcal{H}}_{int}$ as seen, for example, in Eqs. (8.7), (8.9), and (8.10). The average Hamiltonian over full rotor periods thus scales with $P_2(\cos \theta_{RL})$ where θ_{RL} represents the angle between the rotation axis and $\vec{B}_0$ [6]. When $\theta_{RL} = \theta_{MAS}$, this term and thus the average Hamiltonian become zero. Thus, signal sampling at every rotor period leads to liquid-like spectra without anisotropic lineshapes. Sampling at noninteger multiples of the rotor period, however, leads to spinning sidebands [5], if the spinning speed is slow enough, which will be discussed in the next section. Note that the second-order quadrupolar interaction (Eq. (8.13)) also contains other terms as function of θ, which do not average under MAS, leading to anisotropic lineshapes at all spinning rates [25].

8.2.3.2 Lineshapes

The difference between static and MAS spectra are illustrated by numerical simulations of NMR experiments for various interactions (Figure 8.2). These lead to shifts of the energy eigenstates and hence influence position and splitting of the NMR signals [6] (Figure 8.2, first column). In a single crystal, all translation-related spins have the same chemical surrounding and thus the same interaction tensor V as well as the same polar angles θ and ϕ for the orientation of the external field in the PAS. Hence, they give rise to the same signal and the shift of this signal depends on the orientation of the single crystal with respect to $\vec{B}_0$ (Figure 8.2, second column), where the constants σ_{aniso} and η determine the frequency range (cf. Eqs. (8.7), (8.9)/(8.10), and (8.13)). Since a dipolar-coupled spin pair as well as a single quadrupolar nucleus exhibit more than two energy levels with unequal energy differences, they give rise to several signals, which are affected differently by first- and second-order perturbations [25]. For the case of non-integer I, the central transition between $+1/2$ and $-1/2$ energy levels is affected by only the second-order coupling, whereas the satellite transitions are

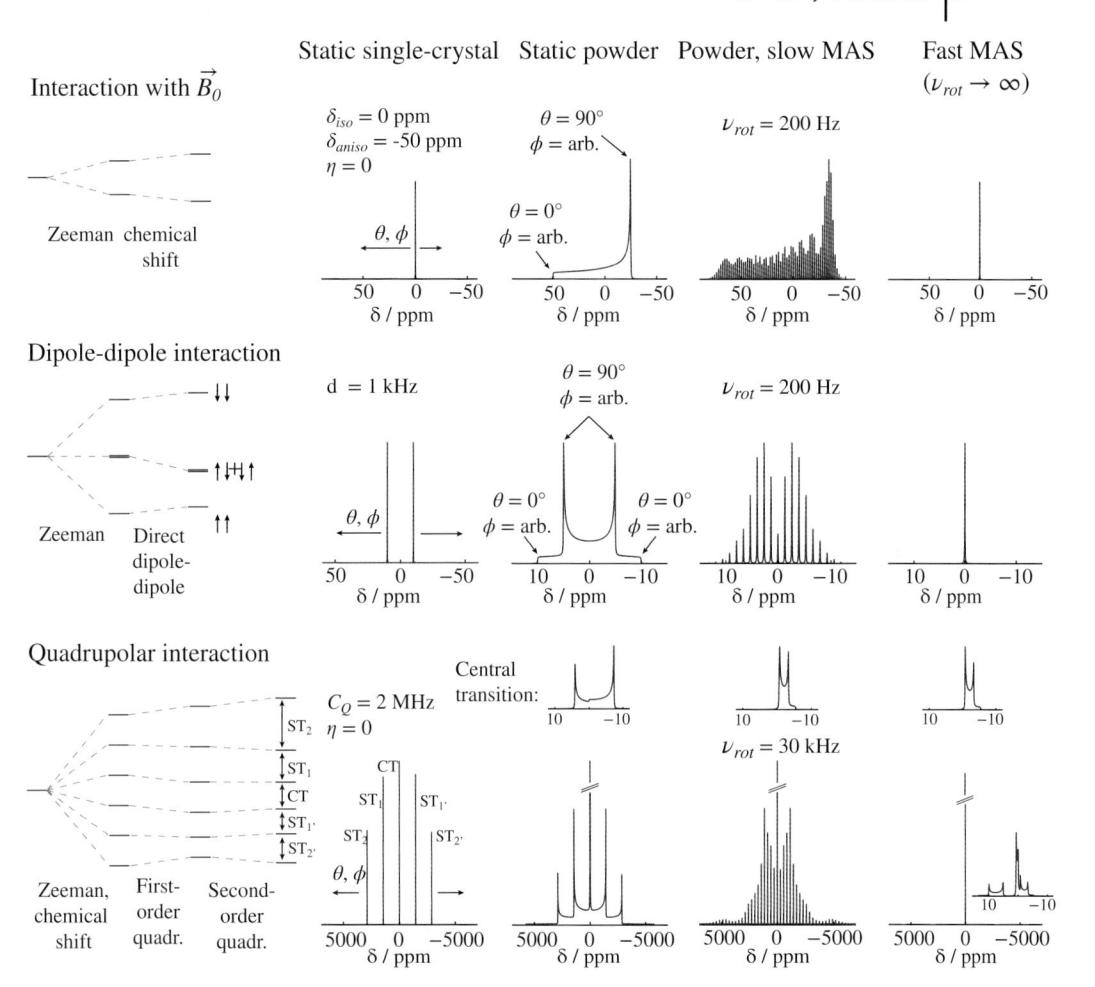

Figure 8.2 Influence of the spin interactions on the energy levels of $I = 1/2$ nuclei (first row, left), a spin-1/2 pair (second row, left), and a spin-5/2 nucleus (third row, left) in a static $\vec{B}_0$ field. The energy shifts are orientation dependent and lead to a variation in the peak positions for single crystals (second column). In a powdered sample, the overlap of the signals from all orientations leads to characteristic lineshapes (third column). Magic-angle spinning decomposes these into sidebands patterns, where the distance between two neighboring lines is the spinning speed (fourth column). If the spinning is significantly faster than the interactions strength ($\omega_{rot} > 10 \cdot \sigma_{aniso}$), only the isotropic parts of the interactions remain visible (last column). For noninteger quadrupolar spins, the central transition is depicted as *insets* above the full spectrum. For the quadrupolar coupling, a magnification of the full spectrum at $\nu_{rot} \to \infty$ illustrates the difference in linewidth to the static case.

affected by both first and second-order shifts. Additionally, every transition exhibits a different excitation behavior (see Ref. [34] for details).

In a static powder sample, uniform orientation distribution of the crystallites causes the whole range of possible frequencies to be present at the same time. The signal is thus the superposition of all these individual signals [4,5], which leads to a distinct shape of the NMR signal, the so-called powder pattern (Figure 8.2, third column). The CSA-type powder shape changes its width with δ_{aniso}, whereas η affects the shape [35]. For a spin-1/2 pair coupled by the direct dipolar interaction (illustrated for the homonuclear case), a so-called Pake doublet [36] is formed, which is the superposition of two CSA-type powder shapes with $\eta = 0$, where both patterns are reflected at the isotropic chemical shift position. For quadrupolar nuclei, the patterns arising from the different possible transitions overlap [25]. Their shapes are mostly affected by first- and second-order quadrupolar coupling as explained above. However, in some rare cases, higher order of perturbation may have a visible effect on the lineshapes.

Slow MAS leads to (partial) averaging of the different interactions and causes spinning sidebands [5]. They appear in multiples of the spinning frequency from the isotropic resonance and resemble the shape of the static powder spectrum for $\nu_{\text{MAS}} \to 0$ (Figure 8.2, fourth column). Since the intensities distributed in a powder pattern are now concentrated on a finite set of signals, the intensities of such sidebands patterns are significantly higher than in a static spectrum. This is used to extract the interaction constants using, for example, the Herzfeld–Berger method [37] or numerical simulations with, for example, SIMPSON [21]. Sufficiently fast spinning eventually leads to full averaging of the interactions, so only the isotropic influences on the NMR lines remain (Figure 8.2, last column). For quadrupolar nuclei, this regime is often not reachable due to the high interaction strength.

8.3
Applications

8.3.1
Wideline Experiments

Since the discovery of MAS [2], most of the ssNMR experiments are done with spinning. However, MAS is not always possible as for *in situ* experiments. For example, Grey and coworkers [38] investigated silicium-based lithium-ion batteries *in situ* using a homemade modified static probe with a flexible plastic battery design. They observed local structural changes for the silicon electrode by following the ^{7}Li NMR signal during charging and discharging the battery (Figure 8.3). Compared to *ex situ* experiments [38], an additional peak at −10 ppm was observed and ascribed to a reaction between

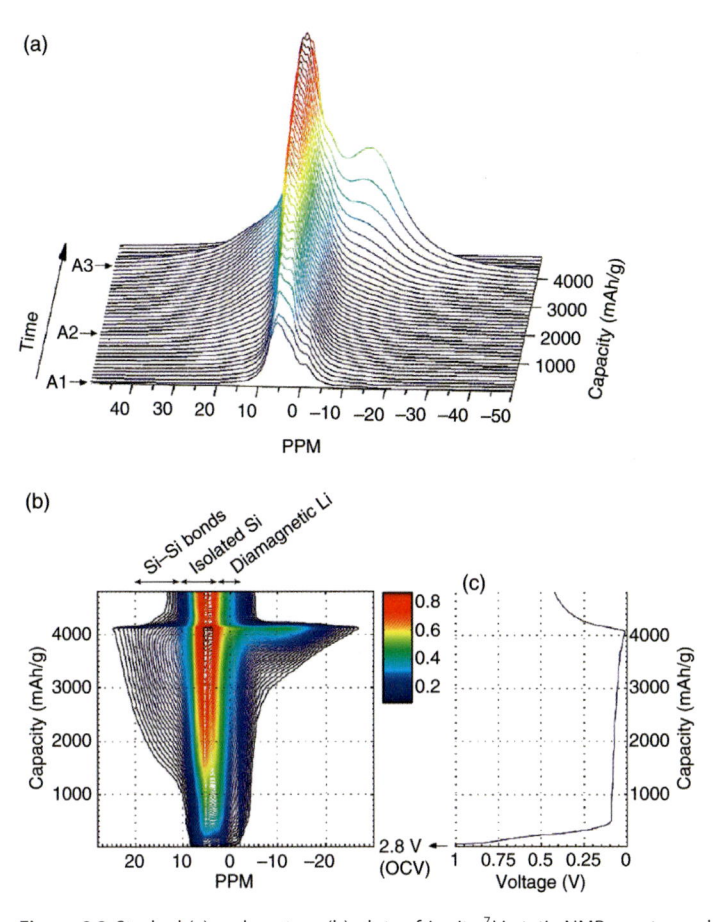

Figure 8.3 Stacked (a) and contour (b) plots of *in situ* ^{7}Li static NMR spectra and electrochemical profile of the first discharge (c) of an actual crystalline Si//Li/Li+ battery. (Adapted from Ref. [38].)

the very reactive, metastable, "Li$_{15}$Si$_4$" phase with the electrolyte, via a "self-charge" mechanism.

In some cases, static experiments have proven to be more useful than MAS, providing more information (e.g., via lineshape analysis) [19]. The chemical shift anisotropy and, for spin >1/2, the quadrupolar interaction can be determined very precisely using static experiments. However, due to the broadness of the signal and hence the short FID, spin or quadrupolar echo experiments [6,39] are required to compensate for the dead time of the probe (often due to acoustic ringing), which can be relatively long compared to the total length of the FID [40,41]. In the case of ^{195}Pt NMR spectra of Pt electrocatalysts, the lineshape became so broad because of the Knight shift, that even a point-by-point acquisition was necessary [42]. Nevertheless, lineshape analyses can still provide information about the surrounding geometry of the observed nuclei as shown for Ga-containing intermetallic phases. In particular, the ^{69}Ga and ^{71}Ga quadrupolar (and thus EFG) tensor turned out to be an excellent probe for the local electronic environment around the Ga atoms [43].

To enhance the usually low sensitivity of static spectra, several groups demonstrated the usefulness of the Carr–Purcell–Meiboom–Gill [44–47] (CPMG or QCPMG for quadrupolar nuclei) experiment that consists of acquiring a train of echoes. The Fourier transform of such an echo train resembles a spikelet spectrum with the overall shape of the regular static one and thus allows to separate the inhomogeneous (e.g., CSA and quadrupolar coupling) from the homogeneous (e.g., homonuclear dipolar coupling) interactions [48] – the latter being represented in the width of each individual spikelet. Broad spectra can be obtained in less time with, however, the drawback of losing some precision in determining the chemical shift and the quadrupolar interaction, respectively. (Q)CPMG experiments were successfully applied on several nuclei too insensitive to be observed, even if acquired with MAS. Examples are ^{207}Pb [49,50], ^{199}Hg [49,50], ^{195}Pt [50], ^{95}Mo [51], ^{14}N [52], and so on.

Pines *et al.* developed a technique called cross-polarization [53–55] (CP), which enables a polarization transfer from abundant spins I with a large γ and hence high polarization (e.g., ^{1}H) to dilute spins S with low γ (e.g., ^{13}C, ^{15}N) via the heteronuclear dipolar interaction. Hereby, continuous rf irradiation on both I and S spins leads to a match of the pseudoenergy level difference in a doubly rotating frame, if the rf amplitudes are chosen in a way that the Hartmann–Hahn condition $\gamma(I)B_1(I) = \gamma(S)B_1(S)$ is fulfilled. CP has several advantages: It enhances the signal of the S spin up to a factor of $\gamma(I)/\gamma(S)$ and only requires to wait for the relaxation of the I spins, which is very often much shorter than that from the S spins, making the overall experimental time much shorter [53–55]. However, CP is not quantitative, as its signal intensities rely on several aspects including the nuclei distances ($\propto 1/r^6$) and their spin–lattice relaxation times in the rotating frame $T_{1\rho}$. Nonetheless, CP is so powerful that it is now routinely used to observe less abundant nuclei such as ^{13}C or ^{15}N but also abundant spins such as ^{31}P since the ^{31}P T_1 are often very long [6,56]. CP can also be applied to quadrupolar nuclei, but it is more troublesome due to the presence of several possible transitions in those cases [57–59].

A crucial wideline NMR application concerns the study of dynamics. Motional processes can be investigated covering more than 14 decades for the jump rates κ by analyzing both spin–spin and spin–lattice relaxation ($10^{-12}s^{-1} \leq \kappa \leq 10^{-3}s^{-1}$), following lineshape changes ($10^{-7}s^{-1} \leq \kappa \leq 10^{-4}s^{-1}$) and performing 2D exchange as well as stimulated echo experiments ($10^{-4}s^{-1} \leq \kappa \leq 10^{2}s^{-1}$) [6,11,60–62]. For example, Spiess *et al.* [63] used temperature-dependent T_1 relaxation measurements and lineshape analyses to identify large-angle jumps as well as their rates about all two- and threefold rotation axes of the tetrahedral P_4 units within the β-phase of white phosphorous at low temperatures.

Similarly, Reimer and coworkers [64] studied the absorption of CO_2 in the Mg_2(dobdc); using ^{13}C lineshapes and the corresponding T_1 data, providing insight into the active CO_2 adsorptions sites and the remaining rotational degrees of freedom for CO_2. More importantly, ^2H ($I = 1$) is ideal to investigate rotational molecular dynamics as its quadrupolar interaction mostly depends on the type of the covalent bond (e.g., CD, ND OD) and thus gives distinct lineshape depending on the rotational processes [62]. In this way, the anisotropic dynamics of liquid crystalline phases was characterized [62]. Also, Hologne and Hirschinger [65] have identified two-site flip motions of dimethyl sulfone in the p-tert-butylcalix[4]arene-nitrobenzene inclusion compounds. Such studies might also be performed under MAS condition [65] or using a CPMG-like echo train to enhance the sensitivity [66].

By making additional use of the anisotropy within relaxation processes (T_{1Z} and T_{1Q}), even for fast motional processes different jump models might be distinguished. This was done, for example, to link the reorientational dynamics of the amide ions within KND_2 salts to the macroscopic phase transitions [67]. In all phases, the amide ion dynamics has to be described as a superposition of thermally activated large-angle jumps and anisotropic librations with characteristic jump geometries within the monoclinic, tetragonal, and cubic phases. In contrast, 2D exchange techniques allow tracing extremely slow motional processes. In this way, the local dynamics of small molecules such as benzene-d_6 or hexamethyl benzene-d_{18} within a rigid polymer matrix could be successfully studied based on ^2H NMR data [68].

Two-dimensional exchange experiments can also provide information about the proximity and interconnectivity of domains. For instance, the formation of two aluminophosphate frameworks, $AlPO_4$-5 and $AlPO_4$-18, has been monitored using hyperpolarized ^{129}Xe gas as a probe demonstrating an intimate contact on mesoscopic length scales between both frameworks [69]. ^{129}Xe NMR spectra [70] are also used to study pore sizes and geometry for a wide variety of porous materials such as zeolites [71], nanotubes [72], or MOFs [14]. Indeed, the ^{129}Xe chemical shift is very sensitive to its surroundings and, following, for example, the Fraissard model [71], one can directly relate the ^{129}Xe isotropic chemical shift with the pore sizes of zeolites.

Finally, 2D exchange wideline spectra can be used to derive connectivities and orientation correlations between neighboring building units if the exchange is driven by spin diffusion. The transfer rates might be enhanced and controlled by applying specific pulse sequences, such as WALTZ17 [73,74] or BLEW12 [75], during the mixing time, where one component of the transverse magnetization is spin-locked in the rotating frame. For the nucleated but still amorphous phase of the molecular glass former triphenyl phosphite, ^{31}P and ^{13}C (Figure 8.4) rf-driven spin diffusion spectra were measured as function of the mixing time to derive average intermolecular molecular distances, the packing density, and the preferred molecular conformer [56,76–78].

8.3.2
High-Resolution Experiments

MAS (Section 2.3) drastically improves the spectral resolution and usually allows distinguishing different chemical environments in a sample. A prominent example is the study of zeolites, where MAS enables to resolve up to five signals in a ^{29}Si NMR spectrum [79]. These signals correspond to the silicon atoms having Q^0, Q^1, Q^2, Q^3, and Q^4 environments, allowing the structures of aluminosilicate networks to be determined. Similarly, ^{31}P MAS NMR spectroscopy has been used to obtain information about the chemical bonding around PO_4 units in various phosphate glasses [80] and the isotropic ^{27}Al NMR chemical shift strongly depends on the Al coordination and nature of bonded atoms. Since ^{27}Al is easy to observe, it has been used to study a plethora of materials [15] and was, for instance, applied to follow hydration processes of alumina cement, where the Al environment changes from tetrahedral in the dry state to octahedral when hydrated [81].

Many other quadrupolar nuclei exhibit, however, broad lineshapes and have a low natural abundance, sometimes coupled with low gyromagnetic ratios, which makes them much harder to measure. Therefore, techniques were developed to increase the sensitivity. Again, QCPMG improves the signal-to-noise ratio [46,47] and population transfer techniques [82] such as RAPT [83], DFS [84], or HS [50] can either saturate or invert the population of the different energy levels in order to achieve the highest possible population differences for the central transition (Figure 8.5). These techniques result in a theoretical signal gain of $2I$, making the experiment up to $(2I)^2$ times faster.

When the spinning speed of MAS is slower than the latitude of the nuclear spin interactions, spinning sidebands arise, which mimic the static powder pattern (Figure 8.2). Fitting these gives access to magnitude and asymmetry of the corresponding interaction and thus provides additional structural information. For example, the analysis of ^{59}Co and ^{6,7}Li MAS spectra of $LiCoO_2$ battery materials allowed for determining quadrupolar coupling constants, which were then correlated to the partial charges of the nuclei using a simple point charge model [85]. The ^{6,7}Li chemical shift anisotropy was shown to arise from the paramagnetism of the sample. Similarly, for dehydrated and water-loaded $Cu_3(BTC)_2$ differences in the ^{1}H–$^{63/65}$Cu hyperfine coupling, manifesting themselves in changes of the ^{1}H NMR spinning sideband intensities, allowed for identifying the active binding sites for water [86].

Many MAS NMR experiments, in particular for low-γ nuclei in a low natural abundance, would still be very difficult to acquire without the use of CP to enhance spectral intensities. For MAS experiments, the CP Hartmann–Hahn condition differs slightly from the one for non-rotating samples to adopt the

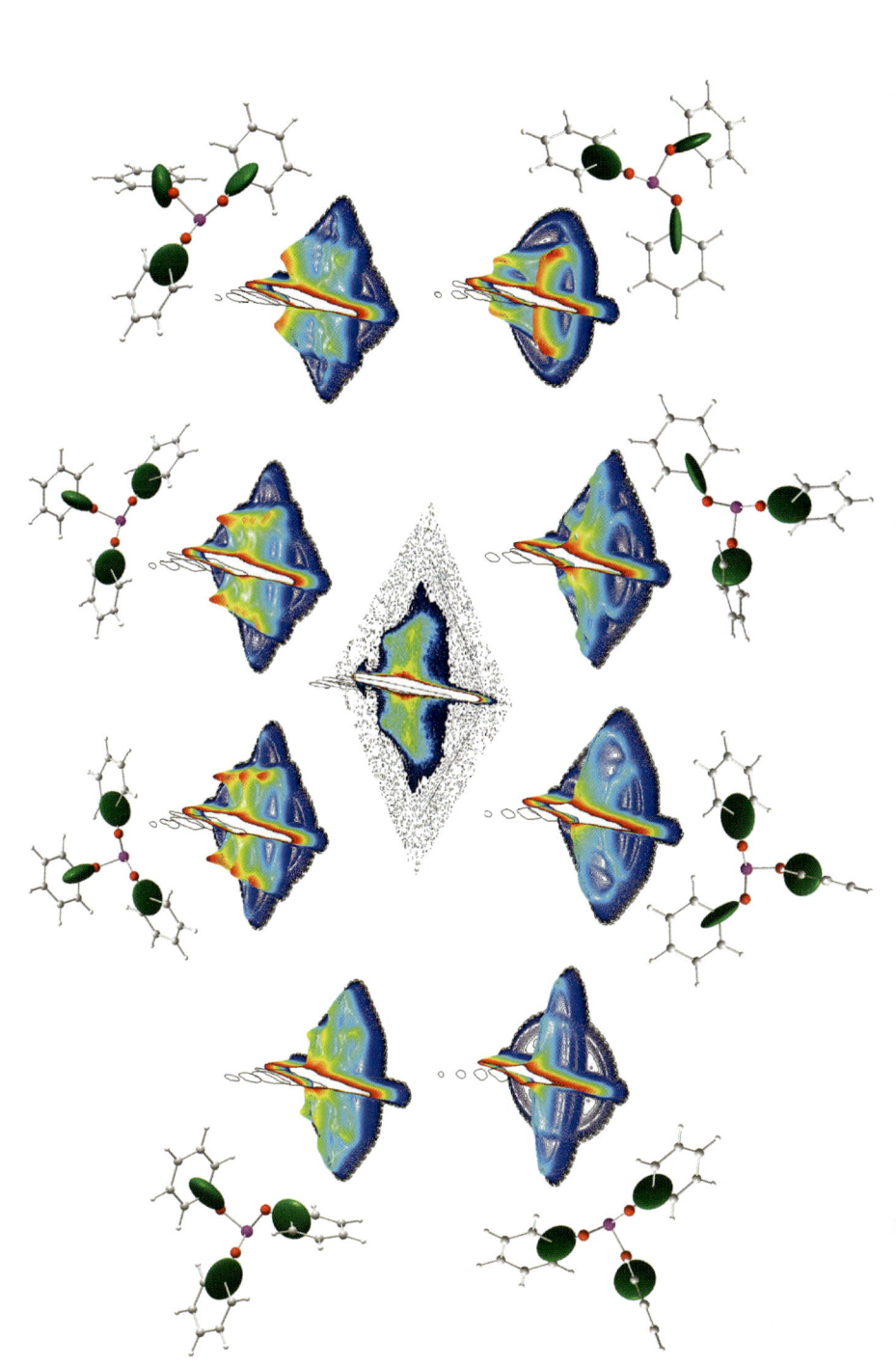

Figure 8.4 2D ^{13}C rf-driven spin diffusion exchange spectra for phase all of triphenyl phosphite. The experimental spectrum ($t_m = 10$ ms) in the center is compared with simulated ones for possible conformers. (Adapted from Ref. [78].)

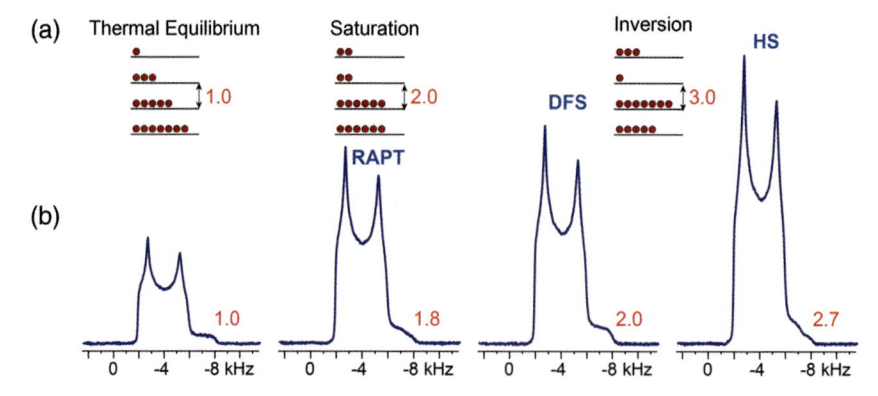

Figure 8.5 *Top:* Schematic representation of the population distribution for spin 3/2 nuclei before and after applying population transfer techniques. *Bottom (left to right):* Comparison of the ^{87}Rb MAS NMR signal intensity of RbClO$_4$ obtained with a Hahn-echo, RAPT, DFS, and HS. The experimental enhancement is shown next to the spectra. (Adapted from Ref. [50].)

additional time dependence due to sample spinning: $\gamma(I)B_1(I) = \gamma(S)B_1(S) \pm n \cdot \nu_{MAS}$ with $(n = 1, 2, \ldots)$ [87]. Only then, ^{15}N CP MAS NMR spectroscopy made it possible to identify the chemical building units for various semicrystalline and amorphous carbon nitride materials, which are promising photocatalysts for water splitting [88–90].

With the increased resolution of MAS, the need for even better resolution arises. For mainly two reasons, interactions might not be averaged out completely using only MAS [10]. First, the spinning speed may not be fast enough, which is often the case for dipolar couplings involving ^{1}H and ^{19}F [6,91–93]. Second, the interaction cannot be canceled by spinning at the magic-angle only, which is the case, for example, for second-order quadrupolar interactions [25,94]. These problems were addressed using a variety of solutions.

When observing nuclei close to protons, ^{1}H heteronuclear decoupling [95,96] needs to be applied during acquisition. While continuous-wave (cw) irradiation is sufficient for non-rotating samples [97], cw appeared to be insufficient in combination with MAS and more refined multiple-pulse schemes for ^{1}H decoupling such as TPPM [98], SPINAL [99], XiX [100], and SW$_f$-TPPM [101] emerged. Their common point is the application of series of pulses with flip angles close to 180° with varying phases, which introduces additional time dependencies, in turn allowing for more efficient averaging of the heteronuclear dipolar interaction. In spite of this, the gain in resolution strongly depends on the magnitude of the dipolar coupling and the internal mobility within the sample [95,96]. The more recent decoupling sequences [101] usually provide narrower resonances, with the drawback, however, of requiring more experimental parameters to optimize. Note that when using a recoupling sequence, which simultaneously requires heteronuclear decoupling, one should only apply cw decoupling to avoid possible interferences with the recoupling [102]. Heteronuclear decoupling is also beneficial for other abundant $I = 1/2$ nuclei with large γ such as ^{31}P and ^{19}F and allowed, for instance, distinguishing pharmaceutical polymorphs with only small differences in the isotropic chemical shifts [103]. For quadrupolar nuclei, specific heteronuclear decoupling sequences were developed [104,105]. Combining ^{1}H and ^{27}Al decoupling, for instance, greatly improved the resolution of ^{31}P NMR spectra for AlPO$_4$-14 [106].

To improve the resolution for spin systems with dominant homonuclear dipole interactions, such as ^{1}H and ^{19}F, is more challenging since detection and decoupling have to be managed on the same channel. To circumvent this problem, homonuclear decoupling was initially used in a 2D fashion with the decoupling sequence being applied in the indirect dimension (t_1 domain) of a 2D separation experiment [107]. The first scheme developed is the Lee–Goldburg (LG) sequence [108]. It relies on an off-resonance cw pulse with a frequency offset $\Delta\omega = \omega_1/\sqrt{2}$, with ω_1 being the rf field strength. This causes the spins to rotate around an effective field $\omega_{\text{eff}} = \sqrt{\omega_1^2 + \Delta\omega^2}$ tilted away from the static field direction by the magic angle. The zero-order average of the dipolar Hamiltonian vanishes in such a case, resulting in an enhanced line narrowing.

When combined with MAS, the so-called CRAMPS (combined rotation and multiple-pulse sequences) experiments arise [5]. The earliest sequence with windows, to enable detection on the same channel, is WHH-4 [109]. It is based on a repetitive cycle of four 90° pulses applied in different directions of the rotating frame by adapting the pulse phases and having two different interpulse delays, which allow for averaging the homonuclear interaction while retaining but scaling the chemical shift [5]. Based on this idea, more stable sequences such as MREV-8 [110,111], BR-24 [112], BLEW12 [75], MSHOT-3 [113], FSLG [114], symmetry-based R and C sequences [115,116], PMLG [117], and DUMBO [118] (a numerically optimized continuous-phase modulation) were developed. They all have their effective field tilted away from the static field direction by a certain angle α, which reduces shift interactions by the so-called scaling factor. When α is equal to the magic angle (e.g., for LG, WHH-4, PMLG, DUMBO), the theoretical scaling factor is $1/\sqrt{3}$. Other sequences, with different α, lead to other scaling factors such as $\sqrt{2/3}$ for MREV-8 [19]. For further details, refer to Refs [107,119,120]. Homonuclear techniques have been broadly applied to organic molecules [121,122], drugs [123–125], inorganic–organic hybrid materials [126], and inorganic compounds [127] where they are often used in combination with 2D recoupling sequences.

The second-order quadrupolar interaction might also significantly broaden the NMR spectrum [94]. However, it is inversely proportional to B_0 and thus is diminished by going to higher fields, as shown by Gan et al. [128]. Nevertheless, in most cases, B_0 is not large enough and additional techniques have to be applied to reach narrow lines. The second-order effect cannot be averaged solely by MAS since it also depends on the fourth-rank Legendre polynom (line 4 in Eq. (8.13)), which requires spinning at angles of 70.12° or 30.56°. The obvious solution is to rotate the sample at two angles at the same time. This is done by DOR [129], which consists of two rotors: an outer one inclined at the magic angle and an inner one inclined at 70.12° or 30.56°. Another solution, DAS [130], uses only one rotor, which toggles between two angles (37.38° and 79.19°) for two equal periods of time with a z-filter in-between. Using ^{23}Na DOR experiments, the structures of two sodium diphosphates ($Na_4P_2O_7$ and $Na_3HP_2O_7 \cdot H_2O$) could be refined based on 2D ^{23}Na spin diffusion correlation spectra as function of the mixing time [131].

In 1995, Frydman and coworkers [132,133] showed that, by manipulating the spin part of the Hamiltonian during MAS, one can obtain high resolution in a so-called 2D MQMAS experiment by correlating a multiple-quantum transition such as 3Q ($-3/2 \leftrightarrow +3/2$) or 5Q ($-5/2 \leftrightarrow +5/2$) [134] with the single-quantum coherence. Later, Gan [135] introduced the STMAS method, which correlates single-quantum satellite transitions with the central one. STMAS has been shown to often have an improved signal-to-noise ratio over MQMAS [136]. However, it is very sensitive to setting the magic angle precisely and is more sensitive to dynamical processes [136]. Both MQMAS and STMAS have been improved over the years and several variants are now available, such as z-filter [137], SPAM-MQMAS [138], or DQF STMAS [139], using shearing [4] or split-t_1 [140] schemes. More details are found in Refs [136,141,142]. Both techniques are meanwhile indispensable for a wide range of materials containing quadrupolar nuclei [15,94]. For example, ^{17}O MAS and MQMAS experiments allowed distinguishing the six polymorphs of $MgSiO_3$ demonstrating the sensitivity of the ^{17}O quadrupole interaction toward its local environment [143]. For orthoenstatite, all six ^{17}O sites could be resolved (Figure 8.6) revealing dependencies between the ^{17}O chemical shift, the quadrupolar coupling constant, and the coordination environment as well as the Si—O bond length.

As mentioned previously, the analysis of spinning sideband patterns (ssbps) provides additional valuable information about structural and dynamical aspects. To derive proper parameters becomes difficult, however, if the ssbps of several signals overlap to heavily form an unambiguous deconvolution of 1D spectra as is often the case for ^{13}C NMR spectra of organic molecules and drugs [103,144,145]. For quadrupolar nuclei with strong second-order effects, additionally, the signal of the central transition becomes too broad and overlaps with the ssbps of the satellites [146,147] In these cases, spinning sidebands separation techniques are helpful. For example, the 2D PASS experiment [148] separates the CSA from isotropic chemical shifts. It is based on a row of π pulses with varying interpulse distances. Changing the timing between the pulses allows the isotropic part of the chemical shift to evolve and the ssbps are sorted according to their order in 1D slices. A similar sequence, QPASS [146], has been developed for quadrupolar nuclei and was used, for example, on ^{71}Ga of $GaPO_4$ quartz [147].

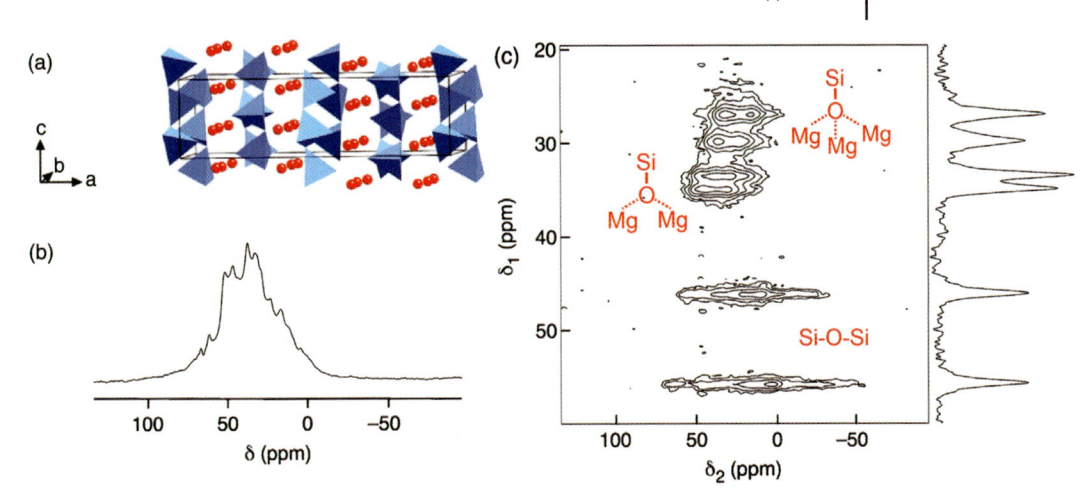

Figure 8.6 (a) Structure of orthoenstatite (MgSiO$_3$) with its corresponding (b) ^{17}O MAS and (c) ^{17}O MQMAS spectra. (Adapted from Ref. [143].)

The CSA can also be recoupled when using fast MAS and two classes of pulse schemes may be differentiated. The first one recouples CSA during the indirect dimension of a 2D MAS experiment, while recording high-resolution MAS spectra in the direct dimension. The earliest sequences in this respect used rotor-synchronized pulses [149,150]. Subsequently, CODEX [151,152] was developed, which is particularly valuable for probing dynamical processes [151,153]. Similar recoupling schemes [154–156] differ mainly in the type of the recoupling pulses. Rotary resonance [157], phase-shifted FSLG [158], and symmetry-based recoupling [159–161] might also be used to recouple the CSA with the advantage that the homonuclear interaction can also be suppressed. The second class also records 2D MAS spectra with high resolution in the direct dimension, but creates ssbps in the indirect dimension with an apparent sample spinning $\nu_{app} = \nu_{MAS}/s$ reduced by an adjustable factor s compared to the actual spinning speed ν_{MAS}. These might alternatively be viewed as a magnification of the CSA in the indirect dimension [162–170].

8.3.3
Dipolar Recoupling Experiments

The gain in resolution obtained through MAS is accompanied by an undesired loss of information, since MAS reduces or cancels the anisotropic parts of interactions, containing valuable information about molecular orientations, distances, and connectivities [12,144]. Extensive efforts were, therefore, made to selectively reintroduce desired interactions by preventing their averaging, while others remain suppressed [9]. This concept is called recoupling and it utilizes the fact that the time dependence of V (Eq. (8.1)) may be accompanied by a periodically time-dependent spin manipulation or interaction that interferes with MAS. Additionally, spin interactions behave differently under spin or space rotations, as characterized by their space and spin rank [115]. For instance, if the space part averaged by MAS is accompanied by a rotation of the spin parts $\vec{I}$ and/or $\vec{J}$ by precisely timed, periodically repeated pulses, both rotations interfere. For defined time intervals, the effective Hamiltonian might then be calculated using Average Hamiltonian theory [171,172] and Floquet theory [173], respectively, allowing for tuning pulse sequences so that only desired interactions are recoupled [22,174].

The majority of recoupling sequences focus on the reintroduction of homo- and heteronuclear dipole–dipole interactions for $I = 1/2$ nuclei. They allow for measuring distances in case of small clusters as well as isolated spin pairs and probing distance sums for dense extended spin systems. Additionally, "through-space" or "through-bond" connectivities might be derived from 2D correlation spectra [12,175], depending on whether the direct dipole interaction or the *J*-coupling is recoupled [12]. Although the principles of recoupling also apply to quadrupolar interactions, its magnitude causes severe perturbations [176] (since already the averaging of it by MAS is incomplete) and hence many sequences may only provide qualitative results [176,177].

8.3.3.1 Homonuclear Dipolar Recoupling

Homonuclear dipolar recoupling (HomDR) experiments [178] are frequently used for structural investigations. They may be applied either to abundant nuclei (e.g., ^{1}H, ^{31}P, and ^{19}F) [102,126,179,180] or rare spins (e.g., ^{13}C, ^{15}N, and ^{29}Si) [181,182]. Due to usually strong signal intensities, the former are easier to record with the drawback that dense arrangements of NMR-active nuclei require large spin systems for proper simulations with the potential to give access to quantitative structural data [102,180]. In contrast, if the sensitivity issue can be overcome, HomDR measurements on rare spins are particularly powerful, since they allow deriving individual distances based on simple two-spins system simulations [182–184].

The simplest HomDR method is rotational resonance (R^2) [185,186]. For MAS experiments, the B-terms within Eq. (8.9) is reintroduced even in the absence of rf fields, when the chemical shift difference between two spins, coupled by the direct homonuclear dipolar interaction, matches a multiple of the spinning speed, which manifests in broadening the resonance lines [185]. The accompanied magnetization exchange between spins by energy-conserving flip-flop processes – also referred to as zero-quantum (ZQ) transfer or spin diffusion – was used to determine intramolecular ^{13}C–^{13}C distances as in zinc acetate [186]. The rotational resonance condition may also be reintroduced by spin echoes as for radio-frequency-driven dipolar recoupling (RFDR) [187,188] and variants such as the finite-pulse RFDR (fpRFDR) [189]. In particular, working with finite real pulses has proven to be significantly more robust and broad-banded than the basic R^2 scheme. As such, RFDR was used to obtain spatial proximities, for example, in the metal–organic framework MIL-121 [190] and phosphate glasses [191]. Usually, when simulating ZQ recoupling schemes, spins within a distance of several tens of angstrom around a central probe nucleus have to be taken into account [192]. This makes density matrix simulations of NMR observables quite demanding and is only possible for special cases exhibiting isolated small clusters as for ^{13}C-labeled p-xylene/Dianin inclusion compounds [193]. Alternatively, a phenomenological rate matrix approach [194], treating the spin exchange as classic diffusion problem, is commonly employed, as, for example, for SrP_2N_4, where ^{31}P fpRFDR exchange spectra measured as function of the mixing time allowed confirming the network topology of corner-sharing PO_4 tetrahedra [192].

Instead of rf pulses, strong heteronuclear dipole interactions such as ^{1}H–X may be used to broaden the X resonance and thus counterbalance chemical shift differences and MAS leading to proton-driven spin diffusion variants of rotational resonance [195–197] (PDSD). Quantitative evaluation of such experiments is difficult [198,199] since the exchange kinetics now depends on both the efficiency of the energy match and the distance sums. For the detection of connectivities, PDSD is a valuable tool and was, for instance, used as restraints in the structure solution of proteins [200], as a cost function for *ab initio* structure solution in combination with quantum chemical modeling [201], and to investigate local order in polymers [194]. Additionally, ^{1}H–X couplings are actively reintroduced by rf fields dipolar-assisted rotational resonance [202] (DARR) to adjust to higher MAS spinning speeds. This sequence is often used for proteins [203] to detect through-space connectivities [9] up to 10 Å.

As alternative to ZQ recoupling, double-quantum (DQ) schemes were developed [10,179]. While the former allowing for treating large spin systems with weak dipolar couplings using kinetic matrices, the latter is advantageous and more accurate for small spin clusters and shorter distances. Most DQ sequences truncate small dipolar interactions in the presence of strong dipole ones, and may not be detected under these circumstances anymore [204]. DQ recoupling was first used in dipolar recovery at the magic angle (DRAMA) [205], where the interference of rf pulses synchronized with the periodic manipulation of the interaction tensor reintroduces DQ homonuclear dipolar interaction terms. This sequence is sensitive to CSA and thus more robust schemes such as homonuclear rotary resonance (HORROR) [206], an adiabatic version, dipolar recoupling enhanced by amplitude modulation (DREAM) [207], and the dipolar recoupling with a windowless sequence (DRAWS) [208] were developed. Here, continuous rf irradiation is used to cause interferences when the rf nutation frequency matches the spinning speed. These sequences were very successfully employed for determination of torsion angles [209] and distances [210] in biological macromolecules, for distance measurements in crystalline organic solids such as nucleic acids [211] and cytidines [212], or to detect Li–Li proximities in lithium over stoichiometric ($x>1$) Li_xCoO_2 battery materials [213].

The back-to-back (BABA) sequence [214,215] uses short 90° pulses instead of a continuous irradiation to achieve HomDR. Therefore, it is well suited for fast MAS and long recoupling times, where high power during continuous irradiation may not be feasible. It was applied to probe through-space proton connectivities in Ba(ClO$_3$)$_2$·H$_2$O salts [216], the local connectivities of PO$_4$ polyhedra in phosphate glasses [217], or to investigate ^{31}P–^{31}P distances in magnesium ultraphosphate [215] MgP$_4$O$_{11}$. Various other sequences employing DQ and ZQ recoupling were identified by applying symmetry arguments [22,183,184,218,219]. This resulted in a variety [220] of R and C sequences, which specifically allowed recoupling selected terms of arbitrary spin interactions, including robust pulse schemes for HomDR. The POST-C7, R14$_2^6$ and SR26$_4^{11}$ sequences proved valuable tools for structure elucidation in inorganic CN materials [221], zeolites [182,222], fluorinated organic materials [102], or phosphate glasses [223]. In particular, the compensation for CSA and pulse imperfections due to supercycling [224] allows for measuring weak dipolar couplings on the order of a few tens of hertz, which was essential for the structure elucidation of supramolecular nano-objects consisting of 1,3,5-tris(2,2-dimethylpropionyl amino)benzene (BTA) additives within the bulk and *i*PP/BTA composites (Figure 8.7) [181,225].

8.3.3.2 Heteronuclear Dipolar Recoupling

Heteronuclear dipolar recoupling (HetDR) covers a range of powerful methods for polarization enhancement and structural investigations. The aforementioned CP technique may be used to detect through-space connectivities and for spectral editing for inorganic and organic materials [12,144,226]. The Lee–Goldburg CP [227] (LG-CP) adds homonuclear decoupling on the *I* channel to suppress spin diffusion, which enables to determine heteronuclear *I–S* distances more accurately [228]. Additionally, it might be used in a spectral-editing fashion by selectively transferring polarization to nearest neighboring spins, for example, to improve the assignment of small organic molecules [229]. Rotary resonance recoupling (R^3) [230,231] and improved versions using amplitude and phase modulations [232–234] recouple the heteronuclear dipolar interaction by matching the sample rotation with the nutation frequency of the spin-locked nuclei of one of the two targeted spin species. This, for instance, allowed for measuring heteronuclear ^{1}H–^{13}C distances of directly bonded pairs in adamantane, ferrocene, and hexamethylbenzene [235].

Echo double-resonance schemes such as spin-echo double resonance (SEDOR) [236,237] for static samples and the MAS experiments rotational-echo double resonance (REDOR) [238,239] as well as transferred-echo double resonance (TEDOR) [240] apply cleverly timed rf pulses to avoid averaging *I–S* dipolar couplings. They can be used to determine *I–S* spin pair distances in selectively enriched materials and provide heteronuclear connectivities, distance sums, and other geometric information for extended spin systems. For instance, ^{23}Na–^{7}Li SEDOR was used to investigate the cation clustering in mixed-alkali disilicate glasses [241] by analyzing second moments. For MAS, REDOR allowed

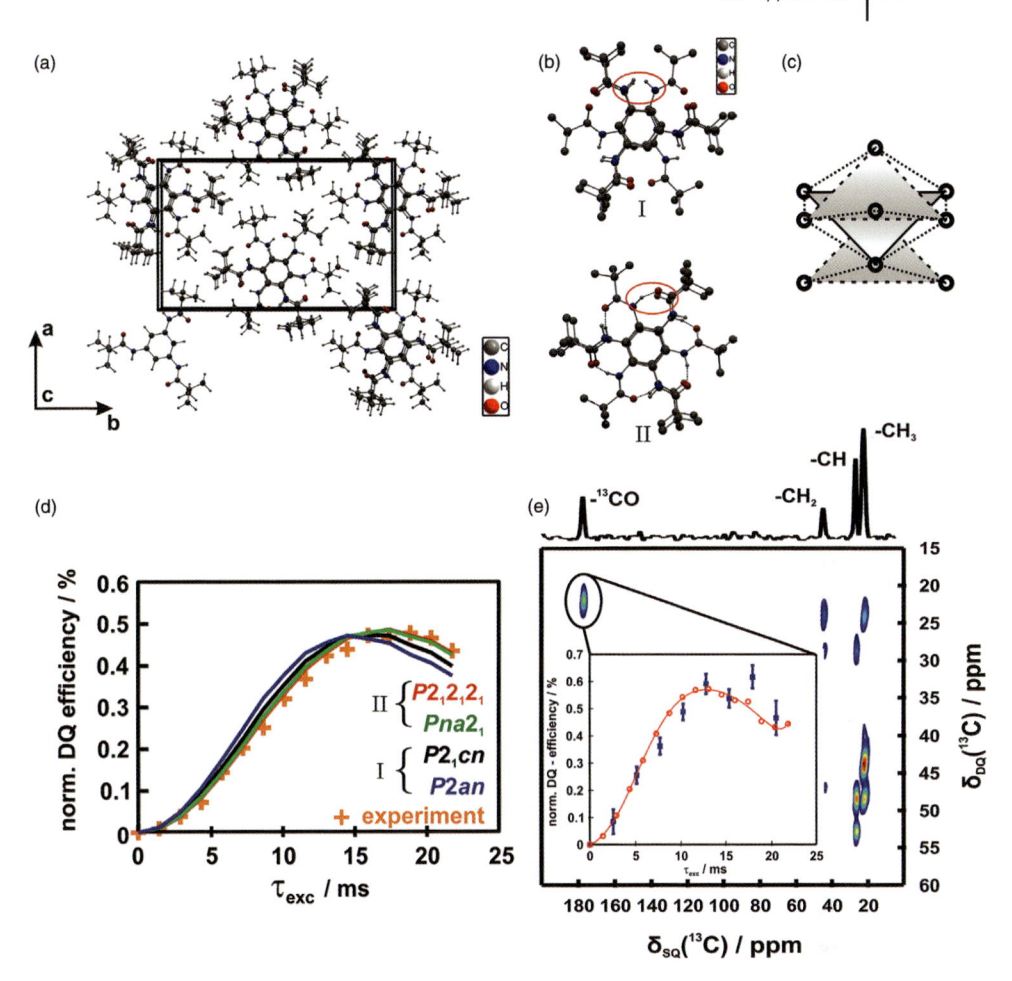

Figure 8.7 (a) Potential structure models derived from powder X-ray diffraction for neat BTA. (b) In all cases, columnar stacks of molecules are formed either without (I) or with (II) intermolecular hydrogen bonds. (c) From these models, nine-spin systems (black circles) were extracted by selecting three subsequent molecules (gray triangles) within each stack for simulating ^{13}C DQ buildup curves. (d) Comparison between the experimental and simulated DQ buildups strongly favor stacks with a helical hydrogen bond network (topology II). (e) Buildup curve for the autocorrelation signal of the additives in the polymer composite at a concentration of 0.1 wt% (blue markers, *inset*) confirms the existence of similar stacks. Red circles (*inset*) depict the neat additive for comparison. The BTA was ^{13}C enriched at the C=O position for these NMR experiments. (Adapted from Refs. [185,225].)

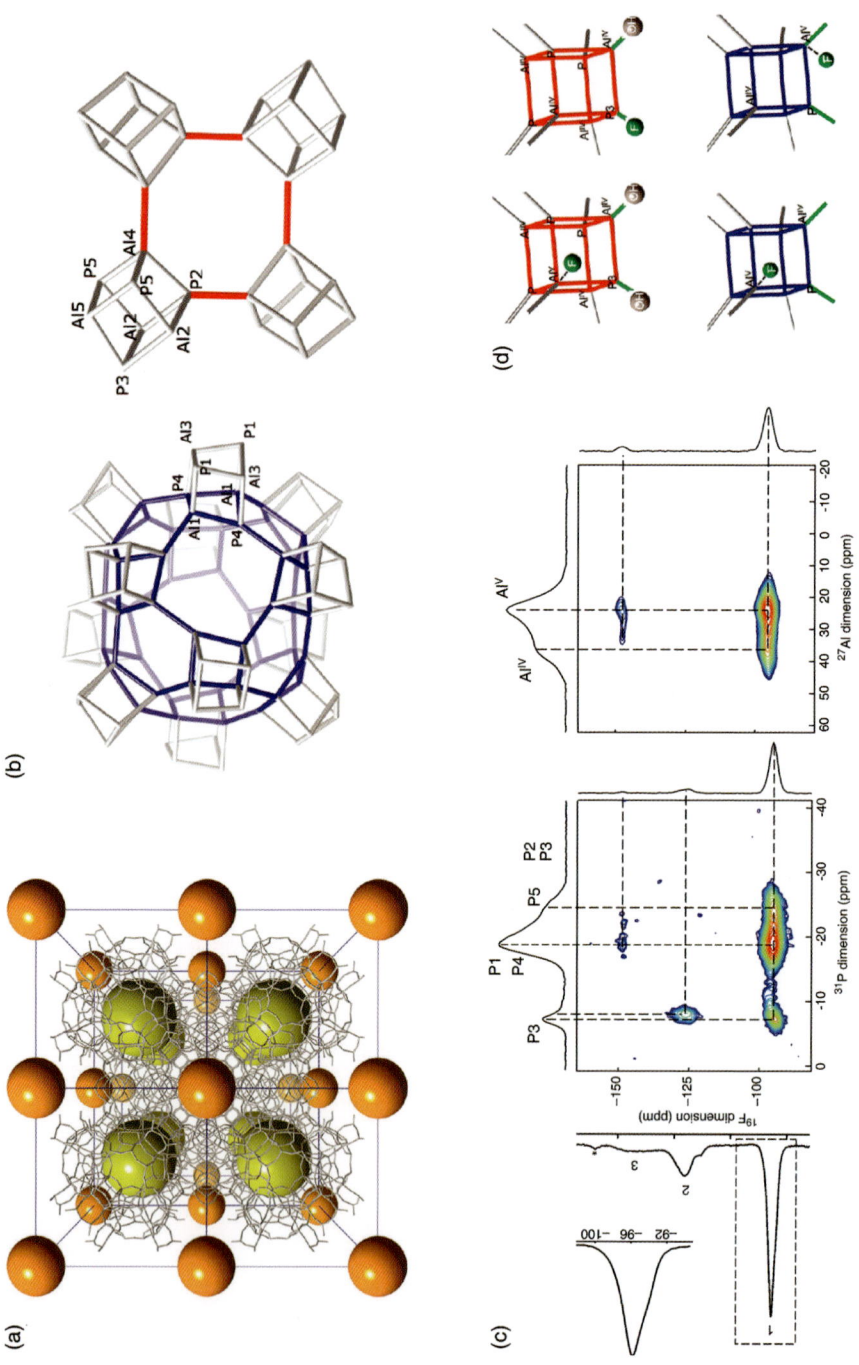

Figure 8.8 (a) The framework of fluorinated aluminophosphate cloverite [180] exhibits pores (yellow spheres) and channels (filled with orange spheres). (b) The basic building units in this structure are double 4-rings (*D4R*), which form cubic cages. Two different *D4R* types exist due to the two different ways of connecting them. (c) The 1D ^{19}F MAS as well as the 2D ^{19}F–^{31}P and ^{19}F–^{27}Al HETCOR spectra (from left to right) reveal three distinct ^{19}F signals. The ^{19}F assigned to peak 1 are in proximity with both the Al and P atoms, whereas the ones for resonances 2 and 3 are either connected to Al or to P. This indicates that F_1 is trapped in the *D4R* cages, while F_2 and F_3 are covalently bonded to either Al or P, thus interrupting the framework. (d) Possible configurations for the two types of *D4R* units (red and blue) with terminal OH and F outside of the cages. (Adapted from Ref. [180])

for analyzing the heterogeneous mesoscale spatial apportionment of functional groups in mixed linker MOFs by comparing the experimental dephasing with spin pair simulations for structural models with ordered and disordered linker arrangements [242]. Furthermore, it is frequently used to derive distance and other geometric restraints in a variety of organic natural products [243]. TEDOR with its better background suppression was employed to unravel the lithium-ion coordination sites in a lithium–glycine–water complex by extracting Li–C distances from 2D experiments [244] and to demonstrate the influence of side group functionalization of poly(arylene vinylene) copolymers on the molecular packing and hence on the hole mobility in organic thin film transistors [245].

The aforementioned symmetry-based R and C sequences [22,183,184,218,219] also allow recoupling the heteronuclear dipolar interaction. The PRESTO scheme [246], such as LG-CP, uses transfer of heteronuclear polarization to estimate distances and connectivities, while suppressing undesired homonuclear dipole–dipole couplings and in particular strong CSA interactions. For example, PRESTO was used to assign chemical groups to supramolecular assemblies [247] and functionalized MOFs [126]. The supercycled symmetry-based $SR\,4_1^2$ sequence [219,248] was applied to measure three-dimensional $^{13}C-^{14}N$ HETCOR experiments [249] in a HMQC-like way [250,251], in spite of the low γ and large C_Q of ^{14}N. In a fluorinated Al cloverite (Figure 8.8a) [180], $^{31}P-^{27}Al$, $^{1}H-^{31}P$, $^{19}F-^{31}P$, and $^{19}F-^{27}Al$ CP HETCOR experiments allowed for linking the periodic framework to a non-periodic subnetwork of fluorine. Two kinds of fluorine atoms have to be distinguished: F^- ions trapped in $D4R$ units (F_1) and F atoms (F_2 and F_3) covalently bonded to terminal Al or P atoms, which interrupt the AlPO network.

References

1 Pake, G.E. (1948) *J. Chem. Phys.*, **16**, 327.
2 Andrew, E.R., Bradbury, A., and Eades, R.G. (1958) *Nature*, **182**, 1659.
3 Ernst, R.R. and Anderson, W.A. (1966) *Rev. Sci. Instrum.*, **37**, 93.
4 Ernst, R.R., Bodenhausen, G., and Wokaun, A. (1991) *Principles of Nuclear Magnetic Resonance in One and Two Dimensions*, Clarendon Press, Oxford.
5 Haeberlen, U. (1976) *High Resolution NMR in Solids: Selective Averaging*, Academic Press, New York.
6 Schmidt-Rohr, K. and Spiess, H.W. (1994) *Multidimensional Solid-State NMR and Polymers*, Academic Press, London.
7 Schaefer, J. and Stejskal, E.O. (1976) *J. Am. Chem. Soc.*, **98**, 1031.
8 Bennett, A.E., Griffin, R.G., and Vega, S. (1994) *NMR Basic Principles and Progress*, vol. **33** (eds P. Diehl, E. Fluck, H. Günther, R. Kosfeld, and J. Seelig), Springer, Berlin, pp. 1–77.
9 de Paepe, G. (2012) *Annu. Rev. Phys. Chem.*, **63**, 661.
10 Duer, M.J. (2004) *Introduction to Solid-State NMR Spectroscopy*, Blackwell, Oxford.
11 Spiess, H.W. (1978) *NMR Basic Principles and Progress*, vol. **15** (eds P. Diehl, E. Fluck, and R. Kosfeld), Springer, Berlin, pp. 55–214.
12 Martineau, C., Senker, J., and Taulelle, F. (2015) *Annu. Rep. NMR Spectrosc.*, **82**, 1.
13 Mesch, M.B., Bärwinkel, K., Krysiak, Y., Martineau, C., Taulelle, F., Neder, R.B.,

Kolb, U., and Senker, J. (2016) *Chem. Eur. J.*, **22**, 16878.

14 Hoffmann, H., Debowski, M., Müller, P., Paasch, S., Senkovska, I., Kaskel, S., and Brunner, E. (2012) *Materials*, **5**, 2537.

15 MacKenzie, K.J.D. and Smith, M.E. (2002) *Multinuclear Solid-State NMR of Inorganic Materials*. Pergamon.

16 Abragam, A. (1994) *The Principles of Nuclear Magnetism*, Clarendon Press, Oxford.

17 Wasylishen, R.E. (2009) Dipolar and indirect coupling: basics, *eMagRes MRI NMR*, John Wiley & Sons, Ltd, Chichester.

18 Slichter, C.P. (1990) *Principles of Magnetic Resonance*, Springer, Berlin.

19 Mehring, M. (1983) *Principles of High Resolution NMR in Solids*, Springer, Berlin.

20 Levitt, M.H. (2008) *Spin Dynamics: Basics of Nuclear Magnetic Resonance*, John Wiley & Sons, Ltd, Chichester.

21 Bak, M., Rasmussen, J.T., and Nielsen, N.C. (2000) *J. Magn. Reson.*, **147**, 296.

22 Levitt, M.H. (2008) *J. Chem. Phys.*, **128**, 52205.

23 Facelli, J.C. (2007) Shielding calculations, *eMagRes MRI NMR*, John Wiley & Sons, Ltd, Chichester.

24 Casabianca, L.B. and de Dios, A.C. (2008) *J. Chem. Phys.*, **128**, 052201.

25 Man, P.P. (2007) Quadrupolar interactions, *eMagRes MRI NMR*, John Wiley & Sons, Ltd, Chichester.

26 Bak, M. and Nielsen, N.C. (1997) *J. Magn. Reson.*, **125**, 132.

27 Eden, M. and Levitt, M.H. (1998) *J. Magn. Reson.*, **132**, 220.

28 Zaremba, S.K. (1966) *Ann. Mat. Pure Appl.*, **73**, 293.

29 Conroy, H. (1967) *J. Chem. Phys.*, **47**, 5307.

30 Cheng, V.B. (1973) *J. Chem. Phys.*, **59**, 3992.

31 Veshtort, M. and Griffin, R.G. (2006) *J. Magn. Reson.*, **178**, 248.

32 Smith, S.A., Levante, T.O., Meier, B.H., and Ernst, R.R. (1994) *J. Magn. Reson. A*, **106**, 75.

33 Tošner, Z., Andersen, R., Stevensson, B., Edén, M., Nielsen, N.C., and Vosegaard, T. (2014) *J. Magn. Reson.*, **246**, 79.

34 Hajjar, R., Millot, Y., and Man, P.P. (2010) *Prog. Nucl. Magn. Reson. Spectrosc.*, **57**, 306.

35 Grant, D.M. (2007) Chemical shift tensors, *eMagRes MRI NMR*, John Wiley & Sons, Ltd, Chichester.

36 Wasylishen, R.E. (2007) Dipolar and indirect coupling tensors in solids, *eMagRes MRI NMR*, John Wiley & Sons, Ltd, Chichester.

37 Herzfeld, J. and Berger, A.E. (1980) *J. Chem. Phys.*, **73**, 6021.

38 Key, B., Bhattacharyya, R., Morcrette, M., Seznec, V., Tarascon, J.-M., and Grey, C.P. (2009) *J. Am. Chem. Soc.*, **131**, 9239.

39 Hahn, E.L. (1950) *Phys. Rev.*, **80**, 580.

40 Buess, M.L. and Petersen, G.L. (1978) *Rev. Sci. Instrum.*, **49**, 1151.

41 Jaeger, C. and Hemmann, F. (2014) *Solid State Nucl. Mag.*, **63**, 13.

42 Tong, Y., Belrose, C., Wieckowski, A., and Oldfield, E. (1997) *J. Am. Chem. Soc.*, **119**, 11709.

43 Haarmann, F., Koch, K., Jeglič, P., Pecher, O., Rosner, H., and Grin, Y. (2011) *Chem. Eur. J.*, **17**, 7560.

44 Carr, H.Y. and Purcell, E.M. (1954) *Phys. Rev.*, **94**, 630.

45 Meiboom, S. and Gill, D. (1958) *Rev. Sci. Instrum.*, **29**, 688.

46 Larsen, F.H., Jakobsen, H.J., Ellis, P.D., and Nielsen, N.C. (1997) *J. Chem. Phys. A*, **101**, 8597.

47 Siegel, R., Nakashima, T.T., and Wasylishen, R.E. (2005) *Concept Magn. Reson. A*, **26**, 62.

48 Maricq, M.M. and Waugh, J.S. (1979) *J. Chem. Phys.*, **70**, 3300.

49 Hung, I., Rossini, A.J., and Schurko, R.W. (2004) *J. Chem. Phys. A*, **108**, 7112.

50 Siegel, R., Nakashima, T.T., and Wasylishen, R.E. (2004) *J. Phys. Chem. B*, **108**, 2218.

51 Forgeron, M.M. and Wasylishen, R.E. (2008) *Phys. Chem. Chem. Phys.*, **10**, 574.

52 O'Dell, L.A. and Schurko, R.W. (2008) *Chem. Phys. Lett.*, **464**, 97.

53 Hartmann, R.S. and Hahn, E.L. (1962) *Phys. Rev.*, **128**, 2042.

54 Pines, A., Gibby, M.G., and Waugh, J.S. (1972) *J. Chem. Phys.*, **56**, 1776.

55 Pines, A., Gibby, M., and Waugh, J.S. (1973) *J. Chem. Phys.*, **59**, 569.

56 Senker, J. and Rössler, E. (2001) *Chem. Geol.*, **174**, 143.
57 Ashbrook, S.E. and Duer, M.J. (2006) *Concept Magn. Reson. A*, **28A**, 183.
58 Vega, A.J. (1992) *J. Magn. Reson.*, **96**, 50.
59 Vega, S. (1981) *Phys. Rev. A*, **23**, 3152.
60 Saalwaechter, K. (2007) *Prog. Nucl. Magn. Reson. Spectrosc.*, **51**, 1.
61 Vogel, M., Medick, P., and Rössler, E. (2005) *Annu. Rep. NMR. Spectrosc*, vol. **56** (ed. G. Webb), Academic Press, pp. 231–299.
62 Hoatson, G.L. and Vold, R.L. (1994) *NMR Basic Principles and Progress*, vol. **32**, Springer, Berlin, pp. 1–67.
63 Spiess, H.W., Grosescu, R., and Haeberlen, H. (1974) *Chem. Phys.*, **6**, 226.
64 Kong, X., Scott, E., Ding, W., Mason, J.A., Long, J.R., and Reimer, J.A. (2012) *J. Am. Chem. Soc.*, **134**, 14341.
65 Hologne, M. and Hirschinger, J. (2004) *Solid State Nucl. Mag.*, **26**, 1.
66 Larsen, F.H., Jakobsen, H.J., Ellis, P.D., and Nielsen, N.C. (1998) *Chem. Phys. Lett.*, **292**, 467.
67 Senker, J. (2004) *Solid State Nucl. Mag.*, **26**, 22.
68 Böhmer, R., Diezemann, G., Hinze, G., and Rössler, E.A. (2001) *Prog. Nucl. Magn. Reson. Spectrosc.*, **39**, 191.
69 Sears, D.N., Demko, B.A., Ooms, K.J., Wasylishen, R.E., and Huang, Y. (2005) *Chem. Mater.*, **17**, 5481.
70 Jameson, C.J. (2009) Xe NMR, *eMagRes MRI NMR*, John Wiley & Sons, Ltd, Chichester.
71 Fraissard, J. (2007) Microporous materials and xenon-129 NMR, *eMagRes MRI NMR*, John Wiley & Sons, Ltd, Chichester.
72 Romanenko, K., Fonseca, A., Dumonteil, S., Nagy, J., d'Espinose de Lacaillerie, J.-B., Lapina, O., and Fraissard, J. (2005) *Solid State Nucl. Mag.*, **28**, 135.
73 Robyr, P., Meier, B., and Ernst, R. (1989) *Chem. Phys. Lett.*, **162**, 417.
74 Robyr, P., Tomaselli, M., Straka, J., Grob-Pisano, C., Suter, U., Meier, B., and Ernst, R. (1995) *Mol. Phys.*, **84**, 995.
75 Burum, D., Linder, M., and Ernst, R. (1981) *J. Magn. Reson.*, **44**, 173.
76 Senker, J., Seyfarth, L., and Voll, J. (2004) *Solid State Sci.*, **6**, 1039.
77 Senker, J., Sehnert, J., and Correll, S. (2005) *J. Am. Chem. Soc.*, **127**, 337.
78 Sehnert, J. and Senker, J. (2007) *Chem. Eur. J.*, **13**, 6339.
79 Klinowski, J., Ramdas, S., Thomas, J.M., Fyfe, C.A., and Hartman, J.S. (1982) *J. Chem. Soc. Farad. Trans. 2*, **78**, 1025.
80 Kirkpatrick, R. and Brow, R.K. (1995) *Solid State Nucl. Mag.*, **5**, 9.
81 Luong, T., Mayer, H., Eckert, H., and Novinson, T.I. (1989) *J. Am. Ceram. Soc.*, **72**, 2136.
82 Siegel, R., Nakashima, T.T., and Wasylishen, R.E. (2005) *Concept Magn. Reson. A*, **26**, 47.
83 Yao, Z., Kwak, H.-T., Sakellariou, D., Emsley, L., and Grandinetti, P.J. (2000) *Chem. Phys. Lett.*, **327**, 85.
84 Kentgens, A. and Verhagen, R. (1999) *Chem. Phys. Lett.*, **300**, 435.
85 Siegel, R., Hirschinger, J., Carlier, D., Matar, S., Ménétrier, M., and Delmas, C. (2001) *J. Chem. Phys. B*, **105**, 4166.
86 Gul-E-Noor, F., Jee, B., Pöppl, A., Hartmann, M., Himsl, D., and Bertmer, M. (2011) *Phys. Chem. Chem. Phys.*, **13**, 7783.
87 Stejskal, E., Schaefer, J., and Waugh, J. (1977) *J. Magn. Reson.*, **28**, 105.
88 Seyfarth, L. and Senker, J. (2009) *Phys. Chem. Chem. Phys.*, **11**, 3522.
89 Seyfarth, L., Seyfarth, J., Lotsch, B.V., Schnick, W., and Senker, J. (2010) *Phys. Chem. Chem. Phys.*, **12**, 2227.
90 Wirnhier, E., Mesch, M.B., Senker, J., and Schnick, W. (2013) *Chem. Eur. J.*, **19**, 2041.
91 Zehe, C.S., Siegel, R., and Senker, J. (2015) *Solid State Nucl. Mag.*, **65**, 122.
92 Chandran, C.V., Madhu, P.K., Wormald, P., and Bräuniger, T. (2010) *J. Magn. Res.*, **206**, 255.
93 Carss, S.A., Scheler, U., Harris, R.K., Holstein, P., and Fletton, R.A. (1996) *Magn. Reson. Chem.*, **34**, 63.
94 Wasylishen, R.E., Ashbrook, S.E., and Wimperis, S. (2012) *NMR of Quadrupolar Nuclei in Solid Materials*, Wiley-VCH Verlag GmbH, Germany.
95 Ernst, M. (2003) *J. Magn. Reson.*, **162**, 1.
96 Hodgkinson, P. (2005) *Prog. Nucl. Magn. Reson. Spectrosc.*, **46**, 197.

97 Bloom, A.L. and Shoolery, J.N. (1955) *Phys. Rev.*, **97**, 1261.

98 Bennett, A.E., Rienstra, C.M., Auger, M., Lakshmi, K.V., and Griffin, R.G. (1995) *J. Chem. Phys.*, **103**, 6951.

99 Fung, B., Khitrin, A., and Ermolaev, K. (2000) *J. Magn. Reson.*, **142**, 97.

100 Detken, A., Hardy, E.H., Ernst, M., and Meier, B.H. (2002) *Chem. Phys. Lett.*, **356**, 298.

101 Thakur, R.S., Kurur, N.D., and Madhu, P.K. (2006) *Chem. Phys. Lett.*, **426**, 459.

102 Zehe, C., Schmidt, M., Siegel, R., Kreger, K., Daebel, V., Ganzleben, S., Schmidt, H.-W., and Senker, J. (2014) *CrystEngComm*, **16**, 9273.

103 Harris, R.K. (2006) *The Analyst*, **131**, 351.

104 Orr, R.M. and Duer, M.J. (2006) *Solid State Nucl. Mag.*, **30**, 130.

105 Delevoye, L., Trébosc, J., Gan, Z., Montagne, L., and Amoureux, J.-P. (2007) *J. Magn. Reson.*, **186**, 94.

106 Delevoye, L., Fernandez, C., Morais, C.M., Amoureux, J.-P., Montouillout, V., and Rocha, J. (2002) *Solid State Nucl. Mag.*, **22**, 501.

107 Brown, S.P. and Spiess, H.W. (2001) *Chem. Rev.*, **101**, 4125.

108 Lee, M. and Goldburg, W.I. (1965) *Phys. Rev*, **140**, A1261.

109 Waugh, J.S., Huber, L.M., and Haeberlen, U. (1968) *Phys. Rev. Lett.*, **20**, 180.

110 Mansfield, P. (1971) *J. Phys. C Solid State*, **4**, 1444.

111 Rhim, W.-K. (1973) *J. Chem. Phys.*, **59**, 3740.

112 Burum, D.P. and Rhim, W.K. (1979) *J. Chem. Phys.*, **71**, 944.

113 Hohwy, M., Bower, P.V., Jakobsen, H.J., and Nielsen, N.C. (1997) *Chem. Phys. Lett.*, **273**, 297.

114 Bielecki, A., Kolbert, A., and Levitt, M. (1989) *Chem. Phys. Lett.*, **155**, 341.

115 Levitt, M.H. (2007) Symmetry-based pulse sequences in magic-angle spinning solid-state NMR, *eMagRes MRI NMR*, John Wiley & Sons, Ltd, Chichester.

116 Madhu, P., Zhao, X., and Levitt, M.H. (2001) *Chem. Phys. Lett.*, **346**, 142.

117 Vinogradov, E., Madhu, P., and Vega, S. (1999) *Chem. Phys. Lett.*, **314**, 443.

118 Sakellariou, D., Lesage, A., Hodgkinson, P., and Emsley, L. (2000) *Chem. Phys. Lett.*, **319**, 253.

119 Coelho, C., Rocha, J., Madhu, P., and Mafra, L. (2008) *J. Magn. Reson.*, **194**, 264.

120 Mote, K.R., Agarwal, V., and Madhu, P.K. (2016) *Prog. Nucl. Magn. Reson. Spectrosc.*, **97**, 1–39.

121 Brown, S.P. (2012) *Solid State Nucl. Mag.*, **41**, 1.

122 Elena, B. and Emsley, L. (2005) *J. Am. Chem. Soc.*, **127**, 9140.

123 Griffin, J.M., Martin, D.R., and Brown, S.P. (2007) *Angew. Chem., Int. Ed.*, **46**, 8036.

124 Mafra, L., Santos, S.M., Siegel, R., Alves, I., Almeida Paz, F.A., Dudenko, D., and Spiess, H.W. (2012) *J. Am. Chem. Soc.*, **134**, 71.

125 Bradley, J.P., Velaga, S.P., Antzutkin, O.N., and Brown, S.P. (2011) *Cryst. Growth Des.*, **11**, 3463.

126 Wack, J., Siegel, R., Ahnfeldt, T., Stock, N., Mafra, L., and Senker, J. (2013) *J. Chem. Phys. C*, **117**, 19991.

127 Trebosc, J., Wiench, J.W., Huh, S., Lin, V. S.-Y., and Pruski, M. (2005) *J. Am. Chem. Soc.*, **127**, 7587.

128 Gan, Z., Gor'kov, P., Cross, T.A., Samoson, A., and Massiot, D. (2002) *J. Am. Chem. Soc.*, **124**, 5634.

129 Samoson, A., Lippmaa, E., and Pines, A. (1988) *Mol. Phys.*, **65**, 1013.

130 Llor, A. and Virlet, J. (1988) *Chem. Phys. Lett.*, **152**, 248.

131 Perras, F.A., Korobkov, I., and Bryce, D.L. (2013) *CrystEngComm*, **15**, 8727.

132 Frydman, L. and Harwood, J.S. (1995) *J. Am. Chem. Soc.*, **117**, 5367.

133 Medek, A., Harwood, J.S., and Frydman, L. (1995) *J. Am. Chem. Soc.*, **117**, 12779.

134 Fernandez, C. and Amoureux, J. (1995) *Chem. Phys. Lett.*, **242**, 449.

135 Gan, Z. (2000) *J. Am. Chem. Soc.*, **122**, 3242.

136 Ashbrook, S.E. and Wimperis, S. (2004) *Prog. Nucl. Magn. Reson. Spectrosc.*, **45**, 53.

137 Amoureux, J.-P., Fernandez, C., and Steuernagel, S. (1996) *J. Magn. Reson. A*, **123**, 116.

138 Gan, Z. and Kwak, H.-T. (2004) *J. Magn. Reson.*, **168**, 346.
139 Kwak, H.-T. and Gan, Z. (2003) *J. Magn. Reson.*, **164**, 369.
140 Brown, S.P., Heyes, S.J., and Wimperis, S. (1996) *J. Magn. Reson. A*, **119**, 280.
141 Goldbourt, A. and Madhu, P.K. (2002) *Monatsh. Chem.*, **133**, 1497.
142 Fernandez, C. and Pruski, M. (2011) *Top. Curr. Chem*, vol. **306**, Springer, Berlin, pp. 119–188.
143 Ashbrook, S.E., Berry, A.J., Frost, D.J., Gregorovic, A., Pickard, C.J., Readman, J.E., and Wimperis, S. (2007) *J. Am. Chem. Soc.*, **129**, 13213.
144 Harris, R.K., Wasylishen, R.E., and Duer, M.J. (eds) (2009) *NMR Crystallography*, John Wiley & Sons, Inc., New York.
145 Harris, R.K. (2004) *Solid State Sci.*, **6**, 1025.
146 Massiot, D., Montouillout, V., Fayon, F., Florian, P., and Bessada, C. (1997) *Chem. Phys. Lett.*, **272**, 295.
147 Massiot, D., Vosegaard, T., Magneron, N., Trumeau, D., Montouillout, V., Berthet, P., Loiseau, T., and Bujoli, B. (1999) *Solid State Nucl. Mag.*, **15**, 159.
148 Antzutkin, O., Shekar, S., and Levitt, M. (1995) *J. Magn. Reson. A*, **115**, 7.
149 Alla, M.A., Kundla, E.I., and Lippmaa, E.T. (1978) *JETP Lett.*, **27**, 194.
150 Yarim-Agaev, Y., Tutunjian, P.N., and Waugh, J.S. (1982) *J. Magn. Reson.*, **47**, 51.
151 DeAzevedo, E.R., Hu, W.-G., Bonagamba, T.J., and Schmidt-Rohr, K. (1999) *J. Am. Chem. Soc.*, **121**, 8411.
152 Hong, M. (2000) *J. Am. Chem. Soc.*, **122**, 3762.
153 Li, W. and McDermott, A. (2012) *J. Magn. Reson.*, **222**, 74.
154 Tycko, R., Dabbagh, G., and Mirau, P.A. (1989) *J. Magn. Reson.*, **85**, 265.
155 Orr, R.M. and Duer, M.J. (2006) *J. Magn. Reson.*, **181**, 1.
156 Liu, S.-F., Mao, J.-D., and Schmidt-Rohr, K. (2002) *J. Magn. Reson.*, **155**, 15.
157 Gan, Z., Grant, D.M., and Ernst, R. (1996) *Chem. Phys. Lett.*, **254**, 349.
158 Gross, J.D., Costa, P.R., and Griffin, R.G. (1998) *J. Chem. Phys.*, **108**, 7286.
159 Chan, J.C.C. and Tycko, R. (2003) *J. Chem. Phys.*, **118**, 8378.
160 Nishiyama, Y., Yamazaki, T., and Terao, T. (2006) *J. Chem. Phys.*, **124**, 064304.
161 Hou, G., Byeon, I.-J.L., Ahn, J., Gronenborn, A.M., and Polenova, T. (2012) *J. Chem. Phys.*, **137**, 134201.
162 Raleigh, D.P., Kolbert, A.C., Oas, T.G., Levitt, M.H., and Griffin, R.G. (1988) *J. Chem. Soc. Farad. Trans. 1*, **84**, 3691.
163 Gullion, T. (1989) *J. Magn. Reson.*, **85**, 614.
164 Kolbert, A. and Griffin, R. (1990) *Chem. Phys. Lett.*, **166**, 87.
165 Crockford, C., Geen, H., and Titman, J.J. (2001) *Chem. Phys. Lett.*, **344**, 367.
166 Eléna, B., Hediger, S., and Emsley, L. (2003) *J. Magn. Reson.*, **160**, 40.
167 Shao, L., Crockford, C., Geen, H., Grasso, G., and Titman, J.J. (2004) *J. Magn. Reson.*, **167**, 75.
168 Strohmeier, M. and Grant, D.M. (2004) *J. Magn. Reson.*, **168**, 296.
169 Orr, R.M., Duer, M.J., and Ashbrook, S.E. (2005) *J. Magn. Reson.*, **174**, 301.
170 Hung, I. and Gan, Z. (2011) *J. Magn. Reson.*, **213**, 196.
171 Haeberlen, U. and Waugh, J.S. (1968) *Phys. Rev.*, **175**, 453.
172 Maricq, M.M. (1982) *Phys. Rev. B*, **25**, 6622.
173 Weintraub, O. and Vega, S. (1993) *J. Magn. Reson. A*, **105**, 245.
174 Tycko, R. and Smith, S.O. (1993) *J. Chem. Phys.*, **98**, 932.
175 Nakashima, T.T. and McClung, R.E.D. (2007) Heteronuclear shift correlation spectroscopy, *eMagRes MRI NMR*, John Wiley & Sons, Ltd, Chichester.
176 Eden, M., Annersten, H., and Zazzi, A. (2005) *Chem. Phys. Lett.*, **410**, 24.
177 Amoureux, J., Trébosc, J., Delevoye, L., Lafon, O., Hu, B., and Wang, Q. (2009) *Solid State Nucl. Mag.*, **35**, 12.
178 Tycko, R. (2009) Dipolar recoupling: homonuclear experiments, *eMagRes MRI NMR*, John Wiley & Sons, Ltd, Chichester.
179 Jäger, C., Feike, M., Born, R., and Spiess, H.-W. (1994) *J. Non-Cryst. Solids*, **180**, 91.
180 Martineau, C., Bouchevreau, B., Tian, Z., Lohmeier, S.-J., Behrens, P., and Taulelle, F. (2011) *Chem. Mater.*, **23**, 4799.

181 Schmidt, M., Wittmann, J., Kress, R., Schneider, D., Steuernagel, S., Schmidt, H.-W., and Senker, J. (2012) *Cryst. Growth Des.*, **12**, 2543.

182 Brouwer, D.H., Kristiansen, P.E., Fyfe, C.A., and Levitt, M.H. (2005) *J. Am. Chem. Soc.*, **127**, 542.

183 Carravetta, M., Eden, M., Johannessen, O.G., Luthman, H., Verdegem, P.J.E., Lugtenburg, J., Sebald, A., and Levitt, M.H. (2001) *J. Am. Chem. Soc.*, **123**, 10628.

184 Kristiansen, P.E., Carravetta, M., Lai, W.C., and Levitt, M.H. (2004) *Chem. Phys. Lett.*, **390**, 1.

185 Andrew, E.R., Clough, S., Farnell, L.F., Gledhill, T.D., and Roberts, I. (1966) *Phys. Lett.*, **21**, 505.

186 Raleigh, D.P., Levitt, M.H., and Griffin, R.G. (1988) *Chem. Phys. Lett.*, **146**, 71.

187 Bennett, A.E., Griffin, R.G., Ok, J.H., and Vega, S. (1992) *J. Chem. Phys.*, **96**, 8624.

188 Sodickson, D.K., Levitt, M.H., Vega, S., and Griffin, R.G. (1993) *J. Chem. Phys.*, **98**, 6742.

189 Ishii, Y. (2001) *J. Chem. Phys.*, **114**, 8473.

190 Volkringer, C., Loiseau, T., Guillou, N., Férey, G., Haouas, M., Taulelle, F., Elkaim, E., and Stock, N. (2010) *Inorg. Chem.*, **49**, 9852.

191 Alam, T.M. and Brow, R.K. (1998) *J. Non-Cryst. Solids*, **223**, 1.

192 Karau, F.W., Seyfarth, L., Oeckler, O., Senker, J., Landskron, K., and Schnick, W. (2007) *Chem. Eur. J.*, **13**, 6841.

193 Zaborowski, E., Zimmermann, H., and Vega, S. (1999) *J. Magn. Reson.*, **136**, 47.

194 Robyr, P., Tomaselli, M., Straka, J., Grob-Pisano, C., Suter, U.W., Meier, B.H., and Ernst, R.R. (1995) *Mol. Phys.*, **84**, 995.

195 Bloembergen, N. (1949) *Physica*, **15**, 386.

196 Suter, D. and Ernst, R.R. (1982) *Phys. Rev. B*, **25**, 6038.

197 Szeverenyi, N.M., Sullivan, M.J., and Maciel, G.E. (1982) *J. Magn. Reson.*, **47**, 462.

198 Grommek, A., Meier, B.H., and Ernst, M. (2006) *Chem. Phys. Lett.*, **427**, 404.

199 Veshtort, M. and Griffin, R.G. (2011) *J. Chem. Phys.*, **135**, 134509-1.

200 Castellani, F., van Rossum, B., Diehl, A., Schubert, M., Rehbein, K., and Oschkinat, H. (2002) *Nature*, **420**, 98.

201 Salager, E., Stein, R.S., Pickard, C.J., Elena, B., and Emsley, L. (2009) *Phys. Chem. Chem. Phys.*, **11**, 2610.

202 Takegoshi, K., Nakamura, S., and Terao, T. (2001) *Chem. Phys. Lett.*, **344**, 631.

203 Crocker, E., Patel, A.B., Eilers, M., Jayaraman, S., Getmanova, E., Reeves, P.J., Ziliox, M., Khorana, H.G., Sheves, M., and Smith, S.O. (2004) *J. Biomol. NMR*, **29**, 11.

204 Bayro, M.J., Huber, M., Ramachandran, R., Davenport, T.C., Meier, B.H., Ernst, M., and Griffin, R.G. (2009) *J. Chem. Phys.*, **130**, 114506.

205 Tycko, R. and Dabbagh, G. (1990) *Chem. Phys. Lett.*, **173**, 461.

206 Nielsen, N.C., Bildsoe, H., Jakobsen, H.J., and Levitt, M.H. (1994) *J. Chem. Phys.*, **101**, 1805.

207 Verel, R., Ernst, M., and Meier, B.H. (2001) *J. Magn. Reson.*, **150**, 81.

208 Gregory, D.M., Mitchell, D.J., Stringer, J.A., Kiihne, S., Shiels, J.C., Callahan, J., Mehta, M.A., and Drobny, G.P. (1995) *Chem. Phys. Lett.*, **246**, 654.

209 Bower, P.V., Oyler, N., Mehta, M.A., Long, J.R., Stayton, P.S., and Drobny, G.P. (1999) *J. Am. Chem. Soc.*, **121**, 8373.

210 Caporini, M.A., Bajaj, V.S., Veshtort, M., Fitzpatrick, A., MacPhee, C.E., Vendruscolo, M., Dobson, C.M., and Griffin, R.G. (2010) *J. Phys. Chem. B*, **114**, 13555.

211 Mehta, M.A., Gregory, D.M., Kiihne, S., Mitchell, D.J., Hatcher, M.E., Shiels, J.C., and Drobny, G.P. (1996) *Solid State Nucl. Mag.*, 7, 211.

212 Kiihne, S.R., Geahigan, K.B., Oyler, N.A., Zebroski, H., Mehta, M.A., and Drobny, G.P. (1999) *J. Phys. Chem. A*, **103**, 3890.

213 Murakami, M., Noda, Y., Koyama, Y., Takegoshi, K., Arai, H., Uchimoto, Y., and Ogumi, Z. (2014) *J. Phys. Chem. C*, **118**, 15375.

214 Graf, R., Demco, D.E., Gottwald, J., Hafner, S., and Spiess, H.W. (1997) *J. Chem. Phys.*, **106**, 885.

215 Saalwächter, K., Lange, F., Matyjaszewski, K., Huang, C.-F., and Graf, R. (2011) *J. Magn. Reson.*, **212**, 204.

216 Gottwald, J., Demco, D.E., Graf, R., and Spiess, H.W. (1995) *Chem. Phys. Lett.*, **243**, 314.

217 Feike, M., Demco, D.E., Graf, R., Gottwald, J., Hafner, S., and Spiess, H.W. (1996) *J. Magn. Reson.*, **122**, 214.
218 Lee, Y.K., Kurur, N.D., Helmle, M., Johannessen, O.G., Nielsen, N.C., and Levitt, M.H. (1995) *Chem. Phys. Lett.*, **242**, 304.
219 Brinkmann, A. and Levitt, M.H. (2001) *J. Chem. Phys.*, **115**, 357.
220 Kristiansen, P.E., Mitchell, D.J., and Evans, J.N.S. (2002) *J. Magn. Reson.*, **157**, 253.
221 Seyfarth, L., Seyfarth, J., Lotsch, B., Schnick, W., and Senker, J. (2010) *Phys. Chem. Chem. Phys.*, **12**, 2227.
222 Brouwer, D.H., Darton, R.J., Morris, R.E., and Levitt, M.H. (2005) *J. Am. Chem. Soc.*, **127**, 10365.
223 Ren, J. and Eckert, H. (2012) *Angew. Chem., Int. Ed.*, **51**, 12888.
224 Kristiansen, P.E., Carravetta, M., van Beek, J.D., Lai, W.C., and Levitt, M.H. (2006) *J. Chem. Phys.*, **124**, 234510.
225 Schmidt, M., Wittmann, J., Kress, R., Schmidt, H.-W., and Senker, J. (2013) *Chem. Commun.*, **49**, 267.
226 Lotsch, B., Döblinger, M., Sehnert, J., Seyfarth, L., Senker, J., Oeckler, O., and Schnick, W. (2007) *Chem. Eur. J.*, **13**, 4969.
227 Ladizhansky, V. and Vega, S. (2000) *J. Am. Chem. Soc.*, **122**, 3465.
228 Seyfarth, L., Sehnert, J., El-Gamel, N., Milius, W., Kroke, E., Breu, J., and Senker, J. (2008) *J. Mol. Struct.*, **889**, 217.
229 Widdifield, C.M., Robson, H., and Hodgkinson, P. (2016) *Chem. Commun.*, **52**, 6685.
230 Oas, T.G., Griffin, R.G., and Levitt, M.H. (1988) *J. Chem. Phys.*, **89**, 692.
231 Levitt, M.H., Oas, T.G., and Griffin, R.G. (1988) *Isr. J. Chem.*, **28**, 271.
232 Costa, P.R., Gross, J.D., Hong, M., and Griffin, R.G. (1997) *Chem. Phys. Lett.*, **280**, 95.
233 Fu, R., Smith, S.A., and Bodenhausen, G. (1997) *Chem. Phys. Lett.*, **272**, 361.
234 Takegoshi, K., Takeda, K., and Terao, T. (1996) *Chem. Phys. Lett.*, **260**, 331.
235 Kitchin, S.J., Harris, K.D., Aliev, A.E., and Apperley, D.C. (2000) *Chem. Phys. Lett.*, **323**, 490.
236 Kaplan, D.E. and Hahn, E.L. (1958) *J. Phys. Radium*, **19**, 821.
237 Shore, S.E., Ansermet, J.-P., Slichter, C.P., and Sinfelt, J.H. (1987) *Phys. Rev. Lett.*, **58**, 953.
238 Gullion, T. and Schaefer, J. (1989) *Adv. Magn. Reson.*, **13**, 57.
239 Gullion, T. and Schaefer, J. (1989) *J. Magn. Reson.*, **81**, 196.
240 Hing, A.W., Vega, S., and Schaefer, J. (1992) *J. Magn. Reson.*, **96**, 205.
241 Yap, A.T.-W., Förster, H., and Elliott, S.R. (1995) *Phys. Rev. Lett.*, **75**, 3946.
242 Kong, X., Deng, H., Yan, F., Kim, J., Swisher, J.A., Smit, B., Yaghi, O.M., and Reimer, J.A. (2013) *Science*, **341**, 882.
243 Matsuoka, S. and Inoue, M. (2009) *Chem. Commun.*, 5664.
244 Haimovich, A. and Goldbourt, A. (2015) *J. Magn. Reson.*, **254**, 131.
245 Jung, I.-S., Lee, Y.J., Jeong, D., Graf, R., Choi, T.-L., Son, W.-J., Bulliard, X., and Spiess, H.W. (2016) *Macromolecules*, **49**, 3061.
246 Zhao, X., Hoffbauer, W., Schmedt auf der Günne, J., and Levitt, M.H. (2004) *Solid State Nucl. Mag.*, **26**, 57.
247 Schmidt, M., Zehe, C.S., Siegel, R., Heigl, J.U., Steinlein, C., Schmidt, H.-W., and Senker, J. (2013) *CrystEngComm*, **15**, 8784.
248 Brinkmann, A. and Kentgens, A.P.M. (2006) *J. Am. Chem. Soc.*, **128**, 14758.
249 Siegel, R., Trébosc, J., Amoureux, J.-P., and Gan, Z. (2008) *J. Magn. Reson.*, **193**, 321.
250 Cavadini, S., Lupulescu, A., Antonijevic, S., and Bodenhausen, G. (2006) *J. Am. Chem. Soc.*, **128**, 7706.
251 Gan, Z. (2006) *J. Am. Chem. Soc.*, **128**, 6040.

9
Modern Electron Paramagnetic Resonance Techniques and Their Applications to Magnetic Systems

Andrej Zorko,[1] Matej Pregelj,[1] and Denis Arčon[1,2]

[1]*Jožef Stefan Institute, Solid State Physics Department, Jamova cesta 39, 1000 Ljubljana, Slovenia*
[2]*University of Ljubljana, Faculty of Mathematics and Physics, Jadranska ulica 19, 1000 Ljubljana, Slovenia*

9.1
Introduction

Electron paramagnetic resonance (EPR) is a well-established spectroscopic technique, where resonant absorption of microwaves yields a unique information about the local structure, dynamics, and distribution of paramagnetic centers. In many aspects, most of the EPR experiments are still conducted in the same way as in the pioneering experiments of Yevgeny Konstantinovich Zavoisky in 1940s [1,2]: at a fixed resonance frequency, usually set to around 9.6 GHz (so-called X-band), a resonance phenomenon of paramagnetic species is observed as a function of sweeping magnetic field. However, especially in the last few decades, we have witnessed a rapid progress in the experimental approaches, extending experiments to ever higher resonance frequencies and magnetic fields, pulsed EPR spectrometers were introduced to overcome some limitations in spectral and time resolution of the conventional continuous wave (c.w.) technique, double resonance techniques explored the specific couplings to extract additional information on the local structure, and attempts to merge microscopy and spectroscopy techniques to detect single paramagnetic spin were witnessed. This rapid progress dramatically expanded the range of applications. While EPR was traditionally in the domain of physics and chemistry of solid state [3–6], it is now, for instance, also indispensable in biology [7–10] or even in medical applications [11].

Over the years, many comprehensive monographs and overviews of the EPR techniques dedicated to different range of applications have been published [3–13]. Below we list some of possible applications and the interested reader is advised to turn to these works directly for more details. The main purpose of this contribution is, however, to review a specific problem of EPR in exchange

---

Handbook of Solid State Chemistry, First Edition. Edited by Richard Dronskowski, Shinichi Kikkawa, and Andreas Stein.
© 2017 Wiley-VCH Verlag GmbH & Co. KGaA. Published 2017 by Wiley-VCH Verlag GmbH & Co. KGaA.

coupled systems. Such systems represent particularly difficult problems for modeling as their response is complicated by the strong exchange interaction between neighboring spins. The presence of such interaction dramatically affects the way the spin system responds to the external magnetic fields and often requires the use of modern theoretical concepts and advanced experimental set-ups.

9.1.1
Relevant Interactions and Their Typical Energy Scales

When a paramagnetic sample is inserted into a strong external magnetic field, B_0, the spin states characterized by the effective electron spin S and the nuclear spin I are determined by a number of various interactions. In general, the static spin Hamiltonian $\mathcal{H}$ will comprise of the following terms [10,14]:

$$\mathcal{H} = \mathcal{H}_{EZ} + \mathcal{H}_{ZFS} + \mathcal{H}_{HF} + \mathcal{H}_{NZ} + \mathcal{H}_{NQ} + \mathcal{H}_{dd} + \mathcal{H}_{e}. \tag{9.1}$$

Here $\mathcal{H}_{EZ}$ is the electron Zeeman interaction, $\mathcal{H}_{ZFS}$ is the electron zero-field splitting term, $\mathcal{H}_{HF}$ is the electron-nuclear hyperfine coupling interaction, $\mathcal{H}_{NZ}$ and $\mathcal{H}_{NQ}$ are the nuclear Zeeman and quadrupole interactions, $\mathcal{H}_{dd}$ is the dipolar interaction between electron spins, and finally $\mathcal{H}_{e}$ is the exchange interaction between neighboring electron spins.

In many cases, the leading term in $\mathcal{H}$ is $\mathcal{H}_{EZ}$, which takes into account the response of the electron spin to the external magnetic field B_0, $\mathcal{H}_{EZ} = \mu_B \vec{B}_0 \cdot \underline{g} \cdot \vec{S}$ (μ_B is the Bohr magneton and g is the g-tensor). In external magnetic field, the energy levels thus split by $\Delta E = \mu_B g B_0$ and the resonance phenomenon at the irradiation frequency ν occurs, when the condition $h\nu = \mu_B g B_0$ is met (h is the Planck constant). The components of g-tensor mirror site-symmetry of the paramagnetic center as the deviations from the isotropic free electron value $g_e = 2.0023$ are introduced through the contribution of the spin–orbit coupling interaction. Therefore, in the cubic environment all three principal values of the g-tensor are identical, $g_x = g_y = g_z$, whereas for the axial symmetry two principal values $g_\perp$ and $g_\parallel$ suffice, that is, one finds that $g_x = g_y \equiv g_\perp$ and $g_z = g_\parallel$. Finally, for lower symmetry, all three principal values are different, $g_x \neq g_y \neq g_z$. In powder samples, the principle values of the g-tensor can be read from the characteristic singularities and shoulders of the inhomogeneously broadened EPR spectra [13].

In cases with $S > 1/2$, the $\mathcal{H}_{ZFS}$ term splits the spin levels already in the absence of external magnetic field. When present, $\mathcal{H}_{ZFS}$ is often the dominant term in the spin Hamiltonian at X-band resonance frequencies. In the lowest order in spin, this interaction takes the form of a fine structure $\mathcal{H}_{ZFS} = \vec{S} \cdot D \cdot \vec{S}$ term and is also known as a single-ion (SI) anisotropy. The second rank zero-field interaction tensor D again reflects the paramagnetic-site symmetry. However, since D is traceless, only two principal values D and E are sufficient. Hence, one can write in the principal system $\mathcal{H}_{ZFS} = D[S_z^2 - \frac{1}{3}S(S+1)] + E(S_x^2 - S_y^2)$. Here D and E are both nonzero in cases of low symmetry, $E = 0$ for the axial symmetry, and both anisotropy parameters vanish for cubic symmetry. In the latter case higher-order terms are usually considered [5,6].

The interaction between the electron and the nuclear spin is contained in the hyperfine term $\mathcal{H}_{HF} = \vec{S} \cdot \underline{A} \cdot \vec{I}$. The hyperfine coupling tensor $\underline{A}$ has in general two contributions: the isotropic Fermi contact interaction $\mathcal{H}_F = a_{iso}\vec{S} \cdot \vec{I}$ (here a_{iso} is the isotropic hyperfine coupling constant that depends on the electron spin density at the nuclear site) and the anisotropic contribution, often dominated by the electron-nuclear dipole–dipole interaction $\mathcal{H}_{N,dd} = \vec{S} \cdot \underline{T}' \cdot \vec{I}$, which is described by the traceless dipolar coupling tensor $\underline{T}'$. Components of $\underline{T}'$ are inversely proportional to the cube of the distance, $1/r^3$, thus providing the basis for the determination of the local environment of the paramagnetic center.

The two interactions, $\mathcal{H}_{NZ}$ and $\mathcal{H}_{NQ}$, describing the splitting of nuclear spin energy levels are by more than three orders of magnitude weaker than all the other terms in $\mathcal{H}$. Nevertheless, in EPR experiments they become highly relevant in electron nuclear double resonance (ENDOR) and electron spin echo envelope modulation (ESEEM) experiments.

The dipolar interaction between electronic spins $\mathcal{H}_{dd} = \sum_{i>j} \vec{S}_i \cdot \underline{T} \cdot \vec{S}_j$ is often the dominant magnetic anisotropy in $S = 1/2$ systems with low magnetic exchange interaction, for example, in diluted magnetic systems.

Finally, $\mathcal{H}_e$ can be separated into the isotropic Heisenberg exchange between the neighboring electron spins $\vec{S}_1$ and $\vec{S}_2$, $\mathcal{H}_{ex} = J\vec{S}_1 \cdot \vec{S}_2$, and the anisotropic exchange interactions $\mathcal{H}'_{ex}$, that is, $\mathcal{H}_e = \mathcal{H}_{ex} + \mathcal{H}'_{ex}$. Although the latter term is usually weaker than the leading isotropic Heisenberg exchange interaction, it is still very important as it generally leads to broadening of the EPR spectra in magnetic systems. $\mathcal{H}'_{ex}$ can be separated [12] into the symmetric traceless part of the form $\mathcal{H}_{AE} = \vec{S}_1 \cdot \underline{\delta}_{12} \cdot \vec{S}_2$, known as the symmetric anisotropy exchange (AE) interaction, and the antisymmetric exchange anisotropy of the form $\mathcal{H}_{DM} = \vec{d}_{12} \cdot \vec{S}_1 \times \vec{S}_2$, known as Dzyaloshinskii–Moriya (DM) interaction [15,16]. Section 9.3 reviews a detailed treatment of the EPR data in the presence of $\mathcal{H}_e$.

9.1.2
Information Extracted from an EPR Experiment

The EPR spectrum $I(\omega, B)$ is given by the imaginary part of the dynamical spin susceptibility for the transverse direction (i.e., perpendicular to the magnetic field), $\chi''_\perp(q, \omega, B)$, calculated at the momentum of the excitation $q = 0$ and at the resonant frequency ω, namely [17],

$$I(\omega, B) = \frac{B_\perp^2}{2\mu_0} \omega \chi''_\perp(q = 0, \omega, B) V. \tag{9.2}$$

Here, $B_\perp$ is the magnitude of the perpendicular (excitation) microwave field, μ_0 the permeability of the vacuum, and V the volume of the sample. It then follows from the Kramers–Kronig relations that the measurements of the integrated EPR signal intensity is directly related to the static uniform spin-only susceptibility [17], χ_0, and thus allows for a quantitative determination of the number of paramagnetic centers in the sample [18].

In principle, all information about the spin Hamiltonian is contained in $\chi''_\perp(q = 0, \omega, B)$. However, it is not always straightforward, how to extract it. The pulsed EPR experiments represent an important step toward this goal. Compared to conventional c.w. EPR methods, pulsed EPR increases the resolution and sensitivity of the EPR experiments by manipulating the electronic spins and thus allows to separate interactions from each other. Reference [10] offers an excellent introduction to pulsed EPR techniques.

9.1.3
Prime Fields of Application

Physicists use EPR techniques mostly to study magnetic systems, to directly probe transition metal, lanthanide, and actinide ions, to measure conduction electrons in conductors and semiconductors or defects in crystals, to list only a few examples. The application of EPR to magnetic systems will be described in detail in the following sections. In metals and semiconductors, the electron paramagnetic resonance is called conducting electron spin resonance (CESR) [19]. CESR is characterized by an asymmetric resonance lineshape, known as a Dysonian lineshape [20]. The asymmetry in the lineshape arises as the itinerant electrons diffuse through the penetration depth of the electromagnetic irradiation. Historically, the Dysononian lineshape has been studied in alkali metals [20,21], but more recently also in alkali-doped fullerides [22–26], graphene or nanostructured graphite and single wall carbon nanotubes [27–30], boron-doped diamond [31], and Bechgaard salts [32,33]. However, as argued in Ref. [29], just the observation of the Dysonian lineshape is not sufficient to unambiguously claim CESR and one should also carefully check the temperature dependence of the CESR linewidth and of the g-factor. The Elliott–Yafet spin-relaxation mechanism applies to metals and predicts that the CESR linewidth is proportional to the resistivity as indeed observed in many archetypal cases [34,35]. An alternative approach to study metallic samples with the EPR technique is to measure magnetic ions embedded in the metallic host. Although such spectra may bare many similarities with a direct detection of CESR, the simultaneous presence of itinerant electrons and localized magnetic-ion states opens many new problems, such as the problem of exchange between the two spin subsystems, ionic and virtual bound states or Kondo effects [36] and is thus, because of high complexity, beyond the scope of this review.

Single-molecule magnets (SMM) emerged in 1990s as a highly intriguing family of metalorganic compounds where strong superexchange interactions between metal ions within the single molecular entity frequently establish a high spin ground state with a large zero-field-splitting. As the magnetic interactions between such molecular entities are usually weak, the superparamagnetic behavior of SMMs thus represent an exciting possibility for the applications in quantum computing, high-density storage, or magnetic refrigeration [37]. EPR was one of the key experimental techniques in this field ever since the first report of the archetypal SMM, [Mn$_{12}$O$_{12}$(O$_2$CCH$_3$)$_{16}$(H$_2$O)$_4$]·2CH$_3$CO$_2$H·4 H$_2$O (abbreviated as Mn$_{12}$Ac) [38]. EPR spectra recorded at high resonance frequencies up to 525 GHz were successfully simulated with the spin Hamiltonian that contained crystal-field terms up to fourth-order [39]. These results and more detailed studies that soon followed [40] enabled the precise elucidation of all terms in effective spin Hamiltonian parameters and the identification of anisotropy terms like $\left(S_+^4 + S_-^4\right)$ that are responsible for the transverse magnetic anisotropy and may play a crucial role in the mechanism of quantum tunneling in Mn$_{12}$Ac. The application of EPR in SMMs was next extended to, for example, detect the excited state in Mn$_{12}$Ac [41] or to many other SMM compounds, including modifications of Mn$_{12}$Ac [42] or Fe$_8$-based SMMs [43]. Interested readers are advised to reach for a several specialized review articles, for example, Ref. [44] provides an extensive and up-to-date review of the subject.

In chemistry, EPR is broadly used for structural and magnetic characterizations [4–6,9], to study the kinetics of radical reactions, numerous polymerization reactions [45] and spin trapping processes [46]. It is also important for the characterization of organometallic [47] and hybrid organic-inorganic compounds [48], studies of catalysis [49,50], in petroleum research, and for the monitoring of oxidation and reduction processes [8,51]. One particular example, where EPR experiments made a substantial contribution to the understanding of chemistry of materials, is the research of zeolites. Zeolites are an important family of materials with periodic arrays of aluminosilicate cages that are widely used in different industrial processes [52]. With more than 200 possible zeolite frameworks known today, they represent a unique playground to control and study the effects of geometry and chemical composition on their chemical and physical properties. One important field of application of zeolites is the metal-nanoparticle catalysis, which is triggered when the transition-metal ions are introduced into the host framework. As an example, manganese species incorporated into such lattices are used as selective catalysts for a wide range of catalytic applications as, for instance, a catalytic disposal of pollutants, that is, ozone decompositions, photocatalytic oxidation of organic pollutants, and so on. Obtaining well-defined and highly dispersed isolated manganese nanoparticles integrated in the framework is a single most important step in the catalytic activation of the material. Not surprisingly, EPR, as a local probe has been extensively used in such studies of the local coordination of manganese ions [53–56] as well as in the studies of atmospheric pollutant adsorption on zeolites [57]. The same approach has been successfully implemented in the numerous studies of nanostructures, for example, incorporation of transition metal oxides or gas adsorption to TiO$_2$-derived nanoparticles [58–61].

EPR technique, being sensitive to free radicals, has been also often used in the studies of the pathogenesis of many diseases, for example, cancer and a variety of inflammatory diseases. In such studies, the most common free radicals and paramagnetic molecules include nitrogen oxide (NO) groups, superoxides (O$_2^-$) and hydroxyls (OH$_2^-$), alky and lipid peroxyl groups, or similar. Moreover, time dependent EPR measurements allow monitoring of a sequence of reactions into which such free radical states enter at various pathological states. One characteristic example is the formation of CO_3^{3-} defects in enamel with a characteristic EPR spectrum [62]. The formation and stability of these centers have been also explored in studies of the human teeth resistance to caries and have provided the basis for oximetry and radiation dosimetry [11]. More detailed survey of EPR in medicine and biology is found elsewhere, for example, in Ref. [7].

9.2
Advanced Electron Paramagnetic Resonance Techniques

9.2.1
Pulsed versus c.w. Electron Paramagnetic Resonance Techniques

The introduction of pulsed EPR methods was for decades limited by difficult experimental and technical requirements imposed by the short electron-spin relaxation times (often on the order of 100 ns) and strong interactions between electronic moments that may exceed the Zeeman interactions. The early work was, therefore, concentrated on magnetically diluted systems with long relaxation times and small linewidth of the respective EPR spectra. Most of the techniques thus used electron spin echo and many modern pulse sequences were already introduced by Mims in 1960s [63–65]. With the appearance of a commercial pulsed EPR spectrometer by Bruker, the pulsed methods rapidly expanded and are now indispensable in numerous fields of research, primarily in chemistry and biochemistry.

Unlike the more conventional c.w. EPR technique, in pulsed EPR experiments excitation microwave fields are switched on only for a very short time (typically, the duration of $\pi/2$-pulse, which is a pulse that rotates the magnetization by 90°, is only a few nanoseconds) to drive the electron magnetization away from the thermal equilibrium and to manipulate it. Broadly speaking, pulsed EPR techniques can be divided into measurements of spin-lattice relaxation and phase-memory relaxation times, Fourier transform EPR, nuclear modulation effect, pulsed ENDOR, and multifrequency EPR. Reference [10] offers an excellent overview for each of these techniques.

9.2.2
Toward Single Spin Electron Paramagnetic Resonance Detection

Many of the recent advances in EPR spectroscopy were stimulated by the desire to manipulate and monitor a single electron spin. An efficient manipulation would open a whole new avenue to the spin electronics – spintronics [66] – with exciting possibilities to process quantum information. The idea is to perform nanoscale magnetic resonance imaging and spectroscopy in a single run. A promising way of development is to modify scanning tunneling microscopes and then implement dopants in crystalline silicon such as phosphorus (Si:P) to use them as EPR active centers [67]. Alternatively, the use of a single nitrogen-vacancy (N-V) defect in the diamond tip of an atomic force microscope cantilever has been proposed as a probe spin [68]. Optical detection of the spin resonance of a nitrogen-vacancy defect centers in diamond is yet another alternative approach that has emerged in the last few years [69,70].

Another novel electron spin resonance technique combined with electric field modulation has been proposed in Ref. [71]. There, the magnetic transitions are induced by an applied electric field that through a magnetoelectric coupling of a material transfers into magnetic modulation. In Cu_2OSeO_3, spin-wave modes were resolved with a high sensitivity via the magnetoelectric coupling. A similar approach has been adopted in Ref. [72] where the high sensitivity for the magnetic resonance spin-flips of a single electron paramagnetic spin center, formed by a defect in the gate oxide of a standard silicon transistor, is achieved when the spin orientation is converted to electric charge. This is then measured as a change in the source/drain channel current. This way, a single paramagnetic center formed by a defect in the gate oxide of a standard silicon transistor has been detected at the resonance frequency of about 45 GHz. The importance of this method then lies in a possibility to directly study the physics of spin decoherence in semiconductors.

9.2.3
High-Frequency EPR Spectroscopy

High-frequency EPR spectroscopy [3,4] has seen astonishing progress in the last two decades, with the development of high-power low-noise microwave sources, effective microwave detectors and high-field superconducting, hybrid, and pulsed magnets [73]. The term high-frequency relates to spectrometers working at frequencies $\nu \gtrsim 50$ GHz, that is above the commercially available and widespread Q-band ($\nu \sim 35$ GHz) spectrometers. Except for a commercial Bruker W-band spectrometer (working at 94 GHz) [74], the high-frequency spectrometers are custom made and typically cover frequency ranges from 50 to several hundred gigahertz. Extending the frequency range beyond a terahertz range appears as the central issue of modern EPR. In extreme cases the EPR spectroscopy can nowadays reach frequencies up to 75 THz with free-electron-laser radiation sources and magnetic fields up to 70 T in pulsed magnetic field facilities [75].

The most obvious reasons why to perform EPR measurements at higher frequencies are increased resolution and sensitivity that such measurements provide. Typically, different paramagnetic species in a sample are characterized by different local environments, so that their resonances are found at different positions, that is, at different g factors. In such a case, the resolution scales linearly with frequency ν, $h\nu = g\mu_B B_0$. Consequently, larger frequencies can resolve spectral structure in the case of small g-factor differences compared to homogeneous line broadening.

The next experimental case that requires the EPR experiments to be performed at higher, and multiple frequencies is a zero-field split EPR spectrum. In the case of non-Kramers, integer-spin systems the zero-field ground state is often a singlet and the excitations to other spin states are gapped. Consequently, the resonance condition is significantly different from the paramagnetic condition $h\nu = g\mu_B B_0$. An illustrative example of a crystal-field driven zero-field splitting was provided by the $S = 1$ one-dimensional (1D) spin system known as DTN, where, additionally, nonlinear two-magnon bound-state excitations could also be observed in a multiple-frequency EPR experiment [76]. Zero-field splitting of the energy levels that requires high-frequency EPR experiments can alternatively originate from collective spin-singlet ground states, as is the case in the two-dimensional (2D) spin system $SrCu_2(BO_3)_2$ [77], or from collective excitations in a magnetically ordered state. This latter scenario is thoroughly explained below in Section 9.4. Last we note, that some exotic magnetic excitations, for example, breather modes in spin-$\frac{1}{2}$ chains, can exhibit highly nonlinear frequency-field dependences and, therefore, require multiple-(high) frequency EPR experiments [78]. Similarly, various materials featuring field-induced phenomena, for example, reflected in field-dependent phase transitions, g factors, or line widths, also require multifrequency EPR experiments

To conclude this section, we briefly touch upon the state-of-the-art high-frequency EPR, namely, the free-electron-laser (FEL) powered multifrequency EPR spectroscopy [75]. In the Dresden High Magnetic Field Laboratory, such experiments are routinely performed with a picosecond pulsed FEL radiation in the frequency range between 1.2 and 75 THz and a spectral resolution better than 1%. Such extreme EPR frequencies require high magnetic fields, which are provided in the range between 0 and 70 T by pulsed magnets with typical pulse duration from 10 to a few hundreds of milliseconds. The use of FELs in pulsed EPR spectroscopy also appears very promising [79]. Since the picosecond-pulse duration imposes serious constrains to electronics, the longer-pulse electrostatic FEL at the University of California, Santa Barbara is more favorable in this respect.

9.3

Electron Paramagnetic Resonance of Exchange Coupled Systems

The magnetically dense insulators represent a vibrant field of research where EPR technique is particularly powerful [12]. A specific strength of this technique in studying electronic properties of a given material stems from the fact that EPR directly detects the electrons that are responsible for these properties, unlike other spectroscopic techniques, for example, like nuclear magnetic resonance (NMR) [80] or muon spectroscopy (μSR) [81] that can only provide indirect information relying on the weak coupling between the electrons and a particular probe. This property of the EPR technique, however, is on the other hand responsible for a relatively elaborate theoretical description of an EPR experiment, which consequently prevents a routine use of this technique.

The foundations of the EPR theory were laid by Kubo and Tomita (KT) more that half a century ago within their General Theory of Magnetic Resonance Absorption [82]. Since then, the theory has been progressively evolving, especially in the modern age with the occurrence of enhanced computational capabilities. Within this section, we overview this progress by summarizing the basic concepts of EPR in exchange coupled systems and providing some illustrative examples.

9.3.1
Kubo–Tomita Theory of the EPR Line Width and Exchange Narrowing

Within the linear-response theory the EPR spectrum $I(\omega)$ is proportional to the imaginary part of the dynamical susceptibility [82]

$$I(\omega) \propto \omega \chi''(q \to 0, \omega) \propto \frac{\omega}{T} \int_{-\infty}^{\infty} \langle S^+(t) S^-(0) \rangle \exp(i\omega t) dt, \qquad (9.3)$$

where $\langle \ \rangle$ denotes canonical averaging and $S^\alpha = \sum_i S_i^\alpha$ is the α-component of the total spin operator. The relation is valid in the high-temperature limit, that is, when $k_B T \gg \mu_B B_0$. Here k_B denotes the Boltzman constant. In the interaction representation, $I(\omega)$ can be expressed as the Fourier transform of the relaxation function $\varphi(t) = \langle \tilde{S}^+(t) S^-(0) \rangle / \langle S^+(0) S^-(0) \rangle$,

$$I(\omega) \propto \int_{-\infty}^{\infty} \varphi(t) e^{i(\omega - \omega_0)t} dt, \qquad (9.4)$$

where $\omega_0 = g\mu_B B_0 / \hbar$ ($\hbar$ is the reduced Planck constant). The calculation of the relaxation function is nontrivial, therefore, various approximation schemes were developed. Within the KT approximation [82] the total spin Hamiltonian given by Eq. (9.1) is divided into two parts, the dominant part $\mathcal{H}_0 = \mathcal{H}_{ex} + \mathcal{H}_{EZ}$ and the magnetic anisotropy part $\mathcal{H}' = \mathcal{H}_{ZFS} + \mathcal{H}_{HF} + \mathcal{H}_{dd} + \mathcal{H}'_{ex}$, which is treated as a perturbation. The nuclear terms $\mathcal{H}_{NZ} + \mathcal{H}_{NQ}$ do not directly affect the EPR spectra and are thus neglected in the following treatment. As the $\mathcal{H}_0$ term is SU(2) invariant, it conserves the total magnetization and therefore leads to a δ-function resonance at the field B_0. The non-commuting anisotropic $\mathcal{H}'$ is thus responsible alone for a finite EPR line width, line shift, or ever line splitting.

In the case of Markovian random processes, the relaxation function is approximated [82] by $\varphi(t) = \exp\left(\int_0^t (t-\tau)\psi(\tau) d\tau\right)$. Furthermore, the correlation function $\psi(\tau) = \psi(0) e^{-\tau^2/2\tau_c}$ is postulated to exhibit a Gaussian decay with the characteristic correlation time given by the exchange interaction J, $\tau_c \approx h/J$. As typically encountered in real exchange-coupled system, the spins fluctuate very rapidly ($\tau_c \omega_0 \ll 1$) and the relaxation function becomes an exponentially decaying function, yielding a Lorentzian spectrum according to Eq. (9.4). This is known as the exchange-narrowing limit [83,84], in which the line width of the ESR absorption line, $\Delta B \propto \tau_c \sqrt{\psi(0)}$, is inversely proportional to J. The zero-time correlation function $\psi(0) = M_2/\hbar^2$ is determined by the second moment of the

ESR line M_2 [83], while the correlation time is well approximated if also the fourth moment M_4 is taken into account, $\tau_c \propto \sqrt{M_2/M_4}$. This yields the full-width-at-half-maximum ESR line width [85]

$$\Delta B = C \frac{k_B}{g\mu_B} \sqrt{\frac{M_2^3}{M_4}}, \tag{9.5}$$

with the corresponding moments expressed as

$$M_2 = \frac{\langle [\mathcal{H}', S^+][S^-, \mathcal{H}']\rangle}{\langle S^+ S^-\rangle},$$
$$M_4 = \frac{\langle [\mathcal{H} - \mathcal{H}_Z, [\mathcal{H}', S^+]][\mathcal{H} - \mathcal{H}_Z, [\mathcal{H}', S^-]]\rangle}{\langle S^+ S^-\rangle}. \tag{9.6}$$

Here [ ] denotes a commutator, and the constant C is of the order of unity. The latter constant somewhat depends on the exact approximation of the ESR line shape. Namely, the experimental line shape is never truly Lorentzian, because the moments of the latter diverge. In systems where isotropic exchange coupling is much stronger than magnetic anisotropy, deviations from the Lorentzian shape occur only in far wings of the resonance and an approximate line shape that is a product of the Lorentzian and a broad Gaussian $\propto e^{-(B-B_0)^2/2B_e^2}$ [85], with $B_e = k_B/g\mu_B \sqrt{M_4/M_2}$ being the exchange field, is applicable. This then yields $C = \sqrt{2\pi}$ [85].

9.3.2
Nagata–Tazuke Theory of the EPR Line Shift

A similar, high-temperature perturbation approximation was derived for the EPR line shift by Nagata and Tazuke (NT) [86]. Within the NT theory the shift of the ESR absorption spectrum from B_0, $\delta B = \frac{1}{g\mu_B}\left(\frac{M_1}{M_0} - B_0\right)$, is given by

$$\delta B = \frac{\langle [S^-, [S^+, \mathcal{H}']]\rangle}{2g\mu_B \langle S^z\rangle}, \tag{9.7}$$

or, equivalently, the g-shift from the free-electron value is $\Delta g = \frac{\delta B}{B_0} g$. The result in Eq. (9.7) is generally exact at any temperature in the first order in anisotropy [87]. The second-order expression is much more cumbersome and can be found in Ref. [88]. For the general form of the magnetic anisotropy in the spin Hamiltonian of a spin-$\frac{1}{2}$ stem,

$$\mathcal{H}' = \frac{1}{2}\sum_{i,j \neq i} S_i \cdot \underline{K}_{ij} \cdot S_j, \tag{9.8}$$

where the coupling tensor $\underline{K}_{ij} = \underline{T}_{ij} + \underline{\delta}_{ij}$ includes both the long-ranged dipolar coupling $\underline{T}_{ij}$ and the short-ranged symmetric anisotropic exchange $\underline{\delta}_{ij}$ between spins at sites i and j. In the paramagnetic regime the NT theory predicts a linear scaling of the line shift with susceptibility χ [86,89,90]

$$\Delta g^z(T) = \frac{\langle S^z \rangle}{2\mu_B B_0} \sum_{j \neq i} \left(2K_{ij}^{zz} - K_{ij}^{xx} - K_{ij}^{yy} \right), \tag{9.9}$$

where $\langle S^z \rangle = \sum_i \langle S_i^z \rangle = \frac{\chi_{mol}(T) B_0}{N_A g \mu_0 \mu_B}$, z denotes the orientation of the applied field and x, y the two perpendicular directions, N_A is the Avogadro number, and μ_0 the vacuum permeability. We emphasize that the antisymmetric anisotropic DM interaction does not contribute to the ESR line shift in the first order [88].

9.3.3
Applications of the Kubo–Tomita and Nagata–Tazuke Theories to Real Materials

Both, the KT and NT theories, were successfully applied to various exchange-coupled systems in the past. Here, we demonstrate their applicability and highlight their limitations with several selected examples from the field of geometrically frustrated 2D spin lattices.

9.3.3.1 SrCu$_2$(BO$_3$)$_2$: An Orthogonal Dimer System
The 2D network of Cu^{2+} spins in SrCu$_2$(BO$_3$)$_2$ forms a network of orthogonal dimers (Figure 9.1), with the dominant intradimer exchange interaction J and somewhat smaller interdimer exchange interaction $J' = 0.63 J$ [91]. As such, it is located in the spin-liquid phase of the corresponding phase diagram, but in a

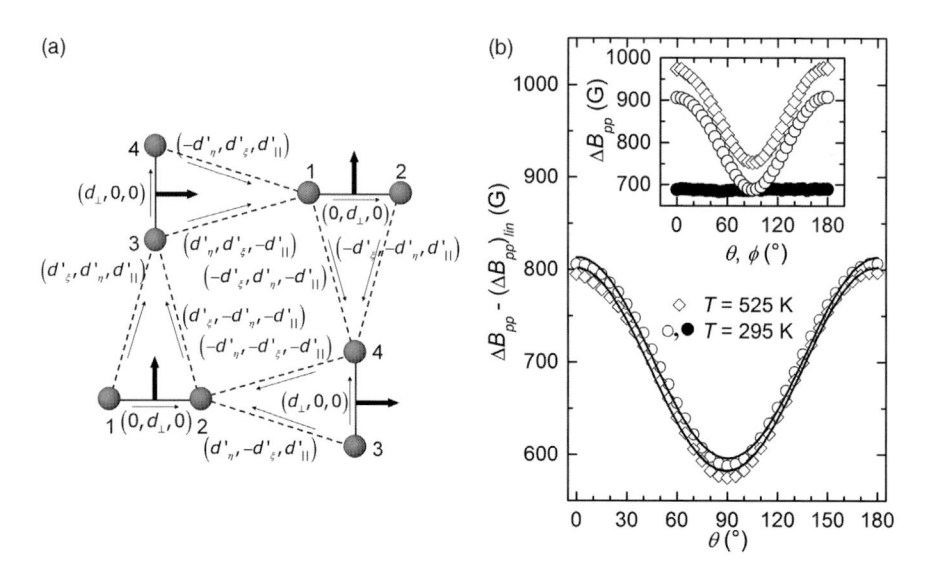

Figure 9.1 (a) The 2D network of orthogonal Cu^{2+} dimers in SrCu$_2$(BO$_3$)$_2$ with corresponding DM vectors. (b) Angular dependent EPR line width in SrCu$_2$(BO$_3$)$_2$, with no dependence in the plane within the 2D network and a pronounced dependence in a perpendicular plane. (Reproduced from Ref. [93].)

close vicinity to a quantum critical point around $J'_c/J \sim 0.68$ that separates this phase from a Néel ordered phase [92]. Therefore, small perturbations to the dominant Heisenberg exchange play an important role in determining the magnetic properties of $SrCu_2(BO_3)_2$. They are responsible for nontrivial properties of this material, such as mixing of the ground singlet and excited states, dispersion of triplon excitations, and their symmetry-breaking superstructures.

Determination of the dominant magnetic anisotropy terms in the spin Hamiltonian of $SrCu_2(BO_3)_2$, therefore, represents one of the milestones in understanding the magnetism of this system. The dominant magnetic anisotropy turns out to be of the DM antisymmetric exchange anisotropy type, $\mathcal{H}'_{ij} = \vec{d}_{ij} \cdot \vec{S}_i \times \vec{S}_j$, with a complex pattern of the intra- and interdimer DM vectors shown in Figure 9.1 [93]. These could be determined quantitatively from an EPR experiment, in which angular-dependent EPR line width was measured (Figure 9.1) [93]. Calculation of the EPR line width [Eq. (9.5)] in the infinite-temperature limit gives an analytical expression for the line-width anisotropy depending on the angle θ between the normal to the Cu planes and the applied magnetic field [93],

$$\Delta B = \frac{\pi}{\sqrt{12}} \frac{k_B}{g\mu_B} \frac{\left(8d_\parallel'^2 + 3d_\perp^2 + \left(8d_\parallel'^2 - d_\perp^2\right)\cos^2\theta\right)^{3/2}}{\left(32d_\parallel'^2 J_1^2 + 3d_\perp^2 J_2^2 + \left(32d_\parallel'^2 J_1^2 + d_\perp^2 J_2^2\right)\cos^2\theta\right)^{1/2}}, \quad (9.10)$$

with $J_1^2 = 3J^2 + 3J'^2 - 2JJ'$ and $J_2^2 = 13J^2 + 6J'^2$. This expression yielded both the intradimer DM anisotropy $d_\perp/J = 4.7\%$ and the interdimer anisotropy $d'_\parallel/J' = 4.6\%$, which values were later confirmed within 20% by ^{11}B NMR-shift measurements [94], by modeling triplet excitations obtained from inelastic neutron scattering [95] and by high-field EPR experiments [96].

9.3.3.2 Magnetic Anisotropy on the Kagome Lattice

Determination of the ground state of a 2D Heisenberg kagome lattice of corner-sharing triangles (Figure 9.2) represent one of the most enduring problems in condensed matter physics. Although a consensus has been reached, that for quantum ($S = \frac{1}{2}$) spins the ground state is not long-range ordered, neither frozen and disordered, the true character of the corresponding quantum spin-liquid (QSL) phase remains elusive. Recent state-of-the-art calculations seem to prefer a spin-gapped QSL [97], while the majority of experiments on known realizations of the kagome lattice speaks in favor of a gapless dynamical state. This discrepancy indicates that the quantum fluctuations that undermine long-range order on the kagome lattice might be highly dependent on smaller perturbations to the dominant isotropic Heisenberg Hamiltonian. Therefore, determination of the magnetic anisotropy terms in real systems is of paramount importance.

The anisotropy of several quantum kagome antiferromagnets (QKA) has successfully been determined by EPR with the use of the KT and NT theories [90,98,99]. The first example is the mineral herbertsmithite, $ZnCu_3(OH)_6Cl_2$, known as the first "structurally perfect $S = \frac{1}{2}$ kagome antiferromagnet" [100].

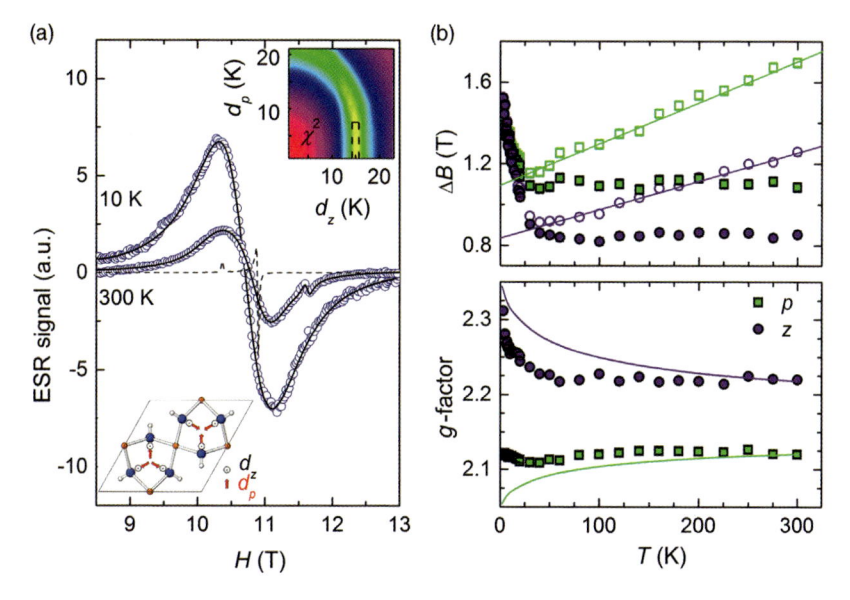

Figure 9.2 (a) EPR spectra of herbertsmithite. The kagome lattice is visualized in the bottom inset. The top inset highlights the quality of the fit for the Dzyaloshinskii–Moriya model. (b) The temperature dependence of the EPR line width (top) and g-factor (bottom) in vesignieite. The solid lines in the top panel are linear fits of the measured high-temperature line width (open symbols), while solid symbols represent intrinsic spin contribution. In the bottom panel the solid lines show the predictions of the symmetric anisotropic-exchange model. (Reproduced from Refs [98,99].)

Based on relatively small g-shifts (of the order of 10%, as typical for Cu^{2+}), it was argued that the antisymmetric part of the exchange anisotropy (DM interaction) dominates over the symmetric part [98]. The DM-vector pattern shown in Figure 9.2 then according to Eq. (9.5) dictates the line-width anisotropy

$$\Delta B(\theta) = \sqrt{2\pi} \frac{k_b}{2g(\theta)\mu_B J} \sqrt{\frac{\left(2d_z^2 + 3d_p^2 + \left(2d_z^2 - d_p^2\right)\cos^2\theta\right)^3}{16d_z^2 + 78d_p^2 + \left(16d_z^2 - 26d_p^2\right)\cos^2\theta}}, \quad (9.11)$$

in the infinite temperature limit, which allowed the authors to determine both the minute in-plane component $d_p/J \sim 0.01(3)$ and the dominant out-of-plane component $d_z/J \sim 0.08(1)$. The value of d_z is within the range $0.06 < d_z < 0.10$, which complies with the exact diagonalization thus explaining a peculiar NMR pattern observed in herbertsmithite [101]. The sizeable DM interaction then raised the fundamental question, to what extent the DM anisotropy affects the ground state of this compound. It was later shown that it places the system to a spin-liquid part of a phase diagram, however, quite close to a quantum critical point determined by the out-of-plane DM component to occur at $d_z^c/J \simeq 0.1$ [102]. This point separates the spin-liquid phase from a Néel ordered phase.

Since quantum criticality induced by the out-of-plane DM anisotropy in the QKA model was theoretically predicted [102], a question whether such a criticality was responsible for magnetic ordering [103] of another QKA representative, vesignieite, $BaCu_3V_2O_8(OH)_2$, arose immediately. The same type of analysis of the EPR line width as described above yielded the DM vector components $d_p/J = 0.19(2)$ and $d_z/J = 0.07(3)$ [99] (Figure 9.2). Alternatively, the same linewidth anisotropy could be modeled also with the symmetric anisotropic-exchange model. However, the extracted symmetric anisotropy parameters that, contrary to the DM interaction, are responsible also for temperature-dependent EPR shifts (Eq. (9.9)) significantly overestimated the measured shift (Figure 9.2). Therefore, the conclusion was reached, that the DM interaction also dominates in vesignieite. While the out-of-plane DM component is the largest anisotropy in herbertsmithite, it is the in-plane DM component that dominates in vesignieite. In the latter compound, the condition $d_p > d_z$ could profoundly affect the quantum critical point because d_p disfavors spin structures from the ground-state manifold of the isotropic J and should therefore be much more efficient in suppressing quantum fluctuations than d_z. This could explain why magnetic ordering in vesignieite occurs at surprisingly high temperature for a highly frustrated system, $T_N/J = 0.17$ [103], despite possessing very similar d_z/J as herbertsmithite.

The scaling of the g-shift with magnetic susceptibility, as predicted by Eq. (9.9) was observed in another QKA representative, kapellasite, $ZnCu_3(OH)_6Cl_2$. This has revealed the presence of the symmetric exchange anisotropy and allowed for its precise determination [90].

9.3.4
Breakdown of the Kubo–Tomita Model and Alternative Approaches

The applicability of the KT approach is limited in several aspects. First, it is a high-temperature approximation that cannot effectively account for developing spin correlations at lower temperatures (compared to J). Next limitation comes from the assumption of Markovian (Gaussian) random processes and exponentially decaying relaxation function [82] $\varphi(t)$ in Eq. (9.4). The KT approximation is, therefore, not always valid, even if the anisotropy can be treated as a perturbation. In low-dimensional systems, it is argued to break down and a diffusional decay of spin correlations may prevail instead, especially at higher temperatures. In this case, significant deviations from Lorentzian profiles and additional line broadening are expected [104]. These effects are particularly important for 1D spin systems [105].

Another potential pitfall of the KT theory relates to the DM interaction. For this type of interaction there exists an issue of a possible hidden symmetry [106]. For linear spin chains with staggered DM vectors, one can effectively transform this anisotropy to the AE term of the magnitude d^2/J by applying a nonuniform spin rotation [107], which makes the DM term of the same order as the AE term and discards the direct applicability of the KT formalism [107,108]. For other lattices, one should distinguish such reducible DM components from irreducible components that cannot be transformed to higher order terms [95]. The components that can be eliminated in the first order are those which sum up to zero within any closed loop on a lattice; for example, for the kagome lattice, the in-plane d_p is reducible while the out-of-plane d_z is irreducible [98].

9.3.4.1 EPR and Exact Calculations

Recently, the EPR line shape was studied numerically by exact diagonalization in 1D and 2D on finite spin clusters [109]. This way the *a priori* assumption of the relaxation function decay was not needed. However, such an approach is limited by the system size and the extrapolation to the thermodynamic limit is difficult [109]. Since the approach is exact, it may still be very interesting for small systems, such as molecular magnets or spin clusters.

In 1D, the exact-diagonalization results showed agreement with the Gaussian decay of spin correlation at early times and a clear slowing down compared to the KT results at longer times was observed, which indicated a crossover into a diffusional regime and an increased line width. Such a broadening process, however can be effectively short-cut by interchain couplings, leading again to narrower lines with the Lorentzian shape [110]. Moreover, an unexpected two-peak line shape was predicted for finite spin clusters, while its appearance in the thermodynamic limit remained questionable [109].

Attempts to calculate EPR spectrum in the limit of 1D Heisenberg chains strongly benefited from the recent theoretical breakthrough in calculating the dynamic spin susceptibility within the Tomonaga–Luttinger-liquid (TLL) formalism. Recently, an analytical expression for $\chi''_\perp(q, \omega, T)$ was used to calculate the corresponding EPR spectrum [111]. Application of this approach to a new molecular-based representative of uniform 1D Heisenberg spin chains CsO_2 allowed for an independent determination of the dimensionless TLL parameters and to demonstrate that in the high-temperature limit the EPR linewidth should be proportional to temperature.

Exact calculations on the 2D kagome lattice, on the other hand, showed that for the irreducible DM component d_z the line width indeed scales like d_z^2/J, as predicted by the KT theory [108]. Moreover, no clear slowing down (such as in 1D) toward the diffusional spin-correlations decay was observed. Interpolation to the spin-diffusional assumption caused a moderate increase of the line width and a slight deviation from the Lorentzian line shape. For herbertsmithite, such an assumption would then slightly decrease the amplitude of the DM vector compared to the above-presented results based on the KT approach, $0.044 \leq d_z \leq 0.08$ [109].

9.3.4.2 Oshikawa–Affleck Theory for 1D Antiferromagnets

Oshikawa and Affleck (OA) have developed a direct approach to EPR spectra calculations in 1D systems by employing an effective-field theory [108,112]. In contrast to the KT theory, this approach works well at intermediate and low temperatures, $T_N \ll T \ll J$, where in general all classical theories break down due to many-body correlation effects. The spin diffusion picture, which predicts a non-Lorentzian line shape in 1D, does not apply to the OA theory. It should be stressed that this approach is still perturbative in magnetic anisotropy.

The OA theory predicts different scaling of the EPR line width for the symmetric and the antisymmetric exchange anisotropy terms. In the former-case, the line width increases linearly with T and is independent of the applied field [108,112,113],

$$\Delta B_{AE}(T) = \frac{2\epsilon k_B (\delta/J)^2}{g\mu_B \pi^3} T, \qquad (9.12)$$

where $\epsilon = 2$ applies when the magnetic field is along the anisotropy axis and $\epsilon = 1$ otherwise. Alternatively, in the case of a staggered field $h_s = c_s B_0$ (c_s is the staggered-field coefficient) that is either due to a staggered DM interaction or a staggered g factor, the line width is given by [108,112,113]

$$\Delta H_{sf}(B_0, T) = 0.69 g\mu_B \frac{k_B J}{(k_B T)^2} h_s^2 \sqrt{\ln\left(\frac{J}{T}\right)}. \qquad (9.13)$$

It thus decreases with increasing T and exhibits a pronounced field dependence.

The OA theory was successfully tested for the first time in EPR experiments on copper benzoate [114] and copper pyrimidine dinitrate [115,116], two prominent examples of the quantum sine-Gordon spin-chain model. In both cases it was found that the term $(B_0/T)^2$, consistent with the dominance of the staggered fields, describes the line width at $T < J$ in the so-called "perturbative spinon" temperature regime. On the other hand, if both the staggered fields and the symmetric exchange anisotropy are important in a real system the overall EPR line width is a sum of terms (9.12) and (9.13), because no cross terms exist. Such is, for instance, the case in the $CuSe_2O_5$ spin-chain compound where simultaneous modeling of the angular, temperature, and frequency-dependent EPR line width with the OA theory yielded both the symmetric anisotropic exchange and the antisymmetric DM anisotropy [113].

9.3.4.3 EPR Moments Approach

The perturbative, classical KT and NT approaches and the *a priori* assumption of the Lorentzian line shape, of course, have limited applicability, as discussed above. However, the above-presented practical examples also demonstrate that these theories are highly useful for extracting magnetic anisotropies in the exchange-coupled spin systems with dimensionality exceeding one. This can usually be done exactly only in the paramagnetic (infinite-temperature) limit, where the canonical averaging, implicitly employing four-spin correlation functions in Eq. (9.6), is feasible. Moreover, also the field-theoretical OA approach is limited, this time to low temperatures and a restricted range of applied fields, while the purely numerical calculations are done on finite spin clusters, and the extrapolation to the thermodynamic limit is challenging.

A more direct, nonperturbative approach, allowing exact calculations of temperature-dependent line shift and line width "in frequency space" has recently been demonstrated for 1D spin chains with Ising, that is, AE, anisotropy of arbitrary strength [117]. In frequency space, the line shift $\delta\omega$ and line width $\Delta\omega$ are both determined by the "shifted moments" $m_n^\omega = \frac{J^{-n}}{2\pi}\int_{-\infty}^{\infty}(\omega - h)^n \chi''(\omega, h)d\omega$,

$$\delta\omega = J\frac{Jm_2^\omega + hm_1^\omega}{Jm_1^\omega + hm_0^\omega},$$

$$\Delta\omega = J^2\frac{Jm_3^\omega + hm_2^\omega}{Jm_2^\omega + hm_1^\omega} - \delta\omega^2, \tag{9.14}$$

where $h = g\mu_B B_0/\hbar$. The moments m_n are static correlations that can be calculated in the case of 1D spin chains to arbitrary precision for any temperature and applied field [117].

In EPR experiments, however, the frequency is kept fixed and the applied field is swept, which is opposite to most of the theoretical treatments. The "shifted moments" defined in the field space, $m_n^h = \frac{J^{-n}}{2\pi}\int_{-\infty}^{\infty}(\omega - h)^n\chi''(\omega, h)dh$ can be expressed by infinitely many terms m_n^ω and their derivatives, therefore, the line shift δB and line width ΔB cannot be calculated exactly at arbitrary T and B_0 [117].

Last, we note that the knowledge of all the moments, $M_n = \int_{-\infty}^{\infty}\omega^n I(\omega)d\omega$, is equivalent to the knowledge of all the derivatives of a particular absorption line and, therefore, exactly determines the line shape. The exact calculation of the moments, however, becomes progressively challenging with increasing n and is in general infeasible for $n > 4$ even in the infinite-temperature limit.

9.4
(Anti)Ferromagnetic Resonance in Magnetically Ordered Phases

In contrast to EPR, which is a response of an ansamble of weakly coupled paramagnetic moments to the applied magnetic field, the antiferromagnetic resonance (AFMR) is by its origin a collective excitation. That is, a correlated precession of a macroscopic number of magnetic moments coupled in the sublattice magnetization that can occur only in (antiferro)magnetically long-range-ordered systems. Consequently, it is dictated by interactions between the magnetic moments and the anisotropies they experience, for example, either of an exchange or of a crystal-field origin. To imitate the AFMR response, it is thus essential to relate the investigated magnetic system with a simplified magnetic structure and exchange lattice model.

The simplest antiferromagnetic model, the so-called "two-sublattice" model, assumes that all magnetic moments in the system can be described as two counterpointing magnetic sublattices coupled by a single antiferromagnetic interaction. Hence, this simplification is meaningful only if all spins ascribed to the same magnetic sublattice experience almost identical magnetic fields and anisotropies. The benefit of such a simple model is that it enables a straightforward calculation of the magnetic eigenstate and eigenmodes (excitations). As a result, tuning of the model (exchange and anisotropy) parameters allows one to reproduce the experimentally derived magnetic structure and AFMR response, for example, angular or magnetic-field dependences. This method, therefore, enables one to determine not only the strength of the dominant exchange interactions but also the magnitude of much weaker, yet very important, magnetic anisotropy terms.

In short, the AFMR is a powerful technique, which can reveal detailed information about the spin Hamiltonian. However, to extract the desired parameters one should have a very good idea about the magnetic ground state and the exchange network, that is, how are the magnetic moments ordered and which interactions are to be consider in the calculations.

9.4.1
Molecular Field Approach

In real magnetic systems, for example, powder or single-crystal samples, there are typically $\sim 10^{20}$ magnetic ions that have finite magnetic moments and similar amount of exchange pathways that connect them. Hence, an exact description (analytical or numerical) of such a huge system is not possible. The idea behind the molecular field approach is thus to describe the system in a similar manner as the crystal structure is described in crystallography, that is, by defining the space group and the unit cell of the crystal. In particular, in magnetically long-range-ordered state the magnetic moments, which have the same size and orientation as well as experience the magnetic fields of the same strength and direction, are ascribed to a single magnetic sublattice [118]. The standard approach is to choose the sublattice magnetizations such that they coincide with the experimentally determined magnetic structure; that is, the number of sublattice magnetizations is given by the number of magnetically inequivalent magnetic ions. Similarly are chosen also the relevant exchange interactions, which are dictated by the crystal structure. However, such a simplification has a cost of losing all information about the interactions between the spins in the individual sublattice and about the differences in their local environments. To gain back some of this information, the unit cell (the number of sublattice magnetizations and interactions between them) can be extended, if the quality of the experimental data is sufficient.

Let us assume a simple magnetic structure with a uniaxial anisotropy, which can be described by a "two-sublattice" model, where each magnetic moment is surrounded by counterpointing moments, corresponding to the opposite magnetic sublattice. The appropriate spin Hamiltonian in the applied magnetic field can thus be written as

$$\mathcal{H} = \sum_{i>j} J_{ij} \mathbf{S}_i \cdot \mathbf{S}_j + \mu_B g \mathbf{B} \cdot \sum_i \mathbf{S}_i + \sum_i D_i S_{zi}^2. \tag{9.15}$$

where indexes i and j run over all spins in the system, J_{ij} is the antiferromagnetic exchange between the nearest-neighboring spins, and D_i is the crystal field anisotropy constant at i-th site. The main simplification of the mean-field approach is the introduction of the magnetic sublattices. Namely, since all magnetic moments within the same sublattice have the same size and orientation, a net-average sublattice magnetization $M_l = -N_l g \mu_B \sum_m S_m$ can be written, where l denotes the sublattice and m runs over the N_l magnetic ions in the l-th sublattice. Considering now that $e^{-F/k_B T} = \sum_n e^{-E_n(B)/k_B T}$, where F is the free energy and the sum runs over the n possible states with the energy $E_n(B)$ depending on the strength of the applied magnetic field B, the free energy of the system in its ground state at $T = 0$ can be written as

$$F = \sum_{k>l}^{N_l} \mathcal{J}_{kl} \mathbf{M}_k \cdot \mathbf{M}_l + \mathbf{B} \cdot \sum_{l=1}^{N_l} \mathbf{M}_l + \sum_{l=1}^{N_l} \mathcal{D}_l M_{zl}^2. \qquad (9.16)$$

Here $\mathcal{J}_{kl} = z_{kl} J_{kl}/N_l(g\mu_B)^2$, $\mathcal{D}_l = D_l/N_l(g\mu_B)^2$ and z_{kl} is the number of the kl neighbors. The obtained expression (Eq. (9.16)) is the starting point for calculations of the magnetic ground state and AFMR modes.

To calculate the AFMR response, the magnetic ground state has to be known. The latter is deduced from minimization of F (Eq. (9.16)) by optimizing the orientations of the sublattice magnetizations. Depending on the magnetic modulation vector, that is, being commensurate or incommensurate, periodic or open boundary conditions are selected, respectively. In principle, this is a straightforward procedure, which, however, can get very complicated when competing interactions are present [119–121]. For further reading on this topic we point the reader to Ref. [122]. Knowing the magnetic ground state, the AFMR modes can then be derived by solving the equations of motion [123,124]

$$\frac{d\mathbf{M}_l}{dt} = \gamma (\mathbf{M}_l \times \mathbf{B}_l). \qquad (9.17)$$

Here $\mathbf{B}_l = -\partial F/\partial \mathbf{M}_l$ is the molecular field acting on the sublattice magnetization $\mathbf{M}_l$ and $\gamma = -g\mu_B/\hbar$. To solve these equations, several further simplifications have to be made [124–126]. First assumption is that sublattice magnetizations oscillate in a simple harmonic motion so that their time dependent part has the $e^{i\omega t}$ form. As a result, the resonant frequencies are simply the eigenvalues of the matrix, with $3l \times 3l$ elements (Eq. (9.17)) (for each sublattice magnetization there are three spatial coordinates). Second, we assume that deviations of sublattice magnetizations from their equilibrium positions are small; hence, only the components perpendicular to them are kept. Consequently, we can write each sublattice magnetization as $\mathbf{M}_l = \mathbf{M}_{0l} + \mathbf{m}_l(t)$, where $\mathbf{M}_{0l}$ represents the equilibrium orientation and $\mathbf{m}_l(t) = \mathbf{m}_l e^{i\omega t}$ is the small oscillating part ($\mathbf{M}_{0l} \cdot \mathbf{m}_l = 0$). Now, the molecular field $\mathbf{B}_l$ can be written as $\mathbf{B}_l(t) = \mathbf{B}_{0l} + \mathbf{b}_l(t)$, where $\mathbf{B}_{0l}$ is the static component parallel to $\mathbf{M}_{0l}$ and $\mathbf{b}_l(t) = \mathbf{b}_l e^{i\omega t}$ is the small ($|\mathbf{b}_l| \propto |\mathbf{m}_l|$) dynamic part perpendicular to it. In the following we can skip $\gamma(\mathbf{M}_{0l} \times \mathbf{B}_{0l})$, which is equal to zero, since $\mathbf{M}_{0l}$ is parallel to $\mathbf{B}_{0l}$. In addition, we neglect $\gamma(\mathbf{m}_l \times \mathbf{b}_l)$, which is proportional to $e^{i2\omega t}$ and is also very small, that is, both $\mathbf{m}_l$ and $\mathbf{b}_l$ are small and perpendicular to the static $\mathbf{M}_{0l}$ and $\mathbf{B}_{0l}$ components. As a result, Eq. (9.17) reduces to:

$$\frac{d\mathbf{m}_l}{dt} = \gamma[(\mathbf{M}_{0l} \times \mathbf{b}_l) + (\mathbf{m}_l \times \mathbf{B}_{0l})]. \tag{9.18}$$

This way the problem is reduced to a linear dependence of an individual magnetic sublattice on the oscillating parts of other sublattices. In other words, the list of $3l$ nonlinear equations is reduced to a set of $2l$ linear equations, which can be solved for a reasonable number of sublettice magnetizations l. The derived expressions (Eq. (9.16) and Eq. (9.18)) thus allow to calculate the magnetic ground state and AFMR modes, respectively, and therefore enable modeling of the experimental results.

Applying this procedure to the "two-sublattice" model with a uniaxial single-ion anisotropy, one obtains the following frequency relation [124,127]:

$$\omega(B) = \omega(0) \pm \gamma B \tag{9.19}$$

Here, $\omega(0) = \gamma\sqrt{B_A(2B_E + B_A)}$, where B_A is the magnetic anisotropy field and B_E is the exchange field, corresponding to the single-ion anisotropy D and the exchange interaction J, respectively. We stress that zero-field degeneracy can be removed by Dzyaloshinsky–Moriya or additional competing exchange interactions.

9.4.2
Application of the Molecular Field Approach to Real Materials

Here we list several examples where the AFMR experiments were essential for determination of weak anisotropy terms that have a profound effect on the magnetic properties of the investigated systems.

9.4.2.1 CuSe$_2$O$_5$: A Spin-1/2 Chain System
The CuSe$_2$O$_5$ system consist of spin-1/2 Cu^{2+} chains running along the crystal c axis, which are bridged by Se$_2$O$_5$ lone-pair dimmers (Figure 9.3a). Due to increased interchain Se–O distances, one may expect interchain exchange, J_{ic}, to be significantly smaller than intrachain one, J. Yet, the system develops long-range magnetic ordering below $T_N = 17$ K, implying that interchain exchange should not be neglected. Indeed, theoretical calculations [128] yield $J_{ic} = 0.12J$, which is thus accounted responsible for the observed magnetic transition. The magnetic ground state is antiferromagnetic and has staggered magnetic moments. The latter is a result of low crystal symmetry and the alternating arrangement of CuO$_4$ plaquettes (Figure 9.3a), which allow for a Dzyaloshinskii–Moriya vector and a staggered g tensor [113].

To quantify the anisotropy terms, an AFMR study was undertaken [129], which involved angular as well as magnetic-field-strength-dependence measurements. The model considered four magnetic sublattices, the dominant intrachain and the weak interchain interactions, antisymmetric Dzyaloshinskii–Moriya and symmetric exchange anisotropies, as well as the anisotropy of the g factor. The

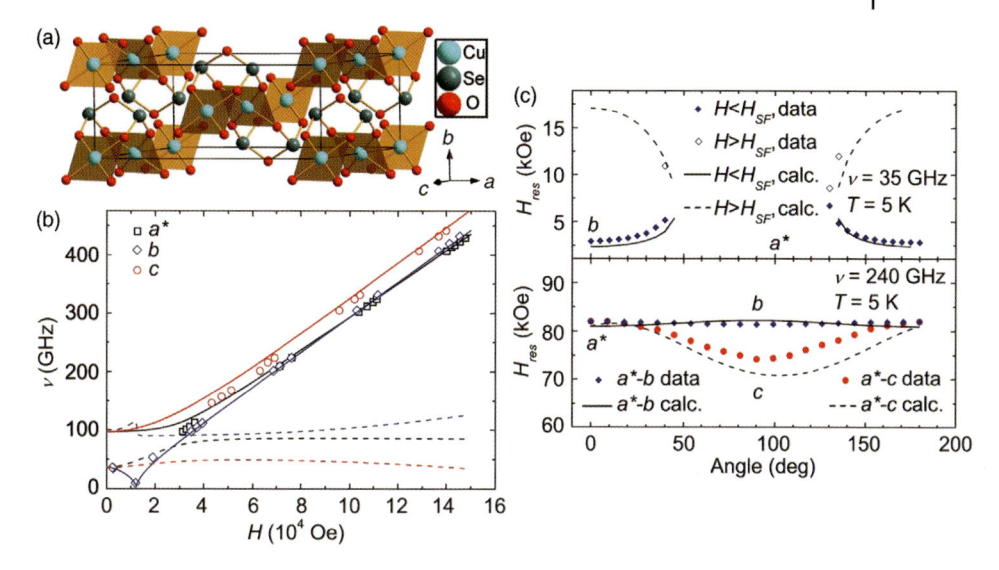

Figure 9.3 (a) Crystal structure of CuSe$_2$O$_5$. (b) The field dependence of the AFMR frequency measured at $T = 5$ K. (c) The angular dependence of the AFMR resonance modes in the a^*c and the a^*b planes measured at different frequencies. Solid and dashed lines show the results of calculations.

derived parameters, which reproduce the experiments very well (Figure 9.3a and b), reveal that the observed spin-flop transition represents a classical spin-flop transition, which originates from the competition of the underlying anisotropies and is not driven by the staggered field. In fact, the observed/calculated field dependences (Figure 9.3a) exhibit a typical response expected for magnetic field applied along the easy (b), intermediate (a^*), and hard (c) magnetic axes.

9.4.2.2 Ba$_3$NbFe$_3$Si$_2$O$_{14}$: A Two-Dimensional Triangular Magnetic Lattice

The Ba$_3$NbFe$_3$Si$_2$O$_{14}$ compound crystallizes in a noncentrosymmetric trigonal unit cell ($P321$ symmetry). The Fe^{3+} ($S = 5/2$) spins reside on vertices of equilateral triangles arranged into a two-dimensional triangular lattice (crystallographic ab planes in Figure 9.4) [130]. The dominant exchange interactions are antiferromagnetic (the Curie–Weiss temperature is -180 K) and thus frustrated. Nevertheless, a long-range-ordered state, realized below $T_N = 26$ K, is characterized by a 120° spin arrangement on each triangle. The moments are bound to the ab planes and form a magnetic helix along the crystallographic c axis, corresponding to the magnetic propagation vector $\mathbf{q} = (0, 0, \tau), \tau \sim 1/7$ [130].

To identify the anisotropy term that is responsible for the chirality of the magnetic ground state, a joint EPR and AFMR study was conducted [131]. The room-temperature EPR signal exhibits a pronounced angular dependence of the resonant field as well as of the signal width. The former reflects a sizable g-tensor anisotropy, whereas the latter is related either to single-ion or exchange

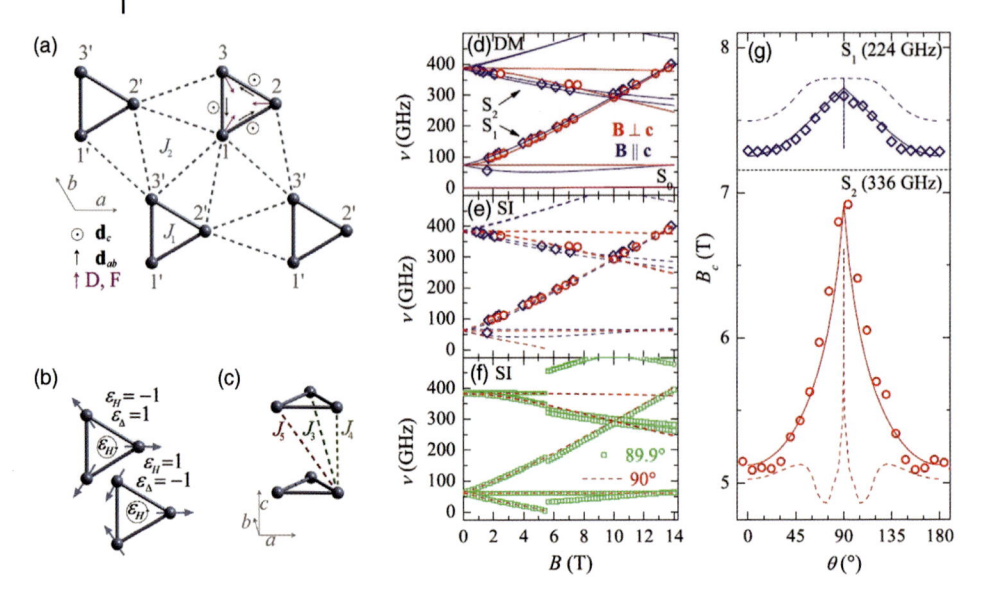

Figure 9.4 The 2D arrangement of Fe^{3+} in $Ba_3NbFe_3Si_2O_{14}$. (a,c) Exchange interactions J_{1-5} and magnetic anisotropies of the Dzyaloshinsky–Moriya (DM) **d** and single-ion (SI) D, F type. The latter vectors are parallel to the twofold rotational c axis. (b) Two possible double chiral ground states, with ϵ_Δ and ϵ_H denoting triangular chirality and helicity (with $\sim 2\pi/7$ pitch along c axis), respectively. Frequency-field diagram of measured (symbols) and simulated (lines) AFMR modes for (d) DM and (e) SI anisotropy at 4 K. (f) Field-induced phase transition predicted by the SI anisotropy model at 5.5 T for $B \perp c$. (g) Angular dependence of the two resonance modes at 4 K; measured (symbols) and simulated for the DM (solid lines) and SI anisotropy (dashed lines).

anisotropy. To distinguish the source of the anisotropy, the combined modeling of the EPR and AFMR data was performed (Figure 9.4). To reproduce the magnetic modulation, the model had to consider 14 pairs of coupled triangles (84 magnetic sublattices and 5 different exchange interactions). Despite the complexity, the modeling enabled the identification of Dzyaloshinskii–Moriya interaction as the dominant source of the anisotropy and thus to be responsible for the observed chiral behavior. Combining AFMR and inelastic neutron scattering allowed also for determination of the SI anisotropy [132].

Similar AFMR analysis allowed estimation of the exchange interactions and magnetic anisotropies in several other layered magnetic systems with triangular topology [126,133].

9.4.3
Linear Spin-Wave Theory

An alternative approach to the resonant-frequency calculations using classical equations of motion is the so-called spin-wave theory. Similarly to the former approach, the starting point is the magnetic ground state, that is, the configuration of classical spins minimizing the magnetic energy (spin Hamiltonian) of the magnetic unit cell (Eq. (9.16)). From this point, however, the two methods start to deviate. In contrast to classical equations of motion, the spin-wave theory considers quantum fluctuations about the equilibrium state and maps them onto a set of a simple harmonic oscillators. For instance, if we denote the equilibrium orientation of a spin as z, such fluctuations will give

$$S_x = 0 \leadsto S_x \neq 0$$
$$S_y = 0 \leadsto S_y \neq 0$$
$$S_z = S \leadsto S_z = S - n.$$

The corresponding energy gain can be treated as a quasi particle a, that is, a boson called a magnon, with the relation $n = a^\dagger a$, where $a^\dagger$ and a are the creation and annihilation operators. The spin algebra is recovered if this changes are rewritten using raising and lowering ladder operators $S^+ = S_x + iS_y$ and $S^- = S_x - iS_y$, respectively, as shown by Holstein and Primakoff [134]:

$$S_z = S - a^\dagger a \tag{9.20}$$

$$S^+ = \sqrt{2S - a^\dagger a}\, a \approx \sqrt{2S}\, a \tag{9.21}$$

$$S^- = a^\dagger \sqrt{2S - a^\dagger a} \approx \sqrt{2S}\, a^\dagger \tag{9.22}$$

$$\mathbf{S}_i \cdot \mathbf{S}_j = S_i^z S_j^z + \frac{1}{2}\left(S_i^+ S_j^- + S_i^- S_j^+\right). \tag{9.23}$$

Here, only the first term of the Taylor series is kept, that is, $1/S$ is considered as being small so higher order terms $1/S^2$ and so on are neglected, and a set of bosons $[a, a^\dagger] = 1$ is introduced, which allows to represent the spin algebra $[S^x, S^y] = iS^z$. This is the foundation of the linear spin-wave theory.

To derive the spin waves, several nontrivial steps have to be made, which we summarize here only briefly, while for detailed explanation and calculation procedure we point the reader to Refs. [135–137]. It is important to be aware that the $a^\dagger$ and a bosonic operators represent excitations localized on magnetic moments, whereas the spin waves (magnons) in a crystal are extended (nonlocalized) periodic harmonic deviations of the magnetic moments from their equilibrium positions. The corresponding (nonlocal) creation and annihilation operators can thus be expressed as the Fourier transformation of the localized bosonic operators. Applying the inverse Fourier transformation to the localized $a^\dagger$ and a, the spin Hamiltonian can be rewritten in a matrix form. The eigenvectors and the eignevalues of this matrix then represent the spin waves written in the basis of non-localized boson operators and their energies, respectively. We note here that in the classical limit the spin-wave theory gives the same results as equations of motion, that is, the latter can be deduced from the former [138]. However, in contrast to the classical equations of motion, the spin-wave theory allows for calculations of dispersion relations over the entire reciprocal space.

9.5 Summary and Outlook

EPR has been for decades one of the key local-probe techniques, owing to its high sensitivity, high resolution, and local-environment insight that it provides. The versatility of this technique for the applications in chemistry, biochemistry, physics, or medicine is hardly met by any other experimental technique. Compared to structural probes, like X-ray or neutron diffraction, the local structural determination from EPR may appear slow and tedious, as only a limited number of structural parameters can be extracted from EPR experiments. However, the real power of EPR as a structural probe emerges in disordered systems, where the structural information over length scales may extend up to ~ 5 nm, something that is difficult to achieve with other complementary methods, even by such methods as the extended X-ray absorption fine structure (EXAFS) technique. Moreover, the EPR spectroscopy yields information regarding the dynamic processes on the time scale ranging between 10^{-10} s and 10^{-3} s, thus significantly extending the time scale offered by the complementary nuclear magnetic resonance (NMR).

Here we have mostly reviewed the application of EPR to magnetic systems. Compared to neutron diffraction technique, EPR is limited only to $q \rightarrow 0$ information. However, we highlight two major advantages that are pertinent to EPR: (i) the very high resolution of resonance spectra allows for a very precise determination of key parameters in the spin Hamiltonian and (ii) the possibility to study magnetic response in extremely high magnetic fields – something that is largely still inaccessible to many other experimental methods. These two unique characteristics allow for comprehensive studies of often very intricate phase diagrams of frustrated or low-dimensional quantum antiferromagnets. Together with the progress in theoretical concepts on quantum magnetic systems, exciting discoveries were made in the past, including fractional spin excitations in one-dimensional antiferromagnets or triplet excitations in two-dimensional lattices. Important instrumental advances that we have witnesses in the last couple of decades extend the range of applicability of EPR to study magnetic systems and will for sure lead to many new discoveries in near future.

References

1. Zavoiskii, E.K. (1967) Electron paramagnetic resonance. *Usp. Fiz. Nauk.*, **93**, 523–527, and references therein.
2. Eaton, G.R. and Eaton, S.S. (1998) *Foundations of Modern EPR*, World Scientific, Singapore.
3. Misra, S.K. (ed.) (2011) *Multifrequency Electron Paramagnetic Resonance: Theory and Applications*, John Wiley & Sons, Weinheim, Weinheim.
4. Lund, A., Shimada, S., and Shiotani, M. (2011) *Principles and Applications of ESR Spectroscopy*, Springer Science & Business Media, Dordrecht.
5. Abragam, A. and Bleaney, B. (1970) *Electron Paramagnetic Resonance of*

Transition Ions, Oxford University Press, Oxford.

6 Pilbrow, J.R. (1990) *Transition Ion Electron Paramagnetic Resonance*, Clarendon Press, USA, Oxford.

7 Hoff, A.J. (ed.) (1989) *Advanced EPR: Applications in Biology and Biochemistry*, 1st edn, Elsevier Science, Amsterdam.

8 Drescher, M. and Jeschke, G. (eds) (2012) *EPR Spectroscopy, Topics in Current Chemistry*, vol. **321**, Springer, Berlin Heidelberg.

9 Brustolon, M. and Giamello, G. (2009) *Electron Paramagnetic Resonance: A Practitioner's Toolkit*, John Wiley & Sons, Inc, Hoboken, N. J.

10 Schweiger, A. and Jeschke, G. (2001) *Principles of Pulse Electron Paramagnetic Resonance*, 1st edn, Oxford University Press.

11 Swartz, H. and Khan, N. (2005) *EPR Spectroscopy of Function In Vivo, Biological Magnetic Resonance*, vol. **23**, Springer, US.

12 Bencini, A. and Gatteschi, D. (1990) *EPR of Exchange Coupled Systems*, Springer-Verlag, Berlin.

13 Poole, C.P. (1996) *Electron Spin Resonance: A Comprehensive Treatise on Experimental Techniques*, Courier Corporation.

14 Weil, J. and Bolton, J. (2007) *Electron Paramagnetic Resonance: Elementary Theory and Practical Applications*, Wiley-Interscience.

15 Dzyaloshinsky, I. (1958) A thermodynamic theory of weak ferromagnetism of antiferromagnetics. *J. Phys. Chem. Solids*, **4** (4), 241–255.

16 Moriya, T. (1960) Anisotropic superexchange interaction and weak ferromagnetism. *Phys. Rev.*, **120** 91–98.

17 Slichter, C.P. (1964) *Principles of Magnetic Resonance with Examples from Solid State Physics*, Harper & Row.

18 Eaton, G., Eaton, S., Barr, D., and Weber, R. (2010) *Quantitative EPR*, Springer.

19 Griswold, T.W., Kip, A.F., and Kittel, C. (1952) Microwave spin resonance absorption by conduction electrons in metallic sodium. *Phys. Rev.*, **88** 951–952.

20 Dyson, F.J. (1955) Electron spin resonance absorption in metals. II. Theory of electron diffusion and the skin effect. *Phys. Rev.*, **98** 349–359.

21 Feher, G. and Kip, A.F. (1955) Electron spin resonance absorption in metals. I. Experimental. *Phys. Rev.*, **98** 337–348.

22 Jánossy, A., Chauvet, O., Pekker, S., Cooper, J.R., and Forró, L. (1993) Conduction electron spin resonance in Rb_3C_{60}. *Phys. Rev. Lett.*, **71**, 1091–1094.

23 Tanigaki, K., Kosaka, M., Manako, T., Kubo, Y., Hirosawa, I., Uchida, K., and Prassides, K. (1995) Alkali effects on the electronic states of K_3C_{60} and Rb_3C_{60}. *Chem. Phys. Lett.*, **240** (5–6), 627–632.

24 Chauvet, O., Oszlànyi, G., Forro, L., Stephens, P.W., Tegze, M., Faigel, G., and Jànossy, A. (1994) Quasi-one-dimensional electronic structure in orthorhombic RbC_{60}. *Phys. Rev. Lett.*, **72**, 2721–2724.

25 Arčon, D., Prassides, K., Margadonna, S., Maniero, A.L., Brunel, L.C., and Tanigaki, K. (1999) Electron spin resonance study of the polymeric phase of Na_2C_{60}. *Phys. Rev. B*, **60**, 3856–3861.

26 Ganin, A.Y., Takabayashi, Y., Pregelj, M., Zorko, A., Arčon, D., Rosseinsky, M.J., and Prassides, K. (2007) EPR analysis of spin susceptibility and line width in the hyperexpanded fulleride (CH_3NH_2) K_3C_{60}. *Chem. Mater.*, **19** (13), 3177–3182.

27 Muranyi, F., Urbanik, G., Kataev, V., and Büchner, B. (2008) Electron spin dynamics of the superconductor CaC_6 probed by ESR. *Phys. Rev. B*, **77**, 024507.

28 Barbon, A. and Brustolon, M. (2012) An EPR study on nanographites. *Appl. Mag. Res.*, **42**, 197–210.

29 Szirmai, P., Fábián, G., Dora, B., Koltai, J., Zolyomi, V., Kürti, J., Nemes, N.M., Forró, L., and Simon, F. (2011) Density of states deduced from ESR measurements on low-dimensional nanostructures; benchmarks to identify the ESR signals of graphene and SWCNTs. *Phys. Status Solidi B*, **248** (11), 2688–2691.

30 Kausteklis, J., Cevc, P., Arčon, D., Nasi, L., Pontiroli, D., Mazzani, M., and Riccò, M. (2011) Electron paramagnetic resonance study of nanostructured graphite. *Phys. Rev. B*, **84**, 125406.

31 Szirmai, P., Fabian, G., Koltai, J., Nafradi, B., Forro, L., Pichler, T., Williams, O.A., Mandal, S., Bäuerle, C., and Simon, F. (2013) Observation of conduction electron spin resonance in boron-doped diamond. *Phys. Rev. B*, **87**, 195132.

32 Torrance, J.B., Pedersen, H.J., and Bechgaard, K. (1982) Observation of antiferromagnetic resonance in an organic superconductor. *Phys. Rev. Lett.*, **49**, 881–884.

33 Dumm, M., Loidl, A., Fravel, B.W., Starkey, K.P., Montgomery, L.K., and Dressel, M. (2000) Electron spin resonance studies on the organic linear-chain compounds (TMTCF)$_2$X (C = S, Se; X = PF$_6$, AsF$_6$, CiO$_4$, Br). *Phys. Rev. B*, **61**, 511–521.

34 Elliott, R.J. (1954) Theory of the effect of spin–orbit coupling on magnetic resonance in some semiconductors. *Phys. Rev.*, **96**, 266–279.

35 Yafet, Y. (1983) Conduction electron spin relaxation in the superconducting state. *Phys. Lett. A*, **98** (5–6), 287–290.

36 Barnes, S.E. (1981) Theory of electron spin resonance of magnetic ions in metals. *Adv. Phys.*, **30** (6), 801–938.

37 Milios, C.J., Piligkos, S., and Brechin, E.K. (2008) Ground state spin-switching via targeted structural distortion: twisted single-molecule magnets from derivatised salicylaldoximes. *Dalton Trans.*, 1809–1817.

38 Caneschi, A., Gatteschi, D., Sessoli, R., Barra, A.L., Brunel, L.C., and Guillot, M. (1991) Alternating current susceptibility, high field magnetization, and millimeter band EPR evidence for a ground S = 10 state in [Mn$_{12}$O$_{12}$(O$_2$CCH$_3$)$_{16}$(H$_2$O)$_4$]·2CH$_3$CO$_2$H·4H$_2$O. *J. Am. Chem. Soc.*, **113**, 5873–5874.

39 Barra, A.L., Gatteschi, D., and Sessoli, R. (1997) High-frequency EPR spectra of a molecular nanomagnet: Understanding quantum tunneling of the magnetization. *Phys. Rev. B*, **56**, 8192–8198.

40 Hill, S., Perenboom, J.A., Dalal, N.S., Hathaway, T., Stalcup, T., and Brooks, J.S. (1998) High-sensitivity electron paramagnetic resonance of Mn$_{12}$-acetate. *Phys. Rev. Lett.*, **80**, 2453–2456.

41 Blinc, R., Cevc, P., Arčon, D., Dalal, N.S., and Achey, R.M. (2001) Excited-state X-band EPR in a molecular cluster nanomagnet. *Phys. Rev. B*, **63** 212401.

42 Redler, G., Lampropoulos, C., Datta, S., Koo, C., Stamatatos, T.C., Chakov, N.E., Christou, G., and Hill, S. (2009) Crystal lattice desolvation effects on the magnetic quantum tunneling of single-molecule magnets. *Phys. Rev. B*, **80**, 094408.

43 Barra, A.L., Debrunner, P., Gatteschi, D., Schulz, C.E., and Sessoli, R. (1996) Superparamagnetic-like behavior in an octanuclear iron cluster. *Europhys Lett.*, **35** 133–138.

44 Baker, M.L., Blundell, S.J., Domingo, N., and Hill, S. (2014) Spectroscopy methods for molecular nanomagnets. *Struct. Bond.*, **164**, 231–291.

45 Emanuel, N.M., Roginskii, V.A., and Buchachenko, A.L. (1982) Some problems of the kinetics of radical reactions in solid polymers. *Russ. Chem. Rev.*, **51** (3), 203.

46 Augusto, O. and Muntz Vaz, S. (2007) EPR spin-trapping of protein radicals to investigate biological oxidative mechanisms. *Amino Acids*, **32** (4), 535–542.

47 Rieger, A.L. and Rieger, P.H. (2004) Chemical insights from EPR spectra of organometallic radicals and radical ions. *Organometallics*, **23** (2), 154–162.

48 Lappas, A., Zorko, A., Wortham, E., Das, R. N., Giannelis, E. P., Cevc, P., and Arčon, D. (2005) Low-Energy Magnetic Excitations and Morphology in Layered Hybrid Perovskite-Poly(dimethylsiloxane) Nanocomposites. *Chem. Mater.*, **17**, 1199–1207.

49 Goswami, M., Chirila, A., Rebreyend, C., and de Bruin, B. (2015) EPR spectroscopy as a tool in homogeneous catalysis research. *Top. Catal.*, **58** (12–13), 719–750.

50 Rhodes, C.J. (2015) The role of ESR spectroscopy in advancing catalytic science: some recent developments. *Prog. React. Kinet. Mec.*, **40** (3), 201–248.

51 Rowlands, C.C. and Murphy, D.M. (1999) Chemical applications of EPR, in *Encyclopedia of Spectroscopy and*

Spectrometry (Second Edition) (ed. J.C. Lindon), Academic Press, Oxford, pp. 221–228.
52 Breck, D.W. (1984) *Zeolite Molecular Sieves: Structure, Chemistry, and Use*, R. E. Krieger.
53 Zhang, J. and Goldfarb, D. (2000) Manganese incorporation into the mesoporous material MCM-41 under acidic conditions as studied by high field pulsed EPR and ENDOR spectroscopies. *J. Am. Chem. Soc.*, **122** (29), 7034–7041.
54 Zhang, J., Luz, Z., and Goldfarb, D. (1997) EPR studies of the formation mechanism of the mesoporous materials MCM-41 and MCM-50. *J. Phys. Chem. B*, **101** (36), 7087–7094.
55 Zabukovec Logar, N., Novak Tušar, N., Mali, G., Mazaj, M., Arčon, I., Arčon, D., Rečnik, A., Ristić, A., and Kaučič, V. (2006) Manganese-modified hexagonal mesoporous aluminophosphate MnHMA: synthesis and characterization. *Microporous Mesoporous Mater.*, **96** (1–3), 386–395.
56 Novak Tušar, N., Ristić, A., Mali, G., Mazaj, M., Arčon, I., Arčon, D., Kaučič, V., and Zabukovec Logar, N. (2010) MnO_x nanoparticles supported on a new mesostructured silicate with textural porosity. *Chem. Eur. J.*, **16** (19), 5783–5793.
57 Yahiro, H., Lund, A., and Shiotani, M. (2004) Nitric oxide adsorbed on zeolites: EPR studies. *Spectrochim. Acta A Mol. Biomol. Spectrosc.*, **60** (6), 1267–1278. EMARDIS-8: Recent Achievements in Fundamental and Practical Aspects of {EPR}.
58 Umek, P., Cevc, P., Jesih, A., Gloter, A., Ewels, C.P., and Arčon, D. (2005) Impact of structure and morphology on gas adsorption of titanate-based nanotubes and nanoribbons. *Chem. Mat.*, **17** (24), 5945–5950.
59 Umek, P., Pregelj, M., Gloter, A., Cevc, P., Jagličič, Z., Čeh, M., Pirnat, U., and Arčon, D. (2008) Coordination of intercalated cu^{2+} sites in copper doped sodium titanate nanotubes and nanoribbons. *J. Phys. Chem. C*, **112** (39), 15311–15319.
60 Umek, P., Bittencourt, C., Gloter, A., Dominko, R., Jaglii, Z., Cevc, P., and Arčon, D. (2012) Local coordination and valence states of cobalt in sodium titanate nanoribbons. *J. Phys. Chem. C*, **116** (20), 11357–11363.
61 Umek, P., Bittencourt, C., Guttmann, P., Gloter, A., Škapin, S.D., and Arčon, D. (2014) Mn^{2+} substitutional doping of TiO_2 nanoribbons: a three-step approach. *J. Phys. Chem. C*, **118** (36), 21250–21257.
62 Cevc, G., Cevc, P., Schara, M., and SkaleriC, U. (1980) The caries resistance of human teeth is determined by the spatial arrangement of hydroxyapatite microcrystals in the enamel. *Nature*, **286** 425–426.
63 Kiel, A. and Mims, W.B. (1967) Paramagnetic relaxation measurements on Ce, Nd, and Yb in $CAWO_4$ by an electron spin-echo method. *Phys. Rev.*, **161**, 386–397.
64 Mims, W.B. (1972) Envelope modulation in spin-echo experiments. *Phys. Rev. B*, **5**, 2409–2419.
65 Mims, W.B. (1972) Amplitudes of superhyperfine frequencies displayed in the electron-spin-echo envelope. *Phys. Rev. B*, **6**, 3543–3545.
66 Žutić, I., Fabian, J., and Das Sarma, S. (2004) Spintronics: fundamentals and applications. *Rev. Mod. Phys.*, **76**, 323–410.
67 Morley, G.W. (2015) Chapter 3, in *Towards Spintronic Quantum Technologies with Dopants in Silicon*, vol. **24**, The Royal Society of Chemistry.
68 Degen, C.L. (2008) Scanning magnetic field microscope with a diamond single-spin sensor. *Appl. Phys. Lett.*, **92** (24), 243111.
69 Häberle, T., Schmid-Lorch, D., Karrai, K., Reinhard, F., and Wrachtrup, J. (2013) High-dynamic-range imaging of nanoscale magnetic fields using optimal control of a single qubit. *Phys. Rev. Lett.*, **111**, 170801.
70 Maurer, P.C., Kucsko, G., Latta, C., Jiang, L., Yao, N.Y., Bennett, S.D., Pastawski, F., Hunger, D., Chisholm, N., Markham, M., Twitchen, D.J., Cirac, J.I., and Lukin, M.D. (2012) Room-temperature quantum

71. Maisuradze, A., Shengelaya, A., Berger, H., Djokić, D.M., and Keller, H. (2012) Magnetoelectric coupling in single crystal Cu_2OSeO_3 studied by a novel electron spin resonance technique. *Phys. Rev. Lett.*, **108** (24), 247211.
72. Xiao, M., Martin, I., Yablonovitch, E., and Jiang, H.W. (2004) Electrical detection of the spin resonance of a single electron in a silicon field-effect transistor. *Nature*, **430**, 435–439.
73. Reijerse, E.J. (2010) High-frequency EPR instrumentation. *Appl. Magn. Reson.*, **37** (1–4), 795–818.
74. Schmalbein, D., Maresch, G., Kamlowski, A., and Höfer, P. (1999) The bruker high-frequency-EPR system. *App. Mag. Res.*, **16** (2), 185–205.
75. Zvyagin, S.A., Ozerov, M., Čižmár, E., Kamenskyi, D., Zherlitsyn, S., Herrmannsdörfer, T., Wosnitza, J., Wünsch, R., and Seidel, W. (2009) Terahertz-range free-electron laser electron spin resonance spectroscopy: techniques and applications in high magnetic fields. *Rev. Sci. Instrum.*, **80** (7), 073102.
76. Zvyagin, S.A., Wosnitza, J., Batista, C.D., Tsukamoto, M., Kawashima, N., Krzystek, J., Zapf, V.S., Jaime, M., Oliveira, N.F. Jr, and Paduan-Filho, A. (2007) Magnetic excitations in the spin-1 anisotropic Heisenberg antiferromagnetic chain system $NiCl_2$-$4SC(NH_2)_2$. *Phys. Rev. Lett.*, **98** (4), 047205.
77. Nojiri, H., Kageyama, H., Ueda, Y., and Motokawa, M. (2003) ESR study on the excited state energy spectrum of $SrCu_2(BO_3)_2$–a central role of multiple-triplet bound states. *J. Phys. Soc. Jpn.*, **72** (12), 3243–3253.
78. Zvyagin, S.A., Cižmár, E., Ozerov, M., Wosnitza, J., Feyerherm, R., Manmana, S.R., and Mila, F. (2011) Field-induced gap in a quantum spin-$\frac{1}{2}$ cain in a strong magnetic field. *Phys. Rev. B*, **83** (6), 060409.
79. Takahashi, S., Allen, D.G., Seifter, J., Ramian, G., Sherwin, M.S., Brunel, L.C., and van Tol, J. (2008) Pulsed EPR spectrometer with injection-locked UCSB free-electron laser. *Infrared Phys. Technol.*, **51** (5), 426–428.
80. Abragam, A. (1961) *The Principles of Nuclear Magnetism*, Oxford University Press, Oxford.
81. Yaouanc, A. and De Réotier, P.D. (2011) *Muon Spin Rotation, Relaxation, and Resonance: Applications to Condensed Matter*, Oxford University Press, Oxford.
82. Kubo, R. and Tomita, K. (1954) A general theory of magnetic resonance absorption. *J. Phys. Soc. Jpn.*, **9** (6), 888–919.
83. Van Vleck, J.H. (1948) The dipolar broadening of magnetic resonance lines in crystals. *Phys. Rev.*, **74** (9), 1168–1183.
84. Anderson, P.W. and Weiss, P.R. (1953) Exchange narrowing in paramagnetic resonance. *Rev. Mod. Phys.*, **25** (1), 269–276.
85. Castner J Jr, T.G. and Seehra, M.S. (1971) Antisymmetric exchange and exchange-narrowed electron-paramagnetic-resonance linewidths. *Phys. Rev. B*, **4** (1), 38–45.
86. Nagata, K. and Tazuke, Y. (1972) Short range order effects on EPR frequencies in Heisenberg linear chain antiferromagnets. *J. Phys. Soc. Jpn.*, **32** (2), 337–345.
87. Maeda, Y., Sakai, K., and Oshikawa, M. (2005) Exact analysis of ESR shift in the spin-1/2 Heisenberg antiferromagnetic chain. *Phys. Rev. Lett.*, **95** (3), 037602.
88. Maeda, Y. and Oshikawa, M. (2005) Direct perturbation theory on the electron spin resonance shift and its applications. *J. Phys. Soc. Jpn.*, **74** (1), 283–286.
89. Nagata, K., Yamamoto, I., Takano, H., and Yokozawa, Y. (1977) EPR g-shift and anisotropic magnetic susceptibility in K_2MnF_4. *J. Phys. Soc. Jpn.*, **43** (3), 857–861.
90. Kermarrec, E., Zorko, A., Bert, F., Colman, R.H., Koteswararao, B., Bouquet, F., Bonville, P., Hillier, A., Amato, A., van Tol, J., Ozarowski, A., Wills, A.S., and Mendels, P. (2014) Spin dynamics and disorder effects in the $S = \frac{1}{2}$ kagome Heisenberg spin-liquid phase of kapellasite. *Phys. Rev. B*, **90** (20), 205103.
91. Kageyama, H., Nishi, M., Aso, N., Onizuka, K., Yosihama, T., Nukui, K.,

Kodama, K., Kakurai, K., and Ueda, Y. (2000) Direct evidence for the localized single-triplet excitations and the dispersive multitriplet excitations in SrCu$_2$(BO$_3$)$_2$. *Phys. Rev. Lett.*, **84** (25), 5876.

92 Koga, A. and Kawakami, N. (2000) Quantum phase transitions in the Shastry-Sutherland model for SrCu$_2$(BO$_3$)$_2$. *Phys. Rev. Lett.*, **84** (19), 4461.

93 Zorko, A., Arčon, D., van Tol, J., Brunel, L.C., and Kageyama, H. (2004) X-band ESR determination of Dzyaloshinsky-Moriya interaction in the two-dimensional SrCu$_2$(BO$_3$)$_2$ system. *Phys. Rev. B*, **69** (17), 174420.

94 Kodama, K., Miyahara, S., Takigawa, M., Berthier, C., Mila, F., Kageyama, H., and Ueda, Y. (2005) Field-induced effects of anisotropic magnetic interactions in SrCu$_2$(BO$_3$)$_2$. *Phys. Condens. Matter*, **17** (4), L61–L68.

95 Cheng, Y.F., Cépas, O., Leung, P.W., and Ziman, T. (2007) Magnon dispersion and anisotropies in SrCu$_2$(BO$_3$)$_2$. *Phys. Rev. B*, **75** (14), 144422.

96 El Shawish, S., Bonča, J., Batista, C.D., and Sega, I. (2005) Electron spin resonance of SrCu$_2$(BO$_3$)$_2$ at high magnetic field. *Phys. Rev. B*, **71** (1), 014413.

97 Yan, S., Huse, D.A., and White, S.R. (2011) Spin-liquid ground state of the $S = 1/2$ Kagome Heisenberg antiferromagnet. *Science*, **332** (6034), 1173–1176.

98 Zorko, A., Nellutla, S., van Tol, J., Brunel, L.C., Bert, F., Duc, F., Trombe, J.C., De Vries, M.A., Harrison, A., and Mendels, P. (2008) Dzyaloshinsky-Moriya anisotropy in the spin-1/2 kagome compound ZnCu$_3$(OH)$_6$Cl$_2$. *Phys. Rev. Lett.*, **101** (2), 026405.

99 Zorko, A., Bert, F., Ozarowski, A., van Tol, J., Boldrin, D., Wills, A.S., and Mendels, P. (2013) Dzyaloshinsky–Moriya interaction in vesignieite: a route to freezing in a quantum kagome antiferromagnet. *Phys. Rev. B*, **88** (14), 144419.

100 Shores, M.P., Nytko, E.A., Bartlett, B.M., and Nocera, D.G. (2005) A structurally perfect $S = \frac{1}{2}$ kagome antiferromagnet. *J. Am. Chem. Soc.*, **127** (39), 13462–13463.

101 Rousochatzakis, I., Manmana, S.R., Läuchli, A.M., Normand, B., and Mila, F. (2009) Dzyaloshinskii–Moriya anisotropy and nonmagnetic impurities in the $S = \frac{1}{2}$ kagome system ZnCu$_3$(OH)$_6$Cl$_2$. *Phys. Rev. B*, **79** (21), 214415.

102 Cépas, O., Fong, C.M., Leung, P.W., and Lhuillier, C. (2008) Quantum phase transition induced by Dzyaloshinskii–Moriya interactions in the kagome antiferromagnet. *Phys. Rev. B*, **78** (14), 140405.

103 Yoshida, M., Okamoto, Y., Takigawa, M., and Hiroi, Z. (2012) Magnetic order in the spin-1/2 kagome antiferromagnet vesignieite. *J. Phys. Soc. Jpn.*, **82** (1), 013702.

104 Richards, P. (1976) Local properties at phase transitions, in *Proc. Intern. School of Phys. E. Fermi, Course LIX* (eds K.A. Müller and A. Rigamonti), North-Holland, Amsterdam, p. 539.

105 Dietz, R.E., Merritt, F.R., Dingle, R., Hone, D., Silbernagel, B.G., and Richards, P.M. (1971) Exchange narrowing in one-dimensional systems. *Phys. Rev. Lett.*, **26** (19), 1186–1188.

106 Shekhtman, L., Entin-Wohlman, O., and Aharony, A. (1992) Moriya's anisotropic superexchange interaction, frustration, and Dzyaloshinsky's weak ferromagnetism. *Phys. Rev. Lett.*, **69** (5), 836.

107 Choukroun, J., Richard, J.L., and Stepanov, A. (2001) High-temperature electron paramagnetic resonance in magnets with the Dzyaloshinskii–Moriya interaction. *Phys. Rev. Lett.*, **87** (12), 127207.

108 Oshikawa, M. and Affleck, I. (2002) Electron spin resonance in $S = \frac{1}{2}$ antiferromagnetic chains. *Phys. Rev. B*, **65** (13), 134410.

109 El Shawish, S., Cépas, O., and Miyashita, S. (2010) Electron spin resonance in $S = \frac{1}{2}$ antiferromagnets at high temperature. *Phys. Rev. B*, **81** (22), 224421.

110 Furuya, S.C. and Sato, M. (2015) Electron spin resonance in quasi-one-dimensional quantum antiferromagnets: relevance of

weak interchain interactions. *J. Phys. Soc. Jpn.*, **84** (3), 033704.

111 Knaflič, T., Klanjšček, M., Sans, A., Adler, P., Jansen, M., Felser, C., and Arčon, D. (2015) One-dimensional quantum antiferromagnetism in the p-orbital CsO_2 compound revealed by electron paramagnetic resonance. *Phys. Rev. B*, **91** 174419.

112 Oshikawa, M. and Affleck, I. (1999) Low-temperature electron spin resonance theory for half-integer spin antiferromagnetic chains. *Phys. Rev. Lett.*, **82** (25), 5136–5139.

113 Herak, M., Zorko, A., Arčon, D., Potočnik, A., Klanjšek, M., van Tol, J., Ozarowski, A., and Berger, H. (2011) Symmetric and antisymmetric exchange anisotropies in quasi-one-dimensional $CuSe_2O_5$ as revealed by ESR. *Phys. Rev. B*, **84** (18), 184436.

114 Asano, T., Nojiri, H., Inagaki, Y., Boucher, J.P., Sakon, T., Ajiro, Y., and Motokawa, M. (2000) ESR investigation on the breather mode and the spinon-breather dynamical crossover in Cu benzoate. *Phys. Rev. Lett.*, **84** 5880.

115 Zvyagin, S.A., Kolezhuk, A.K., Krzystek, J., and Feyerherm, R. (2004) Excitation hierarchy of the quantum sine-Gordon spin chain in a strong magnetic field. *Phys. Rev. Lett.*, **93** 027201.

116 Zvyagin, S.A., Kolezhuk, A.K., Krzystek, J., and Feyerherm, R. (2005) Electron spin resonance in sine-Gordon spin chains in the perturbative spinon regime. *Phys. Rev. Lett.*, **95**, 017207.

117 Brockmann, M., Göhmann, F., Karbach, M., Klümper, A., and Weiße, A. (2011) Theory of microwave absorption by the spin-1/2 Heisenberg-Ising magnet. *Phys. Rev. Lett.*, **107** (1), 017202.

118 Néel, L. (1948) Propértiés magnétique des ferrites; ferrimagnétisme et antiferromagnetisme. *Ann. Phys.*, **3** (2), 137–198.

119 Sachdev, S. (1992) Kagomé-and triangular-lattice heisenberg antiferromagnets: ordering from quantum fluctuations and quantum-disordered ground states with unconfined bosonic spinons. *Phys. Rev. B*, **45** (21), 12377.

120 Mendels, P., Bert, F., De Vries, M.A., Olariu, A., Harrison, A., Duc, F., Trombe, J.C., Lord, J.S., Amato, A., and Baines, C. (2007) Quantum magnetism in the paratacamite family: towards an ideal kagomé lattice. *Phys. Rev. Let.*, **98** (7), 077204.

121 Shastry, B.S. and Sutherland, B. (1981) Exact ground state of a quantum mechanical antiferromagnet. *Phys. B+ C*, **108** (1), 1069–1070.

122 Kaplan, T.A. and Menyuk, N. (2007) Spin ordering in three-dimensional crystals with strong competing exchange interactions. *Philos. Mag.*, **87** (25), 3711–3785.

123 Kittel, C. (1947) Interpretation of anomalous Larmor frequencies in ferromagnetic resonance experiment. *Phys. Rev.*, **71** (4), 270.

124 Keffer, F. and Kittel, C. (1952) Theory of antiferromagnetic resonance. *Phys. Rev.*, **85** (2), 329.

125 Yamada, I. and Kato, N. (1994) Multi-sublattice magnetic structure of $KCuF_3$ caused by the antisymmetric exchange interaction: antiferromagnetic resonance measurements. *J. Phys. Soc. Jpn.*, **63** (1), 289–297.

126 Pregelj, M., Zorko, A., Berger, H., van Tol, J., Brunel, L., Ozarowski, A., Nellutla, S., Jagličič, Z., Zaharko, O., Tregenna-Piggott, P., and Arčon, D. (2007) Magnetic structure of the $S = 1$ $Ni_5(TeO_3)_4Br_2$ layered system governed by magnetic anisotropy. *Phys. Rev. B*, **76** (14), 144408.

127 Nagamiya, T., Yosida, K., and Kubo, R. (1955) Antiferromagnetism. *Adv. Phys.*, **4** (13), 1–112.

128 Janson, O., Schnelle, W., Schmidt, M., Prots, Y., Drechsler, S.L., Filatov, S.K., and Rosner, H. (2009) Electronic structure and magnetic properties of the spin-1/2 Heisenberg system $CuSe_2O_5$. *New J. Phys.*, **11** (11), 113034.

129 Herak, M., Zorko, A., Pregelj, M., Zaharko, O., Posnjak, G., Jagličić, Z., Potočnik, A., Luetkens, H., Van Tol, J., Ozarowski, A., Berger, H., and Arčon, D. (2013) Magnetic order and low-energy excitations in the quasi-one-dimensional

antiferromagnet $CuSe_2O_5$ with staggered fields. *Phys. Rev. B*, **87** (10), 104413.

130 Marty, K., Simonet, V., Ressouche, E., Ballou, R., Lejay, P., and Bordet, P. (2008) Single domain magnetic helicity and triangular chirality in structurally enantiopure $Ba_3NbFe_3SiO_2O_{14}$. *Phys. Rev. Lett.*, **101** (24), 247201.

131 Zorko, A., Pregelj, M., Potočnik, A., Van Tol, J., Ozarowski, A., Simonet, V., Lejay, P., Petit, S., and Ballou, R. (2011) Role of antisymmetric exchange in selecting magnetic chirality in $Ba_3NbFe_3Si_2O_{14}$. *Phys. Rev. Lett.*, **107** (25), 257203.

132 Chaix, L., Ballou, R., Cano, A., Petit, S., de Brion, S., Ollivier, J., Regnault, L.-P., Ressouche, E., Constable, E., Colin, C. V., Zorko, A., Scagnoli, V., Balay, J., Lejay, P., and Simonet, V. (2016) Helical bunching and symmetry lowering inducing multiferroicity in Fe langasites. *Phys. Rev. B*, **93**, 214419.

133 Pregelj, M., Jeschke, H.O., Feldner, H., Valent, R., Honecker, A., Saha-Dasgupta, T., Das, H., Yoshii, S., Morioka, T., Nojiri, H., Berger, H., Zorko, A., Zaharko, O., and Arčon, D. (2012) Multiferroic $FeTe_2O_5Br$: alternating spin chains with frustrated interchain interactions. *Phys. Rev. B*, **86** (5), 054402.

134 Holstein, T. and Primakoff, H. (1940) Field dependence of the intrinsic domain magnetization of a ferromagnet. *Phys. Rev.*, **58** (12), 1098.

135 White, R.M., Sparks, M., and Ortenburger, I. (1965) Diagonalization of the antiferromagnetic magnon–phonon interaction. *Phys. Rev.*, **139** (2A), A450.

136 Chernyshev, A.L. and Zhitomirsky, M.E. (2009) Spin waves in a triangular lattice antiferromagnet: decays, spectrum renormalization, and singularities. *Phys. Rev. B*, **79** (14), 144416.

137 Kowalska, A. and Lindgå rd Mogensen, P.A. (1966) Reports Issued by the Risø National Laboratory. Risø Report. 127.

138 Herring, C. and Kittel, C. (1951) On the theory of spin waves in ferromagnetic media. *Phys. Rev.*, **81** (5), 869.

10
Photoelectron Spectroscopy

Stephan Breuer[1] and Klaus Wandelt[1,2]

[1] University of Bonn, Institute of Physical and Theoretical Chemistry, Wegelerstr. 12, 53115 Bonn, Germany
[2] University of Wroclaw, Institute of Experimental Physics, Maksa Borna 9, 50.204 Wroclaw, Poland

10.1
Introduction

The first versions of the periodic table of chemical elements were incomplete and based on the assumption that the elements are ordered according to the elements' atomic mass, until in 1913 Henry Moseley published his paper "The High Frequency Spectra of the Elements" [1]. In this work he proved a unique relationship between the frequency of the K_α X-ray emission lines $\nu(K_\alpha)$ and the ordinal or proton number Z of the elements:

$$\nu(K_\alpha) \approx R_\infty (Z-1)^2, \quad R_\infty = \text{Rydberg constant,} \tag{10.1}$$

and thereby showed a clear correlation between the chemical identity of the elements and their electronic structure.

Figure 10.1(i) shows schematically the electron structure of an atom in its ground state and Figure 10.1(ii) shows the excitation and X-ray emission process: For example, an electron from the K-shell is excited to an empty valence state or completely removed, and the created core hole is refilled by an electron from a filled higher electron level, which results in the emission of fluorescent X-rays $h\nu'$. The excitation can be done, for example, either by electrons or photons of sufficiently high energy.

X-ray or photon emission in general is not the only decay product following an electronic excitation. If the primary excitation energy is high enough, electrons may even be detached from an atom (Figure 10.1(iii)) as demonstrated by Hertz [2] and Hallwachs [3] through the discovery of the "photoelectric effect." Einstein's interpretation of this effect [4] led to the definition of "light quanta" (photons):

$$E = h\nu, \tag{10.2}$$

Handbook of Solid State Chemistry, First Edition. Edited by Richard Dronskowski, Shinichi Kikkawa, and Andreas Stein.
© 2017 Wiley-VCH Verlag GmbH & Co. KGaA. Published 2017 by Wiley-VCH Verlag GmbH & Co. KGaA.

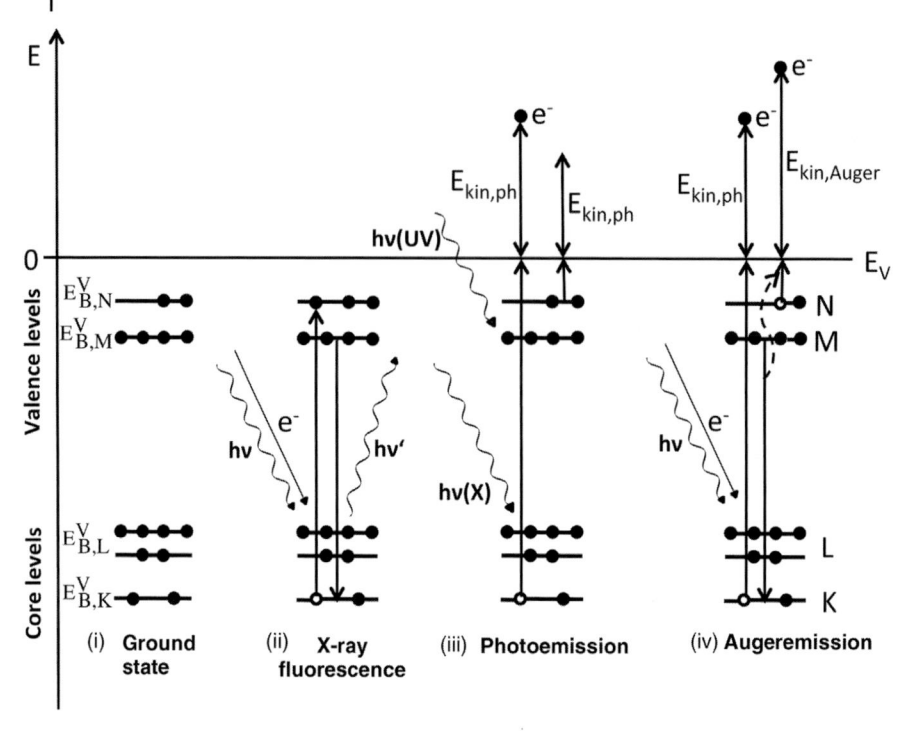

Figure 10.1 Schematic representation of photon- and electron-induced electron excitations: (i) electronic ground state; (ii) X-ray fluorescent de-excitation; (iii) X- or UV-induced photoemission; (iv) Auger electron emission.

and provides an explanation for the kinetic energy $E_{\text{kin,ph}}$ of the emitted electrons (energy conservation):

$$E_{\text{kin,ph}} = h\nu - E_B^V. \tag{10.3}$$

The energy input ($h\nu$) is used to overcome the binding energy (E_B^V) of a particular bound electron, which leaves the emitter atom with the excess energy $E_{\text{kin,ph}}$ (above E_V), as depicted in Figure 10.1(iii) for an isolated atom. E_B^V is defined with respect to the so-called vacuum level E_V, that is, the ionization limit where the core attraction ceases. Thus, measurement of $E_{\text{kin,ph}}$ (above E_V) for the given $h\nu$ in the X or UV range provides E_B^V, which is a clear indicator for the chemical identity of the atomic emitter.

The correlation given by Eq. (10.3) was systematically exploited by Siegbahn since 1960 and led to the development of "electron spectroscopy for chemical analysis" (ESCA) [5–7], nowadays known as X-ray photoelectron spectroscopy (XPS or in short "X-ray photoemission," not *photon*emission!). Using X-ray excitation and detection of *atom-specific* core electron emission makes XPS nowadays the most powerful electron spectroscopy for qualitative and quantitative chemical analyses of surfaces [8–10]. Later, Turner et al. [11] extended the method to UV photoelectron spectroscopy (UPS) using UV and vacuum UV radiation (e.g., HeI radiation of 21.22 eV) for the emission of more weakly bound valence electrons, in particular, from *molecular* orbitals of gaseous or adsorbed atoms/molecules (Figure 10.1(iii)). Further methods taking advantage of zero electron kinetic energy (ZEKE) are discussed in Ref. [12]. All techniques are summarized as photoelectron spectroscopy (PES).

10.2
The Photoemission Experiment

In a photoemission experiment, the sample is irradiated with X or UV photons (Figure 10.1(iii)) and the current of photoelectrons emitted either into the whole half-space above the sample surface or, preferably, into a more or less narrow angular cone around a defined spatial direction is measured and analyzed with respect to

- the kinetic energy of the electrons,
- the photocurrent per kinetic energy interval, and
- the electron spin.

The main ingredients of the experimental setup are the sample, the photon source, and the electron analyzer and detector shown in Figure 10.2. A detailed definition of the geometric parameters of the beam–sample interaction is given in Figure 10.3. Since the photoemitted electrons should not be disturbed by colliding with atoms (as in the Franck–Hertz experiment [13]) or molecules but should travel undisturbed from the sample surface into the analyzer, the sample

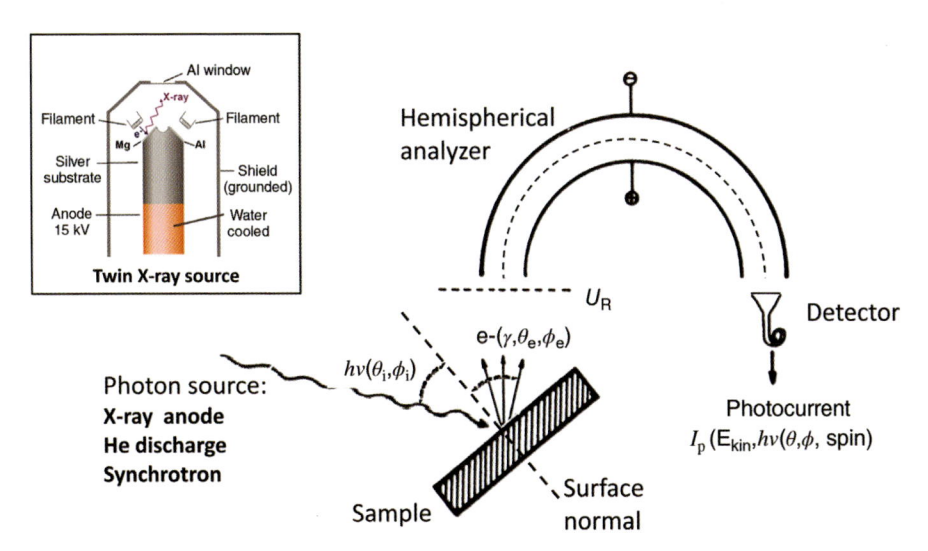

Figure 10.2 Instrumental setup of a photoemission experiment (see text). *Inset*: Mg/Al dual-anode laboratory X-ray source.

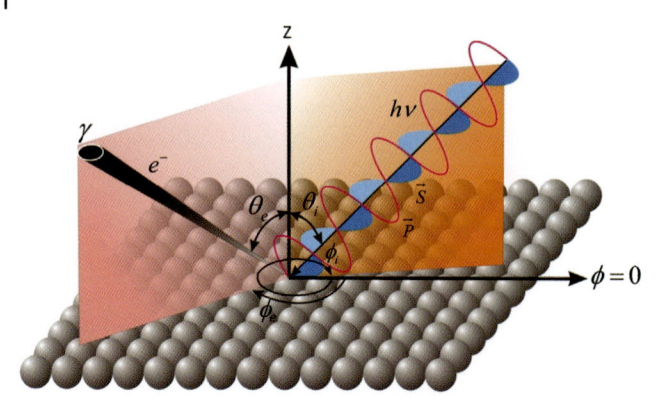

Figure 10.3 Definition of relevant parameters of a photoemission experiment; $z=$ surface normal.

and the analyzer should be kept in vacuum. This in turn puts constraints on the nature of the sample, namely, its vapor pressure, otherwise particular measures have to be taken to make PES measurements possible under "environmental conditions" (see, for example, Ref. [14]).

10.2.1
The Sample

In principle, the sample can be a gas [6,11], a liquid [15], or a solid. In the present context, we concentrate on solid samples.

The sample may be a one- or multicomponent solid of single crystalline, polycrystalline or even amorphous structure, or a powder, and should preferably be conducting in order to avoid charging effects (see Section 10.5). No matter whether the sample is to be analyzed "as received" or after some particular treatment, for example, a cleaning procedure or a controlled reaction step, its status should be stable over the duration of the measurement, which is also best guaranteed under ultrahigh vacuum (UHV) conditions. A "clean" surface can be obtained by vapor deposition of a fresh film or cleavage of a sample of the respective material in UHV. More common is cleaning of a given sample by heating or ion bombardment (sputtering) in UHV or a repeated combination of both. Also, a reactive cleaning is possible, in that the sample is heated in a low-pressure atmosphere (10^{-7}–10^{-5} mbar) of a reactive gas (e.g., oxygen or hydrogen), which oxidizes or reduces surface contaminants such as carbon, sulfur, or oxygen (the most common contaminants) to a volatile compound (CO, CO_2, SO_2, H_2O) that is pumped off. Extended simultaneous annealing and sputtering/reaction also causes continuous diffusion of bulk contaminants to the surface and their removal and, thereby the depletion of a thicker surface-near bulk region. Sophisticated UHV systems possess a separate preparation chamber, and a manipulator permits a transfer of the sample through a valve to the spectrometer in UHV without exposure to atmosphere. Powder samples may be pressed onto a double-sided adhesive tape or a soft and ductile metal foil (Au or In) on the sample holder.

10.2.2
The Photon Source

Common photon sources are divided into laboratory sources and synchrotron light sources.

Laboratory sources provide photons of specific (more or less) discrete energy. UV radiation is obtained from discharge lamps filled with, for example, Hg vapor, Xe, or mostly He. He in its ground state, He($^1S_0$), is excited in a high-voltage discharge to states of higher energy. The discharge takes place in a cell that is connected to the UHV chamber via a differentially pumped capillary, which allows photons to reach the sample but minimizes the passage of He (ground state or excited) gas atoms. Depending on the operation conditions, in particular the He pressure, the excited He atoms or ions emit either the fluorescence radiation He($^1P_1$) → He($^1S_0$) with $h\nu = 21.21$ eV (abbreviated as HeI) or a He$^+$ resonance line with $h\nu = 40.8$ eV (abbreviated as HeII). X-ray tubes directly flanged to the UHV system provide characteristic X-radiation, which is superimposed on a broad (white) background of bremsstrahlung. Typically, light-metal double-anode X-ray tubes with two electrodes (Al and Mg) and two electron filaments, as shown in the *inset* of Figure 10.2, are operated at an acceleration voltage of 15 kV, so that one can switch between AlK$_\alpha$ (1486.6 eV) and MgK$_\alpha$ (1253.6 eV) radiation. The availability of at least two different photon energies is required to discriminate between photoemission and Auger signals (see Figure 10.1(iv) and Section 10.3.6) in the measured spectra, and the range of photon energies (<1500 eV) provided by these Al/Mg double-anode tubes serves optimal surface sensitivity of the measurements (see Figure 10.13).

Apart from the laboratory light sources, synchrotron light covers a wide continuous range of photon energies from IR to hard X-rays (depending on the characteristics of the synchrotron source and the particular beamline), but is only available at specific facilities, for example, BESSY in Berlin.[1] Synchrotron radiation is emitted by electrons that orbit the synchrotron ring and are constantly accelerated toward the center of the orbit. State-of-the-art electron storage ring synchrotrons provide a photon flux magnitudes higher than standard laboratory sources, and the light beam is highly collimated, polarized (Figure 10.3), and – if desired for kinetic measurements – can be provided high-frequency pulsed.

10.2.3
Electron Analyzer and Detector

In a PES experiment, the analyzer separates the photoelectrons primarily according to their kinetic energy E_{kin}, but in dedicated experiments also with respect to

---

1) https://www.helmholtz-berlin.de/quellen/bessy.

their momentum or spin as a function of direction and polarization of the incident light beam. According to Eq. (10.3), E_{kin} is the excess energy of an electron after it escaped the attractive core potential of an *isolated* emitter (gas) atom, that is, after it acquired at least the energy to overcome the "ionization limit" or the "vacuum level E_V" (Figure 10.1).

For convenience, in the case of solid samples (S), E_B^V can be divided into

$$E_B^V = E_B^F + \phi_S, \tag{10.4}$$

so that

$$E_{kin,S} = h\nu - E_B^F - \phi_S, \tag{10.5}$$

where E_B^F is the "binding energy" of an electron with respect to the Fermi level E_F, that is, in metals the top of the valence band as the highest-occupied energy level, and $\phi_S = E_V^S - E_F^S$ is the (surface-specific) work function of the sample (Figure 10.4).

Using $E_F = E_B^F = 0$ as the energy reference level in solid-state PES spectra, on the one hand, has the advantage that E_F^S is directly visible in metal spectra or is defined by the metallic analyzer in the case of a semiconductor or an insulator sample.

On the other hand, the work function ϕ_S of the sample is *per se* not known. However, strictly speaking, the kinetic energy is detected by the analyzer (A) and is therefore different from that outside the sample due to the work function difference $\Delta\phi = \phi_S - \phi_A$ (contact potential) between the sample and the analyzer. In order to scan the whole energy range of the emitted photoelectrons, a variable

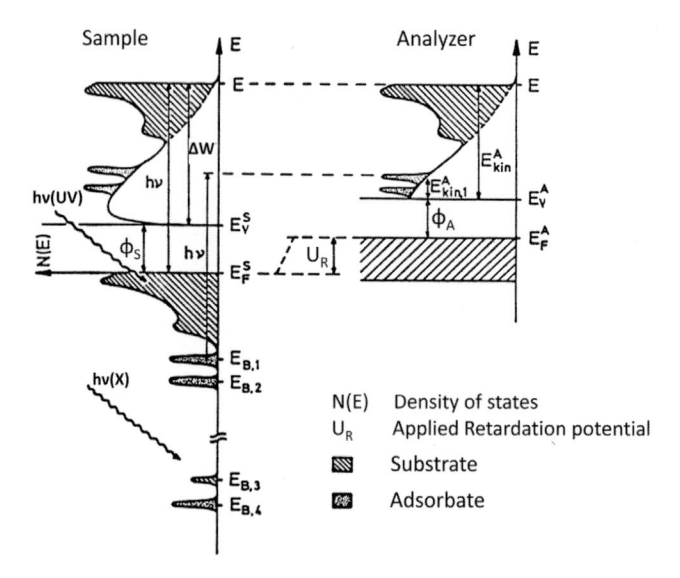

Figure 10.4 Schematic energy diagram of a photoelectron excitation and detection experiment from solid samples.

retardation voltage U_R with $0 < eU_R < E_{kin,max}$ may be applied between the sample and the entrance of the analyzer (Figure 10.4) such that

$$E_{kin,A} = h\nu - E_B^F - \phi_A - eU_R, \tag{10.6}$$

as shown in Figure 10.4, or

$$E_B^F = h\nu - E_{kin,A} - \phi_A - eU_R \tag{10.7}$$

so that with the given work function ϕ_A of the *analyzer*, the desired quantity E_B^F is accessible without the knowledge of ϕ_S. At given photon energy $h\nu$, electrons stemming from the Fermi level ($E_B^F = 0$) have the highest possible kinetic energy $E_{kin,max} = h\nu - \phi_S$, while electrons that just overcome the work function possess a kinetic energy of $E_{kin} = 0$. The total width $\Delta W = E_{kin,max} - 0 = h\nu - \phi_S$ of a spectrum, thus, gives the work function of the sample $\phi_S = h\nu - \Delta W$, as depicted in Figure 10.4 (provided $\phi_S > \phi_A$). Measurement of ΔW, or just shifts of the low kinetic energy "cut-off" ($E_{kin} = 0$) of measured spectra, therefore, yields work function changes $\Delta \phi_S$ due to, for example, adsorption or surface reaction processes. E_B^F (and E_B^V) of localized atomic inner shells depends on Z (compare with Eq. (10.1)) and therefore are atom specific.

The discrimination between charged particles of different kinetic energy, here electrons, can be done by time-of-flight methods, a retardation voltage U_R as shown in Figure 10.4, or the dispersion of the electron trajectories in an electric or a magnetic field. Current PES systems are most widely based on electric field analyzers such as a dispersive "cylindrical mirror analyzer" (CMA) consisting of two coaxial cylindrical electrodes (Figure 10.5) or a "hemispherical analyzer" (HMA), sometimes also called "hemispherical deflection analyzer" (HDA), consisting of two concentric hemispherical electrodes (Figure 10.2). In both cases, the potential on the outer (inner) electrode is controlled to repel (attract) entering electrons in such a way that only electrons of a certain energy ("pass energy" E_{Pass}) can pass the analyzer and reach the detector (Channeltron). Multichannel plate (MCP) detectors allow the simultaneous acquisition of several energy channels at once. This reduces the recording time considerably and thereby possible radiation damage (see Section 10.5).

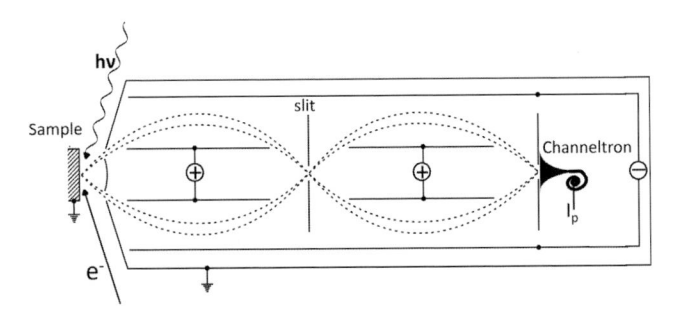

Figure 10.5 Principle of a cylindrical mirror analyzer (CMA).

The energy resolution depends on the analyzer dimensions, the width of the entrance slit, and the ratio E_{Pass}/E_{kin}. In order to improve the resolution, the electrons are therefore often preretarded by an electric field in front of the analyzer entrance to a set constant pass energy of the analyzer. With typical pass energies of, for example, 5 or 50 eV, resolutions of less than 20 meV can be obtained [16].

Additional electron lenses in the analyzer enable focusing of the electron acceptance angle γ for angle-resolved measurements (see Figure 10.3 and Section 10.4.3). Rastering a focused photon beam across the sample surface permits the registration of XPS spectra from micrometer-sized surface spots and thereby a "chemical mapping" of heterogeneous materials ("scanning XPS" or "microprobe"). This, however, is largely restricted to the use of synchrotron light, which still guarantees a high-enough spectral intensity.

The energy-separated electrons that passed the analyzer are counted by the detector, which in the simplest case is a "Faraday cup" as collector. Nowadays electron analyzers use a channeltron as electron multiplier (see Figures 10.2 and 10.5), a twined funnel of semiconducting material with a voltage drop of 1–3 kV between the wide entrance (negative) and the narrow output (positive). Incoming electrons hit the first wall and produce secondary electrons, which are again accelerated, hit the wall, and produce again "secondaries," and so on. With this cascade mechanism, amplifications of 10^9 can be achieved, that is, one incoming photoelectron produces 10^9 secondary electrons, which obviously enhances the sensitivity greatly.

10.2.4
Typical Spectral Features

In this section, let us first take a look at a few typical XPS/UPS spectra in order to illustrate those basic features that convey the most important chemical information.

Figure 10.6 shows XPS spectra of the simplest systems consisting of isolated atoms, namely, of rare gases [6,17]. According to the occupation of their atomic orbitals, we see just one peak for the emission of the 1s electrons of helium, three peaks for the emission of 1s, 2s, and 2p electrons of neon, and so on. The assignment of the respective signal is supported by the known X-ray fluorescence spectra of these atoms; and as emphasized in the introduction (see Ref. [1] and Figure 10.1), the energy positions are a fingerprint of the chemical nature of the particular species. The binding energy of the same electron level, that is, 1s, increases with increasing proton number of the emitter atom.

A close inspection of the spectra in Figure 10.6 reveals important fine structure features: (i) The intensity of peaks with the same second quantum number (s, p, d, or f), although filled with the same number of electrons, that is, 2 for s-levels, is *not* constant (note the intensity axes). This indicates that the probability (cross section) of electron excitation under otherwise constant conditions depends on the atomic number. (ii) While the s-levels show up as just one sharp peak, p-, d-,

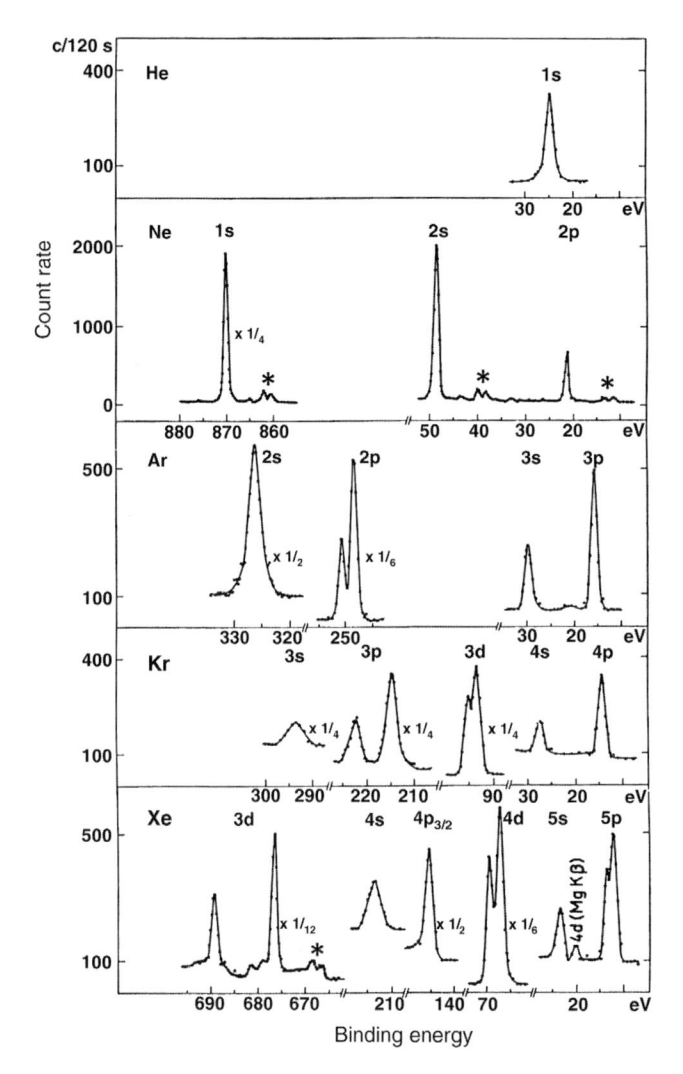

Figure 10.6 MgK$_\alpha$- excited photoemission spectra of rare gases. * denote signals excited by a MgK$_\alpha$ satellite line [17].

and f-levels (the latter not included in Figure 10.6) exhibit "spin–orbit splitting" (or j–j coupling). All electron orbitals except the s-levels are split into two peaks with $j = l + s$ of different energy, l being the angular momentum quantum number with $l = 0, 1, 2, 3$ (or s, p, d, f), and s the spin quantum number with $s = \pm 1/2$; this splitting increases with the ordinal number. (iii) Peaks marked with asterisk (*) are due to MgK$_\alpha$ satellite lines of higher photon energy (see also Figure 10.7).

A closer look at the 1s spectrum of gaseous Ne shown in Figure 10.7, however, indicates that the reality even for this simple atomic system is much more complex. This Ne 1s spectrum is measured with MgK$_\alpha$ = 1253.6 eV radiation,

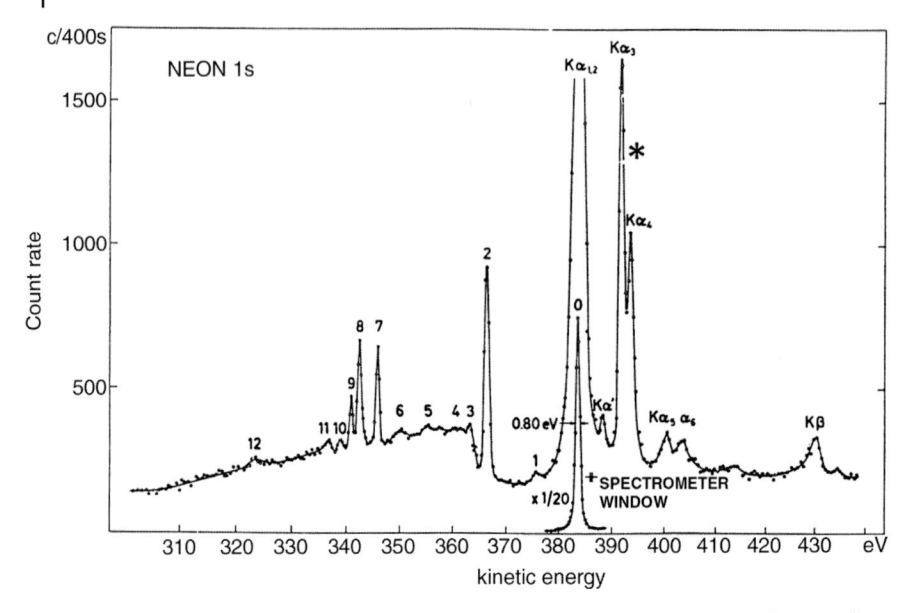

Figure 10.7 Neon 1s electron spectrum excited by MgK$_\alpha$ radiation at a pressure of 0.5 Torr. * denotes signals excited by a MgK$_\alpha$ satellite line as in Figure 10.6 [17].

and plotted against *kinetic* energy of the emitted electrons. Instead of just one sharp line at $E_{kin} = h\nu - E_B^V = 1253.6 - 870.2 = 383.4$ eV, there are several extra lines with higher and lower kinetic energy. All lines with higher kinetic energy are due to the more energetic satellite lines of MgK$_{\alpha,n}$, $n > 1$, radiation (see Section 10.5). Toward lower kinetic energies, a series of 12 peaks is partly superimposed on a continuous background starting at $E_{kin} = 362$ eV. The intensity of this background and the peaks 1–4 is found to be pressure dependent due to secondary collisions between emitted photoelectrons and neutral atoms [17]. The remaining peaks 5–12 are pressure independent and are due to multielectron excitations within one and the same Ne atom; in addition to the main 1s core electron excitation, outer valence electrons are emitted or excited together with the main process, leading to the so-called shake-up (or shake-off) peaks at energies corresponding to the energy sum of the main and shake-up processes [8,9,17].

Besides these multielectron effects within one and the same atom, additional phenomena come into play in condensed phases. Figure 10.8 depicts the UV photoemission 5p spectra of gaseous and adsorbed Xe, respectively. Both the gas-phase spectrum and the spectrum from just one adsorbed Xe monolayer, for example, on a Pd(100) and Ag(111) surface [18], clearly show the splitting into the 5p$_{3/2}$ and 5p$_{1/2}$ spin–orbit components; but in the case of the *adsorbed* Xe atoms, the peaks are not only much broader but the 5p$_{3/2}$ intensity is even split. This broadening and further splitting of the 5p$_{3/2}$ component is due to a "solid-state effect" because the Xe atoms are no longer isolated but are rather in contact with the substrate as well as with each other. First, the individual Xe

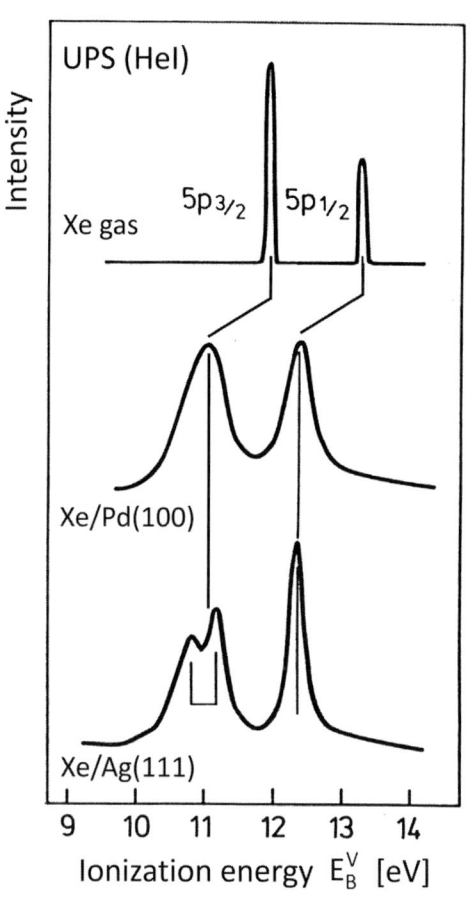

Figure 10.8 UV(HeI)-excited Xe5p spectra of Xe in the gas phase, and adsorbed on a Pd(100) and Ag(111) surface, respectively [18].

atoms (although weakly) interact with the substrate surface that leads to a lifting of the degeneracy of the 5p$_{3/2}$ states into two states with the magnetic quantum number $m_j = \pm 3/2$ and $m_j = \pm 1/2$. In addition, not all Xe atoms may reside in equivalent adsorption sites, that is, in equivalent environment, on the substrate surface, which causes "heterogeneous broadening" of the photoemission peaks. Second, mutual interactions between the adsorbed Xe atoms themselves within the first Xe monolayer lead to broadening due to formation of a valence *band* of the Xe5p states, that is, first to the formation of a 2D valence band within the Xe *monolayer* (see Ref. [18] and references therein). With increasing Xe coverage and built-up of *multilayers*, additional interaction between Xe atoms in *different* layers causes the stepwise formation of the 3D valence band structure of solid Xe, as depicted in Figure 10.9 for the first three Xe layers adsorbed on a Ru(0001) surface [19,20]; note the figure caption concerning the direction of the energy scale. These band structure effects, here demonstrated even for very

Xe/Ru(0001)

He I, normal emission
45 k

Xe dose: 116 L, 100 L, 88 L, 80 L, 68 L, 60 L, 48 L, 40 L, 28 L, 19 L

Binding energy (eV): 5,0 5,5 6,0 6,5 7,0 7,5 8,0

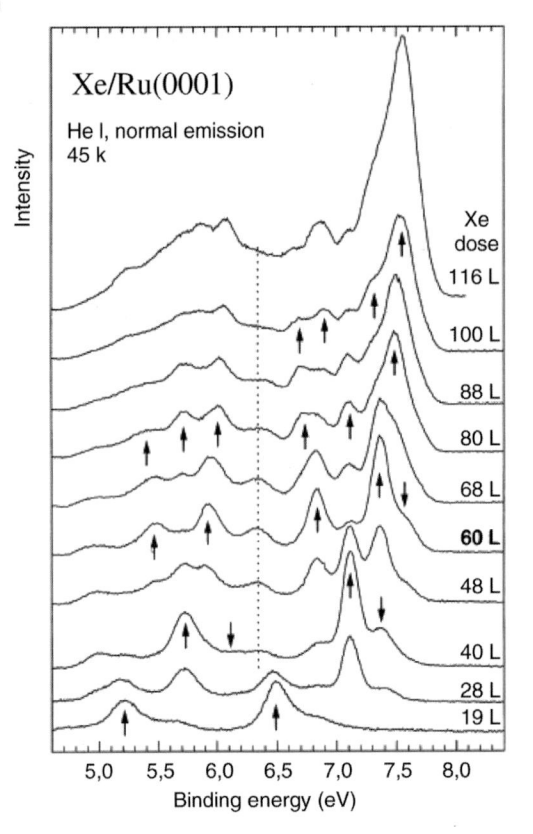

Figure 10.9 UV(HeI) excited Xe5p photoemission spectra of 1 ML (19 L), 2 ML (40 L), 3 ML (88 L), and >3 ML (>100 L) xenon adsorbed on a Ru(0001) surface. The dotted line separates the 5p$_{3/2}$ (E_B^V <6.4 eV) and 5p$_{1/2}$ (E_B^V >6.4 eV) contributions (ML = monolayer, 1 L = 1 Langmuir = 1 Torr·s) [19]. Note that in this figure the energy axis runs in the opposite direction to that in Figure 10.6. With this we want to make the reader explicitly aware of the fact that spectrometers of different producers may give spectra with different directions of the binding energy scale, and that as a consequence the literature is full of spectra with differently oriented energy scales.

weakly interacting rare gases, obviously dominate, above all, the appearance of solid-state *valence* band spectra of solids. This "solid-state effect," that is, the interaction with the substrate and the neighboring Xe atoms, is also responsible for the 1.1. eV shift between the gas and adsorbate spectra in Figure 10.8.

Figure 10.10 shows a survey XPS spectrum of an iron sample. Besides characteristic peaks of Fe3d valence band and Fe3p, Fe2p core-level photoelectrons as well as Fe–Auger electron emission (see Section 10.3.6), we also see carbon and oxygen 1s lines. The sample is obviously carbon and oxygen contaminated (qualitative analysis). All lines are superimposed on a background intensity that continuously increases toward higher *binding* energy, that is, lower *kinetic* energy of the emitted electrons. This increasing background is also a

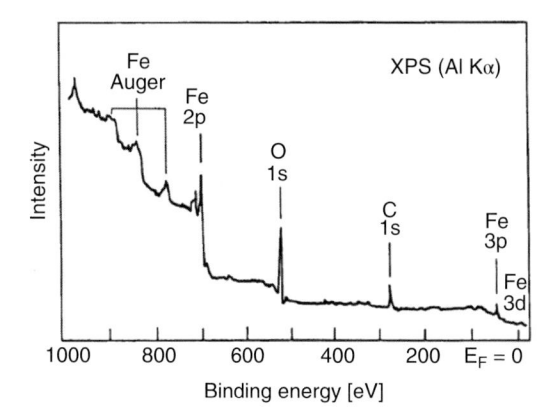

Figure 10.10 AlK$_\alpha$ excited survey XPS spectrum of a carbon (1s)- and oxygen (O1s)-contaminated iron sample.

characteristic solid-state effect due to the limited "mean free path length" of electrons in condensed matter (see Section 10.3.4).

Figure 10.11 shows on an enlarged energy scale the evolution of just the Fe2p$_{3/2}$ spin–orbit component as well as the O1s signal at different states of the oxidation of a *clean* iron sample; both the Fe2p$_{3/2}$ peak of pure iron at 707.0 eV and the O1s signal of atomically adsorbed oxygen at 530 eV binding energy are, as expected, originally not split [21]. However, with increasing oxygen dose (measured in Langmuir, that is, $1\,L = 10^{-6}$ Torr×s), the Fe2p$_{3/2}$ peak remains at 707.0 eV but decreases in intensity and a new broader peak develops at 711.0 eV. By contrast, the O1s signal grows with increasing oxygen exposure, remains unsplit at 530 eV up to very high exposures of *pure oxygen*, but then develops a pronounced shoulder at ~532 eV after exposure to air. Since no iron atoms disappear, and since the Fe2p$_{3/2}$ and the atomic O1s levels are *per se* not further split, and the *new* features at 711.0 and 532 eV develop as a consequence of the exposure to oxygen gas and air, the origin of the new spectral features must be chemistry. The interaction of iron with oxygen leads to a "chemical shift" of the Fe2p$_{3/2}$ core level to 4 eV higher binding energy. Likewise, exposure to air, rather than pure oxygen, results in the occurrence of an additional, chemically different oxygen species on the surface, in this case OH groups [21]. The integral O1s intensity is a measure of the amount of oxygen bound to the iron surface (quantitative analysis). These core-level shifts are actually an expression of the changes of the electronic structure in the *valence* band regime due to chemical bond formation, which may change the partial charge on the respective kind of atoms. As a consequence, the energy needed to remove a core electron from, for instance, a partially positively charged atom is higher than that from the neutral atom (see Section 10.4.2). The change of the electronic structure of the valence band as the cause for the chemical core level shift, however, is more immediately reflected in UPS spectra, as addressed in Section 10.4.2.4, for example, for the oxidation of nickel (see Figure 10.27).

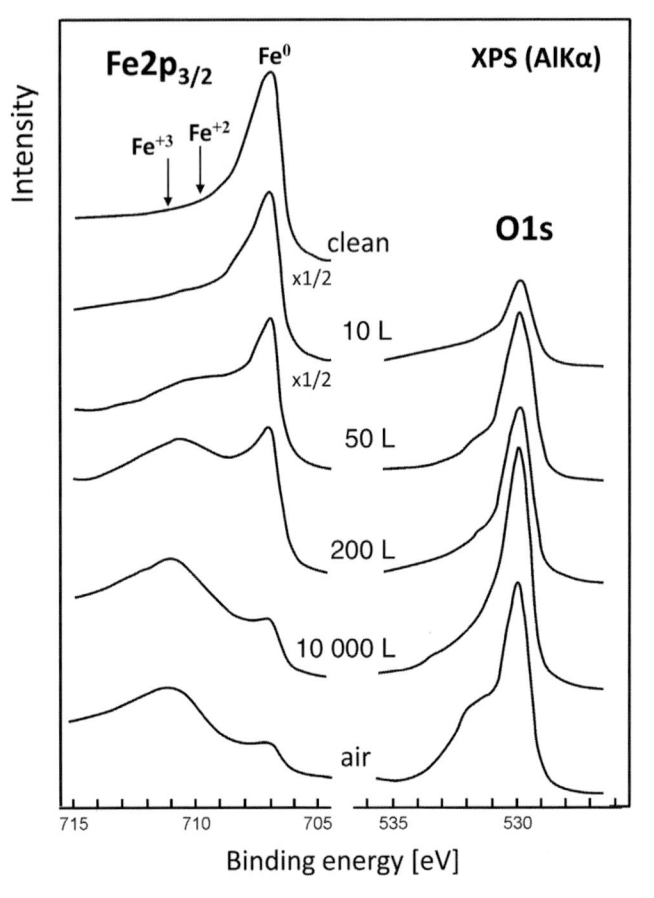

Figure 10.11 AlK$_\alpha$-excited Fe2p$_{3/2}$ and O1s core-level spectra of a clean and oxygen- or air-exposed iron sample. Fe^0, Fe^{2+}, and Fe^{3+} denote the energetic peak positions arising from metallic and oxidized (2+ or 3+) iron [21].

Of course, all the above addressed aspects play a crucial role in the application of photoelectron spectroscopy as a qualitative and quantitative analytical tool, as will be discussed in greater detail in subsequent sections.

10.3
The Photoemission Process

10.3.1
Excitation, Relaxation

As indicated by the occurrence of the Auger process (Figure 10.1) and the complex Ne1s spectrum (Figure 10.7), the photoemission process requires a more detailed contemplation than the simple idea that just one electron is removed, because in reality the removal of an electron from even a single gas atom with N electrons does not leave the remaining $(N-1)$ electron system undisturbed.

Conceptually, three aspects influence the electron emission process: (i) The excitation of one electron to a state where it is no longer attracted by the core potential of its emitter atom, (ii) the time evolution of the $(N-1)$ electron system of the photoionized emitter atom, and (iii) the transport of the liberated electron to the analyzer, which obviously matters only if the emitter atom is embedded in a condensed phase, in which case the electron has to travel through a dense matrix and its surface in order to reach the analyzer in UHV.

The excitation of one electron by absorption of a photon of sufficient energy $h\nu$ (aspect 1) obeys the energy conservation:

$$E_i(N) + h\nu = E_f(N-1,k) + E_{kin}, \tag{10.8}$$

where $E_i(N)$ is the *total* energy of the N-electron system in its initial (i) state and $E_f(N-1,k)$ is the *total* energy of the $(N-1)$ electron system (singly positive ion) in the final (f) state with an "electron hole" in the k-state, respectively. E_{kin} is the kinetic energy of the emitted electron.

In the simplest "one-electron approximation," it is assumed that the removal of this electron does leave all other $(N-1)$ electrons unaffected, in which case the value of the measured binding energy of the emitted electron

$$E_B^V(k) = h\nu - E_{kin} \tag{10.9}$$

is identical to its binding energy in the neutral ground-state emitter atom. This "frozen orbital approximation" is also known as "Koopmans' theorem" in the literature. Based on this first approximation, the energetic position (binding energy) of photoemission peaks provides already valuable information about the identity of the emitter atom as well as its chemical state (see Sections 10.4.1 and 10.4.2).

The total energy of the "frozen orbital" $(N-1)$ electron system $E_f(N-1,k)$, however, is higher by $E_B^V(k)$ than the neutral ground state. As a consequence, the $(N-1)$ electron system rearranges or "relaxes" at best to the *ground state of the $(N-1)$ electron system*, in order to get rid of part of this excess energy (aspect 2). This "relaxation effect" is most drastically reflected by the first and second ionization energy of helium, 24.6 and 54.4 eV, respectively. Even though both electrons of helium are bound in the same ground-state level (1 s) with $E_B^V = 39.5$ eV, the first electron – in the presence of the second electron – can be removed with 24.6 eV photons, while the additional removal of the second electron (from the $(n-1)$ system, namely, the He$^+$ ion) costs 54.4 eV.

Depending on the relative timescale of the emission and the relaxation process, respectively, the outgoing electron may or may not benefit from the released "relaxation energy". During the so-called adiabatic emission process, the photon absorption/electron emission process is slow compared to the relaxation process of the $(N-1)$ electron system, that is, the ionization process is always in equilibrium, so that the outgoing electron fully benefits from the gained relaxation energy:

$$E_{\text{kin}}(k)_{\text{adiab}} = h\nu - E_B^V(k) + E_{\text{relax}}(N-1, k). \tag{10.10}$$

The opposite is true if the photon absorption/electron emission occurs very fast, that is, instantaneously or suddenly. In this case the emitted electron is gone before any relaxation energy is released. This "sudden approximation" represents "Koopmans' theorem," and the lagging relaxation causes subsequent multielectron excitations like "shake-up"- or "shake-off"-processes, in which additional electrons are either just excited to higher bound states or even emitted showing up in spectra as so-called intrinsic satellites [8,9]. In reality, every situation between the adiabatic and sudden approximation is possible depending on the kinetic energy of the outgoing electron; the higher the kinetic energy, that is, the difference between $h\nu$ and E_B^V, the more "sudden" the emission process and the more "Koopmans"-like is the measured binding energy. Vice versa, the slower the electron leaves the emitter atom, the more it profits from the relaxation energy and gains in kinetic energy, which, in turn, according to Eq. (10.10) feigns a lower binding energy. In any case, the emitter atom is left behind as an $(N-n)$ electron system, $n = 1,2,3 \ldots$, that is, as a relaxed or an unrelaxed ion, which eventually de-excites by emission of photons or Auger electrons, as depicted in Figure 10.1. Since the latter are inevitably superimposed on a photoemission spectrum, a short discussion of "Auger spectra" is given in Section 10.3.6.

A general problem of the assessment of photoemission spectra is, therefore, whether spectral features are representative of the system in its "initial state" or in its "final state." "Initial state effects" arise from modifications of the properties of the emitter atom through interactions with its surrounding that existed before the atom was ionized. Namely, bonding of the emitter atom to at least one or many other atoms may not only change its electronic structure but also allow the atom to vibrate, which shows up as "chemical shift," "electronic band formation and dispersion," or "vibrational fine structure" in photoemission spectra. Conversely, "final state effects" are a consequence of the photoelectron emission and hole creation process, because the creation of the electron hole disturbs the $(N-1)$ electron system left behind, and show up in the spectra as "relaxation"-induced peak shifts and multielectron features (as, for example, in Figure 10.7).

10.3.2
Spin–Orbit and Multiplet Splitting

Photoelectron signals from orbitals with an angular momentum unequal 0, namely, p, d or f-orbitals, are split into two peaks, as seen in Figure 10.6, because the electron can be emitted with a favorable ($j = l - s$, j angular momentum, l angular momentum quantum number, s spin quantum number) or less favorable ($j = l + s$) spin–orbit coupling. Consequently, the energy of one peak is higher and the other one lower than the Koopmans energy, which is approximately the center of gravity of the two signals. This "spin–orbit-splitting" is atom and level specific and largely independent of the atom's chemical state.

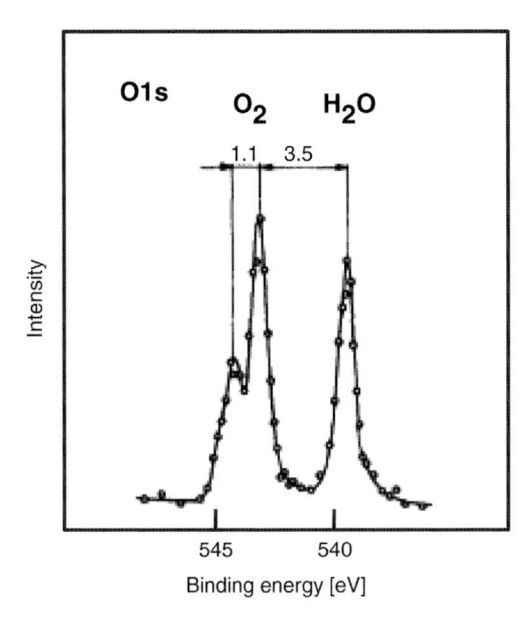

Figure 10.12 O1s photoemission lines of oxygen gas and water vapor [6].

Spectra of atoms, molecules, and solids that possess unpaired electrons, that is, unpaired spins in the initial state, show, in addition, multiplet splitting due to spin–spin coupling effects. The "spin of the electron hole" can interact favorably or unfavorably with the spin of the unpaired electron spins. As an example, Figure 10.12 shows the O1s spectra of gaseous molecular oxygen and water [6,17]. In its ground state the oxygen molecule (and atom) has two unpaired valence electrons (triplet oxygen) whose spins cause the splitting of 1.1 eV seen in Figure 10.12. In the water molecule, the two unpaired electrons of the oxygen *atom* pair with the electrons of the hydrogen atoms that quenches the multiplet splitting. Multiplet splitting is also the reason for the very broad spectrum measured for the valence bands of iron or nickel oxides shown in Figures 10.26 and 10.27, for instance, Fe_x^{II} O whose six 3d valence electrons, due to an octahedral crystal field, are in a high spin state (see Figure 10.26).

10.3.3
Intensity

Decisive for the intensity of PES signals, that is, the photocurrent, is the transition probability P_{fi} of an electron from its initial state $|\psi_i\rangle$ of energy ε_i to a final state $|\psi_f\rangle$ of energy ε_f, which within the dipole approximation of optical transitions is given by Fermi's Golden Rule [8,9]:

$$P_{fi} \approx |\langle\psi_f|\Delta|\psi_i\rangle|^2 \delta(\varepsilon_f - \varepsilon_i - h\nu). \tag{10.11}$$

Δ is the dipole operator that describes the disturbance of the system by the incident light and also accounts for the conservation of momentum; ψ_f and ψ_i

are the wave functions of the final and the initial electronic states, respectively. $\langle \psi_f| \Delta |\psi_i\rangle$ is called the "matrix element." ψ_f, ψ_i, and Δ include all geometric dependencies of the excitation process like the chosen directions of the incident light beam (θ_i, ϕ_i in Figure 10.3) and electron detection (θ_e, ϕ_e in Figure 10.3), and the light's polarization ($\vec{S}$, $\vec{P}$) with respect to the symmetries of the system, as illustrated in Figure 10.3. In the case of "frozen orbital approximation," the wave functions ψ_f and ψ_i remain unaffected during the emission process. In reality, of course, photoexcitation into continuum states (photoionization) is a more complex dynamic process leading to different ionization efficiencies in dependence on photon energy and momentum ("resonances", Cooper minimum, etc.) as described in detail, for example, in Refs [8–10].

A whole spectrum can be regarded as a representation of the electronic density of states (DOS) at specific binding energies. Core-level states result in relatively sharp signals (lines) while the manifold of states in the valence band appears as a broad continuous signal in the spectrum (Figures 10.26 and 10.27). Conclusively, the intensity of a photoelectron emission I_P is determined by

$$I_p(\varepsilon, \nu, \theta, \phi) \approx \sum_k |\langle \psi_f|\Delta|\psi_i\rangle|^2 \delta(\varepsilon_f - \varepsilon_i - h\nu)\delta(\varepsilon - \varepsilon_f) F(\varepsilon_{f'}, \theta, \phi) \qquad (10.12)$$

in which the term $|\langle \psi_f|\Delta|\psi_i\rangle|^2$ arises from the transition probability (see Eq. (10.11)), $\delta(\varepsilon_f - \varepsilon_i - h\nu)$ describes the energy conservation, $\delta(\varepsilon - \varepsilon_f)$ the energy classification (binding energy: the measured kinetic energy fits the final state energy), and finally $F(\varepsilon_{f'}, \theta, \phi)$ accounts for diffraction effects (geometry of the sample). This equation is purely theoretically derived and does not contain specific parameters of the experimental setup or the composition of the sample. Explicit use of Eq. (10.12) is only made in highly sophisticated experiments and not necessarily for routine surface analyses. A more empirical approach for the analysis of photoemission intensities is given in Eq. (10.17).

10.3.4
Photoemission from Solids

Detection of an electron photoexcited within a condensed phase requires its transport through the densely packed matrix of atoms/molecules and its surface into vacuum.

During this transport through the solid, the photoelectrons, due to their dual particle and wave nature depending on their energy, may undergo inelastic collisions with other electrons and lattice atoms and diffraction effects. As one consequence, in condensed matter, electrons as particles have a relatively short "mean free path length" due to collisions with other electrons and atoms. Figure 10.13 displays a plot of the mean fee path length λ of electrons in metals as a function of their kinetic energy [22,23]. The minimum around $E_{kin} \approx 100\,eV$ means two things: (i) Even though the penetration depth of 1–2 keV photons is in the micrometer range, generated photoelectrons of 100–1000 keV kinetic energy have a chance to travel *with undisturbed energy* (elastically) only a

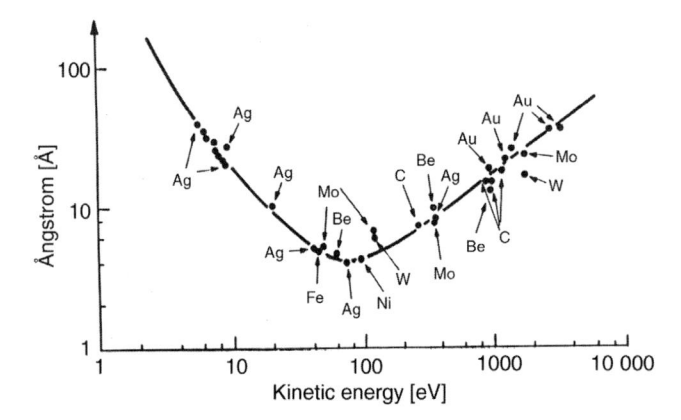

Figure 10.13 Energy dependence of the mean free path length of electrons in metals. (Reproduced with permission from Ref. [23]. Copyright 1987, Wiley-VCH Verlag GmbH & Co. KGaA, Weinheim.)

distance of 1–2 nm within the solid sample (before they suffer an inelastic collision), which makes them reliable probes for only few surface-near atomic layers. On the one hand, this explains the desired "surface sensitivity" of PES as long as *surface* properties are in the focus of interest. On the other hand, it makes XPS (and UPS) less suited for the investigation of bulk properties (e.g., >100 nm below the surface; see however Section 10.4.5). Those electrons that suffer inelastic collisions on their way through the sample create "secondary electrons" of lower energy, which again may collide and so on. This cascade of inelastic collisions ultimately creates a high number of low-energy electrons, as illustrated in Figure 10.14a, which shows the energy distribution of secondary electrons created by primary electrons of an energy E_P. This distribution, of course, holds independent of whether these "primary electrons" of E_p are created inside the solid or whether they impinge on the surface from the outside. Such secondary electrons are also the reason for the stepwise increase of the background on the high binding energy side of each peak in Figure 10.10. Small characteristic signals, which are superimposed on this increasing background of "secondaries," may be brought out by taking the first derivative of the energy distribution (Figure 10.14b).

As an another consequence, depending on their kinetic energy or wave length, electrons may undergo diffraction or "forward scattering." Low-energy electrons ($E \lesssim 200\,eV$), due to their pronounced wave character, may be diffracted; the outgoing electron wave from the emitter may interfere constructively or destructively with waves from other emitters or be scattered from the electron clouds of surrounding atoms, which leads to spatial modulation of the detected intensity (electron flux). Electrons of higher energy ($E \gtrsim 500\,eV$), in turn, may penetrate the electron cloud of an adjacent atom, and as a consequence are attracted by its core and "sucked" through the atom. Figure 10.15 illustrates the angular distribution of electrons ($E \gtrsim 500\,eV$) elastically *forward* scattered from an oxygen ($Z=8$) and Rh ($Z=45$) atom [10]. Both representations (a and b) accentuate a

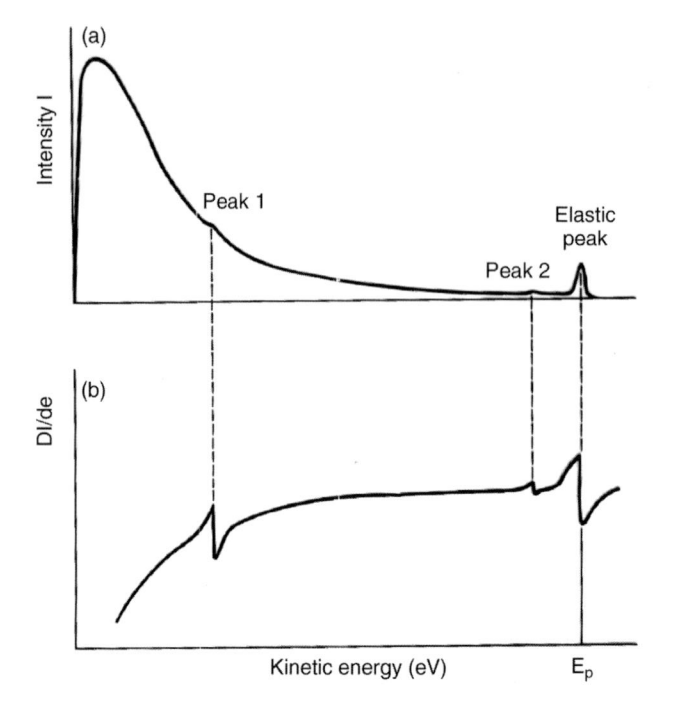

Figure 10.14 (a) Energy distribution of secondary electrons resulting from scattered primary electrons of energy E_p. (b) Derivative of curve (a).

clear intensity enhancement along the trajectory from emitter to scatterer ($\theta_s = 0$); the core potential of the scattering atom focuses the penetrated electrons in forward direction. Hence, variations of the polar angle of detection θ_e or the sample azimuth ϕ_e (see Figure 10.3) at constant $h \cdot \nu$ (i.e., constant E_{kin}, Eq. (10.9)) or variations of the photon energy at constant angles allow the detection of intensity modulations that provide structural information (see Section 10.4.3).

10.3.5
Extrinsic Satellites

The intrinsic satellites mentioned in Section 10.3.1 are due to intra-atomic (or in solids "interband") transitions that accompany the main excitation in multielectron processes. Of quite different nature are "extrinsic satellites."

Photoelectrons leaving the solid still interact electrostatically with the charges in the surface. This interaction decays with increasing distance and sets up a changing Coulomb field that may excite "group oscillations" of the near surface valence band electrons, so-called "surface plasmons." The probability of this process depends on how much time the exiting electrons spend near the surface. Obviously, the more grazing the exit angle, that is, the larger the θ_e in Figure 10.3, the longer the emitted electrons travel in the vicinity of the surface.

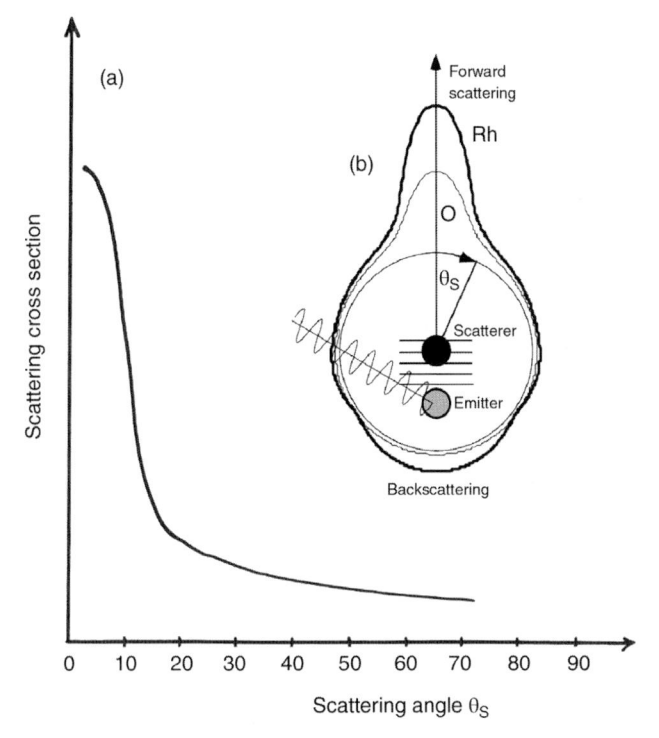

Figure 10.15 (a) Schematic cross section (Reproduced with permission from Ref. [23]. Copyright 2012, Wiley-VCH Verlag GmbH & Co. KGaA, Weinheim.) and (b) angular distribution of photoelectrons of $E_{kin} \geq 500\,eV$ forward scattered from an O and Rh atom [10], respectively.

Figure 10.16 shows the O1s signal of an oxygen-exposed aluminum surface, which is accompanied by a plasmon satellite peak at 10.9 eV higher binding energy. The inset clearly shows the strong dependence of this satellite's intensity on the exit angle θ_e. The analytical importance of this satellite lies in the fact that the energy shift of this peak, $\Delta E_B = 10.9\,eV$, from the main O1s line corresponds to the *surface* plasmon of *metallic* aluminum, which obviously means that the oxygen atoms must be adsorbed *on* the *unperturbed* Al surface.

10.3.6
Auger Spectroscopy

The emission of a photoelectron leaves the emitter atom behind as an ion with one (or more) hole(s). Its de-excitation can proceed either by emission of X-rays as in Figure 10.1(i) or by the so-called Auger mechanism discovered by Pierre Auger in 1925 [23,25]. Instead of emitting X-photons, the de-excitation can also lead to the emission of one (or more) electron(s); the energy gained by the transition of an electron from a higher level into the core hole is transferred (radiationless) to one (or more) bound electron(s) that – in case of sufficient

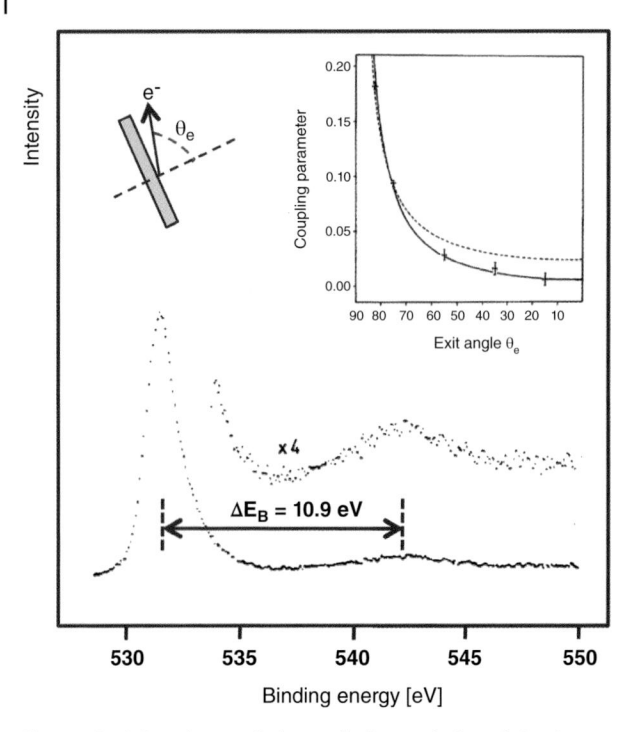

Figure 10.16 O1s photoemission peak plus extrinsic surface plasmon satellite at 10.9 eV higher binding energy. *Inset:* Angular dependence of the plasmon satellite intensity; *dashed line:* theory, *solid line:* best fit to the experimental data points (see Ref. [24] and references therein).

energy – may also leave the emitter atom as a so-called Auger electron(s) (see Figure 10.1 (iv)). As a result, the atom is left behind in a multiply ionized state (Figure 10.1(iv)). In the simplest picture, the kinetic energy of the outgoing Auger electron in Figure 10.1(iv) is given by equating

$$E_K - E_M = E_{\text{kin,Auger}} + E_N, \tag{10.13}$$

which gives

$$E_{\text{kin,Auger}} = E_K - E_M - E_N. \tag{10.14}$$

The kinetic energy of the emitted Auger electron $E_{\text{kin,Auger}}$ depends only on the energies E_K, E_M, and E_N of the three involved levels. The difference $E_K - E_M$ provides the excitation energy to overcome the binding energy E_N. Strictly speaking, the ground-state energies E_M and E_N must meanwhile be replaced by the "relaxed" energies E'_M and E'_N due to the presence of the initially created K-hole. $E_{\text{kin,Auger}}$, thus, is independent of the incident photon energy $h\nu$, in contrast to $E_{\text{kin,ph}}$ of the primary emitted photoelectron leaving the initial core hole in the K-shell behind. While the kinetic energy of a photoelectron emitted from an initial level E_k varies with $h\nu$ (see Eq. (10.9)), an Auger electron of a specific

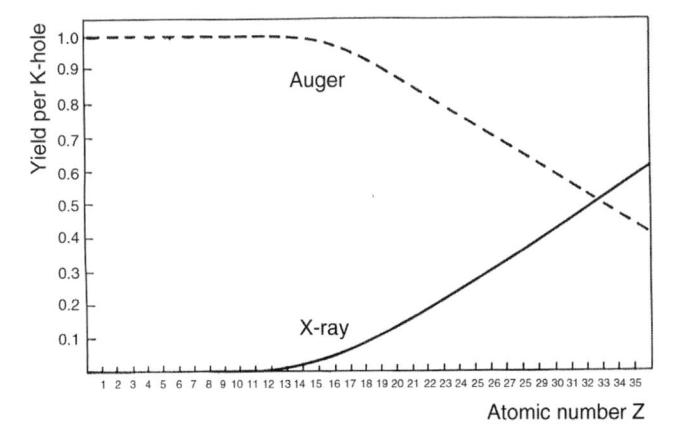

Figure 10.17 X-ray and Auger electron yield after creation of a hole in the K-shell as a function of the atomic number. Z. (Reproduced with permission from Ref. [23]. Copyright 1987, Wiley-VCH Verlag GmbH & Co. KGaA, Weinheim.)

level does not. On the one hand, variation of the incident photon energy $h\nu$ (using, for instance, MgK$_\alpha$ or AlK$_\alpha$ light) enables a distinction between photoemission and Auger emission-related signals in a measured electron spectrum. On the other hand, X-ray emission (Figure 10.1(ii)) and Auger electron emission (Figure 10.1 (iv)) are competing decay processes following the initial core hole creation due to photoelectron emission. While the Auger yield dominates for light elements, the X-ray yield becomes more important for heavy atoms (Figure 10.17) [10,23].

As already pointed out for the photoemission process in Section 10.3.1, it is obvious that in reality a description of the dynamics of the even more involved Auger process leads to a much more complex spectral behavior than the simple Eqs. (10.13) and (10.14) lead to believe; for more theoretical and practical details, the reader must be referred to the specific literature [26,27].

Auger signals are included in the XPS spectrum in Figure 10.10. Their assignment as Auger peaks is based on the fact that their *kinetic* energy is independent of the incident photon energy, as predicted by Eq. (10.14), and on their larger peak width. The larger peak width is a consequence of the convolution over the linewidths of three levels that is reflected in the self-explanatory nomenclature of Auger signals ($X_{l,j}$, $Y_{l,j}$, $Z_{l,j}$), where X denotes the level of the primary core hole, Y the level from which this hole is refilled, and Z the start level of the outgoing Auger electron. The subscripts l, j are the "angular momentum quantum number" and the "total angular momentum quantum number," respectively. As an example, Figure 10.18a shows at the top a derivative Auger spectrum of an as-received oxygen- and sulfur-contaminated iron sample, and below spectra of a clean iron and nickel sample, respectively, as well as of three different iron–nickel alloys. K, L, M specify the core shells, V denotes the valence band, and the subscripts 1, 2, 3 refer to the s and spin–orbit split $p_{1/2}$ and $p_{3/2}$ subshells involved in the respective Auger transition. In this derivative form, Auger

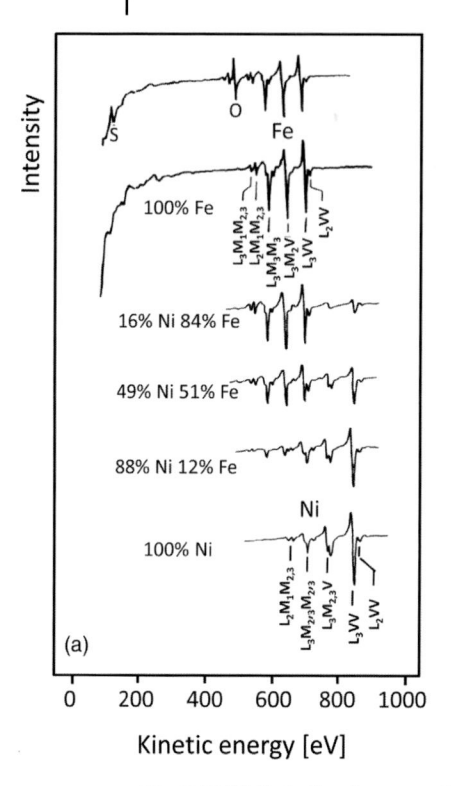

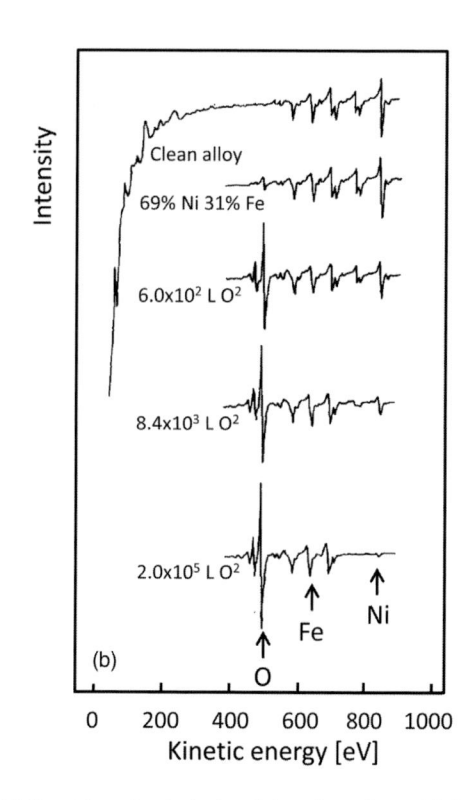

Figure 10.18 Derivative Auger spectra of (a) O and S contaminated and clean iron, clean nickel, and three clean iron–nickel alloys. (b) The successive oxidation of a clean 69%Ni31%Fe alloy.

spectra provide a very clear and powerful fingerprint of the chemical nature of the sample. Figure 10.18b shows at the top again a full and partial spectrum of a clean iron–nickel alloy, which then was exposed to controlled doses of oxygen (lower spectra). Note the increase of the oxygen signal, and the concomitant decrease of the nickel peak, which indicate the surface enrichment of iron oxide (see also Figure 10.36). Based on Eq. (10.14) and the known orbital energies of all atoms, the kinetic energy of all possible Auger transitions of all elements can be calculated and are available in tabulated form [28]. The peaks observed in the range 100–300 eV of the clean iron and alloy surface in Figure 10.18a and b, respectively, are due to diffraction effects; their intensity is temperature dependent.

It shall be mentioned, however, that the spectra shown in Figure 10.18 were actually not excited by photons, but by high-energy (e.g., 3 keV) electrons (like curve a) in Figure 10.14). In the end, however, it does not matter, how the initial core hole, the prerequisite for the occurrence of an Auger transition, was created, either by photoabsorption or by electron impact, except that the cross section for core-hole production by electron impact is higher [23].

10.4
Analytical Applications

10.4.1
Qualitative Analysis

As expressed by Eq. (10.9), measured binding energies E_B are to a first approximation equal to the negative value of the respective orbital energies, and therefore enable an easy identification of the emitter atom, that is, permit a qualitative chemical analysis. Final state relaxation effects and initial state "chemical shifts" (see Section 10.4.2) are relatively small and generally do not impede a discrimination between elements. Tabulated values of electron binding energies of all pure elements can be found, for example, in Ref. [9]. As one example, Figure 10.10 showed already the XPS survey spectrum of a carbon- and oxygen-contaminated iron sample.

As a second example, Figure 10.19 shows a sequence of S2p and Cd3d core-level spectra measured with synchrotron radiation of $h\nu = 650$ eV in a so-called transfer system. These spectra were obtained after the first few steps in an attempt to grow CdS films on a Cu(111) substrate by electrochemical atomic layer epitaxy (ECALE) [29]. The Cu(111) electrode was sequentially immersed and polarized in the following solutions: (1) 5 mM H_2SO_4, (2) 1 mM Na_2S + 5 mM H_2SO_4, (3) 1 mM $CdSO_4$ + 5 mM H_2SO_4, and again (4) 1 mM Na_2S + 5 mM H_2SO_4, and then emersed at the given potential and transferred from the solution through a separately evacuated buffer chamber directly into the UHV

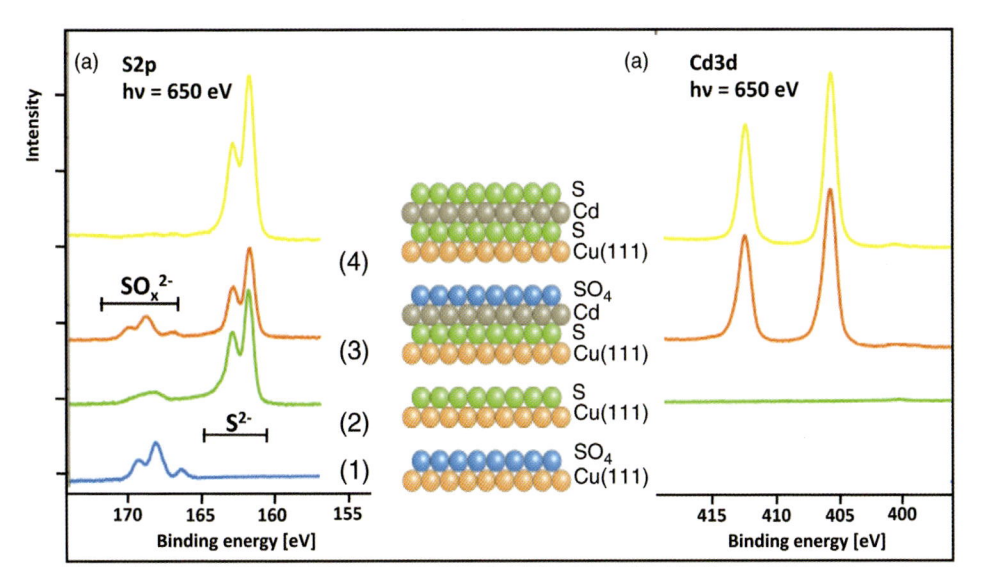

Figure 10.19 (a) S2p and (b) Cd3d XPS spectra taken from a Cu(111) electrode emersed from (1) 5 mM H_2SO_4, (2) 1 mM Na_2S + 5 mM H_2SO_4, (3) 1 mM $CdSO_4$ + H_2SO_4, (4) 1 mM Na_2S + 5 mM H_2SO_4 solution. The spectra were taken with synchrotron radiation of 650 eV (see text).

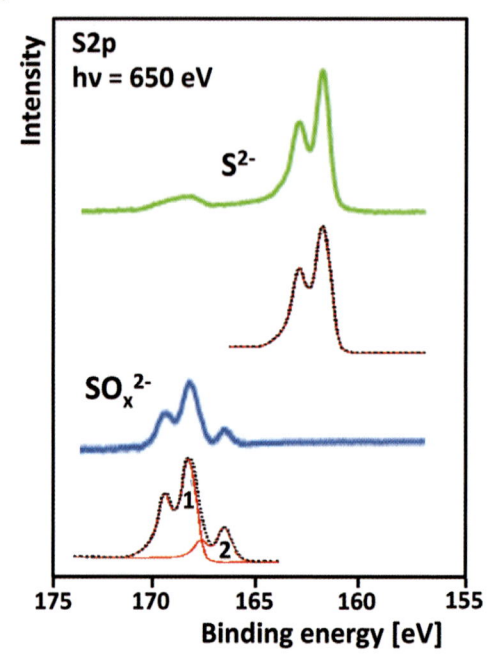

Figure 10.20 S2p spectra (1) and (2) from Figure 10.19. The S^{2-} spin–orbit 2p doublet is used to fit the SO$_x^{2-}$ spectrum with two components.

chamber for XPS measurements. The models (1–4) in Figure 10.19 are only to illustrate the sequence of the reaction steps and do not necessarily imply the structure of the respective deposit. After emersion from the various solutions, two groups of S2p signals are detected denoted SO$_x^{2-}$ and S^{2-}, respectively (copper signals are not shown here). These two groups of signals are shown enlarged in Figure 10.20; the S^{2-} group corresponds to the S2p$_{3/2}$, 2p$_{1/2}$ spin–orbit doublet (at 161.6/162.8 eV) of adsorbed sulfide (S^{2-}) on the Cu surface. Using simply this doublet as a fingerprint of sulfur atoms, the SO$_x^{2-}$ group can be fitted by two such doublets shifted by 1.8 eV with respect to each other and by more than 4.5 eV from the S^{2-} group. Thus, the SO$_x^{2-}$ group represents the coexistence of two sulfur-containing species, namely, (1) SO$_4^{2-}$ (at 167.9/168.1 eV) and (2) SO$_3^{2-}$ (at 166.1/167.3 eV). The SO$_3^{2+}$ anions are a product of radiation damage (see Section 10.5). The drastic "chemical shift" (see Section 10.4.2.1) between the three species containing sulfur in three different oxidation states (S(2−), S(4+), S(6+)) makes their distinction very easy. The Cd3d signal, of course, is seen only after the sample was immersed in the Cd-containing solution (Figure 10.19b).

Besides this chemical identification, important information can be gained from the (relative) intensities of the various spectra. After immersion in pure H$_2$SO$_4$ (1), of course, the Cu(111) is only covered with SO$_4^{2-}$ anions. After immersion in the S^{2-}-containing sulfuric acid solution (2) the S2p spectrum is dominated by the S^{2-} doublet because sulfide interacts with copper much stronger than

SO_4^{2-} (the small SO_4^{2-} signals originates from remnant traces of adhering electrolyte during the transfer process). After immersion in the cadmium-containing sulfuric acid (3), the presence of S^{2-} and SO_4^{2-} plus the decrease of the S^{2-} signal clearly indicate that the sulfide anions cannot be on top of the cadmium layer, because the S^{2-} signal is damped and sulfate is adsorbed again (which can only happen on the sulfide-*free* Cd layer). This reasoning is conclusively supported by the pure S^{2-} spectrum measured after immersing this sample once again in the sulfide-containing solution (4). Sulfide has replaced the terminating SO_4^{2-} layer; the SO_4^{2-} signal has disappeared and the S^{2-} signal has even grown.

10.4.2
Chemical Shift

10.4.2.1 Influence of Ionicity

As described in the previous sections and illustrated in Figures 10.11, 10.12, 10.19, and 10.20, the binding energy of the same electron level, for example, Fe2p, O1s, S2p, may vary with the chemical environment of the respective emitter atom. This is due to a change of the charge density on the emitter atom and, as a consequence, the binding energy of its electrons in the initial state (prior to photo-ionization). As a classical example, Figure 10.21 displays the C1s

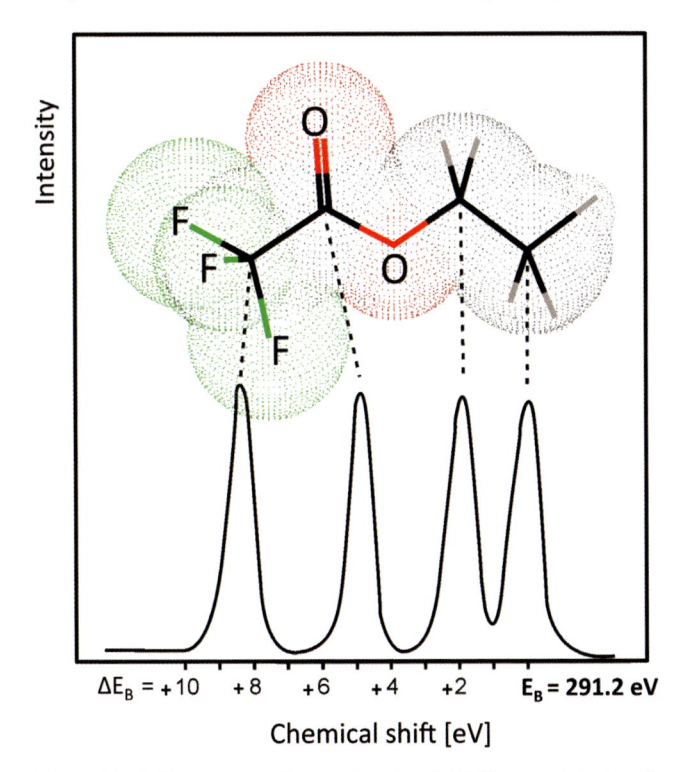

Figure 10.21 C1s spectrum from carbon in ethyl-trifluoroacetate (see, for example, Ref. [17]).

core-level spectrum of gaseous ethyl-trifluoroacetate [17]. Instead of just one C1s line at $E_B^V = 291.2$ eV, we see three further lines at 2, 5, and 8.3 eV higher binding energy. As shown in Figure 10.21, the four signals of equal intensity may be assigned to the four chemically inequivalent carbon atoms, which, from C1 to C4, are bound to an increasing number of electronegative atoms: C1 to none, C2 to one oxygen atom, C3 to two oxygen atoms, and C4 to three fluorine atoms. The increasing withdrawal of negative charge creates an increasing positive partial charge on the respective carbon atom and thereby causes the increase of the C1s binding energy from C1 to C4. This "chemical shift" can be described by a simple "charge potential model" [30]: The charge of a chemically bound atom A is approximated by a point charge whose interaction with the corresponding point charges of neighboring atoms B leads to a shift of the electron binding energies in the initial (i) state:

$$\Delta E_B \approx k_i q_A + \sum_{B \neq A} \frac{q_B}{r_{AB}}, \quad (10.15)$$

with respect to a reference state. q_A and q_B are the partial charges of atoms A and B (with respect to the reference state). The sum accounts for the electrostatic interaction of atom A with all atoms B at distance r_{AB}. It is also obvious that a relationship between the chemical shift and the electronegativity of the bonding partners should hold, for instance:

$$I_A = 1 - \exp[-0.25(X_A - X_B)^2] \propto q_A, \quad (10.16)$$

with I_A = partial ionicity of the emitter atom A bound to an atom B; X_A, X_B = Pauling electronegativities of species A and B, and q_A = partial charge of A that results from I_A [6]. The validity of this relationship is clearly shown in Figure 10.22 [6].

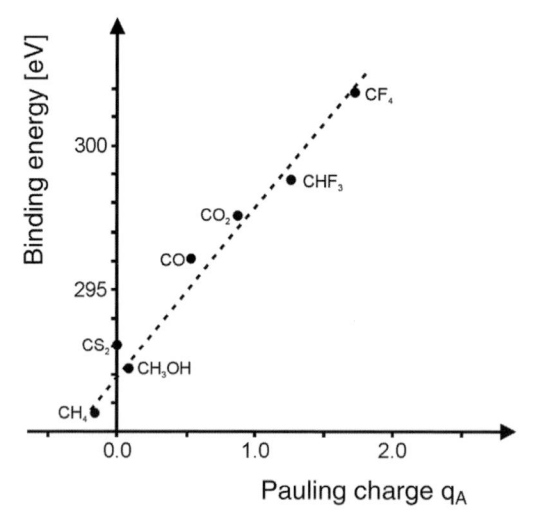

Figure 10.22 Plot of the C1s binding energy of carbon in various compounds versus the Pauling charge of the respective carbon atoms calculated from Eq. (10.16) (see, for example., Ref. [6]).

Likewise, it is this picture of chemical shift that explains the changes of the Fe2p$_{3/2}$ and O1s signals upon oxidation of iron with pure oxygen or air in Figure 10.11, as well as the shifts of the S2p signal from S^{2-}, SO_3^{2-}, and SO_4^{2-} in Figures 10.19 and 10.20.

10.4.2.2 Standard Spectra

Photoelectron spectroscopy is not only able to identify the various elements in composite samples but may also provide information about their distribution in different chemical states. To this end, characteristic "standard spectra" of the different species are required.

As an example, Figure 10.23 displays several Fe2p$_{3/2}$ and O1s spectra from the various nominally stoichiometric iron compounds measured "as received," that is, air exposed [21,24]. All Fe2p$_{3/2}$ signals look disappointingly equal, all spectra have their main maximum at 710 eV and except Fe$_x$O ($x = 0.90–0.95$) a satellite

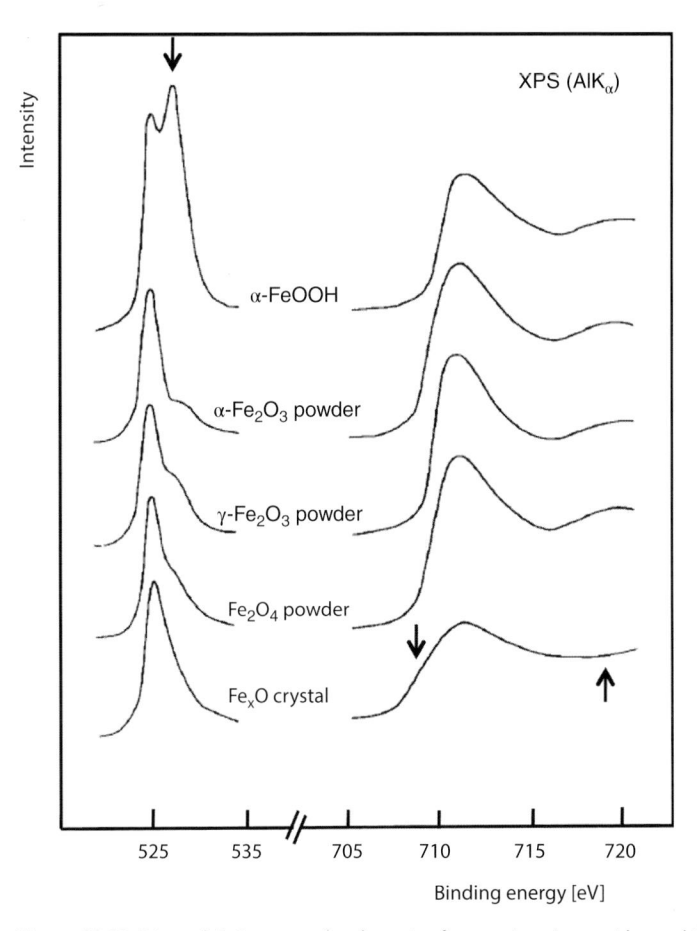

Figure 10.23 O1s and Fe2p$_{3/2}$ core-level spectra from various iron oxides and iron hydroxide samples "as received," that is, air exposed and untreated (see Ref. [24]).

at 719.5 eV. Only the Fe$_x$O spectrum shows a slight broadening toward low binding energies and a lack of the broad satellite at ~719 eV (see arrows). Likewise, the O1s spectra are very similar except the pronounced OH signal (arrow) for FeOOH at higher energy.

The results change completely when corresponding single-crystal samples are cleaved/crushed in UHV and the measurements are performed on these fresh fracture surfaces, which are representative of the bulk composition (before fracture). These results are shown in Figure 10.24. In this case, the spectra of Fe$_3$O$_4$ and Fe$_x$O deviate significantly from those in Figure 10.23. The Fe$_3$O$_4$ spectrum shows a clear shoulder at ~708.5 eV and a (seemingly) weakened satellite at ~719 eV, and the maximum of the Fe$_x$O spectrum occurs at the lower binding energy of 709.7 eV followed by a satellite at 715.7 eV. The O1s signals are largely unaffected.

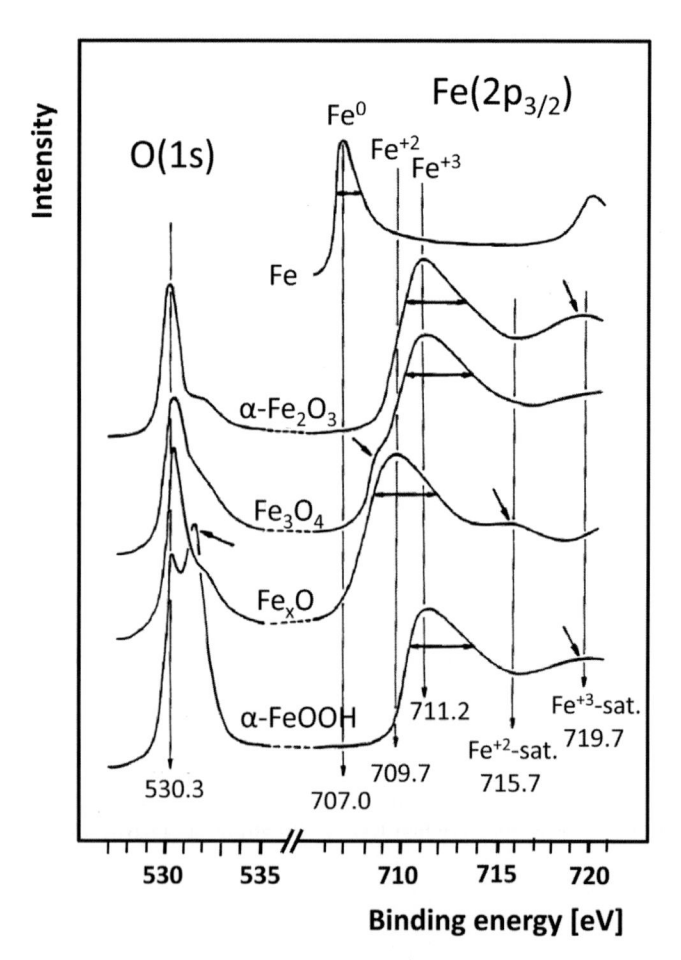

Figure 10.24 O1s and Fe2p$_{3/2}$ "standard spectra" of the indicated iron oxide and hydroxide samples. The stoichiometric compound samples were crushed in UHV. For comparison, the Fe2p$_{3/2}$ spectrum of a clean iron foil is also shown [24].

The spectra in Figure 10.24 may be regarded as the "standard spectra" of the samples in question: The Fe2p$_{3/2}$ signals clearly distinguish between the metallic Fe^0 (707.0 eV), and the chemically shifted Fe^{2+} (709.7 eV) and Fe^{3+} (711.5 eV) cation states. In particular, the Fe$_3$O$_4$ spectrum shows both the Fe^{3+} (maximum) and Fe^{2+} (shoulder) contributions. The seemingly missing satellite peak is only a consequence of a superposition of both the Fe^{3+}-related satellite at 719.7 eV and the Fe^{2+}-related satellite at 715.7 eV, the latter filling the minimum between the former and the main signals in the Fe$_3$O$_4$ spectrum. The difference between these "standard spectra" (Figure 10.24) and those shown in Figure 10.23 is easy to explain: The latter refer to samples that, owing to air exposure, are all superficially fully oxidized to Fe$_2$O$_3$ and therefore all look alike. In summary, XPS core-level spectroscopy permits not only a distinction between metallic and oxidized iron but also between Fe^{2+} and Fe^{3+} cations, even in magnetite, Fe$_3$O$_4$, containing both oxidation states. However, it should not be concealed that the Fe$_3$O$_4$ spectrum in Figure 10.24 does not permit to distinguish between the Fe^{3+} cations with octahedral and tetrahedral coordination, respectively.

For the sake of completeness, Figure 10.26 shows also the complete set of AlK$_\alpha$-excited valence band spectra of the various Fe compounds [21,24]. Note again the distinction between Fe^{2+} and Fe^{3+} cations by the shoulder at ~1 eV binding energy (arrow) as well as the much larger width of the valence bands of the compounds compared to the one of metallic iron. This broadening is largely a consequence of the high-spin configuration of the valence electrons of these species (*inset* in Figure 10.26) and the resultant manifold of multiplet states addressed in Section 10.3.2, as well as the overlap with the O2p states of oxygen and hydroxide anions around 5.5 ev (arrows).

Similar "standard spectra," of course, can be measured and archived for other compounds, as long as pure and stoichiometric samples are available whose surface composition does not deviate from the bulk composition. Sputter cleaning of compound surfaces, however, is not advisable, as discussed in Section 10.5.

10.4.2.3 Surface Core-Level Shifts

It is worth mentioning that even the atoms at the very surface of an atomically clean elemental sample are in a different "chemical environment" than the atoms in the bulk of the same sample. While the bulk atoms are totally surrounded by atoms of the same kind, the very surface atoms have no neighbors on the vacuum side. Already this asymmetry leads to a measurable "surface core-level shift." As an example, Figure 10.25a shows only the d$_{5/2}$ spin–orbit component of a high-resolution 3d core-level spectrum of a Rh(100) surface, split into a bulk component at 307.2 eV and a component from the very first layer of Rh atoms at 306.6 eV [31].

This distinct discrimination between surface and bulk atoms provides a unique opportunity to study the very first steps of a reaction between a solid sample and a surrounding gaseous reactant, for instance, oxygen. It is obvious that impinging oxygen molecules will first interact with the *surface* atoms and initially cause a "chemical shift" of the *surface* component of the core-level spectrum only. Oxygen molecules are known to dissociate spontaneously upon

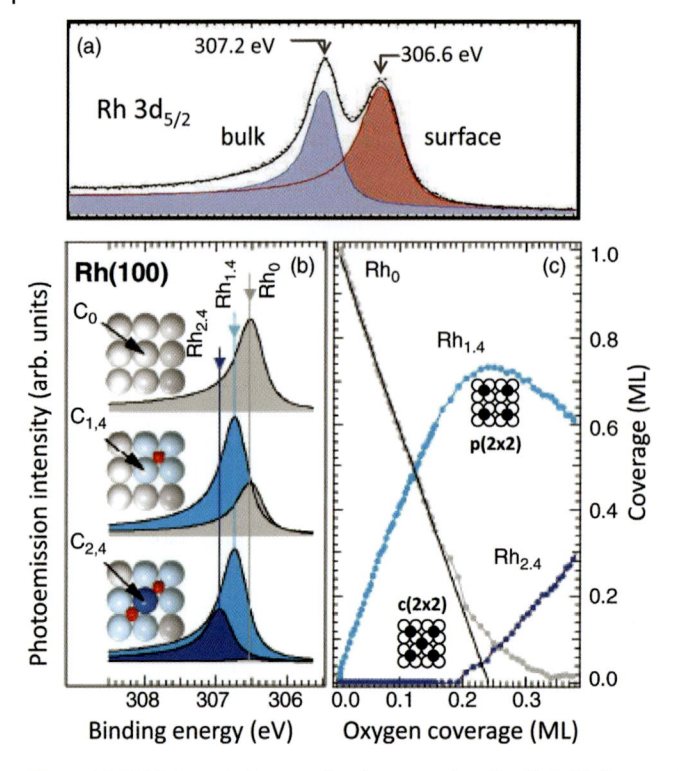

Figure 10.25 High-resolution core-level spectra of the Rh$3d_{5/2}$ spin–orbit component only of (a) a clean [31] and (b) a step-wise oxygen covered Rh(100) surface. Panel (b) shows only the *surface component* of the Rh$3d_{5/2}$ emission for oxygen-free (C_0), the p(2×2)-O ($C_{1,4}$) and the c(2×2)-O ($C_{2,4}$) covered Rh(100) surface. Panel (b) presents the intensity of the Rh$3d_{5/2}$ surface component of Rh atoms in contact with no, one, or two adsorbed oxygen atoms (Rh$_0$, Rh$_{1,4}$, Rh$_{2,4}$) [32].

adsorption on most metal surfaces at room temperature, because the gain of bond energy of two separate oxygen atoms overcompensates the dissociation energy of the oxygen molecule [24].

Figure 10.25b shows only the *surface component* of the Rh($3d_{5/2}$) signal of a bare Rh(100) surface (Rh$_0$) as well as of the same surface covered with $\theta_O = 0.185$ ML (Rh$_{1,4}$) and $\theta_O = 0.375$ ML (Rh$_{2,4}$) of oxygen, respectively (coverage θ_O in ML = monolayer) [32]. At the low coverage ($C_{1,4}$), the oxygen atoms occupy hollow sites between four Rh atoms and form a so-called p(2×2) structure (see *inset*) as known from low energy electron diffraction (LEED) or scanning tunneling microscopy (STM) measurements. In this case, each Rh atom interacts at most with one oxygen atom, which leads to a chemically shifted Rh$3d_{5/2}$ component at 220 meV higher binding energy denoted Rh$_{1,4}$ (one oxygen atom per four rhodium atoms). At the higher oxygen coverage ($C_{2,4}$), a denser c(2×2)-O structure is formed (see *inset*) in which inescapably Rh atoms interact with two oxygen atoms giving rise to a further Rh($3d_{5/2}$) component (Rh$_{2,4}$) at 435 meV higher binding energy than the peak of the bare surface. A plot of the normalized intensity of the three components Rh_0, $Rh_{1,4}$, and $Rh_{2,4}$ as a function of oxygen coverage (Figure 10.25c) illustrates the relative abundance of coexisting Rh surface atoms in contact with no, one, or two oxygen atoms. The low coverage $Rh_{1,4}$ state grows linearly right from the beginning, while the $Rh_{2,4}$ state starts to grow only after the p(2×2) structure is (nearly) saturated. In turn, the Rh_0 state decreases linearly with slope $(dRh/d\theta) = -4.2$ and vanishes once the p(2×2)-O structure is completed. The slope of -4.2 is a clear confirmation that each oxygen atom influences four Rh surface atoms, that is, the oxygen atoms adsorb in *fourfold* hollow sites with no common Rh atom. Likewise, similar adsorption experiments with oxygen on a hexagonal Rh(111) surface resulted, besides the $Rh_0(3d_{5/2})$ peak, in the three signals $Rh_{1,3}$, $Rh_{2,3}$, and $Rh_{3,3}$ chemically shifted by $+345$, $+780$, and $+1120$ meV, respectively [33]. In this case, the slope of the linear decrease of the Rh_0 component as a function of oxygen coverage was found to be -3 in agreement with oxygen adsorption in *threefold* hollow sites with no common Rh atom. Note also that on both the Rh(100) and the Rh(111) surface, the incremental peak shift per O–Rh bond is constant; the shift of $Rh_{2,4}$ (+435 meV) is twice as large as that of $Rh_{1,4}$ (220 meV), and the shifts of $Rh_{2,3}$ and $Rh_{3,3}$ are two and three times larger than that of $Rh_{1,3}$, respectively, indicating a dominantly *local* interaction between the adsorbed oxygen atoms and just the neighboring Rh surface atoms.

In this particular case, the oxygen–metal interaction is first restricted to adsorption *on* the surface, because up to monolayer adsorption only the $3d_{5/2}$ *surface* component is dominantly affected. In turn, such measurements can provide insight into the onset of bulk oxidation, by observing chemical shifts of both the surface and the bulk component. Such measurements as a function of temperature will provide the activation energy of oxygen penetration below the surface.

10.4.2.4 Valence Band Emission

Direct information about chemical interactions between atoms and molecules, that is, bond formation, is reflected by changes of their valence electronic states, which are part of XPS spectra (see Figure 10.26), but can also be measured with UV photoemission. Restricting ourselves to interactions with and within solids, the following two examples give an impression of the specific strength of UPS data.

Figure 10.27 shows HeII ($h\nu = 40.8$ eV) excited valence spectra of a bare and oxygen-exposed Ni(111) surface. Characteristic for the bare metal surface is the rather narrow 3d valence band emission with a high density of states at the Fermi level ($E_F = 0$), as well as an intrinsic satellite peak at ~6 eV binding energy. An exposure of only 6 L oxygen at room temperature suffices to saturate a (2×2)O structure on this surface accompanied with the growth of photoemission intensity at 5.6 eV. From the fact that this exposure does not yet lead to a change of the Ni3d intensity distribution, it must be concluded that the oxygen is *adsorbed only on* the surface [34]. This changes, however, after an exposure of

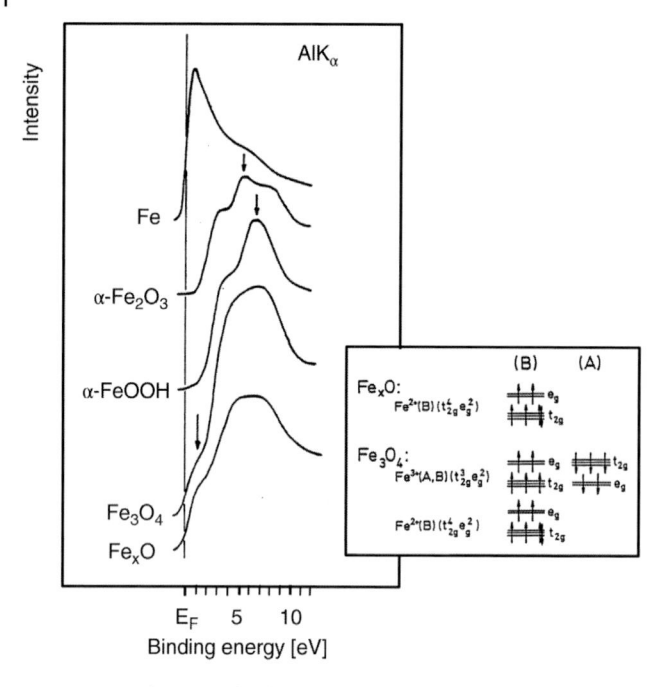

Figure 10.26 AlK$_\alpha$-excited valence band spectra of the indicated iron, iron oxide, and iron hydroxide samples. *Inset:* Schematic representation of the electron occupancies and high-spin configurations of the crystal field split Fe3d states on octahedral (B) and tetrahedral (A) sites in Fe$_x$O and Fe$_3$O$_4$ [24].

100 L O$_2$, the density of states at $E_F = 0$ is reduced and a new peak near ~2 eV binding energy followed by a broad intensity distribution extending up to ~11 eV clearly indicates the nucleation of NiO, a process that is completed after 1000 L O$_2$. This final spectrum indicates the formation of a closed NiO film – within the information depth of these low-energy electrons. The total lack of density of states at E_F is consistent with the insulating character of nickel oxide [34].

Important for the assessment of the reactivity or catalytic activity of surfaces is not only their composition but also the knowledge of their electronic properties. Both the composition and the electronic structure of the *surface* of composite materials are not simply a linear superposition of the contribution of the bulk constituents. The composition at the *surface* may strongly deviate from that in the bulk due to surface segregation [35,36]. Likewise, the electronic interactions between the constituents of, for instance, an alloy cause a mutual modification of the charge density on the constituent metal atoms, known as "ligand effect" [37]. These modifications are immediately recognized in UPS valence band spectra that are roughly a replica of the valence band density of states. As an example, Figure 10.28 shows the valence band spectra of a series of Cu$_x$Pt$_{1-x}$(111) surfaces. While the spectrum of the pure Cu(111) surface shows a very low density of 4s states at the Fermi level and the relatively narrow dominant Cu3d band between

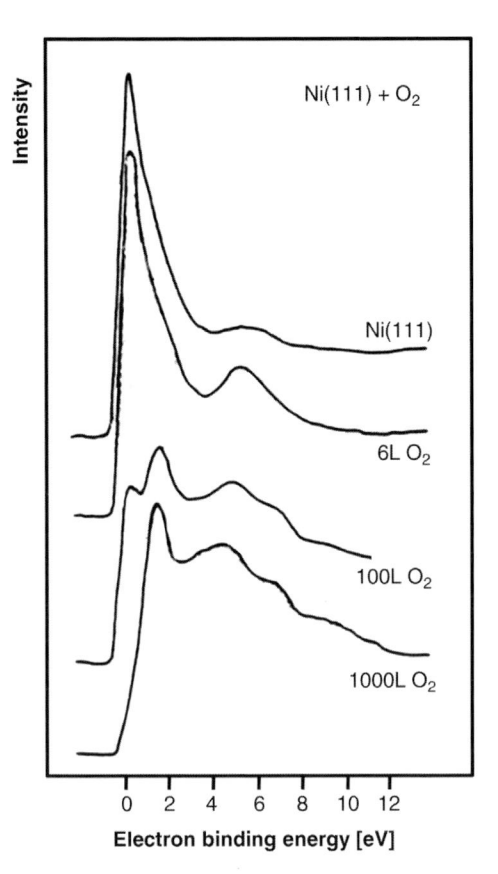

Figure 10.27 HeII (40.8 eV) excited valence band spectra of a clean and oxygen-exposed Ni (111) surface, indicating the transition from chemisorbed oxygen (6 L) to Ni oxide (1000 L).

2 and 4 eV, the valence band of the bare Pt(111) surface is characterized by the 5d states with a high density at the Fermi level as well as two maxima within a broad range of ~8 eV below E_F due to spin–orbit and crystal field splitting.

The spectra of the various Cu_xPt_{1-x} alloys, with $x = 0.15, 0.25, 0.65$, and 0.75, clearly indicates that they are not a simple weighted superposition of the pure metal spectra.

10.4.3
Core-Level Angular Dependence

In Section 10.4.2.3, the adsorption site of the oxygen atoms on a Rh(100) or Rh(111) surface was indirectly concluded from the number of Rh atoms that are chemically affected per oxygen atom. More direct information about the spatial coordination of an adsorbed atom (or molecule) with surface atoms can be obtained by making use of the "forward scattering" effect mentioned in Section 10.3.4. This is nicely illustrated in Figure 10.29 for the case of oxygen adsorption

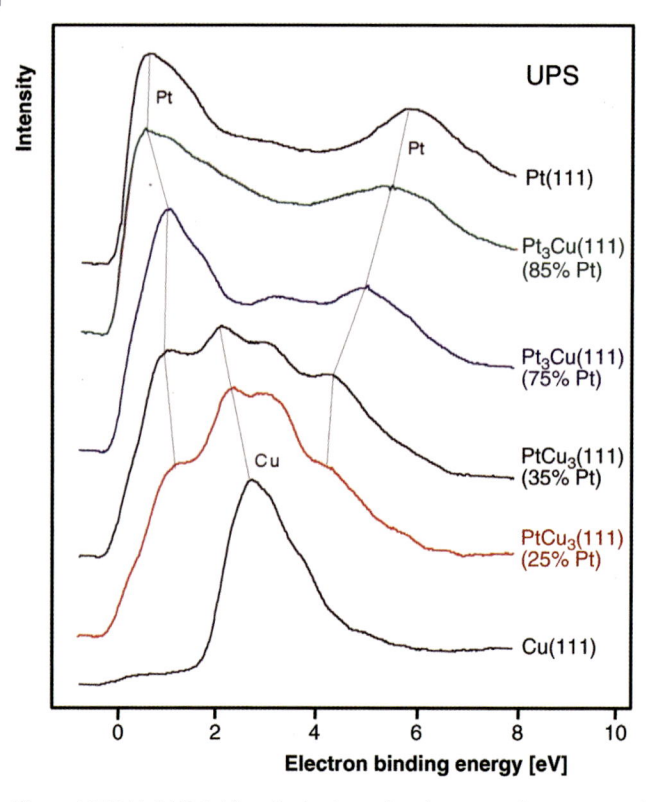

Figure 10.28 HeII (40.8 eV)-excited valence band spectra of a copper, a platinum, and four different Cu–Pt alloy samples.

on a Cu(100) surface, which at room temperature leads also to a c(2×2) superstructure. Pioneering measurements of the azimuthal anisotropy of the O1s photoemission intensity clearly proved that the oxygen atoms occupy fourfold hollow sites between four atoms of the unreconstructed Cu(100) surface, and thereby clearly discarded other proposed structure models [38,39].

Core-level anisotropies may be measured by fixing the polar angle θ_e (Figure 10.3), rotating the sample (stepwise) by full 360° about the surface normal, and registering the photocurrent per angle interval. Figure 10.29a shows the raw azimuthal intensity distribution $I(\phi)$ for the O1s emission obtained for a polar angle of $90° - \theta_e = 13°$ from the c(2×2)O covered Cu(100) surface. The vague impression of a fourfold symmetry of this distribution is clearly verified after subtraction of the isotropic background intensity I_{min} indicated by the dashed circle; the resultant distribution $\Delta I(\phi) = I(\phi) - I_{min}$ clearly shows dominant emission into the four quadrants (Figure 10.29b). Finally, in order to minimize spurious instrumental sources of anisotropy like sample imperfections and electric field inhomogeneities around the sample, the pattern of Figure 10.29b is fourfold averaged by calculating $\Delta \hat{I}(\phi) = \Delta I(\phi) + \Delta I(\phi + 90°) + \Delta I(\phi + 180°) + \Delta I$

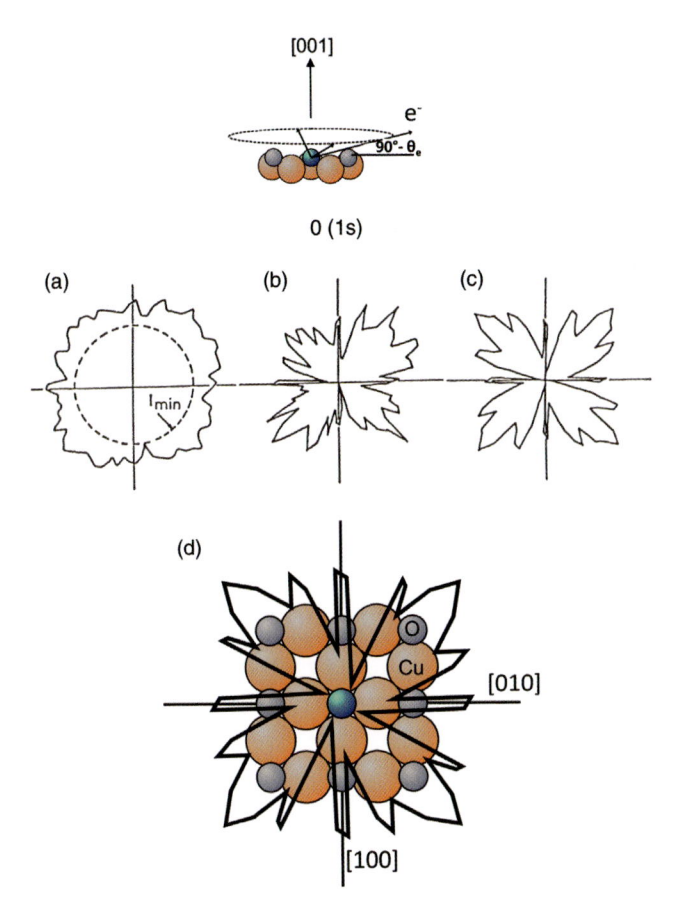

Figure 10.29 Azimuth angle-dependent O1s intensity from oxygen atoms adsorbed on a Cu(100) surface: The raw data (a) were measured at a polar exit angle of $\theta_e = 77°$ (see top). The background corrected (b) and fourfold averaged (c) distribution is superimposed on an O-covered fourfold Cu hollow site in part (d). (Reproduced with permission from Refs. [36, 37]. Copyright 1978, & 1979, American Physical Society.)

($\phi + 270°$), resulting in the appealing "flower pattern" [38,39] of excellent fourfold symmetry in Figure 10.29c. Superposition of this pattern on a fourfold oxygen site on the Cu(100) surface is displayed in Figure 10.29d. All $\Delta \hat{I}(\phi)$ maxima clearly point into the direction of nearest and next-nearest Cu or O atoms of the oxygen emitter atom, in agreement with predictions of the "forward scattering" mechanism addressed in Section 10.3.4 [38,39].

Series of similar $\Delta \hat{I}(\phi)$ results obtained for different polar exit angles θ_e in combination with low-angle single-scattering simulations of the azimuthal intensity anisotropy [38,39] not only clearly supported the notion of oxygen adsorption in fourfold hollow sites but also enabled the determination of the vertical position of the oxygen atoms with respect to the plane of surface copper atoms [39]. While the above presented and discussed data were measured by

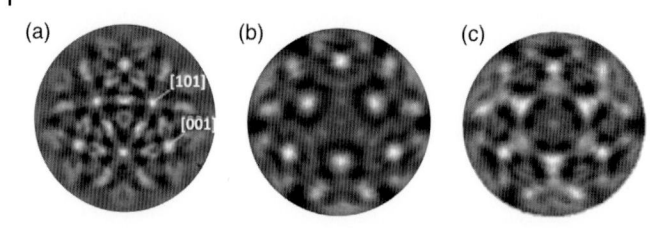

Figure 10.30 High-resolution Cu2p X-ray photoelectron diffraction pattern from (a) clean, (b) 0,8 ML cobalt, and (c) 3.8 ML cobalt-covered Cu(111) surface; E_{kin} (Cu2p) = 808 eV, E_{kin}(Co2p) = 963 eV [10]. (Reproduced with permission from Ref. [23]. Copyright 2012, Wiley-VCH Verlag GmbH & Co. KGaA, Weinheim.)

collecting the photocurrent angular resolved into a narrow acceptance cone (γ in Figure 10.3) and rotating the sample about its surface normal, nowadays the 360° angular distribution of the photoemission intensity can be measured at once by means of a two-dimensional position-sensitive detector placed behind a microchannel plate, each microchannel acting like a channeltron (see Section 10.2.3). As an example, Figure 10.30 compares the 2p photoemission intensity pattern from (a) a bare Cu(111) surface (Cu2p: E_{kin} = 808 eV), (b) a 0.8 ML, and (c) a 3.8 ML Co covered Cu(111) surface (Co2p: E_{kin} = 963 eV). The detailed analysis of these anisotropic intensity distributions provides information about the epitaxial relationship and growth structure of the hcp metal cobalt on the fcc substrate copper [10]. The same technique, of course, can also be applied to investigate the structure of growing surface compounds [40].

10.4.4
Quantitative Analysis

The intensity of PES signals is proportional to the number n_i of contributing emitter atoms i. While, as usual, it is easy to determine the *relative* concentration of the constituents of a multicomponent system, it is more difficult to determine *absolute* concentrations (atoms per cm^3 or atoms per cm^2). The latter requires either a reliable reference value for calibration or a unique relationship between the signal intensities arising from one isolated atom i, on the one hand, and n_i-bound atoms i in a multicomponent system on the other. The latter is hardly ever possible, because the intensity of the n_i-bound atoms deviates from the n_i-fold of the intensity of the isolated atom due to so-called matrix effects, in particular its chemical surrounding and its spatial distribution in the sample. (Note that here "matrix" has a different meaning than in Eq. (10.11)).

The oxygen adsorption on the Rh(100) or Rh(111) surface discussed in the previous section is a clear-cut example for the determination of the absolute number of adsorbed oxygen atoms (in terms of O-atoms per cm^2 within the detected surface area), because (i) all oxygen atoms reside, coverage independent, in equivalent sites, that is, only *on the surface* in fourfold or threefold hollow sites, and (ii) their density (atoms/cm^2) and lateral distribution at monolayer saturation is known from independent experiments (LEED, STM). In this case, the measured O1s photoemission intensity permits a linear interpolation between zero and monolayer coverage. This is no longer strictly true once oxygen atoms start to be incorporated at unknown depth below the surface; their emitted photoelectrons suffer scattering effects causing a damping of the photoelectron current (see below).

In any case, the first task is to determine the intensity of photoemission peaks related to the n_i emitter atoms, that is, the "primary photocurrent" represented by the area of a given peak.

The natural linewidth of photoemission peaks is a Lorentzian function whose linewidth is determined by the lifetime of the excited, that is, the hole, state. Depending on the experimental conditions, however, the measured peak shape suffers modifications, namely, due to the linewidth of the light beam, broadening effects on the part of the spectrometer, and broadening effects due to disorder of the sample (including also thermal noise). As a consequence, the measured peak shape arises from a convolution of Lorentzian and Gaussian contributions, also called Voigt function. Especially in the case of metallic samples, the screening of the created "photo-hole" by the (more or less) free valence electrons calls for the use of an improved so-called Doniach–Sunjic function [41]. Software packages are available for optimal line fitting and the determination of the "true" peak area therefrom.

As displayed in Figure 10.31, however, the background varies across the energy interval of a PES signal; it is always higher on the high binding energy side of each peak, because each primary peak represents an enhanced current of photoelectrons that on their way out of the solid sample produce secondary electrons of lower *kinetic* energy. These lower energy electrons add stepwise to the background intensity toward higher binding energy. A quantitative analysis of XPS spectra, therefore, requires first a careful subtraction of this background intensity in order to determine the "true" intensity of a primary, that is, element-

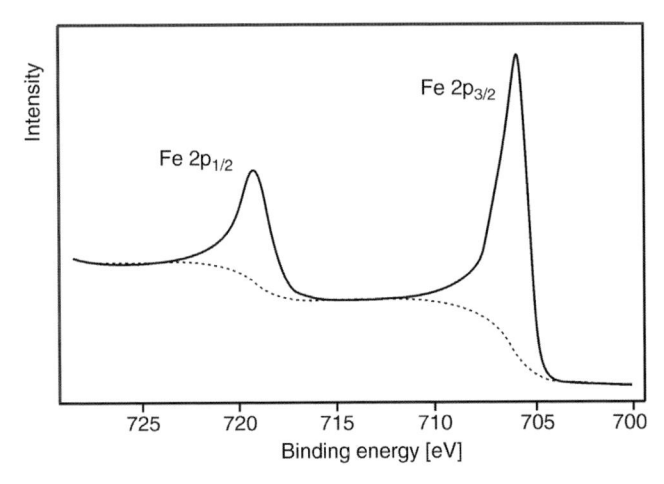

Figure 10.31 Illustration of background correction of an Fe2p$_{3/2}$ spectrum (see text).

specific, peak. Several background correction procedures have been proposed [26,42,43]. One simple approach is to assume that the secondary electron background grows proportional to the increasing photoelectron current across each characteristic peak as the source for secondary electron creation. Thus, the background function is determined by stepwise integrating the peak intensity from the lower binding energy onset up to the high energy limit and normalizing the resulting function to the high binding energy level of the particular peak, as illustrated in Figure 10.31.

From the experimental viewpoint, the resulting intensity of a peak after background subtraction is given by

$$I_i = n_i \sigma_i \lambda_i C, \qquad (10.17)$$

with n_i being the number density of atoms/molecules i in the detected region, σ_i the cross section, and λ_i the inelastic mean free paths length at given energy. C is a constant that includes all other experimental factors like photon flux, angular efficiency, detector yield, and so on.

The assessment of the "detection region" requires particular attention. Besides the surface area illuminated by the incident light beam, λ_i determines the depth z to which i atoms contribute to the signal. This contribution is given by

$$I_i(z) = I_{0,i} \exp\left(-\frac{z}{\lambda \cdot \cos \theta_e}\right). \qquad (10.18)$$

The initial electron flux $I_{0,i}$ generated at a depth z is attenuated by the damping factor $\exp(-z/(\lambda \cdot \cos \theta_e))$. λ is the material-specific inelastic mean free path length (Figure 10.13) and represents the depth from which the initial electron flux $I_{0,i}$ is damped to its one-eth. θ_e is the exit angle, that is, the angle between the surface normal and the direction of detection (see Figure 10.3). The product $\lambda \cdot \cos \theta_e$ represents the effective "escape or information depth" for angles $0<\theta_e<90°$, which can be exploited for the determination of concentration profiles perpendicular to the surface, as detailed in Section 10.4.5. For constant θ_e, Eq. (10.18) is equal to the well-known Lambert–Beer law.

For a given set of experimental parameters, the quantity

$$\text{ASF} = \sigma_i \lambda_i C \qquad (10.19)$$

is a constant "atomic sensitivity factor," which can be determined and tabulated for each species and type of spectrometer. σ and λ are material specific and energy dependent. Values of σ for excitation with K_α radiation have been calculated, for instance, by Scofield for different electronic states [44]. Figure 10.32 shows a plot of the "cross sections" of different electron states as a function of the atomic number for excitation with AlK_α radiation [44]. This cross section obviously enters the quantitative evaluation of XPS intensities (sensitivity factor). According to Figure 10.32, the Ne1s signal of a 50: 50 mixture of Ne and He would be about 30 times more intense than that of the *same* number of He atoms. A plot of λ as a function of kinetic energy is shown in Figure 10.13 and a very detailed discussion of the inelastic mean free path of electrons in solids can be found in Ref. [45].

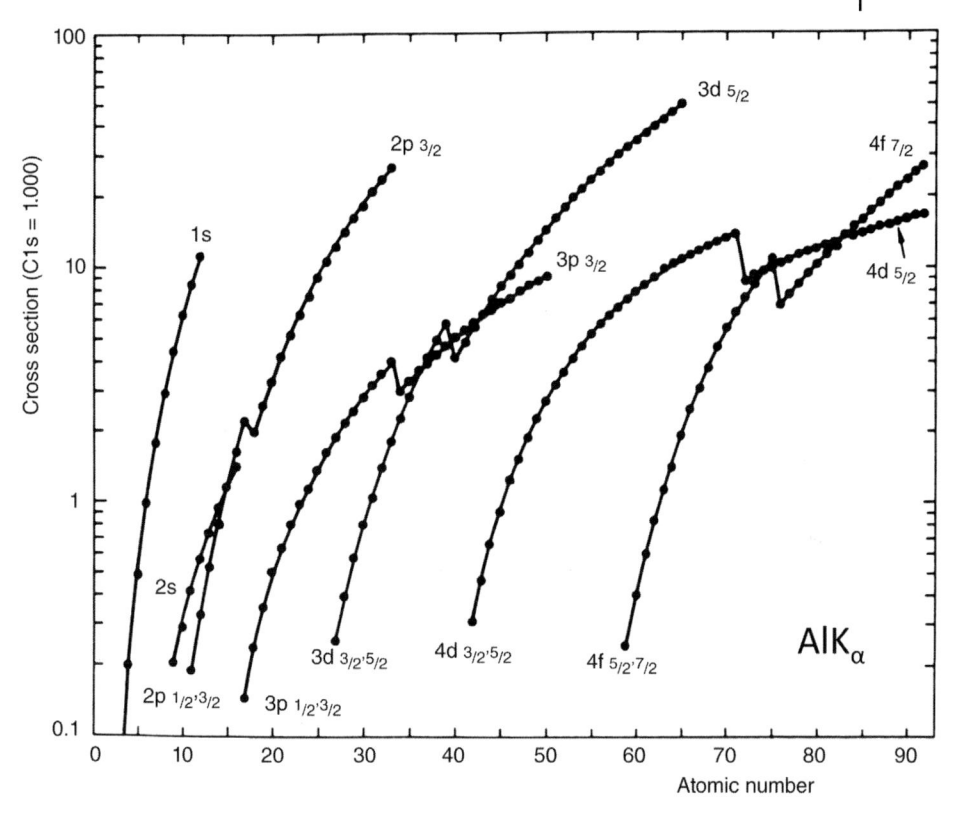

Figure 10.32 Calculated cross section σ for AlK$_\alpha$ radiation relative to the C1s cross section [44]. (Reproduced with permission from Ref. [43]. Copyright 1976, Published by Elsevier B.V.)

According to Eqs. (10.17) and 10.19, the ratio

$$\frac{I_1 \text{ASF}_2}{I_2 \text{ASF}_1} = \frac{n_1}{n_2} \quad (10.20)$$

gives the ratio n_1/n_2 of the atomic density of two species 1 and 2 in the sample and

$$\xi = \frac{n_i}{\sum_j n_j} = \frac{I_i \text{ASF}_i^{-1}}{\sum_j I_j \text{ASF}_j^{-1}} \quad (i \in j) \quad (10.21)$$

the molar fraction of each species if all species j are detected. In the latter two equations, the ASF values still have to be assumed "matrix" independent, that is, independent of the nature and distribution of all other species $j \neq i$ in the sample. This holds true only for homogeneous and similar materials, because especially λ is matrix dependent. Thus, even with most carefully determined ASF factors, the uncertainty of quantitative XPS-derived concentrations is on the order of 10%. A detailed assessment of the quantification of XPS data is given by Jablonski and Powell [45,46].

10.4.5
Depth Profiling

Due to the relatively short mean free length of electrons in condensed matter (see Figure 10.13) compared to X-rays, XPS spectra carry analytical information only about a very thin surface-near layer of at most a few tens of nanometers. This layer, however, may not be representative of the bulk composition of the respective sample due to surface contamination and reaction. But compared to X-ray absorption measurements, the spectral resolution of XPS and, thus, the distinction of species, also in different chemical states, is incomparably better.

There are two options to overcome the limited information depth of XPS data, namely, by "sputter profiling" and angle- or energy-dependent measurements.

"Sputter profiling" is a combination of sputtering and PES (or AES): The sample is stepwise or continuously bombarded with rare gas ions (mostly Ar+) of several hundreds of electrovolts, which erode material from the surface, and an accompanying series of XPS (or AES) spectra yields a concentration profile perpendicular to the surface. On the one hand, sputter profiling can be carried on without any depth limitation – until the whole sample is consumed. Figure 10.33 demonstrates an important example of the application of sputter profiling, namely, the determination of diffusion profiles. An iron substrate was covered with a nickel film by vapor deposition and subsequently annealed at constant temperature for given periods of time. After proper calibration of the depth scale (or sputter rate), the resultant concentration profiles enable the determination of diffusion coefficients and activation energies of diffusion [47].

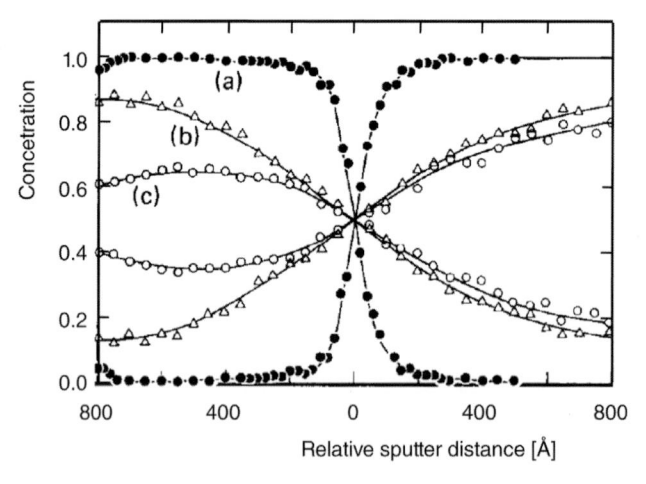

Figure 10.33 Sputter profiles through a nickel film on an iron substrate; (a) annealed at 300 °C for 15 h, (b) annealed at 400 °C for 30 min, (c) annealed at 500 °C for 10 min. Plotted are the FeL$_3$VV (651 eV) and NiL$_3$VV (848 eV) Auger intensity (see Figure 10.18) as a function of the sputter distance (depth).

On the other hand, the destructive nature of the method introduces several problems. First, the sputter rate is atom and matrix dependent, with two consequences: (i) the composition within the detected information depth changes until a steady concentration state is reached, (ii) the surface roughness increases, which worsens the depth resolution. In the light of these problems, sputter profiles have to be taken with due caution. Aspect (i) is convincingly demonstrated in Figure 10.34. The solid lines reproduce the "standard" Fe2p$_{3/2}$ spectra of the stoichiometric single-crystal samples of Fe_2O_4, α-Fe_2O_3, and Fe_xO, as shown in Figure 10.24. After sputtering these samples with Ar+ ions of 1 keV for just a

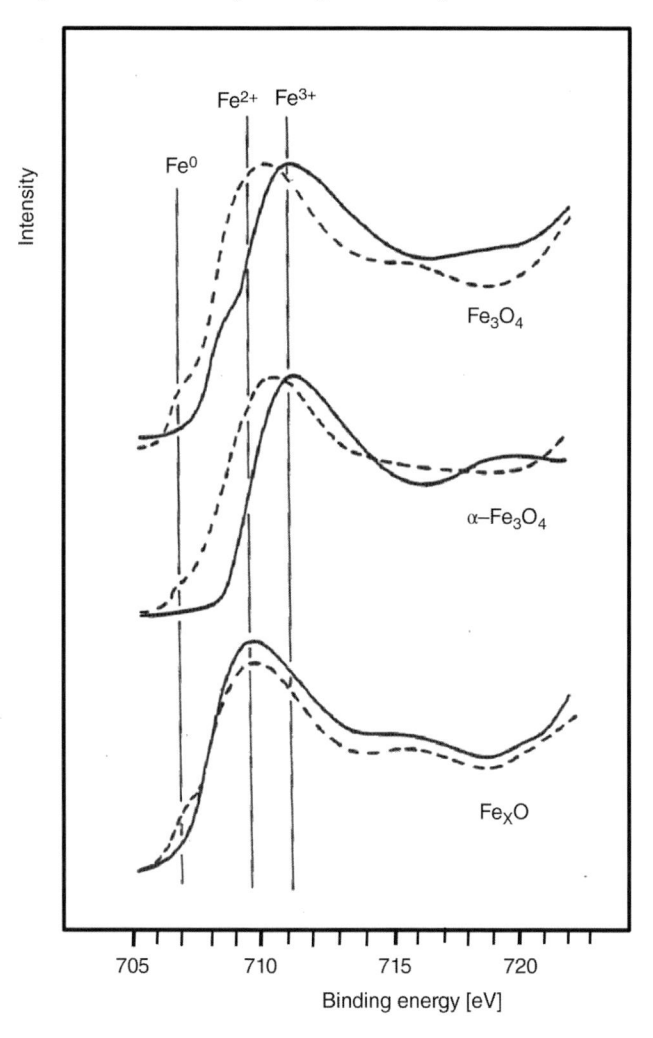

Figure 10.34 Fe2p$_{3/2}$ spectra showing the effect of Ar$^+$ ion bombardment on iron oxide surfaces. *Solid lines:* Single-crystal samples crushed in UHV (see Figure 10.24). *Dashed lines:* Ion-bombarded surfaces of Fe_3O_4 (for 10 min), Fe_2O_3 (for 7 min), and Fe_xO (for 3 min). The ion gun was operated at 5×10^{-5} Torr Ar, 1 kV, 10 μA, and normal incidence.

few minutes as specified in the figure caption, the resultant spectra (dashed lines) show drastic changes in all three cases. Due to preferential sputtering of the lighter component, oxygen, all three materials are severely reduced as indicated by the overall shift of Fe2p$_{3/2}$ intensity to lower binding energies. Most striking is the fact that after sputtering, all three *stoichiometric and single-crystalline* oxide samples are partially reduced even to metallic iron (Fe^0) as verified by the small but unambiguous shoulder at 707.0 eV [48].

The clear lesson from these observations is that surface cleaning and in-depth concentration profiling of multicomponent samples based on sputtering will inevitably cause an adulteration of the composition, and therefore should be handled with due caution.

Alternatively, mere variation of the energy and/or detection angle of the electrons is nondestructive (see Section 10.5) and also provides in-depth information, but only to a rather limited depth. The basis for calculations of the depth distribution of concentrations is Eq. (10.18). As already shown, the mean free path length of photoelectrons in a solid depends on their kinetic energy, which, in turn, depends on the photon energy $h \cdot \nu$ (see Eq. (10.9)). With a tunable light source (usually a synchrotron), one can measure the emission from a certain core level with different photon energies, but the measured intensities must be weighted with the cross sections at the respective energy. While at low photon energies, that is, low kinetic energies (40–80 eV), the photoelectrons from surface-near atoms dominate the signal intensity, with increasing photon energy, and thus increasing kinetic energy, electrons may escape from deeper and deeper layers, as illustrated in Figure 10.35a.

In turn, from Eq. (10.18) one can also see that at *constant* photon energy, the effective attenuation length z depends on the take-off angle θ_e between the surface-normal and the analyzer entrance:

$$z = \lambda \cos \theta_e,$$

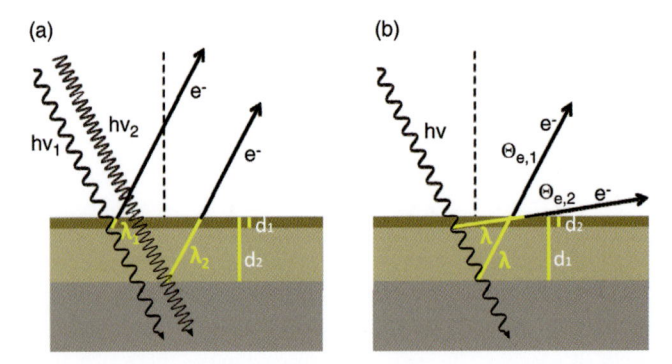

Figure 10.35 Schematic representation of non-destructive "concentration depth profiling", taking advantage of (a) the energy dependence of the mean free path length λ (see Figure 10.13) and (b) the "effective electron escape depth" as a function of the exit angle θ_e.

where is the mean free path length for the studied photoelectrons at the given kinetic energy.

Routinely, XPS spectra are acquired with $\theta_e \approx 0$ (normal emission), because this maximizes the effective electron escape depth and, thus, the detected layer thickness and overall signal intensity. With increasing θ_e, however, the effective escape depth decreases (Figure 10.35b) and the measured signal becomes more sensitive to the surface-near region. This is exploited in Figure 10.36: A $Ni_{0.76}Fe_{0.24}(100)$ alloy surface was exposed to 41, 91, and 141 L of oxygen at room temperature, and the resultant $Ni2p_{3/2}$ and $Fe2p_{3/2}$ XPS signals measured at two different exit angles $\theta_e = 77°$ and $47°$ [24,49]. As can be seen in the spectra of the oxidized sample, at grazing exit ($\theta_e = 77°$) both the $Ni2p_{3/2}$ and the $Fe2p_{3/2}$ signal encompass strong Ni^{2+} and $Fe^{2+/3+}$ contributions from the oxidized metals. At less grazing exit ($\theta_e = 47°$), however, the $Ni^{2+}(2p_{3/2})$ component becomes much less visible than the $Fe^{2+/3+}(2p_{3/2})$ signal. This clearly indicates that room-temperature oxidation of this sample leads to a preferential oxidation

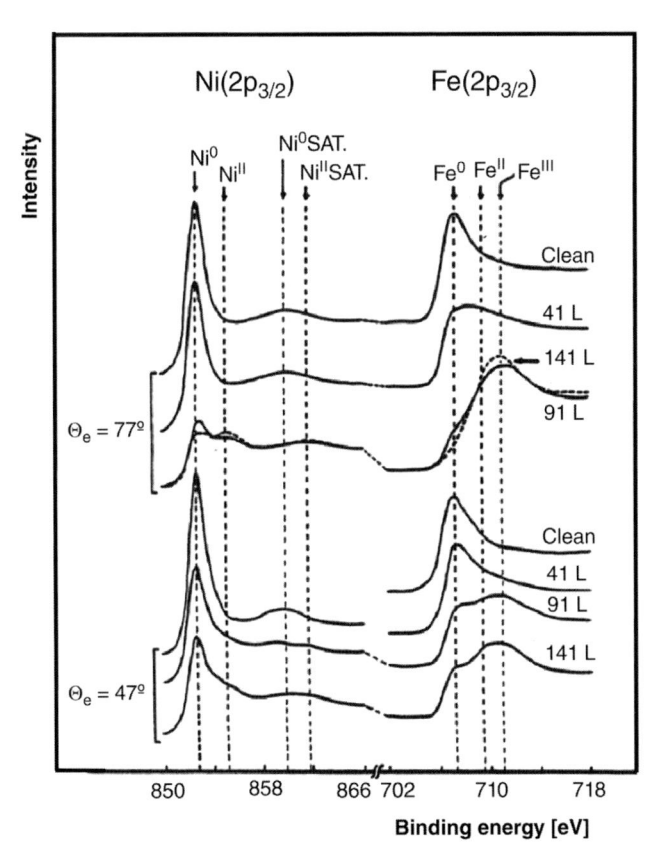

Figure 10.36 $Ni2p_{3/2}$ and $Fe2p_{3/2}$ XPS spectra from a Ni76%Fe24%(100) alloy surface after the indicated oxygen exposures at 293 K. The spectra taken at $\theta_e = 77°$ are more surface sensitive than those measured at $\theta_e = 47°$.

of the iron, in agreement with the difference in heat of formation of iron versus nickel oxide (see also Figure 10.18). The effect is more pronounced at elevated temperatures (300 °C) and leads even to preferential adsorption-induced iron segregation and enrichment of iron oxide at the surface [24,49].

Limited information about the depth of an atom can even be obtained from spectra measured neither with variable energy nor variable angle. Often one spectrum encompasses more than one signal related to one and the same element in the sample. As known from the "universal curve" (Figure 10.13), electrons with lower kinetic energy have a lower escape depth than those with high kinetic energy. If the signals of an element are spread over a wide kinetic energy range, they carry information from different depths and provide a few points of an existing concentration profile, or at least enable a distinction between a predominantly surface, bulk, or uniform distribution of that element.

10.5
Possible Misinterpretations

Some features in PES and AES spectra arising from beam damage, charging effects, ghost lines, and X-ray satellites may lead to misinterpretations.

It is, among others, the high energy of the primary photon beam as well as the excited electrons that calls for the observation of some precautions in the evaluation of PES data. First, UV radiation and even more so X-rays are energetic enough to destroy molecular bonds by virtue of photodissociation. Secondary electrons of ≈ 10 eV are able to overcome the work function, that is, leave the solid, and may also dissociate chemical bonds of adsorbed molecules.

As a consequence, one has to be aware that both the primary photons and the secondary electrons may cause chemical alterations (radiation damage) of the sample during the experiment. As an example, the S2p spectra in Figure 10.19 show the coexistence of SO_4^{2-} and SO_3^{2-} anions on a Cu(111) electrode surface emersed from sulfuric acid solution; the SO_3^{2-} species is only a consequence of radiation damage, as verified by varying the irradiation time. This can only be minimized by reducing the photon flux and/or the registration time t. Measurements of different duration may then enable to extrapolate back to the status of the sample at $t = 0$. Moreover, the high number of secondary electrons (per primary higher energy electron) leaving the sample may cause charging of nonconducting samples, that is, the built-up of a positive charge, which retards outgoing electrons and thereby distorts the spectrum, in particular causing a reduction of the kinetic energy of all emitted electrons, and, as a consequence, feigns a higher binding energy of all peaks.

Two approaches are common to account for or compensate such charging effects. One way is to have a "reference signal" within the measured spectrum, whose correct binding energy is known, but is shifted like all other peaks. The measured shift of this peak may then be used to correct all other binding energies. To this end, insulating samples are often evaporated with traces of gold, and the Au4f signal is used as reference. But also the C1s level (as a ubiquitous contamination peak) can be used (see Figures 10.21 and 10.22). Another way is to "shower" the sample with low-energy electrons from a so-called flood gun in order to compensate the positive charge on the sample.

Quite different in nature are features in the spectra arising from ill-defined primary photon beams. For instance, nonmonochromatized AlK$_\alpha$ radiation actually consists of several satellite lines K_{α_2}, K_{α_2}, K_{α_3}, and so on whose energy is higher than that of the main K_{α_1} – line at 1486.6 eV by up to 23.6 eV. Each of these lines also produces a full photoemission spectrum shifted according to Eq. (10.9) *to lower* binding energy. However, only the satellite line at 1495.9 eV has an appreciable intensity compared to the main K_{α_1} – line (see ∗ in Figures 10.6 and 10.7). In principle, any contribution of this satellite(s) to the original spectrum can be eliminated by subtracting an accordingly shifted and weighted spectrum from the measured one.

Moreover, so-called ghost peaks may arise from contaminations of the X-ray anode material. Like the pure anode material, these contaminations also act as X-ray source but of different photon energy, which according to Eq. (10.9) again produce shifted photoemission signals. Commonly, ghost peaks are weak and arise from copper, silver, or oxygen X-ray lines. Coolable copper or silver tubes are used as supporting material for the Al and Mg coatings (see Figure 10.2). Long-time use of these anodes may partly destroy and oxidize the anode materials. Also, Mg/Al cross-contamination may happen. A possibility to distinguish ghost lines from "real signals" is to compare the spin–orbit splittings of dubious weak signals with those of intense ones.

It is important, however, to emphasize the difference between photoemission signals arising as ghost lines or from X-ray satellites, on the one hand, and shake-up photoemission satellite peaks due to "relaxation effects" discussed in Section 10.3.1 on the other. While the former results from limitations or imperfections of the experimental setup, shake-up peaks are intrinsic features of photoemission spectra due to the response of the remaining $(N-1)$ electron system upon emission of an electron and creation of a hole state. As a thumb rule, X-ray satellite lines produce photoemission peaks at lower binding energy, while shake-up satellite signals occur at higher binding energy than the main peak.

References

1 Moseley, W.G.J. (1913) *Philos. Mag.*, **XCIII-26**, 1024.
2 Hertz, H. (1887) *Ann. Phys.*, **267**, 983.
3 Hallwachs, W. (1888) *Ann. Phys. Chem.*, **269**, 301.
4 Einstein, A. (1905) *Ann. Phys.*, **322**, 132.
5 Siegbahn, K. (1970) *Philos. Trans. R. Soc. Lond.*, **A268**, 33.
6 Siegbahn, K. *et al.* (1969, 1971) *ESCA Applied to Free Molecules*, North-Holland Publishing, Amsterdam.
7 Wendin, G. (1981) *Breakdown of One Electron Pictures in Photoelectron Spectra*, Springer, Berlin.
8 Hüfner, S. (2003) *Photoelectron Spectroscopy: Principles and Applications*, Springer, Berlin.

9. Cardona, M. and Ley, L. (1978) *Photoemission in Solids I + II*, Springer, Berlin.
10. Osterwalder, J. (2012) Photoelectron spectroscopy and diffraction, in *Surface and Interface Science, Vol. 1: Concepts and Methods* (ed. K. Wandelt), Wiley-VCH Verlag GmbH, Germany.
11. Turner, D.W., Baker, C., Baker, A.D., and Brundle, C.R. (1970) *Molecular Photoelectron Spectroscopy*, Wiley-Interscience, New York.
12. Schlag, E.W. (1998) *ZEKE Spectroscopy*, Cambridge Press, ISBN 978–0521675642.
13. Franck, J. and Hertz, G. (1914) *Verh. Dtsch. Phys. Ges.*, **16**, 457.
14. Escudero, C. and Salmeron, M. (2013) *Surf. Sci.*, **607**, 2.
15. Winter, B. and Faubel, M. (2006) *Chem. Rev.*, **106**, 1176.
16. Gorgoi, M., Svensson, S., Schäfers, F., Öhrwall, G., Mertin, M., Bressler, P., Karis, O., Siegbahn, H., Sandell, A., Rensmo, H., Doherly, W., Jung, C., and Eberhardt, W. (2009) *Nucl. Instrum. Methods Phys. Res.*, **A601**, 48.
17. Nordling, C. (1971) *J. Phys. Colloq.*, **32**, C4–254.
18. Wandelt, K. (1987) The local work function of thin metal films: definition and measurement, in *Thin Metal Films and Gas Chemisorption, Studies in Surface Science and Catalysis*, vol. **32** (ed. P. Wissmann), Elsevier, Amsterdam, p. 280.
19. Grüne, M., Pelzer, T., Wandelt, K., and Steinberger, I.T. (1999) *J. Electron Spectrosc. Relat. Phenomena*, **98–99**, 121.
20. Schmitz-Hübsch, T., Oster, K., Radnik, J., and Wandelt, K. (1995) *Phys. Rev. Lett.*, **74**, 2596.
21. Brundle, C.R., Chuang, T.J., and Wandelt, K. (1977) *Surf. Sci.*, **68**, 459.
22. Seah, M.P. and Dench, W.A. (1979) *Surf. Interface Anal.*, **1**, 2.
23. Ertl, G. and Küppers, J. (1985) *Low Energy Electrons and Surface Chemistry*, Wiley-VCH Verlag GmbH, Weinheim.
24. Wandelt, K. (1982) *Surf. Sci. Rep.*, **2**, 1.
25. Auger, P. (1923) *C. R. Acad. Sci.*, **177**, 169.
26. Seah, M.P. (2003) Chapter 13, in *Surface Analysis by Auger and X-ray Photoelectron Spectroscopy* (eds D. Briggs and J.T. Grant), IM Publication, Chichester, p. 345.
27. Varlson, T. (1978) *Photoelectron and Auger Spectroscopy*, Plenum Press, New York, ISBN 978-1-4757-0120-3.
28. Childs, K.D., Carlson, B.A., Vanier, L.A., Moulder, J.F., Paul, D.F., Stickle, W.F., Watson, D.G., and Hedberg, C.L. (eds) (1995) *Handbook of Auger Electron Spectroscopy*, Physical Electronics, Eden Prairie, MN.
29. Gregory, B.W. and Stickney, J.L. (1991) *J. Electroanal. Chem. Interfacial Electrochem.*, **300**, 543.
30. Gelius, U. (1974) *Phys. Scr.*, **9**, 13.
31. Baraldi, A., Bianchettin, L., Vesselli, E., de Gironcoli, S., Lizzit, S., Pellaccia, L., Zampieri, G., Comelli, G., and Rosei, R. (2007) *New J. Phys.*, **9**, 143.
32. Baraldi, A., Lizzit, S., Comelli, G., Kiskinova, M., Rosei, R., Honkala, K., and Norskov, J.K. (2004) *Phys. Rev. Lett.*, **93**, 046101-1.
33. Ganguglia, M.V., Scheffler, M., Baraldi, A., Lizzit, S., Comelli, G., Paolucci, G., and Rosei, R. (2001) *Phys. Rev.*, **B63**, 295415.
34. Conrad, H., Ertl, G., Küppers, J., and Latta, E.E. (1975) *Solid State Commun.*, **17**, 497.
35. Kercher, T.C. and Müller, S. (2014) Surface properties of alloys, in *Surface and Interface Science, Vol. 3: Properties of Composite Surfaces: Alloys, Compounds, Semiconductors* (ed. K. Wandelt), Wiley-VCH Verlag GmbH, Weinheim, Germany.
36. Abraham, F.F. (1981) *Phys. Rev. Lett.*, **46**, 546.
37. Bligaard, T. and Norskov, J.K. (2007) *Electrochim. Acta*, **52**, 5512.
38. Kono, S., Golf Berg, S.M., Hall, N.F.T., and Fadley, C.S. (1979) *Phys. Rev. Lett.*, **41**, 1831.
39. Kono, S., Fadley, C.S., Hall, N.F.T., and Hussain, Z. (1978) *Phys. Rev. Lett.*, **41**, 117.
40. Auwärter, W., Kreutz, T.J., Gerber, T., and Osterwalder, J. (1999) *Surf. Sci.*, **429**, 229.
41. Doniach, S. and Sunjic, M. (1970) *J. Phys. Chem. Solid State Phys.*, **3**, 285.
42. Shirley, D.A. (1972) *Phys. Rev.*, **55**, 4709.
43. Tougaard, S. (1986) *Phys. Rev.*, **B34**, 6779.
44. Scofield, J.H. (1976) *J. Electron Spectrosc.*, **8**, 129.
45. Jablonski, A. and Powell, C. (2012) Spectroscopic data bases and standardization for Auger-electron spectroscopy and X-ray photoelectron

spectroscopy, in *Surface and Interface Science, Vol. 1: Concepts and Methods* (ed. K. Wandelt), Wiley-VCH Verlag GmbH, Weinheim, Germany.

46 Jablonski, A. (2009) *Surf. Sci.*, **603**, 1342.

47 Chuang, T.J. and Wandelt, K. (1979) *Surf. Sci.*, **81**, 355.

48 Chuang, T.J., Brundle, C.R., and Wandelt, K. (1978) *Thin Solid Films*, **53**, 19.

49 Brundle, C.R., Silverman, E., and Madix, R.J. (1979) *J. Vac. Sci. Technol.*, **16**, 474.

11
Recent Developments in Soft X-Ray Absorption Spectroscopy

Alexander Moewes

University of Saskatchewan, Department of Physics & Engineering Physics, 116 Science Place, Saskatoon, SK S7N 5E2, Canada

11.1
Introduction

X-ray absorption spectroscopy (XAS) is one of the oldest and most widely employed spectroscopy techniques today. The principles emerged shortly after Röntgen's discovery of X-rays in 1895 [1,2], while the discovery and naming of the K- and L- lines occurred in 1911 [2,3]. The first tunable synchrotron light source was built in the 1940s, and while their use in the study of materials dates back to roughly 50 years ago, it was only in the 1980s that soft X-ray absorption spectroscopy rose to its current popularity. Today, XAS is routinely applied at all synchrotron sources to solve scientific problems in physics, chemistry, material sciences, life sciences, geology, medicine, and engineering. The technique is powerful because it is element specific, nondestructive, and highly sensitive to the local bonding environment and geometric structure surrounding the absorbing atom as well as to spin, orbital angular momenta, and the polarization of the exciting radiation.

When recording an absorption spectrum, the energy of the incoming photons is typically scanned across one or several binding energy ranges. The number of photons absorbed in a sample is measured as a function of the excitation energy, and the number of transmitted photons N depends only on the number of incoming photons N_0, the thickness of the sample d, and a property of the material being studied, the *linear absorption coefficient* $\mu(E)$, where E is the incident photon energy. This is expressed in the well-known Beer–Lambert law:

$$N(E) = N_0 e^{-\mu(E)d}, \tag{11.1}$$

which was established for visible light long before X-rays were discovered. An absorption spectrum typically displays either the number of transmitted photons $N(E)$ or the linear absorption coefficient $\mu(E)$ as a function of incoming photon energy.

Handbook of Solid State Chemistry, First Edition. Edited by Richard Dronskowski, Shinichi Kikkawa, and Andreas Stein.
© 2017 Wiley-VCH Verlag GmbH & Co. KGaA. Published 2017 by Wiley-VCH Verlag GmbH & Co. KGaA.

Historically, one divides an absorption spectrum into two distinct regions: the immediate vicinity of an absorption edge – labeled X-ray absorption near-edge structure (XANES), which is basically the same as near-edge X-ray absorption fine structure (NEXAFS), and extended X-ray absorption fine structure (EXAFS). An EXAFS spectrum typically extends about 500 eV above the absorption edge under study. The EXAFS process is dominated by single scattering of high-energy photoelectrons off of the potential of the neighboring atoms and the NEXAFS spectrum is caused by multiple scattering of low-energy photoelectrons from the valence potential of the neighboring atoms or excitations into localized bound unoccupied states of the participating atoms (or molecules). Consequently, the two regions probe different parameters of the material. With NEXAFS/XANES, one routinely determines oxidation state, ligation, and local geometry. EXAFS is employed to obtain bonding distances to the nearest neighbors [4,5].

Today, XANES is often referred to when hard (high-energy) X-rays are used, while NEXAFS is the term employed for soft (low-energy) X-rays. EXAFS is almost exclusively limited to hard X-rays, although there have been a few attempts to probe EXAFS with soft X-rays – mostly at the O K-edge (see, for example, Refs [6–8]). The Fourier transformation of the extended absorption range – called the EXAFS oscillation – requires an unperturbed spectrum of up to at least 500 eV above the threshold of the absorption edge. This can be challenging to obtain since, for example, a multielement sample will often have absorption edges from one element overlapping the EXAFS region of another element in the soft X-ray range. It is usually also experimentally difficult to cover such an extended range using only one dispersive optical element in a soft X-ray monochromator where diffraction gratings are employed. Switching gratings during an EXAFS scan is undesirable, as doing so introduces normalization problems associated with the varying transmission of different optical elements.

The different experimental requirements on vacuum, beamline optics, and detectors for soft and hard X-rays have historically established another natural way of categorizing XAS by the energy range of the absorbed photons, therefore distinguishing soft XAS from hard XAS. Nonetheless, it is worth keeping in mind that, due to the different electron binding energies required to excite inner and outer electrons, the different electronic transitions involved as well as the difference in the scattering mechanisms of photoelectrons, soft and hard X-rays probe entirely different parameters.

The interaction of electromagnetic radiation with matter is the general process that allows for the study of materials. Two different aspects of this interaction are especially important for putting any experiment in perspective. First, any given experiment can only represent a small fraction of all the interaction processes associated with an impinging photon. Absorption is only one of the many processes that can occur since impinging radiation can also be inelastically or elastically scattered, refracted, diffracted, and reflected. Second, even when considering only the absorption of photons, these photons cause discrete excitations in the system under study that subsequently decay and lead to the emission of one (or more typically, several) of the following products: electrons (Auger and photoelectrons), photons (fluorescence, luminescence, reflected, scattered, diffracted, and refracted), and other fragments ranging from ions of different ionization levels to desorbed neutral atoms and molecules. Therefore, any one specific technique used to measure absorption will typically only take one of these decay products into account since electrons, ions, and photons are detected differently.

Finally, the energy range of an absorption spectrum only probes one or a few binding energy thresholds and is, for the vast majority of systems, therefore only a small fraction of the energy range in which a material offers transitions. For example, even a relatively light element like Al features excitations ranging from intraband transitions (1–2 eV) to absorption at the K-edge (1560 eV), therefore spanning three orders of magnitude in energy.

When designing any absorption experiment, the key question will be which scientific problem one would like to solve, and therefore which transitions one aims to probe. Other key questions following from this are the spectral resolution required and which measurement technique and detector to employ. Specific questions can also include length and time scales that are to be probed. Table 11.1 shows an overview of the different fundamental excitations and their

Table 11.1 Fundamental excitations and their approximate corresponding energy ranges.

Process	Energy range
Due to free electrons:	
Intraband transitions	≤ Bandwidth
Single-electron oscillations (plasmaoptic)	≤0.5 meV
Collective electron oscillations (plasmons)	≈2–17 eV
Due to bound electrons:	
Inner (core) electron shells	≈4 eV – 98.5 keV
Intrashell excitations (d–d, f–f)	≈100 meV–4 eV
Interband transitions (VB → CB):	≈10 meV–11 eV
Absorption from localized states:	
Excitons	≈E_{Bind} few meV up to few eV
Defects (impurities, vacancies, or interstitials)	Absorption in visible or UV
Spin–orbit splitting (for TM 2p and RE 3d and 4d)	≈3–48 eV (5–23 eV for TM 2p, 19–48 eV for RE 3d, and 2.8–8.8 eV for RE 4d)
Spin waves (magnons)	≈10–400 meV
Charge transfer excitations	≈Up to 12 eV
Correlation effects	≈order of eV
Photon–phonon coupling:	
Photons–acoustic phonons (Brillouin scattering)	≈10^{-6}–0.1 meV
Photons–optical phonons (Polariton scattering)	≈40–70 meV

TM stands for the transition metals (of the fourth period) and RE stands for the rare earth elements.

energy scales. When including low-energy excitations such as phonons (or molecular vibrations), the energies for possible excitations in a solid (or molecule) can extend over up to six orders of magnitude. This again emphasizes that any specific absorption experiment probes only a small subset of all possible processes in terms of both – the excitation processes and the resulting particles that are emitted.

This chapter will focus on the absorption of soft X-rays. The definitions of this energy range vary, but the soft X-ray regime is typically considered as being between 50 and 2000 eV (corresponding to wavelengths of about 0.5–25 nm). The energy range of soft X-rays covers the binding energies of all 1s electrons (K-shell) for the lighter elements ($3 \leq Z \leq 14$) and higher lying electron shells for all heavier elements. The outer electrons and the corresponding electronic states are the ones that govern many different properties of a material, such as chemical bonding, atomic and electronic structure, magnetism, electric and heat conductivity, color, bandgap, and hardness. By probing these unoccupied electronic states with soft XAS, one therefore has access to all these parameters. This is another reason why soft XAS has become so widely distributed.

In the soft X-ray regime, photon absorption occurs almost exclusively through photoexcitation and photoionization with matter. Coherent scattering (Rayleigh) and incoherent scattering (Compton) have about 10^3 and 10^6 times lower cross sections, respectively [9], and are negligible in the soft X-ray range. Due to the inherently low fluorescence yields in the soft X-ray range – for the K-shells (10^{-4}–0.6), L-shells (2×10^{-4}–3×10^{-2}), and M-shells (5×10^{-5} to 2×10^{-2}) – the coherent scattering can contribute to the absorption signal when measuring photon yield and especially when measuring atoms of low concentration in a matrix. But it is generally difficult to distinguish or remove this contribution. This can be attempted with an energy dispersive detector or by positioning of the detector and taking advantage of the spatially different emission characteristics of fluorescence (which are different for K- and L-shells) and coherently scattered photons (which depend on polarization geometry). For the dichroic absorption techniques (see Section 11.2.5), it is impossible to remove the coherent contribution.

Soft X-rays, therefore, interact strongly with matter, which is why this radiation only travels comparatively short distances in materials. Although the attenuation length of the incoming radiation can span orders of magnitude in a given material and can range from tens of nanometers for lower energies up to several micrometers at higher energies, the attenuation length is generally much shorter than for harder X-rays.

Even when considering only soft X-rays, there is already an extraordinary body of work in review (and regular) articles that focus on X-ray absorption spectroscopy of specific research topics such as nanomaterials [10], molecules [11], organic and magnetic compounds [12], transition metal oxide heterostructures [13], C and Si nanostructures [14], polymers [15], minerals and geological systems [16], soils [17], glasses [18], liquids and organic thin films [19], as well as the underlying theory [20], to name but a few. The history of XAS has been described in detail [2]. There are also excellent books that focus on experimental soft X-ray NEXAFS [21] as well as on the corresponding theory [22]. Today, there is almost no research field in materials science and no stable element from H to U where XAS has not been applied.

Finally, it is necessary to realize that since the absorption of a photon and the subsequent photoexcitation serves as the starting point for various other spectroscopy techniques where typically the decay of the excited state is monitored, XAS plays an integral role in other spectroscopy techniques. These include X-ray emission spectroscopy and resonant inelastic X-ray scattering [23–25], as well as resonant elastic X-ray scattering [26], X-ray microscopy [27], and photoelectron and Auger electron spectroscopy [28].

11.1.1
Different Measurement Techniques: Electron Yield and Fluorescence Yield

Outside of any absorption edge, a good approximation for the absorption coefficient $\mu(E)$ as a function of energy E is

$$\mu(E) \approx \frac{\rho Z^4}{AE^3}, \tag{11.2}$$

where ρ is the density of the solid, Z is the atomic number, and A is the atomic mass. Semiempirical values for $\mu(E)$ and the *attenuation length* $d = 1/\mu$ have been obtained for various combinations of parameters (element, energy, incidence angles, densities, etc.) [29], but they are reliable only for energies that are not close to the respective absorption threshold.

If one considers as a simple rule an attenuation length for soft X-rays of about 100 nm as the typical upper limit for many solids, it becomes clear that the relatively short penetration depths of soft X-rays make true absorption experiments – that is, measuring the absorption coefficient in a transmission experiment – difficult for most systems. It is nearly impossible to obtain freestanding samples with uniform thicknesses of this magnitude. This is why soft X-ray absorption in most cases has to rely on secondary processes such as fluorescence or electron emission to be measured. There are a few light elements where the attenuation length is significantly larger above the respective edges, but in any case when an attenuation length for a specific material is needed, one can consult the Web site tool in Ref. [29] since the values fluctuate strongly between materials and photon energy. For example, the attenuation length for Si is 40 nm after the L-edge (or 128 eV), but 1.3 μm above the K-edge (or 1850 eV). The light elements or combinations of them are typically used in materials such as diamond and Si_3N_4 that are used as window material in the soft X-ray range.

In an X-ray absorption transition, an inner electron is excited to another bound state – a localized orbital or a conduction band state – leaving behind a vacant core level often referred to as a *core hole*. This core hole state normally decays almost immediately (with femtosecond lifetimes) and via different processes, all of which can be used to monitor the absorption. The core hole is refilled by an electron from a state higher in energy (either another core level or a valence state) and the energy released in this process can be absorbed by other electrons or (much less likely for lighter elements) emitted as an X-ray. In the former relaxation process, excited electrons near the surface of a sample in ultra-high vacuum (UHV) can be ejected, leaving the surface with a net electric charge. If the sample is grounded, the current drawn from the ground to replenish the charge can be measured with a picoammeter, and the magnitude of this current is then related to the magnitude of the X-ray absorption. This XAS technique is referred to as *total electron yield* (TEY) and it detects all electrons (primary and secondary) removed from the sample. Although in principle relatively simple to monitor, TEY measurements can be inhibited in materials with large bandgaps and poor electron mobility. A large bandgap hampers the movement of charge and can therefore reduce the measured signal strength as the absorption scan proceeds. This phenomenon is referred to as *charging* of the sample. The large number of low-energy electrons emitted can be suppressed by using additional retarding voltages between the sample and electron detector. This method, referred to as partial electron yield (PEY), typically increases the signal to noise ratio but decreases the probing depth to below 5 nm. A modification is the Auger electron yield (AEY) technique, which only detects elastically scattered Auger electrons. AEY requires an electron energy analyzer but provides the highest surface sensitivity of the electron yield measurement techniques.

In the latter of the above relaxation processes, a nonenergy dispersive detector like a channel electron multiplier or several multipliers stacked in a channel plate can measure the emitted X-rays. In this process, the total number of emitted X-rays is proportional to the number of core holes produced and therefore to the X-ray absorption. This XAS technique is called *total fluorescence yield* (TFY). If an energy dispersive detector or a spectrometer is used to integrate over a partial energy range of the fluorescence, a spectrum is measured in the *partial fluorescence yield* (PFY) mode. A sketch of the principal techniques to measure soft XAS is shown in Figure 11.1.

Distortions of the measured spectra can occur when the measured electron yield or fluorescence yield is not proportional to the total (linear) absorption coefficient. This can be for a number of reasons but one generally has to consider two aspects only: the energy-dependent penetration depth for the incoming photons and the escape depth for the electrons in electron yield or the photons in fluorescence yield. Subsequently, the strong variations of the absorption coefficient across an absorption edge effectively lead to very different sample volumes being probed.

In electron yield measurements the penetration depth of the incoming radiation can reach the escape depth of the electrons leading to saturation of the signal. An extreme case occurs when all incoming photons produce photoelectrons that escape the sample, making the signal independent of photon energy.

X-rays have a much longer escape depth than electrons (TFY can probe hundreds of nanometers to micrometers compared to electron escape depths of typically much less than 10 nm), so fluorescence yield measurements are bulk sensitive. There is however the possibility that emitted X-rays will be reabsorbed

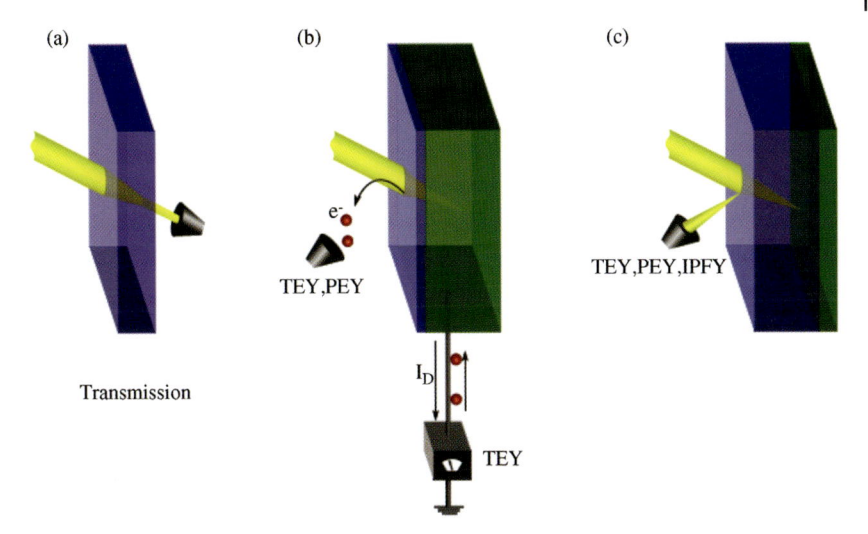

Figure 11.1 Schematics of the different measurement modes for soft X-ray absorption spectra: transmission (a), electron yield (TEY, PEY) (b), and fluorescence yield (TFY, PFY, IPFY) (c). (Reprinted with permission from Ref. [33]. Copyright 2013, University of Saskatchewan.)

by the sample before reaching the detector. This *self-absorption* of X-ray fluorescence photons can suppress the peak height (or contrast) in the measured fluorescence yield spectra of dense systems and the fluorescence yield is not proportional to the linear absorption coefficient any more. In principle, this can be corrected when the absorption spectrum is well known using the known angular dependence of the fluorescence yield. But there are systems where inelastic scattering contributes strongly and this can currently not be accounted for when correcting for self-absorption. However, it is important to realize that the interplay of attenuation length of incoming photons and escape depth of outgoing signal (electrons or photons) is responsible for the distortions discussed above. The strength of this effect depends on concentration and kind of the probed atoms as well as sample thickness and applies to different scenarios for electron and fluorescent yield measurements.

A different problem for electron yield can be *sample charging*, which can occur when the electrical conductivity of the sample is low enough and the incoming photon flux high enough that the photoelectrons cannot be replenished at the rate necessary to maintain an electrically neutral sample. Electron emission in this case will be reduced (with exposure time). Radiation damage can lead to chemical changes with exposure in the sample and therefore can change the measured spectra as well.

Other problems that are typically less important are that both yield techniques depend on the relative probability that the excited atom decays (by emitting electrons or photons). This depends on atom and edge and is usually not exactly known. The detection of the signal can be nonuniform as well. A nonenergy-dispersive fluorescence detector that registers photons of different energy with different efficiency and electric and magnetic fields present can influence how electrons are detected. When different absorption edges present in the sample are close in photon energy (for example, L-edges of Ti, V, and O K-edge), it can be challenging to decouple the different contributions. To summarize, XAS measurements with different yield techniques will not necessarily produce the same result. A comparison of different techniques (to measure the bandgap) is shown in Ref. [31]. For a comparison of various XAS spectra taken with TEY, IPFY (see Section 11.2.2), and TFY, see Ref. [32], and for AEY, PEY, and FY, see Ref. [33].

When a TFY (and/or PFY) spectrum matches the TEY spectrum, the surface of the sample is generally free of contamination and sample charging (in TEY) and self-absorption (in TFY) effects are not significant problems. In this case, the surface electronic states are therefore representative of those in the bulk and the XAS spectra can be trusted to be representative of the true sample. However, agreement of this sort is not often the case. Self-absorption is typically not a significant problem with light nonmetal K-edge spectra (i.e., those from B, C, N, O, and F), so the TFY mode is often used when measuring those edges. Self-absorption and surface oxidation become major problems for transition metal $L_{2,3}$-edge spectra, so great care must be taken to obtain trustworthy XAS measurements. Which absorption technique is applied for a specific sample depends on the requirements of the measurement in terms of probing depth, signal to noise ratio, and sample conductivity.

Electron and fluorescence yield are the most common techniques for measuring absorption spectra. Both take advantage of the fact that the decay products – either the number of electrons (predominantly secondary but also photoelectrons and Auger electrons) or the number of fluorescence photons emitted – are proportional to the number of holes produced in the photoexcitation of the absorption threshold under study. One should keep in mind that in principle other secondary particles such as emitted ions or – if of sufficient intensity – even desorbed atoms or molecules as well as reflected and elastically scattered radiation could be used to monitor the absorption.

Two additional aspects need to be mentioned that are routine but important for XAS. First, XAS requires one to determine the ratio of the number of transmitted photons N to the number of incoming photons N_0 (see Eq. (11.1)). The normalization of the measured signal with respect to the incoming signal is necessary to make the measured spectrum independent of photon number fluctuations that can occur with time, which are typically caused by instabilities in the source. It also eliminates the effect of flux variations with photon energy that occur due to the efficiency characteristics of the monochromator optics (these are diffraction gratings and mirrors in the soft X-ray range). In most cases a measurement of the incoming flux is achieved by detecting the photocurrent of a highly transparent gold mesh at the sample end. Another option is using the photocurrent generated by the incoming photons on an optical element – typically a focusing mirror – close to the sample. Second, the energy calibration of the spectra from a certain beamline might not be exact or reproducible due to backlash in the motion of the optical components. This problem is easily solved by comparing spectra measured from reference samples to excitation energies for the same system found in the literature. For a list of binding energies, fluorescence energies, and relative intensities, see Ref. [34].

11.2
Recent Developments

Taking the number of new publications into account, which make use of XAS each year, it becomes immediately clear that it is not possible to even attempt a review of the entire field. Instead, this chapter focuses on a few important examples that have advanced scientific understanding using soft X-ray absorption (as well as examples that have advanced the understanding and implementation of the technique itself) in the last 10 years. Accordingly, this list cannot be complete.

11.2.1
Calculation of Soft X-Ray Absorption Spectra

11.2.1.1 Notation and Formulas

The notation for X-ray absorption edges is historically named after the participating core electrons of the orbital nl_j the excitation occurs from. When 1 s electrons are excited, a K-edge spectrum is measured and the excitation of 2 s, $2p_{1/2}$, or $2p_{3/2}$ electrons gives to the L_1, L_2, and L_3 edges, respectively, and so on. Momentum conservation together with the fact that the momentum of a soft X-ray photon is very small lead to the well-known *selection rules* for dipole transitions for (one) electron transitions $\Delta l = \pm 1$ (and $\Delta L = 0, \pm 1$, $\Delta S = 0$) but these hold strictly only for *LS* coupling. For the case that spin–orbit interactions are appreciable, the rigid rules are $\Delta J = 0, \pm 1$, $\Delta M_J = 0, \pm 1$, $\pi_f = -\pi_i$ – lowercase (uppercase) symbols refer to single (multiple) electron numbers (the additional rule is that when J (L) is zero, ΔJ (ΔL) cannot be zero).

When it comes to the analysis of XAS, an absorption spectrum rarely provides information in itself. There are two main approaches in the XAS analysis – the measured spectra can be compared either to spectra of reference samples or to calculated *spectra*.

When calculating an absorption spectrum, one typically calculates the absorption transition rate $T^{abs}_{i \to f}$ (the number of transitions per second in a unit volume) for a transition from initial state i to final state f. Since nearly all sources refer to slightly different expressions, in different unit systems and usually only using proportionality expressions, it is worthwhile to show the key expressions in analytical form. This is according to the *Fermi's golden rule* in the single-electron approximation, where only one electron is participating in the transition:

$$T^{abs}_{i \to f} = \frac{2\pi}{\hbar} \left| \langle \psi_f | \widehat{H} | \psi_i \rangle \right|^2 \rho_f^{elec}, \qquad (11.3)$$

where ψ_i and ψ_f are the wave functions of initial and final states, $\hat{H}$ is the Hamiltonian operator, ρ_f^{elec} is the electronic density of states at energy of the final state, and $\hbar$ is the reduced Planck constant.

With the momentum vector operator $\hat{p}$ for the electrons and the vector potential $\vec{A}$ of the electromagnetic wave field (here in SI units), this can be rewritten as

$$T_{i \to f}^{abs} = \frac{2\pi}{\hbar} \left(\frac{q_e}{m}\right)^2 \left|\langle \psi_f | \vec{A} \cdot \hat{p} | \psi_i \rangle\right|^2 \rho_f^{elec}, \quad (11.4)$$

where q_e and m are electron mass and charge, respectively. In the dipole approximation, $\vec{A} \cdot \hat{p} \approx im\omega A_0 e^{\pm i\omega t}\vec{e} \cdot \hat{r}$, this can be rewritten as

$$T_{i \to f}^{abs} = \frac{2\pi}{\hbar} (q_e \omega A_0)^2 \left|\langle \psi_f | \vec{e} \cdot \hat{r} | \psi_i \rangle\right|^2 \rho_f^{elec}. \quad (11.5)$$

Here $\vec{e}$ is the unit vector for the polarization direction of the incoming (monochromatic) wave $A_0 e^{\pm i\omega t}\vec{e}$ with A_0 being the maximum amplitude of the vector potential of the wave and $\hat{r}$ being the position vector operator for the electron.

In order to calculate the linear absorption coefficient μ from Eq. (11.1), one has to consider the density of absorption centers n and the absorption cross section σ, the absorption coefficient is $\mu = n\sigma$. X-ray absorption arises primarily from excitation of core atomic states, that is, the initial states are discrete. The final state may be either discrete (giving rise to absorption lines) or continuous (giving rise to the continuous part of the absorption cross section), such that the total absorption cross section is $\sigma = \sigma_l + \sigma_c$.

The calculation of photon absorption cross sections proceeds through the calculation of the scattering matrix elements that yield the polarized transition rates (for the continuous final states, this yields the golden rule) [35]. The transition rates must then be summed and averaged over photon polarizations and orientations of the absorption centers. The absorption cross section from transition into discrete final states $|\psi_f\rangle$ with Lorentzian line profiles is

$$\sigma_l = \sum_{f,i} \frac{8\pi^2}{3} \alpha \omega_{fi} |\langle \psi_f | \vec{e} \cdot \hat{r} | \psi_i \rangle|^2 \frac{\Gamma}{(\omega_{fi} - \omega)^2 + \Gamma^2}. \quad (11.6)$$

Here α is the Sommerfeld fine structure constant $\alpha = (\mu_0 e^2 c)/4\pi\hbar$ and Γ is the scale parameter (2γ = FWHM) for the Lorentzian distribution. The absorption cross section from transition into continuous final states $|\psi_f\rangle$ is

$$\sigma_l = \sum_{f,i} \frac{8\pi^2}{3} \alpha \omega_{fi} |\langle \psi_f | \vec{e} \cdot \hat{r} | \psi_i \rangle|^2 \rho_f^{elec} \delta(\omega_{fi} - \omega). \quad (11.7)$$

XAS, therefore, allows probing of the *partial* density of *unoccupied* states in the presence of a core hole (or the conduction band states in condensed matter terminology). The selection rule for a change in angular momentum ($\Delta l = \pm 1$) imposes the sensitivity to the orbital momentum. However, we note that the calculation of a *spectrum* involves both, the calculation of the electronic density of states (DOS) and the calculation of the dipole matrix element. With the advent of more computing power and several codes available now for calculations, absorption spectra are now routinely computed with this full approach. Figure 11.2 shows the excitation in the one-electron picture for the N K-edge (excitation of 1 s → CB) for γ-Ge$_3$N$_4$ (top graph) along with measured and calculated absorption spectra.

11.2.1.2 Linewidth and Broadening

When comparing calculated to measured spectra, the calculated spectra are typically broadened to account for broadening effects in the measurements and facilitate a better visual comparison. The *experimental broadening* is of Gaussian shape and reflects the limitations in the optics used. This is nearly constant over the relatively small energy range of any soft X-ray absorption spectrum. There is also broadening due to the finite *lifetime* of the core hole. Linewidth Γ (in eV) and core hole lifetime τ (in s) correlate in the time–energy uncertainty principle ($\Delta E \tau \geq \hbar/2$) [35,36], leading to the following estimation:

$$\Gamma \tau \geq 3.3 \times 10^{-16} \text{ eV s}. \tag{11.8}$$

Typical linewidths are, for example, around 1–2 eV for the transition metal K-edges and about half that value for the L$_3$-edges (Mn: 0.3 eV, Fe: 0.35 eV, Co: 0.4 eV, and Ni: 0.45 eV) in good agreement with the literature [37]. Core holes therefore decay rapidly with typical lifetimes of less than femtoseconds. The core hole lifetime is usually approximated as constant for one specific threshold and it is modeled using a Lorentzian function. Next there is the *final state broadening*, due to the lifetime of the final state of soft XAS; these states typically have an electron in the conduction band – such states have an even shorter lifetime and the associated broadening is larger. This effect is also modeled with a Lorentzian function with a broadening factor ΔE. Since the different final states have different lifetimes, this factor varies with energy; it depends on the energy squared [38]. The contributions to spectral broadening due to lifetime and instrumental broadening are shown in Figure 11.3.

There is in principle also the Doppler broadening due to vibrations of the nuclei in a solid or the molecules in a gas [39], but this is mostly detectable in diatomic gases [40]. For a polar material such as the ionic SiO$_2$, this effect is present but small, but there will be no vibrational contribution to the broadening for a covalently bonded system such as Si$_3$N$_4$. In nearly all solid-state systems, this effect can be neglected in comparison to the core hole broadening. Finally, it should be noted that any isotropy present in the XAS of a crystalline (highly ordered) system will be averaged over – with all possible orientations – when studying the system in amorphous form. In an amorphous system the small changes in bond lengths (compared to a single-crystalline system) will lead to a distribution of electronic levels – occupied as well as unoccupied. This distribution of levels leads to a broadening of spectral features when going from a crystalline to an amorphous system.

It is important to keep in mind that for different elements and for different absorption edges the bound states to which the electron is excited may be

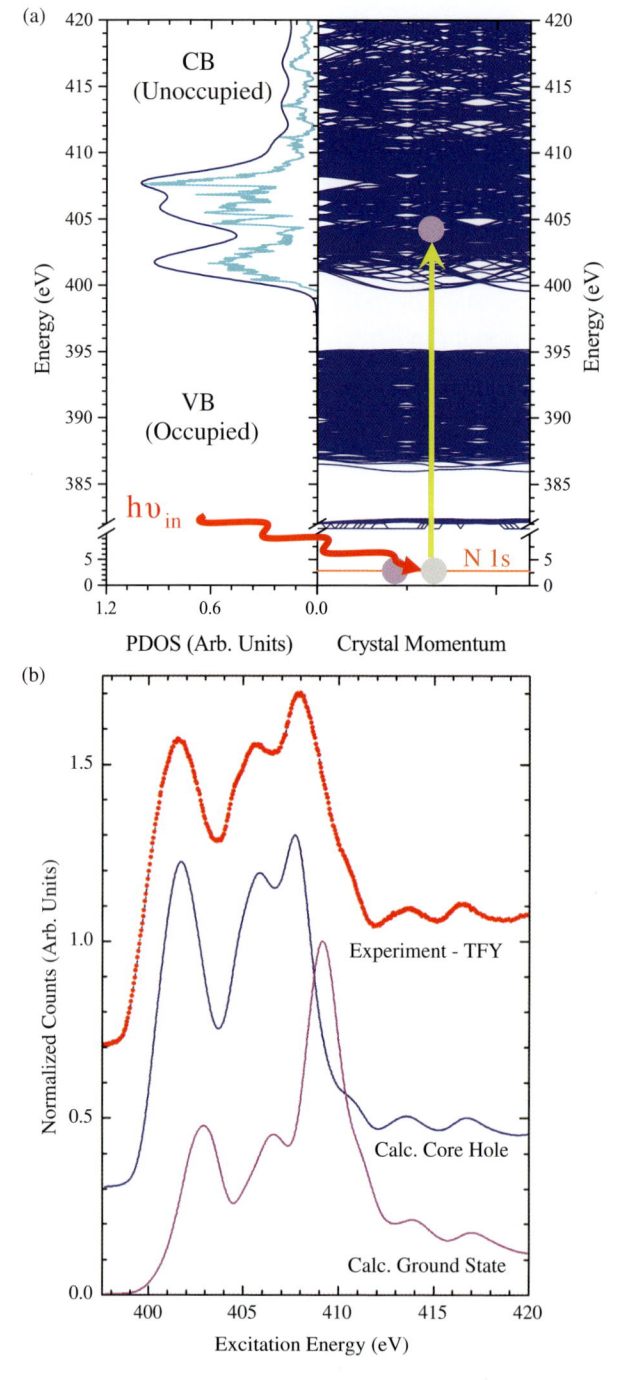

Figure 11.2 (a) The calculated band structure of γ-Ge$_3$N$_4$ (right) with the projected density of states (cyan) along with the broadened spectrum (blue). The yellow arrow depicts the transition of a N 1s electron (magenta) to the previously unoccupied density of states (conduction band) excited by the incoming synchrotron wave (red). The calculated band structure has one N 1s core hole in a 2×2×1 supercell. (b) Comparison of measured NEXAFS (red) in TFY mode and calculated spectra with (blue) and without (magenta) a N 1s core hole.

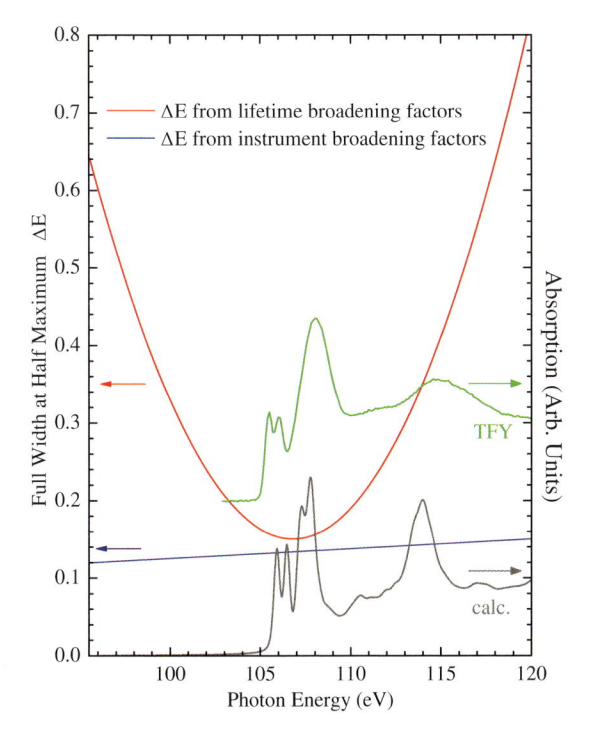

Figure 11.3 Calculated broadening contributions to spectral resolution ΔE of XAS from core hole lifetime hole (red) and instrument (blue). The displayed range of excitation energy is around the Si $L_{2,3}$ absorption threshold. The lifetime broadening is calculated using the typical dependence on photon energy ($\sim E^2$) [38] and the resolution ΔE is calculated from the constant resolving power $E/\Delta E = 800$. It is obvious that (a) with the instrumental broadening available from current monochromator diffraction gratings ($E/\Delta E > 800$), the Si L-edge XAS is dominated by lifetime broadening effects and (b) the energy dependence of the lifetime broadening only allows for sharp spectral features to appear near the minimum of the lifetime broadening curve. Calculated (gray) and measured (green) Si $L_{2,3}$ XAS for SiO_2 are shown for comparison.

dominated by different effects and therefore by either the dipole matrix element or the density of states in Eq. (11.6).

The 3d electrons of transition metals remain relatively localized in most compounds, which has implications for the information that will be obtained from XAS. This localization leads to a strong overlap of the valence 3d and the 2p core wave functions, which causes the spectra in this case to be dominated by multiplet effects [20] whereby the spectra strongly depend on the number of d electrons. This is shown in Figure 11.4. Often electron correlation effects also dominate these spectra and the single-electron picture is inadequate for describing spectra.

The situation is different when the spectra closely relate to a projected unoccupied DOS. When, for example, the transition metal elements discussed

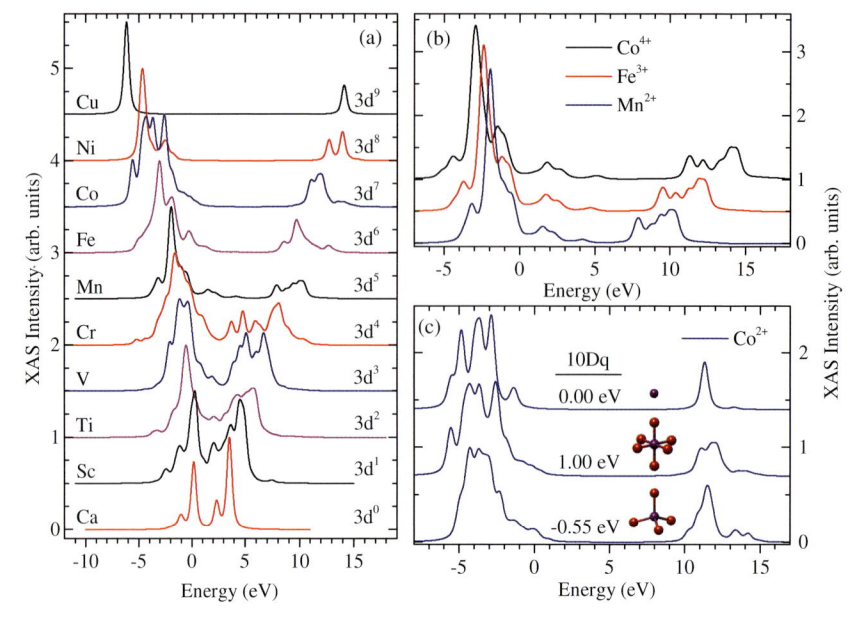

Figure 11.4 Key features of transition metal (from Ca to Cu) $L_{2,3}$ XAS demonstrated using crystal field multiplet calculations. The spectra are plotted with configuration-averaged energies centered at 0 eV for easy comparison (i.e., the different binding energies of the elements are not taken into account). (a) Spectral shapes for divalent, octahedrally coordinated atoms depend strongly on d electron count. (b) Effects of spin–orbit coupling, which gets stronger as Z increases. The general shape is very similar since the number of electrons are the same for each element (with different oxidation state). (c) The spectral shapes are very sensitive to crystal fields imposed by neighboring ligands. The different fields lead to different energy levels of the various multiplets, yielding spectra with markedly different shapes. Thus, one can often deduce the coordination of an atom from the XAS alone. The empirical crystal field splitting parameter is given for each curve. For all calculations shown in this figure, spin–orbit coupling has been included, the empirical crystal field parameter $10Dq = 1.0$ eV unless otherwise given, and intra-atomic Slater integrals are scaled to 80% of Hartree–Fock values. (Reprinted with permission from Ref. [33]. Copyright 2013, University of Saskatchewan.)

previously bond with light element ligands (C, N, O, F, etc.), the ligands are studied by exciting 1 s electrons to unoccupied bound states of predominantly p character. These valence electrons are delocalized in wide bands. In the absence of strong multiplet coupling, such K-edge spectra can be approximated by the unoccupied density of states of p-character (the last term in Eq. (11.5)). These spectra are also described well in the single-electron picture.

In the last decade, it has become common to include the effect of the core hole in the computation of absorption spectra. The final state in the absorption process differs from the ground state in that the promotion of an electron to a previously unoccupied state leaves behind a core hole and the change in potential perturbs the DOS (compared to the ground state). The effect is that the conduction band onset is lowered in energy and the spectral shape itself is changed (for a comparison of ground state and core hole spectra, see the bottom of Figure 11.2 as well [41–43]). For most solid-state systems, the effect of the core hole on the conduction band energy ranges between zero and a few tenths of an electrovolt and is thus small, but there are systems where the core hole effect can be dramatic, having values of up to 1.4 eV for $MnCN_2$ [43] and 2.17 eV for P_3N_5 [41]. When taking the core hole effect into account, a *supercell* is used when calculating spectra, consisting of multiple unit cells of the material. One electron at the core level is then removed from one of the constituent unit cells, which gives a more realistic contribution to the DOS from the core hole than one would obtain by using only one unit cell, as in this case the effect of the core hole would be exaggerated. The upper limit of the size of the supercell is usually imposed by the total number of atoms that is computationally feasible to calculate.

11.2.1.3 Different Approaches, Codes, and Optimization

The transition matrix elements in Eq. (11.6) and the density of electronic states (DOS) are today mostly calculated using density functional theory (DFT). Several software packages are currently available that are tailored to specific problems for solid-state systems such as Wien2k [44], VASP [45], ORCA [46], ADF [47], Gaussian 09 [48], or molecular codes such as StoBe [49], GSCF3 [50], and Quantum Espresso [51]. These codes differ in many ways, such as in which potentials and functionals are used, which approximations are made, and whether or not all electrons are taken into account. For rare earth and transition metal materials, a multiplet crystal field code CTM4XAS [52] will be more fitting but recently progress has also been made in the understanding of the early transition metal $L_{2,3}$ absorption spectra, solving the Bethe–Salpeter equation and taking into account the mixing of the excitation from the $2p_{1/2}$ and $2p_{3/2}$ core levels, which is triggered by the exchange term of the electron–hole Hamiltonian [53]. An interesting comparison of the reproducibility of the different solid-state codes and methods can be found in Ref. [54].

Another interesting aspect concerns the force minimization, often called structure- or geometry-optimization in solid-state codes such as Wien2k and VASP. The input parameter in such codes is the crystal structure (and space group) and this initial structure can be iteratively optimized as the code tries to minimize the forces in the structure. Atomic positions are routinely and accurately measured with X-ray diffraction (XRD). In conventionally grown crystals, XRD provides the experimentally determined crystal structure reflecting the real structure of the sample. Applying a geometry optimization in these cases is unnecessary and typically leads to changes of tenths of an Angstrom in atomic positions. There are, however, many cases where systems are too dilute to be studied with XRD (graphdyine, the functionalization of graphene or silicene, and diluted magnetic semiconductors are some examples) and consequently the exact structure remains unknown. Geometry optimization is an absolute necessity for such problems. In cases where the structure is available from XRD measurements, the difference between measured XRD structure and geometry-optimized

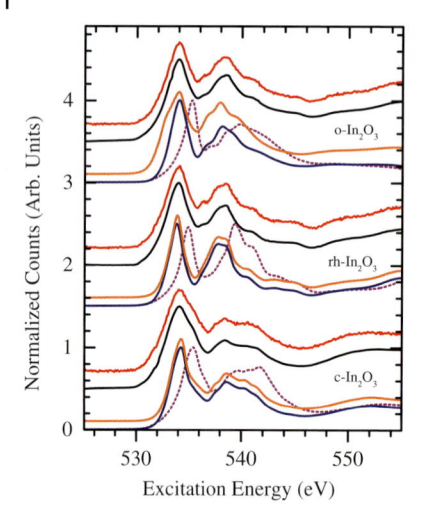

Figure 11.5 Measured and calculated N K-edge XAS for cubic, rhombohedral, and orthorhombic In_2O_3. Experimental TFY (red) and TEY (black) XAS are compared with ground-state calculations (dashed magenta) and core hole calculations, which are calculated for the atomic positions determined experimentally from XRD (blue) and for unit cells with relaxed atomic positions (orange). (Reprinted with permission from Ref. [42]. Copyright 2016, American Physical Society.)

structure can offer a way to evaluate the stress in the crystal. Systems synthesized under exotic conditions such as high pressures and temperature can have severe stress (as well as many defects and impurities) in their crystalline structures. The combined strength of these effects – along with TEY and TFY XAS – is shown in Figure 11.5 [42], where the spectra calculated for the structure experimentally determined from XRD are compared with the slightly different structure obtained from force optimization for a new rhombohedral polymorph of In_2O_3.

Another aspect of XAS is that it can be used to determine structural parameters such as atomic coordinates or the space group when these cannot be obtained by conventional XRD. This can be the case if the system is too dilute for the hard X-rays required for XRD (e.g., silicene [55,56]) or does not consist of elements that are heavy enough to allow for XRD such as graphene.

Another problem can occur in systems that consist of two or more light elements that only differ by one or two in atomic numbers. In this case, the atomic scattering amplitudes are weak and do not differ enough to allow distinguishing the two different elements. XRD will yield the atomic positions but the light atoms cannot be distinguished from one another and hence the space group remains unknown. The experimental XAS can be compared to spectra calculated for the different possible space groups, ultimately allowing determining which model fits the calculation best. This has been demonstrated for GaON, where O and N are indistinguishable using XRD [57].

A new "double cluster" model has been developed recently to understand the XAS (and magnetic diffraction) of perovskite rare earth nickelates $RNiO_3$ [58], which are highly studied materials due to their metal–insulator transition and unique antiferromagnetic ordering, where the transition temperature can be tailored with different sizes of rare earth ions. The understanding of these spectra has been a long-standing problem since they were first measured 25 years ago [59]. This model will also apply to other high oxidation state materials such as manganites and ferrites.

An interesting result concerns the transition metal interaction in Cu_2O_2 species with aromatic substrates and was established by probing the charge on the copper ions and therefore the covalent donor interactions of the peroxo- and oxo-ligands with the Cu centers [60]. Using Cu L-edge XAS, the authors directly quantified the Cu—O bonding interaction of two well-characterized P and O species. It was found that direct electron donation from the aromatic ring to the $Cu(II)_2O_2$ species leads to C—O bond formation with O—O bond cleavage. Separate aromatic hydroxylation mechanisms were also described.

To summarize, with the increases in computing power, software improvements, and the introduction of a variety of codes for electronic structure calculations, the calculation of soft X-ray spectra has blossomed in the last decade and the comparison of measured and calculated spectra has become an important tool for achieving a better understanding of the underlying physics and chemistry of many systems [61,62].

11.2.2
Recent Technical Developments

It has been known for over a quarter of a century that yield measurements often suffer from distortions that can neither be accounted for nor modeled accurately. Electron yield measurements often suffer from distortions due to saturation and sample charging (see Section 11.1.1), while fluorescence yield measurements are known to suffer from self-absorption [63,64]. A systematic study shows that the distortions are especially significant for transition metal elements because of the opening of inelastic scattering channels right below their absorption thresholds [65]. These distortions were also shown to be the strongest in the depolarized measurement geometries.

Among the recent technical innovations, the following two improvements stand out in the sense that they are important steps toward less distorted absorption spectra that will allow for better quantitative analysis. It is highly desirable to measure XAS without distortion effects because more accurate comparisons to calculated spectra will allow for the extraction of more detailed information about the systems under study.

11.2.2.1 Inverse Partial Fluorescence Yield (IPFY)
IPFY is a relatively new method to measure absorption spectra that are free from nonlinear effects such as self-absorption and saturation. It was originally demonstrated for the high-T_C cuprate $La_{1.475}Nd_{0.4}Sr_{0.125}CuO_4$ [66]. The O K_α emission was measured with an energy dispersive photon detector, while the incoming

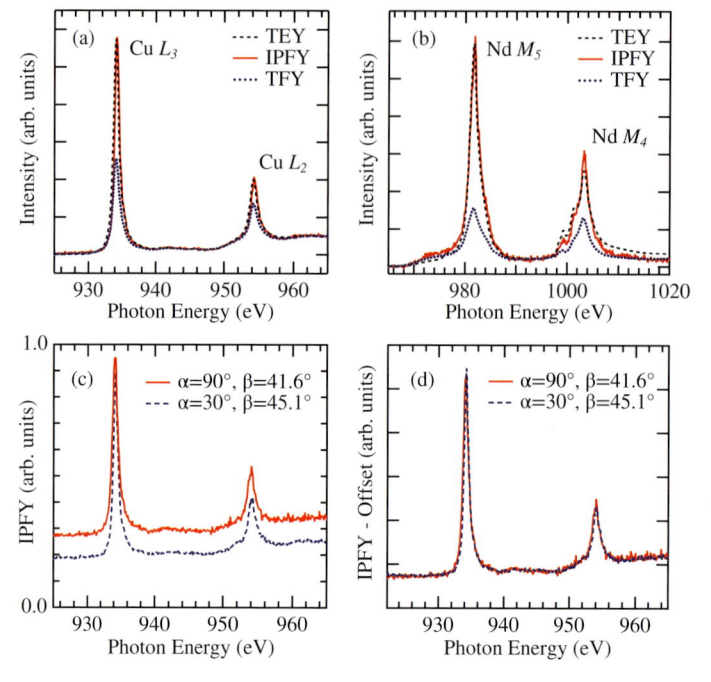

Figure 11.6 Cu $L_{2,3}$ and Nd M_4 XAS for $La_{1.475}Nd_{0.4}Sr_{0.125}CuO_4$. The TEY (dashed black), O K-edge IPFY (red), and TFY (dotted blue) as a function of photon energy through the (a) Cu L-edges and (b) Nd M-edges. The spectra are scaled and offset to match above and below the absorption edges. The IPFY and TEY spectra are in good agreement, while the TFY spectra are distorted due to self-absorption effects and NXES contributions to the TFY. (c) The Cu L IPFY measured for two different geometries. (d) The IPFY spectra offset to match at a point in the pre-edge region. (Reprinted with permission from Ref. [66]. Copyright 2011, American Physical Society.)

photon energy was scanned across the higher binding energy Cu $L_{2,3}$ and Nd $M_{4,5}$ thresholds, respectively.

The advantage of this process is that the fluorescence yield for the detected O K-edge yield is constant across the higher energy absorption edge under study (here Cu and Nd), which is why such measurement effectively probes the linear absorption coefficient $\mu(E)$ and is similar to a transmission experiment. The comparison of some spectra is shown in Figure 11.6. In a second publication, the authors took resonant emission processes into account and established that the angular dependence of the XAS can be used to determine the absorption coefficient $\mu(E)$ in absolute units [67]. The inverse partial fluorescence yield is an important step toward quantitative absorption measurements, and the technique has been used for a number of systems. It is of course limited to systems that contain at least two elements in which both absorption thresholds are accessible by the beamline used. For the remaining systems where the sample composition or an extremely strong reabsorption does not allow IPFY measurements, a special case has been made for measuring the Ta L_3 XAS by integrating over the nonresonant Ta L_3M_5 fluorescence [68]. This formalism is more time consuming and has not become as widely used, but it is remarkable because it allows acquisition of XAS free of self-absorption despite using fluorescence yield measurements.

11.2.2.2 Measuring Soft X-Ray Absorption in Transmission Mode

Green *et al.* [69] use a unique setup to perform quantitative X-ray absorption spectroscopy (the setup is depicted in Figure 11.7c). The system of study is a 200 nm film of Fe-doped In_2O_3 on Al_2O_3 substrate. Instead of monitoring one of the secondary decay products as described above, the substrate luminescence at

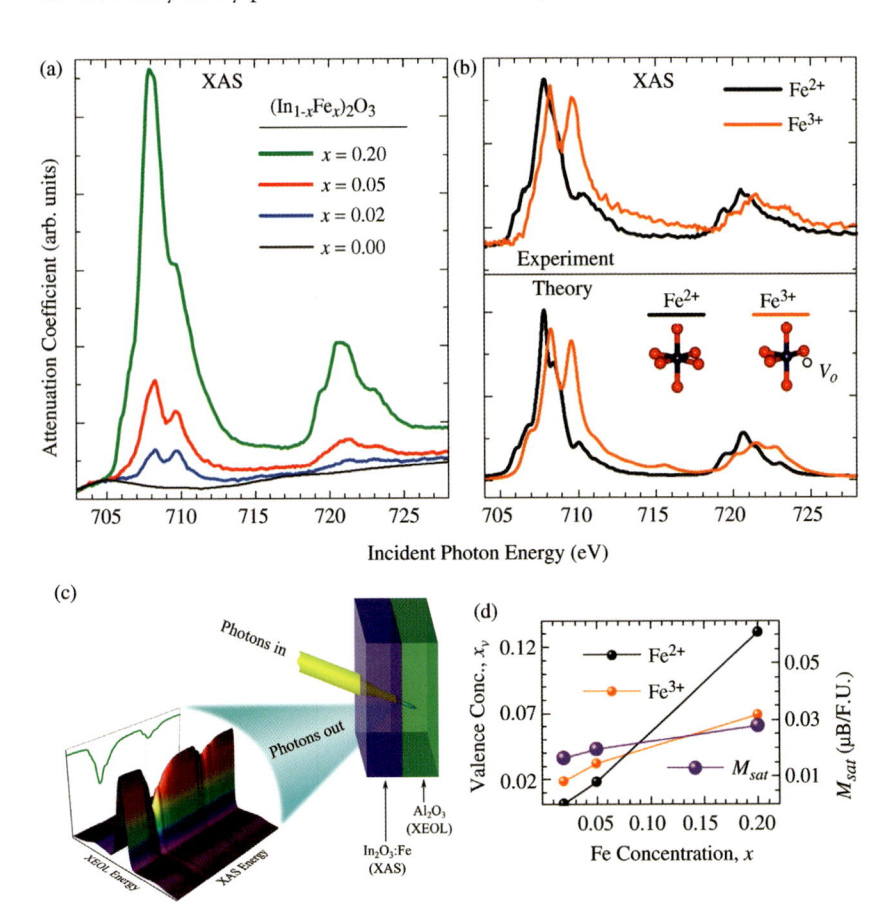

Figure 11.7 Fe $L_{2,3}$-edge XAS of $(In_{1-x}Fe_x)_2O_3$. (a) The raw XAS spectra obtained from the substrate via X-ray excited optical luminescence (XEOL) yield in the setup depicted in part (c). (b) The extracted component XAS spectra (upper panel) with the corresponding calculated spectra (lower panel). (d) The saturation magnetizations (M_{sat}) measured for each sample are correlated with the fractional concentration of Fe^{3+}-V_O centers determined from XAS. (Reprinted with permission from Ref. [69]. Copyright 2015, American Physical Society.)

689 nm was detected with an optical spectrometer while scanning the exciting radiation across the Fe $L_{2,3}$-edge. This has two immediate advantages. First, luminescence photons are generated proportionally to the number of transmitted soft X-ray photons. The constant sensitivity of the material to the luminescence photons passing through the film effectively allows for measurement of the film's transmission coefficient at a given incident X-ray energy. Second (and more importantly), the spectra detected in this transmission mode do not suffer from distortions such as TEY, TFY, and PFY (note that for transmission measurements, the uniform thicknesses of samples and the absence of density fluctuations in these samples are important constraints). This setup allows one to measure the true transmission of the thin film, which is an important advantage when analyzing the XAS quantitatively. Somewhat similar setups were used in earlier works but with different goals of magnetic imaging [70] and circular dichroism [71,72], with the common denominator being the measurement of true absorption spectra.

The in-depth analysis of the Fe $L_{2,3}$ absorption spectra (see Figure 11.7a) in Fe-doped In_2O_3 revealed that there are only two different Fe species in films of different Fe concentration (Figure 11.7b). These are Fe^{2+}, similar but not identical to the coordination of FeO, and Fe^{3+} in a unique octahedral coordination where one of the adjacent O atoms is missing. This was established by comparing the measured spectra with Anderson impurity calculations [23]. It was then possible to link magnetic behavior and the Fe^{3+} component spectra for different Fe concentrations (Figure 11.7d), which revealed that only the Fe atoms that are substituting for indium in the lattice, have 3+ valence, *and* are adjacent to an O vacancy are responsible for the observed room-temperature ferromagnetism in the system. This finding is important because the role of such structures has been speculated about for other DMS systems, but direct experimental evidence of their presence was difficult to establish because vacancies are notoriously difficult to detect [73].

To generalize the above, if the synthesis of a system of interest allows for its deposition or adherence as a sufficiently thin film on a substrate, which exhibits luminescence in an energy range where the film itself is transparent, the absorption spectra will be free of distortions. Combining this result with the angular dependence of absorption spectra, this method could also be advantageous in situations where dopant concentrations need to be determined. This method offers a third way of obtaining undistorted, quantitatively significant absorption spectra that allow for a more detailed analysis when used in conjunction with calculated spectra.

11.2.3
New Sources, Time, and Spatial Resolution

Before the advent of modern synchrotron sources, there was a lack of tunable sources, especially in the soft X-ray regime. It is therefore no surprise that the field of soft XAS has exploded since the third-generation synchrotrons became common during the 1980s. Every modern synchrotron today has one or several beamlines dedicated to X-ray absorption spectroscopy. Today's synchrotrons use insertion devices such as undulators and wigglers and are designed for maximal *brightness* – the number of photons emitted per second, per area of the source, per solid angle the radiation is emitted in, and per spectral bandwidth – which is a property that encompasses the essential performance parameters of modern sources. Today's brightest synchrotrons deliver a peak spectral brightness of up to 10^{25} photons/s/mm^2/mrad^2/0.1%bw, and this value is surpassed by free electron lasers by up to eight orders of magnitude [74]. The pulse length is about 100 times shorter in FELs and these devices offer high spatial and temporal coherence.

The bunch structure of the electrons in a synchrotron results in a natural pulse structure to it that is used to obtain temporal resolution in addition to energy resolution. Progress has also been made in adding temporal resolution to XAS. This is very desirable since the temporal resolution allows for the study of dynamic processes such as chemical and biological reactions as well as electron transfer processes. An interesting niche is an IR-pump, XMCD-probe experiment that allows for measurement of XMCD signals with a tabletop laboratory source used to monitor magnetization dynamics with femtosecond time resolution [75]. Using a laser-driven pulsed high-harmonic generation (HHG) source, it was possible to simultaneously measure the XMCD signal at the $M_{2,3}$-edges (40–75 eV) of Co and Pt in Pt/Co/Pt and Co/Ni/Pt thin films. Magnetization and demagnetization of the Co layer in the films can be altered on a timescale of 92 fs.

The next generation of sources, free electron lasers (FEL), offers increases in brilliance by several orders of magnitude and features also improved time resolution. Unfortunately, stochastic fluctuations in pulse intensity make normalization as well as the tuning of the photon energy difficult. It is no surprise that increased time resolution would come at the expense of spectral resolution, and therefore XAS with FELs has been deemed problematic. However, the La $N_{4,5}$ absorption threshold has been measured in a proof-of-principle experiment detecting an absorption spectrum over a narrow range (101–103 eV) with an FEL source [76].

In another interesting example, an FEL was used to record the Mn $L_{2,3}$ XAS of a dilute Mn aqueous solution [77]. This experiment is especially remarkable because it demonstrates that the ultrashort FEL pulses can outrun the radiation damage – another concern that has been perceived as problematic with the ultrahigh flux FELs – that is otherwise expected with the ultrahigh photon flux associated with these sources. This could open up FELs for use in the study of biological systems possibly where chemical dynamics under functional conditions are of great interest and radiation damage has been a notorious problem (even with the less brilliant third-generation synchrotron sources).

A number of publications use the O K-edge absorption to study real-time breaking of bonds of CO to a Ru surface [78] and the dynamics of laser-induced collisions of CO with O and a Ru substrate, which give rise to new features in the O K-absorption on a few 100 fs timescale [79].

The spot size of the synchrotron beam on a sample ranges typically between tens to hundreds of micrometers depending on the focusing optics of the beamline. Such macroscopic beam spots always average the signal over the illuminated sample area. A much smaller spot size can be obtained with X-ray microscopes where a smaller beam spot is achieved by using special X-ray optics such as Fresnel zone plates. Typically, the dramatically improved spatial resolution comes at the expense of some energy resolution or photon flux on the sample, but many important problems, especially in polymers and magnetism, have nonetheless been studied with X-ray microscopy. Spatial resolutions of a few nanometers are achievable today [27]. The following two features of XRM are important, and XAS without spatial resolution cannot probe these: In *spectromicroscopy*, spectra can be taken from an area as small as the beam spot; in *microspectroscopy*, a particular signal (such as emitted electrons or fluorescence photons) can be measured, while the sample is raster-scanned across the beam spot allowing generation of an image showing the spatial distribution of the measured signal. A study of (Na,H)Ti nanoribbons has shown that transmission X-ray microscopy today can achieve excellent spectral resolving power ($E/\Delta E = 10^4$), spatial resolution (25 nm), and fast acquisition (seconds per image) [80].

11.2.4
Detectors

Fluorescence yields are inherently low as it takes about 10^3–10^4 incoming photons to produce a fluorescence photon because the excitations relax predominantly via Auger electron emission [34,81]. Fluorescence yields can therefore be difficult (but not impossible) to measure in strongly diluted systems. When diodes are used for TFY detection, the relatively weak TFY signal can be distorted by the dominating Auger electron signal. This is why a combination of photodiode and repelling meshes to suppress the emitted electrons has been suggested as a means for obtaining better TFY measurements [82].

Energy dispersive solid-state detectors (SSD) have been used for about a decade for PFY and have become more popular recently with the advent of the IPFY measurements. While the energy dispersive solid-state detectors required for IPFY (see Section 11.2.1) allow for XAS free of distortions, they do have the disadvantage of having an inherent dead time that many nondispersive detectors do not suffer from. It has been recently shown that measuring with an array of filtered nondispersive TFY detectors allows for measurement of PFY and IPFY spectra without dead time [83]. Demonstrated for the example of Fe_2O_3, this option is less expensive and offers improved signal-to-noise ratios when compared to conventional energy-dispersive solid-state detectors.

11.2.5
Polarization Effects

Synchrotron radiation from a bending magnet is horizontally polarized. Special insertion devices allow for complete selection of the polarization direction including circular and elliptical polarization. The scalar product in Eq. (11.6) introduces a directional dependence proportional to $\cos^2(\theta)$ with θ being the angle between the electric field vector and the direction of the orbital (such as a π-bonded orbital). The absorption is maximal when the direction of the electric field vector of the exciting radiation is parallel to the orientation of the unoccupied final state orbital surrounding the excited atom. Taking spectra at two (or more) different incidence angles allows for determination of the orientation of, for example, the highly oriented π-bonds of atoms or molecules in the sample. This technique is referred to as X-ray linear dichroism (XLD) and was pioneered by Stöhr et al. for CO and NO on Ni [21].

Recently, NEXAFS at the C K-edge (in combination with polarization modulated infrared absorption spectroscopy) revealed the molecular orientations of self-assembled monolayers (SAM). The SAMs studied turn out to be surprisingly well oriented, more specifically the phenyl ring of model phenylphosphonic acid on indium zinc oxide is oriented with a defined tilt angle of 12°–16° from the surface normal [84]. Given that SAMs find applications in improving device performances of organic light-emitting diodes (OLEDs), organic field-effect transistors (OFETs), and organic photovoltaics (OPVs), NEXAFS is a key technique for organic electronics research.

The polarization of X-rays can also couple to the spin of an electron. X-ray magnetic linear dichroism (XMLD) is used to study antiferromagnetic materials such as many of the transition metal oxides [85]. When using circularly polarized X-rays, the different sum rules allow one to measure spin and orbital moments. Pioneered by Schütz et al. [86], this X-ray magnetic circular dichroism (XMCD) is ideal for studying ferromagnetics. The chemical selectivity of NEXAFS makes it an ideal tool to study the role of different elements, particularly dopants, in various matrices.

For example, the magnetic phase transition with temperature in $LaCoO_3$ has been a topic of interest for about the last 50 years. Recent research using XMCD and NEXAFS has shown that it is an inhomogeneously mixed-spin system, which means it is described by a low-spin and a triply degenerate high-spin excited state [87]. XMCD measurements at the O K-edge as a function of Sr-doping of $LaCoO_3$ reveal, for small Sr concentrations, that ferromagnetic clusters form with a nonmagnetic matrix, while, with higher Sr-content, Sr increasingly induces the formation of magnetic O-hole states [88], which are responsible for the magnetism.

An interesting attempt has been made to combine the chemical and magnetic contrast of NEXAFS with the high spatial resolution of scanning tunneling microscopy (STM). Magnetic contrast between left and right circular polarized soft X-rays at the Fe $L_{2,3}$-edges of an Fe film was achieved by illuminating a sample under an STM tip with a monochromatic soft X-ray beam (tens of microns size) [89], and Cu L_3 absorption was measured by monitoring sample and tip currents simultaneously in far-field and near-field geometry [90].

11.2.6
Liquids and in Operando XAS

Arguably, one of the areas of greatest progress in X-ray absorption spectroscopy in the last 15 years has been the study of liquids with soft XAS. The improved photon flux of modern synchrotrons aside, this development was triggered when soft X-ray absorption (and emission) spectroscopy were employed for exploring the different forms of H_2O. This became possible with improved signal statistics that allowed more accurate fitting of the otherwise relatively plain O K-edge spectrum of water. Starting in 2002, a large number of studies were devoted to the adsorption of water [91] and its local bonding structure – see, for example, Refs [92,93] and later Refs [94,95]. More studies followed that looked at interface mixing of water and alcohol [96] as well as isotope and temperature effects [97]. What these studies have in common is that they are aimed at understanding the hydrogen bond network. Although the H atoms cannot be studied directly with soft XAS, measurements at the O K-edge have proven to be sensitive to the fluctuations in the H-bonds of the water molecules. It is worth noting that the excitations in the absorption are orders of magnitude faster ($<10^{-16}$ s) than the molecular vibrations in the liquid ($\approx 10^{-10}$ s) [98]. Therefore, NEXAFS provides snapshots in time of the H-bonding configuration, allowing one to distinguish broken or distorted bonds (liquid form) and fully coordinated bonds (ice).

These studies were carried out with specially designed liquid cells that faced the challenge of containing the liquid behind a window that would be thin enough to allow for a sufficient signal from the liquid, yet thick enough to avoid ruptures. It was also realized that in order to minimize radiation damage, it would not be enough to simply contain the liquid. Instead, it would have to be continuously exchanged, leading to the concept of flow cells. Absorption spectroscopy with liquid flow cells uses exclusively fluorescence yield mode because the attenuation length of electrons in liquids is small. An alternative concept with additional challenges is the spectroscopy of liquids using liquid microjets [99]. In recent years, there has been a multitude of new cell designs – and projects well beyond the study of water – that allow studying liquids *in situ* [100], at different temperatures [101,102], and at liquid–solid interfaces [103], as well as reactions in electrodepositions [104] and catalysis [105]. Some cells also allow the probing of gases (and liquids) [106] and gas–solid interfaces [107].

Noteworthy advances have also been made in the ability to study active electrochemical interfaces at extremely close range, and in one particular case the role of water on gold electrodes [62]. Simulations by the authors indicate that the electron yield at the gold–water interface is sensitive to only two molecular layers of liquid water, and the combination of theory and experiment was vital in this study.

Finally, when dealing with very dilute liquid or even gaseous sample systems, signals are much weaker than in solids. Normalizing and analyzing the absorption spectra is more difficult – especially when measuring in yield modes. It is worth noting that despite being one of the most widely used synchrotron-based spectroscopy techniques, the interpretation of XAS can still be challenging, controversial, and sometimes leading to misinterpretations (see, for example, the range of comments in Refs [61,108–111]). A particularly lively discussion has taken place in articles around the interpretation of the O K-edge spectra of water – see, for example, Refs [112,113] and the reference cited therein. Underlying this discussion is the fact that the O K-edge absorption spectra of water and ice are relatively featureless with the typical π^* and σ^* peaks. In order to evaluate and even deconvolute such spectra to obtain contributions from different H-bonds, it becomes crucial that the spectra are measured with high accuracy, reproducibility, and minimal distortions [33].

NEXAFS has become an important tool in battery research, an area that especially benefits from the understanding of new materials. Particularly, the interface and interaction of liquid and solid are of interest in this field. The advancement here is owed to the fact that battery and electrode materials and charge dynamics can now be studied while the battery is being operated. This so-called *in operando* mode is another important advance from *in situ* measurements in the last decade.

It has been stressed that NEXAFS is an ideal tool because its element selectivity allows for the identification of substrates, reactants, solvents, and catalysts in solid–liquid interfaces. Yuzawa *et al.*, for example, studied the time and temperature dependence of *in situ* solid–liquid heterogeneous catalytic reactions – in this case cyanopyrazine (PzCN) hydration to produce pyrazinamide (PzCONH$_2$) on a TiO$_2$ catalyst [114].

One research goal for Li-ion batteries is to increase the capacity of the batteries, and one avenue for achieving this is the optimization of the materials involved. The structural and dynamic effects of applying a potential to the electrolyte material have been studied extensively, and more recently the focus has shifted to the less studied effects on the electrode materials during operation. Liu *et al.* [115] revealed the charge dynamics of Li ions and electrons in two Li-ion battery cathodes, Li(Co$_{1/3}$Ni$_{1/3}$Mn$_{1/3}$)O$_2$ and LiFePO$_4$. The difference in performance is discussed in terms of phase transformation, mesoscale morphology, and the conductivity of the electrodes. More recently, Luo *et al.* have shown that oxygen loss and Li$^+$ removal for the similar Li$_{1.2}$(Co$_{0.3}$Ni$_{0.3}$Mn$_{0.544}$)O$_2$ are compensated by the formation of electron holes localized on the O atoms [116]. The Mn^{4+} and Li^{1+} ions facilitate the localization of O$_2^{2-}$ species. In a study of a model electrode material that consists of interconnected graphene sheets, the field-induced changes in electronic structure were monitored when a bias was applied and the charge was stored in the double layer [117] and the role of specific Cl$^-$ and OH$^-$ anions from the electrolyte was elucidated through this work: It was found that it is not sufficient to simply take the isolated properties of the electrolyte and electrode into account, rather one has to consider polarization-induced and electrolyte-mediated modifications that are made to the electrode as well.

As for gases, the Ar L$_{2,3}$ absorption for Ar$_2$, ArNe dimers, and Ar clusters was measured at a gas phase beamline where two different drift chambers for the gas allowed for differentiation of electron and ion signals in one, and fast ions and total ion yield plus photons in the other [118]. Detecting one electron and two ions in a coincidence experiment in the former chamber and discriminating by kinetic energy of the ions in the latter setup enabled the measurement of chemical shifts between Ar atoms and the different Ar dimers well below the natural linewidth.

11.2.7
Other

It has been stressed in this chapter that the immense wealth of soft X-ray absorption spectroscopy research performed in the last decade makes a comprehensive listing of all progress made impossible, let alone one that does justice to research areas where NEXAFS has been employed. The goal of the following paragraphs is to highlight additional examples to emphasize that there is literally no research area in the natural sciences where NEXAFS has not been applied.

One of the most important unsolved problems in the physics of correlated electron systems is the high temperature superconductivity observed in cuprates. Peets *et al.* [119] studied two overdoped high-temperature superconductors ($La_{2-x}Sr_xCuO_{4\pm\delta}$ for different Sr concentrations and $Tl_2Ba_2CuO_{6+\delta}$). It was found that electron correlation decreases in the overdoped cuprates and no upper Hubbard band is observed in the O K-edge XAS. The unexpected inapplicability of the single-band Hubbard model – frequently employed for describing the transition from conducting to insulating systems – may lead to the future development of new theories for describing high-temperature superconducting cuprates.

NEXAFS is particularly suited to nondestructively study buried layers and interfaces. A study of the Ti $L_{2,3}$ absorption spectrum at the $LaAlO_3$–$SrTiO_3$ interface found that orbital reconstruction at the interface is related to the generation of an electron gas [120]. In particular, the Ti 3d degeneracy is lifted and the $3d_{xy}$ states become the first states available for conducting electrons.

Soft XAS has also made many contributions to the study of two-dimensional systems. The fine structure of the graphene absorption has been debated [121], and the role of core hole effects [122] as well as the rippling of graphene and its interaction with Cu and Ni [123] and diamond [124] substrates have been studied. Today, functionalization and doping [125–129] of graphene have been extensively studied with NEXAFS. NEXAFS and XMCD have been applied to 3d transition metals [130] and the orientation of dipolar molecules on graphene [131]. It is noteworthy that the spectral XAS features in functionalized graphene oxide (GO) were originally assigned by comparing them with spectra of reference systems with similar hydroxyl (C—OH), epoxide (C—O—C) and carbonyl (C=O) functional groups [132]. When the spectra were compared with calculated GO spectra, it was realized that the bonding situation was far more complex [61]. A new effect has now been studied wherein the electronic structure of graphene oxide changes reversibly with temperature (<160 K) [133]. This remarkable effect is hypothesized to result from the formation of an epoxide in which an O atom is uniquely bonded to three carbon atoms accompanied by increased buckling of the GO lattice.

When the hexagonally arranged carbon atoms in graphene are replaced with Si atoms, the resulting monolayer is referred to as *silicene*. This material was first synthesized on an Ag(111) substrate in 2012 [55]. The motivation behind replacing C with Si is that silicene might have semiconducting properties and this could find applications in micro- and nanoelectronic devices that require a bandgap. Silicene does not yet exist as a freestanding layer, but it has been studied experimentally on substrates Ag(111), Ir(111), and ZrB_2(001). Interaction with substrate materials are always important in the study of monolayers, and the question here was whether Si would retain its semiconducting properties on Ag(111) or become electrically conducting as a result of these interactions. Soft X-ray spectroscopy has shown that silicene on Ag(111) is conducting [56] and has allowed the orientation and structural organization of the Si atoms on this substrate to be determined. Furthermore, speculation that stacked layers of silicene might exhibit semiconducting behavior has been addressed: Silicene on Ag (111) remains conducting until up to three layers, but beyond this the material turns into polycrystalline Si [134].

A relatively new area of research is the employment of NEXAFS for the understanding of materials for phosphor-converted light emitting diodes (pc-LEDs). This field of research is not an obvious place for soft XAS because the processes governing to the emission of visible light take place at energies much lower than the soft X-ray range. Nonetheless, it has been shown that NEXAFS can contribute to a better understanding of the correlation between structural and luminescence properties as well as to explaining bandgap and narrowband emission [135–137]. The new generation pc-LEDs use Eu^{2+} as the luminescing element, which is unique in that the energy of the Eu^{2+} 5d level decreases when placed in a semiconducting lattice in such a way that the decay from the Eu^{2+} 5d to 4f ground state leads to an unusually broad luminescence emission, making it the most technologically important emitter today. Unlike the energy levels of the very localized rare earth 4f orbitals, the energies of the 5d orbitals strongly depend on the host lattice. This makes a prediction of the energies of the 5d band (and hence the luminescence energies of Eu^{2+}) in a host lattice difficult.

This chapter has not addressed much in the way of NEXAFS studies on polymers, but any review of XAS should at least touch on the topic of radiation damage. Organic materials – especially polymers – are particularly sensitive to radiation damage by soft X-rays [138], which often presents a problem for researchers. Chemical modifications of organic molecular structures occur through bond-breaking, aromatic ring opening, and cross-linking processes. The dominating damage mechanism in polymers is the removal of carbonyl groups [139]. Radiation damage has been studied for different edges, for example, the carbon edge [140] and for different systems, such as glycine at the N and C K-edges [141] and the Mn and Fe L-edges [142]. There are often clear trends in spectral features that emerge with increasing exposure time and dosage.

Sample cooling and the minimization of dosage by rapid scanning of the sample to constantly change the location of the incoming beam are now standard measures taken to avoid damaging samples. No matter whether these measures are taken or not, it remains wise to compare early low-exposure spectra with the spectra taken later with a higher accumulated radiation dosage to check for radiation-induced variations in spectral features. Radiation damage was expected to become much more dramatic with the advent of extremely bright free electron laser sources, but in certain conditions it seems possible to conduct measurements before damage occurs [77], meaning FEL sources offer exciting possibilities for time-resolved measurements.

11.3
Conclusions and Outlook

NEXAFS studies have become increasingly popular since the 1980s with the advent of powerful and tunable third-generation synchrotron sources. Although only a small fraction of the research performed with NEXAFS could be covered in this chapter, it should have become clear that great advances have been made particularly in the last decade, in both scientific understanding and in the development of new instrumentation. A subjective summary and outlook of the recent advances in NEXFS is the following:

- New techniques such as IPFY have emerged that enable measurement of true, undistorted, and sometimes even absolute soft X-ray absorption spectra, opening the door for improved quantitative analyses that provide deeper insight into systems under study.
- The advent of liquid flow cells has facilitated the study of a wide array of systems from different phases of H_2O to liquid–solid interfaces. This technology is especially relevant for battery materials research where *in situ* and *in operando* NEXAFS will allow for new insights and, hopefully, the development of higher capacity, longer lasting batteries.
- Concentration levels for species in solution of <1 mM will be achieved, making NEXAFS of liquids relevant for studying many processes in biology and the life sciences.
- With improved computing power and the plethora of codes available today, calculating NEXAFS *spectra* by taking into account partial DOS and transition matrix elements as well as the core hole effect has become more common.
- XAS is still a powerful technique in itself, but it has gained importance as a complimentary tool for other techniques such as microscopy and elastic/inelastic scattering spectroscopy.
- Spatially resolved spectroscopic measurements seem to be possible in modern X-ray microscopes, which offer resolutions of a few nm.
- Many new areas are waiting to be explored, especially with time-resolved XAS, which will assist in unraveling the dynamics of chemical reactions and charge, transfer processes.

- Free electron laser sources with unprecedented brightness and time resolution promise to offer a window into this dynamic behavior, enabling scientists to move beyond the traditional realms of ordered structures and equilibrium into the world of disordered and transient phenomena.

References

1 Sagnac, G. (1901) *Ann. Chim. Phys.*, **23**, 145–198.
2 Stumm von Bordwehr, R. (1989) *Ann. Phys. Fr.*, **14**, 377–466.
3 Inokuti, M. and Noguchi, T. (1974) *Am. J. Phys.*, **42**, 1118–1119.
4 Penner-Hahn, J.E. (2005) *Coord. Chem. Rev.*, **249**, 161.
5 Solomon, E.I. et al. (2005) *Coord. Chem. Rev.*, **249**, 97.
6 Stöhr, J. (1992) *NEXAFS Spectroscopy*, vol. **25**, Springer Series in Surface Sciences, Springer, Heidelberg.
7 Yang, B.X., Kirz, J., and Sham, T.K. (1987) *Phys. Rev. A*, **36**, 4298–4310.
8 Wikfeldt, K.T. et al. (2010) *J. Chem. Phys.*, **132**, 104513.
9 Hubbell, J.L. et al. (1980) *J. Phys. Chem. Ref. Data*, **9**, 1023–1147.
10 Hemraj-Benny, T. et al. (2006) *Small*, **2**, 26–35.
11 Adachi, J. et al. (2005) *J. Phys. B*, **38**, R127.
12 Ade, H. et al. (2009) *Nat. Mater.*, **8**, 281.
13 Chakhalian, J. et al. (2014) *Rev. Mod. Phys.*, **86**, 1189.
14 Zhong, J. et al. (2014) *Adv. Mater.*, **26**, 7786.
15 Ade, H. and Urquhart, S.G. (2002) *Chemical Applications of Synchrotron Radiation* (ed. T.K. Sham), World Scientific Publishing, Singapore, pp. 285–355.
16 Henderson, G.S. et al. (2014) *Rev. Mineral. Geochem.*, **78**, 75.
17 Gillespie, A.W. et al. (2013) *Adv. Agron.*, **133**, 1–32.
18 Moulton, B.J.A. et al. (2016) *Chem. Geol.*, **420**, 213.
19 Hähner, G. (2006) *Chem. Soc. Rev.*, **35**, 1244.
20 de Groot, F.M.F. (2005) *Coord. Chem. Rev.*, **249**, 31–63.
21 Stöhr, J. et al. (1981) *Phys. Rev. Lett.*, **47**, 381–384.
22 van Veenendahl, M. (2015) *Theory of Inelastic Scattering and Absorption of X-Rays*, Cambridge University Press.
23 Kotani, A. and Shin, S. (2001) *Rev. Mod. Phys.*, **73**, 203–246.
24 M. Simon and T. Schmitt (2013) *J. Electron Spectros. Relat. Phenomena*, **188**, 1.
25 de Groot, F. and Kotani, A. (2008) *Core Level Spectroscopy of Solids*, Taylor & Francis/CRC Press, London.
26 Fink, J. et al. (2013) *Rep. Prog. Phys.*, **76**, 056502.
27 Hitchcock, A.P. (2015) *J. Electron Spectros. Relat. Phenomena*, **200**, 49–63.
28 Damascelli, A., Hussain, Z., and Shen, Z.X. (2003) *Rev. Mod. Phys.*, **75**, 473–541.
29 Henke, B.L. et al. (1993) *At. Data Nucl. Data Tables*, **54**, 181–342 (a good Web site resource is http://henke.lbl.gov/optical_constants/).
30 Green, R.J. (2013) Transition metal impurities in semiconductors: induced magnetism and band gap engineering. Ph.D. thesis, University of Saskatchewan.
31 Baer, M. et al. (2008) *Appl. Phys. Lett.*, **93**, 244103.
32 Leedahl, B. et al. (2014) *J. Phys. Chem. C*, **118**, 28143–28151.
33 Näslund, L.A. et al. (2005) *J. Phys. Chem. B*, **109**, 13835–13839.
34 Thompson, A.C. et al. (2009) *X-Ray Data Booklet*, 3rd edn, Lawrence Berkeley National Laboratory, Berkeley, CA.
35 Dick, R. (2016) *Advanced Quantum Mechanics Materials and Photons*, 2nd edn, Springer.
36 Heisenberg, W. (1927) *Z. Phys.*, **43**, 172.
37 Krause, M.O. and Oliver, J.H. (1979) *J. Phys. Chem. Ref. Data*, **8**, 329–338.
38 Goodings, D.A. and Harris, R. (1969) *J. Phys. C Solid State Phys.*, **2**, 1808–1816.
39 Asplund, L. et al. (1985) *J. Phys. B*, **18**, 1569–1579.

40 Braun, C. et al. (2011) *J. Am. Chem. Soc.*, **133**, 4307–4315.

41 Tolhurst, T.M. et al. (2016) *Chem. Eur. J.*, **22**, 10475–10483.

42 de Boer, T. et al. (2016) *Phys. Rev. B*, **93**, 155205.

43 Boyko, T.D. et al. (2013) *J. Phys. Chem. C*, **117**, 12754–12761.

44 Tran, F. and Blaha, P. (2009) *Phys. Rev. Lett.*, **102**, 226401.

45 Kresse, G. and Furthmuller, J. (1996) *Phys. Rev. B*, **54**, 11169–11186.

46 Neese, F. (2008) ORCA: An Ab Initio, DFT, and Semiempirical Electronic Structure Package, Version 2.6 Revision 35, Universität Bonn, Bonn, Germany.

47 Guerra, C.F. et al. (1998) *J. Theor. Chem. Acc.*, **99**, 391.

48 Frisch, M.J. et al. (2009) Gaussian 09 Revision A.1, Gaussian, Inc., Wallingford, CT.

49 Hermann, K. et al. (2014) StoBe-deMon, Version 3.3, StoBe Software, Berlin.

50 Kosugi, N. (1987) *Theor. Chim. Acta*, **72**, 149.

51 Gianozzi, P. et al. (2009) *J. Phys. Condens. Matter*, **21**, 395502.

52 Stavitski, E. and de Groot, F.M.F. (2010) *Micron*, **41**, 687–694.

53 Laskowski, R. et al. (2010) *Phys. Rev. B*, **82**, 205104.

54 Lejaeghere, K. et al. (2016) *Supramol. Sci.*, **351**, 1415.

55 Vogt, P. et al. (2012) *Phys. Rev. Lett.*, **108**, 155501.

56 Johnson, N.W. et al. (2014) *Adv. Funct. Mater.*, **24**, 5253–5259.

57 Boyko, T.D. et al. (2011) *Phys. Rev. B*, **84**, 085203.

58 Green, R.J., Haverkort, M.W., and Sawatzky, G.A. (2016) *Phys. Rev. B*, **94**, 195127.

59 Medarde, M. et al. (1992) *Phys. Rev. B*, **46**, 14975–14984.

60 Quayyum, M.F. et al. (2013) *J. Am. Chem. Soc.*, **135**, 17417.

61 Hunt, A. et al. (2014) *Adv. Mater.*, **26**, 4870–4874.

62 Velasco-Velez, J.-J. et al. (2014) *Supramol. Sci.*, **346**, 831.

63 Tröger, L. et al. (1992) *Phys. Rev. B*, **46**, 3283.

64 Eisebitt, S. et al. (1993) *Phys. Rev. B*, **47**, 14103.

65 Kurian, R. et al. (2012) *J. Phys. Condens. Matter*, **24**, 452201.

66 Achkar, A.J. et al. (2011) *Phys. Rev. B*, **83**, 081106.

67 Achkar, A.J. et al. (2011) *Sci. Rep.*, **1**, 182.

68 Blachucki, W. et al. (2014) *Phys. Rev. Lett.*, **112**, 173003.

69 Green, R.J. et al. (2015) *Phys. Rev. Lett.*, **115**, 167401.

70 Vaz, C.A.F. et al. (2012) *Appl. Phys. Lett.*, **101**, 083114.

71 Jakob, G. et al. (2007) *Phys. Rev. B*, **76**, 174407.

72 Meinert, M. et al. (2011) *J. Phys. D Appl. Phys.*, **44**, 215003.

73 Ciatto, A. et al. (2011) *Phys. Rev. Lett.*, **107**, 127206.

74 Huang, Z. (2013) Proceedings of the 4th International Particle Accelerator Conference (IPAC 2013), Shanghai, China. See also SLAC-Pub-15449 http://slac.stanford.edu/pubs/slacpubs/15250/slac-pub-15449.pdf.

75 Willems, F. et al. (2015) *Phys. Rev. B*, **92**, 220405.

76 Bernstein, D.P. et al. (2009) *Appl. Phys. Lett.*, **95**, 134102.

77 Mitzner, R. et al. (2013) *J. Phys. Chem. Lett.*, **4**, 3641–3647.

78 Dell'Angela, M. et al. (2013) *Supramol. Sci.*, **339**, 1302–1305.

79 Öström, H. et al. (2015) *Supramol. Sci.*, **347**, 978–982.

80 Guttmann, P. et al. (2012) *Nat. Photonics*, **6**, 25–29.

81 Hubbell, J.L. et al. (1994) *J. Phys. Chem. Ref. Data*, **23**, 339–364.

82 Thielemann, N., Hoffmann, P., and Föhlisch, A. (2012) *Rev. Sci. Instrum.*, **83**, 093105.

83 Boyko, T.D. et al. (2014) *J. Synchrotron Radiat.*, **21**, 716–721.

84 Gliboff, M. et al. (2013) *Langmuir*, **29**, 2166–2174.

85 van der Laan, G. et al. (1986) *Phys. Rev. B*, **34**, 6529–6531.

86 Schütz, G. et al. (1987) *Phys. Rev. Lett.*, **58**, 737–740.

87 Haverkort, M.W. et al. (2006) *Phys. Rev. Lett.*, **97**, 176405.

88 Medling, S. et al. (2012) *Phys. Rev. Lett.*, **109**, 157204.

89 Cummings, M.L. et al. (2012) *Ultramicrosocpy*, **112**, 22.

90 Rose, V. et al. (2013) *Adv. Funct. Mater.*, **23**, 2646.
91 Ogasawara, H. et al. (2002) *Phys. Rev. Lett.*, **89**, 276102.
92 Myneni, S. et al. (2002) *J. Phys. Condens. Matter*, **14**, L213–L219.
93 Guo, J.-H. et al. (2002) *Phys. Rev. Lett.*, **89**, 137402.
94 Wernet, Ph. et al. (2004) *Supramol. Sci.*, **304**, 995–999.
95 Smith, J.D. et al. (2004) *Supramol. Sci.*, **306**, 851–853.
96 Guo, J.-H. et al. (2003) *Phys. Rev. Lett.*, **91**, 157401.
97 Fuchs, O. et al. (2008) *Phys. Rev. Lett.*, **100**, 027801.
98 Dantus, M., Bowman, R.M., and Zehwail, A.I. (1990) *Nature*, **343**, 737–739.
99 Wilson, K.R. et al. (2004) *Rev. Sci. Instrum.*, **75**, 725–736.
100 Bora, D.K. et al. (2014) *Rev. Sci. Instrum.*, **85**, 043106.
101 Fuchs, O. et al. (2008) *Nucl. Instrum. Methods A*, **585**, 172–177.
102 Meibohm, J. et al. (2014) *Rev. Sci. Instrum.*, **85**, 103102.
103 Schwanke, C. et al. (2014) *Rev. Sci. Instrum.*, **85**, 103120.
104 Bozzini, B. et al. (2015) *J. Vac. Sci. Technol. A*, **33**, 031102.
105 Kristiansen, P.T. et al. (2013) *Rev. Sci. Instrum.*, **84**, 113107.
106 Brown, M.A. et al. (2013) *Rev. Sci. Instrum.*, **84**, 073904.
107 Benkert, A. et al. (2014) *Rev. Sci. Instrum.*, **85**, 015119.
108 Green, R.J. et al. (2014) *Phys. Rev. Lett.*, **112**, 129301.
109 de Groot, F.M.F. (2012) *Nat. Chem.*, **4**, 766–768.
110 Regier, T.Z. et al. (2012) *Nat. Chem.*, **4**, 765–766.
111 Pacile, D. et al. (2009) *Phys. Rev. Lett.*, **102**, 099702.
112 Fuchs, O. et al. (2008) *Rev. Lett.*, **100**, 249802.
113 Nilsson, A. et al. (2005) *Science*, **308**, 793.
114 Yuzawa, H., Nagasaka, M., and Kosugi, N. (2015) *J. Phys. Chem. C*, **119**, 7738–7745.
115 Liu, X. et al. (2013) *Nat. Commun.*, **4**, 2568.
116 Luo, K. et al. (2016) *Nat. Chem.*, **8**, 684–691.
117 Bagge-Hansen, M. et al. (2015) *Adv. Mater.*, **277**, 1512.
118 Jabbari, G. et al. (2015) *Phys. Chem. Chem. Phys.*, **17**, 22160.
119 Peets, D.C. et al. (2009) *Phys. Rev. Lett.*, **103**, 087402.
120 Salluzzo, M. et al. (2009) *Phys. Rev. Lett.*, **102**, 166804.
121 Pacile, D. et al. (2008) *Phys. Rev. Lett.*, **101**, 066806.
122 Zhang, L. et al. (2012) *Phys. Rev. B*, **86**, 245430.
123 Lee, V. et al. (2010) *J. Phys. Chem. Lett.*, **1**, 1247–1253.
124 Yu, J. et al. (2012) *Nano Lett.*, **12**, 1603–1608.
125 Usachov, D. et al. (2011) *Nano Lett.*, **11**, 5401–5407.
126 Schiros, T. et al. (2012) *Nano Lett.*, **12**, 4025–4031.
127 Schultz, B.J. et al. (2011) *Nat. Commun.*, **2**, 1–8.
128 Chang, C.-K. et al. (2013) *ACS Nano*, **2**, 1333–1341.
129 Zhao, L. et al. (2011) *Supramol. Sci.*, **333**, 999–1003.
130 Eelbo, T. et al. (2013) *Phys. Rev. Lett.*, **110**, 136804.
131 Johnson, P.S. et al. (2014) *Langmuir*, **30**, 2559–2565.
132 Christl, I. and Kretzschmar, R. (2007) *Environ. Sci. Technol.*, **41**, 1915–1920.
133 Hunt, A. et al. (2015) *J. Phys. Chem. Lett.*, **6**, 3163–3169.
134 Johnson, N.W. et al. (2014) *Adv. Funct. Mater.*, **24**, 5253–5259.
135 Tolhurst, T. et al. (2016) *Adv. Opt. Mater.*, **4**, 584–591.
136 Tolhurst, T. et al. (2015) *Adv. Opt. Mater.*, **3**, 546–550.
137 Boyko, T.D. et al. (2013) *Phys. Rev. Lett.*, **111**, 097402.
138 Howells, M.R. (ed.) (2009) *J. Electron Spectros. Relat. Phenomena*, **170** (1–3), 1–68.
139 Wang, J. et al. (2009) *J. Electron Spectros. Relat. Phenomena*, **170**, 25–36.
140 Leontowitch, A.F.G., Hitchcock, A.P., and Egerton, R.F. (2016) *J. Electron Spectros. Relat. Phenomena*, **206**, 58–64.
141 Wilks, R.G. et al. (2009) *J. Phys. Chem. A*, **113**, 5360–5366.
142 Schooneveld, M.M. and DeBeer, S. (2015) *J. Electron Spectros. Relat. Phenomena*, **198**, 31–56.

12
Vibrational Spectroscopy

Götz Eckold[1] and Helmut Schober[2,3]

[1]*University of Göttingen, Institute of Physical Chemistry, Tammannstr. 6, D-37077, Göttingen, Germany*
[2]*Institut Laue Langevin, 71 avenue des Martyrs, F-38000 Grenoble, France*
[3]*Université Grenoble–Alpes, F-38000 Grenoble, France*

12.1
Why Spectroscopy

The most remarkable microscopic property of a material is certainly its atomic structure. It is described by the correlation between the positions of atoms averaged over time. Since the structure is a direct reflection of the interaction between atoms, it is in some cases sufficient to correctly predict physical properties of a material. In general, such predictions, however, require the knowledge of the dynamics of the atoms. This holds in particular for functional materials that are the backbone of new high-tech devices. A functional material is very different from a structural material. Its physical and chemical properties have to be sensitive to environmental changes such as temperature, pressure, electromagnetic fields, pH, the presence of adsorbed molecules. A major challenge of functional architecture is the need for a control that is at the same time highly sensitive and stable. Sensitivity and stability are a function of the energy that is required to trigger changes of properties. These energies are directly encoded in the fluctuations of the system, that is, in its dynamics. Studying the fluctuations does not only allow determining the thermodynamic equilibrium properties of a material, but via the fluctuation–dissipation theorem equally gives access to transport properties in the linear regime. In this article, we will concentrate on studying structural excitations in fully relaxed crystalline materials where the atoms vibrate about well-defined equilibrium positions.

Regarding a solid as a huge molecule consisting of N atoms (N being of the order 10^{23}), there are $3N$ individual vibrational states that characterize the variety of interatomic interactions. If the material is in a crystalline state with N_c atoms in the primitive cell, then these vibrations can be classified according to 3

Handbook of Solid State Chemistry, First Edition. Edited by Richard Dronskowski, Shinichi Kikkawa, and Andreas Stein.
© 2017 Wiley-VCH Verlag GmbH & Co. KGaA. Published 2017 by Wiley-VCH Verlag GmbH & Co. KGaA.

N_c phonon dispersion sheets. The full knowledge of a dispersion sheet requires determining the frequency and eigenvector of the corresponding phonon as a function of the wave vector throughout the Brillouin zone. Electron–phonon coupling and anharmonicities introduce finite life times of the phonon excitations that have to be determined via the line broadening. There are theoretical methods to calculate all the phonon modes: Phenomenological treatments on the basis of pair potentials may be used to obtain a flavor of the different vibrations, while *ab initio* calculations are frequently used to predict the phonon frequencies and eigenvectors without or with only few adjustable parameters. Even though, the experimental determination of phonon modes and their wave vector dependence – the phonon dispersion – is crucial for the understanding of chemical bonding in solids. Inelastic scattering of neutrons and synchrotron X-rays allow the detection of arbitrary modes with wave vectors all over the Brillouin zone of a crystalline solid and with frequencies from zero to some 100 THz. While these techniques require large-scale facilities as high-brilliance sources of neutrons or X-rays, some of the vibrational states are also accessible by optical methods such as infrared(IR) or Raman spectroscopy. Due to the small momentum of photons in the visible or infrared spectral range, however, only phonon modes with very large wavelengths, that is, near the center of the Brillouin zone, can be detected with these methods. If there are N_c atoms within a primitive cell of a crystalline solid (N_c being on the order of 10 to some 100), a maximum of $3N_c$ modes are thus accessible by optical spectroscopy. This information obtained by home-lab experiments is, however, extremely valuable since it provides evidence for correlated motions where equivalent atoms in different unit cells are vibrating in phase. Moreover, local vibrations of impurity atoms or other defects that are without dispersion are easily accessible by these methods.

In the following, we will illustrate the power of neutron spectroscopy using selected examples of crystalline materials. We put the neutrons in context with optical spectroscopy as the main laboratory supplements to neutron spectroscopy investigations. It is, however, not intended to provide a comprehensive review on vibrational properties of solids. Rather we want to show, how different phenomena are reflected in the phonon spectra. In many cases, we will use material from research that we were personally involved in for the simple reason that it is these subjects that we are best familiar with.

12.2
Fundamentals of Lattice Dynamics

12.2.1
Phonons in the Harmonic Approximation

We will, in the following section, briefly introduce the concept of phonons in the harmonic approximation. Details can be found in the respective textbooks and dedicated articles [1,2].

12.2 Fundamentals of Lattice Dynamics

In perfect crystals, the vibrational states can most conveniently be described on the basis of the smallest unit needed to generate the whole (infinite) lattice by translation, that is, the *primitive cell*. Each individual primitive cell containing N_c atoms is characterized by a running index ℓ and a vector r_ℓ pointing to its origin (Figure 12.1). The equilibrium positions of all atoms in the crystals are given by

$$r^o_{\kappa\ell} = r_\ell + r^o_\kappa, \quad \kappa = 1,\ldots,N_c, \quad \ell = 1,2,\ldots, \tag{12.1}$$

r^o_κ being the vector of the κth atom with respect to the origin of the primitive cell.

The vibrating lattice is described by the time-dependent displacements of all individual atoms

$$u_{\kappa\ell}(t) = r_{\kappa\ell}(t) - r^o_{\kappa\ell} = r_{\kappa\ell}(t) - r_\ell - r^o_\kappa. \tag{12.2}$$

Within the adiabatic approximation, the motions of electrons and ions are decoupled. The forces acting on the ions are described via a potential V. This potential V, which contains in particular the electronic energy of the whole crystal, depends on the position vectors of all atoms,

$$V = V(r_1, r_2, \ldots, r_{\kappa\ell}, \ldots), \tag{12.3}$$

and exhibits a minimum for a completely relaxed system, that is, if all atoms occupy their equilibrium positions. For small relative atomic displacements, V can be expanded in a Taylor series with respect to $u_{\kappa\ell}(t)$:

$$V = V^o + \frac{1}{2} \sum_{\kappa\ell} \sum_{\kappa'\ell'} \sum_{\alpha=1}^{3} \sum_{\beta=1}^{3} u^\alpha_{\kappa\ell}(t) \cdot V_{\alpha\beta}(\kappa\ell, \kappa'\ell') \cdot u^\beta_{\kappa'\ell'}(t) + \cdots, \tag{12.4}$$

if $u^\alpha_{\kappa\ell}(t)$ denotes the cartesian co-ordinate of $u_{\kappa\ell}(t)$ in direction α. In the *harmonic approximation*, third and higher order terms are neglected.

The expansion coefficients in Eq. (12.4) are the partial derivatives of the potential energy with respect to the atomic displacements taken at the equilibrium positions:

$$V_{\alpha\beta}(\kappa\ell, \kappa'\ell') = \left.\frac{\partial^2 V}{\partial u^\alpha_{\kappa\ell} \partial u^\beta_{\kappa'\ell'}}\right|_o. \tag{12.5}$$

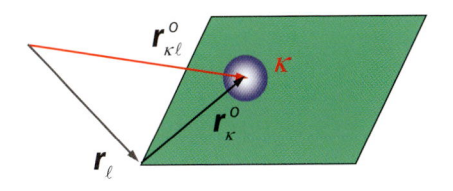

Figure 12.1 Notation of the position vector of atom κ within primitive cell ℓ.

Using the matrix notation

$$V(\kappa\ell,\kappa'\ell') = \begin{pmatrix} V_{11}(\kappa\ell,\kappa'\ell') & V_{12}(\kappa\ell,\kappa'\ell') & V_{13}(\kappa\ell,\kappa'\ell') \\ V_{21}(\kappa\ell,\kappa'\ell') & V_{22}(\kappa\ell,\kappa'\ell') & V_{23}(\kappa\ell,\kappa'\ell') \\ V_{31}(\kappa\ell,\kappa'\ell') & V_{32}(\kappa\ell,\kappa'\ell') & V_{33}(\kappa\ell,\kappa'\ell') \end{pmatrix}, \quad (12.6)$$

and dropping the constant V°, Eq. (12.4) reads

$$V = \frac{1}{2}\sum_{\kappa\ell}\sum_{\kappa'\ell'} \boldsymbol{u}_{\kappa\ell} \cdot \boldsymbol{V}(\kappa\ell,\kappa'\ell') \cdot \boldsymbol{u}_{\kappa'\ell'} + \cdots . \quad (12.7)$$

The product $-\boldsymbol{V}(\kappa\ell,\kappa'\ell')\boldsymbol{u}_{\kappa'\ell'}$ is just the force $\boldsymbol{f}(\kappa\ell)$ acting upon atom $(\kappa\ell)$ if the atom $(\kappa'\ell')$ is displaced by $\boldsymbol{u}_{\kappa'\ell'}$ (Figure 12.2).

Hence, the matrix $\boldsymbol{V}(\kappa\ell,\kappa'\ell')$ may be regarded as a *force constant matrix* and its elements $V_{\alpha\beta}(\kappa\ell,\kappa'\ell')$ as *force constants*. These parameters may be calculated with the help of specific interaction models such as pair potentials, tensor force models, or more complicated many body interactions. Alternatively, they can also be obtained from *ab initio* calculations in terms of the well-known Hellmann–Feynman forces.

The Hamiltonian of the perfect harmonic crystal can now be written in the form

$$H = \sum_{\kappa l} \frac{\boldsymbol{p}_{\kappa\ell}^2}{2m_\kappa} + \frac{1}{2}\sum_{\kappa\ell}\sum_{\kappa'\ell'} \boldsymbol{u}_{\kappa\ell} \cdot \boldsymbol{V}(\kappa\ell,\kappa'\ell') \cdot \boldsymbol{u}_{\kappa'\ell'}, \quad (12.8)$$

if $\boldsymbol{p}_{\kappa\ell}$ and m_κ are the momentum and the mass of atom $(\kappa\ell)$, respectively.

Consequently, the equations of motion for all individual atoms are given by

$$m_\kappa \frac{d^2\boldsymbol{u}_{\kappa\ell}}{dt^2} = -\sum_{\kappa'\ell'} \boldsymbol{V}(\kappa\ell,\kappa'\ell') \cdot \boldsymbol{u}_{\kappa'\ell'} . \quad (12.9)$$

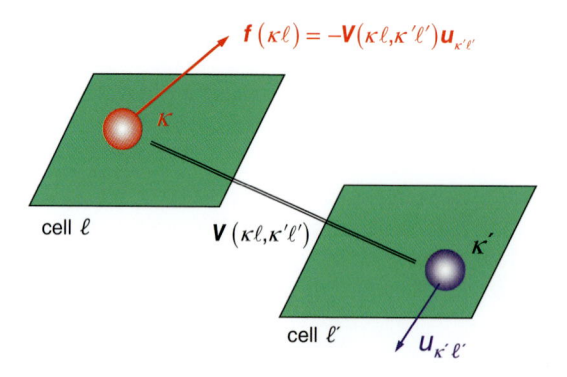

Figure 12.2 Interaction between atoms in different unit cells.

12.2 Fundamentals of Lattice Dynamics

Solutions of this set of coupled differential equations are of the form

$$u_{\kappa\ell}^{\pm} = \frac{1}{\sqrt{Nm_\kappa}} \cdot e_\kappa \cdot e^{i(q\, r_\ell \pm \omega t)}, \qquad (12.10)$$

which are plane waves with wave vector q and polarization vector e_κ. The upper index $\pm$ distinguishes two waves with identical frequencies that are traveling in opposite directions. If a finite crystal is considered or if periodic boundary conditions are applied, the wave vector is restricted to a sequence of discrete and equidistant values that are, however, very close to each other. Thus, for practical work q can be treated as a continuous variable. The polarization vectors e_κ are, in general, different for every atom κ. Moreover, they depend on q and for each wave vector there are $3N_c$ different modes of vibration characterized not only by different e_κ-vectors but also by different vibrational frequencies ω. Hence, Eq. (12.10) can be written more specifically as

$$u_{\kappa\ell}^{\pm}(q,j) = \frac{1}{\sqrt{Nm_\kappa}} \cdot e_\kappa(q,j) \cdot e^{i(q\, r_\ell \pm \omega_q j t)}, \qquad (12.11)$$

where the running index $j = 1, \ldots, 3N_c$ labels the different fundamental vibrations or *phonons*.

If the ansatz (Eq. (12.11)) is inserted into the equation of motion (Eq. (12.9)), the following eigenvalue equation is obtained:

$$\begin{aligned}
\omega_{q,j}^2 \, e_\kappa(q,j) &= \sum_{\kappa'\ell'} \sqrt{\frac{1}{m_\kappa m_{\kappa'}}} \cdot V(\kappa\ell, \kappa'\ell') \cdot e^{iq(r_{\ell'} - r_\ell)} \cdot e_{\kappa'}(q,j) \\
&= \sum_{\kappa'} \sqrt{\frac{1}{m_\kappa m_{\kappa'}}} \left[\sum_{\ell'} V(\kappa l, \kappa'\ell') \cdot e^{iq(r_{\ell'} - r_\ell)} \right] \cdot e_{\kappa'}(q,j).
\end{aligned} \qquad (12.12)$$

The summation over all primitive cells on the right-hand side of Eq. (12.12) yields the Fourier-transformed force constant matrix

$$F_{\kappa\kappa'}(q) = \sum_{\ell'} V(\kappa\ell, \kappa'\ell') \cdot e^{iq(r_{\ell'} - r_\ell)} \qquad (12.13)$$

that is independent of ℓ for infinite crystals. $F_{\kappa\kappa'}(q)$ contains all interactions of type κ atoms with type κ' atoms. Using this notation Eq. (12.12) reduces to

$$\omega_{q,j}^2 \, e_\kappa(q,j) = \sum_{\kappa'} \sqrt{\frac{1}{m_\kappa m_{\kappa'}}} \cdot F_{\kappa\kappa'}(q) \cdot e_{\kappa'}(q,j). \qquad (12.14)$$

If for a given vibration characterized by (q,j), we combine the 3D polarization vectors $e_\kappa(q,j)$ of all atoms within a primitive cell to a $3N_c$-dimensional polarization vector $e(q,j)$

$$e(q,j) = \begin{pmatrix} e_1(q,j) \\ \vdots \\ e_N(q,j) \end{pmatrix} = \begin{pmatrix} e_1^x(q,j) \\ e_1^y(q,j) \\ e_1^z(q,j) \\ \vdots \\ e_N^x(q,j) \\ e_N^y(q,j) \\ e_N^z(q,j) \end{pmatrix} \quad (12.15)$$

and simultaneously the 3×3 matrices $F_{\kappa\kappa'}(q)$ to a $3N_c \times 3N_c$ matrix $F(q)$

$$F(q) = \begin{pmatrix} F_{11}^{xx} & F_{11}^{xy} & F_{11}^{xz} & \cdots & \cdots & \cdots & \cdots & F_{1N}^{xx} & F_{1N}^{xy} & F_{1N}^{xz} \\ F_{11}^{yx} & F_{11}^{yy} & F_{11}^{yz} & \cdots & \cdots & \cdots & \cdots & F_{1N}^{yx} & F_{1N}^{yy} & F_{1N}^{yz} \\ F_{11}^{zx} & F_{11}^{zy} & F_{11}^{zz} & \cdots & \cdots & \cdots & \cdots & F_{1N}^{zx} & F_{1N}^{zy} & F_{1N}^{zz} \\ \vdots & \vdots & \vdots & \cdots & F_{\kappa\kappa'}^{xx} & F_{\kappa\kappa'}^{xy} & F_{\kappa\kappa'}^{xz} & \cdots & \vdots & \vdots & \vdots \\ \vdots & \vdots & \vdots & \cdots & F_{\kappa\kappa'}^{yx} & F_{\kappa\kappa'}^{yy} & F_{\kappa\kappa'}^{yz} & \cdots & \vdots & \vdots & \vdots \\ \vdots & \vdots & \vdots & \cdots & F_{\kappa\kappa'}^{zx} & F_{\kappa\kappa'}^{zy} & F_{\kappa\kappa'}^{zz} & \cdots & \vdots & \vdots & \vdots \\ F_{N1}^{xx} & F_{N1}^{xy} & F_{N1}^{xz} & \cdots & \cdots & \cdots & \cdots & F_{NN}^{xx} & F_{NN}^{xy} & F_{NN}^{xz} \\ F_{N1}^{yx} & F_{N1}^{yy} & F_{N1}^{yz} & \cdots & \cdots & \cdots & \cdots & F_{NN}^{yx} & F_{NN}^{yy} & F_{NN}^{yz} \\ F_{N1}^{zx} & F_{N1}^{zy} & F_{N1}^{zz} & \cdots & \cdots & \cdots & \cdots & F_{NN}^{zx} & F_{NN}^{zy} & F_{NN}^{zz} \end{pmatrix}$$
(12.16)

Equation (12.14) can be written in matrix notation and takes the simple form

$$\omega_{q,j}^2 \cdot e(q,j) = [M \cdot F(q) \cdot M] \cdot e(q,j) = D(q) \cdot e(q,j), \quad (12.17)$$

where the diagonal matrix

$$M = \begin{pmatrix} \frac{1}{\sqrt{m_1}} & 0 & 0 & & & & \\ 0 & \frac{1}{\sqrt{m_1}} & 0 & & & & \\ 0 & 0 & \frac{1}{\sqrt{m_1}} & & & & \\ & & & \ddots & & & \\ & & & & \frac{1}{\sqrt{m_N}} & 0 & 0 \\ & & & & 0 & \frac{1}{\sqrt{m_N}} & 0 \\ & & & & 0 & 0 & \frac{1}{\sqrt{m_N}} \end{pmatrix} \quad (12.18)$$

contains the inverse square root masses of the atoms.

The $3N_c \times 3N_c$ matrix

$$D(q) = M \cdot F(q) \cdot M \tag{12.19}$$

is called the *dynamical matrix*. It is Hermitian and contains all information about the dynamical behavior of the crystal. The squares of the vibrational frequency $\omega_{q,j}$ and the polarization vectors $e(q,j)$ are eigenvalues and corresponding eigenvectors of the dynamical matrix.

The wave vector dependence of the vibrational frequencies is called *phonon dispersion*. For each wave vector q there are $3N_c$ fundamental frequencies yielding $3N_c$ phonon *branches* when $\omega_{q,j}$ is plotted versus q. In most cases, the phonon dispersion is displayed for wave vectors along high-symmetry directions. These dispersion curves are, however, only special projections of the dispersion hypersurface in the four-dimensional q-ω-space. As a simple example, the phonon dispersion of bcc-hafnium [3] is displayed in Figure 12.3. The wave vectors are restricted to the first Brillouin zone and the phonon dispersion for different directions of the wave vector are combined in one single diagram making use of the fact that different high-symmetry directions meet at the Brillouin zone boundary. Note, that in Figure 12.3 the moduli of the wave vectors are scaled by the Brillouin zone boundary values and represented by the reduced coordinates ξ. Due to the simple bcc structure of Hafnium with one atom per primitive cell there are only three phonon branches. Moreover, for all wave vectors along the directions $[0\ 0\ \xi]$ and $[\xi\ \xi\ \xi]$ two of them exhibit the same frequencies – they are said to be *degenerate*. Hence, in the corresponding parts of Figure 12.3 only two branches can be distinguished.

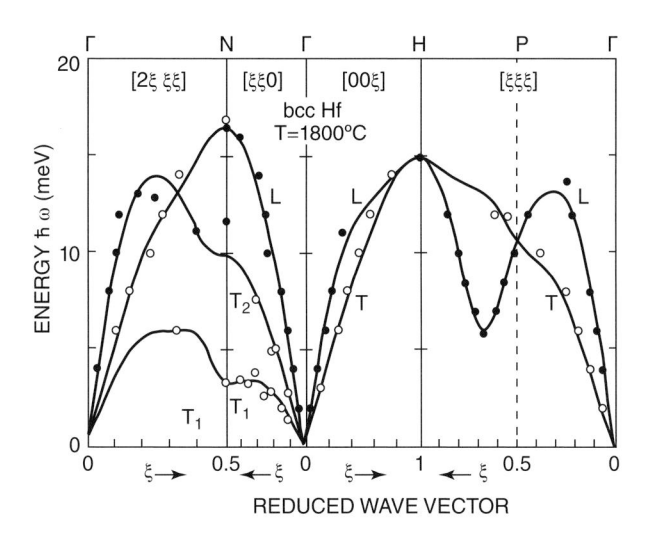

Figure 12.3 Phonon dispersion of bcc-hafnium for wave vectors along the main symmetry directions of the cubic structure (after Ref. [3]). The symbols represent experimental data obtained by inelastic neutron scattering and the full lines are results of model calculations. (Reproduced with permission from Ref. [61]. Copyright 2010, John Wiley & Sons.)

Whereas in this simple example, the different branches can be separated quite easily, this is no longer true for more complicated crystal structures. For illustration, the phonon dispersion of the high-T_c superconductor material Nd_2CuO_4 is shown in Figure 12.4 for the main symmetry directions of the tetragonal structure (space group $I4/mmm$, 7 atoms per primitive cell) [4]. Note, that in many

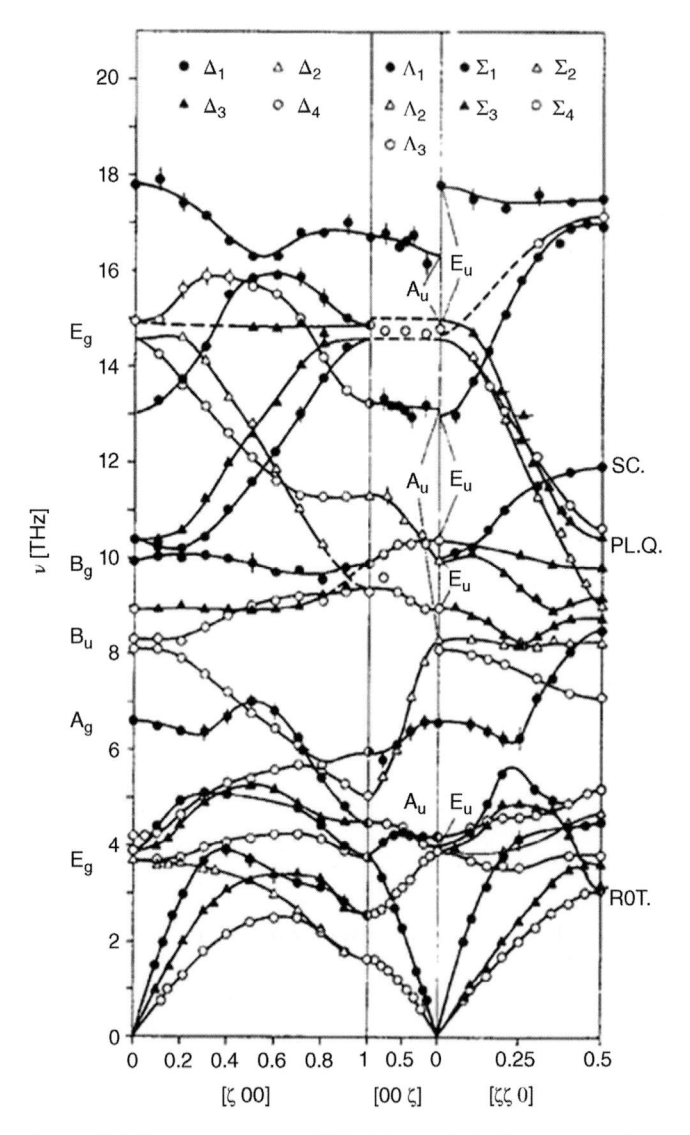

Figure 12.4 Phonon dispersion of Nd_2CuO_4 along the main symmetry directions of the tetragonal structure (after Ref. [4]). (The symbols represent experimental data obtained by inelastic neutron scattering and the full lines are drawn to guide the eye.) (Reproduced with permission from Ref. [4]. Copyright 1995, American Chemical Society.)

publications on lattice dynamics the frequency $\nu = \omega/2\pi$ is used rather than the angular frequency ω.

The 21 phonon branches of Nd_2CuO_4 with their intricate dispersion reflect the details of the interatomic interactions between all atoms of the structure. The phonon frequencies ν cover a range from 0 to 18 THz. In crystals with strongly bonded molecular groups, such as SiO_4 tetrahedra in quartz or SO_4 tetrahedra in sulphates, for example, the highest frequencies are found near 35 THz and correspond to bond stretching vibrations. Soft materials like organic molecular crystals, on the other hand, exhibit a large number of phonon branches within a rather small frequency range. It is, therefore, not always trivial to separate them. Deuterated naphthalene ($C_{10}D_8$) is a well-investigated example. The low frequency part of its phonon dispersion is shown in Figure 12.5 [5].

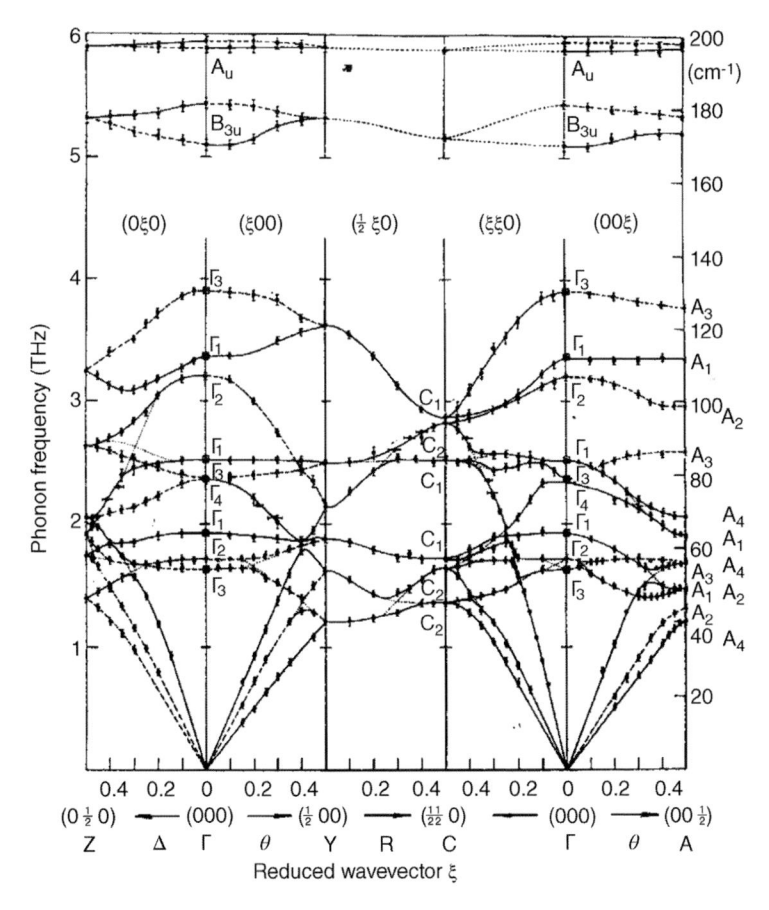

Figure 12.5 Low-frequency part of the phonon dispersion of deuterated naphthalene at 6 K (after Ref. [5]). (The symbols represent experimental data obtained by inelastic neutron scattering and the full lines are drawn to guide the eye). (Reproduced with permission from Ref. [5]. Copyright 2005, American Chemical Society.)

In the limit of long wavelengths, close to the center of the Brillouin zone, there are always three particular modes with identical polarization vectors for all atoms and zero frequency at the Γ-point ($q=0$). These are called *acoustic modes* and exhibit a linear dispersion for long wavelength that correspond to sound waves. The slopes of the acoustic dispersion curves are related to the elastic constants.

12.2.2
Amplitudes of Lattice Vibrations and Normal Coordinates

The plane-wave solutions (Eq. (12.10)) of the equations of motion form a complete set of orthogonal functions if q is restricted to the first Brillouin zone. Hence, the actual displacement of an atom ($\kappa\ell$) can be represented by a linear combination of the $u_{\kappa\ell}^{\pm}(q,j)$

$$u_{\kappa\ell} = \sum_q \sum_j \left[A_{q,j} \cdot u_{\kappa\ell}^{+}(q,j) + A'_{q,j} \cdot u_{\kappa\ell}^{-}(q,j) \right]. \tag{12.20}$$

Making use of the fact that the displacements are real quantities, this equation reduces to

$$u_{\kappa\ell} = \frac{1}{\sqrt{Nm_\kappa}} \sum_q \sum_j Q_{q,j} \cdot e_\kappa(q,j) \cdot e^{iqr_\ell}, \tag{12.21}$$

where

$$Q_{q,j} = A_{q,j} \cdot e^{-i\omega_{q,j}t} + A^{*}_{-q,j} \cdot e^{i\omega_{q,j}t} \tag{12.22}$$

are called *normal coordinates*. They reflect the relative weight and amplitude of a particular vibrational mode (q,j) that is temperature dependent.

In terms of these normal coordinates, the Hamiltonian of the lattice (Eq. (8)) is reduced to a sum of independent harmonic oscillators.

$$H = \frac{N_Z}{2N_c} \sum_q \sum_j \left[\left|\frac{dQ_{q,j}}{dt}\right|^2 + \omega_{q,j}^2 \cdot |Q_{q,j}|^2 \right] \tag{12.23}$$

(N_Z is the number of primitive cells within the crystal).

Using the well-known quantization method for harmonic oscillators, we can immediately go from the classical to the quantum picture. Each oscillator is characterized by its frequency $\omega_{q,j}$ and normal coordinates $Q_{q,j}$. The quantum levels are equidistant.

$$E_n = \left(n + \frac{1}{2}\right) \cdot \hbar\omega_{q,j}. \tag{12.24}$$

The excitations that lead from one level to the next are called phonons.

As long as anharmonic effects are neglected the oscillators are decoupled and, therefore, there are no interactions between the individual phonons. The respective amplitudes are directly coupled to the excitation level of the oscillators that can be determined by quantum statistical methods.

In thermal equilibrium the average number of phonons is given by the Bose factor:

$$n_{q,j} = \frac{1}{\exp(\hbar\omega_{q,j}/k_B T) - 1}, \qquad (12.25)$$

and the corresponding contribution of these phonons to the lattice energy is

$$E_{q,j} = \left(n_{q,j} + \frac{1}{2}\right)\hbar\omega_{q,j}. \qquad (12.26)$$

The mean square amplitude of the normal oscillator coordinate is obtained as

$$\langle |Q_{q,j}|^2 \rangle = \frac{\hbar}{\omega_{q,j}}\left(n_{q,j} + \frac{1}{2}\right). \qquad (12.27)$$

The total energy that is stored in the harmonic phonon system is given by the sum over all phonon states (q,j):

$$E_{\text{Ph}} = \sum_q \sum_j \hbar\omega_{q,j}\left(n_{q,j} + \frac{1}{2}\right). \qquad (12.28)$$

12.2.3
Phonon Density of States and Lattice Heat Capacity

Related thermodynamic quantities, such as the internal energy or the heat capacity, are determined by the frequency distribution of the lattice vibrations rather than by details of the phonon dispersion. Hence, it is useful to introduce the so-called *phonon density of states* $G(\omega)$ in such a way that $G(\omega)d\omega$ is the number of phonons with frequencies between ω and $\omega + d\omega$. Using this quantity, the sum in (Eq. (12.28)) may be replaced by an integral expression:

$$E_{\text{Ph}} - E_o = \int_0^\infty \frac{\hbar\omega}{\exp(\hbar\omega/k_B T) - 1} \cdot G(\omega) \cdot d\omega. \qquad (12.29)$$

Here, E_o is the energy at $T=0$. The derivative with respect to temperature provides the lattice heat capacity at constant volume:

$$c_V = k_B \cdot \int_0^\infty \left(\frac{\hbar\omega}{k_B T}\right)^2 \cdot \frac{e^{(\hbar\omega/k_B T)}}{\left(e^{(\hbar\omega/k_B T)} - 1\right)^2} \cdot G(\omega) \cdot d\omega. \qquad (12.30)$$

As an example, Figure 12.6 displays the phonon dispersion of GaAs as determined by inelastic neutron scattering along with the phonon density of states [6,7].

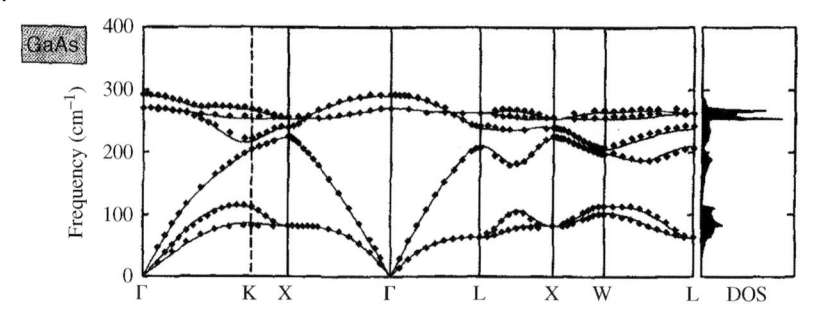

Figure 12.6 Phonon dispersion and density of states for GaAs. The experimental data are from Ref. [6], the full lines and the density of states (DOS) are results of *ab initio* model calculations. (Reproduced with permission from Ref. [7]. Copyright 1988, American Chemical Society.)

Obviously, even in this relative simple substance $G(\omega)$ (DOS) exhibits a rather complicated multipeak structure. Integral properties like the heat capacity are, however, not extremely sensitive on details of $G(\omega)$. For low temperatures, only low frequency vibrations are excited that are dominated by the acoustic modes. This is the reason, why the well-known Debye model that is based on a simple quadratic approximation of $G(\omega)$ is able to reproduce the T^3-behavior of the lattice heat capacity near $T=0$.

12.3
Spectroscopic Studies of Selected Systems

In chemistry, IR spectroscopy and Raman spectroscopy are widely used for the identification of particular bonds within molecules. The basic principle is the energy exchange between the probe (electromagnetic radiation in the case of optical spectroscopy) and the sample under consideration. Hence, spectroscopy is always governed by the law of energy conservation that allows the determination of energy levels. Characteristic frequencies of O—H stretching vibrations, for example, are found in the 3000 cm^{-1} regime (which corresponds to a frequency of $\nu \approx 100$ THz or an energy of $E \approx 400$ meV)[1] and typical C—C single bonds exhibit excitation frequencies that are lower by about a factor of three. If the molecular structure becomes more complex, frequency shifts are observed that provide important information about specific bonding situations such as the formation of hydrogen bonds.

In solids, the vibrational properties are determined by the collective nature of lattice phonons where a large number of atoms are involved. Unlike single molecules, the excitations are described by propagating waves that carry not only energy but also momentum. The momentum of a phonon with a wave vector q is given by $\hbar q$. Hence, spectroscopy needs to take into account the conservation

---

[1] In optical spectroscopy, it is convenient to use the wave number scale while scattering methods usually adopt the frequency or energy scale: $\bar{\nu} = \frac{1}{\lambda} = \frac{\nu}{c} = \frac{\omega}{2\pi c} = \frac{E}{hc}$

of momentum as well. Due to the fact that photons in the infrared or visible spectral range exhibit wavelengths, which are orders of magnitude larger than characteristic lattice parameters of solids, they are not able to transfer considerable amounts of momentum to the sample and, hence, optical spectroscopy is sensitive to a small subset of lattice vibrations – namely, those with almost vanishing wave vectors close to the Γ-point. Figure 12.7 shows in the frequency–wavelength representation, the interesting regime for atom dynamics in solids around 1 THz and 0.1 nm along with the traces of electromagnetic radiation and neutrons. This clearly demonstrates why neutrons are particularly suited for spectroscopic studies in solids. Nevertheless, optical spectroscopy as a home-lab technique provides important information about specific phonons with accuracies that are hardly achievable with other methods. Due to the enormous increase in brilliance of modern synchrotron sources and the development of new beam optical devices over the last years, very high resolution can nowadays be achieved that allows inelastic X-ray scattering studies over a wide range of energy and momentum transfers. Moreover, the combination of different techniques allows the assignment of excitations to specific phonon modes due to their particular selection rules.

The crystalline structure imposes constraints on the possible displacement patterns of vibrations. These eigenvectors, that is, the collection of 3D displacement vectors of all N_c atoms within a primitive cell, are subject to transformation properties of the point group of the crystal if phonons of zero wave vector are concerned, or the subgroup of the point group that leaves the wave vector q invariant for arbitrary directions. If a corresponding symmetry operation is applied, not only the atoms but also their respective displacement vectors are transformed. Hence, each of the phonon modes can be assigned to one of the

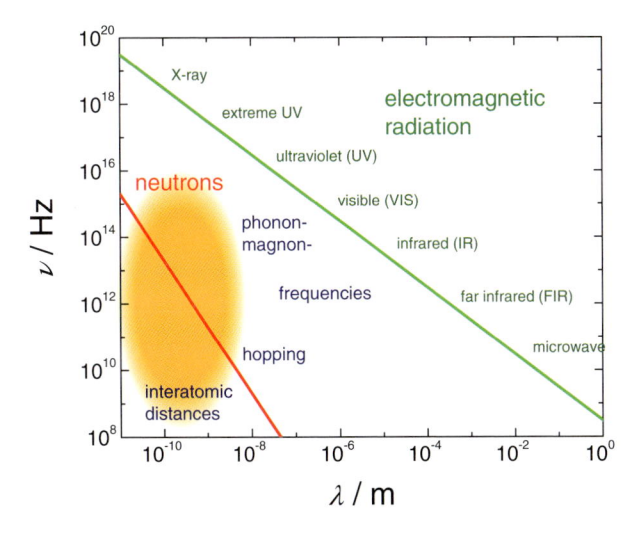

Figure 12.7 Frequency–wavelength relationship for electromagnetic radiation and neutrons.

irreducible representations of the point group. A comprehensive description of the symmetry properties is given in Ref. [2].

There are a variety of different experimental techniques that are frequently used for optical spectroscopy or inelastic neutron and X-ray scattering. In most cases it is the competition between high intensity and high resolution that determines the most appropriate method. While Fourier transform IR spectroscopy is almost exclusively used at least in the far-infrared regime, high-resolution Raman spectrometers use up to three gratings for the monochromatisation of the scattered beam. In X-ray and neutron spectroscopy, crystal monochromators are employed whenever detailed information about particular phonon modes is required. In contrast, neutron time-of-flight techniques allow access to a broader range of (Q,ω) space and are particularly efficient for the study of, for example, the phonon density of states.

A detailed description of different techniques is beyond the scope of this chapter. For more information the reader is referred to Refs [8,9].

12.3.1
IR Spectroscopy

In IR spectroscopy, the electromagnetic wave is used to excite a lattice vibration. This can only be performed if the vibration is associated with an electric dipole moment that interacts with the electric field of the IR radiation. Hence, IR spectroscopy is restricted to ionic solids and vibrations that belong to irreducible representations that are contained in the vector representation of the electric dipole moment.

IR spectroscopy is a technique that is frequently used in chemistry since it allows the assignment of individual modes as fingerprints of specific bonds within molecular or functional groups. A compilation of characteristic frequencies is given in Ref. [10]. Usually, the wave number regime between some hundred (bending vibrations) and several thousand (stretching vibrations of CH- or OH-groups) cm^{-1} is studied. In solids, however, significant variations of those frequencies are frequently observed that reflect the strong interatomic interactions and the modification of the electronic density. In systems with strong hydrogen bonds, this effect is very pronounced and can even lead to structural phase transitions like in KH_2PO_4 where the hydrogen bonds form a three-dimensional network that becomes ferroelectric at low temperatures. For the study of collective lattice vibrations, the low frequency, far-infrared regime (FIR) below, say, 500 cm^{-1} is particularly interesting. Here, the many-body interactions are responsible for the vibrational properties.

12.3.1.1 Thin Films versus Bulk Samples
Fourier transform spectroscopy is almost exclusively used for FIR studies, since it provides high intensities at reasonable resolution. This is achieved by the simultaneous use of a broad wavelength spectrum that is modulated by a (Michelson-) interferometer.

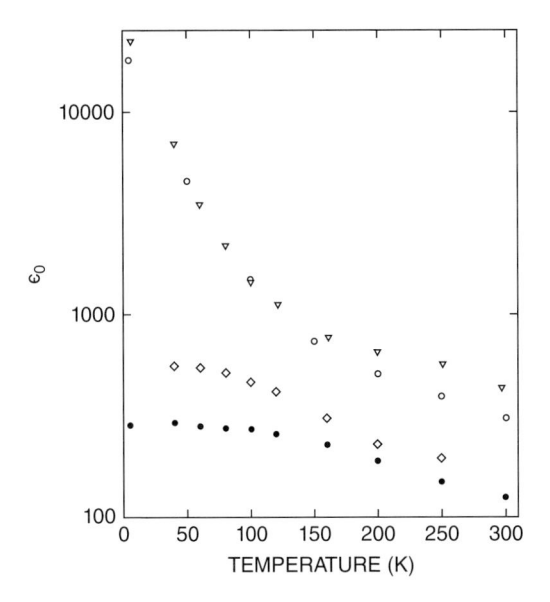

Figure 12.8 Temperature dependence of the low-frequency limit of the permittivity for a thin film (full circles) and for bulk SrTiO$_3$ (empty circles) as obtained from FIR experiments compared with data from macroscopic dielectric experiments (diamonds and triangles) (after Ref. [11] and references therein). (Reproduced with permission from Ref. [11]. Copyright 1999, American Chemical Society.)

For a large variety of samples, the absorption of FIR radiation is rather strong and transmission experiments, which allow the direct determination of the frequency-dependent absorption coefficient, are difficult and may require the preparation of thin films. As shown in SrTiO$_3$, however, the vibrational properties of thin films may be considerably different to those of bulk samples [11]. In Figure 12.8, the low-frequency permittivity as deduced from FIR-spectra is compared for thin films and bulk samples. It is clearly seen that the permittivity of thin films is reduced by more than an order of magnitude at low temperatures. Internal strains within the film are supposed to be responsible for this striking difference.

12.3.1.2 IR Reflectivity of Multiferroic Systems

Due to the difficulties with IR transmission experiments, many FIR studies focus on reflectivity measurements that can be transformed into absorption spectra using Kramers–Kronig analysis (see Box 12.1) or that can be fitted by some microscopic models for the individual vibrational modes. A most recent example is the study of the multiferroic material MnWO$_4$ [12]. The reflectivity spectra in Figure 12.9a show a number of different features varying with the reflection angle. The shape can be explained on the basis of specific models such as the generalized Drude–Lorentz model employed by these authors allowing the determination of the dispersion of the absorption coefficient or, equivalently, the imaginary part of the permittivity. Alternatively, the Kramers–Kronig analysis,

> **Box 12.1**
>
> The interaction of infrared radiation with matter is usually described by the complex, frequency-dependent permittivity $\epsilon = \epsilon' - i\epsilon''$ and the complex refractive index
>
> $$n = n' - in'' = \sqrt{\epsilon}. \tag{12.31}$$
>
> If the beam hits the sample under an angle φ, the reflection coefficient is given by
>
> $$r = \frac{\cos\varphi - \sqrt{n^2 - \sin^2\varphi}}{\cos\varphi + \sqrt{n^2 - \sin^2\varphi}}, \tag{12.32}$$
>
> and the reflectivity by
>
> $$R = |r|^2. \tag{12.33}$$
>
> Similarly, the transmission coefficient is
>
> $$t = (1-r)(1+r)\exp\left(-\frac{\omega}{c}n''d\right), \tag{12.34}$$
>
> if ω is the angular frequency of the radiation and d is the thickness of the sample under consideration. The fraction of transmitted intensity is
>
> $$T = |t|^2. \tag{12.35}$$
>
> Since real and imaginary parts of the permittivity and the refractive index are related by Kramers–Kronig relations,
>
> $$\epsilon'(\omega) - \epsilon'(\infty) = \frac{2}{\pi}\int_0^\infty \frac{x\epsilon''(x)}{x^2 - \omega^2}dx,$$
>
> $$\epsilon''(\omega) = -\frac{2\omega}{\pi}\int_0^\infty \frac{\epsilon'(x) - \epsilon'(\infty)}{x^2 - \omega^2}dx + \frac{\sigma_0}{\epsilon_0\omega}, \tag{12.36}$$
>
> where $\epsilon'(\infty)$ and σ_0 are the permittivity at high frequencies, far beyond the range of atomic vibrations, and the DC conductivity, respectively, and the integrals are evaluated by their principal values.
>
> Using these relations, it is possible to obtain the complex permittivity from a single reflectivity or transmission experiment.

which, as a result from linear response theory, is independent of any particular model, yields similar results as demonstrated in Figure 12.9b. The well-defined peaks enable the accurate determination of phonon eigenfrequencies and damping constants even from the broad reflectivity spectra.

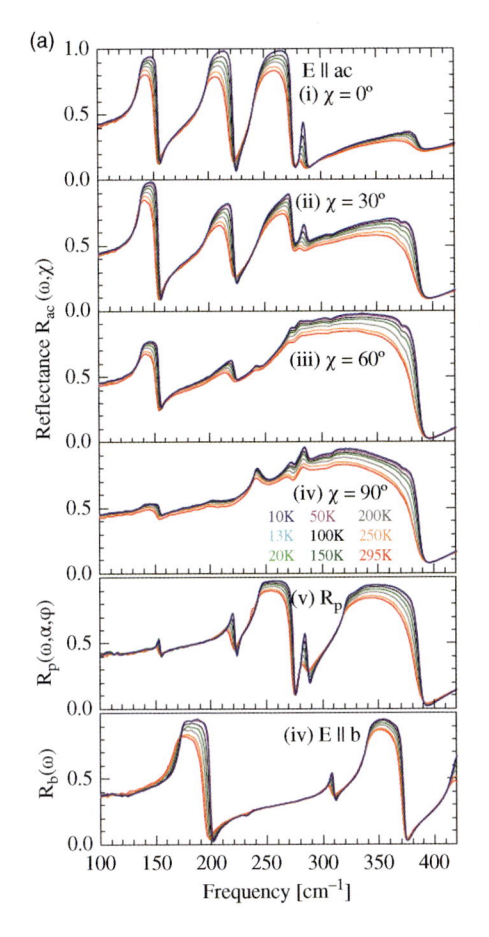

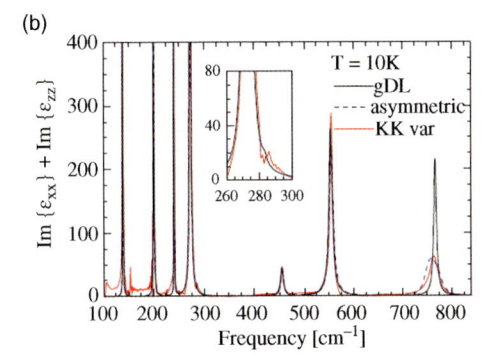

Figure 12.9 (a) Temperature dependence of the reflectance in the low-frequency range. (i)–(iv): $R_{ac}(\omega,\chi)$, (v) $R_p(\omega,11°,80°)$, (vi) $R_b(\omega)$. (b) The sum $\text{Im}\{\varepsilon_{xx}(\omega)\} + \text{Im}\{\varepsilon_{zz}(\omega)\}$ shows all eight B_u phonon modes. A Kramers–Kronig constrained variational analysis of $R_{ac}(\omega,\chi)$ (KKvar, red line) confirms the results of a generalized Drude–Lorentz fit [gDL, black line] with the exception of the line shape of the highest mode at 767 cm^{-1}. (Reproduced with permission from Ref. [12]. Copyright 2001, American Chemical Society.)

12.3.2
Raman Spectroscopy

Raman spectroscopy is also called inelastic light scattering and involves the polarization of solids by the incident laser light. If lattice vibrations are associated with a variation of the polarizability, the emission of dipole radiation is induced, that is shifted in frequency with respect to the incident beam by the vibration frequency $\pm \nu_s$. If the frequency decreases, the respective quantum oscillator is excited, that is, a phonon is created and the corresponding spectral line is called the Stokes signal. If the frequency of the scattered beam is higher than that of the incident beam, a phonon is absorbed and the anti-Stokes spectrum is observed. Due to the fact that the scattered intensity is proportional to a tensor quantity, namely, the phonon induced polarizability, the only vibrations that can contribute belong to an irreducible representation that is also contained in the tensor representation. In polarized Raman experiments, the polarization of both, the incident and the scattered beam is controlled and particular selection rules apply, which help to assign phonon modes to individual irreducible representations. In Porto notation, the setting of a spectrometer with the incident beam along x with polarization along a and the scattered beam along y with polarization along b is denoted as $y(b,a)x$.

Quite frequently, Raman spectra are complementary to IR spectra due to the different selection rules. In centrosymmetric systems, in particular, all IR active modes belong to odd (u) representations, while all Raman active ones need to be even (g).

Box 12.2

In the classical picture, inelastic light scattering is described on the basis of the polarization of the sample induced by the incident electromagnetic wave. The scattered intensity is consequently obtained as the dipole radiation emitted by the vibrating sample.

If $E = E_0 \exp(i(k_i r - \omega t))$ is the electric field of the incident wave of wave vector k_i and frequency ω_0, the induced dipole density (polarization) is determined by the polarizability tensor α:

$$P = \alpha E. \tag{12.37}$$

The excitation of a phonon that is described by a normal coordinate Q_{qj} might lead to a variation of this polarization according to

$$\alpha = \alpha_0 + \frac{\partial \alpha}{\partial Q_{qj}} Q_{qj}. \tag{12.38}$$

The quantity $(\partial \alpha / \partial Q_{qj})$ is called the Raman tensor.

Since Q_{qj} is varying periodically in time with the phonon frequency ω_{qj}, the polarization is modulated and consists of three individual contributions with frequencies ω_0, $\omega_0 + \omega_{qj}$, and $\omega_0 - \omega_{qj}$:

$$P = \alpha_0 E_0 e^{i(k_i r - \omega_0 t)} + \frac{\partial \alpha}{\partial Q_{qj}} \left[A_{qj} e^{ik_i r} E_0 e^{-i(\omega_0 + \omega_{qj})t} + A^*_{-qj} e^{ik_i r} E_0 e^{-i(\omega_0 - \omega_{qj})t} \right]$$

$$\tag{12.39}$$

Consequently, the scattered light consists of three components: Rayleigh, anti-Stokes, and Stokes scattering corresponding to elastic scattering, the absorption, and the creation of a phonon, respectively. Modulus and direction of the polarization is clearly determined by the elements of the Raman tensor and, hence, polarized Raman scattering can be used to distinguish between different phonon modes: If a polarization filter is used to select scattered light with the

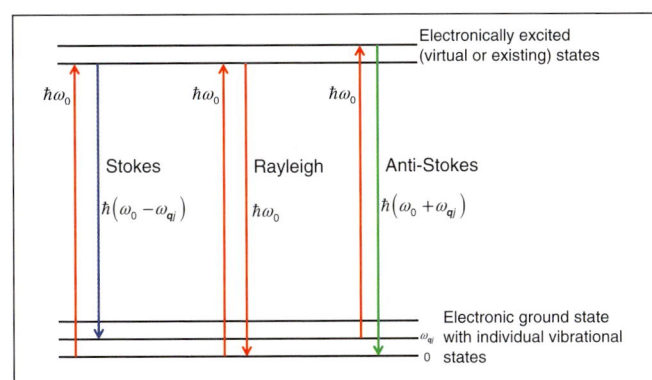

Figure 12.10 Electronic states involved in the Raman process.

electric field along E_1, the intensity is proportional to the expression $|E_1(\partial\alpha/\partial Q_{qj})E_0|^2$ that is nonzero only if the mode (qj) belongs to an irreducible representation of the point group that is also contained in the usual three-dimensional (reducible) tensor representation.

In a quantum mechanical description, the Raman process is described by the excitation of an (virtual or existing) electronic state by the incident photon followed by a transition to a different vibrational state within the electronic ground state as shown in Figure 12.10. Strongly enhanced intensity is obtained if the electronically excited state is an actually existing state of the crystal. This process is called resonant Raman scattering.

Different experimental techniques are employed in Raman spectroscopy: Lasers with different and sometimes even continuously variable wavelengths are used as brilliant and coherent light sources. Resonance Raman spectra with enhanced intensity can be obtained if the energy of the incident light corresponds to a transition between different electronic states of the sample. While Fourier transform Raman spectrometers use an interferometer to analyze the spectrum of scattered light, conventional instruments use gratings as monochromators and either single photomultiplier detectors or CCD cameras for a multichannel operation. If spectra at small Raman shifts of a few cm^{-1} are of interest, the suppression of the elastically scattered intensity is crucial in order to separate the inelastic signals from the much stronger elastic Rayleigh line. Hence, high-resolution spectrometers use up to three gratings that can be operated in different modes in order to improve Rayleigh suppression and/or energy resolution.

12.3.2.1 Lattice Anomalies in Multiferroics

An example of a recent high-resolution study with a triple Raman spectrometer is shown in Figure 12.11 [13]. Here, extremely small anomalies of Raman active lattice modes of MnWO$_4$ are observed close to the multiferroic phase transition. Even if the electric polarization due to a displacement of the oxygen ions is only very small, the corresponding changes of mode frequencies of the order of some 0.1 cm^{-1} could be determined with sufficient accuracy along with the damping reflected by the temperature-dependent line width. In Figure 12.11, the temperature variation of the all A$_g$ and B$_g$ modes, selected by the polarization of incident and scattered beam, is displayed. There are individual modes that are almost not affected by the phase transition (e.g., A$_g$2), while others show either a softening (e.g., A$_g$0) or a stiffening (e.g., B$_g$5) when entering the magnetically ordered phase. This study shows that variations of less than 0.1 cm^{-1} can be identified. Care must, however, be taken to correct the spectra for any drift of the detector or grating performance that usually lead to shifts of several 0.1 cm^{-1}. To this end, the authors applied a sophisticated calibration method using the well-known emission lines from neon as a standard.

12.3.2.2 Raman Spectra of Multidomain Samples

The power of polarized Raman spectroscopy can also help to understand structural features such as domain formation at phase transitions. SrTiO$_3$, a well-studied perovskite exhibits an antiferrodistortive phase transition at 105 K leading to a tetragonal phase, where lattice modes become Raman active. There are three tetragonal domains with their symmetry axis along x, y, or z. For different polarization settings of the Raman spectrometer, the visibility of modes in the individual domains varies strongly as shown by Gibhardt et al. [14]. Moreover, the domain distribution can be modified by the application of an electric field. Hence, the Raman spectra directly reflect the domain structure. As an example, data taken at 45 K and different electric fields are shown in Figure 12.12. While the mode at about 44 cm^{-1} increases with electric field in the b(cc)a polarization, the opposite behavior is observed in c(a,b)a polarization. This is a clear signature of a redistribution of domains under the influence of an electric field.

12.3.2.3 Local Distortions Revealed by Raman Scattering

Raman active modes at higher energies can often be used to characterize the local bonding configurations in complex solids since the correlation length of phonons is usually limited to a few unit cells. Su et al. [15] have shown that the different local structure of various nanoporous silicotitanates gives rise to well-defined shifts of individual Raman lines as shown in Figure 12.13 for the sequence BaTiSi$_3$O$_9$, K$_2$TiSi$_3$O$_9$, and Cs$_2$TiSi$_6$O$_{15}$. Of particular interest is the mode close to 960 cm^{-1} that corresponds to stretching vibrations of the Si—O—Ti network. It is found to soften when Cs is replaced by K and finally by Ba and reflect the local distortions of the building blocks consisting of TiO$_6$ octahedra and SiO$_4$ tetrahedra.

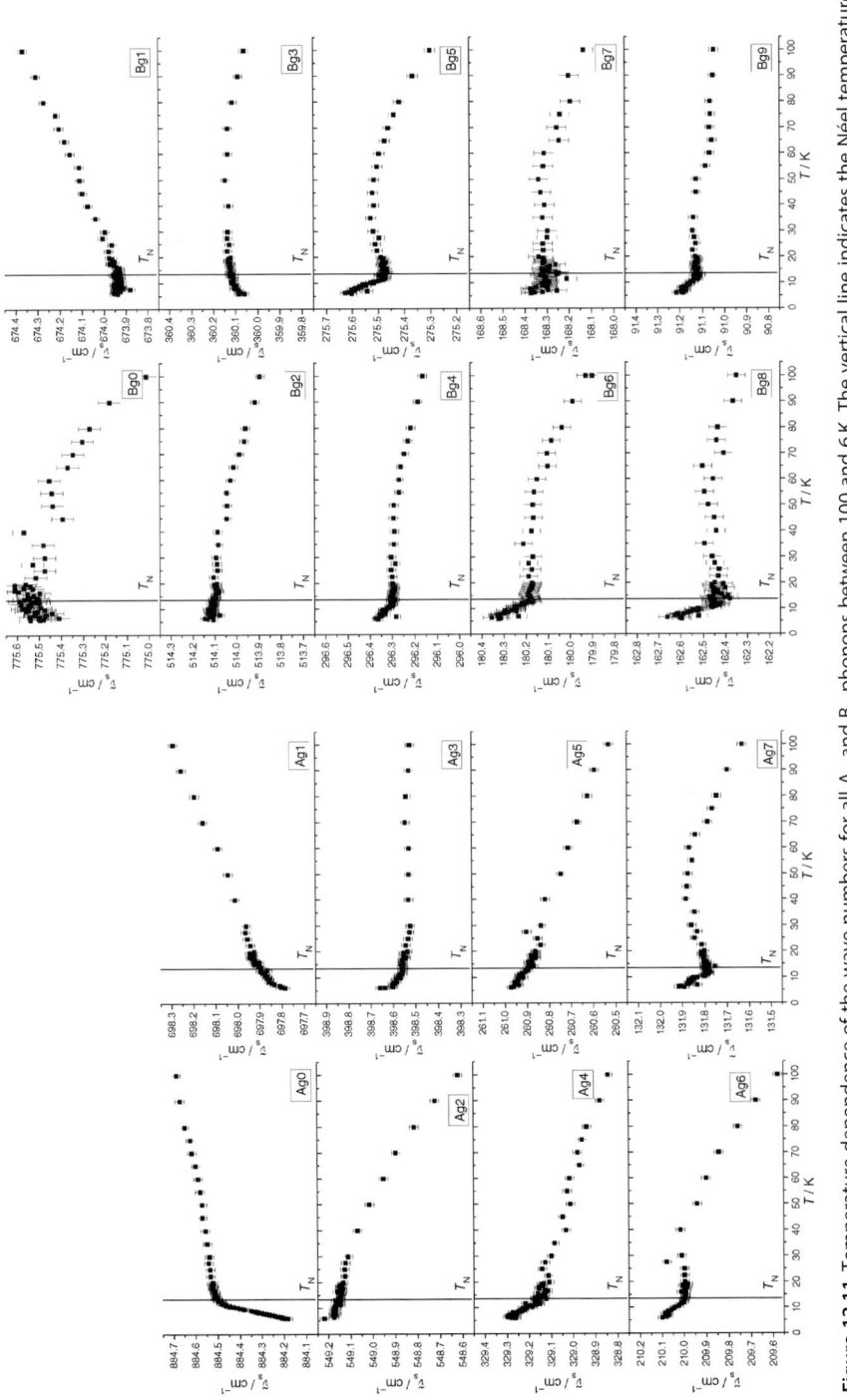

Figure 12.11 Temperature dependence of the wave numbers for all A_g and B_g phonons between 100 and 6 K. The vertical line indicates the Néel temperature $T_N = 13.5$ K. (Reproduced with permission from Ref. [13]. Copyright 2015, http://iopscience.iop.org/article/10.1088/2053-1591/2/9/096103/meta. Used under CC BY 3.0 https://creativecommons.org/licenses/by/3.0/.)

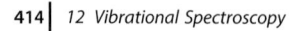

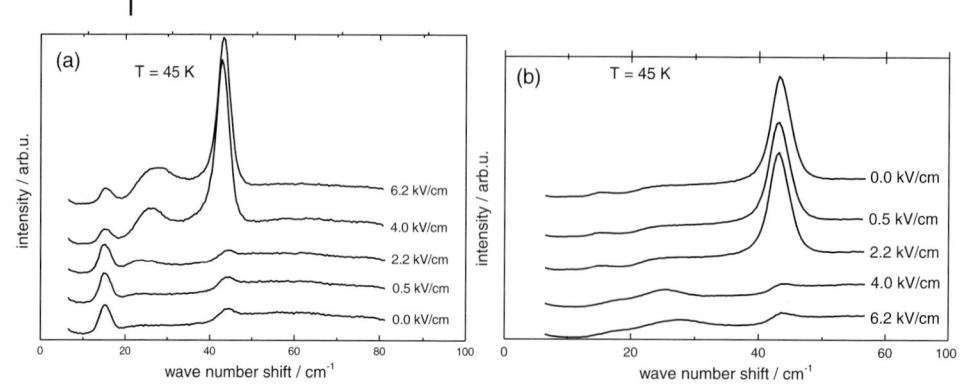

Figure 12.12 Raman spectra of SrTiO$_3$ at 45 K for configuration b(cc)a (a) and b(ab)a (b) as a function of the applied electric field (after Ref. [14]).

12.3.3
Inelastic Neutron Scattering

Since thermal neutrons exhibit both energies and momenta on the order of typical phonon quantities, they are able to interact with almost every phonon mode. This is an advantage, on the one hand, since it allows in principle a complete determination of phonon dispersion curves just as shown in Figures 12.1–12.3. On the other hand, it is not at all clear how to assign individual phonon modes to the experimental signals. The fact, however, that each phonon mode can be detected in arbitrary Brillouin zones with specific intensity provides in principle sufficient information to identify the modes.

In a typical scattering experiment, neutrons with a wave vector k_i are prepared and hit the sample where they are scattered into a solid angle $d\Omega$ around some wave vector k_f as illustrated in Figure 12.14. The momentum transferred to the

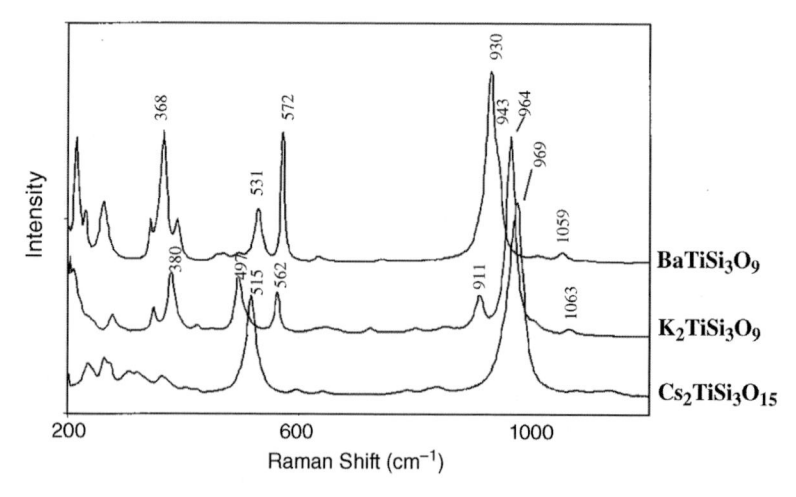

Figure 12.13 Raman spectra of Cs$_2$TiSi$_6$O$_{15}$, BaTiSiO$_3$, and K$_2$TiSi$_3$O$_9$. (Reproduced with permission from Ref. [15]. Copyright 1987, American Chemical Society.)

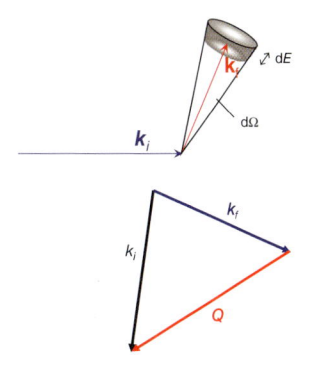

Figure 12.14 Layout of a typical scattering experiment.

sample is given by the scattering vector $\mathbf{Q}$:

$$\hbar \mathbf{Q} = \hbar(\mathbf{k}_i - \mathbf{k}_f), \tag{12.40}$$

and the energy transfer is

$$\hbar\omega = \frac{\hbar^2}{2m_n}\left(k_i^2 - k_f^2\right). \tag{12.41}$$

If all neutrons, that exhibit energy transfers within an interval dE, are counted then the scattered intensity is given by the double differential cross section:

$$\frac{d^2\sigma}{d\Omega dE} = \frac{k_f}{k_i} S(\mathbf{Q}, \omega), \tag{12.42}$$

where the scattering function $S(\mathbf{Q},\omega)$ depends only on the energy and momentum transfer, but not on the specific choice of the energy of the incident neutrons.

Box 12.3

As shown in textbooks on scattering theory [16], the (coherent) scattering from phonons with wave vector $\mathbf{q}$ and branch index j is given by the expression

$$S_{\text{coh}}(\mathbf{Q}, \omega) = \frac{\hbar}{4\pi N_c \omega_{qj}} \left| \sum_\kappa b_\kappa \frac{\mathbf{Q} \cdot \mathbf{e}_\kappa(qj)}{\sqrt{m_\kappa}} \exp(-W_\kappa)\exp(-i\mathbf{q}\mathbf{r}_\kappa) \right|^2$$

$$\times \left\{ n_{qj}\, \delta(\omega + \omega_{qj}) \sum_g \delta_{\mathbf{Q},\mathbf{g}-\mathbf{q}} + (n_{qj} + 1)\delta(\omega - \omega_{qj}) \sum_g \delta_{\mathbf{Q},\mathbf{g}+\mathbf{q}} \right\}. \tag{12.43}$$

The δ-functions within the curly brackets guarantee the energy and momentum conservation. The sum runs over all reciprocal lattice vectors $\mathbf{g}$ and considers that part of the momentum is transferred onto the rigid lattice, while the rest is used to create or absorb a phonon. Hence, a phonon can in principle be detected in arbitrary Brillouin zones. The different prefactors, n_{qj} and $(n_{qj} + 1)$, reflect the principle of detailed balance stating that the absorption of a phonon ($\omega = -\omega_{qj}$) requires the thermal occupation of excited energy states.

The quadratic expression in the first line of Eq. (12.43) is somewhat similar to the atomic form factor used to describe Bragg reflections. The sum runs over all atoms

(continued)

(continued)
within the primitive unit cell; b_κ, m_κ, and $\exp(-W_\kappa)$ are the (coherent) scattering length, mass, and Debye–Waller factor of atom κ, respectively. The dot product $Q \cdot e_\kappa$ indicates that only those phonons can be observed that exhibit at least one displacement component in the direction of the overall momentum transfer. Since Q varies with the Brillouin zone (g), this allows the distinction between different modes and eventually also the determination of the entire polarization vector.

Since neutrons are scattered by atomic nuclei, the scattering power depends on the nuclear structure, the number of nucleons, or the nuclear spin. Hence, even chemically identical atoms may exhibit different scattering behavior. As a consequence, there is another contribution to the scattered intensity, which is called incoherent scattering and that does not experience interference effects from distinct scattering particles. It is simply the sum of the individual contribution of all atoms:

$$S_{\text{inc}}(Q, \omega) = \frac{\hbar}{8\pi} \sum_\kappa \sigma_\kappa^{\text{inc}} \frac{|Q \cdot e_\kappa(qj)|^2}{m_\kappa \omega_{qj}} \exp(-2W_\kappa) \\ \times \left\{ n_{qj} \delta(\omega + \omega_{qj}) + (n_{qj} + 1) \delta(\omega - \omega_{qj}) \right\} \quad (12.44)$$

($\sigma_\kappa^{\text{inc}}$ being the incoherent scattering cross section of atom κ).

Summing up, the contributions from all phonon states yield the total incoherent scattering. The sum of δ-functions can be replaced by the phonon density of states $G(\omega)$ if $\langle |Q \cdot e_\kappa(qj)|^2 \rangle$ is used to denote the averaged polarization factor of modes with frequencies between ω and $\omega + d\omega$:

$$S_{\text{inc}}(Q, \omega) = \frac{\hbar}{8\pi} \sum_\kappa \sigma_\kappa^{\text{inc}} \frac{\langle |Q \cdot e_\kappa(qj)|^2 \rangle}{m_\kappa} \exp(-2W_k) \frac{2n(\omega) + 1}{\omega} G(\omega). \quad (12.45)$$

Hence, the incoherent scattering is proportional to the (weighted) phonon density of states. For powder samples, the orientational average of Qe_κ guarantees that the scattered intensity is isotropic.

The extraordinary large incoherent scattering cross section of hydrogen – being almost an order of magnitude larger than that of any other nucleus – allows the very sensitive detection of vibrations associated with the displacement of protons.

The inelastic structure factor (Eq. (12.43)) describes how the intensity of a particular phonon mode varies if measured within different Brillouin zones as characterized by a reciprocal lattice vector g. Hence, data taken for a good number of Brillouin zones allow the determination of phonon eigenvectors. This method has been applied to different systems like semiconductors (Si [17,18], GaAs [19]), minerals (SiO_2 [20]), ferroelectrics ($KNbO_3$ [21]), or molecular crystals (naphthalene [22]) using three-axes spectrometers that allow the detection of scattered intensities at (almost) arbitrary points in Q–ω space.

Moreover, the experimental data can be arranged according to their eigenvectors and the corresponding symmetries in order to obtain dispersion curves like those that have been displayed in Figures 12.3–12.5. These represent profiles along main symmetry directions of the 3D dispersion surface and reflect the variety of chemical bonds between all individual atoms of the solid. The comparison with model calculations based on phenomenological potentials or with *ab initio* calculations based on the solution of Schrödinger's equation allows the identification of interatomic interactions that are responsible for the individual phonon modes. Phenomenological treatments need the specification of pair or three-body potentials and can also account for polarization (shell model [23]) or bond charge effects [24]. Computer software such as UNISOFT[2] is available that does not only calculate the phonon frequencies and eigenvectors but is also able to visualize the lattice vibrations [25]. DFT *ab initio* calculations are based on programs such as VASP[3] or ABINIT[4] that evaluate the electronic ground state for an adequate supercell of the crystal and apply periodic boundary conditions to consider the infinite crystal. While the ground state corresponds to the static equilibrium structure, effective force constants can be determined when shifting individual atoms from their equilibrium position and calculating the increase in energy. Based on these Hellmann–Feynman forces, the dynamical matrix, Eq. (12.19), can be collected in just the same way as for phenomenological models – now, however, with no or only few free parameters (e.g., electron correlation parameter). PHONON is a powerful software package that calculates phonon dispersion from first principles[5].

Based on the detailed description of lattice excitations by computational methods, also averaged quantities like the phonon density of states or the lattice heat capacity can be obtained.

When dealing with a powder we average over all phonon signals possessing the same $|Q|$. This does not imply that we cannot access certain aspects of the dispersion relations. In particular, when combined with *ab initio* calculations, powder-averaged coherent scattering can give access to the lattice dynamics as was recently successfully demonstrated on skutterudites [26]. We will come back to this work in Section 12.3.3.6. Despite this possibility, most powder investigations aim at extracting the phonon density of states from the data. This is particularly attractive when dealing with incoherent scattering. In that case and for cubic monoatomic samples, there is a direct connection between the scattered intensity and the density of states. However, practically all systems of interest are complex and thus do not fulfill these extremely stringent requirements. Fortunately, an excellent idea of the phonon density of states can be obtained even for systems composed of several coherent scatterers under what is called the incoherent approximation [27]. Time-of-flight spectroscopy is the most suitable technique to obtain data from polycrystalline samples since it simultaneously covers a broad range of the relevant Q–ω space. We will show several examples later in the chapter.

---

2) http://www.uni-pc.gwdg.de/eckold/unisoft.html.
3) http://www.vasp.at/.
4) http://www.abinit.org/.
5) http://www.computingformaterials.com/.

The determination of bond energies within solids and the characterization of chemical bonding is one aspect of lattice dynamical investigations. Another topic is the variation with external thermodynamic parameters such as temperature, pressure, or electric and magnetic field.

12.3.3.1 Elastic Constants and Mechanical Anisotropy

The three acoustic phonon modes with linear dispersion close to the center of the Brillouin zone correspond to sound waves and the slope is given by the respective sound velocity. Hence, all individual elements of the matrix of elastic constants can be obtained if these acoustic phonons are determined for a number of propagation directions and for different polarization vectors. In a recent study on multiferroic $Bi_2Mn_4O_{10}$, the mechanical anisotropy and all nine elastic constants of the orthorhombic system could be determined with good accuracy [28]. This required the determination of phonon branches not only in high-symmetry but also in low-symmetry directions. In a three-dimensional representation, the transverse and longitudinal sound velocities are illustrated in

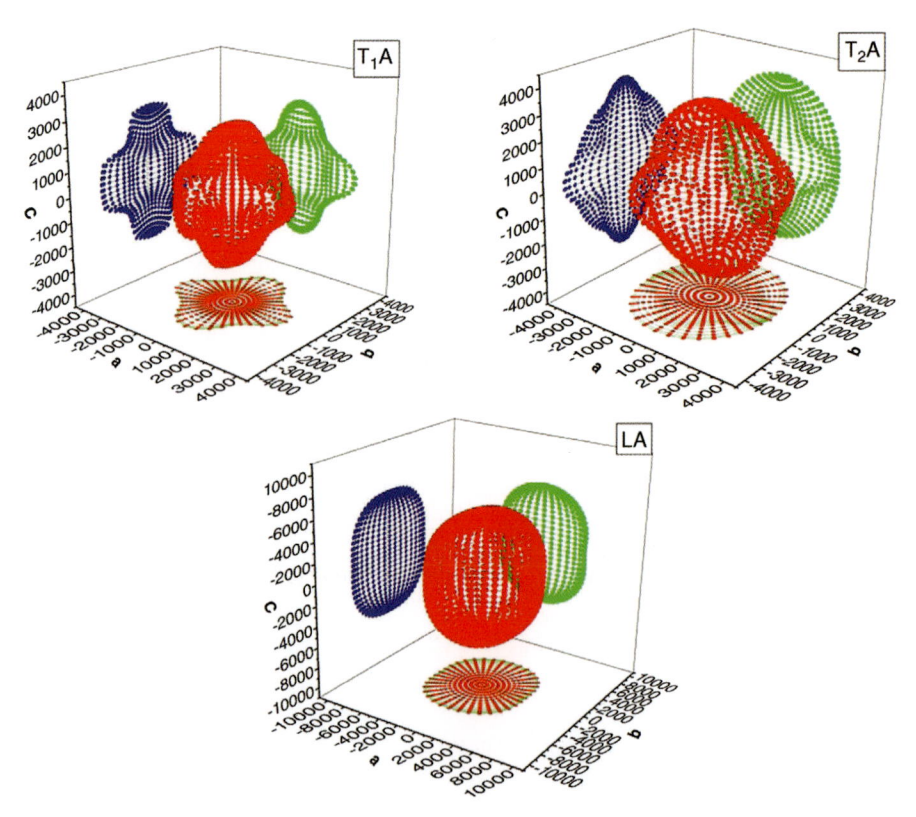

Figure 12.15 Three-dimensional representation of the anisotropic behavior of sound velocities (in ms^{-1}) in multiferroic $Bi_2Mn_4O_{10}$. T_1A and T_2A and LA denote the transverse and longitudinal modes, respectively. (Reproduced with permission from Ref. [28]. Copyright 2008, Elsevier.)

Figure 12.15. Obviously, there are soft directions along [110], [101], and [011]. These results may be important for the understanding of the multiferroic transition that is related to the interaction between electric, magnetic, and elastic properties.

12.3.3.2 Softmode Phase Transitions

Softmode phase transitions are associated with a particular phonon mode that induces instability of the lattice. Frequently, a high-symmetry phase is dynamically stabilized at high temperatures. If on cooling, a lattice excitation softens, its amplitude becomes larger and larger and finally, a displacive transition into a new crystallographic phase can be observed at a critical temperature T_c where the atoms are frozen at positions that correspond to the eigenvector of the softmode. Usually, Landau theory is used to describe this phase transition that may be of first or second order and the normal coordinate of the softmode serves as the order parameter. During a first-order transition, the phonon frequency does not vanish but exhibits a discontinuity at T_c, while a second-order transition is associated with a condensation of the softmode at zero frequency as illustrated in Figure 12.16 for $BaTiO_3$ and $SrTiO_3$ (after Refs [29,30]). Both compounds are cubic perovskites at high temperatures. $BaTiO_3$ exhibits a sequence of three first-order transitions into a tetragonal, orthorhombic, and rhombohedral low-temperature phase. Different behavior is found in the Sr compound that yields a second-order transition at 105 K into a tetragonal phase with almost complete condensation of the phonon mode.

Both compounds differ not only in the order of the transition but also in the wave vector of the softmode. While in $BaTiO_3$ the softening takes place at the Γ-point with zero wave vector, the transition in $SrTiO_3$ is associated with a zone boundary mode at $q = (½\ ½\ ½)$. In the former case, the atomic displacements are identical in all unit cells corresponding to a ferro-transition while the latter case leads to a superstructure with opposite displacements in adjacent unit cells that is characteristic for antiferro-transition. A third type is found when a phonon mode gets soft at arbitrary points of the Brillouin zone. Here, the transition leads to an incommensurate phase that is associated with the formation of satellite reflections which are, in general, varying in temperature or pressure. This variation is independent of the behavior of the main Bragg reflections that characterize the average structure. The three scenarios are depicted in Figure 12.17. In $BaTiO_3$, the softmode carries an electric dipole moment and, hence, the low-temperature phase is ferroelectric. In $SrTiO_3$, a similar phonon mode exists but there is a competition with the zone boundary mode involving a rotation of TiO_6 octaedra in opposite direction. The softening of this antiferrodistortive mode is more pronounced and, consequently, the ferroelectric transition is suppressed. A polar phase can only be induced in $SrTiO_3$ at temperatures below about 40 K if an external electric field is applied. Incommensurate phases are frequently found as the result of a delicate balance of competing interactions in solids. Usually, this modulated structure is formed as an intermediate phase between a normal (*para-*)high-temperature phase and a low-temperature superstructure phase.

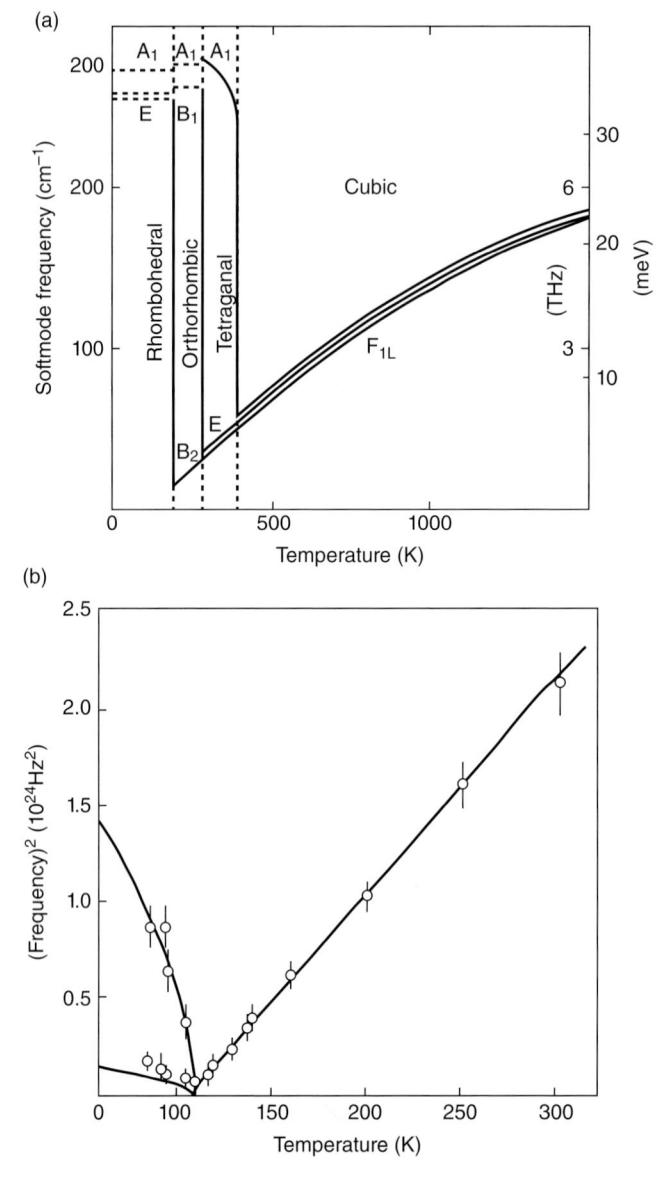

Figure 12.16 Temperature dependence of the softmode frequency of (a) BaTiO$_3$ (first-order transition [29]) and (b) SrTiO$_3$ (second-order transition [30]). (Reproduced with permission from Ref. [30]. Copyright 1969, Elsevier.)

The incommensurate structure can be regarded as a superposition of the normal structure of the *para*-phase and the displacement field of the softmode that is no longer dynamic in character but static. In most cases, the wave vector of this modulation varies with temperature until it locks in at a superlattice position and the satellites become usual superlattice reflections. This transition, therefore,

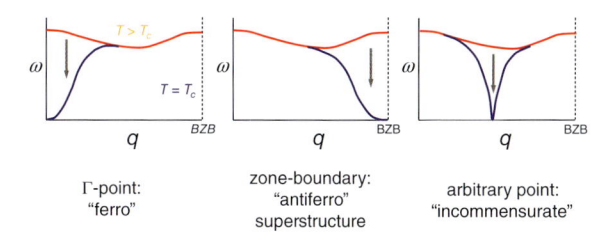

Figure 12.17 Scenarios for softmode transitions at different points of the Brillouin zone.

came to be called a lock-in transition. One of the most prominent examples of these systems is K_2SeO_4 [31], a member of the so-called A_2BX_4 family with building blocks of rather rigid BX_4^{2-} units with strong covalent character. Here, the softmode is the collective and propagating librational mode of BX_4 tetrahedra associated with a small displacement of the A^+ ions. While the probabilities for positive and negative displacements are equal within the incommensurate phase, there is a preference of one direction in the lock-in phase, which is ferroelectric. Figure 12.18 shows the soft mode dispersion of K_2SeO_4 for various temperatures along the a^*-direction of the orthorhombic crystal. The phonon condenses near $2/3\ a^*$ at 128 K and the lock-in transition to a threefold superstructure is observed at 93 K.

The α–β transition in quartz is another famous example of a softmode transition. Since the lattice consists of a framework of edge sharing SiO_4 tetrahedra, a tilt of a single and almost rigid unit gives rise to a collective excitation of the entire crystal. There are two modes that soften when the phase transition is approached, a zone boundary mode at the M-point of the hexagonal lattice and

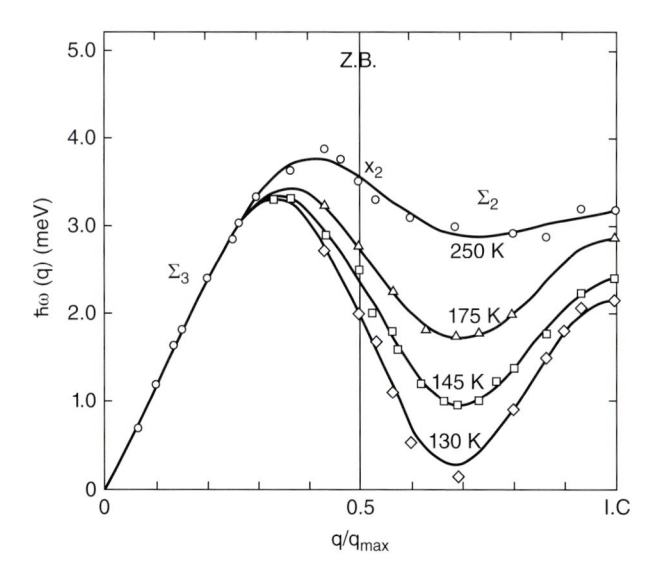

Figure 12.18 Softmode dispersion of K_2SeO_4 along a^* at various temperatures (after Ref. [31]). (Reproduced with permission from Ref. [31]. Copyright 2016, American Chemical Society.)

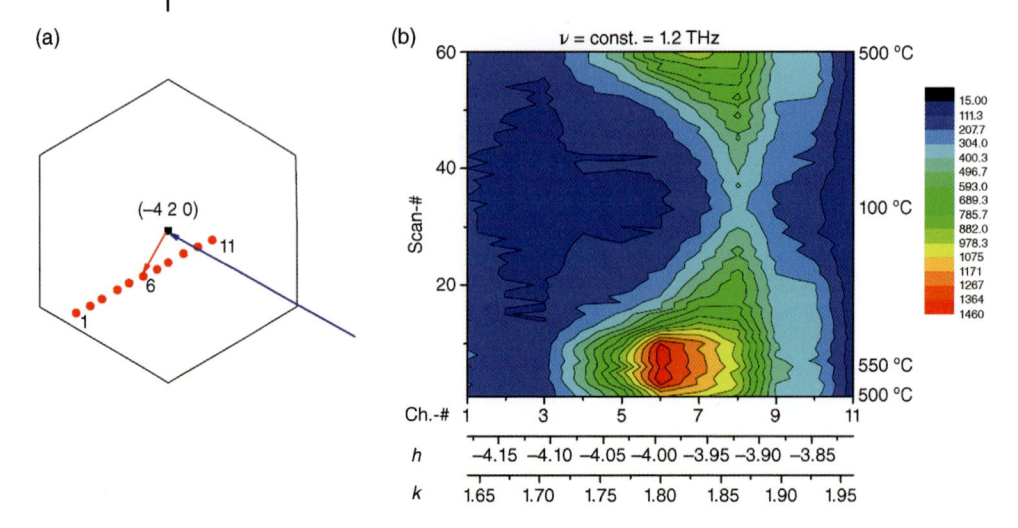

Figure 12.19 Temperature dependence of a constant energy scan at 1.2 THz across the transverse acoustic phonon in quartz (after Ref. [32]). (a) Trace of the scan realized using the multianalyzer technique. (b) Contour map during continuous heating and cooling between 100 and 550 °C (close to the phase transition). (Reproduced with permission from Ref. [35]. Copyright 1993, American Chemical Society.)

a zone center mode. While this general behavior is well known from earlier experiments [20], a new multiplexing neutron technique has recently been used to study the gradual softening of the entire phonon dispersion during continuous heating and cooling [32]. Figure 12.19 shows the evolution of the scattered intensity for a constant energy scan at 1.2 THz along a selected path within the Brillouin zone. It can easily be seen that the phonon intensity varies strongly with temperature and the maximum is shifted in q-space. For two selected temperatures, an intensity map is displayed in Figure 12.20 for a wide range of the Brillouin zone. While at 200 °C the intensity contour is almost symmetric

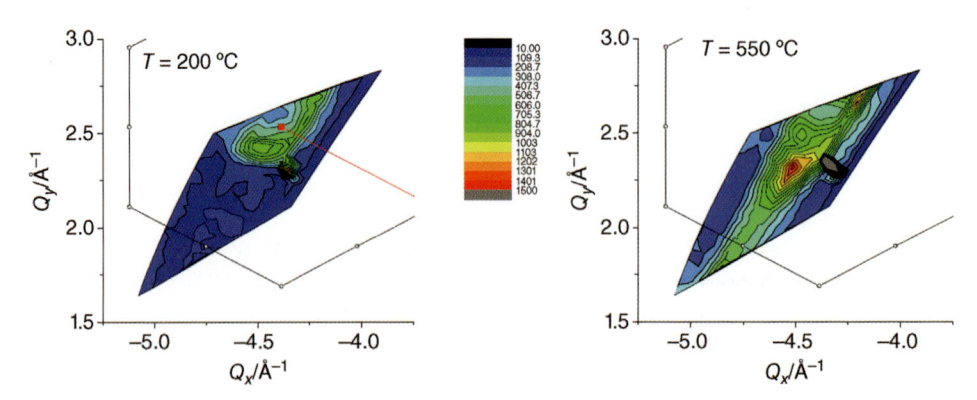

Figure 12.20 Constant energy maps taken at 1.2 THz of quartz within the Brillouin zone (−4 2 0) for two different temperatures. (Reproduced with permission from Ref. [32]. Copyright 1993, American Chemical Society.)

around the Bragg peak (red dot) being characteristic for an ordinary acoustic mode, a well-defined anisotropic intensity distribution is observed when the phase transition is approached.

Even this phase transition in quartz is accompanied by an intermediate incommensurate phase: Due to the anharmonic interaction between the zone center softmode and an acoustic mode, the condensation is not exactly at $q=0$. Instead, it is observed at 0.03 a^* within a temperature interval of about 1.5 K between the low-temperature α-phase and the high-temperature β-phase [33]. This corresponds to a static modulation with very long wavelength of tens of unit cells and is a direct consequence of the long-range correlation of the soft tilt-mode.

12.3.3.3 Anharmonicity

In the harmonic approximation of vibrational properties as briefly described in Section 12.2.1, there is no thermal expansion of the lattice that clearly contradicts experience. There are essentially two approaches to overcome this difficulty. Either one keeps higher order terms in the energy expansion (Eqs (12.4) or (12.8)) and solve the equations of motion using perturbation theory or – in the quasiharmonic approximation – one uses the fact that the interatomic interactions vary implicitly with the volume, that is, with the atomic separation. In the former case, the phonons are no longer eigenstates of the crystal and the detailed consideration of intrinsic anharmonicity requires the description of phonon–phonon interactions. This approach has been used, for example, for organic crystals like naphthalene or anthracene [34]. As a result, the temperature dependence of phonon frequencies as well as line widths is obtained. Figure 12.21 shows characteristic data of anthracene with a very pronounced anharmonic softening and broadening of phonon modes.

More frequently, the quasiharmonic approximation is used to understand the origin of thermal expansion in terms of vibrational properties. Based on thermodynamic relations, the components of the tensor of thermal expansion

$$\alpha_{kl} = \left(\frac{\partial \epsilon_{kl}}{\partial T}\right) \tag{12.46}$$

can be related to the so-called mode Grüneisen parameters:

$$\gamma_{qj,kl} = -\frac{\partial \ln \omega_{qj}}{\partial \epsilon_{kl}}, \tag{12.47}$$

that describe the variation of a particular mode frequency ω_{qj} with respect to a lattice deformation ϵ_{kl} being an element of the (dimensionless) tensor of mechanical deformation. These quantities can be determined experimentally if external mechanical stress is applied since this does not only lead to a deformation as described by the tensor of elastic constants, but also induces a significant shift in phonon frequencies. Systems with negative thermal volume expansion that are important materials for practical applications need to exhibit negative Grüneisen parameters at least for specific phonon modes. This was nicely shown

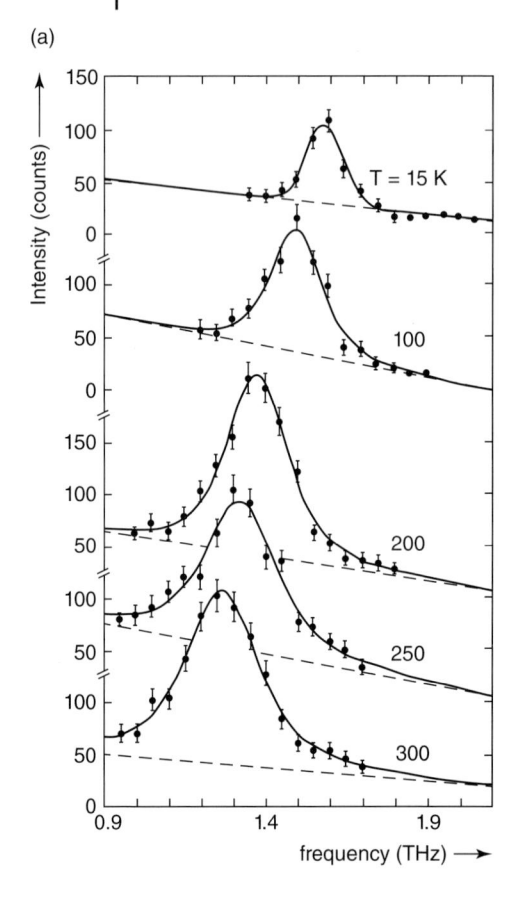

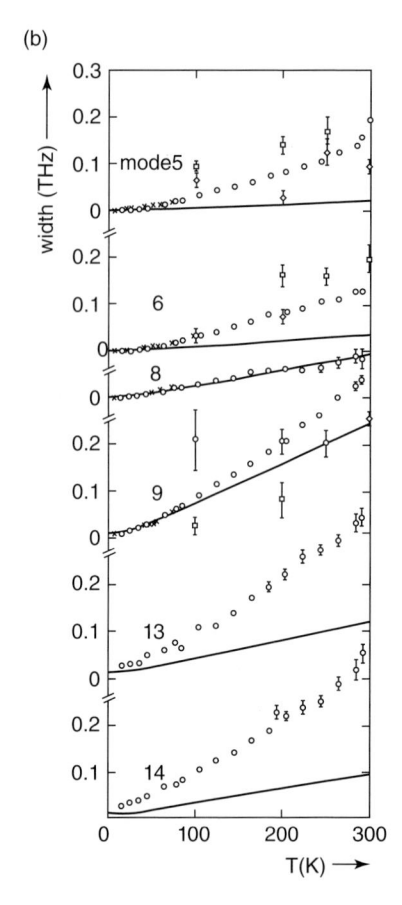

Figure 12.21 Temperature dependence of phonon frequencies and line widths in deuterated anthracene (after Ref. [34]). (a) Selected phonon spectra at different temperatures. (b) Line widths of several phonons compared with predictions from model calculations. (Reproduced with permission from Ref. [37]. Copyright 1995, American Chemical Society.)

in a high-pressure study of GaSb [35] where zone boundary modes could be identified as the main drivers for negative thermal expansion. In Figure 12.22, characteristic spectra for ambient pressure and for 7 GPa are displayed along with the dispersion of Grüneisen parameters for transverse acoustic modes. These data clearly show that the zone boundary modes soften considerably if pressure is applied and the volume is reduced.

To tailor materials for specific applications, it is often required to control thermal expansion. This triggered lately a strong interest in systems that show negative thermal expansion in at least one direction. Spectroscopic studies help to understand and to predict expansion coefficients quantitatively. Advanced materials that show a strong volume contraction with temperature are open frameworks with stiff structural units. A typical example is the simple three-dimensionally connected Zn–cyanide $Zn(CN)_2$ (see Figure 12.23). It features a

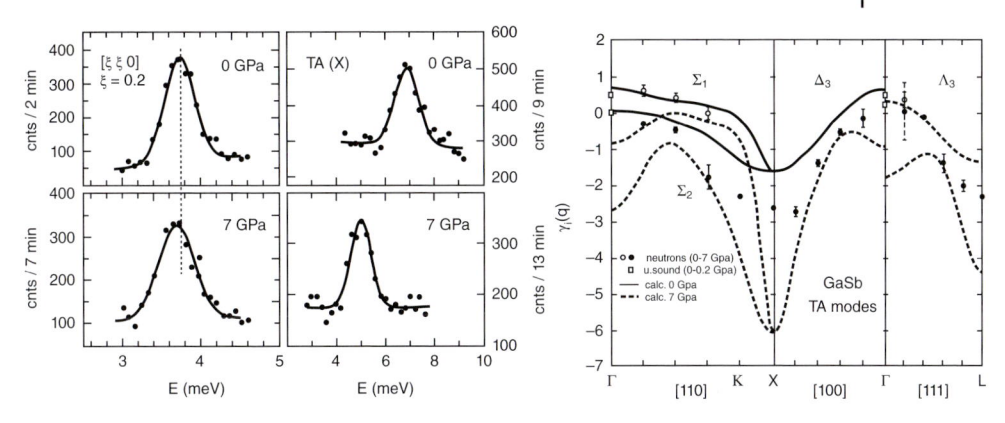

Figure 12.22 Pressure dependence of transverse acoustic modes (after Ref. [35]). (a) Softening of the zone boundary mode TA(X). (b) Dispersion of the Grüneisen parameter for TA modes.

strong negative thermal expansion coefficient of $\alpha_V = -51 \times 10^{-6}\,\mathrm{K}^{-1}$. Neutron powder investigations as a function of pressure give access to the powder-averaged mode Grüneisen parameters that in turn allow determining the contribution of the phonon bands to the thermal expansion (see Figure 12.24) [36].

Using *ab initio* calculations, which have been verified by the spectroscopic data, it is possible to identify the specific vibrational patterns, that is, phonon eigenvectors that are responsible for the negative thermal expansion behavior. As can be seen in Figure 12.25, transverse vibrations of the stiff cyanide units induce a contraction of the lattice planes as their vibrational amplitude increases [37].

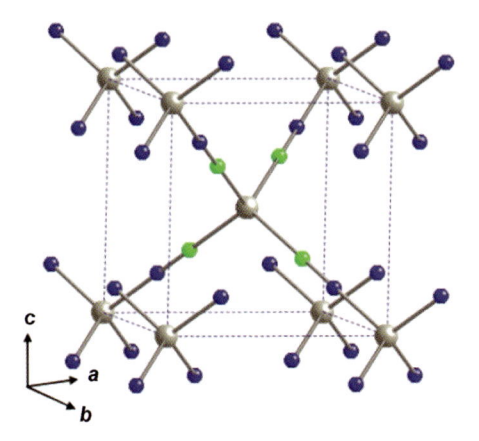

Figure 12.23 The structure of $Zn(CN)_2$ in the space group *P43m*. Color coding: Zn, gray spheres; C, green spheres; N, blue spheres. (Reproduced with permission from Ref. [37]. Copyright 1993, Elsevier.)

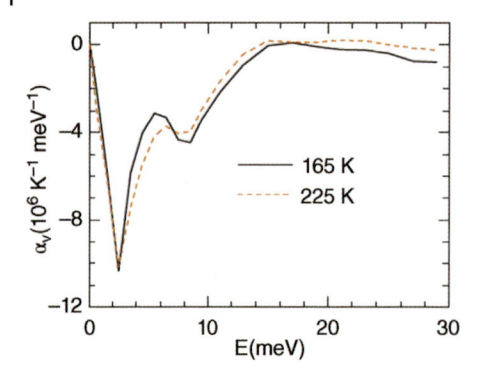

Figure 12.24 The contribution of phonon bands on the thermal expansion coefficient of $Zn(CN)_2$ as a function of energy as obtained from the pressure dependence of the phonon density of states. (Reproduced with permission from Ref. [36]. Copyright 1999, American Chemical Society.)

12.3.3.4 Nonequilibrium Studies

Since phonons directly reflect the interatomic interactions in solids, they are particularly sensitive for any changes of chemical bonds not only during phase transitions but also during chemical processes taking place in the solid phase. Demixing reactions have been investigated in great detail in recent

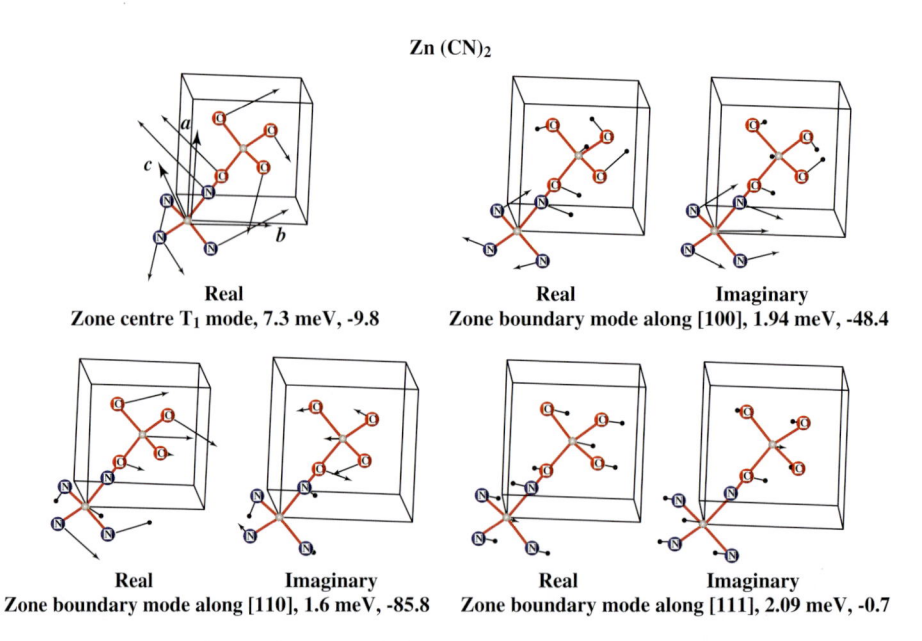

Figure 12.25 Vibrational patterns of a few selected modes. The frequencies and mode Grüneisen parameters are given below the figures. In particular, low-frequency modes around 2 meV contribute substantially to the negative thermal expansion. (Reproduced with permission from Ref. [40]. Copyright 1999, American Chemical Society.)

years. It could be shown that phonons provide unique information about the kinetics and the intermediate states that are passed during the phase separation. As a model system with a simple rock salt structure, mixed silver alkali halides were studied that exhibit a pronounced miscibility gap at low temperatures. If a sample is quenched from a homogeneous state to temperatures well within the miscibility gap, it starts to decompose and finally two product phases with different concentrations are formed. This process is diffusion limited and in general two different mechanisms can be distinguished: Nucleation and growth, on the one hand, is associated with the formation of small nuclei of the product phase that can subsequently grow in volume. Spinodal decomposition, on the other hand, is driven by small fluctuations of the concentration that are thermodynamically stable and grow in amplitude until the final concentrations of the product phases are reached. Usually, it is assumed that this process can be monitored by the splitting of Bragg reflections that is due to the fact that parent and product phases exhibit different lattice parameters. However, if mechanical strains play a dominant role, the shapes of the Bragg peaks and the average structure depends on the relaxation of residual stresses and do not directly reflect the concentration distribution. In AgCl–NaCl crystals, even one year of aging at room temperature did not yield well-defined and sharp Bragg reflections. In contrast, phonons are much better suited to monitor the variation of concentration, since they are highly sensitive to changes of the interatomic interaction. It could be shown [38] that the basic process of the chemical demixing, that is, the separation into silver-rich and alkali-rich regions, in fact takes place on a time scale of seconds or minutes. As an example, Figure 12.26 displays the time evolution of acoustic phonon spectra of a AgBr–NaBr single-crystalline sample that was quenched from 350 (homogeneous phase) to 100 °C [39]. While in the beginning, the well-defined acoustic phonon of the parent, homogeneous phase is observed, the separation into two components corresponding to the silver- and sodium-rich phases is clearly seen already after about 1 min. At the same time, almost no changes in the Bragg peaks are observed. Figure 12.27 shows the time-correlation function both for Bragg reflections and for phonons. Obviously, the time scales differ by more than an order of magnitude. Hence, chemical demixing takes place at constant lattice parameter resulting in large coherence strains. The adaptation of the equilibrium structure is a second step of the process.

While these results have been obtained for temperatures where the demixing reaction is governed by spinodal decomposition, similar experiments have been performed within the nucleation regime. The time evolution of phonon spectra is qualitatively different, since in this case instead of a continuous concentration distribution there is a coexistence of three well-defined phases – the parent phase and the two product phases – with varying volume fractions. Consequently, the phonon spectrum can be described as the superposition of three components during the phase separation as illustrated in Figure 12.28 [39].

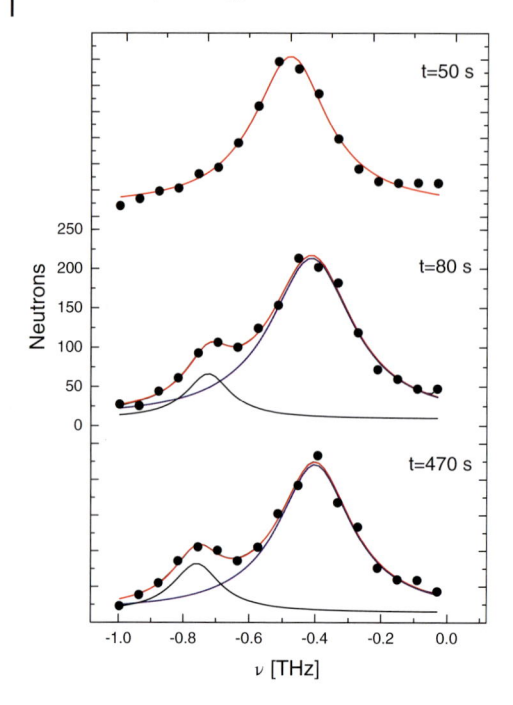

Figure 12.26 Time evolution of a transverse acoustic phonon in $Ag_{0.45}Na_{0.55}Br$ at $q = (0.2\ 0.2\ 0)$ and 373 K (after Ref. [39]).

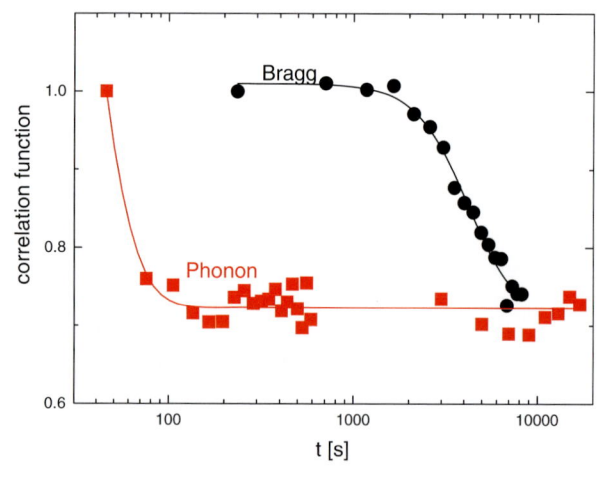

Figure 12.27 Time self-correlation function for acoustic phonons and Bragg reflections in a AgBr–NaBr mixed crystal after quenching into the miscibility gap (after Ref. [39]).

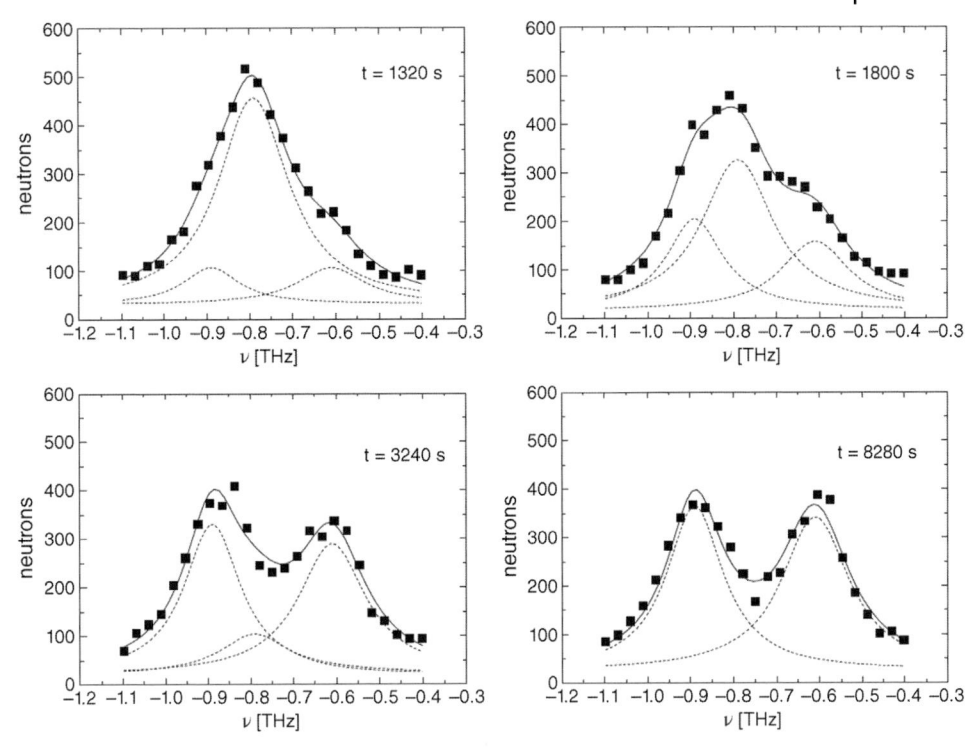

Figure 12.28 Time evolution of the transverse acoustic phonon in $Ag_{0.32}Na_{0.68}Br$ during demixing at 473 K within the nucleation regime (after Ref. [39]).

This example demonstrates that phonons are particularly suited to elucidate the intrinsic mechanisms during solid-state processes, and provide complementary information as compared to structural studies.

12.3.3.5 Molecular Crystals

In the following section, we will briefly show how neutron powder spectroscopy can contribute important information on the dynamics of molecular crystals. It is the specificity of molecular crystals that they are characterized by two clearly separated energy scales. Strong bonds within the molecular units are confronted with rather weak bonds acting between the molecules. This leads to intra- and intermolecular bands separated by a gap. The size of the gap can be taken as a direct indicator of the molecular crystal character of a material.

We use the well-known fullerenes to illustrate the information that can be gained by neutron powder spectroscopy on such systems. Fullerenes form crystalline structures with the molecules spinning nearly unhindered at high temperatures (rotator phase) before evaporating into the gas phase. In the neutron data, the rotational diffusion shows up as quasielastic intensity that allows determining the character of the rotational motion, including rotational diffusion constants and activation energies. At lower temperatures, the cages lock into fixed orientations with respect to each other. Compared to the stiffness of the cages, the intermolecular bonding is weak making C_{60} and C_{70} close-to-ideal representatives of molecular crystals. However, due to the complex electronic structure characterized by short and long carbon–carbon bonds, fullerenes will develop intermolecular covalent bonds under specific conditions, for example, exposure to light, high pressure at elevated temperatures, or doping with an appropriate amount of alkali ions (A_1C_{60}, A = K, Rb, Cs, Rb_1C_{70}). These covalent bonds lead to the formation of a large variety of polymer topologies.

When the cages form covalent networks the clear separation of external and internal energy scales present in the monomer system breaks down. Intercage bonding influences the spectra of molecular crystals in essentially two ways. On one hand, the bonding deforms the cages, lowering the symmetry. This leads to changes of the intramolecular vibration frequencies. Often accompanied by a change in the selection rules, these changes are best investigated with optical spectroscopy that, as we have already shown, possesses excellent resolution and is not concerned by the powder averaging. On the other hand, the bond formation leads to a more or less pronounced breakdown of the energy separation and thus to the filling of the gap. The form of the spectra in the original gap region allows the extraction of quantitative information on the bonds that have formed between the molecular cages in a rather straightforward way.

Figure 12.29 shows concrete examples of fullerene systems in the monomer, dimer, and various polymer phases. The separation of energy scales in the monomer leads to a pronounced gap in the density of states $G(\omega)$ that extends from 6 to 30 meV. Upon the creation of $(C_{60})_2$ dimers via the formation of single bonds between cages we observe sharp bands in the gap region. These bands originate from dispersionless phonons equally called Einstein modes. These modes correspond to the vibrations of the $(C_{60})_2$ dumbbells. Only when long-range connectivity is established within the crystal via polymerization in the form of chains or sheets do we observe broad bands in the gap region indicating the presence of dispersive phonon branches in this region.

Via the spectroscopic signature of the various phases it is possible to follow transformation processes like the polymerization in real time.

12.3.3.6 Guest–Host Systems

Crystalline systems that feature cages within their structure offer the possibility of hosting guest ions or molecules. These systems have received sustained attention as the inclusion of the guests considerably modifies the physical properties of the material. Spectroscopy is ideally suited for studying the dynamics of the guest within the host structure.

The importance of neutron spectroscopy to investigate guest dynamics has recently been underlined by studies of endohedrally doped fullerenes. Due to the advances of molecular surgery, it has become possible to introduce a large

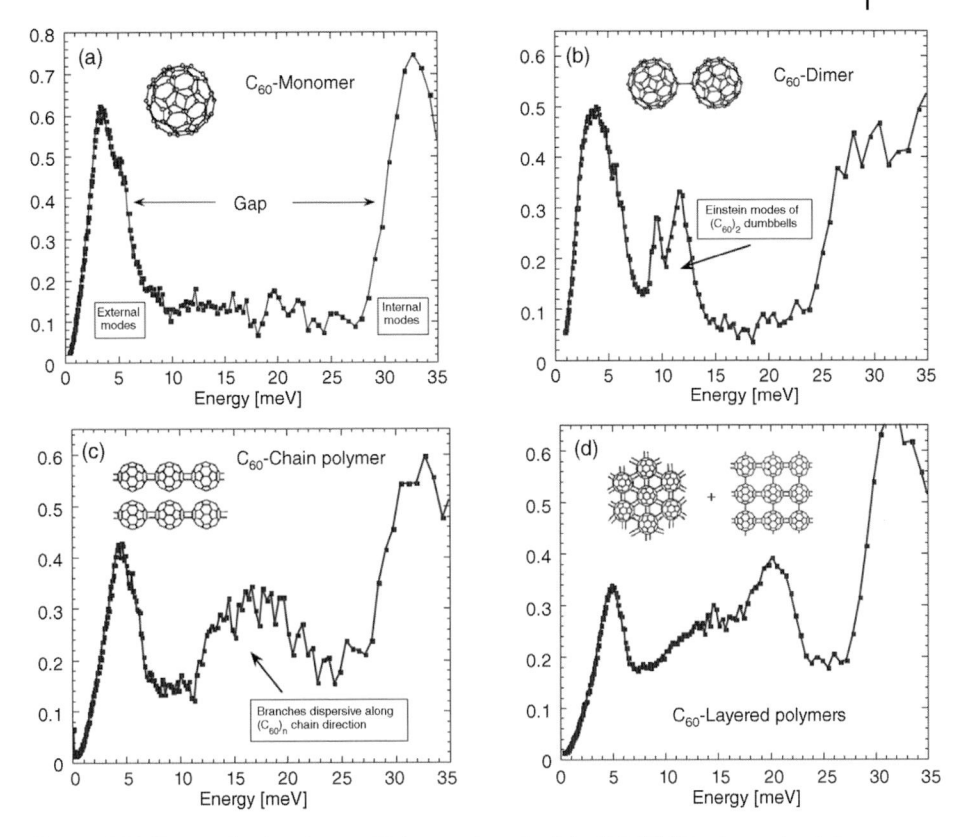

Figure 12.29 Phonon density-of-states $G(\omega)$ as obtained from the inelastic neutron scattering spectra. (a) C_{60} at 300 K in the monomer plastic crystal phase [40]. (b) $(C_{60})_2$ dimer of Rb_1C_{60} [40]. (c) Linear chains in pressure polymerized C_{60} [41]. (d) 2D polymer sheets in pressure polymerized C_{60} (the sample is a mixture of 2D rhombohedral and tetragonal networks) [42].

variety of small molecules into the fullerene cages. It was in particular possible to produce high-purity samples of $H_2@C_{60}$ and $H_2O@C_{60}$. These systems offer the possibility to study the dynamics of the molecules in a highly symmetric local "nanolaboratory" environment. The emphasis of the neutron investigation lies clearly with the study of the quantum rotors. Contrary to photons, neutrons can induce direct transitions between *para-* and *ortho-*species. Figure 12.30 shows measurements of $H_2O@C_{60}$ at higher energies in both spin states. Combining data obtained from several instruments it is possible to map out the dynamical states of an entrapped quantum rotor with good resolution [43]. A splitting of the *ortho-*H_2O ground state is interpreted as a clear indication for a symmetry-breaking interaction. In contrast to $H_2@C_{60}$, the ground state of $H_2O@C_{60}$ appears to be sensitive to the environment outside the cages.

In our second example of guest–host systems, we would like to demonstrate how a combination of advanced *ab initio* lattice dynamics calculations and

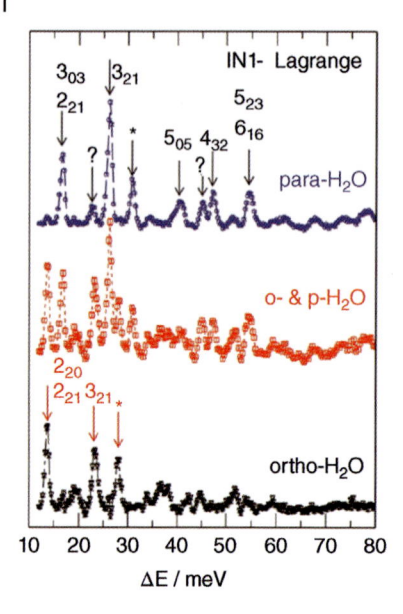

Figure 12.30 INS spectra of $H_2O@C_{60}$ recorded at 2.5 K on the instrument IN1-Lagrange. The blue spectrum corresponds to the initial *para*-H_2O, the red spectrum to a mixture of *para*- and *ortho*- species during the transformation process, and the black spectrum extrapolates toward *ortho*-H_2O. The labels indicate the final rotational states. The transitions originating in 1_{01} are denoted in red. The transitions originating in 0_{00} are denoted in black. (Reproduced with permission from Ref. [43]. Copyright 1980, Elsevier.)

powder-averaged coherent neutron data allows gaining deep insight into the dynamics of thermoelectric materials. To reach a high efficiency, thermoelectric material should combine good electric conductivity with poor thermal conductivity. Many guest–host compounds combine these properties. It was argued that the poor thermal conductivity could be attributed to resonant scattering of the heat carrying phonons of the host lattice by the encaged guest ions, so-called rattlers. This concept, which gained appreciable attention under the name "phonon glass," was recently challenged by spectroscopic investigations on a number of thermoelectric materials. Prominent examples are the skutterudites $LaFe_4Sb_{12}$ and $CeFe_4Sb_{12}$. Neutron spectroscopy experiments and *ab initio* computational work clearly show that in the fully filled skutterudite lattices the La/Ce guest ions couple coherently to the vibrations of the cages [44]. The powder-averaged dispersion relations are compared with the spectroscopic data in Figure 12.31. There is no indication of resonant scattering as an energy dissipating mechanism. The low thermal conductivity in these guest–host systems is traced back to the dispersion relations themselves. Only few branches show dispersion and, therefore, can contribute to the heat transport with finite group velocities. The population of heat-carrying phonons is thus rather small explaining the low thermal conductivity.

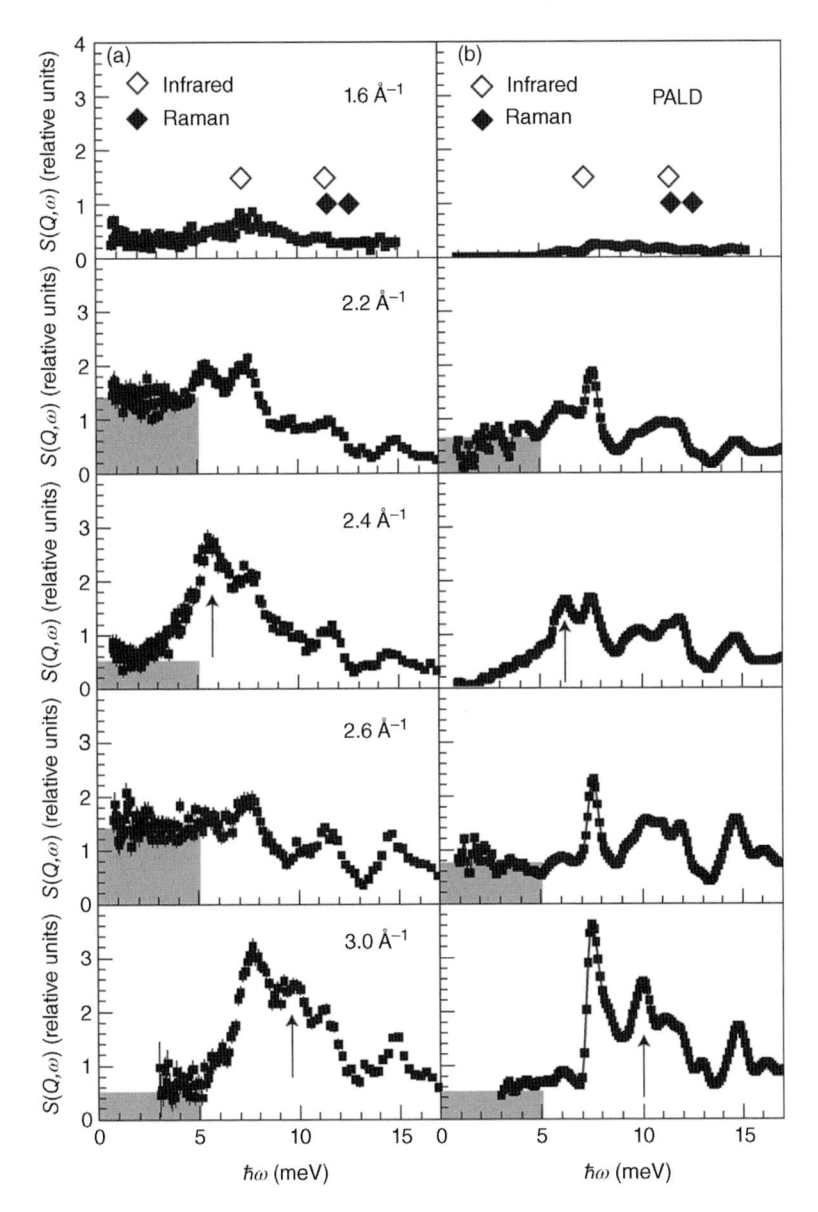

Figure 12.31 (a) Measured neutron spectra of LaFe$_4$Sb$_{12}$ at constant momentum transfer for the indicated Q-values. (b) Spectra as obtained by powder averaging the *ab initio* lattice dynamics calculations (PALD). Raman and infrared frequencies are indicated by open and filled symbols. (Reproduced with permission from Ref. [44]. Copyright 2008, Nature Publishing.)

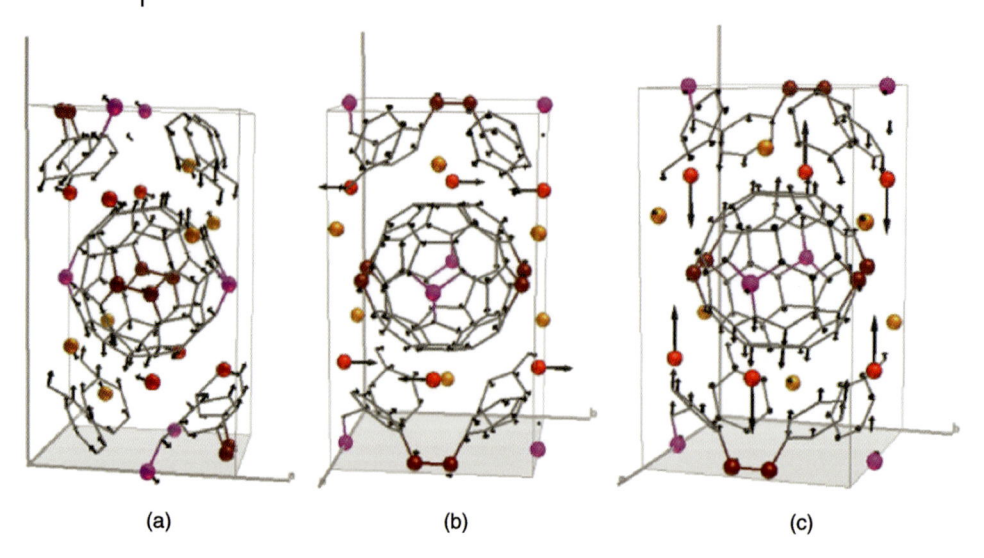

Figure 12.32 Vibrational modes in Li_4C_{60} as obtained by *ab initio* calculations verified against neutron powder spectroscopy data. (a) Mode with H_g character at 16.9 meV [$A_g(3)$]. (b) Hybrid mode at 18 meV [$B_g(1)$]. (c) Hybrid mode at 30.2 meV [$A_g(4)$]. (Reproduced with permission from Ref. [45]. Copyright 2015, American Chemical Society.)

12.3.3.7 Phonon-Enhanced Ionic Transport

Lattice vibrations may play an important role in ionic transport via the so-called phonon-enhanced diffusion process. Specific phonon modes lead to deformations of the structure that lower the energy barriers for ionic jumps. Such a mechanism has recently been proposed for Li_4C_{60} [45]. Li_4C_{60} is a particularly interesting material. The charge transfer induced by the intercalation of Li-ions leads to the formation of two-dimensional C_{60} polymer sheets featuring alternating sequences of single and [2+2] cycloaddition bonds. In addition, Li_4C_{60} possesses already at room temperature a very high ionic mobility of $\sigma \sim 10^{-2}$ S/cm comparable to liquid electrolytes making it potentially very interesting for electrode applications. Neutron spectroscopic studies have allowed gaining a very complete picture of the dynamics of both the Li-ions and the carbon cages. Due to the polymerization, the low-lying cage modes are spread out over the gap region of the monomeric compound. As can be seen from Figure 12.32, some of these modes have a clear hybrid character. Large amplitude Li motions accompany the deformation of the cages. These hybrid modes are ideal vectors to help transport a Li-ion to a new equilibrium position, thus enhancing ionic transport.

12.3.4
Inelastic X-Ray Scattering

The enormous increase in brilliance achieved by modern synchrotron sources combined with beam optics allowing for very high resolution has made it possible to investigate inelastic signals over a very extended energy and momentum transfer range. Inelastic X-ray scattering (INXS) has become a powerful spectroscopic tool in all areas of solid-state physics and chemistry. As INXS measures the same density correlation functions as neutrons, the theoretical formulism describing the scattering is very similar. The main difference resides in the fact that X-rays are scattered by the electrons while neutrons interact with the nuclei. Therefore, the nuclear scattering lengths have to be replaced by the form factors of the electronic cloud in the formula for the scattering law (Eq. (12.43)). Details can be found in recent reviews on the technique [46]. The main advantage of INXS resides in the possibility to go to very small momentum transfers even for high excitation energies as well as in the small sample volumes needed. In conventional INXS, the energy of the X-ray beam is determined by the reflection of the high-resolution monochromator and can be as low as a few millielectron volts. Recently, large progress has been achieved in preserving good resolution over an extended continuous range of incident energies. This allows tuning the X-ray energy to various absorption edges making the observed scattering element specific while at the same time strongly enhancing the intensities via the resonance effect. An excellent review of resonant inelastic X-ray scattering (RIXS) is given by Ament *et al.* [47]. We will illustrate both conventional INXS and RIXS with a specific example.

12.3.4.1 INXS Probing Superconductivity

For many years it was thought that conventional phonon-mediated superconductivity would not hold any surprises. In addition, the discovery of high-temperature superconductivity had shifted the attention to the so-called unconventional coupling mechanisms. The discovery of superconducting transition temperatures close to 30 K in doped fullerenes and MgB_2, therefore, came as a surprise. It immediately triggered renewed interest in the study of electron–phonon coupling. This interest has in the meantime spread to unconventional superconductivity. An increasing number of scientists are convinced that electron–phonon interactions cannot be completely discarded when it comes to give a complete picture of superconductivity in high-T_c materials.

Let us briefly expose how spectroscopy rapidly established the mechanisms underlying superconductivity in MgB_2. This will allow us, in particular, to highlight the important role that high-resolution X-ray inelastic scattering plays in modern spectroscopic investigations.

MgB_2 is a strongly two-dimensional material with hexagonal layers of Mg ions interspaced between graphite-like boron sheets. The in-plane vibrations feature the two main ingredients required for high transition temperatures: high phonon frequencies and elevated electron–phonon coupling. An almost complete picture of the phonon dispersion could be obtained using INXS. The high intensity of the X-ray beam allows working with rather small samples. The data shown in Figure 12.33 have been obtained on a sample with dimensions of $400 \times 470 \times 40\,\mu m^3$ [48].

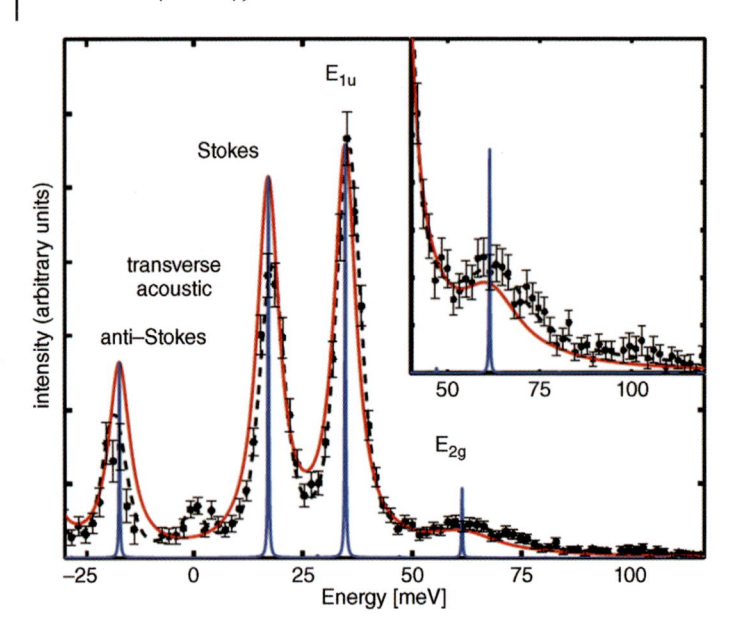

Figure 12.33 INXS spectra for a particular scattering geometry in MgB$_2$. The blue spectrum corresponds to *ab initio* calculations. The red spectrum shows the same calculations but taking experimental broadening and electron–phonon coupling into account. The insert zooms into the region of the strongly damped in-plane E$_{2g}$-mode. The peak at zero is due to diffuse scattering. (Reproduced with permission from Ref. [48]. Copyright 1980, Elsevier.)

From the full set of measured spectra, it is possible to extract both the frequencies and widths of the phonons for all the main symmetry directions. In particular the high-frequency E$_{2g}$ modes show very strong damping indicative of strong electron–phonon coupling.

An excellent way of experimentally demonstrating electron-phonon coupling consists in ionic substitution. Replacing Mg by Al in MgB$_2$ has only minor influence on the structure. In particular, the bond lengths change very little. One may, therefore, expect that the vibrational spectra, in particular of the boron sheets, remain almost unchanged. This is not the case as demonstrated by the powder neutron data shown in Figure 12.34 [49]. The replacement of Mg by Al fills the holes within the band structure of the boron sheets. This reduces the renormalization of the phonon frequencies. The effects are absolutely impressive, for example, the nearly dispersionless branch starting at the Raman active E$_{2g}$ mode of 125 meV in AlB$_2$ drops to 70 meV in MgB$_2$. It is this strong coupling of high-frequency modes that allows obtaining very high transition temperatures within a conventional framework.

12.3.4.2 RIXS Probing Electron–Phonon Coupling in Semiconductors

Electron–phonon coupling has a strong influence on the transport properties of polar semiconductors as it modifies the effective mass and lifetime of the charge

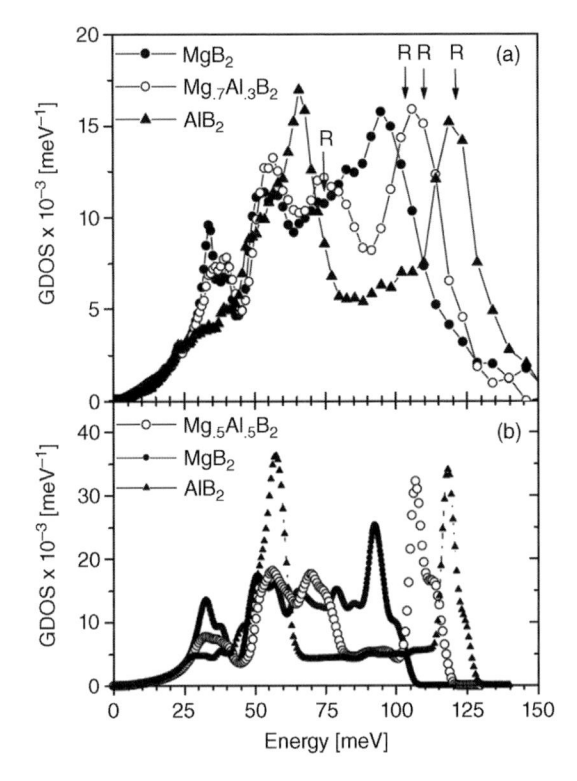

Figure 12.34 The generalized phonon density of states $G(\omega)$ for a series of samples of $Mg_{1-x}Al_xB_2$ with $x = 0, 0.3, 1$. The arrows indicate Raman lines. As can be seen there is a very strong renormalization of the frequency bands with x. (Reproduced with permission from Ref. [49]. Copyright 2012, American Chemical Society.)

carriers. A material that has wide-ranging application in photovoltaics and photocatalysis is TiO_2. Using RIXS in combination with other experimental probes, it has been possible to identify those phonons that are involved in the formation of the polarization cloud of the polarons in these materials [50]. In addition, it was possible to determine the electron–phonon coupling quantitatively. Figure 12.35 shows the principle of the measurement. The X-ray energy is tuned to the L_3 edge of Ti. In the case where the X-ray absorption leads to the excitation of the conduction electron into the t_{2g} crystal field orbital, a local exciton is produced as the electron is strongly bound via Coulomb interactions to the hole. The hole is, therefore, well screened and the lattice as a consequence remains in its normal state. If the excitation involves the more delocalized e_g orbital, then the screening is less effective leading to a concomitant lattice distortion. This creation of phonons induces a shift of the energy of the emitted phonon leading to an inelastic signal (see Figure 12.36). In addition to the momentum transfer, selectivity for specific excitations can be achieved by using the polarization of the light as an experimental variable. These signals allow identifying not only the

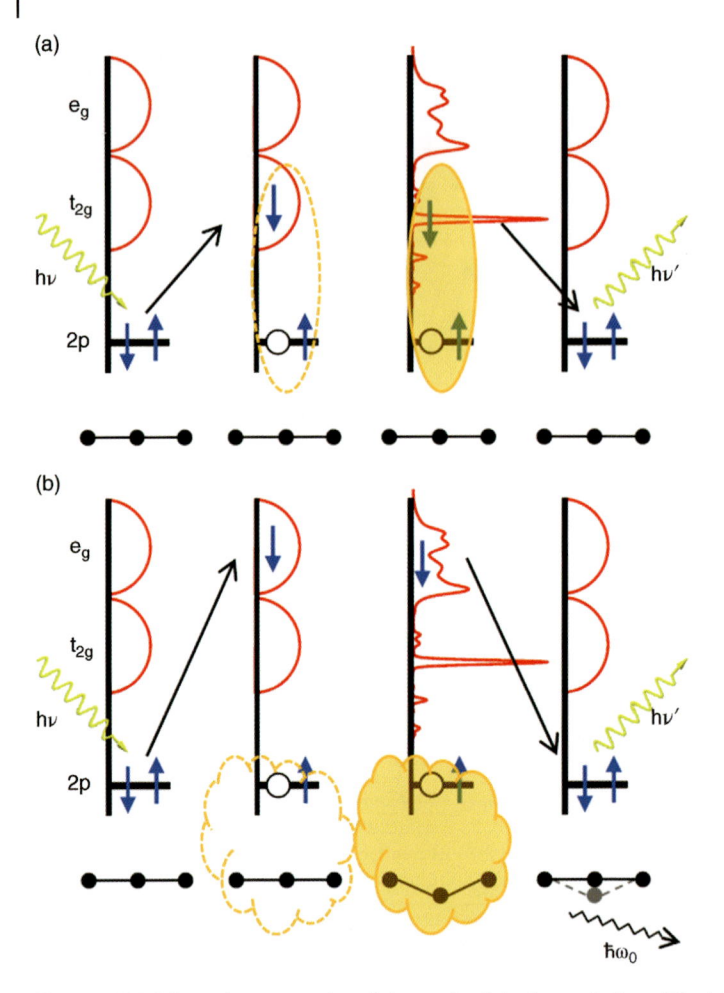

Figure 12.35 Schematic presentation of the RIXS excitations along the Ti L_3 edge in TiO_2 (a) via the localized t_{2g} crystal field orbital and (b) via the more extended e_g orbital. In the latter case, the incomplete screening of the hole leads to the excitation of the lattice reminiscent of strong electron–phonon coupling. (Reproduced with permission from Ref. [50]. Copyright 2015, American Chemical Society.)

type and energy of the vibrations involved but equally determining the electron–phonon coupling constant.

12.4
Concluding Remarks

With this contribution, we have tried to cover a broad range of applications of vibrational spectroscopy using different techniques with special emphasis on neutron scattering. We have collected examples from various fields of solid-state

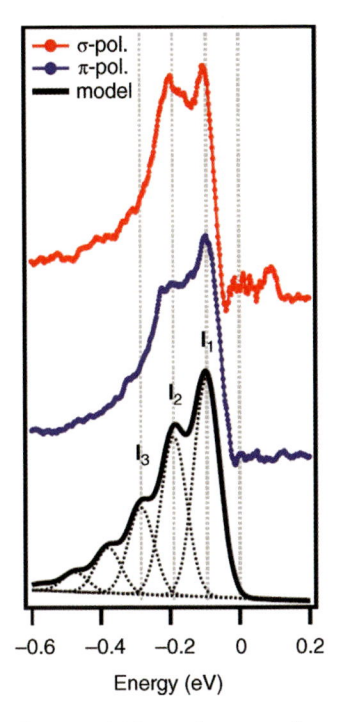

Figure 12.36 Spectral response for two polarizations in TiO$_2$ compared to model calculations using a single phonon mode with an energy of 95 meV and a width of 375 meV as input. I$_n$ indicate the respective inelastic satellites. The elastic peak has been subtracted. (Reproduced with permission from Ref. [50]. Copyright 2015, American Chemical Society.)

research that illustrate the information content of spectroscopic studies. This demonstrates how properties of materials or microscopic mechanisms of solid-state processes are reflected by the underlying dynamics. This chapter is not intended to serve as an overall review of vibrational spectroscopic studies. But even if the selection of examples is biased by our own research we hope that the reader has got an impression of the power of spectroscopic investigations using photons and neutrons as very efficient probes for the understanding of solid-state properties and processes on a microscopic scale.

Modern materials such as multiferroics, ionic conductors, or superconductors are studied using all types of spectroscopies and molecular crystals as well as guest–host systems are examples for the detailed investigation of chemical bonding in solids. Moreover, it is shown how anisotropy, domain structures, and the difference between bulk samples and thin films determine the vibrational properties. Beyond the harmonic approximation, anharmonic effects and even dynamical instabilities of crystal lattices are exciting topics that help to elucidate the underlying phenomena, and new types of time-resolved techniques allow the monitoring of transformation processes in real time.

References

1 Maradudin, A.A. (1963) *Ann. Phys.*, **14**, 89.
2 Eckold, G. (2003) *International Tables for Crystallography*, vol **D**, (ed A. Authier), Springer, The Netherlands, (ISBN 978-1-4020-0714-9) p. 266.
3 Trampenau, J., Heiming, A., Petry, W., Alba, M., Herzig, C., Miekeley, W., and Schober, H.R. (1991) *Phys. Rev. B*, **43**, 10963–10969.
4 Pintschovius, L., Pyka, N., Reichardt, W., Rumiantsev, A.Y., Mitrofanov, N.L., Ivanov, A.S., Collin, G., and Bourges, P. (1991) *Physica C*, **185–189**, 156–161.
5 Natkaniec, I., Bokhenkov, E.L., Dorner, B., Kalus, J., Mackenzie, G.A., Pawley, G.S., Schmelzer, U., and Sheka, E.F. (1980) *J. Phys. C*, **13**, 4265–4283.
6 Strauch, D. and Dorner, B. (1990) *J. Phys. Condens. Matter*, **2**, 1457–1474.
7 Giannozzi, P., De Gironcoli, S., Pavone, P., and Baroni, S. (1991) *Phys. Rev. B*, **43**, 7231–7242.
8 Sablinskas, V. (2014) Instrumentation, in *Handbook of Spectroscopy*, vol. **1** (eds G. Gauglitz and D.S. Moore), Wiley, Weinheim, (ISBN 978-3-527-32150-6) pp. 39–70.
9 Schober, H. (2009). Neutron scattering instrumentation, in *Neutron Application in Earth, Energy and Environmental Sciences* (eds L. Lang, R. Rinaldi, and H. Schober), Springer, Berlin, (ISBN 978-0-387-09415-1) pp. 37–104.
10 Socrates, G. (2001) *Infrared and Raman Characteristic Group Frequencies: Tables and Charts*, John Wiley & Sons, Inc. (ISBN 0-471-85298-8).
11 Fedorov, I., Zelezny, V., Petzelt, J., Trepakov, V., Jelinek, M., Trtik, V., Cernansky, M., and Studnicka, V. (1998) *Ferroelectrics*, **208–209**, 413–427.
12 Möller, T., Becker, P., Bohatý, L., Hemberger, J., and Grüninger, M. (2014) *Phys. Rev. B*, **90**, 155105.
13 Ziegler, F., Gibhardt, H., Leist, J., and Eckold, G. (2015) *Mater. Res. Express*, **2**, 096103.
14 Gibhardt, H., Leist, J., and Eckold, G. (2015) *Mater. Res. Express*, **2**, 015005.
15 Su, Y., Balmer, M.L., and Bunker, B.C. (2000) *J. Phys. Chem. B*, **104**, 8160–8169.
16 Lovesey, S.W. (1984) *Theory of Neutron Scattering from Condensed Matter* (International Series of Monographs on Physics 72), vol. **1**, Clarendon Press, Oxford, (ISBN 0-19-852015-8); Squires, G.L. (2012) *Introduction to the Theory of Thermal Neutron Scattering*, 3rd edn, Cambridge University Press, Cambridge (ISBN 978-1-107-64406-9); Schober, H. (2014) *J. Neutron Res.*, **17**, 109–357.
17 Strauch, D., Mayer, A.P., and Dorner, B. (1990) *Z. Phys. B*, **78**, 405–410.
18 Kulda, J., Strauch, D., Pavone, P., and Ishii, Y. (1994) *Phys. Rev. B*, **50**, 13347.
19 Strauch, D. and Dorner, B. (1986) *J. Phys. C: Solid State Phys.*, **19**, 2853.
20 Boysen, H., Dorner, B., Frey, F., and Grimm, H. (1980) *J. Phys. C: Solid State Phys.*, **13**, 6127.
21 Currat, R., Comès, R., Dorner, B., and Wiesendanger, E. (1974) *J. Phys. C: Solid State Phys.*, **7**, 2521.
22 Pawley, G.S., Mackenzie, G.A., Bokhenkov, E.L., Sheka, E.F., Dorner, B., Kalus, J., Schmelzer, U., and Natkaniec, I. (1980) *Mol. Phys.*, **39**, 251–260.
23 Dick, B.J. and Overhauser, A.W. (1958) *Phys. Rev.*, **112**, 90.
24 Johnson, F.A. (1974) *Proc. R. Soc. Lond. A*, **339**, 73.
25 Elter, P. and Eckold, G. (2000) *Physica B*, **276–278**, 268. (Eckold G., JÜL-2639, Jülich 1992, ISSN 0366–0885). Available at http://www.uni-pc.gwdg.de/eckold/unisoft.html.
26 Koza, M.M., Johnson, M.R., Viennois, R., Mutka, H., Girard, L., and Ravot, D. (2008) *Nat. Mater.*, **7**, 805.
27 Schober, H. (2014) *J. Neutron Res.*, **17**, 109.
28 Ziegler, F., Murshed, M.M., Gibhardt, H., Sobolev, O., Gesing, T.M., and Eckold, G. (2016) *Phys. Status Solidi B*, 1–7. doi: 10.1002/pssb.201552670
29 Luspin, Y., Servoin, J.L., and Gervais, F. (1980) *J. Phys. C: Solid State Phys.*, **13**, 3761.
30 Cowley, R.A., Buyers, W.J.L., and Dolling, G. (1969) *Solid State Commun.*, **7**, 181.
31 Axe, J.D., Iizumi, M., and Shirane, G. (1980) *Phys. Rev. B*, **22**, 3408.

32 Sobolev, O., Hoffmann, R., Gibhardt, H., Jünke, N., Knorr, A., Meyer, V., and Eckold, G. (2015) *Nucl. Instrum. Methods Phys. Res. A*, **772**, 63–71.

33 Aslanyan, T.A. and Levanyuk, A.P. (1979) *Solid State Commun.*, **31**, 547; Dolino, G., Bachheimer, J.P., Berge, B., and Zeyen, C.M.E., (1984) *J. Phys. France*, **45**, 361.

34 Jordan, J., Kalus, J., Schmelzer, U., and Eckold, G. (1989) *Phys. Status Solidi (B)*, **155**, 89.

35 Klotz, S., Braden, M., Kulda, J., Pavone, P., and Steininger, B. (2001) *Phys. Status Solidi (B)*, **223**, 441.

36 Mittal, R., Chaplot, S.L., and Schober, H. (2009) *Appl. Phys. Lett.*, **95**, 201901.

37 Mittal, R., Zbiri, M., Schober, H., Marelli, E., Hibble, S.J., Chippindale, A.M., and Chaplot, S.L. (2011) *Phys. Rev. B*, **83**, 024301.

38 Eckold, G., Caspary, D., Gibhardt, H., Schmidt, W., and Hoser, A. (2004) *J. Phys. Condens. Matter*, **16**, 5945–5954.

39 Elter, P., Eckold, G., Gibhardt, H., Schmidt, W., and Hoser, A. (2005) *J. Phys. Condens. Matter*, **17**, 6559–6573.

40 Schober, H., Tölle, A., Renker, B., and Heid, R., and Gompf, F. (1997) *Phys. Rev. B*, **56**, 5937.

41 Renker, B., Schober, H., Heid, R., and v. Stein, P. (1997) *Solid State Commun.*, **104**, 527.

42 Schober, H., Renker, B., and Heid, R. (1999) *Phys. Rev. B*, **60**, 998.

43 Goh, K.S.K., Jiménez-Ruiz, M., Johnson, M.R., Rols, S., Ollivier, J., Denning, M.S., Mamone, S., Levitt, M.H., Xuegong, Lei, Yongjun, Li, Turro, N.J., Murata, Y., and Horsewill, A.J. (2014) *Phys. Chem. Chem. Phys.*, **16**, 21330.

44 Koza, M.M., Johnson, M.R., Viennois, R., Mutka, H., Girard, L., and Ravot, D. (2008) *Nat. Mater.*, **7**, 805.

45 Rols, S., Pontiroli, D., Cavallari, C., Gaboardi, M., Aramini, M., Richard, D., Johnson, M.R., Zanotti, J.M., Suard, E., Maccarini, M., and Riccó, M. (2015) *Phys. Rev. B*, **92**, 014305.

46 Baron, A.Q.R. (2015) doi: arXiv.org 1504.01098.

47 Ament, L.J.P., van Veenendaal, M., Devereaux, T.P., Hill, J.P., and van den Brink, J. (2011) *Rev. Mod. Phys.*, **83**, 705.

48 Shukla, A., Calandra, M., d'Astuto, M., Lazzeri, M., Mauri, F., Bellin, Ch., Krisch, M., Karpinski, J., Kazakov, S.M., Jun, J., Daghero, D., and Parlinski, K. (2003) *Phys. Rev. Lett.*, **90**, 095506.

49 Renker, B., Bohnen, K.B., Heid, R., Ernst, D., Schober, H., Koza, M., Adelmann, P., Schweiss, P., and Wolf, T. (2002) *Phys. Rev. Lett.*, **88**, 067001.

50 Moser, S., and Fatale, S., Krüger, P., Berger, H., Bugnon, P., Magrez, A., Niwa, H., Miyawaki, J., Harada, Y., and Grioni, M. (2015) *Phys. Rev. Lett.*, **115**, 096404.

13
Mößbauer Spectroscopy

Hermann Raphael

Oak Ridge National Laboratory, Materials Science & Technology Division, 1 Bethel Valley Rd, Oak Ridge, TN 37831-6064, USA

13.1
Introduction

Mössbauer spectroscopy and related synchrotron radiation-based nuclear resonance techniques are powerful albeit specialized methods that provide unique insights into the dynamics and coordination of atoms by probing transitions between a nuclear excited state and a nuclear ground state. A continued growth of nuclear resonance scattering capabilities can be expected with the ongoing trend in increasing synchrotron source brightness.

The electronic density is central to the current modeling [1,2] and understanding of most materials properties. Using nuclear spectroscopic techniques in the suite of analytical methods for solid-state chemistry might thus be surprising, but the tremendous success of nuclear magnetic resonance, NMR, and Mössbauer spectroscopy well justify this approach. In probing nuclei, we start with an *a priori* handicap with respect to techniques that directly probe the electron cloud because the gathered information must be obtained from hyperfine interactions between the electrons and the nuclei. Whereas a typical energy range for Coulomb interactions is a few electronvolts and that for fine structure splitting is 10^{-4} eV, hyperfine splitting ranges around 10^{-8} eV. High resolution is thus essential. Nuclear magnetic resonance (NMR) spectroscopies [3] take a direct approach to the problem in probing how the ground-state nuclear magnetic moment interacts with the electromagnetic field and neighboring electrons. In contrast, Mössbauer spectroscopy takes a more convoluted route in probing how neighboring electrons influence transitions between the nuclear ground state and a nuclear excited state in order to obtain information on the electron configuration, magnetism, and the lattice dynamics. As energies larger than 10^4 eV are typical for the first nuclear excited state, a resolution power of $\Delta E/E \sim 10^{-12}$ is needed. To set a scale, this is equivalent to resolving a millimeter-sized object at

---

Handbook of Solid State Chemistry, First Edition. Edited by Richard Dronskowski, Shinichi Kikkawa, and Andreas Stein.
© 2017 Wiley-VCH Verlag GmbH & Co. KGaA. Published 2017 by Wiley-VCH Verlag GmbH & Co. KGaA.

the surface of the moon. It is thus not surprising that this approach came as a rather late – although brilliant – afterthought of fundamental nuclear physics research.

Resonance spectroscopic techniques rely on the overlap of an absorption and emission line or signal. The corresponding observation of resonance absorption for γ-ray photons [4] was elusive and had been sought for 20 years until it was finally first observed by Rudolf Mössbauer in 1957 [5]. The essential problem lies in that nuclear absorption or emission of γ-rays involves recoil energy, E_R, as a consequence of the conservation of linear momentum. Whereas for visible light photons, for example, for solar light spectroscopy, the recoil energy is much smaller than the bandwidth determined by the Heisenberg uncertainty principle, for nuclei the opposite is typically the case (see Figure 13.1). It was reasoned that either shifting the emission line by the Doppler effect at high velocity [6] or broadening the emission line by increasing the atomic thermal motion with increasing temperature [7] would yield the required overlap between absorption and emission line. Surprisingly, upon *cooling* a source, Mössbauer did not observe the expected *decrease* in resonance absorption in the sample, but a strong

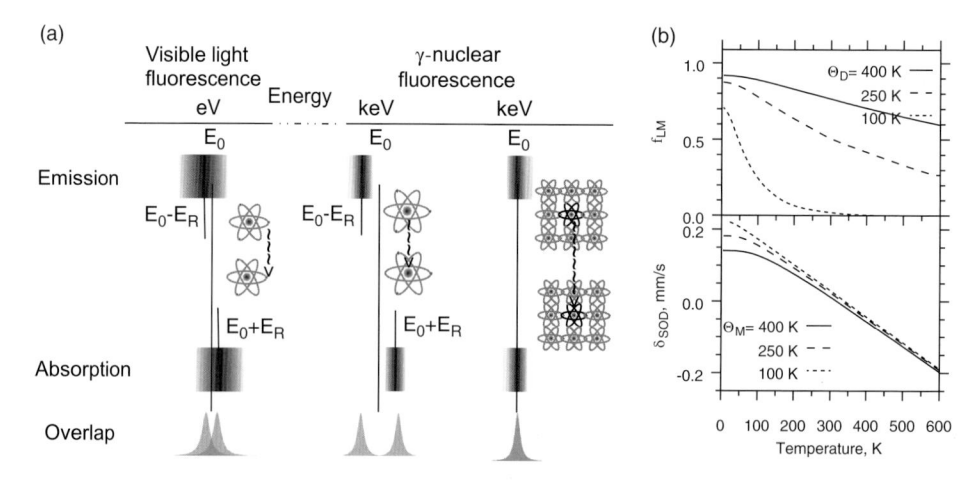

Figure 13.1 (a) A photon emitted by a free atom or nucleus has energy $E_0 - E_R$, where E_0 and E_R are the excited level and the recoil energy, respectively. Upon absorption, the necessary energy is $E_0 + E_R$. For visible light fluorescence, E_R is small in comparison to the Heisenberg bandwidth (width of the shaded bands) $\Delta E = \hbar/\tau$, with τ being the level lifetime. For nuclear fluorescence, E_R is typically much larger than ΔE and the absence of overlap between the emission and absorption lines prevents resonant absorption. This situation changes when the nucleus is bound in a solid and undergoes moderate mean square displacements. A significant fraction of photons f_{LM} can then undergo resonant emission and absorption *without recoil*, that is, the Mössbauer effect. (b) Temperature dependence of the Lamb–Mössbauer factor f_{LM} and second-order Doppler shift δ_{SOD} for different Debye Θ_D and Lamb–Mössbauer Θ_M temperatures (see text). The data are shown for the iron-57 nuclear resonance. δ_{SOD} is displayed normalized to the same high temperature limit.

increase [5]. He correctly surmised that this increase is a consequence of the recoil-free absorption or emission of γ-rays that can be achieved whenever the average thermal motion energy of the nuclei is much smaller than the recoil energy as can be the case if the nucleus is bound in a solid (see Figure 13.1). In a follow-up experiment, Mössbauer then demonstrated that the absorption spectrum can be recorded by Doppler-shifting the source relative to the sample with velocities of a mere few centimeters per second [8]. Mössbauer spectroscopy was born, and its discoverer was awarded the Nobel Prize in Physics in 1961 [9,10]. Though it primarily probes the lattice dynamics, which defines the magnitude of the resonant absorption and quantifies the atomic displacements, the technique became more widely used because the spectral distribution of the narrow recoil-free absorption probes the hyperfine interaction Hamiltonian. Thus, besides quantifying atomic motion, a Mössbauer spectrum reveals priceless information on electronic, structural, and magnetic properties.

The availability of chemical elements that possess isotopes with suitable nuclear properties is a crucial prerequisite. The primary criterion that decides upon the feasibility of Mössbauer spectroscopy for a specific isotope is related to the resolution and magnitude of the effect. The typical hyperfine interaction energy scale of 10^{-8} eV corresponds to a lifetime of the nuclear excited state of ~30 ns. Nuclear excited states with a lifetime significantly shorter than 0.2 ns yield not very useful broad emission lines, whereas for lifetimes significantly larger than 20 μs, broadening effects by, for example, impurities or vibrations make any spectroscopy extremely challenging. A second limitation is the recoil energy, $E_R = E_\gamma^2/2Mc^2$, where E_γ is the γ photon energy, M is the mass of the emitting nucleus, and c is the speed of light. If this energy becomes too large, as compared to the Debye energy of the solid, the magnitude of the Mössbauer effect becomes very small, even when both the source and the sample are cooled to liquid helium temperature. Note that, in liquid and gases, as there is no elastic scattering, there is also no recoil-free emission or absorption of γ-radiation, and thus, no Mössbauer effect. Pragmatically, only nuclear excited states with energies $E_\gamma < 200$ keV are considered. Further limiting factors are the isotopic abundance, the lifetime of the source and/or of the sample, transparency of the material to the radiation ($E_\gamma > 5$ keV), and the nuclear resonance cross sections:

$$\sigma_n = \frac{2\pi}{k^2} \frac{1}{1+\alpha} \frac{2I_e + 1}{2I_g + 1}, \tag{13.1}$$

where $k = E_\gamma/(\hbar c)$ is the wave vector for the resonant photons and other quantities are defined in Table 13.1. Combining all these factors with the potential scientific interest in the physics and chemistry of an element has historically led to iron-57, tin-119, europium-151, and antimony-121 being among the most popular isotopes. The essential nuclear parameters for these transitions are listed in Table 13.1. The reader is referred to Ref. [11] for a complete list of possible Mössbauer transitions.

Table 13.1 Selected common Mössbauer active nuclides and their nuclear parameters: resonance energy E_0; lifetime of the excited state τ_0; natural isotopic abundance a; nuclear spin in the ground and excited states: I_g and I_e, respectively; internal conversion coefficient α; nuclear resonance absorption cross section σ_n; and ratio of the nuclear to photoelectric absorption cross section, σ_n/σ_{ph}. Values are from Ref. [11], with updated energies from Ref. [12–16]. Useful information on Mössbauer active isotopes is also available at the Mössbauer Effect Data Center.[1]

Nuclide	E_0 (keV)	τ_0 (ns)	a(%)	I_g	I_e	α	σ_n/kbarn	σ_n/σ_{ph}
^{187}Os	9.777	3.43	1.6	$1/2^-$	$3/2^-$	264.	194.4	5.84
^{57}Fe	14.4129	141.	2.14	$1/2^-$	$3/2^-$	8.18	2464.0	428.58
^{151}Eu	21.5412	14.0	47.8	$5/2^+$	$7/2^+$	28.0	242.6	29.06
^{149}Sm	22.5015	10.3	13.8	$7/2^-$	$5/2^-$	29.2	120.1	17.29
^{119}Sn	23.8793	25.6	8.58	$1/2^+$	$3/2^+$	5.22	1380.5	562.59
^{125}Te	35.4920	2.14	6.99	$1/2^+$	$3/2^+$	14.0	259.0	44.11
^{121}Sb	37.1292	4.99	57.3	$5/2^+$	$7/2^+$	11.11	195.4	40.26
^{129}Xe	39.5813	1.47	26.4	$1/2^+$	$3/2^+$	12.31	234.7	47.24
^{61}Ni	67.408	7.60	1.19	$3/2^-$	$5/2^-$	0.139	709.1	7046.
^{73}Ge	68.752	2.51	7.76	$9/2^+$	$7/2^+$	0.227	337.5	2121.
^{197}Au	77.351	2.76	100.	$3/2^+$	$1/2^+$	4.36	38.1	56.22
^{191}Ir	82.407	5.89	37.3	$3/2^+$	$1/2^+$	10.9	15.1	6.20
^{155}Gd	86.546	9.13	14.7	$3/2^-$	$5/2^-$	0.434	341.7	304.61
^{99}Ru	89.571	28.8	12.7	$5/2^+$	$3/2^+$	1.498	81.2	315.04

13.2
Spectral Parameters

The simplest approach for describing and modeling a spectrum is to first calculate the positions of the resonance lines, second determine the intensities based on the Clebsch–Gordan coefficient for the nuclear transition and material-specific parameters, such as preferred crystallographic or magnetic orientation, and third to decorate the lines with a Lorentzian function with a width determined by the natural lifetime of the excited state – and possible spectrometer broadening. Although not entirely valid, it is often tempting, and frequently used, to add an additional line broadening that absorbs unknown material parameters, such as distributions of local environments. The type of information that can be obtained from Mössbauer spectroscopy is defined by the interaction Hamiltonian of the electrons and the probed nucleus (see Figure 13.2). These interactions are (a) the Coulomb interaction between the electron and nuclear charge, including dipolar, quadrupolar, or higher order terms, (b) the Zeeman interaction between electronic and nuclear spins, and (c) the coupling of the orbital electronic moment with the nuclear spin.

---

1) Mössbauer Effect Data Center. http://www.medc.dicp.ac.cn/.

$$\delta = E_S - E_A$$
$$= (\Delta E_e^S - \Delta E_g^S)$$
$$-(\Delta E_e^A - \Delta E_g^A)$$

Isomer shift   Quadrupole splitting   Magnetic hyperfine splitting

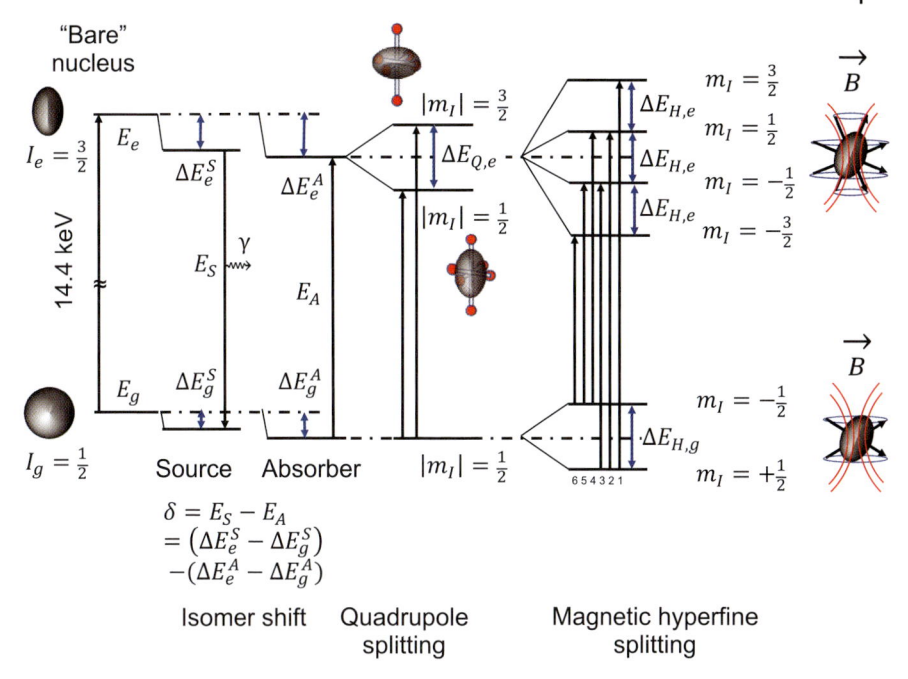

Figure 13.2 The Coulomb interactions of the nucleus with the surrounding electronic environment yield an isomer shift of the levels relative to a source δ (left). The interaction of a nonspherical nucleus with the electronic electric field gradient gives rise to the quadrupole splitting ΔE_Q (center). The precession of the nuclear spin in the externally applied or internal magnetic induction yields a Zeeman magnetic hyperfine splitting of the nuclear levels $\Delta E_{H,g}$ and $\Delta E_{H,e}$ (right). These interactions are illustrated here for the 14.4 keV $I_g = 1/2$ to $I_e = 3/2$ transition of iron-57 (see text).

13.2.1
Area and Recoil-Free Fraction

The intensity of the effect gives access to the Lamb–Mössbauer factor f_{LM}, which quantifies the recoil-free fraction and is related to the atomic displacement parameter [17] $\langle x^2 \rangle$ by

$$f_{LM} = \exp(-k^2 \cdot \langle x^2 \rangle). \tag{13.2}$$

In simple terms, k^2, exactly as E_R, is related to the strength of the knock of the absorbed photon, whereas $\langle x^2 \rangle$ is related to the lattice rigidity. In an ideal case, the recoil-free fraction in a material can be determined from the total integrated nuclear absorption – the area of a Mössbauer spectrum – if the sample thickness, isotopic abundance, source strength, and so on are known. Strictly speaking such a determination is very difficult, and also requires knowledge of the recoil-free fraction in the source and precise knowledge of the detector efficiency and discrimination.

In practice, f_{LM} is obtained through modeling the temperature dependence of the spectral absorption area, which will always decrease with temperature as the atomic displacement increases. The most commonly used approximation is that of a Debye solid, in which the distribution of vibrational modes, phonons, is assumed to be quadratic up to a cutoff frequency or energy, the Debye energy, E_D. This approximation provides means to model the atomic displacement parameters and finally yields

$$\log(f_{LM}(T)) = -\frac{6E_R}{k_B\Theta_D}\left[\frac{1}{4} + \left(\frac{T}{\Theta_D}\right)^2 \int_0^{\frac{\Theta_D}{T}} xe^{-x}dx\right], \tag{13.3}$$

where $\Theta_D = E_D/k_B$ is the Debye temperature, k_B is the Boltzmann constant, and T is the temperature. The temperature dependence of the logarithm of the absorption area can usually be fitted with this equation, up to an additive constant that absorbs all the unknown parameters. Note that a more precise and less model-dependent determination of the recoil-free fraction is nowadays possible by means of a direct measurement of the element-specific density of phonon states, $g(E)$, as can be achieved by nuclear resonant inelastic X-ray scattering [18,19], see section 13.3.7.

As the material under investigation can bear different species or crystallographic sites for the utilized Mössbauer element, the spectrum can comprise several subspectra pertaining to each configuration. A priori, the ratio of the number of nuclei in each configuration will contribute to determining the relative signal strength or absorption area for each subspectrum. However, special care must be taken for the f_{LM} of each site or oxidation state, an f_{LM} that provides an additional weighting factor that requires correcting for. This correction is notably important for investigations of compounds that contain divalent and trivalent iron ions, because their f_{LM} differ significantly at room temperature and above.

13.2.2
Isomer Shift

A most informative quantity that can be obtained from a Mössbauer spectroscopy experiment is the isomer shift δ. Empirically, this quantity is obtained from the center of mass of a particular component or subspectrum. The isomer shift is a relative quantity determined by the difference in the energy of the nuclear ground state in the source and the sample – which we will call the absorber – minus the same difference for the nuclear excited state. Because the electronic density around the nucleus, in general, does not change during the Mössbauer transition, the only factors to consider are the difference in electronic density, ρ, at the nucleus in the source and in the absorber $\rho_A(0) - \rho_S(0)$ and the transition-specific nuclear parameters such as the charge, and the difference in size and shape of the nucleus in both states. Quantum mechanics indicates that in the absence of strong relativistic effects only s electrons have a significant probability of being located in the nucleus. However, p, d, and to a lesser extent f electrons still have a strong influence on the isomer shift because they strongly interact with and contribute to screening the s electrons. In the absence of quadrupolar or magnetic hyperfine interactions, the isomer shift is the relative (Doppler) velocity that is required in order to restore the source–absorber resonance between the lines shifted by the Coulomb interaction of the nucleus with its surrounding electrons. Formally, see Figure 13.2,

$$\delta = E_A - E_S = \frac{2\pi}{5} Z e^2 (\rho_A(0) - \rho_S(0)) \left[R_e^2 - R_g^2 \right], \tag{13.4}$$

where E_A and E_S denote the resonant photon energy for the absorber and the source, respectively, Z is the atomic number and e is the electron charge, and R_e and R_g are the nuclear radii in the nuclear excited and ground states, respectively.

In compounds that contain different sites or valence states for the Mössbauer element under consideration, several resonance lines could be present. In the more general case of valence fluctuations related to electron hopping, these resonance lines could be separated or collapsed to an average position depending on the valence fluctuation frequency; see the relaxation phenomena discussion below.

For almost all elements investigated by Mössbauer spectroscopy, empirical relations have been gathered that relate an observed isomer shift – relative to a standard sample or a specific source – to an electronic configuration or an oxidation state.[2] Unfortunately, although a specific electronic configuration or density uniquely determines the isomer shift, the converse is not true. Notably, for iron-57 Mössbauer spectroscopy, an isomer shift in the range of 0.1–0.4 mm/s relative to ambient temperature α-iron is not very informative without further insights, as it can be related to iron(I), iron(II), iron(III), or iron(IV) in various spin states. Based on this observation alone, high-spin iron(II) is the only common configuration that can be ruled out, as its isomer shift ranges from 0.8 to 1.4 mm/s. Isomer shift tables or diagrams have been compiled for the most commonly used elements and are accessible at the Mössbauer Effect Data Center[1] or can be found in the literature [20,21]. Data for iron-57, tin-119, antimony-121, and europium-151 are shown in Figure 13.3.

13.2.3
Second-Order Doppler Shift

A further effect that influences the relative velocity (position) of a resonance line arises from a nonnegligible relativistic correction. This influence was pointed out in a brilliant article by Josephson [22] and published back to back with the article by Pound and Rebka, which established the gravitational redshift and apparent weight of γ photons by the Mössbauer effect [23]. The basis of this effect originates in the nonnegligible change in mass of the nucleus as it transitions from

---

2) A notable exception is hafnium where the isomer shift bears little information on the electronic configuration but, instead, the quadrupole interaction provides this information [20].

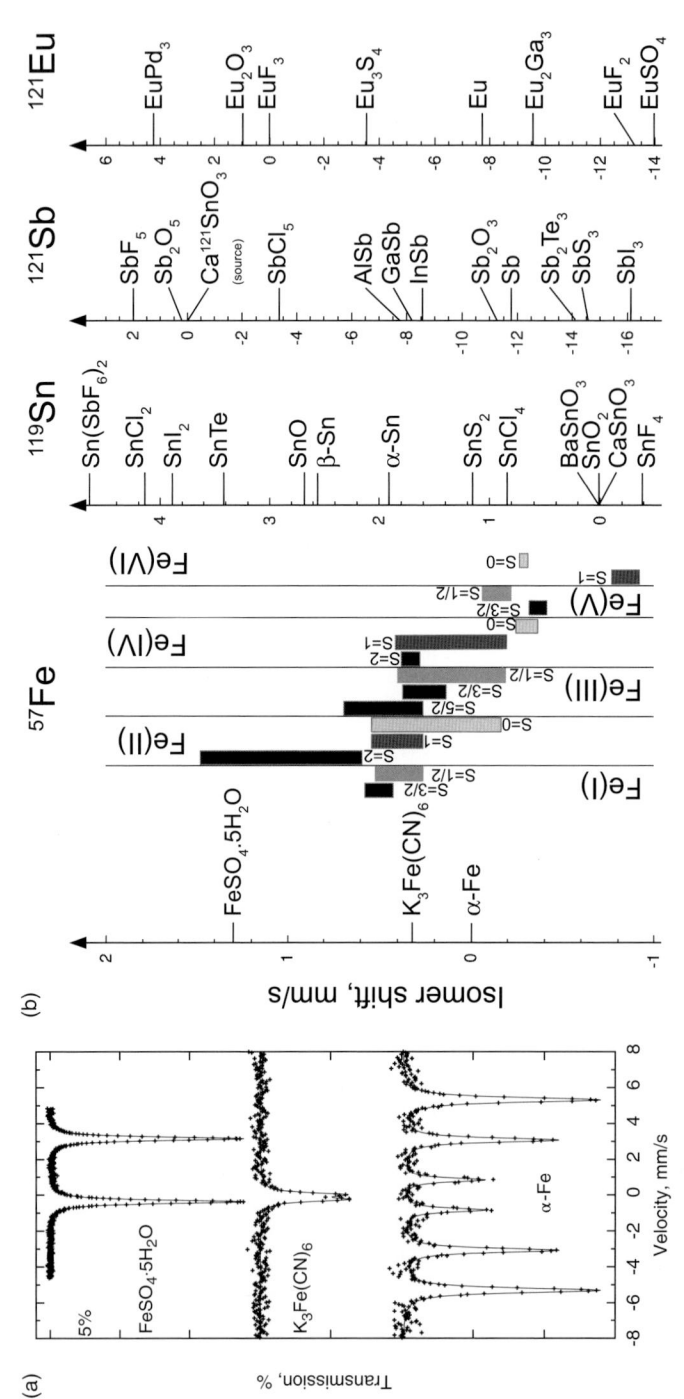

Figure 13.3 (a) Room temperature iron-57 transmission Mössbauer spectra for the indicated samples. Hydrated iron sulfate exhibits the large quadrupole splitting and isomer shift characteristic of high-spin ($S = 2$) Fe(II). Potassium iron hexacyanide exhibits a small isomer shift and quadrupole splitting compatible with Fe(III). The magnetic sextet of α-iron is utilized as both calibrant and reference for the isomer shift. (b) Typical isomer shifts for selected substances and isotopes. For iron-57, typical ranges in isomer shifts are given for different oxidation states and spin states. Calcium stannate is a typical source for both tin-119 and antimony-121 Mössbauer spectroscopy. (Data adapted from Refs [20].)

the ground state to the nuclear excited state, or vice versa. This mass difference given by $\delta m = E_\gamma/c^2$ leads to a change in mean-square velocity for the nucleus because its kinetic energy is conserved. As a consequence, both the source and the absorber temperature – that determine the kinetic energy – have an influence on the resonance energy. It is thus crucial, when providing a value for the isomer shift of a substance, to indicate the reference sample or source as well as its temperature.[3] The effect can be substantial, typically 0.073 mm/s per 100 K at high temperatures, for iron-57. More generally, the second-order Doppler shift (SOD) is – in velocity units – $\delta_{SOD} = -\langle v^2\rangle/2c$, where $\langle v^2\rangle$ is the mean-square velocity of the nucleus.

The SOD must be considered and corrected for when determining an oxidation state. In a similar way as for the Lamb–Mössbauer factor, the temperature dependence of the SOD also provides material-specific thermodynamic information, namely, the mean kinetic energy of the nuclei [24]. In the Debye solid approximation,

$$\delta_{SOD} = -\frac{9k_B}{16M_{eff}c}\left[\Theta_M + 8T\left(\frac{T}{\Theta_M}\right)^3\int_0^{\frac{\Theta_M}{T}} x^3/(e^x-1)dx\right] \quad (13.5)$$

$$\simeq \frac{3k_BT}{2M_{eff}c}\left\{1 + \frac{1}{20}\left(\frac{\Theta_M}{T}\right)^2 + \cdots\right\},$$

where M_{eff} is the effective mass and Θ_M is the Mössbauer temperature. The second line provides a series expansion useful at high temperatures. Typically, the effective mass should be the mass of the nucleus, however deviations can be observed in practice depending on the particular vibrational density of states in the material. In the same way, because different parts of the phonon spectrum have different weights for the $\langle v^2\rangle$ and $\langle u^2\rangle$ quantities, the Debye temperature and Mössbauer temperature can be very different if the density of phonon states differs strongly from that of a Debye solid [25–27].

Note that strictly speaking neither the isomer shift nor the second-order Doppler shift belong to the hyperfine interactions that we will discuss below, as they do not lead to a hyperfine level splitting.

13.2.4
Quadrupole Splitting

Whereas the isomer shift provides a localized measure of the electron density at the nucleus – a Coulomb-type charge monopole interaction – the quadrupole splitting of the nuclear levels measures the distortion of the spatial distribution

---

3) In a first approximation, SOD is not observed when the Mössbauer spectral measurement is carried out when the source and the sample temperature are modified simultaneously. Any residual SOD observed would point to a difference in the rate of kinetic energy change in the source relative to the sample/absorber.

of the electronic charge around the nucleus.[4] In the simple case of a transition between an $I_g = 1/2$ nuclear ground and $I_e = 3/2$ nuclear excited state, where I denotes the spin of the nucleus, the quadrupole splitting is pragmatically determined by the splitting between the two lines of the symmetric doublet observed in the Mössbauer spectrum of a random powder.

The quadrupole splitting originates in the asymmetric shape of both the electronic environment, quantified by the electric field gradient, EFG, and the nucleus – in at least one of the ground or excited states – quantified by the nuclear quadrupole moments Q_g and Q_e. With an adequate choice of coordinates, and in absence of other hyperfine interactions, the second-rank EFG tensor $V_{ij} \equiv \partial^2 V/\partial i \partial j$, where $i = x, y, z$ and V is the electric potential, takes a diagonal form with only V_{xx}, V_{yy}, $V_{zz} \neq 0$. Because the EFG does not involve a net charge, Laplace's equation further states that $\nabla\nabla V = 0$, that is, the trace of the EFG tensor $V_{xx} + V_{yy} + V_{zz} = 0$ [20]. Thus, only two parameters are sufficient to fully characterize the EFG: V_{zz}, the principal element of the EFG, and $0 \leq \eta \leq 1$, the asymmetry parameter, with the conventions that $|V_{zz}| \geq |V_{yy}| \geq |V_{xx}|$ and $\eta \equiv (V_{xx} - V_{yy})/V_{zz}$. A convenient visualization of the EFG is an ellipsoid having V_{zz} as the longest axis and η quantifying the relative difference in the shorter axes.

The interaction Hamiltonian between the nuclear quadrupole moment and the EFG takes a simple form:

$$\hat{H}^Q_{(g,e)} = \sum_{i,j} Q^{ij} V_{ij} = \frac{eQ_{(g,e)}}{I \cdot (2I-1)} \left(V_{zz} \hat{I}_z^2 + V_{yy} \hat{I}_y^2 + V_{xx} \hat{I}_x^2 \right), \quad (13.6)$$

where I is the nuclear spin and $\hat{I}_{x,y,z}$ are the components of the nuclear spin operator along the Cartesian directions. In the absence of further interactions, the matrix elements are diagonal in the $|I, m_I\rangle$ states, where m_I is the nuclear spin magnetic number and the level energies are simply

$$E^Q_{g,e}(I, m_I) = \frac{eQ_{g,e} V_{zz}}{4I_{g,e} \cdot (2I_{g,e}-1)} \left[3m_I^2 - I.(I+1) \right] \sqrt{(1+\eta^2/3)}. \quad (13.7)$$

A detailed derivation of Eq. (13.7) can be found in, for example, Ref. [20].

The absorption line positions are then obtained by calculating the splitting in the ground and excited nuclear states and considering which transitions are allowed, depending on whether the nuclear transition is of the M1 magnetic dipolar, E2 electric quadrupolar, or mixed type. In general, quadrupole split spectra can be quite complicated with, for example, eight different lines for the 5/2–7/2 (mostly) M1 dipolar transition of europium-151 or antimony-121 with

---

4) There is no measurable static dipole for the nuclei which would enable a direct probe of the electronic dipole moment.

allowed transitions for $\Delta m = -1, 0, 1$; however, in practice, these lines usually overlap and are thus poorly resolved.

In the rather simple and most useful case of $I_g = 1/2$ to $I_e = 3/2$ transition for iron-57 and tin-119, the level splitting observed in the spectrum is simply $\Delta E = \pm 1/4 e Q_e \cdot V_{zz} \sqrt{(1 + \eta/3)}$. Note that there is no splitting for $I = 0$ or $I = 1/2$ nuclear states.

As seen in Eq. (13.7), the quadrupole interaction is only sensitive to the absolute value of the magnetic quantum number. The sign of the quadrupole interaction is thus usually not resolved and can be obtained only in the presence of an additional splitting by a magnetic hyperfine interaction, or, in rare cases, relative signs of the quadrupole splittings for two sites can be determined for relaxation spectra.

Two terms contribute to the EFG, namely, the charge distribution on the neighboring atoms that defines the lattice contribution and the anisotropy in charge distribution in the molecular or atomic orbitals of the probed atom itself, called the valence contribution. In the axial symmetry approximation, $\eta = 0$, the EFG is given by $V_{zz} = (1 - g_\infty) V_{zz}^{lat} + (1 - R) V_{zz}^{val}$. The two scaling factors are the Sternheimer antishielding factor, $1 - g_\infty \sim 7 - 10$ for iron-57, which describes how the crystal lattice charge distribution EFG is enhanced by the inner electron shell and the Sternheimer shielding factor $(1 - R) \sim 1/4 - 1/3$ for iron-57, which describes how the valence contribution is decreased [20]. Note that if the probed atom exhibits an orbital moment, an additional valence contribution to the EFG is expected.

A simple analysis of the point symmetry provides ample information on the EFG parameters: A three- or fourfold axis is sufficient to guarantee axial symmetry that is, $\eta = 0$, whereas three- and fourfold axes, that is, cubic point symmetry, lead to $V_{zz} = 0$ and thus the absence of quadrupole interaction. The presence of a quadrupole splitting and its temperature dependence is a particularly powerful probe for the crystal field and possible Jahn–Teller distortions, see section 13.4.3.

Providing a general perspective of the relative intensities of the absorption lines is beyond the scope of this chapter and we refer the reader to more specialized literature [11,20]. For 1/2 to 3/2 transitions, in the absence of magnetic hyperfine splitting, a quadrupole interaction leads to a simple doublet of two Lorentzian absorption lines with the same area, if the sample is a randomly oriented powder. However, occasionally the observed spectrum reveals an asymmetry in the intensity. This asymmetry can be related to the presence of two absorption sites with similar quadrupole interaction but slightly different isomer shift, the signature of preferential orientation or texture, the existence of a small magnetic hyperfine splitting, or, rather rarely, a Goldanskyi–Karyagin effect [28] related to an anisotropy in the recoil-free fraction. By combining measurements at different temperatures and a magic-angle spinning [29] experiment at $\theta = 54.7°$ between the sample normal and the radiation, a definite answer on the cause of the asymmetry can be obtained [30].

13.2.5
Magnetic Hyperfine Splitting

Most Mössbauer active nuclei have a nonzero spin both in the ground state and the excited state.[5] In interacting with unpaired electronic spins in a material, a magnetic hyperfine splitting of the nuclear levels occurs. The magnitude of this splitting is determined both by the effective magnetic hyperfine field H_{eff} experienced by the nucleus and the coupling constant for the ground state g and/or excited state e. The interaction Hamiltonian has a rather simple form:

$$\hat{\mathcal{H}}^{mag}_{(g,e)} = -\hat{\mu} \cdot \hat{\mathbf{B}} = -g_{N,(g,e)}\mu_N \hat{\mathbf{I}} \cdot \hat{\mathbf{B}}, \tag{13.8}$$

where $\hat{\mu}$ and $\hat{\mathbf{B}}$ are the magnetic dipole moment and magnetic induction operators, respectively, g_N is the nuclear Landé factor, and μ_N is the nuclear magneton. The matrix elements are diagonal in the $|I, m_I\rangle$ states and the level energies are

$$E^{mag}_{g,e}(I, m_I) = -g_{N,(g,e)}\mu_N B m_I. \tag{13.9}$$

Thus, in the absence of further perturbation by a quadrupole interaction, the splitting between levels is proportional to the nuclear magnetic quantum number. The magnetic hyperfine interaction completely removes the degeneracy of the states as it depends on m_I rather than m_I^2, as was the case for the quadrupole interaction.

Behind this simple expression, a rather complicated sum of phenomena [31] contributes to the effective induction $B_{eff} = \mu_0 H_{eff}$ that the nucleus experiences. The effective field is the sum $H_{eff} = H_{mag} + H_{dip} + H_{orb} + H_{FC}$. The first term H_{mag} includes the lattice magnetization, the demagnetizing factor, and possibly an externally applied magnetic field. The second term is the dipolar magnetic field generated at the nucleus by neighboring magnetic ions, a term that typically leads to the observation of a so-called transferred magnetic field on nonmagnetic ions. The third term arises from any orbital magnetic moment that the Mössbauer atom could possess if the crystal field does not fully quench it. The sum of the first three terms is typically small, on the order of 2–3 T for iron-57. The last term is the Fermi contact term, that is, in most cases, the dominating term for magnetic ions. As only the s electrons are literally in "contact" with the nucleus and because s electrons are mostly spin pairs that carry no moment, an *a priori* small contribution could be expected. However, similar to the case of the isomer shift, a small imbalance in spin-up and spin-down s electron density at the nucleus arises by their interaction with unpaired d or f electrons. The contribution of the outer s electronic shells to the hyperfine magnetic field is also sensitive to magnetic moments at adjacent atoms, and provides a particularly useful probe into local chemistry and magnetism, notably for tin-119 and antimony-121 that experience large so-called transferred fields. For iron-57, the largest

---

5) With a notable exception of a series of $I_g = 0$ to $I_e = 2$ transitions of rare earth elements, see for example, the Appendix A in Ref. [11].

term is a very large negative contribution to the effective field caused by the 2 s shell, smaller contributions from the 1 s and 3 s shells, and a positive contribution from the 4 s shell that only partly mitigates the 2 s contribution [17]. Consequently, in α-Fe, the measured hyperfine field is quite large, at −33 T, where the negative sign was verified by spectroscopy in external applied fields. This value is much larger than expected for the 2.2 μ_B moment of iron and is indicative of the huge Fermi contact term.

In general, the nuclei experience a combination of quadrupole and magnetic hyperfine interactions. If the quadrupolar interaction is small relative to the magnetic interaction, it can be treated as a perturbation and an approximate simplified linearized solution for the absorption energies can be obtained. In the more general case, the solution of this problem requires the diagonalization of the Hamiltonian in both the ground state and excited state, a diagonalization that is trivial for a spin state of 0 or 1/2 but increases in complexity for larger spin values.

13.2.6
Relaxation Phenomena

An interesting question arises when the environment that the Mössbauer nucleus experiences is not static. First, the relevant timescale τ must be considered. How long does it take for the splitting of the nuclear levels to be established? Semiclassically, for magnetic hyperfine splitting, this timescale is the inverse Larmor precession frequency obtained from the hyperfine field and the nuclear magnetic moment. More generally, the timescale to consider is given by the hyperfine splitting energy $\tau = \hbar/\Delta E$ [17].

If the nuclear environment is fluctuating, that is, when a time-dependent electric field gradient or magnetic field [32] changes very slowly or very rapidly compared to the nuclear timescale, the spectra take a simple form described by either a weighted sum of the spectra for each configuration or by an effective time averaged Hamiltonian, respectively. In the intermediate timescale, the spectrum is a more complex relaxation spectrum. A major complication arises because such a relaxation spectrum is no longer described as a discrete sum of Lorentzians with positions determined by the Hamiltonian eigenvalues. The detailed treatment is complex and involves the use of the Liouville matrices. These matrices operate on the nuclear transition matrix in the Hilbert space of product wave functions of the ground and excited states $|\psi_g\rangle \cdot |\psi_e\rangle$. The interested reader is referred to Ref. [32–35]. Examples of the information that can be obtained are provided below. Note that the lifetime of the nuclear excited state plays only a secondary role in that it provides an estimate of the slowest fluctuations that can be resolved.

In summary, through the hyperfine parameters – quadrupole splitting and hyperfine splitting – and the other spectral parameters, isomer shift and Lamb–Mössbauer factor, Mössbauer spectroscopy provides partial access to a large collection of material-specific quantities related to coordination, bonding, oxidation state, magnetism, and dynamics. It is thus not surprising that, rapidly after the discovery of the Mössbauer effect and in parallel with nuclear physics-specific investigations, applications were found in most scientific fields. These applications span archeology, art, currency forgery detection, geology, geophysics and planetary physics, condensed matter, inorganic and organic chemistry, biology, medicine, and fundamental physics, at the basic research, applied research, and industrial application levels. Practical aspects and selected examples related to solid-state chemistry are described in the following sections.

13.3
Practical Aspects

13.3.1
Experimental Implementation

A radiation source, an absorber or scatterer, and a detector are the fundamental components of a nuclear resonance experiment. Beyond these elements, many variations are possible depending on the specific application. In the vast majority of experiments, a conventional transmission Mössbauer spectrum is the desired outcome. Such a spectrum is obtained by mounting the radiation source on a drive that modifies the emitted radiation energy by the Doppler effect and by recording the radiation transmitted through a sample as a function of the source velocity. Nowadays the most popular scheme involves a constant acceleration drive providing a periodic and linear velocity as a function of time, with a triangular waveform. The number of γ photons detected per time interval is stored with help of a multichannel analyzer, and the spectrum is gradually acquired by acquisition cycles (see Figure 13.4). More exotic variations discussed below involve the use of other detection schemes, operate in emission mode, or involve synchrotron radiation as a source.

13.3.2
Radiation Sources

The availability of a suitable radiation source is one of the most limiting aspects for Mössbauer spectroscopy, and it is worthwhile to discuss the fundamental aspects with the example of iron-57. From a nuclear physics standpoint, the parent isotope, its availability, lifetime, and finally the lifetime of the final excited state are the crucial factors. For iron-57, the parent nucleus is cobalt-57 or manganese-57 with half-life $t_{1/2}$ of 270 days or 85 s, respectively (see decay schemes in Figure 13.4). Obviously, for conventional laboratory-based experiments, the former is preferable, whereas the latter is invaluable for accelerator-driven emission spectroscopy following the implantation of manganese-57 [36]. The cobalt-57 decay cascade that follows an electron capture by the nucleus populates to

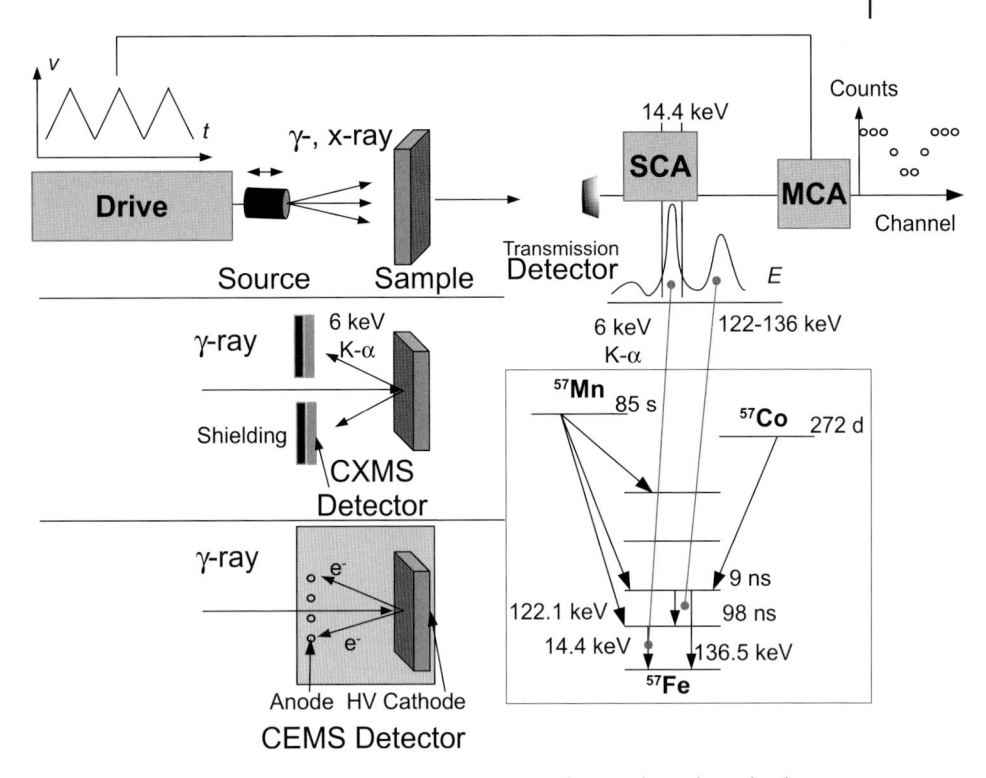

Figure 13.4 Schematic setup for transmission Mössbauer spectroscopy and alternative detection schemes for conversion X-rays (CXMS) and electrons (CEMS) (see text). The single-channel analyzer (SCA) discriminates the detector signal and feeds the multichannel analyzer (MCA), which registers counts as a function of the drive velocity. The energy spectrum in the SCA depends on the decay cascade for the specific isotope – see bottom right for iron-57. For CXMS X-ray and γ sample, fluorescence is detected, the detector is shielded from direct source radiation, for CEMS, and high voltage (HV) is applied within a sealed gas flow chamber in order to collect conversion electrons at the anode.

nearly 100% a 136 keV nuclear excited state with $t_{1/2} = 9$ ns, which in turn decays with 90% probability to the 14.4 keV nuclear excited state with $t_{1/2} = 98.3$ ns half-life. This 14.4 keV state is used for iron-57 Mössbauer spectroscopy. Note that, first, the other 10% photons with 136 keV energy have also been used for Mössbauer spectroscopy, and second, the 122 keV photons emitted in the transition from 136 to 14.4 keV provides an interesting internal time standard for specialized time-window experiments [37,38].

The production of cobalt-57 from a nickel-58 target occurs by ^{58}Ni(p, 2p)^{57}Co reaction in a cyclotron. Besides cyclotron or accelerator-based production, a neutron capture reaction in a high-flux isotope reactor is the second route, which is however not applicable for cobalt-57. For short half-life parents, such as cobalt-61 ($t_{1/2} = 1.65$ h) for nickel-61 spectroscopy or platinum-197 ($t_{1/2} = 18.3$ h) for gold-197 spectroscopy, a close proximity of the source preparation and the spectroscopy laboratory is required. Note that in-neutron-beam spectroscopy, where the γ-ray decay following a neutron capture is used, is possible and has been successfully demonstrated at RIKEN [39]. A combination of all the factors above with the scientific interest leads to iron-57 and tin-119 being the most investigated isotopes, and currently the only commercially available sources. Next popular, arguably, are europium-151 and antimony-121 due to the very long source (half-life $t_{1/2} > 50$ years), large nuclear resonant absorption cross section, and large natural isotopic abundance.

From a materials science point of view, the challenge consists in embedding the radionuclide in a suitable matrix. The aspects to optimize are the stiffness, hyperfine interactions, homogeneity, and transparency. Currently, cobalt-57 alloyed in rhodium is the most popular choice because (a) the strong bonding of cobalt in rhodium provides a large recoil-free fraction for the emitted 14.4 keV radiation even at room temperature, (b) rhodium is a nonmagnetic cubic solid where substitutional cobalt or iron does not experience hyperfine interactions and thus the emitted radiation is a narrow single line, and (c) rhodium is sufficiently transparent to 14.4 keV radiation, with a 20 μm mean-free path. Note that as a source ages, nuclear resonant self-absorption and defects caused by the presence of iron-57 and radiation damage lead to gradual broadening of the emitted radiation.

A further important parameter to consider is the internal conversion tied to the transition between the nuclear excited and ground states, which is quantified by the internal conversion coefficient α, the ratio between the number of decays by electron versus γ emission. The emission of an electron, typically from the inner shells, is followed by X-ray fluorescence as the core hole is filled. For the 14.4 keV state of iron-57, $\alpha = 8.18$. Thus, for every 14.4 keV γ-ray emitted by the source, eight times more high-energy conversion electrons and X-rays are emitted. The magnitude of this conversion coefficient is determined by the electronic structure and the multipolarity of the nuclear transition. Beyond a reduction in the number of emitted useful photons, the internal conversion process also reduces the nuclear resonance absorption cross section, see Eq (13.1), and enables the detection of nuclear resonance absorption by detecting the emitted conversion electrons or X-rays in CEMS or CXMS Mössbauer spectroscopy experiments (see Figure 13.4).

Finally, the natural isotopic abundance of the studied nucleus is a further limitation to consider (see Table 13.1). For iron, although iron-57 makes up only 2% of the natural isotopic abundance, the rich interest in iron chemistry combined with the huge resonance cross section has contributed to making iron the most studied element. In contrast, more stringent experimental conditions make, for example, gold-197 with 100% isotopic abundance, much less studied. Note that isotopic enrichment of the sample is often used if the acquisition time must be reduced or if a very thin sample must be investigated. For iron-57, an overall enhancement by a factor 50 is possible, but at a cost of typically 1–5 $/mg isotope. Isotopic labeling can further be used for contrast or in diffusion experiments.

13.3.3
Detectors

From a detection point of view, the radiation cascade, X-ray fluorescence from the source and sample, and detection efficiency are the major concerns. An optimal signal-to-noise-ratio (SNR) can be achieved only if the Mössbauer γ-radiation can be discriminated efficiently from any parasitic radiation. For iron-57, the 6 keV iron K-edge fluorescence and the 122 keV nuclear fluorescence are easily discriminated by pulse-height analyzing electronics for most proportional detectors. The choice between a Kr-gas proportional counter, sodium iodide scintillator or yttrium aluminum garnets, or perovskites scintillators [38] is typically guided by a balance between source strength, spectrometer geometry, and acquisition time.

Detectors for transmission Mössbauer spectroscopy should be optimized in order to discriminate the γ photons from any noise and cope with the usually high photon flux in excess of 10^5/s. For relatively low energy γ-rays, $E < 40$ keV, NaI or other scintillators, gas proportional counters, or new silicon drift detectors, with improved energy resolution, are utilized. For high energies, Ge or CdTe detectors can be employed. The detection can sometimes be enhanced by a simple filtering technique, such as by placing a thin palladium foil, ~20 μm thick, between the sample and the detector for ^{119}Sn Mössbauer spectroscopy. By efficiently absorbing unwanted X-ray fluorescence from the source and the sample, while sufficiently transmitting the 23.88 keV γ photons, a strong improvement in the SNR can be obtained. For other geometries, alternative detection schemes can be utilized. If transmission geometry is not possible, for example, when investigating art, solid rocks, surfaces, or metal sheets, the so-called backscattering geometry can be utilized. In that case, the γ and conversion X-rays or electrons produced by the nuclear decay that follows the nuclear resonance absorption are detected [40,41] (see Figure 13.4). This scheme was particularly successful for the Mössbauer spectroscopy instrument MIMOS II featured into Mars landing missions Spirit and Opportunity [42,43].

Information in terms of depth contrasting is gained from the different mean free paths upon exit of the 6 keV X-ray and the 14.4 keV γ-fluorescence. When investigating thin films or surfaces, detection of the electrons emitted through the internal conversion process, with an exit depth of ~200 nm, is utilized [44]. Typically, the sample is placed inside a small chamber that acts as a gas proportional counter and an optimized flowing gas mixture is chosen in order to enable efficient multiplication, detection, and discrimination of the emitted electrons. This method is very successful for investigating, for example, surface reactions [45], corrosion processes [46,47], or thin film magnetism [48].

Finally, the use of nuclear resonant detectors, NRD, is another possibility that is currently being improved. These detectors are comprised of a known single-line absorber with optimized thickness and an isomer shift matched to the source and placed in a conversion electron detection cell. This arrangement strongly increases the sensitivity and discrimination of γ photons with precisely the energy emitted by the source. Either by keeping source and NRD fixed and moving the sample or by moving source and NRD together and keeping the sample fixed, faster data acquisition and linewidth reduction is possible [49,50] in a way similar to a lock-in detection mechanism.

13.3.4
Sample Thickness and Sample Environment

The sample thickness is an important parameter to optimize whenever possible. The natural sample thickness t is the inverse of the product of the recoil-free fraction, the density of absorbing nuclei n_a, – taking into account the natural isotopic abundance a – and the cross section for each nucleus σ_n (see Table 13.1):

$$t^{-1} = f_{LM} \cdot n_a \cdot \sigma_n = f_{LM} \cdot a \cdot \rho/m_M \cdot f_{at} \cdot N_A \cdot \sigma_n, \tag{13.10}$$

where ρ is the density, m_M is the molar mass, and f_{at} is the atomic fraction contributed by the element considered (i.e. atomic percent/100), and N_A is the Avogadro number. For bcc-iron at ambient temperature, $f_{LM} \sim 0.8$ and $t \sim 2.7\,\mu m$. This thickness must further be multiplied by a factor 4, because the strongest absorption line of the iron magnetic sextet accounts for only 1/4 of the absorption intensity (see Figure 13.3a) and overall $t \sim 11\,\mu m$ [17]. As a rule of thumb, ~20 μm of natural abundance iron is plenty. Note that if the X-ray absorption of the sample is large, a thicker sample will lead to slower acquisition time, and the optimal thickness is then 2 times the x-ray absorption length. Details for the optimization involve balancing the overall sample absorption with the effective thickness of Mössbauer nuclides [51]. Because often only a very small amount of sample is required, a few tens of milligram for an absorber of 1–2 cm², the sample is usually mixed with a low-absorption matrix material, such as boron nitride, in order to uniformly fill the sample holder.

Another aspect to consider is the broadening and distortion of the absorption lines for samples with large natural thickness, both in transmission [52,53] and backscattering geometries [54]. A practical approximation for the broadening is that the effective linewidth increases by a factor $(1 + 0.135t)$ [17]. Because typical conventional Mössbauer experiments utilize samples with ~10% absorption, such line broadening is subtle and a treatment by the so-called transmission integral approach [55,56] is usually not required. It is nevertheless advisable to use samples thinner than t for quantitative evaluation.

According to the question do be answered, the sample is placed in rather compact sample environments, such cryostats, ovens, magnets, with a typical source-to-detector distance of 10–15 cm. The sample environment will also determine the detector type; in particular, stray magnetic fields prevent the use of photomultiplier-based detectors. High-pressure experiments utilizing diamond anvil cells are also possible if a specialized point source with high specific activity is available.

13.3.5
Calibration and Data Treatment

In order to guarantee reproducibility, a calibration of the instrument is highly advisable every time the velocity range of the Doppler drive is changed, and an optimization of the pulse height analysis must be carried out after each sample change, as the detector load changes with the sample transmission and fluorescence.

The acquired data obtained as counts per channel, for example, 1024 channels, are reduced to a calibrated Doppler velocity scale by utilizing a known absorber, typically α-Fe with its characteristic sextet. For fitting purposes, either the measured counts can be utilized or the data can be reduced to percent transmission or emission in order to avoid numerical pitfalls associated with large numbers. Due to the triangular shape for the velocity versus time, the spectrum is acquired twice, on velocity increase and decrease, respectively, and should also be folded (the velocity decrease part is inverted and added the the increase part) for presentation purposes.

In order to prepare a fit, a reasonable model based on the material structure must be prepared, where a hyperfine interaction Hamiltonian, in practice, an isomer shift, an EFG, and a magnetic hyperfine field, is attributed to each site occupied by the Mössbauer element. The Hamiltonian defines the position and intensity of the absorption lines that are decorated by a Lorentzian line shape in most cases, see Section 13.2. The thus generated subspectrum is then added to those for other sites in order to generate the overall profile. Most spectra and fits shown in Section 13.10 illustrate this approach. The fitting of the spectral parameters is then most frequently carried out by a χ^2 minimization routine. This routine must be provided with the standard deviation for the obtained data in order to prevent overfitting and obtaining error bars on the fitted parameters. As for some complex spectra 20–50 fitting parameters can be necessary for a 500 data point spectrum, this aspect is crucial. Various fitting software, such as MossWinn, Normos or WinNormos, Recoil, Confit, MossFit, and WMoss, are provided by equipment vendors or expert experimental teams. A list of software, and much more information, is hosted on the Mössbauer Information eXchange server MIX, http://www.kfki.hu/mixhp/mossba.htm#soft. A simple code for modeling iron-57 data in the gnuplot software is provided here:

```
# The isomer shift, is, and quadrupole splitting, qs, and line-
width, g, are in mm/s.
# The hyperfine field is in Tesla/mu0.
# The qs is treated as perturbation for the sextet and yields the
quadrupole shift.
# gama is a magnetic texture parameter. 0<gama<4, with gama=2
for 3D averaged magnetism,
# and gama=4 for a thin iron foil with in-plane magnetization.
# This script plots the spectrum for iron powder at room
temperature.
```

```
set samples 500
e16=10.6244/66.16; e25=6.156/66.16; e34=1.6793/66.16
set xlabel "Velocity, mm/s"; set ylabel "Transmission, %"
bs=100.;area=10.;is=0.;hf=33.08;g=0.25;gama=2

ltz(x,e,g)=g/6.28319/((x-e)**2+(g/2)**2)

doublet(x,is,qs,g)=0.5*ltz(x,is-qs/2.,g)+0.5*ltz(x,is+qs/
2.,g)

sextet(x,is,qs,hf,g)=(1./4*(ltz(x,is+qs/2-hf*e16,g)+ltz(x,
is+qs/2+hf*e16,g))\
             +1./6*gama/2.*(ltz(x,is-qs/2-hf*e25,g)+
             ltz(x,is-qs/2+hf*e25,g))\
             +1./12*(ltz(x,is-qs/2-hf*e34,g)+ltz
             (x,is-qs/2+hf*e34,g))
      )    /(1+(gama/2.-1)/3)

ironspectrum(x)=bs-area*sextet(x,is,0.,hf,g)

plot ironspectrum(x)
```

13.3.6
Synchrotron Radiation

With the advent of high brilliance synchrotron radiation sources, the idea to use these as sources for nuclear resonance spectroscopy arose in the 1970s. This idea is attributed to Stan Ruby, at Argonne National Laboratory [57,58]. The extremely broad energy band of the synchrotron radiation beam, on the order of 1–100 eV, in comparison to the bandwidth of a Mössbauer source, on the order of 1–1000 neV, was – and still is – a major concern for synchrotron-based nuclear resonance scattering or absorption experiments. The filtering of the photons involved in the nuclear resonant processes from the electronic process and the transmitted beam is the primary issue. The first observation of a nuclear resonance effect by exploiting the time structure of a purely nuclear diffraction peak in an yttrium iron garnet crystal dates back to 1985 [59], after extensive effort by the group of Erich Gerdau. Since then, the technique has evolved and became available to the general user community by submission of experimental proposals to one of the four major facilities that feature a nuclear resonance beamline (the European Synchrotron Radiation Facility (ESRF), SPring-8, the Advanced Photon Source, APS, and Petra III). The first major breakthrough was arguably the development of superior crystal optics [60] that can now provide millelectrovolts photon bandwidth or better by means of high-resolution monochromators for tens of different isotopes [61–63], a bandwidth reduction that is crucial for avoiding detector overload or destruction. An alternative approach, based on high-speed choppers, is currently under development [64]. The second major breakthrough was the development of avalanche photodiodes (APD) (see below).

A typical synchrotron Mössbauer spectroscopy experiment is nowadays carried out in the time domain by using the time structure of the synchrotron radiation in the so-called nuclear forward scattering (NFS) technique (Figure 13.5). The so-called *prompt* radiation impinging on and transmitted through the sample is separated by the time of arrival in the detector from the *delayed* radiation that undergoes a nuclear resonance transition in the sample [65]. The nuclear decay will occur with the lifetime of the nuclear excited state as typical timescale. Thus, in any such experiment, fast detectors are required, typically with <1 ns time resolution, and APD are typically utilized [66]. Nuclear resonant photons emitted by the sample in the beam direction, in so-called forward scattering, are delayed with respect to the prompt radiation. If a very thin sample presenting a single nuclear absorption line is used, a signal exponentially decaying with the transition lifetime is expected. For samples with finite thickness, this signal is modulated by dynamical beats, originating in multiple scattering processes [60,65]. The presence of the dynamical beat in the time-domain expression of thickness induced line-broadening in Mössbauer spectroscopy.

If a sample contains either different environments for the probed nuclei or hyperfine interactions, the spectrum of the radiation emitted in forward direction will carry a superposition of energies of all the absorption lines. In the detector, which records the intensity and not the amplitude of the beam, a coherent "quantum beat" pattern of the superposed waves will be recorded, exactly as expected for the sound emitted by a pair of slightly misadjusted tuning forks. Though for simple cases the hyperfine splitting in time domain spectra can be extracted from the beating period, in general advanced modeling and fitting is required. Software suites are available [67,68]. Note that because by essence this technique uses a very broad band – compared to a Mössbauer radiation source – NFS does not provide the isomer shift but only the absolute value of relative isomer shifts if several components are present. Obtaining absolute isomer shift values from an NFS experiment requires measuring spectra of a combination of known reference samples with the sample under investigation [16].

Even with the tremendous increase in brightness of synchrotron radiation beams, measuring a typical transmission Mössbauer spectrum should be carried with conventional instrumentation. First, for a typical $1\,cm^2$ sample, the photon flux through the sample per bandwidth for a conventional experiment is similar to or sufficiently close to that for a synchrotron radiation experiment [11]. Second, the cost, overbooking factor, and logistics involved must be considered as well as the isotopic enrichment often required in order to optimize beam time usage. The true strength of synchrotron-based experiments lies in the small beam size, the intrinsic polarization [69,70], and the tunable energy. The small beam size made possible investigations of minute amounts of material in complex sample environments, such as high-pressure cells, now at pressures surpassing 200 GPa [71], extremely thin films down to monolayer thickness grown *in situ* [72,73], or pulsed magnetic fields above 30 T [74]. The intrinsic polarization provides unique insights in magnetic phenomena, such as exchange bias coupling [75,76], spin canting in nanomaterials [77], and magnetization dynamics [78]. The tunable energy provides access to nuclear resonance experiments for several isotopes that are challenging for laboratory experiments, such as nickel-61 [79] or ruthenium-99 [12] or simply impossible due to the lack of a source, such as for osmium-187 [13].

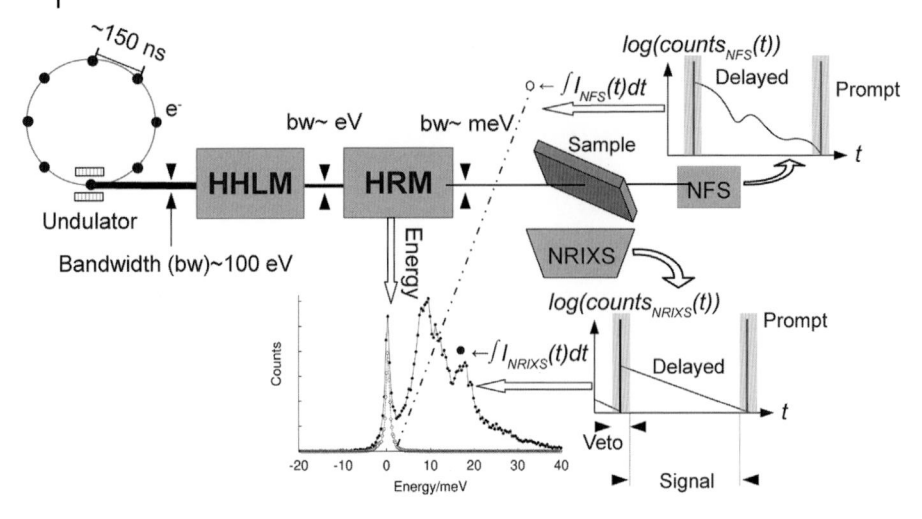

Figure 13.5 In a synchrotron radiation experiment, the bandwidth (bw) of *pulsed* synchrotron radiation beam, with ~150 ns as typical interbunch spacing, is reduced by a standard high-heat load monochromator (HHLM) and then by a high-resolution monochromator (HRM). When the beam energy matches the nuclear resonant energy, delayed nuclear forward scattering (NFS) is observed in the forward detector (top right) and delayed nuclear resonant inelastic X-ray scattering (NRIXS) is observed in the transverse detector (bottom right). The desired delayed signals are discriminated in time domain with a so-called veto that eliminates the prompt response, that is, diffraction or X-ray fluorescence. The HRM is scanned in energy around the nuclear resonance energy defined as zero in the bottom left inset. The time-integrated NRIXS intensity (closed symbols) as a function of energy provides the inelastic scattering function and the time-integrated NFS signal (open symbol) provides the resolution function of the HRM (open symbols) (see bottom left).

A noteworthy recent advance is the development of synchrotron radiation-based energy domain Mössbauer spectroscopy [80]. At the ESRF, an energy domain synchrotron Mössbauer source based on pure nuclear diffraction in iron borate is available for users and provides beam between 10^4 and 10^6 nuclear resonant photons per second in a narrow beam, depending on the desired bandwidth [81]. At SPring-8 several experiments with more exotic transitions, such as ytterbium-174 [82], have demonstrated the possibility for energy domain spectroscopy by utilizing a velocity transducer and a reference absorber in order to record a spectrum of an unknown sample [83].

13.3.7
Phonon-Assisted Nuclear Resonance Absorption

Whereas Mössbauer spectroscopy is concerned with the resonant recoil-free absorption or emission of nuclear fluorescence, as quantified by the recoil-free fraction, that is, Lamb–Mössbauer factor f_{LM}, the absorption with recoil is more common, and certainly dominant in most cases. However, in a solid, the recoil is tied to the presence of atomic displacements and vibrational modes, as evidenced by Eq. (13.2). This relation was recognized very early and experimentally harnessed in 1979 by Weiss and Langhoff [84]. By utilizing the Mössbauer transition in terbium-159 and extremely large Doppler shifts, ~100 m/s, they were able to reveal the phonon sidebands around the Mössbauer spectrum and thus obtain the density of phonon states for terbium in terbium oxide. This experiment is feasible essentially for two reasons: The high energy of the transition, 59 keV, leads to a Doppler energy shift of 20 meV for a 100 m/s velocity, and, second, the very short nuclear excited state lifetime of 0.2 ns leads to a rather broad band for the emitted γ photons, 3 µeV, more favorably matched to the typical millielectronvolt phonon energies than the 5 neV bandwidth, for example, in iron-57. Though a real breakthrough for phonon spectroscopy, the tremendous experimental challenges have prevented this technique from becoming mainstream.

The availability of synchrotron radiation has resurrected the possibility for phonon spectroscopy based on nuclear resonance absorption by providing the required combination of the ideal millielectronvolt radiation bandwidth transmitted through the high-resolution monochromator, HRM, with the tunability of the energy. In the first experiments by Makoto Seto and coworkers, in 1995, the inelastic scattering [85] by iron was obtained and the associated element-specific density of phonon states [86] for iron and iron oxide was then extracted by Wolfgang Sturhahn and coworkers. These data have been obtained by recording the nuclear resonance fluorescence that follows a phonon-assisted nuclear resonance absorption process as a function of the radiation energy transmitted through the HRM (Figure 13.5). Since 1995, this technique has been expanded to a dozen more isotopes and utilized with a wealth of available sample environments, notably high-pressure cells, in, for example, geophysical application [87], iron-based superconductors [88,89], or thermoelectrics [90].

The applicability of the above technique, called "nuclear inelastic scattering" (NIS), or "nuclear resonant inelastic X-ray scattering" (NRIXS), to a series of interesting Mössbauer transitions with high energy is limited by two major problems. First, suitable monochromators with millielectronvolt bandwidth and sufficient spectral transmission cannot be based on Si crystals for nuclear resonances with energies above ~30 keV. Thus, other schemes, notably backscattering monochromatization with a sapphire single crystal [14,62,91], have been utilized, for example, for the ^{121}Sb, ^{125}Te, and ^{129}Xe transitions at 37.1, 35.4, and 39.9 keV, respectively [14,15,92,93]. Second, the recoil energy increases quadratically with increasing transition energy and in turn strongly increases the multiphonon contribution to the inelastic scattering, an increase that eventually prevents the extraction of a density of phonon states [15].

A preliminary visual analysis of the inelastic scattering data is rather easy and reveals the energy of the phonon bands. The detailed analysis needed in order to extract the thermodynamic quantities related to the element-specific density of phonon states is however more complex and requires specialized software provided by the synchrotron radiation facilities [67,94].

13.4
Examples and Applications

In the following, examples that bear a relation to solid-state chemistry of how Mössbauer spectroscopy has contributed to our understanding of the structure and properties of materials are provided. These examples represent only a small selection, somewhat partial to the author's experience, of a host of excellent literature. The reader is referred to specialized literature for further information concerning key concepts [21,31,95–100] and examples [20,101–104].

13.4.1
Valence Determination

The probably most frequent request received in a Mössbauer spectroscopy laboratory is related to the valence determination in inorganic powder compounds. This question arises, for example, in mineralogy, in synthetic chemistry, or even in industrial settings and aims at understanding a phase formation mechanism or determining whether a synthesis route has generated a phase pure material. In principle, the valence state is determined by considering the isomer shift of the identified spectral components; however, in particular for iron, these data alone can yield an ambiguous result that needs to be confirmed by considering the quadrupole splitting or crystallography and magnetometry data. In natural or synthetic phosphates, iron is often in the high-spin configuration and it is then rather easy to identify Fe(II) that exhibits a uniquely high isomer shift, between 0.6 and 1.2 mm/s, see Figure 13.3 for Fe(II) and $S = 2$, and a typically large quadrupole splitting, $2 < \Delta E_Q < 4$ mm/s. The ratio of Fe(II) to Fe(III) in mixed valence alluaudites [105,106] or the Fe(III) impurity content in $NaFePO_4F$ [107] were, for example, determined by Mössbauer spectrometry (see Figure 13.6a). For an intermediate valence state, intermediate isomer shifts can be observed, for example, for europium transition metal silicides [108,109]. Note that it is necessary to correct the measured spectral areas for a possible difference in f_{LM}. In particular, Fe(II) and Fe(III) exhibit different bonding and coordination and, in a room-temperature spectrum, the Fe(II) content is usually underrepresented in the Fe(II) relative spectral area.

13.4.2
Site Occupancy/Distribution

In complex structured materials, the site occupation by the Mössbauer element can be partial or a near neighbor site can host randomly distributed ions. In this case, the diversity in the local environment of the probed nucleus typically leads to a superposition of spectral components and the relative area is representative of the relative probability of each configuration. A typical approach utilized for analyzing such spectra is the use of a binomial distribution function in a fitting

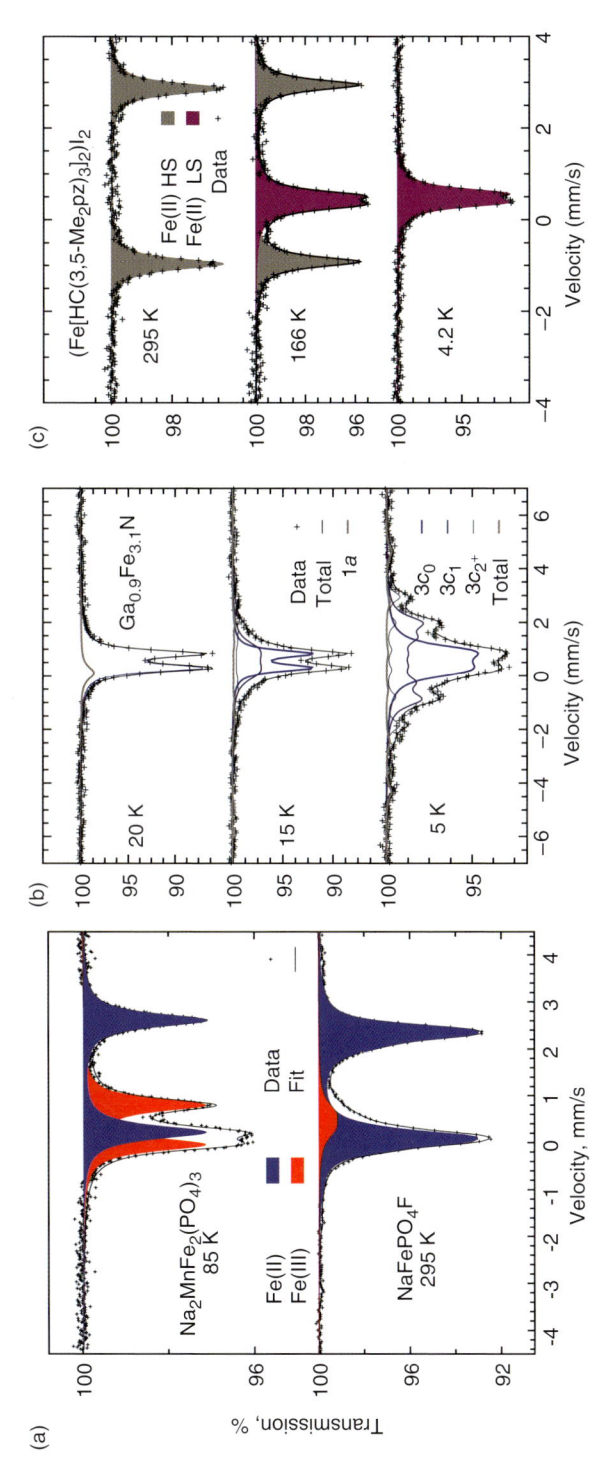

Figure 13.6 (a) A quantitative Fe(II) and Fe(III) content can be determined from the relative spectral areas, after correcting for the different f_{LM}. In the Na$_2$Mn-Fe$_2$(PO$_4$)$_3$ alluaudite, the 50–50% Fe(II)–Fe(III) assumption could be validated (data courtesy of Ref. [106]). In NaFePO$_4$F, the optimal calcination temperature of 873 K, which minimizes Fe(III) content, could be determined [107]. In the data for the sample calcined at 893 K that is shown, 9% Fe(III) impurities are observed. (b) A binomial distribution model is required for the spectral analysis of Ga$_{0.9}$Fe$_{3.1}$N. Fe on the 3c site exhibits a different subspectrum depending on the number (0, 1, 2+) of Fe near neighbors on 1a. Fe on the Ga 1a site has 12 Fe near neighbors and exhibits a larger hyperfine field [110]. Note the temperature dependence that reveals the onset of magnetism at 15 K. (c) The large reduction in quadrupole splitting when Fe(II) changes from the high-spin (HS) $S = 2$ to the low-spin (LS) $S = 0$ state enables easy identification of the two spin states.

attempt. Unless the site distribution is nonstochastic, this approach yields good results. Preferential near neighbor configurations or site order models can also be tested, in particular in conjunction with diffraction data. This approach was successful in modeling, for example, $Ga_{0.9}Fe_{3.1}N$ spectra [110] (see Figure 13.6b). Iron on the gallium-site, $1a$, has 12 iron near neighbors. In contrast, iron on the $3c$ site can have 0, 1, and 2 or more (2+) gallium near neighbors. Each configuration exhibits a different subspectrum. A detailed modeling revealed that the magnetism is driven by the iron on the gallium site that forms a 13-atom iron cluster. Many spectra, for example, alluaudites discussed above [106,112] and the FeMn(P,As) phases discussed below [113,114] can be successfully modeled by such a discrete distribution model. In more complex cases, a continuous distribution of environments can be encountered and a specialized fitting approach has been developed [115–117].

13.4.3
Spin Crossover

A domain in which Mössbauer spectroscopy has been particularly successful is the investigation of spin transition or crossover in iron-containing compounds [118,119]. In such materials, a temperature- or pressure-driven modification in the crystal field leads to a transition in the electronic configuration. The probably most common example is the transition from a $3d^6$ $S = 2$ high-spin iron(II) configuration observed at high temperatures to a $S = 0$ low-spin configuration at low temperatures, that is, when the crystal field becomes sufficiently large, it becomes energetically more favorable to violate Hund's first rule. The combination of Mössbauer spectroscopy and magnetometry is particularly powerful in order to understand the transition mechanism and clarify whether the transition is purely electronic or lattice driven. The model utilized for an iron pyrazolyl [111] is provided as an example in Figure 13.6c. Recently, Mössbauer spectral investigations of light-induced spin transitions have also been carried out by utilizing a specialized cryostat with light-feeding glass-fiber (see, for example, Ref. [20]).

13.4.4
Charge Order and Relaxation

Similar way to spin crossover transitions, temperature or pressure can drive a metal-to-insulator transition, a charge-order transition, or a combination thereof. Not surprisingly, Mössbauer spectroscopy was instrumental in clarifying the Verwey transition in Fe_3O_4 by providing evidence that the intermediate valence iron on the octahedral sites present at high temperatures undergoes a valence separation from $Fe^{"+2.5"}$ to Fe(II) and Fe(III) below the Verwey transition [120–122]. In this particular example, only the charge-ordered state and the fast fluctuation regime are observed. In contrast, in Fe_2OBO_3 and other charge-order compounds [123–125], the valence transition is more gradual, and the rate

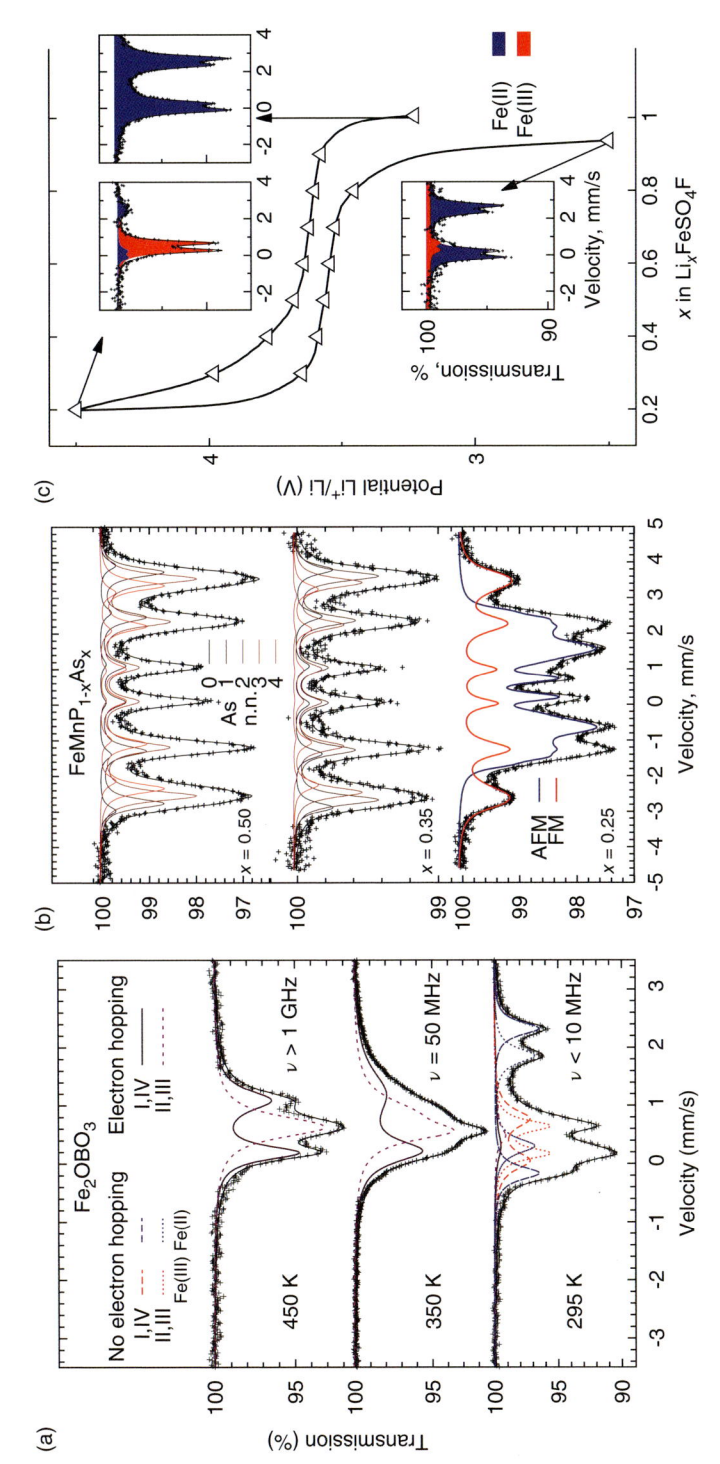

Figure 13.7 (a) Iron oxoborate Fe$_2$OBO$_3$ is charge ordered below 285 K, with sites I–IV bearing Fe(II) and Fe(III). Upon heating, electron hopping sets in. The data analysis requires a relaxation profile fitting that yields the hopping frequency ν [124]. Above 450 K, iron is seen as Fe$^{2.5+}$ in the fast relaxation limit. (b) In the magnetocaloric FeMnP$_{1-x}$As$_x$, a binomial distribution model for iron with 0–4 arsenic near neighbors is needed. The components with 0–4 As near neighbors are displayed in shades of red. For $x = 0.25$, a further incommensurate antiferromagnetic component (in blue, AFM) is observed (the binomial ferromagnetic components with 0–4 arsenic near neighbors are summed up for visibility, in red, FM) [113]. (c) Using an *operando* electrochemical cell [129], Mössbauer spectra, *inset*, were acquired as a function of time at different points of the potential versus lithium content curve, starting at $x = 1$, during the charge and discharge of LiFeSO$_4$ (data courtesy of Ref. [130]).

for electron hopping from Fe(II) to Fe(III) can be measured (see Figure 13.7a). Strong evidence for this charge-order and charge-hopping scenario was first provided by Mössbauer spectroscopy [123] and later confirmed by detailed crystallographic investigation [124]. Although the case of $FeOBO_3$ is somewhat unique in the sense that the charge-order involves one full electronic charge, several other systems have been investigated in similar manner and involve a less pronounced Fe(II)–Fe(III) charge separation [126,127]. A further material class in which Mössbauer spectroscopy provides a detailed insight into phase transitions and electronic instabilities are the iron perovskites $CaFeO_3$ and $SrFeO_3$, which undergo a charge disproportionation from formally iron(IV) at high temperatures to iron(III) and iron(V) at low temperatures in an ordered or fast relaxation state, respectively [128].

13.4.5
Magnetic Phase Transitions

Mössbauer spectral investigations are particularly important in complement to magnetometry measurements or in preparation for neutron diffraction experiments and should be used whenever possible. Although the information provided by Mössbauer spectroscopy is purely local, a detailed temperature- and composition-dependent study often helps to draw a detailed phase diagram with boundaries between ferro-, ferri-, or antiferromagnetic phases, in particular when combined with magnetometric data. For example, in the Fe_2P-type magnetocaloric materials [131,132], detailed phase diagrams and microscopic insight into the magnetic degree of freedom have been obtained on the Fe–Mn–P material with arsenic, germanium, or silicon substituting phosphorus. In the case of $FeMnP_{1-x}(As,Ge)_x$, the influence of As or Ge on the ordering temperature and magnetic coupling was investigated [113,114,133] (see Figure 13.7b), and the magnetostrictive phase transition could be analyzed in the framework of the magnetostriction exchange coupling model developed by Bean and Rodbell [134]. It is noteworthy that, first, the Mössbauer spectral data show clear evidence for a magnetic triple point at the boundary between the ferro-, para-, and antiferromagnetic phases. Second, *a priori* quite surprisingly, the data also provide direct evidence for incommensurate antiferromagnetic order (see $FeMnP_{0.75}As_{0.25}$ spectrum in Figure 13.7b). The subtlety resides in the modulation of the magnitude of the hyperfine field or in the modulation of the relative orientation between the hyperfine field, that is, the magnetic structure, and the electric field gradient, that is, the crystalline structure [135,136]. In $FeMnP_{1-x}As_x$, the modeling of the incommensurate magnetic Mössbauer spectra complements neutron diffraction data and provides evidence that the iron moments although strictly aligned along the *c*-axis are coupled to and align in the "up" or "down" direction of the spiraling Mn moments [113]. A further important example of incommensurate magnetism is found in the magnetoelectric multiferroic $BiFeO_3$. Therein, the cycloidal magnetic order of the iron moments can be directly probed through a detailed analysis of the Mössbauer [48,137] or nuclear forward scattering data [138], for example, as a function of temperature and strain for films deposited on substrates.

13.4.6
Superparamagnetism

A topic in which Mössbauer spectroscopy has played an important role is the characterization, description, and understanding of magnetic nanomaterials, in particular iron-bearing nanoparticles [139–142]. Below a critical size of ~30 nm, these nanoparticles no longer sustain magnetic domain walls, and the magnetization direction in the single-domain particle is fluctuating with a temperature-dependent frequency described by an Arrhenius law. Through a detailed analysis of the temperature dependence of the Mössbauer spectra and in particular their relaxation timescale, information on the magnetic energy barrier is obtained. Furthermore, by analyzing the Mössbauer spectrum at the lowest temperature, in the so-called magnetically blocked state, information about spin-canting can be retrieved [143]. The extensive data and literature on this topic have led to a description of the particles with a core–shell model with prominent spin-canting at the surface and a more uniformly magnetized core. This model is however still evolving as new Mössbauer spectral, small-angle neutron scattering [144,145] and nuclear forward scattering data become available [77].

13.4.7
Transferred Field

Even though many Mössbauer isotopes correspond to elements that do not have any magnetic moment, information on the magnetism can be obtained through the polarizing effect that nearby magnetic ions have on the probed nucleus, through the so-called transferred hyperfine field (see Section 13.7). A striking example is the high-temperature ferromagnetic material MnSb in which a rather large hyperfine magnetic field is observed by ^{121}Sb Mössbauer spectroscopy [146].

In general, the temperature dependence of the hyperfine field – transferred or proper – tracks, in first approximation, the order parameter of the magnetic transition. Note however that quite generally there is no strict correspondence between a measured hyperfine field and the magnetic moment of the probed ion. At best, a rough proportionality can be qualitatively used when comparing materials belonging to the same material class.

13.4.8
Superconductivity

With the discovery of high-T_c superconductors, Mössbauer spectral measurements have been extensively used for probing the electronic system, the magnetism, and the Meissner effect [147] in superconductors and parent compounds. Doping has been utilized in cuprates with mixed success. More recently, with the advent of iron-based superconductors [148], this field has known a revival and Mössbauer spectral investigations as a function of temperature, pressure, and applied magnetic field have been carried out [149–152]. Furthermore, because of the unique means of phonon spectroscopy by NRIXS, the effect of temperature and pressure on the iron vibrational modes has been clarified and evidence for a coupling between spin and lattice degrees of freedom has been shown [88,89]. Attempts to utilize a nuclear resonant nucleus in order to observe the Meissner effect date back to the 1960s, for example, tin-119 in niobium–tin superconductors, but they were not successful because of the below detection limit diamagnetic fraction, whenever a measurable external field is applied [147]. This situation changed very recently with the discovery of superconductivity in H_2S at high pressures [153], where a tin-119 probe foil was successfully utilized as an internal magnetometer [154]. This experiment, carried out by synchrotron nuclear forward scattering (NFS) provided direct evidence for a full Meissner shielding at 153 GPa below 60 K, with an onset of the shielding at 120 K [154].

13.4.9
In Situ, Operando, Remote Operation

A strong trend in recent years has been to use data acquisition either for *in situ* or automated investigation where the spectrometer is adapted to the sample and its environment or *operando* data acquisition was a process is followed in real time in a working device, modified just sufficiently to enable spectroscopy. Historically, the investigation of catalysis process has been extensively pursued [155–157], notably by emission Mössbauer spectroscopy, in which, for example, a cobalt catalyst is prepared with some activity of cobalt-57, and the catalytic process is followed by recording the iron-57 emission Mössbauer spectrum [158].

The identification of electrochemical reaction mechanisms in new battery materials has also been greatly facilitated, at least for iron-bearing materials, by the availability of *operando* cells, which enable Mössbauer data acquisition at different points of a charge and discharge cycle. This *in situ* acquisition can also be combined with X-ray diffraction. Such a cell is described in detail in Ref. [129] and was used, for example, in order to elucidate the reaction mechanism in the high-voltage Li(Fe$_{1-\delta}$Mn$_\delta$)SO$_4$F fluorosulfate cathodes [130,159], see Figure 13.7c. NaFePO$_4$F [160], or in FeNCN [161], a representative of the new class of carbodiimide-based anode materials [162].

When an absorber preparation is not possible, backscattering geometry CXMS is the best option, and a specific detector and spectrometer, MIMOS, has been developed [42]. Such a system has been very successful in investigating martian soil and rocks [43], in field expeditions [163], or for *in situ* monitoring of biomineralization [164].

13.4.10
Nuclear Resonance Scattering

The applications of nuclear resonance scattering of synchrotron radiation span the same diversity as Mössbauer spectroscopy and some examples have already been provided. An extensive review of nuclear resonance scattering methods and applications of the first 15 years was published in *Hyperfine Interactions*, 123–125 (1999–2000). Since then, these methods have conquered many more isotopes, for resonance energy from 6 keV to the now highest accessed energy of 89.6 keV for nuclear forward scattering by the ruthenium-99 resonance [12]. A recent example is the direct *in situ* study of the thermal decomposition reaction of K_2FeO_4 to $KFeO_2$, a decomposition during which reaction intermediates could be observed [165]. Another promising avenue is the *in situ* characterization of new materials synthesized at high pressures, in the gigapascal range, and high temperatures [166–169].

High-resolution phonon spectroscopy, with $\Delta E \lesssim 1$ meV, can routinely be carried out for iron-57, tin-119, samarium-149, europium-151, or dysprosium-161. An example of the inelastic scattering function and the instrumental resolution function obtained for $FeSb_2$ [170] is shown in Figure 13.8a. For isotopes with a resonance energy above 30 keV, such as antimony-121 or tellurium-125, inelastic measurements are more difficult and utilize a backscattering monochromatization scheme [14]. This approach permits the experimental determination of the element-specific density of phonon states, for example, in Sb_2Te_3 [93] or in

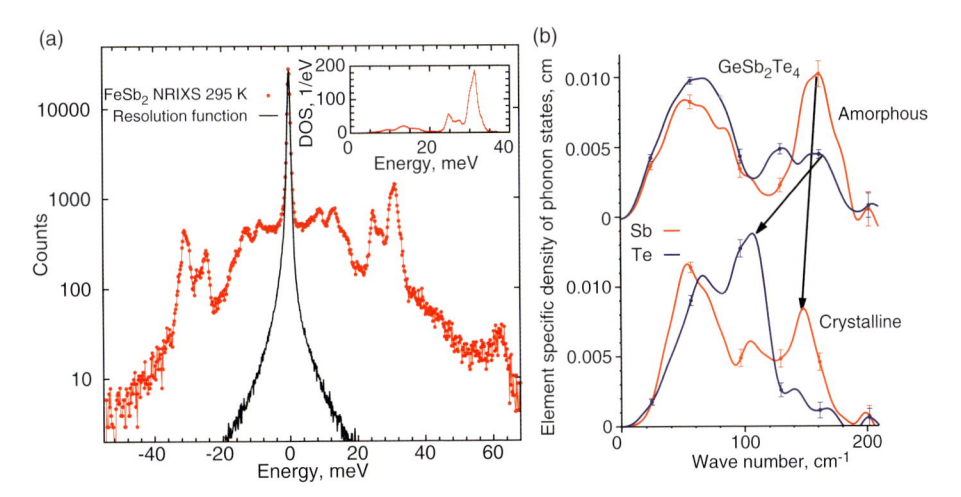

Figure 13.8 (a) Example of the iron-specific inelastic scattering function (NIS) and the associated resolution function measured on $FeSb_2$ at 295 K. These data enable the extraction of the corresponding density of phonon states, see inset [170]. (b) By means of antimony-121 and tellurium-125 nuclear inelastic scattering, the antimony and tellurium-specific density of phonon states in crystalline and amorphous $GeSb_2Te_4$ could be determined and it indicates a strong softening of optical phonon modes upon crystallization.

$GeSb_2Te_4$, where a crystallization-induced phonon softening is observed (see Figure 13.8b) [92].

13.5 Conclusion

Mössbauer spectroscopy and nuclear resonance scattering techniques nowadays provide ample opportunity for probing a large diversity of materials. Fortunately, the limited availability of sources for some isotopes is now compensated by the advent of synchrotron radiation-based techniques. With the development of new detection schemes and sample environments, more effective measurements adapted to current scientific problems have become possible, to the point of, for example, reaching new extremes in pressure or even operating on Mars. It is the hope of the author to have provided a general overview of the possible applications and to have enticed the reader to reach out to the members of the international Mössbauer spectroscopy community for further information or collaboration.

Acknowledgments

The author acknowledges his various coauthors who participated in many of the cited work, foremost his PhD students and postdoctoral fellows, as well as Prof. Fernande Grandjean and Dres. John Budai, Tom Watkins, and Alexander Chumakov for many helpful suggestions to improve this chapter.

This work was supported by the U.S. Department of Energy, Office of Science, Basic Energy Sciences, Materials Sciences and Engineering Division.

Abbreviations

AFM	antiferromagnetic
APD	avalanche photodiodes
APS	Advanced Photon Source, at Argonne National Laboratory, Chicago, USA
bw	bandwidth
CEMS	conversion electron Mössbauer spectroscopy
CXMS	conversion X-ray Mössbauer spectroscopy
EFG	electric field gradient
ESRF	European Synchrotron Radiation Facility, Grenoble, France
f_{LM}	Lamb–Mössbauer factor
FM	ferromagnetic
HHLM	high-heat load monochromator
HRM	high-resolution monochromator
HS	high spin

LS	low spin
MCA	multichannel analyzer
MIMOS	miniaturized Mössbauer spectrometer
NFS	nuclear forward scattering
NIS	nuclear inelastic scattering, used at the ESRF and PETRA III, same as NRIXS
NMR	nuclear magnetic resonance
NRD	nuclear resonant detector
NRIXS	nuclear resonance inelastic X-ray scattering, used at the APS, same as NIS
PETRA III	Positron-Electron Tandem Ring Accelerator III, Hamburg, Germany
SCA	single-channel analyzer
SPRing-8	Super Photon Ring – 8 GeV, Hyōgo Prefecture, Japan
SNR	signal-to-noise ratio
SOD	second-order Doppler shift

References

1 Hohenberg, P. and Kohn, W. (1964) Inhomogeneous electron gas. *Phys. Rev.*, **136** (3B), B864.

2 Kohn, W. and Sham, L.J. (1965) Self-consistent equations including exchange and correlation effects. *Phys. Rev.*, **140** (4A), A1133–A1138.

3 Gerald, R.E. (2006) Solid-state nuclear magnetic resonance, in *Encyclopedia of Analytical Chemistry*, John Wiley & Sons, Ltd, Chichester.

4 Kuhn, W. (1929) Scattering of thorium C" γ-radiation by radium G and ordinary lead. *Philos. Mag.*, **8**, 625.

5 Mössbauer, R.L. (1958) Kernresonanzfluoreszenz von Gammastrahlung in Ir191. *Z. Phys.*, **151**, 124–143.

6 Moon, P.B. (1951) Resonant nuclear scattering of gamma-rays: theory and preliminary experiments. *Proc. Phys. Soc. A*, **64**, 76.

7 Malmfors, K. (1953) Nuclear resonance scattering of gamma-rays. *Ark. Fys.*, **6**, 49.

8 Mössbauer, R.L. (1958) Kernresonanzabsorption von Gammastrahlung in Ir191. *Naturwissenschaften*, **45**, 538.

9 Mössbauer, R.L. (1961) *Nobel Lecture: Recoilless Nuclear Resonance Absorption of Gamma Radiation*, The Nobel Foundation.

10 Kalvius, M. and Kienle, P. (2012) *The Rudolf Mössbauer Story: His Scientific Work and Its Impact on Science and History*, Springer, Berlin.

11 Röhlsberger, R. (2004) *Nuclear Condensed Matter Physics with Synchrotron Radiation: Basic Principles, Methodology and Applications*, in Springer Tracts in Modern Physics, Springer, Berlin.

12 Bessas, D., Merkel, D., Chumakov, A., Rüffer, R., Hermann, R., Sergueev, I., Mahmoud, A., Klobes, B., McGuire, M., Sougrati, M., and Stievano, L. (2014) Nuclear forward scattering of synchrotron radiation by ^{99}Ru. *Phys. Rev. Lett.*, **113** (14), 147601.

13 Bessas, D., Sergueev, I., Merkel, D.G., Chumakov, A.I., Rüffer, R., Jafari, A., Kishimoto, S., Wolny, J.A., Schünemann, V., Needham, R.J., Sadler, P.J., and Hermann, R.P. (2015) Nuclear resonant scattering of synchrotron radiation by ^{187}Os. *Phys. Rev. B*, **91** (22), 224102.

14 Sergueev, I., Wille, H.-C., Hermann, R.P., Bessas, D., Shvyd'ko, Y.V., Zając, M., and Rüffer, R. (2011) Milli-electronvolt monochromatization of hard X-rays with a sapphire backscattering monochromator. *J. Synchrotron Radiat.*, **18**, 802–810.

15 Klobes, B., Desmedt, A., Sergueev, I., Schmalzl, K., and Hermann, R.P. (2013) ^{129}Xe nuclear resonance scattering on solid Xe and ^{129}Xe clathrate hydrate. *EPL*, **103** (3), 36001.

16 Simon, R.E., Sergueev, I., Persson, J., McCammon, C.A., Hatert, F., and Hermann, R.P. (2013) Nuclear forward scattering by the 68.7 keV state of 73Ge in $CaGeO_3$ and GeO_2. *EPL*, **104** (1), 17006.

17 Fultz, B. (2011) *Mössbauer Spectrometry*, Wiley Online Library.

18 Rüffer, R. and Chumakov, A. (2000) Nuclear inelastic scattering. *Hyperfine Interact.*, **128** (1–3), 255–272.

19 Chumakov, A. and Sturhahn, W. (1999) Experimental aspects of inelastic nuclear resonance scattering. *Hyperfine Interact.*, **123** (1–4), 781–808.

20 Gütlich, P., Bill, E., and Trautwein, A.X. (2011) *Mössbauer Spectroscopy and Transition Metal Chemistry: Fundamentals and Applications*, Springer, Berlin.

21 Shenoy, G.K. and Wagner, F.E. (1978) *Mössbauer Isomer Shifts*, North-Holland Pub. Co.

22 Josephson, B.D. (1960) Temperature-dependent shift of γ-rays emitted by a solid. *Phys. Rev. Lett.*, **4** (7), 341–342.

23 Pound, R.V. and Rebka, G.A., Jr. (1960) Apparent weight of photons. *Phys. Rev. Lett.*, **4** (7), 337.

24 Sturhahn, W. and Chumakov, A. (1999) Lamb–Mössbauer factor and second-order Doppler shift from inelastic nuclear resonant absorption. *Hyperfine Interact.*, **123** (1–4), 809–824.

25 Rechenberg, H. (2000) Comment on "Mössbauer effect study of filled antimonide skutterudites". *Phys. Rev. B*, **62** (10), 6827.

26 Long, G.J., Hautot, D., Grandjean, F., Morelli, D.T., and Meisner, G.P. (2000) Reply to 'Comment on 'Mössbauer effect study of filled antimonide skutterudites''. *Phys. Rev. B*, **62** (10), 6829.

27 Möchel, A., Sergueev, I., Wille, H.C., Voigt, J., Prager, M., Stone, M.B., Sales, B.C., Guguchia, Z., Shengelaya, A., Keppens, V., and Hermann, R.P. (2011) Lattice dynamics and anomalous softening in the $YbFe_4Sb_{12}$ skutterudite. *Phys. Rev. B*, **84** (18), 184306.

28 Goldanskii, V.I., Makarov, E.F., Suzdalev, I.P., and Vinogradov, I.A. (1968) Quantitative test of the vibrational anisotropy origin of the asymmetry of quadrupole Mössbauer doublets. *Phys. Rev. Lett.*, **20** (4), 137–140.

29 Greneche, J. and Varret, F. (1982) A new method of general use to obtain random powder spectra in ^{57}Fe Mössbauer spectroscopy: the rotating-sample recording. *J. Phys. Lett. (Paris)*, **43** (7), 233–237.

30 Pfannes, H.D. and Gonser, U. (1973) Goldanskii–Karyagin effect versus preferred orientations (texture). *Appl. Phys.*, **1** (2), 93–102.

31 Greenwood, N.N. (1971) *Mössbauer Spectroscopy*, Springer, The Netherlands.

32 Blume, M. and Tjon, J. (1968) Mössbauer spectra in a fluctuating environment. *Phys. Rev.*, **165** (2), 446.

33 Tjon, J. and Blume, M. (1968) Mössbauer spectra in a fluctuating environment: II. Randomly varying electric field gradients. *Phys. Rev.*, **165** (2), 456.

34 Shenoy, G. and Dunlap, B. (1976) Mössbauer relaxation line shapes in the presence of complex hyperfine interactions. *Phys. Rev. B*, **13** (3), 1353.

35 Litterst, F. and Gorobchenko, V. (1982) Mössbauer relaxation spectra for nuclear motion in an octahedral cage with reorienting electric field gradient. *Phys. Status Solidi B*, **113** (2), K135–K138.

36 Gunnlaugsson, H.P., Weyer, G., Dietrich, M., Fanciulli, M., Bharuth-Ram, K., and Sielemann, R. (2002) Charge state dependence of the diffusivity of interstitial Fe in silicon detected by Mössbauer spectroscopy. *Appl. Phys. Lett.*, **80** (15), 2657.

37 Lynch, F.J., Holland, R.E., and Hamermesh, M. (1960) Time dependence of resonantly filtered gamma rays from Fe57. *Phys. Rev.*, **120** (2), 513–520.

38 Novak, P., Pechousek, J., Prochazka, V., Navarik, J., Kouril, L., Kohout, P., Vrba, V., and Machala, L. (2016) Time differential 57Fe Mössbauer spectrometer with unique 4π YAP:Ce 122.06keV

gamma-photon detector. *Nucl. Instrum. Methods Phys. Res. A*, **832**, 292–296.
39 Kobayashi, Y. (2010) 99Ru and 61Ni Mössbauer spectroscopic studies using the accelerator at RIKEN. *J. Phys.*, **217**, 012023–012028.
40 Swanson, K.R. and Spijkerman, J.J. (1970) Analysis of thin surface layers by Fe-57 Mössbauer backscattering spectrometry. *J. Appl. Phys.*, **41** (7), 3155–3158.
41 Bonchev, Z., Jordanov, A., and Minkova, A. (1969) Method of analysis of thin surface layers by the Mössbauer effect. *Nucl. Instrum. Methods*, **70** (1), 36–40.
42 Fleischer, I., Klingelhöfer, G., Morris, R.V., Schröder, C., Rodionov, D., and de Souza, P.A. (2012) In-situ Mössbauer spectroscopy with MIMOS II. *Hyperfine Interact.*, **207** (1–3), 97–105.
43 Morris, R.V. (2004) Mineralogy at Gusev crater from the Mossbauer spectrometer on the spirit rover. *Science*, **305** (5685), 833–836.
44 Tricker, M.J., Thomas, J.M., and Winterbottom, A. (1974) Conversion-electron Mössbauer spectroscopy for the study of solid surfaces. *Surf. Sci.*, **45** (2), 601–608.
45 Costa, R., Lelis, M., Oliveira, L., Fabris, J., Ardisson, J., Rios, R., Silva, C., and Lago, R. (2006) Novel active heterogeneous Fenton system based on $Fe_{3-x}M_xO_4$ (Fe, Co, Mn, Ni): the role of M2+ species on the reactivity towards H_2O_2 reactions. *J. Hazard. Mater.*, **129** (1–3), 171–178.
46 Oh, S.J., Cook, D., and Townsend, H. (1998) Characterization of iron oxides commonly formed as corrosion products on steel. *Hyperfine Interact.*, **112** (1/4), 59–66.
47 Abdelmoula, M., Refait, P., Drissi, S.H., Mihe, J., and Génin, J. (1996) Conversion electron Mössbauer spectroscopy and X-ray diffraction studies of the formation of carbonate-containing green rust one by corrosion of metallic iron in $NaHCO_3$ and ($NaHCO_3$+NaCl) solutions. *Corros. Sci.*, **38** (4), 623–633.
48 Agbelele, A., Sando, D., Infante, I., Carrétéro, C., Jouen, S., Le Breton, J.M., Barthélémy, A., Dkhil, B., Bibes, M., and Juraszek, J. (2016) Insight into magnetic, ferroelectric and elastic properties of strained $BiFeO_3$ thin films through Mössbauer spectroscopy. *Appl. Phys. Lett.*, **109** (4), 042902.
49 Odeurs, J., Hoy, G.R., and L'abbé, C. (2000) Enhanced resolution in Mössbauer spectroscopy. *J. Phys. Condens. Matter*, **12** (5), 637–642.
50 Belyaev, A.A., Volodin, V.S., Irkaev, S.M., Panchuk, V.V., and Semenov, V.G. (2010) Application of resonant detectors in Mössbauer spectroscopy. *Bull. Russ. Acad. Sci., Phys.*, **74** (3), 412–415.
51 Long, G.J., Cranshaw, T., and Longworth, G. (1983) The ideal Mössbauer effect absorber thickness. *Mössbauer Eff. Ref. Data J.*, **6** (2), 42–49.
52 Margulies, S. and Ehrman, J.R. (1961) Transmission and line broadening of resonance radiation incident on a resonance absorber. *Nucl. Instrum. Methods*, **12**, 131–137.
53 Margulies, S., Debrunner, P., and Frauenfelder, H. (1963) Transmission and line broadening in the Mössbauer effect. II. *Nucl. Instrum. Methods*, **21**, 217–231.
54 Fultz, B. and Morris, J.W. (1981) The thickness distortion of 57Fe backscatter Mössbauer spectra. *Nucl. Instrum. Methods*, **188** (1), 197–201.
55 Shenoy, G. and Friedt, J. (1974) Influence of absorber thickness on the Mössbauer quadrupole spectrum of 121Sb. *Nucl. Instrum. Methods*, **116** (3), 573–578.
56 Chen, Y.L. and Yang, D.P. (2007) *Mössbauer Effect in Lattice Dynamics: Experimental Techniques and Applications*, Wiley-VCH Verlag GmbH, Weinheim, Germany.
57 Ruby, S.L. (1974) Mössbauer experiments without conventional sources. *J. Phys. Colloq.*, **35** (C6), C6-209–C6-211.
58 Shenoy, G.K. (2007) Scientific legacy of Stanley Ruby. *Hyperfine Interact.*, **170** (1–3), 5–13.
59 Gerdau, E., Rüffer, R., Winkler, H., Tolksdorf, W., Klages, C.P., and Hannon, J.P. (1985) Nuclear Bragg diffraction of synchrotron radiation in yttrium iron garnet. *Phys. Rev. Lett.*, **54** (8), 835–838.
60 Hastings, J.B., Siddons, D.P., van Bürck, U., Hollatz, R., and Bergmann, U. (1991) Mössbauer spectroscopy using

synchrotron radiation. *Phys. Rev. Lett.*, **66** (6), 770–773.

61 Toellner, T. (2000) Monochromatization of synchrotron radiation for nuclear resonant scattering experiments. *Hyperfine Interact.*, **125** (1/4), 3–28.

62 Shvyd'ko, Y. (2004) *X-Ray Optics*, Springer Series in Optical Sciences, vol. **98**, Springer, Berlin.

63 Ishikawa, T., Tamasaku, K., and Yabashi, M. (2005) High-resolution X-ray monochromators. *Nucl. Instrum. Methods Phys. Res. A*, **547** (1), 42–49.

64 Toellner, T., Alp, E., Graber, T., Henning, R., Shastri, S., Shenoy, G., and Sturhahn, W. (2011) Synchrotron Mössbauer spectroscopy using high-speed shutters. *J. Synchrotron Radiat.*, **18** (2), 183–188.

65 Kagan, Y., Afanas' ev, A.M., and Kohn, V.G. (1978) Time delay in the resonance scattering of synchrotron radiation by nuclei in a crystal. *Phys. Lett. A*, **68** (3–4), 339–341.

66 Baron, A.Q. (2000) Detectors for nuclear resonant scattering experiments. *Hyperfine Interact.*, **125** (1–4), 29–42.

67 Sturhahn, W. (2000) CONUSS and PHOENIX: evaluation of nuclear resonant scattering data. *Hyperfine Interact.*, **125** (1–4), 149–172.

68 Shvyd'ko, Y.V. (2000) MOTIF: evaluation of time spectra for nuclear forward scattering. *Hyperfine Interact.*, **125** (1–4), 173–188.

69 Bergmann, U. (1995) Mössbauer spectroscopy with synchrotron radiation. *Appl. Radiat. Isotopes*, **46** (6–7), 525–530.

70 Röhlsberger, R., Bansmann, J., Senz, V., Jonas, K.L., Bettac, A., Meiwes-Broer, K.H., and Leupold, O. (2003) Nanoscale magnetism probed by nuclear resonant scattering of synchrotron radiation. *Phys. Rev. B*, **67** (24), 245412.

71 Potapkin, V., Dubrovinsky, L., Sergueev, I., Ekholm, M., Kantor, I., Bessas, D., Bykova, E., Prakapenka, V., Hermann, R., Rüffer, R. et al. (2016) Magnetic interactions in NiO at ultrahigh pressure. *Phys. Rev. B*, **93** (20), 201110.

72 Partyka-Jankowska, E., Sepiol, B., Sladecek, M., Kmiec, D., Korecki, J., Slezak, T., Zajac, M., Stankov, S., Rüffer, R., and Vogl, G. (2008) Nuclear resonant scattering studies of electric field gradient in Fe monolayer on W(110). *Surf. Sci.*, **602** (7), 1453–1457.

73 Stankov, S., Rüffer, R., Sladecek, M., Rennhofer, M., Sepiol, B., Vogl, G., Spiridis, N., Slezak, T., and Korecki, J. (2008) An ultrahigh vacuum system for *in situ* studies of thin films and nanostructures by nuclear resonance scattering of synchrotron radiation. *Rev. Sci. Instrum.*, **79** (4), 045108.

74 Strohm, C., van der Linden, P., and Rüffer, R. (2010) Nuclear forward scattering of synchrotron radiation in pulsed high magnetic fields. *Phys. Rev. Lett.*, **104** (8), 087601.

75 L'abbé, C., Meerschaut, J., Sturhahn, W., Jiang, J., Toellner, T., Alp, E., and Bader, S. (2004) Nuclear resonant magnetometry and its application to Fe/Cr multilayers. *Phys. Rev. Lett.*, **93** (3), 037201.

76 Röhlsberger, R., Thomas, H., Schlage, K., Burkel, E., Leupold, O., and Rüffer, R. (2002) Imaging the magnetic spin structure of exchange-coupled thin films. *Phys. Rev. Lett.*, **89** (23), 237201.

77 Herlitschke, M., Disch, S., Sergueev, I., Schlage, K., Wetterskog, E., Bergström, L., and Hermann, R.P. (2016) Spin disorder in maghemite nanoparticles investigated using polarized neutrons and nuclear resonant scattering. *J. Phys. Conf. Ser.*, **711**, 012002.

78 Bocklage, L., Swoboda, C., Schlage, K., Wille, H.C., Dzemiantsova, L., Bajt, S., Meier, G., and Röhlsberger, R. (2015) Spin precession mapping at ferromagnetic resonance via nuclear resonant scattering of synchrotron radiation. *Phys. Rev. Lett.*, **114** (14), 147601.

79 Sergueev, I., Dubrovinsky, L., Ekholm, M., Vekilova, O.Y., Chumakov, A., Zając, M., Potapkin, V., Kantor, I., Bornemann, S., Ebert, H. et al. (2013) Hyperfine splitting and room-temperature ferromagnetism of Ni at multimegabar pressure. *Phys. Rev. Lett.*, **111** (15), 157601.

80 Mitsui, T., Seto, M., Kikuta, S., Hirao, N., Ohishi, Y., Takei, H., Kobayashi, Y.,

Kitao, S., Higashitaniguchi, S., and Masuda, R. (2007) Generation and application of ultrahigh monochromatic X-ray using high-quality ^{57}FeBO$_3$ single crystal. *Jpn. J. Appl. Phys.*, **46** (2R), 821.

81 Potapkin, V., Chumakov, A.I., Smirnov, G.V., Celse, J.P., Rüffer, R., McCammon, C., and Dubrovinsky, L. (2012) The ^{57}Fe synchrotron Mössbauer source at the ESRF. *J. Synchrotron Radiat.*, **19** (4), 559–569.

82 Masuda, R., Kobayashi, Y., Kitao, S., Kurokuzu, M., Saito, M., Yoda, Y., Mitsui, T., Iga, F., and Seto, M. (2014) Synchrotron radiation-based Mössbauer spectra of 174Yb measured with internal conversion electrons. *Appl. Phys. Lett.*, **104** (8), 082411.

83 Seto, M., Masuda, R., Higashitaniguchi, S., Kitao, S., Kobayashi, Y., Inaba, C., Mitsui, T., and Yoda, Y. (2009) Synchrotron-radiation-based Mössbauer spectroscopy. *Phys. Rev. Lett.*, **102** (21), 217602.

84 Weiss, H. and Langhoff, H. (1979) Observation of one phonon transitions in terbium by nuclear resonance fluorescence. *Phys. Lett.*, **69A**, 448.

85 Seto, M., Yoda, Y., Kikuta, S., Zhang, X.W., and Ando, M. (1995) Observation of nuclear resonant scattering accompanied by phonon excitation using synchrotron radiation. *Phys. Rev. Lett.*, **74** (19), 3828–3831.

86 Sturhahn, W., Toellner, T.S., Alp, E.E., Zhang, X., Ando, M., Yoda, Y., Kikuta, S., Seto, M., Kimball, C.W., and Dabrowski, B. (1995) Phonon density of states measured by inelastic nuclear resonant scattering. *Phys. Rev. Lett.*, **74** (19), 3832–3835.

87 Mao, H., Xu, J., Struzhkin, V., Shu, J., Hemley, R., Sturhahn, W., Hu, M., Alp, E., Vocadlo, L., Alfè, D. *et al.* (2001) Phonon density of states of iron up to 153 gigapascals. *Science*, **292** (5518), 914–916.

88 Ksenofontov, V., Wortmann, G., Chumakov, A., Gasi, T., Medvedev, S., McQueen, T., Cava, R., and Felser, C. (2010) Density of phonon states in superconducting FeSe as a function of temperature and pressure. *Phys. Rev. B*, **81** (18), 184510.

89 Sergueev, I., Hermann, R., Bessas, D., Pelzer, U., Angst, M., Schweika, W., McGuire, M.A., Sefat, A., Sales, B.C., Mandrus, D. *et al.* (2013) Effect of pressure, temperature, fluorine doping, and rare earth elements on the phonon density of states of LFeAsO studied by nuclear inelastic scattering. *Phys. Rev. B*, **87** (6), 064302.

90 Sergueev, I., Glazyrin, K., Kantor, I., McGuire, M., Chumakov, A., Klobes, B., Sales, B., and Hermann, R. (2015) Quenching rattling modes in skutterudites with pressure. *Phys. Rev. B*, **91** (22), 224304.

91 Shvyd'ko, Y.V. and Gerdau, E. (1999) Backscattering mirrors for X-rays and Mössbauer radiation. *Hyperfine Interact.*, **123** (1–4), 741–776.

92 Matsunaga, T., Yamada, N., Kojima, R., Shamoto, S., Sato, M., Tanida, H., Uruga, T., Kohara, S., Takata, M., Zalden, P., Bruns, G., Sergueev, I., Wille, H.C., Hermann, R.P., and Wuttig, M. (2011) Phase-change materials: vibrational softening upon crystallization and its impact on thermal properties. *Adv. Funct. Mater.*, **21** (12), 2232–2239.

93 Bessas, D., Sergueev, I., Wille, H.C., Perßon, J., Ebling, D., and Hermann, R.P. (2012) Lattice dynamics in Bi$_2$Te$_3$ and Sb$_2$Te$_3$: Te and Sb density of phonon states. *Phys. Rev. B*, **86** (22), 224301.

94 Kohn, V. and Chumakov, A. (2000) DOS: evaluation of phonon density of states from nuclear resonant inelastic absorption. *Hyperfine Interact.*, **125** (1–4), 205–221.

95 Gonser, U. (1975) *Mössbauer Spectroscopy*, Springer, Berlin.

96 Bancroft, G.M. (1973) *Mössbauer Spectroscopy: An Introduction for Inorganic Chemists and Geochemists*, McGraw Hill.

97 Wertheim, G.K. (1964) *Mössbauer Effect: Principles and Applications*, Academic Press, New York.

98 Frauenfelder, H. (1962) *The Mossbauer Effect*, Benjamin, New York.

99 Dickson, D.P. and Berry, F.J. (1986) *Mossbauer Spectroscopy*, Cambridge University Press, Cambridge.

100 Gonser, U. (1981) *Mössbauer Spectroscopy II: The Exotic Side of the Method*, Topics in Current Physics, vol **25**, Springer, Heidelberg.

101 Goldanskii, V.I. and Herber, R.H. (1968) *Chemical Applications of Mössbauer Spectroscopy*, vol. **66**, Academic Press, New York.

102 Long, G.J. *Mössbauer Spectroscopy Applied to Inorganic Chemistry*, Modern Inorganic Chemistry, vols **1–2**, Springer Science+Business Media, New York, pp. 1984–1987.

103 Long, G.J. and Grandjean, F. (1989) *Mössbauer Spectroscopy Applied to Inorganic Chemistry*, vol. **3**, Springer Science+Business Media, New York.

104 Long, G.J. and Stevens, J.G. (1986) *Industrial Applications of the Mössbauer Effect*, Plenum Press, New York.

105 Hatert, F., Rebbouh, L., Hermann, R.P., Fransolet, A.M., Long, G.J., and Grandjean, F. (2005) Crystal chemistry of the hydrothermally synthesized $Na_2(Mn_{1-x}Fe_x^{2+})_2Fe^{3+}(PO_4)_3$ alluaudite-type solid solution. *Am. Mineral.*, **90** (4), 653–662.

106 Hatert, F., Long, G.J., Hautot, D., Fransolet, A.M., Delwiche, J., Hubin-Franskin, M.J., and Grandjean, F. (2004) A structural, magnetic, and Mössbauer spectral study of several Na-Mn-Fe-bearing alluaudites. *Phys. Chem. Miner.*, **31** (8), 487–506.

107 Brisbois, M., Krins, N., Hermann, R.P., Schrijnemakers, A., Cloots, R., Vertruyen, B., and Boschini, F. (2014) Spray-drying synthesis of Na_2FePO_4F/carbon powders for lithium-ion batteries. *Mater. Lett.*, **130**, 263–266.

108 Sampathkumaran, E., Gupta, L., Vijayaraghavan, R., Gopalakrishnan, K., Pillay, R., and Devare, H. (1981) A new and unique Eu-based mixed valence system: $EuPd_2Si_2$. *J. Phys. C Solid State Phys.*, **14** (9), L237.

109 Nemkovski, K.S., Kozlenko, D.P., Alekseev, P.A., Mignot, J.M., Menushenkov, A.P., Yaroslavtsev, A.A., Clementyev, E.S., Ivanov, A.S., Rols, S., Klobes, B. et al. (2016) Europium mixed-valence, long-range magnetic order, and dynamic magnetic response in $EuCu_2(Si_xGe_{1-x})_2$. *Phys. Rev. B*, **94**, 195101.

110 Burghaus, J., Sougrati, M.T., Möchel, A., Houben, A., Hermann, R.P., and Dronskowski, R. (2011) Local ordering and magnetism in $Ga_{0.9}Fe_{3.1}N$. *J. Solid State Chem.*, **184** (9), 2315–2321.

111 Reger, D., Little, C., Smith, M., Rheingold, A., Lam, K.C., Concolino, T., Long, G., Hermann, R., and Grandjean, F. (2002) Synthetic, structural, magnetic, and Mössbauer spectral study of {Fe[HC(3,5-Me2pz)3]2}I2 and its spin-state crossover behavior. *Eur. J. Inorg. Chem.*, **2002** (5), 1190–1197.

112 Hermann, R.P., Hatert, F., Fransolet, A.M., Long, G.J., and Grandjean, F. (2002) Mössbauer spectral evidence for next-nearest neighbor interactions within the alluaudite structure of $Na_{1-x}Li_xMnFe_2(PO_4)^3$. *Solid State Sci.*, **4** (4), 507–513.

113 Hermann, R.P., Tegus, O., Brück, E., Buschow, K.H.J., de Boer, F.R., Long, G.J., and Grandjean, F. (2004) Mössbauer spectral study of the magnetocaloric $FeMnP_{1-x}As_x$ compounds. *Phys. Rev. B*, **70** (21), 214425.

114 Sougrati, M.T., Hermann, R.P., Grandjean, F., Long, G.J., Brück, E., Tegus, O., Trung, N.T., and Buschow, K.H.J. (2008) A structural, magnetic and Mössbauer spectral study of the magnetocaloric $Mn_{1.1}Fe_{0.9}P_{1-x}Ge_x$ compounds. *J. Phys. Condens. Matter*, **20** (47), 475206.

115 Le Caër, G. and Dubois, J. (1979) Evaluation of hyperfine parameter distributions from overlapped Mossbauer spectra of amorphous alloys. *J. Phys. E Sci. Instrum.*, **12** (11), 1083.

116 Le Caër, G. and Brand, R. (1998) General models for the distributions of electric field gradients in disordered solids. *J. Phys. Condens. Matter.*, **10** (47), 10715.

117 Varret, F., Gerard, A., and Imbert, P. (1971) Magnetic field distribution analysis of the broadened Mössbauer spectra of zinc ferrite. *Phys. Status Solidi B*, **43** (2), 723–730.

118 Gütlich, P., Hauser, A., and Spiering, H. (1994) Thermal and optical switching of

iron (II) complexes. *Angew. Chem., Int. Ed. Engl.*, **33** (20), 2024–2054.
119 Breuning, E., Ruben, M., Lehn, J., Renz, F., Garcia, Y., Ksenofontov, V., Gütlich, P., Wegelius, E., and Rissanen, K. (2000) Spin crossover in a supramolecular Fe4II [2×2] grid triggered by temperature, pressure, and light. *Angew. Chem., Int. Ed.*, **39** (14), 2504–2507.
120 Bauminger, R., Cohen, S., Marinov, A., Ofer, S., and Segal, E. (1961) Study of the low-temperature transition in magnetite and the internal fields acting on iron nuclei in some spinel ferrites, using Mössbauer absorption. *Phys. Rev.*, **122** (5), 1447.
121 Hargrove, R.S. and Kündig, W. (1970) Mössbauer measurements of magnetite below the Verwey transition. *Solid State Commun.*, **8** (5), 303–308.
122 Coey, J., Morrish, A., and Sawatzky, G. (1971) A Mössbauer study of conduction in magnetite. *J. Phys. Colloq.*, **32** (C1), C1–C271.
123 Attfield, J., Bell, A., Rodriguez-Martinez, L., Greneche, J., Cernik, R., Clarke, J., and Perkins, D. (1998) Electrostatically driven charge-ordering in Fe_2OBO_3. *Nature*, **396** (6712), 655–658.
124 Angst, M., Hermann, R.P., Schweika, W., Kim, J.W., Khalifah, P., Xiang, H.J., Whangbo, M.H., Kim, D.H., Sales, B.C., and Mandrus, D. (2007) Incommensurate charge order phase in Fe_2OBO_3 due to geometrical frustration. *Phys. Rev. Lett.*, **99** (25), 256402.
125 Angst, M., Hermann, R.P., Christianson, A.D., Lumsden, M.D., Lee, C., Whangbo, M.H., Kim, J.W., Ryan, P.J., Nagler, S.E., Tian, W., Jin, R., Sales, B.C., and Mandrus, D. (2008) Charge order in $LuFe_2O_4$: antiferroelectric ground state and coupling to magnetism. *Phys. Rev. Lett.*, **101** (22), 227601.
126 Karen, P., Woodward, P.M., Lindén, J., Vogt, T., Studer, A., and Fischer, P. (2001) Verwey transition in mixed-valence $TbBaFe_2O_5$: two attempts to order charges. *Phys. Rev. B*, **64** (21), 214405.
127 Karen, P., Gustafsson, K., and Lindén, J. (2007) $EuBaFe_2O_5$: extent of charge ordering by Mössbauer spectroscopy and high-intensity high-resolution powder diffraction. *J. Solid State Chem.*, **180** (1), 138–147.
128 Takano, M., Nakanishi, N., Takeda, Y., Naka, S., and Takada, T. (1977) Charge disproportionation in $CaFeO_3$ studied with the Mössbauer effect. *Mater. Res. Bull.*, **12** (9), 923–928.
129 Leriche, J., Hamelet, S., Shu, J., Morcrette, M., Masquelier, C., Ouvrard, G., Zerrouki, M., Soudan, P., Belin, S., Elkam, E. et al. (2010) An electrochemical cell for operando study of lithium batteries using synchrotron radiation. *J. Electrochem. Soc.*, **157** (5), A606–A610.
130 Ati, M., Sougrati, M.T., Recham, N., Barpanda, P., Leriche, J.B., Courty, M., Armand, M., Jumas, J.C., and Tarascon, J.M. (2010) Fluorosulfate positive electrodes for Li-ion batteries made via a solid-state dry process. *J. Electrochem. Soc.*, **157** (9), A1007–A1015.
131 Tegus, O., Brück, E., Buschow, K., and De Boer, F. (2002) Transition-metal-based magnetic refrigerants for room-temperature applications. *Nature*, **415** (6868), 150–152.
132 Caron, L., Hudl, M., Höglin, V., Dung, N., Gomez, C.P., Sahlberg, M., Brück, E., Andersson, Y., and Nordblad, P. (2013) Magnetocrystalline anisotropy and the magnetocaloric effect in Fe_2P. *Phys. Rev. B*, **88** (9), 094440.
133 Malaman, B., Le Caër, G., Delcroix, P., Fruchart, D., Bacmann, M., and Fruchart, R. (1996) Magneto-elastic transition and magnetic couplings: a Mössbauer spectroscopy study of the system. *J. Phys. Condens. Matter*, **8** (44), 8653.
134 Bean, C. and Rodbell, D. (1962) Magnetic disorder as a first-order phase transformation. *Phys. Rev.*, **126** (1), 104.
135 Häggström, L., Gustavsson-Seidel, A., and Fjellvåg, H. (1989) A Mössbauer study of helimagnetic FeAs. *Europhys. Lett.*, **9** (1), 87.
136 Le Caër, G. and Dubiel, S. (1990) Influence of spin-density-wave parameters on 119Sn-site Mössbauer spectra of chromium: theoretical calculations. *J. Magn. Magn. Mater.*, **92** (2), 251–260.

137 Lebeugle, D., Colson, D., Forget, A., Viret, M., Bonville, P., Marucco, J.F., and Fusil, S. (2007) Room-temperature coexistence of large electric polarization and magnetic order in $BiFeO_3$ single crystals. *Phys. Rev. B*, **76** (2), 024116.

138 Lazenka, V., Lorenz, M., Modarresi, H., Bisht, M., Rüffer, R., Bonholzer, M., Grundmann, M., Van Bael, M.J., Vantomme, A., and Temst, K. (2015) Magnetic spin structure and magnetoelectric coupling in $BiFeO_3$-$BaTiO_3$ multilayer. *Appl. Phys. Lett.*, **106** (8), 082904.

139 Mørup, S. and Tronc, E. (1994) Superparamagnetic relaxation of weakly interacting particles. *Phys. Rev. Lett.*, **72** (20), 3278.

140 Chinnasamy, C., Narayanasamy, A., Ponpandian, N., Chattopadhyay, K., Guerault, H., and Greneche, J. (2000) Magnetic properties of nanostructured ferrimagnetic zinc ferrite. *J. Phys. Condens. Matter*, **12** (35), 7795.

141 Tartaj, P., Gonzalez-Carreno, T., Bomati-Miguel, O., Serna, C., and Bonville, P. (2004) Magnetic behavior of superparamagnetic Fe nanocrystals confined inside submicron-sized spherical silica particles. *Phys. Rev. B*, **69** (9), 094401.

142 Rebbouh, L., Hermann, R.P., Grandjean, F., Hyeon, T., An, K., Amato, A., and Long, G.J. (2007) Fe_{57} Mössbauer spectral and muon spin relaxation study of the magnetodynamics of monodispersed γ-Fe_2O_3 nanoparticles. *Phys. Rev. B*, **76** (17), 174422.

143 Linderoth, S., Hendriksen, P.V., Bo, F., Wells, S., Davies, K., Charles, S., Mo, S. et al. (1994) On spin-canting in maghemite particles. *J. Appl. Phys.*, **75** (10), 6583–6585.

144 Disch, S., Wetterskog, E., Hermann, R.P., Wiedenmann, A., Vainio, U., Salazar-Alvarez, G., Bergström, L., and Brückel, T. (2012) Quantitative spatial magnetization distribution in iron oxide nanocubes and nanospheres by polarized small-angle neutron scattering. *New J. Phys.*, **14** (1), 013025.

145 Krycka, K., Booth, R.A., Hogg, C., Ijiri, Y., Borchers, J.A., Chen, W., Watson, S., Laver, M., Gentile, T.R., Dedon, L. et al. (2010) Core–shell magnetic morphology of structurally uniform magnetite nanoparticles. *Phys. Rev. Lett.*, **104** (20), 207203.

146 Ruby, S. and Kalvius, G. (1967) Magnetic hyperfine interaction in Sb_{121} using the mössbauer effect. *Phys. Rev.*, **155** (2), 353.

147 Heberle, J. (1966) Some applications of superconducting magnets, in *Mössbauer Effect Methodology*, Springer, pp. 95–109.

148 Kamihara, Y., Watanabe, T., Hirano, M., and Hosono, H. (2008) Iron-based layered superconductor $La[O_{1-x}F_x]FeAs$ (x=0.05–0.12) with T_c=26K. *J. Am. Chem. Soc.*, **130** (11), 3296–3297.

149 Kitao, S., Kobayashi, Y., Higashitaniguchi, S., Saito, M., Kamihara, Y., Hirano, M., Mitsui, T., Hosono, H., and Seto, M. (2008) Spin ordering in LaFeAsO and its suppression in superconductor $LaFeAsO_{0.89}F_{0.11}$ probed by Mössbauer spectroscopy. *J. Phys. Soc. Jpn.*, **77** (10), 103706.

150 McGuire, M.A., Christianson, A.D., Sefat, A.S., Sales, B.C., Lumsden, M.D., Jin, R., Payzant, E.A., Mandrus, D., Luan, Y., Keppens, V. et al. (2008) Phase transitions in LaFeAsO: structural, magnetic, elastic, and transport properties, heat capacity and Mössbauer spectra. *Phys. Rev. B*, **78** (9), 094517.

151 McGuire, M.A., Hermann, R.P., Sefat, A.S., Sales, B.C., Jin, R., Mandrus, D., Grandjean, F., and Long, G.J. (2009) Influence of the rare-earth element on the effects of the structural and magnetic phase transitions in CeFeAsO, PrFeAsO and NdFeAsO. *New J. Phys.*, **11** (2), 025011.

152 Medvedev, S., McQueen, T.M., Troyan, I., Palasyuk, T., Eremets, M., Cava, R., Naghavi, S., Casper, F., Ksenofontov, V., Wortmann, G. et al. (2009) Electronic and magnetic phase diagram of β-$Fe_{1.01}Se$ with superconductivity at 36.7K under pressure. *Nat. Mater.*, **8** (8), 630–633.

153 Drozdov, A., Eremets, M., Troyan, I., Ksenofontov, V., and Shylin, S. (2015) Conventional superconductivity at 203 Kelvin at high pressures in the sulfur hydride system. *Nature*, **525** (7567), 73–76.

154 Troyan, I., Gavriliuk, A., Rüffer, R., Chumakov, A., Mironovich, A., Lyubutin, I., Perekalin, D., Drozdov, A.P., and Eremets, M.I. (2016) Observation of superconductivity in hydrogen sulfide from nuclear resonant scattering. *Science*, **351** (6279), 1303–1306.

155 Dumesic, J.A. and Topsøe, H. (1977) Mössbauer spectroscopy applications to heterogeneous catalysis. *Adv. Catal.*, **26**, 121–246.

156 Berry, F.J. (1984) Mossbauer spectroscopy in heterogeneous catalysis. *Mössbauer Appl. Spectrosc. Inorg. Chem.*, **1**, 391.

157 Lázár, K. (2013) Mössbauer spectroscopy in catalysis. *Hyperfine Interact.*, **217** (1–3), 57–65.

158 Topsøe, H., Clausen, B.S., Candia, R., Wivel, C., and Mørup, S. (1981) In situ Mössbauer emission spectroscopy studies of unsupported and supported sulfided Co–Mo hydrodesulfurization catalysts: evidence for and nature of a Co–M–S phase. *J. Catal.*, **68** (2), 433–452.

159 Barpanda, P., Ati, M., Melot, B.C., Rousse, G., Chotard, J.N., Doublet, M.L., Sougrati, M.T., Corr, S., Jumas, J.C., and Tarascon, J.M. (2011) A 3.90V iron-based fluorosulphate material for lithium-ion batteries crystallizing in the triplite structure. *Nat. Mater.*, **10** (10), 772–779.

160 Brisbois, M., Caes, S., Sougrati, M.T., Vertruyen, B., Schrijnemakers, A., Cloots, R., Eshraghi, N., Hermann, R.P., Mahmoud, A., and Boschini, F. (2016) Na_2FePO_4F/multi-walled carbon nanotubes for lithium-ion batteries: operando Mössbauer study of spray-dried composites. *Sol. Energy Mater. Sol. C*, **148**, 67–72.

161 Herlitschke, M., Tchougréeff, A.L., Soudackov, A.V., Klobes, B., Stork, L., Dronskowski, R., and Hermann, R.P. (2014) Magnetism and lattice dynamics of FeNCN compared to FeO. *New J. Chem.*, **38** (10), 4670–4677.

162 Sougrati, M.T., Darwiche, A., Liu, X., Mahmoud, A., Hermann, R.P., Jouen, S., Monconduit, L., Dronskowski, R., and Stievano, L. (2016) Transition-metal carbodiimides as molecular negative electrode materials for lithium- and sodium-ion batteries with excellent cycling properties. *Angew. Chem., Int. Ed.*, **55** (16), 5090–5095.

163 Fleischer, I., Klingelhoefer, G., Rull, F., Wehrheim, S., Ebert, S., Panthöfer, M., Blumers, M., Schmanke, D., Maul, J., and Schröder, C. (2010) In-situ Mössbauer spectroscopy with MIMOS II at Rio Tinto, Spain. *J. Phys. Conf. Ser.*, **217**, 012062.

164 Zegeye, A., Abdelmoula, M., Usman, M., Hanna, K., and Ruby, C. (2011) In situ monitoring of lepidocrocite bioreduction and magnetite formation by reflection Mössbauer spectroscopy. *Am. Mineral.*, **96** (8–9), 1410–1413.

165 Machala, L., Procházka, V., Miglierini, M., Sharma, V.K., Marušák, Z., Wille, H.C., and Zbořil, R. (2015) Direct evidence of Fe(V) and Fe(IV) intermediates during reduction of Fe(VI) to Fe(III): a nuclear forward scattering of synchrotron radiation approach. *Phys. Chem. Chem. Phys.*, **17** (34), 21787–21790.

166 Lavina, B. and Meng, Y. (2015) Unraveling the complexity of iron oxides at high pressure and temperature: synthesis of Fe_5O_6. *Sci. Adv.*, **1** (5), e1400260–e1400260.

167 Bykova, E., Dubrovinsky, L., Dubrovinskaia, N., Bykov, M., McCammon, C., Ovsyannikov, S.V., Liermann, H.P., Kupenko, I., Chumakov, A.I., Rüffer, R., Hanfland, M., and Prakapenka, V. (2016) Structural complexity of simple Fe_2O_3 at high pressures and temperatures. *Nat. Commun.*, **7**, 10661.

168 Ismailova, L., Bobrov, A., Bykov, M., Bykova, E., Cerantola, V., Kantor, I., Kupenko, I., McCammon, C., Dyadkin, V., Chernyshov, D., Pascarelli, S., Chumakov, A., Dubrovinskaia, N., and Dubrovinsky, L. (2015) High-pressure synthesis of skiagite-majorite garnet and investigation of its crystal structure. *Am. Mineral.*, **100** (11–12), 2650–2654.

169 Kothapalli, K., Kim, E., Kolodziej, T., Weck, P.F., Alp, E.E., Xiao, Y., Chow, P., Kenney-Benson, C., Meng, Y., Tkachev, S., Kozlowski, A., Lavina, B., and Zhao, Y. (2014) Nuclear forward scattering and

first-principles studies of the iron oxide phase Fe$_4$O$_5$. *Phys. Rev. B*, **90** (2), 024430.

170 Diakhate, M., Hermann, R., Möchel, A., Sergueev, I., Søndergaard, M., Christensen, M., and Verstraete, M. (2011) Thermodynamic, thermoelectric, and magnetic properties of FeSb$_2$: a combined first-principles and experimental study. *Phys. Rev. B*, **84** (12), 125–210.

14
Macroscopic Magnetic Behavior: Spontaneous Magnetic Ordering

Heiko Lueken and Manfred Speldrich

RWTH Aachen University, Institute of Inorganic Chemistry, Landoltweg 1, 52074 Aachen, Germany

14.1
Introduction

Automated, high-accuracy measurements are state of the art in today's magnetic materials sciences. The measurements serve to determine electronic configurations of magnetically active ions, interatomic exchange interactions, diamagnetic contributions, metallic character, magnetic anisotropy, superparamagnetism, single-molecule magnetism, delocalization of electrons in mixed valence systems, and multiferroic properties.

Systems with localized valence electrons have the advantage that the free ion behavior is a convenient starting point to clarify the magnetism of the condensed phase. Systems with delocalized valence electrons, for example, 3d, 4d, and 5d intermetallics, defy this magnetic treatment due to band magnetism with a non-integer number of magnetic electrons per atom. For intermetallics of the lanthanide (actinide) series, the magnetic behavior is generally characterized by delocalized 5d(6d) and 6s(7s) electrons of the outer subshells as well as localized 4f(5f) electrons of the inner subshell.

To take full advantage of experimental data, special attention should be given to the purity of the samples, measurement conditions, graphical presentation of the results, and adequate models. In practical operation, the magnetic properties of compounds are examined as function of temperature (T) and the applied magnetic field (B).

Two principal systems of electromagnetic units are utilized in magnetometry, which can be called the *rational* (the SI system) and *irrational* (the CGS-emu system) [1]. In this chapter, SI units are applied throughout. Conversion factors are compiled in Table 14.1.

Handbook of Solid State Chemistry, First Edition. Edited by Richard Dronskowski, Shinichi Kikkawa, and Andreas Stein.
© 2017 Wiley-VCH Verlag GmbH & Co. KGaA. Published 2017 by Wiley-VCH Verlag GmbH & Co. KGaA.

14 Macroscopic Magnetic Behavior: Spontaneous Magnetic Ordering

Table 14.1 Magnetic quantities; definitions, units, and conversion factors in the SI and in the CGS-emu system [1].

Quantity		SI	CGS-emu	Factor[a]	
μ_0	Permeability of vacuum	$\frac{4\pi}{10^7}$ H/m [b]	1		
B	Magnetic induction	T	G	10^{-4}	$\frac{T}{G}$
H	Magnetic field strength	A/m	Oe	$\frac{10^3}{4\pi}$	$\frac{A/m}{Oe}$
M	Magnetization	A/m	G	10^3	$\frac{A/m}{G}$
M_m	Molar magnetization	A m^2/mol	G cm^3/mol	10^{-3}	$\frac{A\,m^2}{G\,cm^3}$
μ_a [c]	Atomic magnetic dipole moment	$\mu_a = M_m/N_A$ A m^2	$\mu_a = M_m/N_A$ G cm^3		
m	Magnetic dipole moment	A m$^2 \equiv$ J/T	G cm$^3 \equiv$ erg/G	10^{-3}	$\frac{A\,m^2}{G\,cm^3}$
μ_B	Bohr magneton	A m^2	G cm^2	10^{-3}	$\frac{A\,m^2}{G\,cm^3}$
χ	Magnetic volume susceptibility	$M = \chi_1 H$	$M = \chi_1^{(ir)} H^{(ir)}$	4π	
χ_g	Magnetic mass susceptibility	$\chi_g = \chi/\rho$ m^3/kg	$\chi_g^{(ir)} = \chi^{(ir)}/\rho$ cm^3/g	$\frac{4\pi}{10^3}$	$\frac{m^3/kg}{cm^3=g}$
χ_m	Molar magnetic susceptibility	m^3/mol	cm^3/mol	$\frac{4\pi}{10^6}$	$\frac{m^3}{cm^3}$
μ_{eff}	Effective Bohr magneton number [2]	1 [d]	1^e	1	
$N_A\mu_B$ [f]		5.58 494 A m^2/mol	5.58 494 × 10^3 erg/(G mol)		
$\frac{\mu_B}{k_B}$		0.67 171 K/T	0.67 171 × 10^{-4} K/G		
$\mu_0 \frac{N_A\mu_B^2}{3k_B}$		1.57 141 × 10^{-6} m^3 K/mol	1.25 049 × 10^{-1} cm^3 K/mol		

a) Factor applied to the value in CGS-emu units to obtain the value in SI units.
b) H = Henry; H/m = Vs/Am.
c) μ_a/μ_B is the atomic magnetization in Bohr magnetons, independent of the unit systems SI and CGS-emu.
d,e) $\mu_{eff} = \sqrt{3k_B/\mu_0 N_A\mu_B^2}\sqrt{\chi_m T} = 797.74 \sqrt{\chi_m T/(m^3\,K/mol)}$ (SI) = 2.8279 $\sqrt{\chi_m T/(cm^3\,K/mol)}$ (CGS-emu).
f) $\mu_0 N_A\mu_B = 7.018 \times 24 \times 10^{-6}$ T m^3/mol.

14.2
Basic Prinziples

14.2.1
Magnetic Quantities and Units [3,4]

14.2.1.1 Magnetic Dipole Moment m

The orbital motion of an electron with charge $-e$ and mass m_e generates a magnetic dipole moment m, which is related to an angular momentum l:

$$m = \gamma_e l, \quad \text{where} \quad \gamma_e = -\frac{e}{2m_e} \quad (\gamma_e : \text{magnetogyric ratio}). \tag{14.1}$$

In an atom, m is specified by the quantum numbers of the orbital angular momentum l and m_l. The magnitude of the angular momentum is given by $\sqrt{l(l+1)}\hbar$, and the component along the z-axis by $m_l\hbar$. In units of Bohr magnetons, $\mu_B = e\hbar/(2m_e)$, the magnetic dipole moment induced by orbital angular momentum is thus $\sqrt{l(l+1)}\mu_B$ with $-m_l\mu_B$, the component along z-axis. In addition to the orbital contribution, an electron carries a spin characterized by the quantum numbers $s = 1/2$ and $m_s = \pm 1/2$. The magnitude of the spin is $\sqrt{s(s+1)}\hbar = \sqrt{3}\hbar/2$. The spin is related to the magnetic moment in a similar way as the orbital angular momentum, but with a proportionality constant $g_e \neq 1$ (*electron g-factor*). Therefore, the spin magnetic moment is characterized by a magnitude equal to $\sqrt{3}g_e\mu_B/2$, and a component along the z-axis equal to $-g_e\mu_B m_s$, which is approximately $\mp\mu_B$ since $g_e \approx 2$.

In the classical picture, the energy of a magnetic dipole in a magnetic field is $E = -\boldsymbol{m} \cdot \boldsymbol{B}$. Due to space quantization, the energy for a free single electron reads $E = g\mu_B m_s B$, and the energy difference between the two states is $\Delta E = g\mu_B B$. Electrons in atoms have, in general, both orbital angular momentum and spin, which can moreover interact (*spin–orbit coupling*). The corresponding g-factor thus differs from g_e.

14.2.1.2 Magnetic Field, Magnetization, and Magnetic Susceptibility

A magnetic field can be described by the vector fields $\boldsymbol{B}$ (*magnetic induction*, measured in Tesla (T)) and $\boldsymbol{H}$ (*magnetic field strength*, measured in amperes per meter (A/m)). In free space (vacuum), they are related by $\boldsymbol{B} = \mu_0\boldsymbol{H}$, where the factor $\mu_0 = 4\pi \times 10^{-7}$ H/m is the permeability of free space (Table 14.1). Moving charges are the origin of $\boldsymbol{H}$, while $\boldsymbol{B}$ describes the effect of the field upon materials.

In a magnetic solid, the vector relationship is $\boldsymbol{B} = \mu_0(\boldsymbol{H} + \boldsymbol{M})$, where the magnetization is defined as $\boldsymbol{M} = \boldsymbol{m}\,V^{-1}$ (magnetic dipole moment per volume). Provided magnetization is not saturated, $\boldsymbol{M}$ and $\boldsymbol{H}$ are linearly related: $\boldsymbol{M} = \chi\boldsymbol{H}$, where χ is the dimensionless magnetic volume susceptibility. In this case, $\boldsymbol{B}$ and $\boldsymbol{H}$ are also linearly related: $\boldsymbol{B} = \mu_0(1+\chi)\boldsymbol{H} = \mu_0\mu_r\boldsymbol{H}$, where $\mu_r = 1 + \chi$ is the relative permeability. Many materials exhibit μ_r values deviating slightly from 1, that is, $|\chi| \ll 1$. Furthermore, χ serves to classify materials: paramagnets $\chi > 0$, vacuum $\chi = 0$, diamagnets $\chi < 0$. In general, the magnetic susceptibility is the sum of all potential contributions $\left(\chi = \chi_{\text{dia}} + \chi_{\text{para}} + \cdots\right)$. Since the paramagnetic properties are normally of interest, χ denotes in the following sections solely the paramagnetic part of the total magnetic susceptibility. In practice, the molar susceptibility $\chi_m = \chi M_r \rho$ (M_r: molar mass; ρ: mass density) is most frequently used.

Diamagnetism

In presence of an external magnetic field, the *diamagnetic* contribution to the magnetic moment is based in a classical explanation on the electromagnetic inductive effect. The orbital angular momenta of the electrons are perturbed by the applied field inducing currents that generate a magnetic field whose direction opposes, according to Lenz's law, that of the applied field. Magnetism of solids is according to the Bohr–van Leeuwen theorem, however, a quantum mechanical effect but the corresponding results agree with most classical results, for example, Langevin diamagnetism: In the one-electron approximation, the diamagnetic susceptibility per mole of an atom is given by

$$\chi_{m,\text{dia}} = -\frac{\mu_0 N_A e^2}{6 m_e} \sum_{i=1}^{n} \langle r_i^2 \rangle, \tag{14.2}$$

where the radial integrals $\langle r_i^2 \rangle$ describe the extension of the orbitals occupied by the n electrons. N_A, e, and m_e are the Avogadro constant, elementary charge, and mass of the electron, respectively.

For free atoms, the diamagnetic susceptibility is independent of both T and B. In general, diamagnetic contributions are present for both, closed- and open-shell electron configurations. Thus, experimental magnetic susceptibility data have to be corrected for diamagnetic contributions [5].

Paramagnetism

A requirement for paramagnetism is the presence of unpaired electrons (open-shell atoms). Net spins and orbital angular momenta are polarized in the direction of the applied field. Paramagnetism is observed in various forms, differing in magnitude and dependence on T and B, for example,

i) *Curie paramagnetism*: Inverse dependence of χ_m on T, independence of χ_m on B at weak applied fields, but inverse dependence at strong fields. Magnetic dipoles tend to line up by increasing B, while increasing T favors a random orientation. Therefore, the magnetization of a Curie paramagnet depends on the ratio B/T.

ii) *Temperature-independent paramagnetism* (TIP) or *Van Vleck paramagnetism*: Second-order effect involving mixing of the ground-state and unoccupied, excited multiplets induced by the applied field.

iii) *Pauli paramagnetism* (of conduction electrons): Observed for metallic systems.

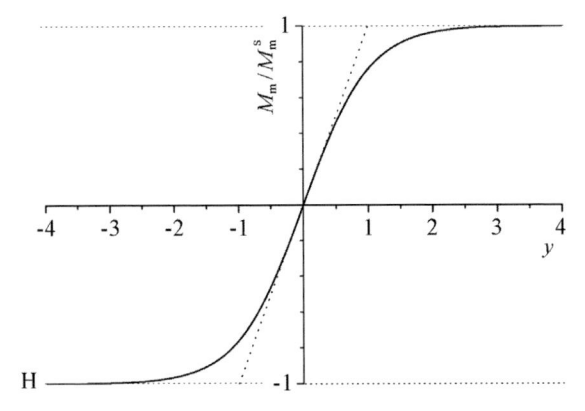

Figure 14.1 Brillouin function $B_J(\gamma)$ for $J = 1/2$ and $g = 2$ ($\gamma = \mu_B B/k_B T$). The dotted line passing the origin corresponds to Curie law behavior (see text).

14.2.1.3 Brillouin Function, Curie Law

The magnetic moment of a free ion is characterized by its total angular momentum composed of the orbital angular momentum and the spin: $\boldsymbol{J} = \boldsymbol{L} + \boldsymbol{S}$, measured in units of $\hbar$. The total angular momentum quantum number J can take an integer or a half-integer value according to an even or odd number, respectively, of unpaired electrons. In Eq. (14.3), the magnetization as a function of $B = |\boldsymbol{B}|$ and T is given by the Brillouin function $B_J(\gamma) = M_m/M_m^s$, where M_m^s is the molar saturation magnetization at $T = 0$ K and $\gamma = g_J J \mu_B B/(k_B T)$.[1)]

$$B_J(\gamma) = \frac{M_m}{M_m^s} = \frac{2J+1}{2J} \coth\left[\left(\frac{2J+1}{2J}\right)\gamma\right] - \frac{1}{2J}\coth\left(\frac{\gamma}{2J}\right), \qquad (14.3)$$

where

$$\gamma = \frac{g_J J \mu_B B}{k_B T} \quad \text{and} \quad M_m^s = N_A g_J J \mu_B.$$

For $J = 1/2$ and $g=2$, the Brillouin function reduces to $M_m/M_m^s = \tanh \gamma$, depicted in Figure 14.1. Figure 14.2 displays the atomic magnetic dipole moment divided by the Bohr magneton, μ_a/μ_b, as a function of B/T for compounds[2)] of $Cr^{3+}(S = 3/2)$, $Mn^{2+}(S = 5/2)$, and $Gd^{3+}(S = 7/2)$ [6].

For $\gamma \ll 1$, the Brillouin function approximates to $B_J(\gamma \ll 1) \approx (J+1)\gamma/(3J)$ (see dotted line in Figure 14.1.). The corresponding magnetization and magnetic susceptibility are $M_m = (N_A\mu_B^2 B/3k_B T)g_J^2 J(J+1)$ and $\chi_m = \mu_0(N_A\mu_B^2/3k_B T)g_J^2 J(J+1)$, respectively, reproducing the *Curie law*:

$$\chi_m = \frac{C}{T}, \quad \text{where } C = \mu_0 \underbrace{\frac{N_A\mu_B^2}{3k_B}g_J^2 J(J+1)}_{\mu_{\text{eff}}^2}, \quad g_J = \frac{3}{2} + \frac{S(S+1) - L(L+1)}{2J(J+1)}. \qquad (14.4)$$

---

1) For a pure spin system, the argument of the Brillouin function $B_s(\gamma)$ reads $\gamma = gS\mu_B B = (k_B T)$.
2) $KCr(SO_4)^2 \cdot 12H_2O$, $NH_4Fe(SO_4)_2 \cdot 12H_2O$, $Gd_2(SO_4)_3 \cdot 8H_2O$.

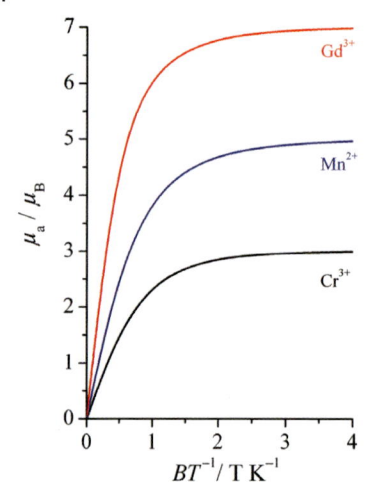

Figure 14.2 Variation μ_a versus B/T for $S = 3/2\,(Cr^{3+})$, $5/2\,(Mn^{2+})$, and $7/2\,(Gd^{3+})$. (Adapted from Ref. [6].)

Written as *term symbol* $^{2S+1}L_J$, the quantum numbers S, L, and J specify the ground state of the free ion and are determined using Hund's rules.[3] μ_{eff} is the *effective Bohr magneton number* [2], and $\mu_a^s = M_m^s / N_A = g_J J \mu_B$ is the *atomic magnetic saturation moment* measured at low T and large B.

14.2.2
Classification of Magnetic Materials

14.2.2.1 Magnetically Dilute Systems

In insulators, a magnetically isolated d^N or f^N metal ion is surrounded by ligands. This leads to a splitting of the degenerate electronic energy states of the free ion due to combination of interelectronic repulsion (H_{ee}), spin–orbit coupling (H_{so}), the electrostatic field produced by the ligands (H_{lf}), and the applied magnetic field H_{mag} (Zeeman effect). Depending on the relative strengths of H_{ee}, H_{so}, and H_{lf}, the following scenarios can be identified from experiment for $3d^N$ [7] and $4f^N$ [8] ions (Table 14.2).

Energy splittings of d and f electron states caused by H_{ee} are about[4] 10^4 cm^{-1}, and splittings by H_{so} up to 10^3 cm^{-1}. The effect of H_{lf} on d ions extends to[5] 2×10^4 cm^{-1} (high-spin configurations) or 4×10^4 cm^{-1} (low-spin configurations), and is much weaker on 4f ions (10^2 cm^{-1}) and 5f ions (10^3 cm^{-1}). Energies

---

3) Hund's rules: (1) Considering Pauli's exclusion principle, the spins of the electrons are arranged in a way that $S = |m_{s1} + m_{s2} + \cdots + m_{sN}| = |M_S|$. (2) Given the first rule, $L = |M_L|$ is maximized by summing up the m_l values. (3) J is obtained using $J = |L - S|$ if the shell is less than half full, and $J = |L + S|$ if it is more than half full.
4) Instead of cm^{-1}, other energy equivalents are in use, for example, K, kJ/mol, s^{-1} = Hz [9].
5) The splitting factor $\Delta = f(L) \cdot g(M)$ for octahedral complexes [ML$_6$] may be estimated based on the spectrochemical series for ligands (L) and metals (M) to a good approximation [10].

Table 14.2 Ranking of energetic effects acting on a 3d and 4f metal center and the associated coupling schemes.

System	Ranking of energetic effects	Coupling scheme
$3d^N$	$H_{ee} > H_{lf} > H_{so}$	Weak crystal field
	$H_{lf} > H_{ee} > H_{so}$	Strong crystal field
	$H_{lf} \approx H_{ee} > H_{so}$	Intermediate crystal field
$4f^N$	$H_{ee} > H_{so} > H_{lf}$	Ln strong field
	$H_{ee} > H_{so} \gg H_{lf}$	Ln weak field

regarding the interatomic exchange interactions (H_{ex}) have been observed up to a few 10^2 cm^{-1} for nd systems. In contrast to the very small splittings detected for 4f ions (less than 1 cm^{-1}), which is comparable to H_{mag} corresponding to an energy equivalent of about 0.5 cm^{-1} or $B = 1 \text{ T}$.[6]

The magnetic properties of d and f metal compounds arise from the ground state of the metal ion as well as from higher, thermally populated states. A special electronic situation represents $3d^5$ and $4f^7$ ions with ground states $^6S_{5/2}$ (Mn^{2+}, Fe^{3+}) and $^8S_{7/2}$ (Eu^{2+}, Gd^{3+}), respectively. H_{lf} and H_{so} virtually do not affect the ground state since orbital contributions are negligible, leading to the observation of spin-only paramagnetism ($S = 5/2$ and $7/2$, respectively).[7] Therefore, compounds as $NH_4Fe(SO_4)_2 \cdot 12\, H_2O$ and $Gd_2(SO_4)_3 \cdot 8\, H_2O$ obey the Curie law (Eq. (14.4)) with *effective Bohr magneton number* $\mu_{eff} \approx 2\sqrt{S(S+1)} = \mu_{so}$ (spin-only formula), that is, $\mu_{eff} = 5.9$ and 7.9 for Fe^{3+} and Gd^{3+}, respectively. Another case to highlight is the orbital singlet ground state ^1A of $3d^6$ low-spin systems in a strong octahedral crystal field generating $\chi_m = const.$ (TIP), corresponding to a square root variation of $\mu_{eff} \sim \sqrt{T}$ (refer to footnotes d and e in Table 14.1).

In case of a nonhalf-filled 3d subshell, the ground state can be A, E, or T [12]. For the former two, χ_m is often given as a sum of a temperature-dependent Curie term C/T and a TIP contribution χ_0, where μ_{eff} is close to the spin-only value and varies only little with temperature. The μ_{eff} behavior of systems with T ground terms, however, is strongly temperature dependent [13,14]. As an example, the variation μ_{eff} versus T and χ_m^{-1} versus T of a $3d^1$ ion, for example, Ti^{3+}, V^{4+}, is shown in Figure 14.3 for crystal fields of octahedral and orthorhombic symmetry. For the octahedral system (ground term $^2T_{2g}$), the unquenched orbital contribution to the magnetic moment and the spin part produce via H_{so} a nonmagnetic ground state ($\mu_{eff} \to 0$ for $T \to 0$) [13]. A linear increase of χ_m^{-1} versus T is not observed until $T > 150$ K. Reducing the crystal

---

6) For a reliable detection of such weak exchange interactions, B must be one or more orders of magnitude weaker than 1 T. While for a more detailed characterization of nd systems, magnets generating high continuous fields up to 40 T and pulsed fields up to 100 T are available nowadays.
7) Spectroscopic studies have confirmed that the ground term of the Gd^{3+} free ion is composed of 97% $^8S_{7/2}$ and 3% $^6P_{7/2}$, $^6P_{5/2}$, $^6P_{3/2}$ corresponding to $g = 1.993(2)$ [11] instead of $g = 2.0023 = g_e$.

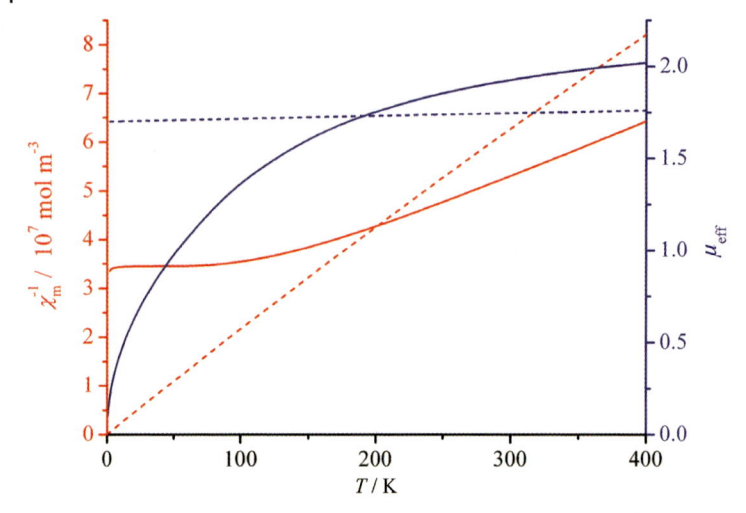

Figure 14.3 Calculated variation μ_{eff} versus T and χ_m^{-1} versus T of a 3d^1 ion in octahedral (solid curves) and orthorhombic crystal fields (dashed curves).

field symmetry, for example, to C_{2v}, the quenching of orbital contributions to μ_{eff} is (nearly) complete so that H_{so} has no effect. Pure spin magnetism results with a temperature-independent $\mu_{eff} = 1.73$.

In 4d and 5d electron systems, H_{so} increases and H_{ee} decreases compared to the 3d case. No simple formula is available to describe the variation μ_{eff} versus T. As rule of thumb, the μ_{eff} values are lower than those of comparable 3d systems.

The magnetic behavior of 4f systems (insulating as well as metallic systems)[8] is accurately predictable using, for example, computer programs [15,16]. Equation (22.4) applies for temperatures well above 150 K, that is, neglect of the influence of the crystal field due to a sufficient thermal population of all m_J substates of the specific J ground multiplet. Because of Kramers' theorem [17], centers with an odd number of 4f electrons exhibit a degenerate ground state leading to Curie behavior at temperatures below 10–20 K in the absence of magnetic exchange interactions. In case of an even number of f electrons, degenerate as well as non-degenerate ground states can be observed generating Curie behavior and TIP, respectively, at low temperatures.

In contrast to lanthanides, H_{so} and H_{lf} increase for actinides due to the larger effective nuclear charge and the fact that 5f electrons are more accessible for ligands. The challenge in modeling actinide complexes and solids originates from the observation that interelectronic repulsion ($H_{ee} \approx 10^4$ cm^{-1}), spin–orbit coupling ($H_{so} \approx 10^3$ cm^{-1}), and crystal field potential ($H_{lf} \approx 10^3$ cm^{-1}) are roughly of the same order. The situation is, therefore, more complicated than

---

8) Exceptions are Ln intermetallics with intermediate valence, for example, Ce$^{3+/4+}$, Eu$^{2+/3+}$, and Yb$^{2+/3+}$ [4].

for most 3d^N and 4f^N systems, since a simple approximation is not available. Thus, numerical (computational) methods are necessary to predict the magnetic behavior for such ions. So far, only the program package CONDON [15,18–20] is capable of modeling homo- and polynuclear systems due to the numerical approach.[9]

14.2.2.2 Local Magnetic Ordering within a Dinuclear Molecule

One of the simplest concentrated magnetic systems is a dinuclear molecule where each metal ion has a single electron ($S_1 = S_2 = 1/2$) as, for example, a dinuclear Cu(II) complex without intermolecular interactions. Introducing isotropic spin–spin coupling described by the Heisenberg operator[10] $\hat{H}_{ex} = -2J_{ex}\hat{S}_1 \cdot \hat{S}_2$, the mononuclear centers interact either with total spins S_i opposed (antiferromagnetic nature, $J_{ex} < 0$, ground state $S' = S_1 - S_2 = 0$) or aligned (ferromagnetic nature, $J_{ex} > 0$, ground state $S' = S_1 + S_2 = 1$). The magnetic behavior of such systems is described in terms of Bleaney–Bowers equation [22]:

$$\chi_m = \mu_0 \frac{N_A \mu_B^2 g^2}{3k_B T}\left[1 + \frac{1}{3}\exp\left(\frac{-2J_{ex}}{k_B T}\right)\right]^{-1} + \chi_0. \quad (14.5)$$

The results derived from this equation are presented as χ_m versus T, χ_m^{-1} versus T, and μ_{eff} versus T curves for $g = 2$ and various J_{ex} values in the range -50 to $+50\,\mathrm{cm}^{-1}$ in Figure 14.4. The χ_m versus T plot is well suited to reveal antiferromagnetic spin–spin couplings ($J_{ex} < 0$) due to the display of the typical maximum of χ_m, whereas couplings of ferromagnetic nature are readily identified by the μ_{eff} versus T representation.[11] The χ_m^{-1} versus T plot shows Curie behavior for $J_{ex} = 0$, while distinct deviations from straight lines are observed for the other cases at lower temperatures (see Section 14.2.2.3 and Eq. (14.6)).

14.2.2.3 Magnetically Condensed Systems

Materials with high magnetic dipole concentration tend to order magnetically on cooling below the Curie temperature T_C (ferro and ferrimagnets) and the Néel temperature T_N (antiferromagnets), respectively. If the magnetism of a metal ion is solely caused by the spin, the magnetic behavior in the paramagnetic regime is described by the *Curie–Weiss law*:

$$\chi_m = \frac{C}{T - \theta}, \quad \text{with} \quad \theta = \frac{2S(S+1)}{3k_B}\sum_i^n z_i J_{ex,i}, \quad (14.6)$$

---

9) The program package CONDON treats single-ion effects (H_{ee}, H_{so}, H_{lf}), interatomic exchange interactions, and in addition the impact of applied magnetic field. This key feature enables CONDON to describe the wide range of (s, p,) 3–5d and 4/5f homo- and heteropolynuclear systems, whereas effective models are inapplicable, for example, actinide compounds.
10) In literature, different definitions of J_{ex} exist. Instead of the version $\hat{H}_{ex} = -2J_{ex}\hat{S}_1 \cdot \hat{S}_2$, preferred here, $\hat{H}_{ex} = +2J_{ex}\hat{S}_1 \cdot \hat{S}_2$ and $\hat{H}_{ex} = +J_{ex}\hat{S}_1 \cdot \hat{S}_2$ are often used [21].
11) Instead of the μ_{eff} versus T curve, a $\chi_m T$ versus T plot is alternatively used in literature, where $\chi_m T \sim \mu_{eff}^2$.

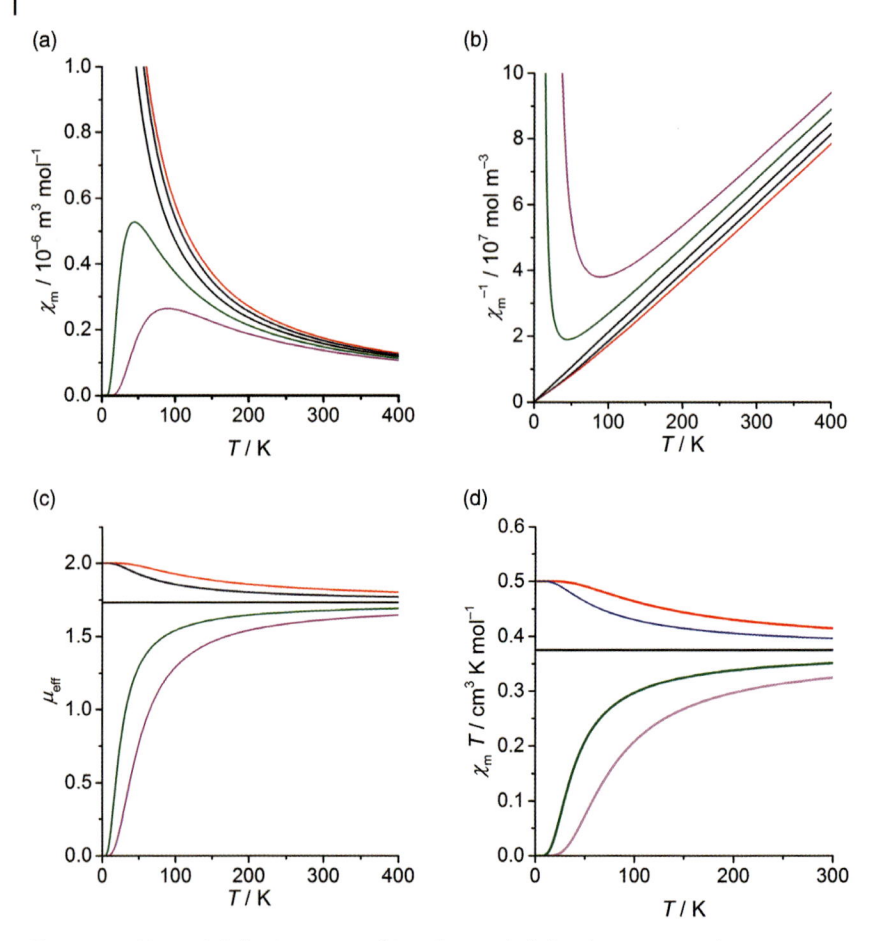

Figure 14.4 Magnetic behavior per metal ion of a coupled dinuclear system with $S_1 = S_2 = 1/2$. (a) χ_m versus T. (b) χ_m^{-1} versus T. (c) μ_{eff} versus T. (d) $\chi_m T$ versus T. $J_{\text{ex}}/\text{cm}^{-1}$: -50 (pink solid line), -25 (green), 0 (black), $+25$ (blue), $+50$ (red).

where θ is the Weiss constant and z_i is the number of ith nearest magnetic neighbors of a given magnetic center. $J_{\text{ex},i}$ stands for the mean exchange interaction between the center and the ith neighbors, and n is the number of sets of neighbors for which $J_{\text{ex},i} \neq 0$ [23]. Positive and negative θ values refer to predominant ferro and antiferromagnetic interactions, respectively. For nonhalf-filled subshells where both exchange and crystal field effects play a major role, and the single-ion behavior therefore deviates from the Curie law, the variation of the magnetic susceptibility with temperature above the ordering temperature may be described by the molecular field (mf) model:

$$\chi_m^{-1} = \chi_m'^{-1} - \lambda_{\text{mf}}, \quad \text{with} \quad \lambda_{\text{mf}} = \frac{2\sum_{i=1}^{n} z_i J_{\text{ex},i}}{\mu_0 N_A \mu_B^2 g^2}, \tag{14.7}$$

where χ'_m refers to the single-ion susceptibility. The mf parameter λ_{mf} thus generates a parallel shift of χ_m^{-1} compared to the χ'^{-1}_m versus T curve, similar to θ for the Curie–Weiss law. The g factor may deviate more or less from 2.0, depending on orbital contributions.

14.3 Crystal Field Splittings

14.3.1 Introduction of the Crystal Field

Crystal field theory (CFT) was developed by physicists Hans Bethe and John Hasbrouck van Vleck in the 1930s [2]. It was the first theoretical model to describe the spectroscopic properties of transition metal complexes. CFT describes the interaction of, for example, d orbital electrons of a transition metal and the electrostatic field originating from the ligands that form the corresponding coordination complex.

In free atoms and ions, "outer" or non-core electrons are subject to three main constraints: (a) they possess kinetic energy, (b) they are attracted to the nucleus, and (c) they repel each other. Within the environment of other ions – as, for example, within the lattice of a crystal – those electrons are subject to another constraint. Namely, they are additionally affected by the nonspherical electric field caused by the surrounding ions, which is also known as the *crystal field*.

In the simplest form, an arrangement of negative point charges is the source of the crystal field. This simplification is not essential but perfectly adequate for our introduction.[12]

14.3.2 Splitting of d Orbitals in Octahedral Symmetry

In general, the results of crystal field theory depend on the number electrons of the central ion, and on the spatial arrangements as well as the type of the

---

12) Suggestions for further reading: References [1–4] below give a detailed introduction to the subject on a technical level, while Refs [5,6] provide profound knowledge insights:

1) Figgis, B.N. (1966) *Introduction to Ligand Fields*, John Wiley & Sons, Inc., New York.
2) Cotton, R.A. (1990) *Chemical Applications of Group Theory*, 3rd ed., John Wiley & Sons, Inc., New York.
3) Schläfer, H.L. and Gliemann, G. (1967) *Einführung in die Ligandenfeldtheorie*, Akademische Verlagsgesellschaft, Frankfurt.
4) Mabbs, F.E. and Machin, D. J. (1973) *Magnetism and Transition Metal Complexes*, Chapman and Hall, London.
5) Griffith, J.S. (1971) *The Theory of Transition-Metal Ions*, Cambridge University Press, Cambridge.
6) Ballhausen, C.J. (1962) *Introduction to Ligand-Field Theory*, McGraw-Hill, New York.

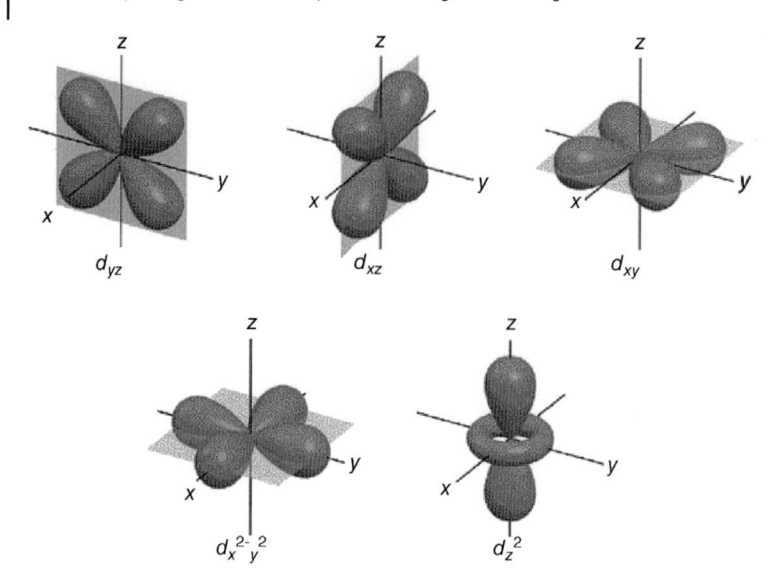

Figure 14.5 The real, single-electron 3d orbitals.

ligands. The most common coordination geometry for first series transition metals is octahedral; we therefore discuss an octahedral coordination of a 3d transition metal by point charges. The corresponding set of five single-electron d orbitals are shown as real orbitals in Figure 14.5. Next, six point charges are equidistantly put on the positive and negative x-, y-, and z-axes of the same coordinate system to form an octahedral coordination complex. This is conveniently drawn by placing the charges at the centers of each face of a cube, which itself is centered on the metal atom as depicted in Figure 14.6.

Some d orbitals (Figure 14.5) are more directed toward the point charges (Figure 14.6) than others, the d_{z^2} and $d_{x^2-y^2}$ orbitals are directed exactly toward

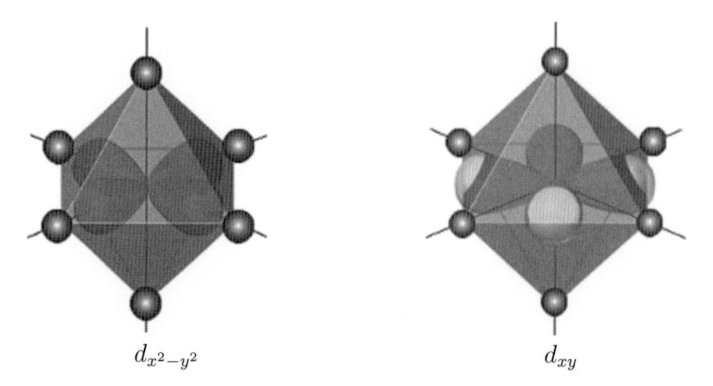

Figure 14.6 The $d_{x^2-y^2}$ orbital points toward (four of) the ligands and d_{xy} points between these ligands.

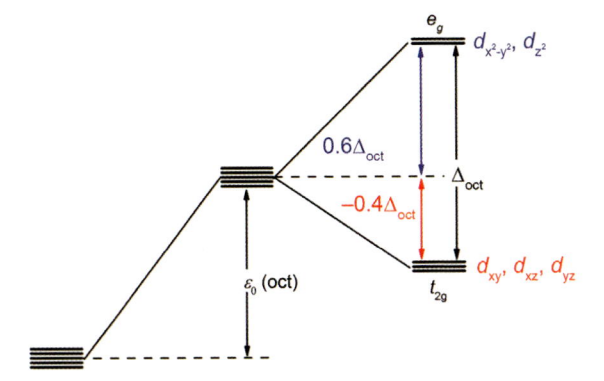

Figure 14.7 The d_{xy}, d_{xz}, and d_{yz} orbitals lie at lower energies than d_{z^2} and $d_{x^2-y^2}$ since the latter point toward the orbitals of the ligands, thus experiencing larger repulsion.

the six charges, while each of the lobes of the d_{xy}, d_{xz}, and d_{yz} orbitals point between two of the x-, y-, and z-axes on which the charges are situated. A single electron placed in the d_{xy} orbital is therefore less repelled by the crystal field than being placed in the $d_{x^2-y^2}$ orbital. The d_{xy}, d_{yz}, and d_{xz} orbitals are energetically equivalent as evident from the geometry of the orbitals. The repulsion suffered by an electron situated in the $d_{x^2-y^2}$ orbital is, although not immediately obvious but also due to symmetry reasons, the same as in the d_{z^2} orbital.

The energies of the five d orbitals in octahedral symmetry are divided into two groups, as shown in Figure 14.7. In general, all d orbital energies of a metal complex increase compared to the free ion due to the electron repulsion in the crystal field. In an octahedral crystal field, two orbitals have higher energies than the other three. Since spectroscopic and most other d electron properties of interest are characterized by relative energies, or splittings, rather than absolute energies, a more common representation of the differential crystal field effect upon d orbitals is shown in Figure 14.8. The energy levels are arranged relative to the mean energy of the whole d orbital set. This so-called barycenter rule implies that the higher pair of orbitals lies at an energy $+0.6\Delta_{oct}$ and the lower trio at $-0.4\Delta_{oct}$

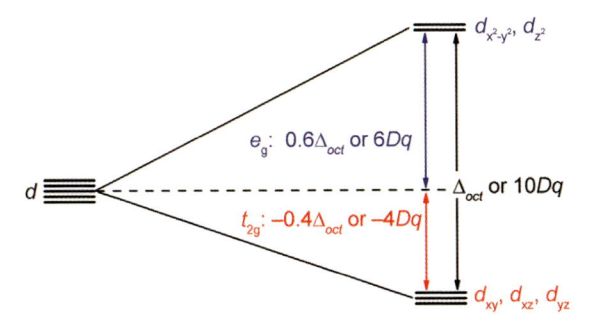

Figure 14.8 Barycenter splitting of 3d orbitals in an octahedral crystal field.

provided that the splitting energy between these two subsets is Δ_{oct} ("barycenter" means "center of mass").[13]

The subsets of d orbitals in Figure 14.8 may also be labeled according to their symmetry properties: d_{z^2} and $d_{x^2-y^2}$ as e_g, and d_{xy}, d_{xz}, d_{yz} as t_{2g}. These labels are derived from group theory, and indicate the transformation properties of the wave functions under various symmetry operations. For *our* purposes, the labels *a* and *b* refer to an orbital (i.e., spatial) singlet, *e* to an orbital doublet, and *t* to an orbital triplet. Lower case letters are used for one-electron wave functions (i.e., the orbitals). Since the inversion center *i* is an element of the octahedral point group, the subscript g (from German *gerade*, even, whereas u, *ungerade*, odd) denotes the behavior of these functions under central inversion. Since all d orbitals are centrosymmetric, they are thus all labeled g. The subscript 2 in t_{2g} denotes that two of the corresponding orbitals are symmetric and one is antisymmetric with respect to each of the rotation operations by 180° around the C_2 axes perpendicular to the principal axes.

A transition metal ion with a single d electron (electron configuration d^1) in an octahedral environment is in its ground state, if the electron is located in the t_{2g} orbitals. This electron may be promoted to the higher energy e_g orbitals by absorption of the appropriate energy Δ_{oct}. A redistribution of the electron within the t_{2g} or e_g orbitals, respectively, does not involve a change of energy due to the equivalence of the three Cartesian directions in an octahedron, and therefore takes place spontaneously and continuously. Thus, only a single change of energy may occur within the d orbitals corresponding to the *transition* $t_{2g} \to e_g$. Illuminating such d^1 complexes with light of varying frequency ν induces a single electronic transition characterized by the resonance frequency $\nu_0 = \Delta_{oct}/h$.

14.3.3
Splitting of d Orbitals in Tetrahedral and Other Symmetries

Tetrahedral Symmetry

A convenient placement of the four ligands in tetrahedral symmetry is shown in Figure 14.9. Represented by point charges, they occupy alternate corners of a cube centered around the metal. The reference frame is therefore the same as in Figure 14.6, that is, the *x*, *y*, and *z* axes go through the centers of the cube faces.

The lobes of the d_{xy} orbital are directed toward the midpoints of the cube edges, whereas those of the $d_{x^2-y^2}$ orbital point toward the centers of the cube faces. The lobes of the d_{xy} orbital are thus closer to the point charges than those of the $d_{x^2-y^2}$ orbital or, in other words, the angle between the lobe of the $d_{x^2-y^2}$ orbital and metal–ligand (M–L) axis is larger than between the d_{xy} orbital lobes and the M–L axis. An electron in the d_{xy} orbital thus experiences a larger repulsion than in the $d_{x^2-y^2}$ orbital. For symmetry reasons, the physical situation of an

---

13) An older alternative label for the octahedral field splitting is 10 Dq (= Δ_{oct}) where, in the literal crystal field theory, q is the charge on each ligand and D is a quantity related to the geometry. Both Δ_{oct} and 10 Dq are commonly used.

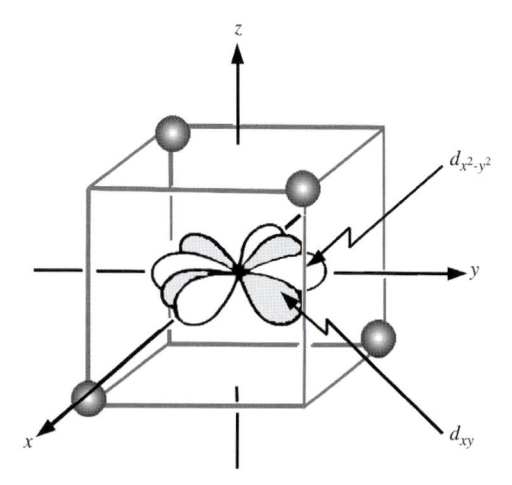

Figure 14.9 Lobes of the d_{xy} orbital are closer to the point charges than those of the $d_{x^2-y^2}$ orbital.

electron is the same in the d_{xy}, d_{xz}, or d_{yz} orbital, which therefore form a set of three degenerate orbitals. Less obvious but also due to symmetry reasons, the d_{z^2} and $d_{x^2-y^2}$ are equivalent and form another set of two degenerate orbitals. The energy diagram for the five d orbitals in a tetrahedral crystal field is shown in Figure 14.10.

As for octahedral symmetries, the orbital energies obey the barycenter rule but their order is reversed. The subsets of the d orbital subsets are labeled t_2 and e, and the splitting Δ_{tet} or $10\,Dq$. In this case, the subscript 2 in t_2 denotes that two of the corresponding orbitals are symmetric and one is antisymmetric with respect to each of the reflection operations about the dihedral mirror planes σ_d. Note the absence of the subscript g (which does *not* imply the opposite, u): Although the d orbitals are centrosymmetric, the tetrahedral symmetry lacks a center of inversion, and the corresponding symmetry operation is thus not possible.

The d orbitals are split in inverse order in the tetrahedral symmetry compared to octahedral symmetry. In an octahedral crystal field, the orbitals are oriented

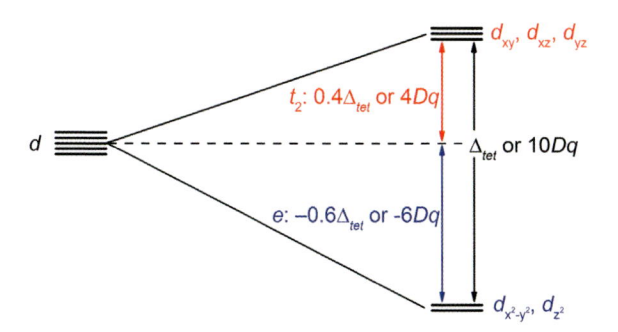

Figure 14.10 Barycenter splitting of 3d orbitals in a tetrahedral crystal field.

either directly at or between the point charges. For a tetrahedron, however, all d orbitals point between the ligands. The magnitude of the tetrahedral splitting is thus less than in the octahedron. Simple geometrical calculations show that these splittings are related as given in Eq. (14.8), if the central metal (same d orbital radial functions) and the bond lengths are the *same*.[14]

$$\Delta_{tet} = 4/9 \Delta_{oct} \tag{14.8}$$

Other Environments

The splitting patterns in crystal fields of symmetries other than octahedral or tetrahedral can be similarly determined. In general, the degeneracy of the d orbitals may be lifted even more up to the removal of degeneracy resulting in five energetically separated orbitals. In turn, the physical description gets more complex as, for example, further splitting parameters such as Δ_{oct} have to be introduced (see Box 14.1). We briefly look at some of such scenarios later on. At this point, we restrict our attention to the *cubic* symmetries – octahedral and tetrahedral symmetries – as most 3d metal complexes exhibit these symmetries, at least approximately.

14.3.4
Hole Formalism: d^1 and d^9 Electron Configurations

Consider a d^9 electron configuration of an octahedral complex in the so-called *strong-field* configuration $t_{2g}^6 e_g^3$, as shown in Figure 14.11 on the left side. Six electrons are in the t_{2g} orbitals, while three are in the e_g orbitals, which are therefore incompletely filled. An electron of the t_{2g} orbitals is promoted to the e_g orbitals by absorption of a photon of energy Δ_{oct}, as shown in Figure 14.11 on the right side. Because the e_g orbitals are now completely occupied, no further electron promotions are possible, and the $t_{2g}^5 e_g^4$ configuration represents the single excited state of the octahedral d^9 configuration. Therefore, only a single absorption band is observed in the d–d spectrum. Note that this is a very simplified description of the real physical scenario since the interelectronic repulsions of the d electrons are entirely neglected.

A description reflecting the observed multiplicities of spectra is that the excitation $t_{2g}^6 e_g^3 \rightarrow t_{2g}^5 e_g^4$ corresponds to a transfer of an *electron hole* (lack of an electron) of the e_g orbitals to the t_{2g} orbitals. The excitation of a d^9 complex can thus be understood as a redistribution of a single electron hole within a completely occupied d shell. While the excitation of octahedral d^1 complexes implies the electron transition $t_{2g} \rightarrow e_g$, the transition $e_g \rightarrow t_{2g}$ of an electron *hole*, therefore, formally describes the excitation of an octahedral d^9 configuration. This *hole formalism* may also explain the corresponding excitations of d^1 and d^9 tetrahedral complexes. Note that octahedral and tetrahedral complexes of d^1 or d^9

---

14) Some authors prefer to write $\Delta_{tet} = -4/9 \Delta_{oct}$ to emphasize the inversion of t_2 and e orbital subsets. However, if the Δ's are defined as the orbital *splittings*, it is better to omit the sign.

Box 14.1

Impact of a lower symmetry on the crystal field splitting:
Splitting of the d orbital energies in (1) a regular octahedron and (2) a tetragonally elongated octahedron.

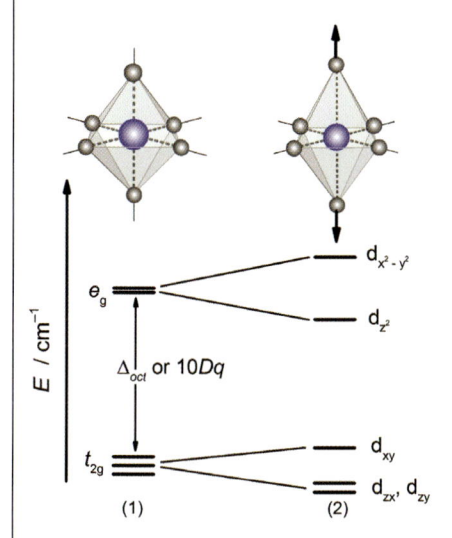

The latter may be achieved by either an asymmetric arrangement of ligands of the same kind (four short and two long *trans*-contacts (bond lengths)) or by different ligands as in a *trans*-MA$_4$B$_2$ complex. The weaker field along the z-direction repels an electron less in the d$_{z^2}$, d$_{xz}$, or d$_{yz}$ orbitals than in the d$_{x^2-y^2}$ and d$_{xy}$ orbitals (A$_4$ plane), thus splitting the e_g and t_{2g} orbitals as shown in the diagram to the right. In case of a tetragonally compressed octahedron, the order of this orbital splitting is inverted (d$_{z^2}$ ↔ d$_{x^2-y^2}$/d$_{xz}$, d$_{yz}$ ↔ d$_{xy}$).

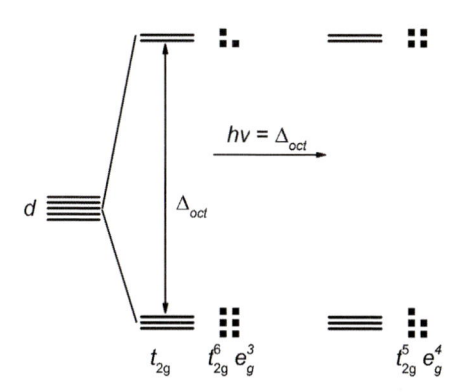

Figure 14.11 The electronic transition of an octahedral d^9 complex.

electron configurations exhibit just a *single* electronic transition in their d–d spectra. Unfortunately, explaining the larger numbers of transitions in, for example, $3d^2$ complexes involves a more sophisticated discussion. However, we cannot avoid this discussion and therefore delve into crystal field theory in the following sections to understand the observed spectra of d electrons in transition metal complexes.

14.3.5
Multiple Transitions for $3d^2$ Electron Configurations

We start from the orbital splitting of a d^1 metal center in an octahedral crystal field as shown in Figure 14.8. Obviously, there are more combinations for two electrons to be distributed over the t_{2g} and e_g orbitals than for a single electron: Both electrons may occupy either the t_{2g} orbitals or the e_g orbitals, or each set of these orbitals is occupied by one electron as indicated on the left side of Figure 14.12. Due to interelectronic repulsion, however, the degeneracy is partially lifted giving rise to a larger number of less degenerate energy states. For first series transition metals, the total orbital angular momentum L and total spin S are *good quantum numbers,* thus allowing the characterization of those energy states with respect to L and S.

The spins of two electrons can either be aligned or opposed. For aligned spins, the total spin is $S = s_1 + s_2 = 1/2 + 1/2 = 1$, and for opposed spins $S = s_1 + s_2 = 1/2 - 1/2 = 0$. The spin multiplicity, that is, the number of projections of S along the specified axis, is $2S + 1$. The two-electron states associated with these spin quantum numbers are therefore called (spin) triplets ($S = 1$) or singlets ($S = 0$), respectively. To simplify the discussion at this point, we only consider the electronic states of maximum total spin since the ground term of a free ion has maximum spin multiplicity according to Hund's first rule.

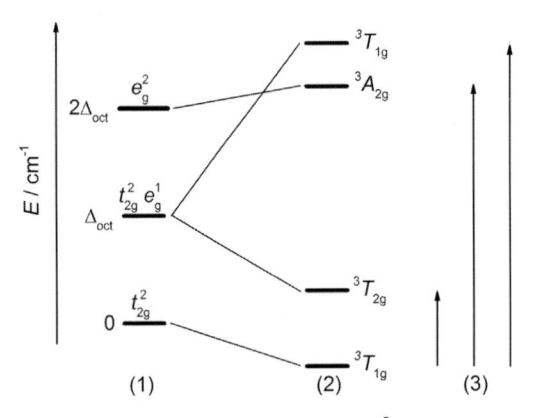

Figure 14.12 The four-spin triplets of a $3d^2$ electron configuration: (1) strong field configuration, (2) strong field and interelectronic repulsion denoted by term symbols, and (3) electronic transitions to excited states.

Neglecting thus all spin singlet states, four degenerate spin triplets arise as shown below.

Due to Pauli's exclusion principle, aligned spins must not occupy the same orbital, while opposed spins may. Since there are multiple ways to realize this rule for the electron configurations t_{2g}^2, $t_{2g}^1 e_g^1$, and e_g^2, further *spatial*[15] degeneracies may arise besides the spin triplet degeneracy.

First, we start with the t_{2g}^2 electron configuration. The spin triplet states characterized by $|M_S| = 1$ are readily identified as the possible distributions of two aligned spin electrons over the three t_{2g} orbitals. There are three different arrangements of both electrons being in spin-up state, and another three of both being spin-down electrons (this conclusion may also result from noting that the empty orbital can be any of the three). For $M_S = 0$, there are additionally three linear combinations of two different t_{2g} orbitals occupied by a spin-up and a spin-down electron, respectively. Since there are three distributions for each projection of S along the specified axis, the states also form a spatial or orbital triplet. In total, there are nine degenerate states, which we denote by the crystal field *term* symbol $^3T_{1g}$.[16]

Next, we consider the electron configuration e_g^2. There are only two combinations to distribute two aligned spin electrons over both e_g orbitals – either both are spin-up or both are spin-down. A third state, a linear combination of the two e_g orbitals occupied by a spin-up and a spin-down electron, respectively, completes the spatial singlet, which is therefore denoted by $^3A_{2g}$.[17] As the e_g orbitals of a single-electron configuration are higher in energy than the t_{2g} orbitals by Δ_{oct}, the states of the e_g^2 electron configuration are at energies that lie $2\Delta_{oct}$ higher than those of t_{2g}^2 (i.e., Δ_{oct} per electron). Note that while the energy of the $^3A_{2g}$ term is larger than that of the $^3T_{1g}$ (derived from the t_{2g}^2 electron configuration), the energy gap between both terms is, however, different from $2\Delta_{oct}$, as we will see below.

So far, we skipped the $t_{2g}^1 e_g^1$ electron configuration. For this configuration, one electron occupies the t_{2g} orbitals, while the other one occupies the energetically different e_g orbitals. For a single electron, there are as many spin-up and spin-down configurations as the degeneracy of the associated orbital set. In total, there are thus 12 states of aligned spins and 12 states of opposed spins. As for

---

15) The acronym *spatial* is often associated with the orbital angular momentum since both the radial and the angular distributions of a one-electron spatial orbital depend on this physical quantity.

16) The superscript represents the spin multiplicity $2S + 1 = 3$, that is, a spin triplet. As discussed in Section 14.3.2, the label T_{1g} for the spatial triplet is derived from group theory, but we use capital letters because the symbol denotes *many*-electron wave functions instead of orbitals, that is, one-electron wave functions. The subscript g means the wavefunctions are even under inversion through the center of symmetry possessed by the octahedron (since each d orbital is of g symmetry, so also is any product of them), and the right subscript 1 describes other symmetry properties we need not discuss here.

17) The character A means that it is orbitally onefold degenerate (singlet state) and it is uppercase because we describe many-electron wave functions. The subscript is g because the product of d orbitals is even under the octahedral center of inversion, and the right subscript 2 describes other symmetry properties we need not discuss here.

t_{2g}^2 and e_g^2, we start with the alined spin states to identify the spin triplet states. Six of the corresponding 12 states are spin-up configurations, while the remaining six are spin-down configurations. However, they do not belong to a spatial sextet but to two spatial triplets since three spin-up states act differently with respect to the symmetry operations of the octahedral group than the other three. Though not explicitly mentioned, all 45 states of the d^2 electron configuration are *two*-electron wave functions that are approximately characterized as a *product* of two orbitals. The same symmetry operations acting on two-electron wave functions may thus yield different results than from acting on the individual one-electron wave functions. Finally, the two spatial as well as spin triplets of the $t_{2g}^1 e_g^1$ electron configuration are completed by linear combinations (in total six) of the opposed spin states, and denoted by the term symbols $^3T_{1g}$ and $^3T_{2g}$.

However, the $^3T_{1g}$ and $^3T_{2g}$ terms are not degenerate but differ in energy. Moreover, this difference is about twice the magnitude of the crystal field splitting Δ_{oct} for many complexes. Also, remember our remark about a shift of the $^3T_{1g}$ and $^3A_{2g}$ terms relative to the energies of the t_{2g}^2 and e_g^2, respectively, electron configurations. So far, we focused on *crystal field* energies, that is, the interaction of the d electrons and the electric field caused be the surrounding ions. We did not consider the interactions between both electrons, therefore neglecting *interelectronic repulsion*, which is obviously an inadequate approximation for our purposes. However, the electron–electron repulsion energies are important, and differ from term to term.

A qualitative explanation is as follows: Regarding the t_{2g}^2 and e_g^2 electron configurations, the shifts are different since the spatial "proximities" of the t_{2g} orbitals are different from those of the e_g orbitals. Similar considerations regarding the spin triplet terms $^3T_{1g}$ and $^3T_{2g}$ derived from the $t_{2g}^1 e_g^1$ electron configuration may explain the split of both terms: Higher and lower energy triplets are formed dependent on the particular product of two occupied d orbitals as shown in Eq. (14.9).

$$^3T_{1g} : d_{xy}d_{x^2-y^2} d_{yz}d_{z^2} d_{xz}d_{z^2},$$
$$^3T_{2g} : d_{xy}d_{z^2} d_{yz}d_{x^2-y^2} d_{xz}d_{x^2-y^2}.$$
(14.9)

The states of the $^3T_{2g}$ term are lower in energy than those of the $^3T_{1g}$ term. Examining the involved orbitals reveals that the orbitals of the $^3T_{1g}$ states are closer to each other than those of the $^3T_{2g}$ states as depicted by two examples in Figure 14.13; the pair of d_{xy} and $d_{x^2-y^2}$ orbitals are more compact than the pair of d_{xy} and d_{z^2} orbitals. Bearing in mind that electrons repel each other, the interelectronic repulsions for states of the $^3T_{1g}$ term are thus larger than for states of $^3T_{2g}$. Please note that this is a very rough explanation since the two-electron wave functions are approximately based on the *products* of the two orbitals, and not on their *sums*.

To summarize, we again consider Figure 14.12: The relative energy levels of the two $^3T_{1g}$ terms, the $^3T_{2g}$ term and the $^3A_{2g}$ term are shown in the central column of the diagram. The ground term is the $^3T_{1g}$ term derived from the t_{2g}^2

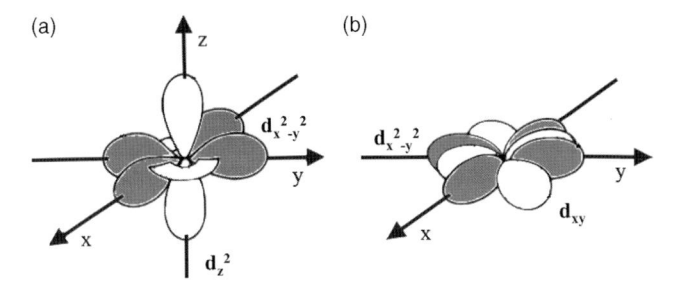

Figure 14.13 Relative electron crowding for different orbital pair densities. (a) d_{xy} and d_{z^2} orbitals. (b) d_{xy} and $d_{x^2-y^2}$.

electron configuration. Spin-allowed electronic transitions, that is, transitions between terms of the same spin multiplicity, are excitations from the ground term to the respective excited terms: $^3T_{1g} \rightarrow {}^3T_{2g}$, $^3T_{1g} \rightarrow {}^3A_{2g}$, and $^3T_{1g} \rightarrow {}^3T_{1g}$. Since the terms are separated by different energies, the d–d spectra of octahedrally coordinated d^2 ions thus exhibit *three* bands (instead of two). Similarly, spectra of tetrahedrally coordinated d^2 ions also exhibit three bands. However, the energetic order of the terms is partially reversed. The ground term is the ^3A term, and the transitions are $^3A_2 \rightarrow {}^3T_2$, $^3A_2 \rightarrow {}^3T_1$, and $^3A_2 \rightarrow {}^3T_1$. In addition, the behavior of d^8 metal centers can now be readily understood by applying the hole formalism, that is, describing a d^8 electron configuration as two electron holes in a full d shell.

Note that we adopted for the description of a d^2 metal center the so-called strong-field limit, that is, the contributions of the crystal field are much larger than those of the electron–electron repulsion. In the opposite limit, the "weak-field" limit, the interelectronic repulsion dominates the contributions of the crystal field.

As already mentioned, CFT is based primarily on symmetry of ligands around a transition metal center and how this anisotropic (properties depending on direction) crystal field affects the atomic orbitals of the metal. The energies of which may increase, decrease, or not be affected at all. Once the ligands' electrons interact with the electrons of the d orbitals, the electrostatic interactions cause the energy levels of the d orbital to fluctuate depending on the orientation and the nature of the ligands. Ligands are classified as strong or weak based on the spectrochemical series and discussed in the next section.

14.3.6
The Spectrochemical Series

Another factor that plays a key role in a transition metal complex is the nature of the ligands. The d orbital energy splitting or crystal field splitting is influenced by how strongly the ligand interacts with the metal. Ligands that interact only weakly produce little change in the d orbital energy levels, whereas ligands that interact strongly produce a larger change in d orbital energy levels. The

Box 14.2

Strong and Weak-Field Approximation (Octahedral Symmetry)

We assume in the "strong field" approximation that the crystal field splitting of the 3d electrons is larger than the splitting due to the interelectronic repulsion. The energy levels in this approximation for an ion of $3d^2$ electron configuration in an octahedral crystal field are shown at the right side of the figure below.

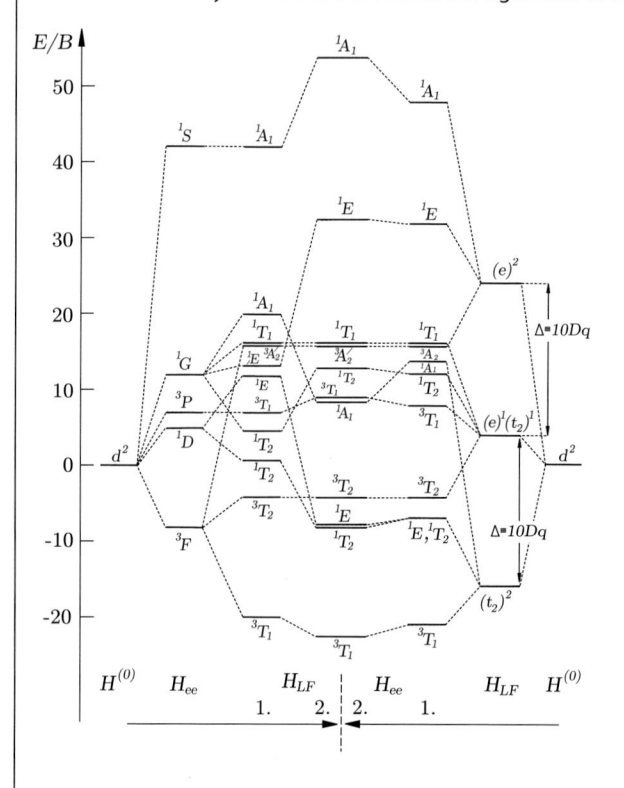

The degenerated free ion energy level ($H^{(0)}$) is primarily split by the crystal field (H_{LF}), and secondarily by interelectronic repulsion (1.).

The "weak" field approximation is illustrated at the left side of the figure. In this case, the interelectronic repulsion (H_{ee}) primarily splits the degenerate free ion energy level ($H^{(0)}$), which is secondarily split by the crystal field (1.). The ground state of a high-spin $3d^2$ electron configuration taking into account only interelectronic repulsion is a ^3F term (LS term) determined by Hund's rules. As a perturbation, the crystal field splits this term into further states from which the $^3T_1$ term forms the ground state.

Note that these approximations are two boundary cases of the CFT, but often suffice to, at least, approximately describe the energy splittings of 3d metal centers. However, there are also cases in which interelectronic repulsion and crystal field effect are of the same order. The splitting of such a scenario is depicted in the middle of the figure.

magnitude of Dq in a given complex is a direct measure of the interaction between the "spectral" metal d electrons and their molecular environment.[18]

$$Dq \approx 100 f(\text{ligands}) \times g(\text{metal}) \, \text{cm}^{-1}. \quad (14.11)$$

The ligands may be ordered according to the magnitude of Dq almost independent of the central metal (see Eq. (14.12)).

$$I^- < Br^- < SCN^- < Cl^- < F^- < O^{2-} \approx OH^- \lesssim H_2O < NCS^- < NH_3 < CN^- < PR_3 < CO$$
$$Dq \quad \text{increasing} \quad \Rightarrow$$
$$(14.12)$$

And metal centers can be ordered almost independent of the ligands (see Eq. (14.13)).

$$Mn^{II} < Ni^{II} < Co^{II} < Fe^{III} < Cr^{III} < Co^{III} < Ru^{III} < Mo^{III} < Rh^{III} < Pd^{II} < Ir^{III} < Pt^{IV}$$
$$Dq \quad \text{increasing} \quad \Rightarrow$$
$$(14.13)$$

These lists are called the *spectrochemical series*. A selection of f and g values addressed in Eq. (14.11) are presented in Table 14.3. At an empirical level, these numbers roughly predict Dq for various metal–ligand combinations. In addition, these values are also used in the empirical relationship known as the *law of average environment*. This law implies that the splitting parameter Dq of a metal complex composed of a mixed set of ligands is given by the appropriately weighted average of the corresponding complexes composed of a single type of ligand [10,27]. For example, from Table 14.3, $\Delta_{\text{oct}} = 10 \, Dq$ for $[NiF_6]^{4-}$ is

Table 14.3 Some typical f and g values for octahedral transition metal complexes.

Ion	g	Ligand	f
Co(II)	9.3	6 Br$^-$	0.76
Co(III)	19.0	6 Cl$^-$	0.8
Cr(III)	17.0	6 CN$^-$	1.7
Fe(III)	14.0	3 en	1.28
Mn(II)	8.5	6 F$^-$	0.9
Ni(II)	8.9	6 H$_2$O	1.0
V(II)	12.3	6 NH$_3$	1.25

18) Suggestions for further reading: The references give a detailed introduction into aspects of the nephelauxetic and spectrochemical series.
   1) Gerloch, M. (1973) *Ligand Field Parameters*, Cambridge University Press, Cambridge.
   2) Jørgensen, C.K. (1962) *Absorption Spectra and Chemical Bonding in Complexes*, Pergamon Press, Oxford.
   3) Jørgensen, C.K. (1970) *Modern Aspects of Ligand-Field Theory*, North Holland.

8010 cm^{-1}, and for $[Ni(H_2O)_6]^{2+}$ is 8900 cm^{-1}. The law of average environment predicts that Δ_{oct} for $[NiF_4(H_2O)_2]^{2-}$ is $10 \times 100 \times (4/6 \times 0.9 + 2/6 \times 1.0)$ $\times 8.9$ cm^{-1} = 8307 cm^{-1}.

Note that the law of average environment is, however, *strictly* invalid concerning chemistry and physics, because the symmetry of a six-coordinate complex with dissimilar ligands cannot be exactly octahedral. In this case, further splittings of the energy states occur introducing further splitting parameters than a single one such as Δ_{oct}. However, if the deviation from octahedral symmetry is small enough, and the spectral bands are thus broadened instead of split, then the law of average environment approximately yields the correct splitting energies.

An explanation for the nephelauxetic series came readily to hand (see Box 14.3). Let us see how successfully we can provide one for the spectrochemical series. First, note that Dq values increase with decreasing sizes of the donor halides:

$$I^- < Br^- < Cl^- < F^-$$
Ionic radii decreasing $\Rightarrow$ (14.14)
Dq increasing $\Rightarrow$

This seems reasonable in terms of the crystal field theory set out in Section 14.3.1, for which shorter bonds indeed imply larger values of Dq. However, the

---

Box 14.3

Nephelauxetic Effect

The interelectronic repulsion (H_{ee}) may be characterized in terms of *Racah parameters A, B,* and *C* [24] from which *B* is the most relevant for electronic transitions. The interelectronic repulsion is found to be weaker in metal complexes than in free ions, meaning that the value of *B* is smaller for a complex than for the corresponding free ion. Due to the chemical bonding, the electrons are partially "drawn" from the metal to the ligands, which separates the electrons, and hence reduces repulsion at the metal ion [25,26].

The reduction of *B* from its free ion is normally reported in terms of

$$\beta = \frac{B^*_{complex}}{B_{free\ ion}},$$

where β is the nephelauxetic parameter, which refers to the electron *cloud expansion*. The values of β are dependent on the ligand.

The ligands can be arranged in the *nephelauxetic series* as follows:

$$I^- < Br^- < Cl^- < CN^- < NH_3 < H_2O < F^-$$
β increasing $\Rightarrow$ (14.10)

A small value of β represents a larger expansion of the d electrons at the metal ion and a greater covalent character of the metal–ligand bonds in the complex. Although parts of this series seem quite similar to the spectrochemical series of ligands – for example, iodide, bromide, and chloride occupy similar positions in both series – others, such as cyanide, water, and fluoride, occupy very different positions. The ordering roughly reflects the ability of the ligands to form covalent bonds with metals. Those that are characterized by a small parameter β, and are at the start of the series, are the softer ligands. The nephelauxetic character of a ligand is usually different for electrons in the t_{2g} and the e_g orbitals.

ability of the crystal field theory to rationalize the spectrochemical series stops right here: We might also expect the splitting energies to vary according to the charge of the ligands. As shown in Eq. (14.12), the negatively charged halides generate, however, smaller splittings than neutral water or ammonia ligands. Additionally, note that very similar Dq values are observed for H_2O, OH^-, and O^{2-} ligands. We also find from Eq. (14.13) that Dq increases with higher metal oxidation states as abstracted in Eq. (14.15).

$$\begin{array}{c} M(II) < M(III) < M(IV) \\ Dq \quad \text{increasing} \quad \Rightarrow \end{array} \qquad (14.15)$$

Based on the expectations of the crystal field theory, the d orbitals contract with increasing positive charge of the ion and hence interact less with the ligand "point charges." The observed modest decreases in bond lengths as one goes through the series from left to right (Eq. (14.15)) are unlikely to compensate for, let alone override, the effects of such orbital contractions. Finally, to add another discrepancy, we also note from Eq. (14.13) that Dq values increase as we go down the periodic table of elements (Eq. (14.16)).

$$\begin{array}{c} 3d < 4d < 5d \\ Dq \quad \text{increasing} \quad \Rightarrow \end{array} \qquad (14.16)$$

While this observation is still compatible with the crystal field theory in that the radial extensions of 4d and 5d orbitals are larger than those of 3d orbitals, the diffuseness of these orbitals, however, increases along the series in Eq. (14.16), which tends to decrease the Dq values.

These and various further observations made over the years prove that the assumption of the crystal field theory describing ligands as electrostatic point charges fails to provide even a qualitative explanation of the spectrochemical series. This is mainly due to the description of chemical "bonding" of metal and ligands in terms of the *crystal* field theory as electrostatic phenomenon. Note that – notwithstanding crystal field theory failing to explain the *magnitudes* of the splitting energy parameters – the splitting pattern of the energy levels and all corresponding properties are accounted for with extraordinary success by this model.

14.4
Magnetic Materials

The macroscopic behavior of magnetic materials is classified using a few magnetic parameters. The most significant magnets, the ferromagnets, can be classified on this basis. The main uses of the ferromagnets indicate how the macroscopic properties determine the suitability of materials for a given application.

14.4.1
Magnetic Properties of Ferromagnets

The most important class of the magnetic materials are the ferromagnets. The applications that these materials find are very diverse. They are used because of their high permeabilities, which enable high magnetic inductions to be obtained with only modest magnetic fields, their ability to retain magnetization, and thereby act as a source of field, and, of course, the torque on a magnetic dipole in a field can be used in electric motors. It is not really surprising that the few ferromagnetic elements in the periodic table – iron, cobalt, nickel, and several of the lanthanides – are so technologically vital.

14.4.1.1 Permeability

By far the most important single property of ferromagnets is their high relative permeability. The permeability of a ferromagnet is not a constant function of the magnetic field. Instead, in order to characterize the properties of a given ferromagnetic material, it is necessary to measure the magnetic flux density B as a function of the magnetic field strength H over a continuous range of H to obtain a hysteresis curve.

For ferromagnets, the initial relative permeabilities μ_r usually lie in the range 10–10^5. The highest values occur for special alloys such as permalloy and supermalloy, which are nickel–iron alloys. These materials find use in, for example, magnetic flux concentrators. Permanent magnet materials do not have such high permeabilities, but their applications depend on their remanence, which is the next most important property.

When considering magnetic properties, it is first necessary to define quantities that represent the response of these materials to the field. These quantities are magnetic moment m and magnetization M. We define the quantity M as the magnetic moment per unit volume of a solid: $M = m/V$.

14.4.1.2 Hysteresis

The most common representation of the bulk magnetic properties of a ferromagnetic material is by plotting the magnetic flux density B as a function of the magnetic field strength H. Alternatively, plots of the magnetization M against H are used, which contain the same degree of information since $B = \mu_0(H + M)$. The term "hysteresis," meaning to lag behind, was introduced by Ewing, who

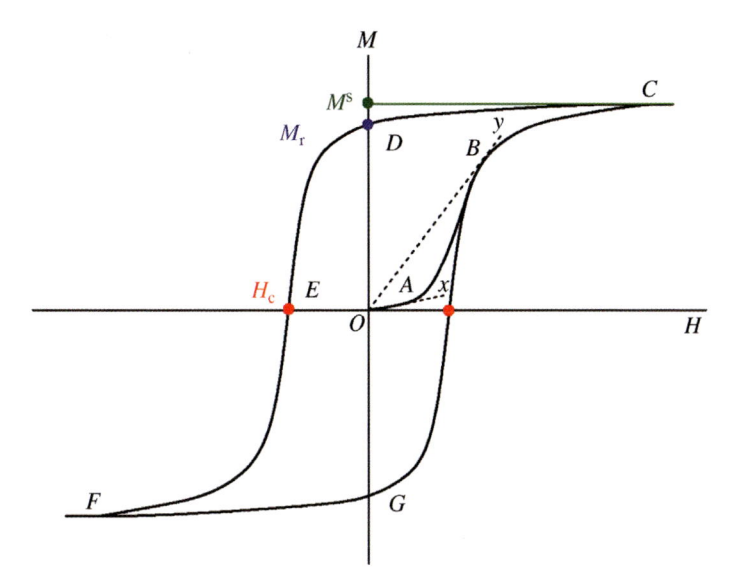

Figure 14.14 A typical hysteresis loop of a ferromagnetic material: magnetization M against magnetic field strength H. Starting at the origin, the upward curve is the initial magnetization curve. The downward curve after saturation, along with the lower return curve, forms the main loop. The intercepts H_c and M_r are the coercivity and saturation remanence, respectively. The saturation magnetization (M^s) is the maximum induced magnetic moment that can be obtained in an applied magnetic field, beyond this field no further increase in magnetization occurs.

was the first to systematically investigate it. A typical hysteresis loop is shown in Figure 14.14.

The suitability of ferromagnetic materials for applications is determined from the characteristics shown by their hysteresis loops. Materials for transformer applications need to have high permeability and low hysteresis losses because of the need for efficient conversion of electrical energy. Materials for electromagnets need to have low remanence and coercivity in order to ensure that the magnetization can easily be reduced to zero as needed. Permanent magnetic materials need high remanence and coercivity in order to retain the magnetization as much as possible.

14.4.1.3 Saturation Magnetization

It can be seen from the hysteresis plot (Figure 14.14) that the ferromagnet in its initial state is not magnetized. Application of a field H causes the magnetization to increase in the field direction. By continuously increasing H, the magnetization eventually reaches saturation at a value that we designate M^s. Under this condition, all the magnetic dipoles within the material are aligned in the direction of the magnetic field H. The saturation magnetization only depends only on the magnitude of the (atomic) magnetic moments m and the number of atoms

Table 14.4 Saturation magnetization of various ferromagnets.

Material	10^6 A/m
Iron	1.71
Cobalt	1.42
Nickel	0.48
78 Permalloy (78% Ni, 22% Fe)	0.86
Supermalloy (80% Ni, 15% Fe, 5% Mo)	0.63
Metglas 2605 ($Fe_{80}B_2O$)	1.27
Metglas 2615 ($Fe_{80}P_{16}C_3B_1$)	1.36
Permendur (50% Co, 50% Fe)	1.91

per unit volume n:

$$M^S = nm.$$

M^S, therefore, depends on the elements present in a specimen, it is not sensitive to structure. Some typical values of saturation magnetization for different materials are shown in Table 14.4.

14.4.1.4 Remanence

When the magnetic field is reduced to zero after magnetizing a magnetic material, the remaining magnetic flux density is called the remanent flux density B_R, and the remaining magnetization is called the remanent magnetization M_R:

$$B_R = \mu_0 M_R.$$

The term "remanence" is used to describe the value of either the remaining flux density or magnetization when the field has been removed after the magnetic material has been magnetized to *saturation*. The remanence, therefore, becomes the upper limit for all remanent flux densities or magnetizations, respectively.

14.4.1.5 Coercivity

The remanent flux density can be reduced to zero by applying a reverse magnetic field of strength H_c. The magnetic field strength that is required for removing remanence (in terms of *flux density*) is called "coercivity." The magnetic field strength needed to reduce the remanent flux density from an arbitrary level to zero is denoted as the coercive field or coercive force. The "intrinsic coercivity," denoted as H_{ci}, is defined as the magnetic field strength at which the remanence (in terms of *magnetization*) is reduced to zero.

In soft magnetic materials, H_c and H_{ci} are so close in value that usually no distinction is made. However, in hard magnetic materials there is a clear difference between them, with H_{ci} always being larger than H_c. In addition, these quantities are strongly dependent on the condition of the sample, being affected by such factors as heat treatment or deformation.

14.4.1.6 Differential Permeability

As a sidenote, the permeability μ is not a particularly useful parameter to characterize ferromagnets as may be realized from the hysteresis loop: Almost any value of μ can be obtained, including $\mu = \infty$ at the remanence ($B = B_R$, $H = 0$), and $\mu = 0$ at the coercivity ($B = 0$, $H = H_c$).

The differential permeability $\mu' = dB/dH$ is a more useful quantity, although it is, in general, also field dependent. The maximum differential permeability μ'_{max}, which usually occurs at the coercivity, and the initial differential permeability μ'_{max}, which represents the slope of the initial magnetic flux density at the origin, are more useful since they are related to other material properties such as the number and strength of pinning sites.

14.4.1.7 Curie Temperature

All ferromagnets when heated to sufficiently high temperatures become paramagnetic. The transition temperature or critical temperature from ferromagnetic to paramagnetic behavior is called the Curie temperature T_C. At this temperature, the permeability of the material drops suddenly and both coercivity and remanence become zero. This property of ferromagnets was known long before the work of Curie (Table 14.5).

14.4.1.8 Classification of Ferromagnetic Materials

A simple classification of ferromagnetic materials is based on their coercivity as it is a structure-sensitive magnetic property in contrast to, for example, saturation magnetization. Historically, it was observed that iron and steel specimens that were mechanically hard also had high coercivity, while those that were soft had low coercivity. Therefore, the terms "hard" and "soft" were used to distinguish ferromagnets on the basis of their coercivity. Generally, hard magnetic

Table 14.5 Curie temperatures of various materials.

Material	Curie temperature T_C(°C)
Iron	770
Nickel	358
Cobalt	1130
Gadolinium	20
$Nd_2Fe_{14}B$	312
Alnico	850
$SmCo_5$	720
Hard ferrites	400–700
Barium ferrite	450

Table 14.6 Ferromagnetic materials: properties and application (mr: magnetic data recording; sm: soft magnetic material; hm: hard magnetic material).

Material	Structure	T_C	Permeability/μ	Coercivity (A/m)	Appl.
Metglas	Amorphous	395	7 000 000	0.3	sm
Supermalloy[a]	ccp	673	100 000	0.4	sm
Alnico[b]	[c]	>800	8	70 000	hm
$SmCo_5$	$CaCu_5$	700–850	1	680 000	hm
Sm_2Co_{17}	Th_2Zn_{17}	700–850	1	720 000	hm
$Nd_2Fe_{14}B$	$Nd_2Fe_{14}B$	585	1	2 000 000	hm
CrO_2	Rutil	192		50 000	mr

a) Typical composition (mass%): Ni (79), Fe (15.5), Mo (5.0), Mn (0.5).
b) Alnico 2 (mass%): Ni (18–21), Al (8–10), Co (17–20), Cu (2–4), Nb (0–1), Fe (rest).
c) Heterogeneous system: ferromagnetic segregations (Fe/Co) in a (Ni/Al) matrix.

materials are defined as materials with coercivities larger than 10 kA/m (125 Oe), while "soft" magnetic materials are those with coercivities below 1 kA/m (12.5 Oe). The properties of some ferromagneitic materials are given in Table 14.6.

14.4.1.9 Permanent Magnets

Permanent magnets are one of the three most important classes of magnetic materials, the others being electrical steels and magnetic recording media. Permanent magnets find applications in electrical motors and generators, loudspeakers, television tubes, moving-coil meters, magnetic suspension devices, and clamps.

The application determines the choice of the magnetic material based on its hysteresis characteristics. The properties of these materials are usually represented by the "demagnetization curve," which is the part of the hysteresis loop in the second quadrant, where the magnetization is reduced from remanence to zero. The demagnetization curves for some permanent magnet materials depend as much on the metallurgical treatment and processing of the material as on its chemical composition.

In recent years, a permanent magnet material based on neodymium–iron–boron has been discovered. This material has superior magnetic properties for many applications when compared with its predecessor samarium–cobalt. For example, its coercivity can be as high as 1.12×10^6 A/m (14 000 Oe) compared to 0.72×10^6 A/m (9000 Oe) for samarium–cobalt.

In addition to the coercivity, another parameter of prime importance to permanent magnet users is the maximum energy product BH_{max}. It is determined by the maximum value of the product $|BH|$ in the second, or demagnetizing, quadrant of the hysteresis loop. It represents the magnetic energy stored in a permanent magnet material. The maximum energy product by itself, however, does not give sufficient information for permanent magnet users to decide on the suitability of a material for a particular application, but it is one parameter that is widely quoted when comparing various permanent magnet materials. For many years, the maximum energy product was about 50×10^3 J m^{-3} (a few MG Oe). The development of samarium–cobalt permanent magnets raised this value to about 160×10^3 J/m^3 (20 MG Oe), and in the neodyme–iron–boron magnets energy products of typically 320×10^3 J/m^3 (40 MG Oe) have been achieved.

In most applications, the stability of the permanent magnet is of highest importance, and the material must be thus operated sufficiently far below its Curie temperature since the spontaneous magnetization decreases rapidly with temperatures above 75% of T_C. This is one of the main issues that arose with neodymium–iron–boron magnets for higher temperature applications.

14.4.2
Paramagnetism and Diamagnetism

Paramagnets do not find nearly as many applications as ferromagnets. The description of paramagnetism is, however, of essential importance in the understanding of magnetism. Paramagnetism is a simpler phenomenon than ferromagnetism, and quite reasonable theories have been developed, which are in good agreement with experimental observations.

Diamagnets generally do not find many applications that depend on their magnetic properties either, except for the special case of the superconductors, which are perfect diamagnets with $\chi = -1$.

14.4.2.1 Paramagnets

The study of paramagnetism allows for the investigation of the magnetic moments of almost isolated atoms, since unlike ferromagnetism, paramagnetism is not a cooperative phenomenon. Solid-state chemists and physicists are, therefore, more familiar with the underlying theories of paramagnetism such as the temperature dependence of paramagnetic susceptibility, and its description using the classical expression, the Langevin function [28], or its quantum mechanical analog, the Brillouin function [29]. Materials exhibiting paramagnetism are usually atoms and molecules characterized by unpaired electrons yielding a net spin, giving rise to a net magnetic moment. These include atoms and ions with partially filled inner shells, such as transition metals, but also the elemental aluminum and β-tin as well as O_2.

Examples of paramagnetic materials are elemental platinum, aluminum, oxygen, and various salts of the transition metals such as chlorides, sulfates and carbonates of manganese, chromium, iron, and copper, in which the paramagnetic moments reside on the Cr^{3+}, Mn^{2+}, Fe^{2+}, and Cu^{2+} center, respectively. These salts obey at higher temperatures the Curie law, which states that the molar magnetic susceptibility χ_m is inversely proportional to the temperature T, because the magnetic moments are localized on (almost) isolated metal ions.

Salts and oxides of rare earth (lanthanide) elements are strongly paramagnetic at ambience temperature. In these solids, the magnetic properties are determined by highly localized 4f electrons. These are closely bound to the nucleus, and are effectively shielded by the outer electrons from the crystal field. Rare earth metals are also paramagnetic for the same reason. However, if the temperature is decreased, many of the lanthanide compounds exhibit ordered states such as ferromagnetism.

All ferromagnetic metals such as cobalt, iron, and nickel become paramagnetic above their Curie temperatures, as do the antiferromagnetic metals chromium and manganese above their transition temperatures of 35 and −173 °C, respectively. Paramagnetic metals that do not exhibit a ferromagnetic state include all alkali metals (sodium series), and the alkaline earth metals (calcium series) with the exception of beryllium. The 3d, 4d, and 5d transition metals are all paramagnetic with the exception of copper, zinc, silver, cadmium, gold, and mercury that are diamagnetic.

14.4.2.2 Temperature Dependence of Paramagnetic Susceptibility

Only in the limiting case of pure spin paramagnets, the magnetic susceptibility is inversely proportional to temperature (see Box 14.4). This dependence is known

---

Box 14.4

Estimated Magnetic Behavior of d Ions in Octahedral Crystal Fields

Only in the case of d^5 centers (first row) the magnetic behavior of the complex can be described by the Curie law. In all other cases, the quantum mechanical effects acting on the metal center yield more complex temperature dependencies.

System	Order of energetic effects	N [a]	Ground state	$\chi_m(T)$ [b]
$3d^N$	$H_{ee} \approx H_{lf} > H_{so}$	5(hs)	$^6A_1$	C/T
	$H_{lf} > H_{ee} > H_{so}$	1, 2, 6 (hs), 7 (hs)	T	$f(T)$
		4 (hs), 9	E	$C'/T + \chi_0$
		3, 8	A_2	$C'/T + \chi_0$
		6 (ls)	$^1A_1$	χ_0
$4d^N$	$H_{lf} > H_{ee} > H_{so}$			$f(T)$
$5d^N$	$H_{lf} > H_{ee} \approx H_{so}$			$f(T)$

a) hs and ls denote high-spin and low-spin configuration, respectively.
b) $\chi_m = f(T)$ stands for complex dependency of χ_m on T; C' is a constant that deviates more or less from C; χ_0 is a temperature-independent parameter.

as the *Curie law* $\chi = C/T$, where T is the temperature in Kelvin and C is the Curie constant. This law is obeyed by half-filled subshell systems as Mn^{2+}(hs), Fe^{3+}(hs), Eu^{2+}, or Gd^{3+} centers where crystal field effects are negligible. Note, however, that the relation of susceptibility and temperature is usually more complex for many compounds. Additionally, there are paramagnets whose susceptibility is (almost) independent of temperature. Two theories have evolved to deal with these two types of paramagnetism: the localized magnetic moment model, which leads to the Curie law, and the conduction band electron model according to Pauli, which leads to (an almost) temperature-independent and rather weaker susceptibility [29]. The dependence of the inverse susceptibility on temperature of some selected paramagnetic solids is shown in Figure 14.15.

If the magnetic susceptibilities of spin-paramagnets, recorded at low applied magnetic fields, cannot be explained by the Curie law, and if corresponding structural reasons are given, cooperative magnetic phenomena such as ferro-, antiferro-, and ferrimagnetism at low temperatures should be taken into consideration. For such systems, the *Curie–Weiss law* (see Eq. (14.6)) can be applied for the interpretation of the magnetic behavior at sufficiently high temperatures. The sign of θ indicates the nature of the ordering that dominates below a sufficiently low temperature (see Figure 14.16). Positive values of θ are typical for substances that order ferromagnetically below the Curie temperature T_C, that is, that adopt magnetic ordering with parallel-oriented magnetic moments. Negative values of θ occur for compounds with antiferromagnetic and ferrimagnetic ordering below a distinct critical temperature (Néel temperature T_N and Curie temperature T_C, respectively).

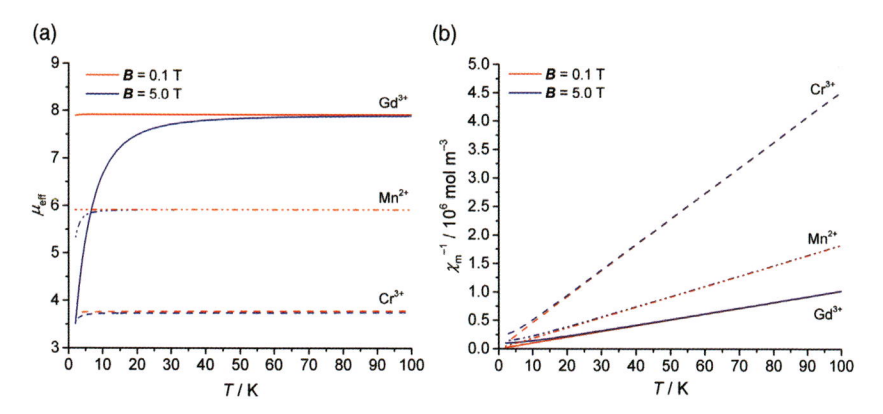

Figure 14.15 $\mu_{eff} - T$ (a) and $\chi_m^{-1} - T$ diagrams (b) for (dashed line) Cr^{3+} (d^3), (dashed–dotted line) Mn^{2+} (d^5), and (solid line) Gd^{3+} (f^7), calculated with spectroscopically determined data and different applied magnetic fields. All compounds exhibit Curie behavior and the deviations in the low-temperature region are caused by the applied magnetic field $B = 0.1, 35.0$ T.

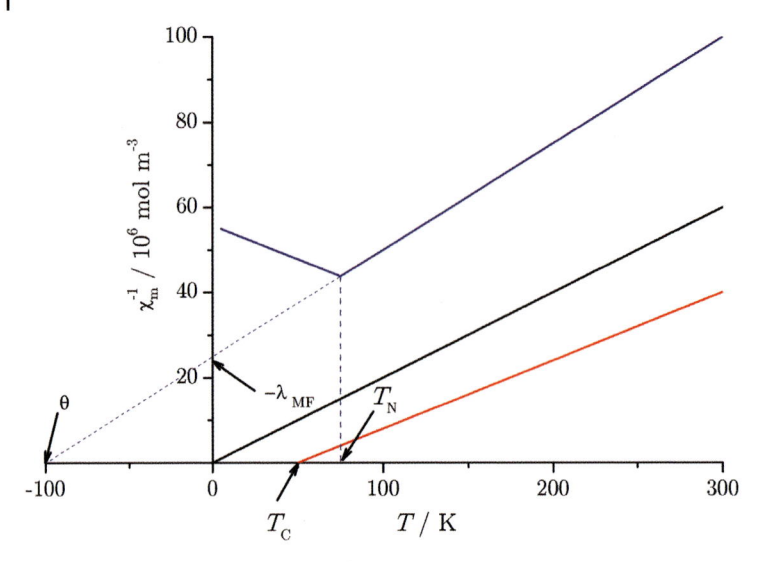

Figure 14.16 Inverse susceptibility χ_m^{-1} as a function of T for Curie law (blue solid curve), antiferromagnetic spin–spin coupling (black solid curve) below T_N, and ferromagnetic coupling (red solid curve) below T_C.

14.4.2.3 Applications of Paramagnets

There are very few applications of paramagnetic materials on account of their magnetic properties. They are primarily object to scientific studies of magnetism. They allow electronic properties studies of materials with net atomic magnetic moments in the absence of strong cooperative effects. Eventually, these studies will help in understanding the phenomenon of ferromagnetism.

One other application is cooling to very low temperatures. A paramagnetic salt is isothermally magnetized, and then cooled to as low a temperature as possible by conventional cryogenic means, for example, by using liquid helium at reduced pressure. Afterward, it is thermally isolated and adiabatically demagnetized whereupon the temperature drops further. By this procedure, temperatures down to the millikelvin scale can be achieved.

14.4.2.4 Diamagnets

Compounds without permanent atomic magnetic moments are incapable of exhibiting paramagnetism or ferromagnetism. However, each compound exhibits in the presence of an external magnetic field a contribution to the magnetic susceptibility that forces the material out of the field. This diamagnetic contribution may be classically explained by arising from the change in orbital angular momentum due to the applied field which is, however, quantum mechanically incorrect. The quantum mechanical treatment of magnetism shows that the diamagnetic susceptibility per mole of an atom is given by Eq. (14.2). For closed-shell compounds, the diamagnetic susceptibility is to a very good approximation independent of both T and B. Note that all diamagnetic substances shown $\chi < 0$

and $M < 0$. The application of diamagnetic materials is limited (see, for example, Refs [30,31]).

14.4.2.5 Superconductors

A superconductor is a material that can conduct electricity or transport electrons from one atom to another without electrical resistance. This means no heat, sound, or any other form of energy would be released from the material below its "critical temperature" T_C, the temperature at which the material becomes superconductive. In normal conductors, the electrical resistance decreases as the temperature is lowered but does not disappear completely (see Figure 14.17a). But for superconductors the resistance is truly zero. Unfortunately, most materials must be in an extremely low energy state in order to become superconductive [32]. A second feature of a superconductor is occurrence of the Meissner–Ochsenfeld effect [33], that is, the complete expulsion of all magnetic fields from the interior of the superconductor independent of the chosen path to arrive at superconductivity, in contrast to the ideal conductor, which is path dependent. Such a state represents *perfect* diamagnetism, with volume magnetic susceptibility $\chi = -1$ (see Figure 14.17b) below the critical temperature. A hot topic of scientific research is the development of compounds that become superconductive at higher temperatures.

Superconductors are classified into two different types. *Type I* superconductors consist of basic conductive elements that are used everywhere from electrical wiring to computer microchips. At present, type I superconductors are known to have T_C between 0.000 325 and 7.8 K at standard pressure. Some type I superconductors require enormous pressures in order to reach the superconductive state. *Type II* superconductors are usually composed of metallic components such as copper or lead as well as nonmetallic components as oxides. They reach a superconductive state at (much) higher temperatures compared to type I superconductors. The cause of the dramatic increase of the critical

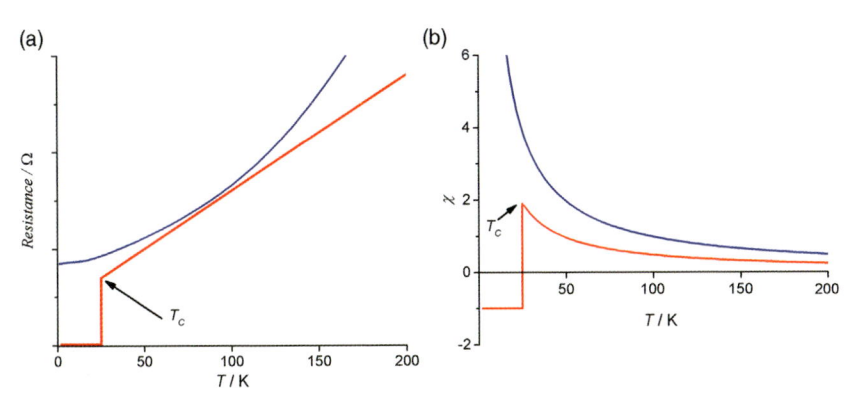

Figure 14.17 Schematic electrical resistance as a function of temperature for a (blue curve) normal conductor and (red curve) superconductive compound (a); volume magnetic susceptibility χ versus T for a (blue curve) paramagnetic and (red curve) a superconductive compound (b).

temperature is not fully understood. The highest T_C reached at standard pressure to date is 135 K or −138 °C by the compound $HgBa_2Ca_2Cu_3O_8$, a cuprate perovskite.

An important application of superconductivity are superconducting magnets, in which the magnitudes of the magnetic field are about 10 times larger than those produced by the best conventional electromagnets. Superconducting magnets are currently used in medical magnetic resonance imaging (MRI) units, which produce high-quality images of internal organs without the need for excessive exposure of patients to X-rays or other harmful radiation. Such magnets are also considered as energy storages.

References

1 Mills, I., Cvitaš, T., Homann, K., Kallay, N., and Kuchitsu, K. (1993) *Quantities, Units and Symbols in Physical Chemistry*, 2nd edn, Blackwell Science, Oxford, Sections 2.3, 7.2, 7.3, 7.4.
2 Blundell, S. (2001) *Magnetism in Condensed Matter*, Oxford University Press, Oxford.
3 Lueken, H. (1999) *Magnetochemie*, Teubner, Stuttgart, Germany.
4 Haberditzl, W. (1964) *Über ein neues Diamagnetismus-Inkrementsystem*, Akademie-Verlag, Berlin.
5 Henry, W.E. (1952) *Phys. Rev.*, **88**, 559–652.
6 Van Vleck, J.H. (1932) *The Theory of Electric and Magnetic Susceptibilities*, Oxford University Press, Oxford.
7 Williams, A.F. (1979) *A Theoretical Approach to Inorganic Chemistry*, Springer, Berlin.
8 Görller-Walrand, C. and Binnemans, K. (1996) Chapter 155, in *Handbook on the Physics and Chemistry of Rare Earths*, vol. 23 (eds K.A. Gschneidner and L. Eyring), Elsevier, Amsterdam.
9 Hellwege, K.-H. (1988) *Einfüuhrung in die Festkörperphysik*, Springer, Berlin.
10 Jørgensen, C.K. (1962) *Absorption Spectra and Chemical Bonding in Complexes*, Pergamon Press, Oxford.
11 Sytsma, J., Murdoch, K.M., Edelstein, N.M., Boatner, L.A., and Abraham, M.M. (1995) *Phys. Rev. B*, **52**, 12668–12676.
12 Schläfer, H.L. and Gliemann, G. (1967) *Einführung in die Ligandenfeldtheorie*, Akadademische Verlagsgesellschaft, Frankfurt (English version: *Basic Priciples in Ligand Field Theory*, Wiley-Interscience, New York, 1969).
13 Mabbs, F.E. and Machin, D.J. (1973) *Magnetism and Transition Metal Complexes*, Chapman and Hall, London.
14 Griffth, J.S. (1971) *The Theory of Transition-Metal Ions*, Cambridge University Press, Cambridge.
15 Schilder, H. and Lueken, H. (2004) *J. Magn. Magn. Mater.*, **281**, 17–26.
16 (a) Urland, W. (1976) *Chem. Phys.*, **14**, 393–401; (b) Urland, W. (1977) *Chem. Phys. Lett.*, **46**, 457–460.
17 Abragam, A. and Bleaney, B. (1970) *Electron Paramagnetic Resonance of Transition Ions*, Clarendon Press, Oxford.
18 Schilder, H., Speldrich, M., Lueken, H., Sutorik, A.C., and Kanatzidis, M.G. (2004) *J. Alloy Compd.*, **374**, 249–252.
19 Speldrich, M., Schilder, H., Lueken, H., and Kögerler, P. (2011) *Isr. J. Chem.*, **51** (2), 215–227.
20 van Leusen, J., Speldrich, M., Schilder, H., and Kögerler, P. (2015) *Coord. Chem. Rev.*, **289**, 137–148.
21 Hatscher, S., Schilder, H., Lueken, H., and Urland, W. (2005) *Pure Appl. Chem.*, **77**, 497–511.
22 Bleaney, B. and Bowers, K.D. (1952) *Proc. R. Soc. Lond. USA*, **A214**, 451–465.
23 Smart, J.S. (1966) *Effective Field Theories of Magnetism*, Saunders, Philadelphia, PA.
24 Racah, G. (1942) *Phys. Rev.*, **62** (9–10), 438.

25 Schäffer, C.E. and Jørgensen, C.K. (1958) *J. Inorg. Nucl. Chem.*, **8**, 143–148.
26 Tchougréeff, A. and Dronskowski, R. (2009) *Int. J. Quantum Chem.*, **109** (11), 2606–2621.
27 Figgis, B.N. (1987) in *Comprehensive Coordination Chemistry*, vol. **1** (eds G. Wilkinson, R.D. Gillard, and J.A. McCleverty), Pergamon, Oxford, p. 213.
28 Brandt, S. and Dahmen, H.D. (2005) *Elektrodynamik: Eine Einführung in Experiment und Theorie*, Springer, Berlin, p. 239.
29 Kittel, C. (2004) *Introduction to Solid State Physics*, 8th ed., John Wiley & Sons, Inc., New York, pp. 303–304.
30 Mulay, L.L. and Boudreaux, E.A. (eds) (1976) *Theory and Applications of Molecular Diamagnetism*, John Wiley & Sons, Inc., New York.
31 Spaldin, N. (2003) *Magnetic Materials, Fundamentals and Device Applications*, Cambridge University Press.
32 Bardeen, J., Cooper, L.N., and Schrieffer, J.R. (1957) *Phys. Rev.*, **108** (5), 1175–1204.
33 Meissner, W. and Ochsenfeld, R. (1933). *Naturwissenschaften*, **21** (44), 787–788.

15
Dielectric Properties

Rainer Waser[1,2] and Susanne Hoffmann-Eifert[1]

[1]*Forschungszentrum Jülich GmbH, Peter Grünberg Institute, Electronic Materials, Wilhelm-Johnen-Straße, 52428 Jülich, Germany*
[2]*RWTH Aachen University, Institut für Werkstoffe des Elektrotechnik II, Sommerfeldstraße 24, 52074 Aachen, Germany*

15.1
Applications of Dielectrics

Dielectric materials are key components in all fields of electronic devices spanning the range from high-voltage insulators for power electronics to high-permittivity dielectric gate oxides in state-of-the-art field effect transistors for next-generation information technology. This broad range of applications emphasizes the strong influence of dielectrics on the evolution of today's information society. The material classes involved range from crystalline and amorphous oxides to organic compounds and polymers. The integration of nanometer dielectric thin films into semiconductor technology is a key for enabling the continuation of size and power scaling of integrated circuits. Even the applications of dielectric materials in the field of information technology are extremely broad. Very low losses and a specific temperature dependence of the dielectric properties are required for new microwave dielectrics for oscillator and filter applications. Ultralow-permittivity dielectrics are used as insulators between the metallization layers on advanced CMOS circuits, whereas high-permittivity dielectrics are being used for the cell capacitors in ultrahigh-density integrated dynamic random access memories (DRAM) as well as for high-permittivity gate dielectrics in various modern field effect transistors (FETs). Even thinner dielectrics are applied as tunneling barriers, for example, for magnetoresistive random access memory (MRAM) devices and in superconducting quantum computing systems. Additional interesting properties arise when the insulators obey an induced or a remanent polarization, which leads to the classes of pyroelectric, piezoelectric, and ferroelectric materials. The basic understanding of the effects arising in the interaction of the electromagnetic field with matter is one of the key issues in understanding the principles of the design and operation of the new electronic as well as photonic devices. Therefore, the present chapter will discuss the basics of the solid-state physics of dielectrics with respect to the properties in the frequency range of electronic devices, for example, megahertz to gigahertz. In particular, we will focus on the solid-state properties caused by microscopic and macroscopic charge inhomogeneity generated by electrical fields. For further details or a broader view of the topic, the reader is referred to comprehensive textbooks on solid-state physics [1–7], electronic materials [8–10], or specifically on dielectric properties [11–14].

15.2
Polarization of Condensed Matter

Before we start with the specific description of dielectric phenomena and their characterization in the range of technical frequencies, some definitions might be useful:

Dielectrics are insulating materials that are used technically because of their property of polarization to modify the dielectric function of the vacuum, for example, to increase the capacity (i.e., the ability to store charge) of capacitors. If a vacuum capacitor with the capacitance C_0 is filled with a dielectric insulator, its capacitance will increase by the amount $C = \varepsilon_r \cdot C_0$. The relative permittivity ε_r (also called the "*k*-value") is a characteristic property that defines the polarizability of the insulating material. Ideal dielectrics are perfect insulators, which means that they do not conduct electricity. In other words, the number of free charge carriers, that is, ionic or electronic, with a significant mean free path is negligible.

Polarization is the separation of positive and negative charge barycenters of bound charges. If this separation is induced by an applied electric field E, it is called dielectric polarization and the respective materials' property is the dielectric permittivity ε_r. There are several mechanisms for the dielectric polarization determined by the polarizable unit, that is, atom, cation–anion pair, orientable permanent dipole, and so on.

Dielectric properties of matter describe the interaction of the polarizable material with an electromagnetic wave propagating through this material. Therefore, dielectrics have to the characterized according to the frequency range. For technical electronic applications, this range yields from millihertz to gigahertz, covering, for example, devices such as decoupling capacitors and storage capacitors up to high-frequency filters for mobile communication. For even higher frequencies the material is characterized by its *optical* properties, such as the refractive index n, determined by free and bound charges.

The chapter will start with a brief review of the macroscopic and microscopic description of dielectric polarization. Subsequently, resonance and relaxation phenomena will be outlined, followed by a discussion of phonons and their interaction with electromagnetic waves.

15.2.1
Dielectrics in an Electrostatic Field

According to the *Poisson equation*, each *free* charge acts as a source for the dielectric displacement D:

$$\text{div } \boldsymbol{D} = \rho_{\text{free}}, \tag{15.1}$$

where ρ_{free} denotes the density of free (conducting) charges. Based on this relation, the overall charge neutrality of matter in an external field is described by

$$\boldsymbol{D} = \varepsilon_0 \boldsymbol{E} + \boldsymbol{P}. \tag{15.2}$$

The term $\varepsilon_0 E$ describes the vacuum contribution to the displacement D caused by an externally applied electric field E with the vacuum permittivity $\varepsilon_0 = 8.854 \times 10^{-12}$ As/Vm. P represents the electrical polarization of the matter in the system. Equation (15.2) is independent of the cause of the polarization. The polarization may exist spontaneously (pyroelectric polarization), it may be generated by mechanical stress (piezoelectric polarization), or it may be induced by an external electric field (dielectric polarization).

Figure 15.1 illustrates a parallel plate capacitor before (cf. Figure 15.1a) and after insertion of material with polarization P. If the insertion is performed at a constant applied voltage (cf. Figure 15.1b), that is, $E =$ constant, additional free charges need to flow into the system to increase D according to Eq. (15.2). If the insertion is performed for constant charges on the plates, that is, $D =$ constant, the electric field E and, hence, the voltage between the plates will decrease according to Eq. (15.2).

The dielectric polarization of a linear dielectric material is given by

$$\boldsymbol{P} = \varepsilon_0 \chi_e \boldsymbol{E}, \tag{15.3}$$

which, in turn, yields a simple linear term for the dielectric displacement

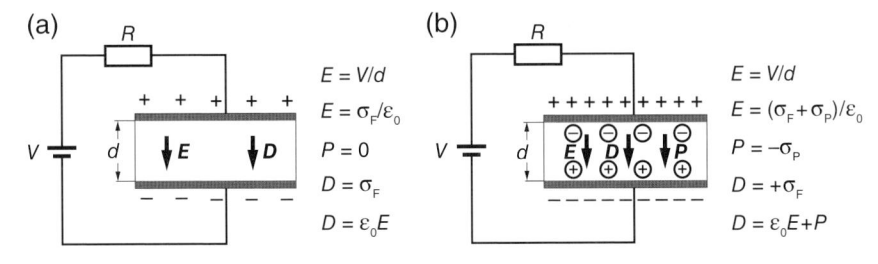

$$\boldsymbol{D} = \varepsilon_0 (1 + \chi_e) \boldsymbol{E} = \varepsilon_0 \varepsilon_r \boldsymbol{E}. \tag{15.4}$$

Figure 15.1 Insertion of a dielectric material with polarization P into a parallel plate capacitor at constant voltage. (a) Vacuum capacitor with free charges "+" and "−" on the plates. (b) Plate capacitor with the dielectric material inside showing the free charges "+" and "−" on the plates as well as the polarization $\oplus$ and $\ominus$ inside the dielectric.

The material properties that are used to describe linear dielectric materials are the electrical susceptibility, χ_e, and the relative dielectric permittivity ε_r. In general, the polarizability of a dielectric material depends on the direction of the applied electrical field and the direction the displacement is measured. Therefore, generally both quantities, χ_e and ε_r, are tensors, that is, direction dependent in all spatial dimensions.

Under high electrical fields, the displacement of the bound charges might not be linear anymore, due to, for example, electrostriction effects. Then, χ_e or ε_r themselves become field-dependent, that is, $\varepsilon_r = \varepsilon_r(E)$. This enables a tuning of the material dielectric properties by an externally applied electric field as, for example, utilized in tunable filters.

Table 15.1 gives an overview of typical values of the dielectric permittivity for selected materials at room temperature.

15.2.2
Electrostatic Field Energy

A capacitor with the capacitance C stores an energy of

$$W = \frac{1}{2} C \cdot V^2 \tag{15.5}$$

if a voltage V is applied.

For a parallel plate capacitor defined by the volume $V = A \cdot d$, the capacitance $C = Q/V$, the voltage $V = E \cdot d$, and the dielectric displacement $D = Q/A$, the energy density w of the dielectric material is given:

$$w = \frac{1}{2} DE = \frac{1}{2} \varepsilon_0 E^2 + \frac{1}{2} PE. \tag{15.6}$$

The first term describes the energy density of the vacuum, while the second term gives the contribution due to the polarization of the dielectric material. For

Table 15.1 Values of the dielectric permittivity of selected materials at room temperature.

Material	ε_r
Vacuum	1
Dry air, 1 bar	1.00055
Water	80.3
Mineral oil	2.2
Quartz glass	3.78
Mica	6.8
Polyethylene	2.3
Titanium oxide	≈100
Capacitor ceramics	≈1000

linear dielectrics that are characterized by the dielectric permittivity, the second term can be written as

$$w = \frac{1}{2}\varepsilon_0\varepsilon_r E^2. \tag{15.7}$$

15.2.3
Dipoles and Polarization

When a neutral entity (Figure 15.2a), for example, a molecule, is put into an electric field, the positive and negative charges of this entity will be separated due to the electrical forces as shown in Figure 15.2b. The charge separation creates a dipole moment p that is proportional to the displacement δs of the charges:

$$\boldsymbol{p} = q \times \delta s. \tag{15.8}$$

The total dipole moment of a material results as the sum of the individual dipole moments, that is, $\boldsymbol{p}_{total} = \sum_i \boldsymbol{p}_i$. With this, the polarization of the material defined as the total dipole moment per volume is given: $\boldsymbol{P} = \boldsymbol{p}_{total}/V$. The change of the individual dipole moment is proportional to the applied electric field:

$$\boldsymbol{p} = \alpha \cdot \boldsymbol{E}, \tag{15.9}$$

with the proportionality factor α defining the polarizability of a single entity. Equation (15.9) holds for independent polarizable species, not for lattices. If the density of dipoles per volume is defined as n_{dipole}, the total polarization is given by

$$\boldsymbol{P} = n_{dipole} \cdot \alpha \cdot \boldsymbol{E}. \tag{15.10}$$

For the more general case of different dipoles (i) contributing to the total polarization, Eq. (15.10) becomes

$$\boldsymbol{P} = \sum_i n_i \cdot \alpha_i \cdot \boldsymbol{E}. \tag{15.11}$$

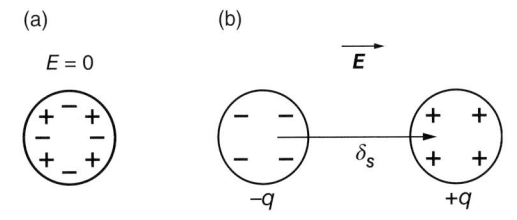

Figure 15.2 Neural entity containing positive and negative charges: (a) with a common barycenter and (b) displaced by δs forming a local dipole $\boldsymbol{p}$.

For a dielectric material of perfectly diluted, independent dipoles, the macroscopic susceptibility is directly related to the microscopic polarizability via

$$\chi = \frac{\sum_i n_i \cdot \alpha}{\varepsilon_0}. \tag{15.12}$$

This is because the field at the position of each dipole equals the macroscopic field inside the dielectric material. The scenario changes for polarizable matter where each dipole is affected by the field of its neighboring dipoles. In this case, the electric field E has to be replaced by the local electric field E_{loc}, which leads to modification of Eq. (15.11).

The energy of a dipole in an electric field is calculated from the energy that has to be brought up in order to separate the positive and negative charges in the electric field. With $F = q \cdot E$, it follows

$$W = \int_0^{\delta s} F\,dx = \int_0^{\delta s} \frac{q^2 x}{\alpha}\,dx = \frac{1}{2}pE. \tag{15.13}$$

Adding the field energy yields the total energy of a permanent dipole in an electric field

$$W_{\text{permanent}} = -pE. \tag{15.14}$$

For the case of an induced dipole, additional energy is required to separate the positive and negative charges. Therefore, this energy is lower:

$$W_{\text{induced}} = -\frac{1}{2}pE. \tag{15.15}$$

15.3
Mechanisms of Polarization

In order to describe macroscopic and microscopic polarization mechanisms in ensembles that might contain induced as well as permanent dipoles, one has to discuss the different physical mechanisms of polarization in the solid state. In general, four different types of polarization can be distinguished. Details are described in textbooks such as Ref. [3].

15.3.1
Electronic Polarization

The electronic polarization describes the displacement of the negatively charged electron shell against the positively charged nucleus. Figure 15.3 shows the simple model of a spherical atom consisting of a core with the positive charge ($+Q$) and of a negatively charged electron shell. For the model, the negative charge ($-Q$) is homogeneously distributed within a sphere of radius R.

15.3 Mechanisms of Polarization

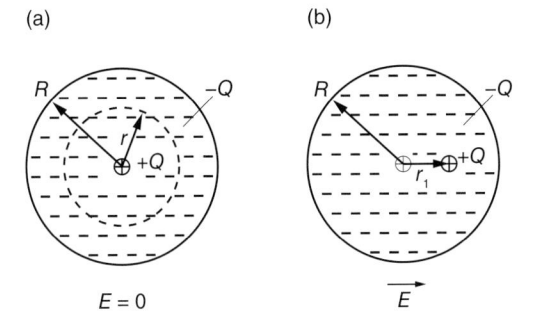

(a)   (b)

Figure 15.3 Model of a spherical, nondeformable atom with the charge +Q in the core and −Q in the electron shell of radius R. (a) Field-free case with common barycenters. (b) Separation of the positive and negative barycenters due to an applied electric field.

For this scenario, the electronic space charge density reads

$$\rho = -\frac{Q}{4/3 \times \pi \times R^3}. \tag{15.16}$$

Without an external applied electric field, the barycenters of the positive charge of the nucleus and of the negative charge of the electron shell fall into identical positions. However, in an electric field E, the negatively charged electron shell (here assumed to be stiff) is displaced against the positively charged nucleus by the distance r_1 (see Figure 15.3b). Here the nucleus is considered as a point charge, whereas the field of the electron shell derives from

$$\text{div } \boldsymbol{D} = \text{div } \varepsilon_0 \boldsymbol{E} = \rho. \tag{15.17}$$

Due to the system's symmetry, the electric field of the shell is fully characterized by its radial component E_r, which is calculated from the volume integral for the dashed sphere of radius r in Figure 15.3a:

$$\varepsilon_0 \int \text{div } \boldsymbol{E} dV = \int \rho dV = -\frac{Qr^3}{R^3}. \tag{15.18}$$

Rearrangement of the left-hand side of Eq. (15.18) by means of Gauss' law yields

$$\varepsilon_0 \oint E_r dA = 4\varepsilon_0 \pi r^2 E_r = -\frac{Qr^3}{R^3}. \tag{15.19}$$

Thus, the electric field inside the sphere of the electron shell reads:

$$E_r = -\frac{Qr}{4\pi\varepsilon_0 R^3}. \tag{15.20}$$

In the center of the sphere, that is, at $r=0$, the electric field is zero. This is the position of the positively charged nucleus in field-free equilibrium. If now an external electric field E is superimposed, the position r_1, where the total electric field inside the sphere adds to zero follows from $E_{tot} = E + E_r(r_1) = 0$. Insertion into Eq. (15.20) yields

$$Q r_1 = 4\pi \varepsilon_0 R^3 E. \tag{15.21}$$

This equation defines the equilibrium position of the positive nucleus charge under application of an external electric field. Even for high electric fields, the relative displacement r_1/R is small ($<10^{-5}$), justifying the assumptions of the model. This displacement of the positive nucleus against the barycenter of the electron shell induces a microscopic dipole moment $p = Q/r_1$ that is proportional to the strength of the applied electric field:

$$p = \alpha_{el} E, \tag{15.22}$$

with the electronic polarizability

$$\alpha_{el} = 4\pi \varepsilon_0 R^3. \tag{15.23}$$

This very simple model is quite rough and therefore Eq. (15.23) should only be taken as an approximation. However, two important properties of the electronic polarizability already become visible. First, α_{el} is approximately proportional to the volume of the electron shell. Therefore, bigger atoms give a larger contribution to the total polarization of a material. Second, the temperature dependence of α_{el}, that is, $TK_{\alpha,el}$, can generally be neglected since the atomic radius is nearly temperature-independent up to the ionization temperature.

If the electric field is switched off, the positively charged nucleus will be pushed back into its equilibrium position by the force $|F| = |Q \cdot E|$. The undamped vibration of the negative electron shell with the mass m_{shell} against the positive core is described in terms of a forced vibration:

$$m_{shell} \frac{d^2 r}{dt^2} - Q E_r = 0. \tag{15.24}$$

The differential equation can be solved by the Ansatz $r = r_1 \cdot \exp(j\omega t)$ resulting in the resonance frequency of the electronic polarization:

$$\omega_{el}^2 = \frac{Q^2}{m_{shell} \, \alpha_{el}}. \tag{15.25}$$

The values of the electronic resonance frequency are typically in the ultraviolet regime, that is, at about 10^{15} Hz. Corresponding phenomena are described by the topic of *Optics*.

15.3.2
Ionic Polarization

Ionic polarization is observed in materials with ionic bonds, but also in crystals with predominantly covalent bonding if there is more than one atom in the unit

15.3 Mechanisms of Polarization

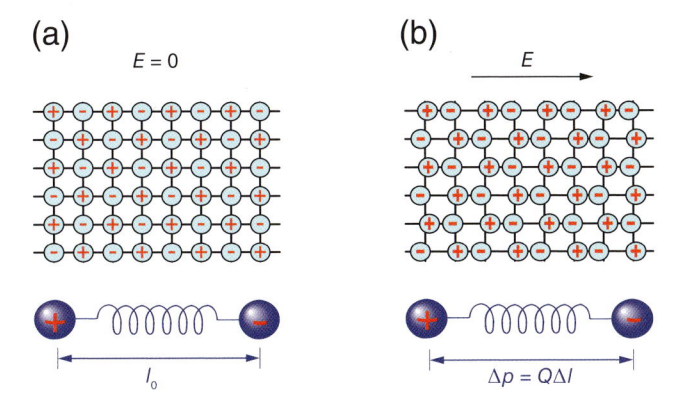

Figure 15.4 Ionic lattice (a) without, $E=0$, and (b) with, $E \neq 0$, an applied electric field. The lower pictures show isolated pairs of ions, respectively.

cell. Ionic polarization describes the mutual displacement of the positive and negative sublattices under the influence of an applied electric field. The principle is schematically shown in Figure 15.4.

In a simple approach for describing the response of an ionic lattice consisting of positively and negatively charged ions on an electrical field, the complex binding interactions might be represented by springs. The distance l_0 of positively and negatively charged ions in equilibrium is described by a relaxed spring with the spring constant k, which summarizes the complex binding conditions in the particular lattice (see Figure 15.4a). With this definition, we can calculate the elongation Δl of a particular lattice as the response to an external applied electric field E from the equivalence of the Coulomb force and the spring force:

$$Q \cdot E = k \cdot \Delta l. \tag{15.26}$$

The change in the dipole moment $\Delta p = Q \cdot \Delta l$ is as follows:

$$\Delta p = \frac{Q^2 E}{k}. \tag{15.27}$$

From this, the ionic polarizability of an isolated dipole (see Figure 15.4b, lower graph) is derived as

$$\alpha_{\text{ion}} = \frac{Q^2}{k}. \tag{15.28}$$

An accurate model of the lattice vibrations requires a detailed description of the coupling coefficients in the different directions of the crystal lattice. This is dealt with in the chapter on phonon spectroscopy. The resonance frequency of the ionic vibrations is typically in the infrared regime between 10^{12} and 10^{13} Hz. In general, the temperature coefficient of the ionic polarizability $TK_{\alpha,\text{ion}}$ is weakly positive:

$$TK_{\alpha,\text{ion}} \approx +10^{-4} \cdots + 10^{-3} \text{ K}^{-1}. \tag{15.29}$$

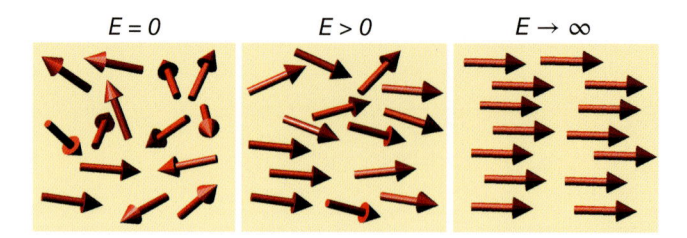

Figure 15.5 Permanent dipoles symbolized by arrows: complete thermodynamic disorder in the field free situation ($E=0$), partial ordering in a moderate electric field ($E>0$), and complete ordering in an "infinite high" electric field ($E \to \infty$).

This is because of the thermal expansion of the lattice and the slight decrease of the spring constant with increasing temperature.

15.3.3
Orientation Polarization

Many substances contain molecules – either regular constituents or impurities – that carry a permanent electric dipole moment. If these dipoles are able to reorient themselves by motion or by rotation, their contribution to the dielectric polarization is defined as the *orientation polarization*. At ambient temperatures, usually all dipole moments are mutually compensated because of thermodynamic orientation disorder. In contrast an electric field causes a preferential orientation of the dipoles, while the thermal movement of the atoms or molecules perturbs the alignment. This is illustrated in Figure 15.5.

The energy of a permanent dipole in an electric field is given as

$$W = -\boldsymbol{pE} = -pE \cos \Theta, \tag{15.30}$$

where Θ is the angle between p and E, as is shown in Figure 15.6.

The average degree of orientation, described by the mean value of $\cos\Theta$, that is, $\langle\cos\Theta\rangle$, is a function of the applied field E and the temperature T. Therefore, the total polarization of an entity of the volume V with a density n_{or} of permanent dipoles is calculated from

$$P = n_{or} V \cdot p \langle \cos \Theta \rangle. \tag{15.31}$$

Every dipole contributes by $p\langle\cos\Theta\rangle$ to the total polarization.

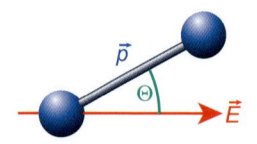

Figure 15.6 Permanent dipole in an electric field.

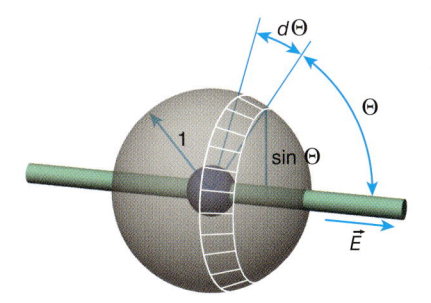

Figure 15.7 Calculation of the density of states of dipoles pointing toward the shaded area in field-free case.

The mean value is calculated according to

$$\langle \cos \Theta \rangle = \frac{\int_0^\pi n_{or}(\Theta) \cos \Theta \, d\Theta}{\int_0^\pi n_{or}(\Theta) \, d\Theta}. \tag{15.32}$$

For the calculation of the mean value of dipoles pointing into a certain solid angle Θ, we will first discuss the field-free situation according to the picture shown in Figure 15.7.

The density of states $z(\Theta)d\Theta$ of the dipoles that point toward the shaded area on the sphere is given as

$$z(\Theta)d\Theta \sim 2\pi \sin \Theta \, d\Theta. \tag{15.33}$$

Under the influence of an electric field E, the energy of the dipoles (see Eq. (15.30)) changes as a function on the direction. The Boltzmann distribution $f(\Theta) \sim \exp(-W(\Theta)/k_B T)$ becomes

$$f(\Theta) \sim \exp \frac{pE \cos \Theta}{k_B T}, \tag{15.34}$$

where k_B denotes the Boltzmann constant and T the absolute temperature measured in Kelvin.

The density of occupied states follows as

$$n_{or}(\Theta)d\Theta = z(\Theta) \cdot f(\Theta)d\Theta \sim 2\pi \sin \Theta \cdot \exp\left(\frac{pE \cos \Theta}{k_B T}\right) d\Theta. \tag{15.35}$$

With this, the mean value $\langle \cos \Theta \rangle$ and therefore the total polarization P can be calculated. The solution of the integral in Eq. (15.32) gives

$$\langle \cos \Theta \rangle = \coth \frac{pE}{k_B T} - \frac{k_B T}{pE} \equiv L\left(\frac{pE}{k_B T}\right). \tag{15.36}$$

$L(x) = (\coth x - 1/x)$ is the so-called *Langevin function* [6], which is plotted in Figure 15.8. It can be approximated by the following expressions:

$$L(x) \approx \frac{x}{3}, \quad \text{for} \quad x \ll 1, \quad \text{and} \quad L(x) \approx 1 - \frac{1}{x}, \quad \text{for} \quad x \gg 1.$$

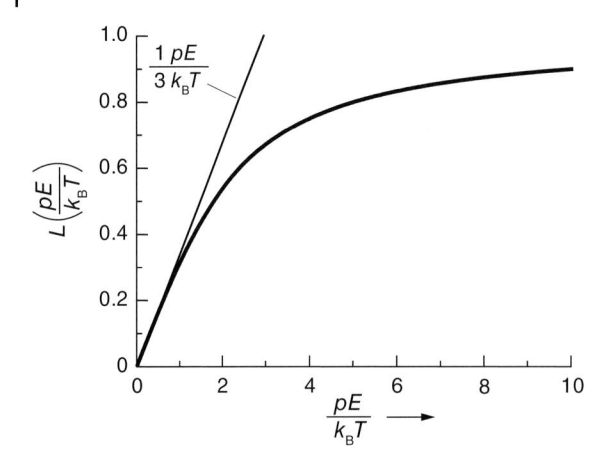

Figure 15.8 Plot of the Langevin function including the approximation for low electric fields.

For all technically applicable cases, the polarization is far from saturation and is proportional to the applied field:

$$P \sim \langle \cos \Theta \rangle \approx \frac{pE}{3k_B T}, \quad \text{for} \quad \frac{pE}{k_B T} \ll 1. \tag{15.37}$$

In other words, the disorder of the dipoles due to thermal vibrations is only slightly affected by the electric field. For these conditions, the average polarizability originating from permanent dipole moments p is given as follows:

$$\langle \alpha_{or} \rangle \simeq \frac{p^2}{3 k_B T}. \tag{15.38}$$

Noticeable is the strong temperature dependence ($\sim T^{-1}$), which is one of the main characteristics of the orientation polarization. This temperature dependence is basically used to differentiate between contributions of orientation polarization and those of ionic and electronic polarization. Another consequence of the thermal disorder is a certain probability of collisions of the dipoles that delays the orientation after application of an electric field. Consequently, the establishing of the equilibrium condition defined by $\langle \cos\Theta \rangle$ is retarded, and the degree of retardation is characterized by the relaxation time τ_{or}. In alternating electric fields of high frequencies, $\omega\tau \gg 1$, the permanent dipoles are unable to respond to the changes of the electric field and the contribution from α_{or} becomes negligible.

In contrast to the free orientation of dipoles in gases, the rotation of permanent dipoles in solids is limited by the number of available regular lattice sites. The simplest case is described by two possible positions where the dipole points toward $+x$ or $-x$ direction. Here the total density of dipoles is simply the sum of the two entities: $n_{or} = n_+ + n_-$. For the field-free case ($E = 0$), equal quantities of dipoles point toward each direction: $\Delta n_{or} = n_+ - n_-$. If now an electric field is applied in +x direction, the density n_+ of the dipoles in the direction of the field increases, whereas n_- decreases accordingly. With

$$n_+ = \frac{n_{\text{or}}}{2} \exp\left(\frac{pE}{k_B T}\right) \quad \text{and} \quad n_- = \frac{n_{\text{or}}}{2} \exp\left(-\frac{pE}{k_B T}\right), \tag{15.39}$$

the amount of dipoles contributing to the net polarization is calculated as follows:

$$\frac{\Delta n_{\text{or}}}{n_{\text{or}}} = \tanh\frac{pE}{kT} \approx \frac{pE}{kT}, \quad \text{for} \quad \frac{pE}{k_B T} \ll 1. \tag{15.40}$$

Therefore, for the specific case of a solid with permanent dipole moments p that can only *orient along a defined axis*, the mean polarizability is

$$\langle \alpha_{\text{or,1-axis}} \rangle \simeq \frac{p^2}{k_B T}. \tag{15.41}$$

As a consequence of the limitation in the number of possible orientations, this value is by a factor of 3 larger compared to the purely random orientation (compare Eq. (15.38)). Apart from the different prefactors, the temperature dependence is still the same.

15.3.4
Maxwell–Wagner Polarization

Space charge polarization describes a polarization effect in a dielectric material that shows spatial inhomogeneity of charge carrier densities. Space charge polarization effects are not only of importance in semiconductor field-effect devices (see Ref. [7]) and composite materials from metallic nanoparticles enclosed in polymers or glass matrices but they also occur in inhomogeneous ceramic materials. A technically important class of ceramics with electrically conducting grains and insulating grain boundaries are the so-called varistors with extremely high polarizability. The different regimes, generally built from electrical (slightly) conducting grains, representing the bulk phase, "b," and insulating grain boundaries, "gb," lead to the so-called *Maxwell–Wagner polarization* (see Ref. [9]). A good approximation of the polycrystalline material is given by the brick wall model, which is depicted in Figure 15.9. Utilizing the dimensions shown in the

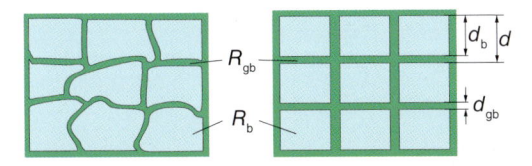

Figure 15.9 Microstructure of a polycrystalline ceramic material. R_b is the mean resistance of the slightly conducting grains, and R_{gb} is the value for the insulating grain boundaries. In the simple brick-wall model, the grains are approximated by cubes of equal size (d_b) and distance (d_{gb}).

figure, the polarizability of a certain volume V_c of a ceramic material is approximated by

$$\alpha_{sc} \cong \varepsilon_0 \varepsilon_r \frac{d}{d_{gb}} V_c, \quad \text{with} \quad d = d_b + d_{gb}. \tag{15.42}$$

15.3.5
Material Classification by the Polarization Mechanism

Technically relevant materials for electronic devices can be classified by the dominating polarization mechanism. The class of materials with *sole electronic polarization* comprises covalently bonded materials such as carbon, silicon, and germanium, that is, materials with sp³ hybridization, as well as nonpolar plastic material such as polyethylene (PE), polytetrafluoroethylene (PTFE), and polystyrene (PS). In polyethylene, for example, the four C—H dipoles of each C-atom perfectly compensate each other resulting in a total dipole moment of zero. Materials with *ionic bonds* but without permanent dipoles, such as sodium chloride (NaCl) as well as oxides and sulfides, typically exhibit electronic and ionic polarization. *Polar liquids*, such as water (H_2O) or dichloromethane (CH_2Cl_2), as well as polar long-chain polymers such as polyvinyl chloride (PVC) and polyester, exhibit permanent dipole moments and therefore show orientation polarization in addition to electronic and ionic polarization. Table 15.2 gives a compact summary of different polarization mechanisms. One important characteristic is the type of the dipole. Electronic and ionic polarization are based on induced dipoles, whereas in case of the orientation polarization, permanent dipoles are oriented in an electric field. The energy of the specific dipole in the electric field is twice as much for the permanent dipole compared to the induced dipole. All polarization mechanisms contribute to the dispersion, that is, frequency dependence, of the dielectric properties of materials in a characteristic manner that is sketched in Figure 15.10. The frequency dependence will be discussed in more detail in the subsequent chapters.

Table 15.2 Mechanisms of polarization.

Polarization mechanism	Type of the dipole	Energy in the field	Polarizability α	Temperature coefficient TK	Cutoff mechanism
Electronic (el)	Induced	$-1/2\,pE$	$4\pi\varepsilon_0 R^3$	≈ 0	Resonance in the ultraviolet (UV)
Ionic (ion)	Induced	$-1/2pE$	Q^2/k	$\approx 10^{-4}\,K^{-1}$	Resonance in the infrared (IR)
Orientation (or)	Permanent	$-pE$	$p^2/3kT$	$\propto T^{-1}$	Relaxation

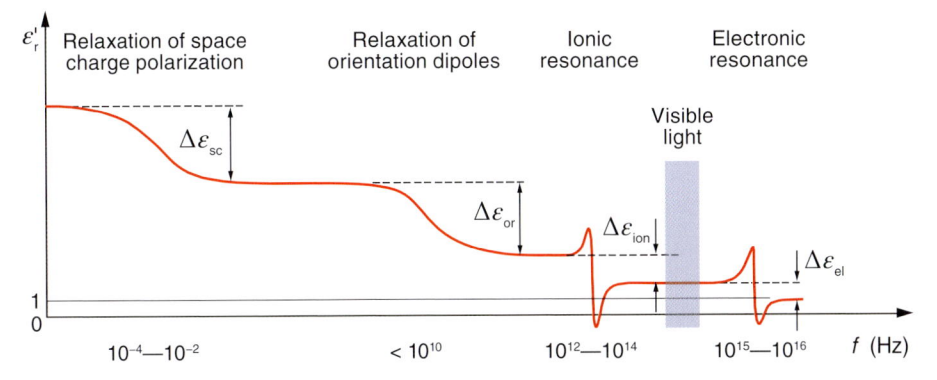

Figure 15.10 Cutoff characteristics of different polarization mechanisms as a function of the frequency of the applied alternating electric field.

15.4
Electric Fields in Matter

15.4.1
Macroscopic Description

The macroscopic electric field E_m inside a dielectric material, which is inserted into an external applied electric field E_a, follows directly from Maxwell's theory. For an easy understanding, we consider different forms of a dielectric inserted into a parallel plate capacitor in a way that the space between the plates is not completely filled by the dielectric. From Figure 15.11 we can derive the following statements: The external applied field E_a originates from the charges on the capacitor's plates and equals the field in vacuum. The applied field causes a polarization inside the dielectric. The polarization charges at the surface of the

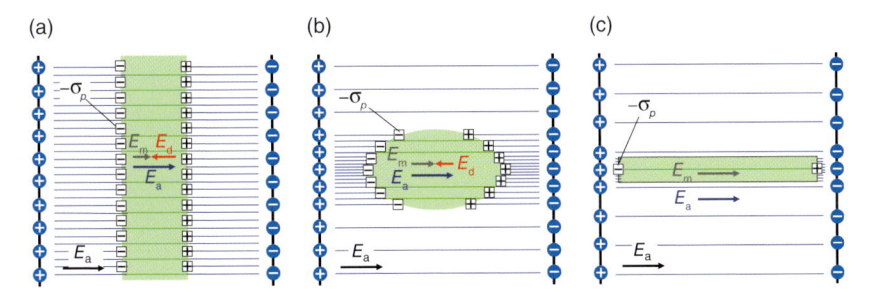

Figure 15.11 Electric fields E_m inside dielectric bodies of different shapes, which are placed inside a homogeneous electric field E_a of a parallel plate capacitor. E_d defines the depolarizing field caused by uncompensated polarization charges on the surface of the dielectric. The value of the depolarizing field depends strongly on the shape of the dielectric body. The pictures show the cases of (a) a disk, (b) an equipotential ellipsoid, and (c) a bar.

dielectric material induce a depolarizing field E_d in the dielectric, which is opposite to the applied field. In total, the electric field inside the dielectric material follows from the superposition:

$$E_m = E_a + E_d. \tag{15.43}$$

Intuitively, Figure 15.11a–c shows that the depolarizing field strongly depends on the shape of the dielectric body.

The depolarizing field is higher for bodies with a larger surface area oriented perpendicular to the lines of the electric flux. This shape anisotropy is expressed by the depolarizing factor or shape factor $\mathcal{N}$:

$$E_m = E_a - \mathcal{N}\frac{P}{\varepsilon_0}. \tag{15.44}$$

Calculation yields characteristic values of the shape factor for certain bodies, that is, $\mathcal{N}=1$ for an infinitely thin disk (cf. Figure 15.11a), $\mathcal{N}=1/3$ for a sphere, and $\mathcal{N}=0$ for an infinitely slim and long bar (cf. Figure 15.11c). This follows directly from the requirement of Maxwell's theory that the tangential component of the electric field is uniformly continuous.

15.4.2
Microscopic Description and the Lorentz Field

The microscopic description of polarization in a dielectric material performed so far utilized the polarizability α of a microscopic dipole induced by a local electric field E_{loc} at the site of the dipole:

$$p = \alpha \cdot E_{loc}. \tag{15.45}$$

In the absence of any interaction between the polarized particles, that is, in diluted matter, the local electric field is identical to the applied electric field, that is, $E_{loc} = E_a$.

However, in condensed matter, the density of microscopic dipoles is high and their electrostatic interaction is not negligible anymore. Therefore, the local field E_{loc} at the position of a particular dipole is given by the superposition of the macroscopic field E_a and the fields of all other dipoles:

$$E_{loc} = E_a + \sum E_{dipole}. \tag{15.46}$$

Figure 15.12 gives an illustration of the feedback loop utilized for the calculation of the local electric field.

For the explicit calculation of the local electric field, the contributions of all the other dipoles in Eq. (15.46) have to be explicitly analyzed, that is,

$$E_{loc} = E_a + \underbrace{E_d + E_L + E_{nf}}_{\sum E_{dipole}}. \tag{15.47}$$

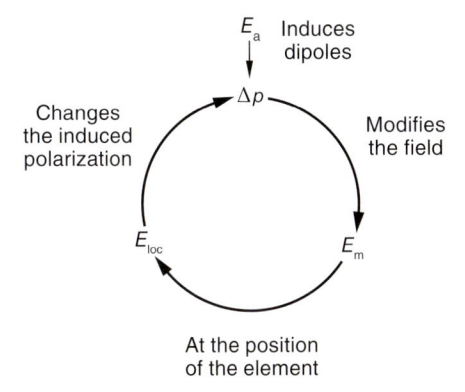

Figure 15.12 Feedback loop for calculation of the local electric field.

Here the electrical fields describe the depolarizing field E_d (with $E_m = E_a + E_d$), the Lorentz field E_L, and the near field term E_{nf} inside the material. Lorentz determined the electric fields E_L and E_{nf} by means of a "Gedankenexperiment," which is illustrated in Figure 15.13. For this, a sphere is cut out of the material at the position of a fixed point where the local field should be calculated (see Figure 15.13a). In consequence the polarization charges of the (homogeneous) dielectric material at the inner shell of the hollow sphere generate an electric field E_L at the position of the test point, called the Lorentz field. The calculation yields

$$E_L = \frac{P}{3\,\varepsilon_0}. \tag{15.48}$$

The near field response of the cut sphere results from the superposition of the fields of all individual dipoles indicated for the zoomed picture in Figure 15.13b.

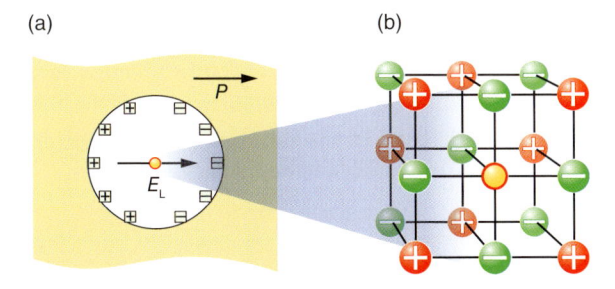

Figure 15.13 Illustration of the Gedankenexperiment of Lorentz for determining the contributions of (a) the Lorentz field E_L and (b) the near field E_{nf} to the local field. The yellow point marks the position where the local field is considered. For cubic crystal structures, the near field (at the position of the yellow sphere) equals zero, that is, $E_{nf} = 0$, due to symmetry considerations.

From this it can be intuitively understood that for cubic crystal structures, the near field term equals zero, that is, $E_{nf}=0$, for the reason of symmetry.

With this, the local electric field in a dielectric material with cubic structure is obtained as

$$E_{loc} = E_m + \frac{P}{3\varepsilon_0}. \tag{15.49}$$

Considering the microscopic picture in Figure 15.13, the local electric field acting on an atomic dipole inside a dielectric material is increased by the term $P/3\varepsilon_0$ compared to the macroscopic electric field inside the dielectric. Equation (15.49) is derived for cubic crystal structures but is often also applied to other crystal structures as a first approximation.

15.4.3
Clausius–Mosotti Equation

Clausius and Mossotti (see, for example, Ref. [6]) derived a method to calculate the local electric field E_{loc} quantitatively, which lead to a relation between the atomic polarizability α and the macroscopic permittivity ε_r. The *Clausius–Mosotti equation* formulates the result for cubic crystal structures, which obey induced dipoles:

$$\frac{n_{dipole} \cdot \alpha}{3\varepsilon_0} = \frac{\varepsilon_r - 1}{\varepsilon_r + 2}. \tag{15.50}$$

For a superposition of different polarization mechanisms and various species, the term $(n_{dipole} \cdot \alpha)$ is replaced by $\Sigma_i \, (n_i \cdot \alpha_i)$. From Eq. (15.50), the local field follows:

$$E_{loc} = \frac{\varepsilon_r + 2}{3} \cdot E_m. \tag{15.51}$$

In dielectric materials with electronic and ionic polarizability, both contributions, that is, α_{ion} and α_{el}, can easily be separated by means of the Clausius–Mosotti equation. The measurement of the dielectric permittivity ε_r at radio frequencies yields $n_{dipole} \cdot (\alpha_{ion} + \alpha_{el})/3\varepsilon_0$, whereas the measurement at optical frequencies results in $n_{dipole} \cdot \alpha_{el}/3\varepsilon_0$ (compare Figure 15.10). The two approaches, namely, the Lorentz field and the Clausius–Mosotti equation, are strictly derived for induced dipoles only. However, for values of $n_{dipole}\alpha/3\varepsilon_0 \ll 1$, the formalism can also be applied to materials that exhibit orientation polarization.

15.4.4
Temperature Coefficient of the Dielectric Permittivity

For materials that exhibit only electronic and ionic polarization,, the effective temperature coefficient of the permittivity TK_ε results from the superposition of the individual polarization contributions summarized as TK_α, plus a contribution TK_n, which results from the feedback of the lattice expansion on the polarizability. Differentiating the Clausius–Mosotti equation yields

$$TK_\varepsilon = \frac{1}{\varepsilon_r}\frac{d\varepsilon_r}{dT} = \frac{(\varepsilon_r - 1)\cdot(\varepsilon_r + 2)}{3\varepsilon_r}(TK_\alpha + TK_n). \quad (15.52)$$

The density of polarizable species n is inversely proportional to the volume V. Therefore, it holds $TK_n = -TK_V$ with values of the volume expansion coefficient being in the range of 1×10^{-5}–6×10^{-5} K^{-1}. Taking into account the values for the electronic and ionic polarizability, that is, $TK_{\alpha,\text{el}} \approx 0$ and $TK_{\alpha,\text{ion}} \approx +10^{-4}$ to 10^{-3} K^{-1}, it follows that materials with electronic polarization exhibit a slightly negative value of TK_ε, while for materials with ionic polarization the positive value of $TK_{\alpha,\text{ion}}$ often dominates the total TK_ε.

For materials with orientation polarization, $\alpha_{\text{or}} \sim T^{-1}$, a negative temperature coefficient is expected. However, the decrease is supercompensated by the temperature dependence of the relaxation time. This results in an increase of ε_r with increasing temperature at a constant frequency, which equals a positive temperature coefficient TK_ε. Figure 15.14 shows characteristic TK_ε and ε_r values of different classes of dielectric materials.

15.5
The Frequency Dependence of the Dielectric Behavior of Matter

15.5.1
Dielectrics in Alternating Electrical Fields

For a material that contains atomic or molecular entities carrying a charge q that can be set into vibration by application of an electrical field, the following

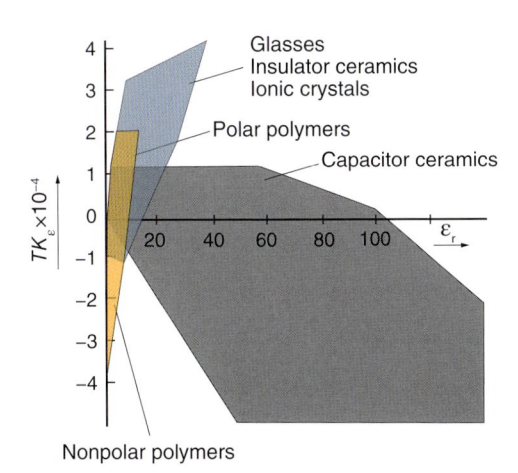

Figure 15.14 Temperature coefficient TK_ε and dielectric permittivity ε_r of different classes of dielectric materials.

differential equation of a forced damped vibration is derived:

$$m\,\delta\ddot{s}(t) + \gamma\,\delta\dot{s}(t) + k\,\delta s(t) = q\,E_{\text{loc}}(t). \tag{15.53}$$

Here, δs defines the deflexion of the charged entity, m is its mass, γ is the friction coefficient, and k is the restoring constant. Figure 15.15 illustrates the analogy between different vibrating systems: (a) the schematic corresponding to Eq. (15.53), (b) the mechanical, (c) the electrical equivalents, and (d) an equivalent LCR resonator circuit.

In the mechanical analogue, the charge is replaced by the mass, the electrostatic restoring force is replaced by a spring, and the damping is represented by a frictional cylinder. Multiplying Eq. (15.53) by the charge q of the vibrating entity results in a differential equation containing the contribution of the polarization that participates in the vibration. Applying the Clausius–Mosotti equation finally yields a differential equation for the dielectric displacement D as a function of the macroscopically applied electric field E_m:

$$\frac{1}{\omega_0^2}\ddot{D}(t) + \tau\dot{D}(t) + D(t) = \varepsilon_0\varepsilon_s E_m(t) + \varepsilon_0\varepsilon_\infty\left(\tau\dot{E}_m(t) + \frac{1}{\omega_0^2}\ddot{E}_m(t)\right), \tag{15.54}$$

with the resonance circular frequency $\omega_0 = (k/m)^{1/2}$ and the relaxation time $\tau = \gamma/k$. The contribution of the studied polarization mechanism is given by the difference of the dielectric permittivity values before (ε_s, "s" = static) and after (ε_∞, "∞" = infinite frequency) the cutoff of this polarization event, that is, $\Delta\varepsilon = \varepsilon_s - \varepsilon_\infty$.

Before we solve Eq. (15.54), two other equivalent differential equations should be mentioned that relate to the alternative circuits displayed in Figure 15.15. The field quantity-related differential equation is obtained by restricting to the

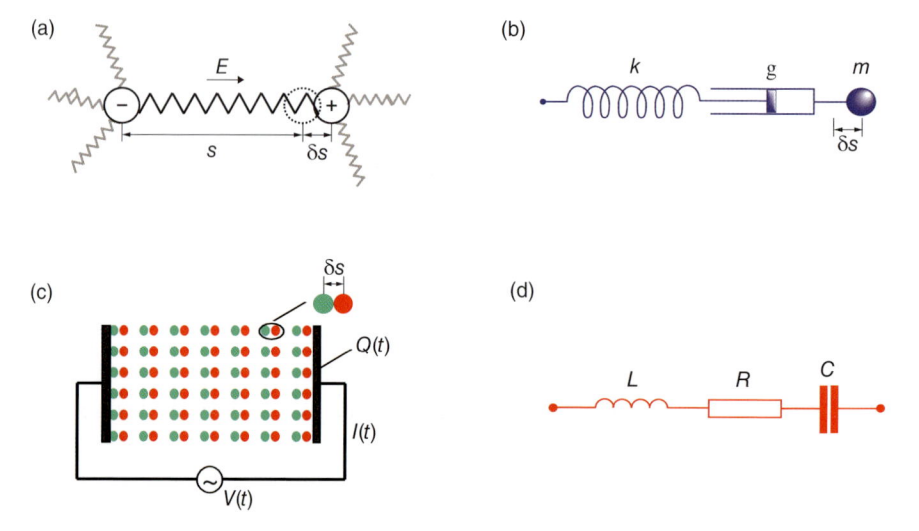

Figure 15.15 (a) Scheme of a vibrating entity and its corresponding representations in (b) purely mechanical and (c) purely electrical presentation. (d) Equivalent LCR resonator circuit.

vibration-related part, that is, $\varepsilon_\infty \equiv 0$, and by substituting $E_m = V/d$ and $Q = D \cdot A$. This yields

$$\frac{1}{\omega_0^2} \ddot{Q}(t) + \tau \dot{Q}(t) + Q(t) = C \cdot V(t). \tag{15.55}$$

Insertion of circuit parameters, such as the inductance L, the capacitance C, and the resistance R as well as $\omega_0^{-2} = LC$ and $\tau = RC$, results in the differential equation of an *LCR* resonance circuit:

$$L \cdot \dot{I}(t) + R \cdot I(t) + \frac{1}{C} \int I(t)\, dt = V(t). \tag{15.56}$$

This describes the current response $I(t)$ on an applied voltage signal $V(t)$.

15.5.2
The Complex Dielectric Function

The discussion of the dielectric behavior of condensed matter in alternating electric fields, that is, the solution of the differential Eq. (15.54), requires a description of the relative dielectric permittivity as a complex function:

$$\underline{\varepsilon}_r = \varepsilon'_r + i\varepsilon''_r. \tag{15.57}$$

This is because moving charges cause a frequency-dependent phase shift between applied field and charge displacement. The real part ε'_r (formerly ε_r) characterizes the displacement of the charges, and the imaginary part ε''_r the dielectric losses.

Analogue, the complex electrical susceptibility, reads

$$\underline{\chi}_e = \chi'_e + i\chi''_e. \tag{15.58}$$

The loss tangent and the quality factor, respectively, are defined as

$$\tan \delta := \frac{\varepsilon''_r}{\varepsilon'_r} \quad \text{and} \quad Q := (\tan \delta)^{-1}. \tag{15.59}$$

In addition to the losses caused by dipole reorientation $(\tan\delta)_{\text{dipole}}$, the residual leakage current $(\tan\delta)_{\text{cond}}$ of the nonperfect insulator is added so that in general $\tan\delta$ and hence ε''_r, respectively, become the sum of both contributions:

$$\tan \delta = (\tan \delta)_{\text{dipole}} + (\tan \delta)_{\text{cond}}. \tag{15.60}$$

The term $(\tan\delta)_{\text{cond}}$ can be separated experimentally because of its distinct frequency dependence.

The total polarization of a dielectric material is the sum of different contributions that apply to a certain material in a given frequency range. Each contribution stems from a short-range movement of charges that respond to an electric field on different timescales and, hence, through a Fourier transform, in different frequency regimes. This leads to the characteristic dispersion of the real part of the dielectric function, which is observed from Figure 15.10. Basically, two different phenomena regarding the frequency dependence of the dielectric function can be distinguished, that is, *resonance* effects and *relaxation* effects, depending on whether the oscillating masses experience a restoring force or not, respectively.

15.5.3
Resonance Phenomena

Resonance phenomena are associated with electronic and ionic polarization. In an electric field, the charged species (either electrons and nuclei or positively and negatively charged ions) are displaced from their equilibrium position. The displacement as a function of time is obtained from the solution of the differential Eq. (15.54). The terms on the left-hand side of Eq. (15.54) describe the force of inertia, the friction force, and the restoring force due to electrostatic Coulomb interaction. The term on the right-hand side represents the driving force caused by the local electric field. If E_{loc} is a DC field and this is switched off at a given moment, the electric charges are pulled back into their original position driven by the Coulomb force. For a negligible friction force, the limiting case of undamped resonance at frequency ω_0 is obtained. In contrast, a supercritical damped oscillation with a time constant $\tau = \gamma/\omega_0^2$ is observed for the case that the Coulomb force is significantly smaller than the friction force. Hence, for the complete scenario where the induced dipoles are excited to damped oscillations by the application of an alternating electric field of the angular frequency $\omega = 2\pi \cdot f$, the local electric field evolves like

$$\underline{E}_{loc} = E_{loc,0} \cdot e^{i(kr-\omega t)}, \tag{15.61}$$

where k denotes the wave vector and r the position.

For the frequency-dependent complex amplitude, the ansatz is chosen:

$$\underline{\delta s} = \delta s_0 \cdot e^{i(kr-\omega t)}. \tag{15.62}$$

With this, the frequency-dependent complex amplitude of the damped oscillation for species i with the charge q_i and the effective mass m_i^* follows:

$$\underline{\delta s_0} = \frac{(q_i/m_i^*) \cdot E_{loc,0}}{\omega_{0,i}^2 - \omega^2 + i\gamma_i \omega}. \tag{15.63}$$

The frequency-dependent dipole field associated with the oscillation is given as

$$\underline{p}_i = q_i \, \underline{\delta s_0} \cdot e^{i(kr-\omega t)} = \underline{\alpha}_i \, E_{loc,0} \cdot e^{i(kr-\omega t)}. \tag{15.64}$$

Therefore, the complex dynamic polarizability reads:

$$\underline{\alpha}_i(\omega) = \frac{q_i^2/m_i^*}{\omega_{0,i}^2 - \omega^2 + i\gamma_i \omega} = \alpha_i'(\omega) + i\alpha_i''(\omega), \tag{15.65}$$

with the real part

$$\alpha'_i(\omega) = \frac{q_i^2}{m_i^*} \frac{\omega_{0,i}^2 - \omega^2}{\left(\omega_{0,i}^2 - \omega^2\right)^2 + \gamma_i^2 \omega^2}, \tag{15.66}$$

and the imaginary part

$$\alpha''_i(\omega) = \frac{q_i^2}{m_i^*} \frac{\gamma_i \omega}{\left(\omega_{0,i}^2 - \omega^2\right)^2 + \gamma_i^2 \omega^2}. \tag{15.67}$$

For $\omega = 0$, we obtain the static polarizability as

$$\alpha_{i,s} := \alpha'_i(0) = \frac{q_i^2}{m_i^* \omega_{0,i}^2} \quad \text{and} \quad \alpha''_i(0) = 0. \tag{15.68}$$

This frequency dependence holds for the electronic α_{el} as well as for the ionic polarization α_{ion}. The resonance frequency of the electronic polarization $\omega_{0,el}$ typically lies in the ultraviolet region at about 10^{15} Hz. The ionic polarization is correlated to the lattice vibrations corresponding to the polar optical phonon branches. Therefore, the resonance frequency of the ionic polarization $\omega_{0,ion}$ lies in the infrared region between 10^{12} and 10^{13} Hz.

The frequency dependence of the complex dielectric function in the frequency range of $f > 10^{11}$ Hz follows directly from Eqs. (15.66) and (15.67) and the equation of Clausius and Mossotti. From this, we obtain

$$\varepsilon_r(\omega) = \varepsilon'_r(\omega_\infty) + \frac{\varepsilon'_r(\omega_s) - \varepsilon'_r(\omega_\infty)}{1 - (\omega/\omega_0)^2 + i\gamma\omega/\omega_0^2}, \tag{15.69}$$

where $\varepsilon'_r(\omega_\infty)$ and $\varepsilon'_r(\omega_s)$ denote values of the relative permittivity at frequencies significantly above (index ∞) and below (index s), respectively, the resonance frequency ω_0 of the oscillator. The corresponding dispersion curves, that is, $\varepsilon'_r(\omega)$ and $\varepsilon''_r(\omega)$, for a damped oscillation are plotted in Figure 15.16.

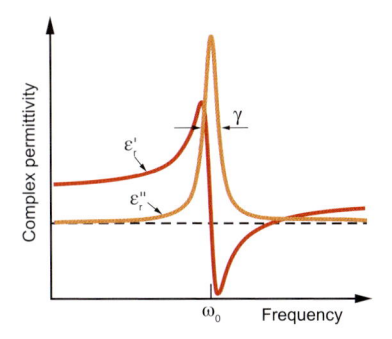

Figure 15.16 Frequency dependence of the real and imaginary parts of the dielectric function for a damped resonance phenomenon.

Analogous to $\underline{\varepsilon}_r$, the optical refractive index $\underline{n}(\omega)$ is complex and frequency dependent. For an insulator ($\sigma = 0$) with negligible magnetization ($\mu_r = 0$), the refractive index follows from Maxwell's law of dispersion:

$$\underline{n}(\omega) = \sqrt{\underline{\varepsilon}_r(\omega)}. \tag{15.70}$$

Hence, the optical properties of matter could be described as electric properties under the influence of alternating fields of high frequency $10^{12}\,\text{Hz} < f < 10^{18}\,\text{Hz}$.

15.5.4
Relaxation Phenomena

15.5.4.1 Debye Relaxation

Relaxation phenomena might result from two different physical origins. The first case is described by a supercritical damped oscillation (so called overdamped oscillator), where the relaxation time τ is significantly smaller than the reciprocal resonance frequency ω_0^{-1}. In this case, the relaxation time is defined by the relation of friction and spring constant, $\tau = \gamma/k$. The second situation is found for vibrations without a restoring force, such as the orientation polarization. After switching off the electrical field, permanent dipoles relax into the equilibrium position with zero net polarization due to thermal disorder. Explicitly the Debye relaxation describes a system with a single relaxation time τ. One example is the orientation polarization in a material with only *one* type of permanent dipole, which can be oriented by an external electric field:

$$\tau\dot{D}(t) + D(t) = \varepsilon_0\varepsilon_s E_m(t) + \varepsilon_0\varepsilon_\infty \tau \dot{E}_m(t). \tag{15.71}$$

The velocity by which the system approaches toward its equilibrium position is proportional to its displacement from equilibrium. The complex permittivity is immediately obtained from the general solution by omitting the term due to the driving force:

$$\underline{\varepsilon}_r(\omega) = \varepsilon'_r(\omega_\infty) + \frac{\varepsilon'_r(\omega_s) - \varepsilon'_r(\omega_\infty)}{1 + i\gamma\omega/\omega_0^2} = \varepsilon'_r(\omega_\infty) + \frac{\Delta\varepsilon'_r}{1 + i\omega\tau},$$

$$\text{with}\quad \Delta\varepsilon'_r := \varepsilon'_r(\omega_s) - \varepsilon'_r(\omega_\infty) \quad\text{and}\quad \tau := \gamma/\omega_0^2. \tag{15.72}$$

The *relaxation step* $\Delta\varepsilon'_r$ is defined as the maximum value of the gain in permittivity originating from a relaxation process. For the case of orientation polarization, the magnitude of the relaxation step is temperature dependent in the same way as the mean value of the polarizability:

$$\Delta\varepsilon'_{r,\text{or}} \propto \langle\alpha_\text{or}\rangle \propto T^{-1}. \tag{15.73}$$

From Eq. (15.72), the real and imaginary parts of the dielectric permittivity are calculated:

$$\varepsilon'_r(\omega) = \varepsilon'_r(\omega_\infty) + \frac{\Delta\varepsilon'_r}{1 + \omega^2\tau^2} \quad\text{and}\quad \varepsilon''_r(\omega) = \frac{\omega\tau\Delta\varepsilon'_r}{1 + \omega^2\tau^2}. \tag{15.74}$$

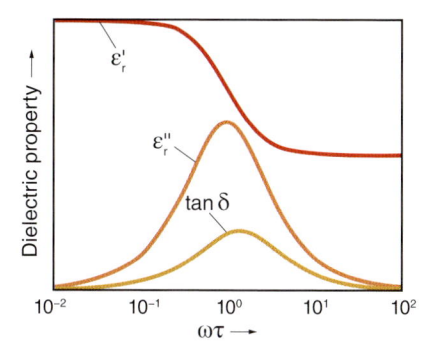

Figure 15.17 Frequency dependence of the real and imaginary parts of the dielectric function and of the loss tangent for a Debye-type relaxation process.

The corresponding frequency dependencies of the dielectric quantities are shown in Figure 15.17.

The thermal activation of the resulting Debye-type relaxation process can be easily understood from a bistable relaxation model shown in Figure 15.18.

The *relaxation time* τ of this thermally activated process is given as

$$\tau = \tau_0 \cdot e^{W_0/k_B T}, \tag{15.75}$$

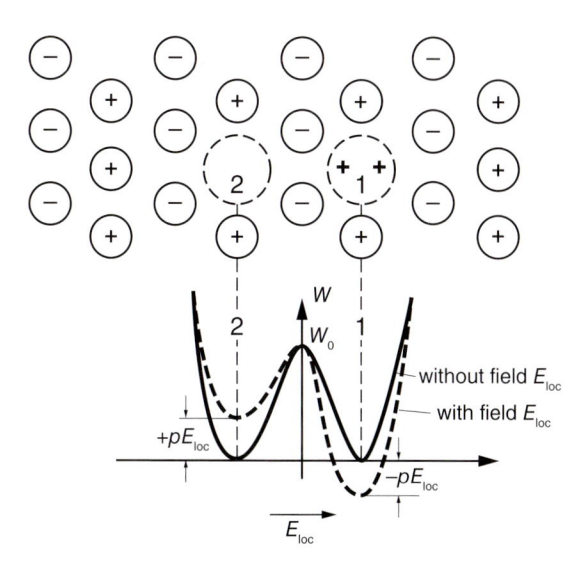

Figure 15.18 Bistable relaxation model for a Debye-type process: the doping ion (1) and the vacancy (2) form a dipole. The polarization of the entity switches if the doping ion changes the site with the vacancy. Therefore, the ion has to overcome the energy barrier W_0, defined for the field-free case. In an applied electric field, one position is lower in energy than the other.

15 Dielectric Properties

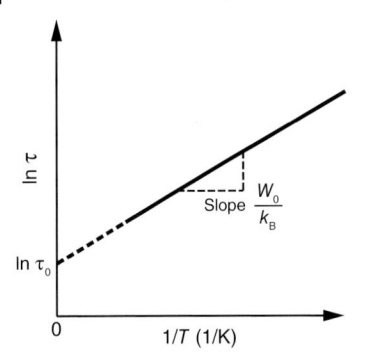

Figure 15.19 Arrhenius plot for calculation of the energy barrier W_0 of the relaxation process.

which leads to an exponential shift of the relaxation frequency with temperature. The activation energy is determined from a semilogarithmic plot of the relaxation time versus the reciprocal temperature shown in Figure 15.19.

The Debye-type relaxation process can be described by an equivalent circuit shown in Figure 15.20. The differential equation for this circuit is derived as

$$\tau \dot{Q}(t) + Q(t) = (\Delta C + C_\infty)V(t) + \tau C_\infty \dot{V}(t). \tag{15.76}$$

The relaxation time is defined as $\tau = R \cdot \Delta C$. Equation (15.76) is equivalent to Eq. (15.71) setting $C_\infty = \varepsilon_0 \varepsilon_\infty \cdot A/d$, $\Delta C = \varepsilon_0 \Delta \varepsilon \cdot A/d$, $D = Q/A$, and $E = V/d$.

The differential equation can be solved in the time or in the frequency domain. In the *time domain*, the solution is obtained by applying a voltage pulse represented by a step function. The response of the dielectric displacement $D(t)$ and the displacement current $\dot{D}(t)$, respectively, on the abrupt change of the electrical field $E(t)$ represented by a step function is derived as the solution of the differential equation. The dielectric displacement follows as

$$D(t) = \varepsilon_0 \varepsilon_\infty E_0 + \varepsilon_0 \Delta \varepsilon E_0 \left(1 - \exp\left(-\frac{t}{\tau}\right)\right). \tag{15.77}$$

And the displacement current is given as

$$\dot{D}(t) = \frac{\varepsilon_0 \Delta \varepsilon}{\tau} E_0 \exp\left(-\frac{t}{\tau}\right). \tag{15.78}$$

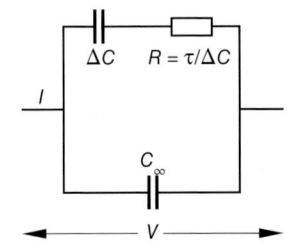

Figure 15.20 Equivalent circuit for Debye-type relaxation.

15.5 The Frequency Dependence of the Dielectric Behavior of Matter

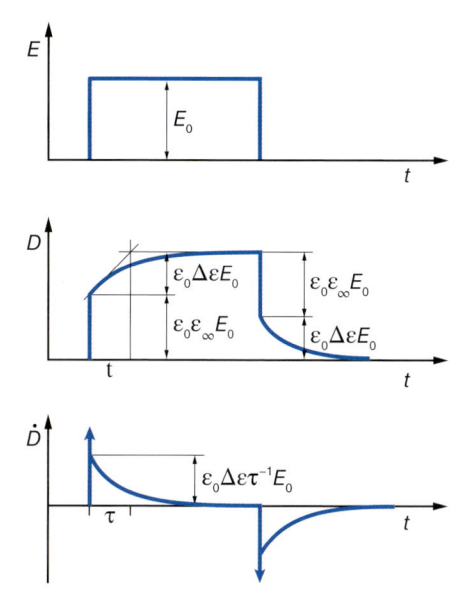

Figure 15.21 Dielectric displacement $D(t)$ and corresponding displacement current $\dot{D}(t)$ upon an electric field pulse of amplitude E_0.

The time dependencies of the characteristic measures for the case of a Debye-type relaxation process with one relaxation time τ are schematically plotted in Figure 15.21.

In the *frequency domain*, Eq. (15.76) is solved for an harmonic alternating electrical field, that is, $E_m = \hat{E}_m \exp(i\omega t)$. Defining the dielectric displacement as $D = \hat{D} \exp(i\omega t)$, the solution yields the frequency-dependent complex dielectric function for a Debye-type relaxation process:

$$\varepsilon_r = \frac{\hat{D}}{\varepsilon_0 \hat{E}} = \frac{\Delta \varepsilon}{1 + i\omega\tau} + \varepsilon_\infty, \tag{15.79}$$

with the real and imaginary part of the relative permittivity following as

$$\varepsilon'_r = \frac{\Delta \varepsilon}{1 + (\omega\tau)^2} + \varepsilon_\infty \quad \text{and} \quad \varepsilon''_r = \frac{\Delta \varepsilon \, \omega \tau}{1 + (\omega \tau)^2}. \tag{15.80}$$

Here $\varepsilon_r, \varepsilon_\infty$, and $\Delta\varepsilon$ represent relative dielectric permittivity quantities.

For the case of a Debye-type relaxation process, the Kramers–Kronig integral relationship between the real and imaginary parts of the permittivity simplifies to

$$\varepsilon'_r(\omega) = \varepsilon'_r(\omega_\infty) + \frac{\varepsilon''_r(\omega)}{\omega \tau} \quad \text{and} \quad \varepsilon''_r(\omega) = \left(\varepsilon'_r(\omega) - \varepsilon'_r(\omega_\infty)\right) \omega \tau. \tag{15.81}$$

Therefore, the plot "ε''_r versus ε'_r," the so-called *Cole–Cole diagram*, results in a semicircle for a single Debye-type relaxation process (see Figure 15.22).

Some dielectric materials show a nearly ideal Debye-type relaxation behavior. One example is clophen, which is used as an insulating oil in transformers. The

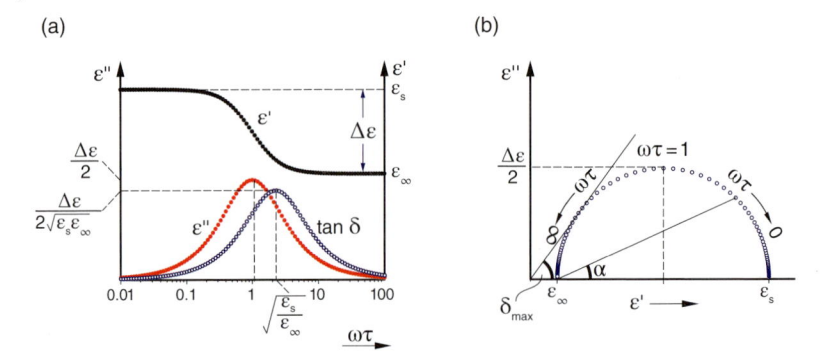

Figure 15.22 Debye relaxation. (a) Frequency dependence of the real and imaginary part of the complex dielectric function and of the loss tangent tan δ. (b) Cole–Cole diagram for a single Debye relaxation.

clophen-dipoles orient in an electric field with a well-defined relaxation time. This becomes apparent from the clear semicircle of the Cole–Cole plot in Figure 15.23a. However, it is more likely that the relaxation of the permanent dipoles in an electric field is described by a certain distribution of relaxation times. This results in a broadening of the relaxation step in the dispersion plot and, in consequence, in a superposition of different semicircles. In the Cole–Cole diagram, this superposition appears as a part of a semicircle with the center being shifted below the ε_r'-axis. Such a behavior is obtained for polyurethane (PUR) as plotted in Figure 15.23b. In addition, Figure 15.23b shows the contribution from the DC leakage current represented by the strong increase of ε_r'' in the low-frequency regime.

15.5.4.2 Maxwell–Wagner Relaxation

Maxwell–Wagner relaxation processes occur in inhomogeneous dielectrics with regions of different conductivity (see Figure 15.9, Maxwell–Wagner polarization).

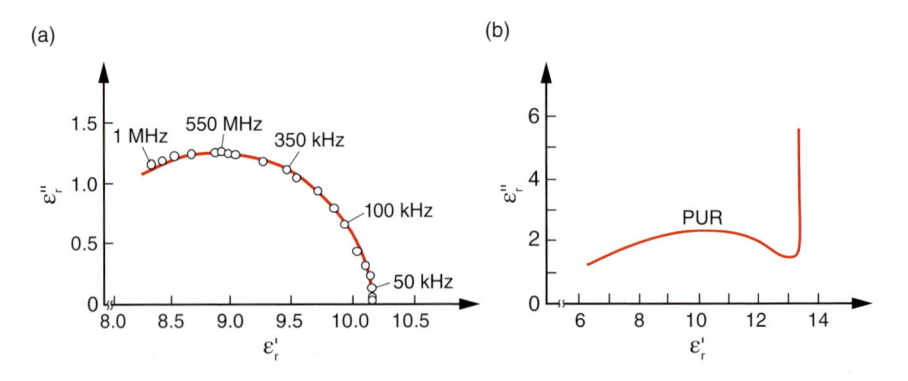

Figure 15.23 Cole–Cole plots of (a) clophen that shows a nearly ideal Debye relaxation and (b) polyurethane that is characterized by several orientation polarization events resulting in a distribution of relaxation times as well as by direct leakage current.

This is the case for polycrystalline ceramics with slightly conducting grains and highly insulating grain boundary regions (see Ref. [9]). A corresponding equivalent circuit of such a two-phase dielectric material can be transformed into one that is identical with the equivalent circuit of the Debye relaxation (see Figure 15.20). Therefore, impedance analysis of polycrystalline technical ceramics provides microscopic information about relevant properties of grain (bulk) and grain boundary regions of a polycrystalline material and thus enables to design requested dielectric properties and insulation strength of capacitor ceramics.

15.5.4.3 Dielectrics with a Distribution of Relaxation Times

In a more general case of a heterogeneous dielectric, often observed for high-k dielectric thin films, a variety of polarization events is obtained with a distribution of relaxation times τ_k. The resulting relaxation current of such a system can be interpreted as the sum of exponential decays according to Debye-type processes yielding a nonexponential curve as is shown in Figure 15.24. The corresponding equivalent circuit of such an inhomogeneous dielectric material with a distribution of relaxation times is shown in Figure 15.25.

The voltage-step response of such a heterogeneous dielectric with a distribution of relaxation times is described by the *Curie-von Schweidler* behavior

$$j_R(t) = \beta \cdot t^{-\alpha} + \delta(t), \tag{15.82}$$

where $j_R(t) \triangleq \dot{D}(t)$ denotes the polarization charging current of the dielectric material and the δ-function represents the high-frequency ionic and electronic polarization events. The Curie-von Schweidler law describes a class of dielectrics with a power-law dependence of the dielectric relaxation current on time. It is typically observed for partially or completely disordered materials such as glasses and polymers [12], but it has been observed in high-k dielectric thin film capacitors as well [15]. Regarding the application, the relaxation current sets a limit for

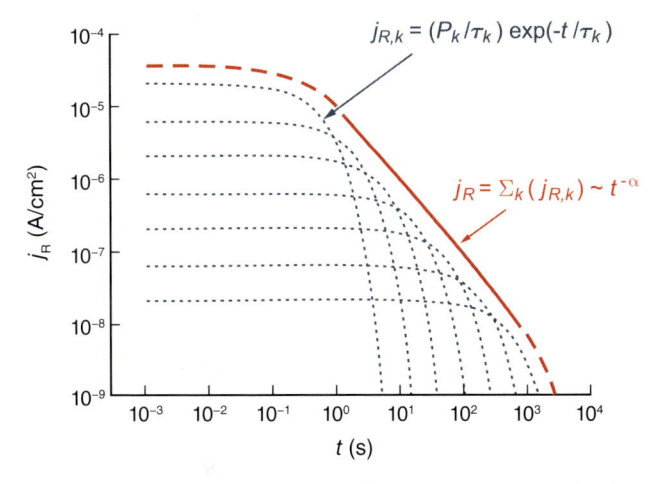

Figure 15.24 A nonexponential decay of relaxation currents can be described by a superposition of exponential decays having different time constants.

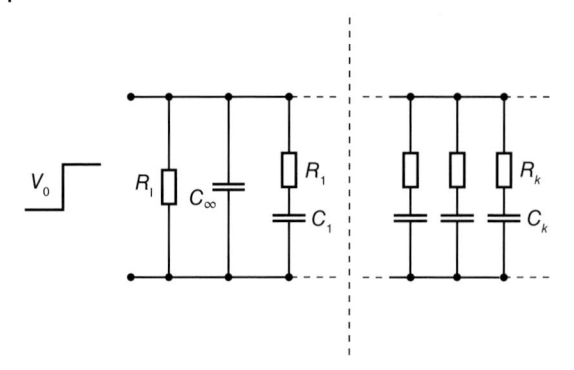

Figure 15.25 A network representing an inhomogeneous dielectric material with a distribution of relaxation times. R_1 and C_1 represent a relaxation element with a single relaxation time, R_l a leakage resistance, C_∞ the high-frequency capacitance, and a series of R_k and C_k elements with different $R_k \cdot C_k = \tau_k$ products representing a distribution of relaxation times.

the high-k dielectric thin films in future integrated capacitors as discussed in the following.

In general, the exponent α in the Curie-von Schweidler dependence (Eq. (15.82)) is close to unity, whereas the deviation from a value of one causes a dispersion in the frequency dependence of the susceptibility. For time-invariant linear systems, the frequency domain can be derived from the time domain by a *Fourier transform*. This yields

$$\underline{\chi}_e(\omega) = \frac{1}{\varepsilon_0 E(t)} \cdot \mathcal{F}\{j_R(t)\} = \frac{1}{\varepsilon_0 E_0} \cdot \left(\beta \int_0^\infty t^{-\alpha} \cdot e^{i\omega t} dt \right), \qquad (15.83)$$

with

$$E(t) = \begin{cases} 0, & \text{for } -\infty \leq t < 0, \\ E_0, & \text{for } 0 \leq t \leq \infty. \end{cases} \qquad (15.84)$$

Solving the integral yields

$$\underline{\chi}_e(\omega) = \chi'_e(\omega) + i\chi''_e(\omega) = \frac{\beta \Gamma(1-\alpha)}{\varepsilon_0 E_0} \omega^{(\alpha-1)} \left(\sin\left(\frac{\alpha\pi}{2}\right) + i \cos\left(\frac{\alpha\pi}{2}\right) \right) + \chi_\infty, \qquad (15.85)$$

with the gamma function

$$\Gamma(x) = \int_0^\infty e^{-z} z^{x-1} dz. \qquad (15.86)$$

From Eq. (15.85), the frequency dependence of the real part of the relative permittivity follows:

$$\varepsilon'_r(\omega) \approx \chi'_e(\omega) = \frac{\beta \Gamma(1-\alpha)}{\varepsilon_0 E_0} \cdot \omega^{(\alpha-1)} \cdot \sin\left(\frac{\alpha\pi}{2}\right) + \chi_\infty. \qquad (15.87)$$

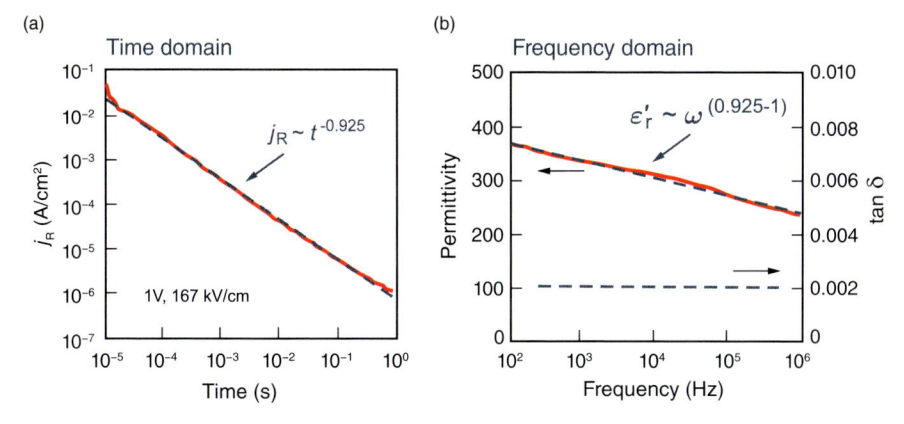

Figure 15.26 Dielectric response of $(Ba_{0.7}Sr_{0.3})TiO_3$ thin films in (a) the time and (b) the frequency domains, respectively. (After Ref. [15].)

The $\omega^{(\alpha-1)}$ term leads to a dispersion of $\varepsilon'_r(\omega)$, which is increasingly pronounced as α deviates from the value of 1. The loss factor of a system that shows a Curie–von Schweidler behavior is constant over the entire frequency range:

$$\tan \delta(\omega) = \left|\frac{\varepsilon''_r(\omega)}{\varepsilon'_r(\omega)}\right| \approx \left|\frac{\chi''_e(\omega)}{\chi'_e(\omega)}\right| = \cot\left(\frac{\alpha\pi}{2}\right). \qquad (15.88)$$

As an example, Figure 15.26 shows the Curie–von Schweidler-type relaxation behavior found in a thin film of barium strontium titanate prepared by MOCVD [15].

15.6
Polarization Waves in Ionic Crystals

So far we have focused on the dielectric properties of matter at low frequencies that apply for dielectric capacitors and insulators in electronic circuits. This frequency regime is dominated by relaxation phenomena originating from orientation polarization of permanent dipoles or space charge polarization effects in inhomogeneous materials. Significant contributions to the total permittivity might be added from ionic polarization effects and further smaller contributions come from electronic polarization phenomena. However, resonance frequencies related to ionic and electronic vibrations appear at higher frequencies, in particular in microwave to infrared regime and at optical frequencies, respectively. The description of the propagation of high-frequency waves in ionic crystals is covered by the theory on acoustic and optical phonons. Classically, these are described as harmonic oscillators. All optical lattice vibrations that coincide with an electric dipole moment are called polar optical phonons. The concept of polar optical phonons is comprehensively described in textbooks (see, for example, Refs [13,14]) and is also discussed in the chapter on "infrared characterization." Here, we will focus on important consequences of the high-frequency phonon theory that reach into the low-frequency regime, that is, affects the dielectric properties of polar materials. The chapter will close with an example of the low- and high-frequency dielectric properties of $SrTiO_3$, representing a material with a displacive phase transition at about 105 K [16].

15.6.1
The Lyddane–Sachs–Teller Relation

In 1941, Lyddane, Sachs, and Teller derived a formula for the dependence of the phonon–polariton frequencies and the dielectric properties in the case of ionic polarization [17]. For structures with a single optical branch, equivalent to a unit cell consisting of two atoms, the *Lyddane–Sachs–Teller (LST)* relation reads

$$\frac{\varepsilon_s}{\varepsilon_\infty} = \frac{\omega_{LO}^2}{\omega_{TO}^2}. \tag{15.89}$$

Here, ε_s denotes the "static" dielectric permittivity ($\omega \to 0$) and $\varepsilon_\infty = 1 + \chi_\infty = n^2$ is the high-frequency optical permittivity, which takes into account the contributions of the vibrations of the valence electrons. ω_{LO} and ω_{TO} denote the eigenfrequencies of the longitudinal and transverse optical phonon modes, which are determined, for example, from neutron scattering experiments. The *LST* relation can also be derived for complex structures on the basis of a tensor description (see Ref. [18] and references cited therein).

The *LST* relation gives a correlation between the ratio of the square of the longitudinal and transverse optical mode frequencies at wave vector $q = 0$ and the ratio of the values of the real part of the relative dielectric permittivity at frequencies much lower and higher than the resonance frequency of the ionic relaxation. The *LST* relation shows that a large value of the permittivity in ionic crystals is connected with a wide gap between the frequencies ω_{LO} and ω_{TO}, also called the LO–TO splitting, and especially with a low value of the resonance frequency of the transverse optical phonon ω_{TO}. Remember that in nonpolar crystals phonon–polaritons do not exist, for example, in germanium.

15.6.2
Softening of the Transverse Optical Phonon

A further consequence of the retardation ($E_{loc} \neq E_a$) in systems where polar optical phonons exist is that the polarization fields act in different ways on the longitudinal and transverse modes of the vibrating system. Solving, again for the simplest structure with one optical branch, the set of differential equations assuming a classical damped oscillator dispersion model for the polar phonon contribution and considering the screening of the dipole by its polarized environment yields for the total susceptibility:

$$\chi_e = \frac{(-\omega^2 - i\omega\gamma_{TO} + \omega_0^2)3N_a\alpha + 3\omega_p^2}{(-\omega^2 - i\omega\gamma_{TO} + \omega_0^2)(3 - N_a\alpha) - \omega_p^2}, \tag{15.90}$$

where γ_{TO} denotes the damping frequency of the transverse optical phonon, ω_0 the resonance frequency of the optical phonon, ω_p the mode plasma frequency (see Ref. [19]), N_a the concentration, and α the polarizability. For the limit of very high frequencies ($\omega \to \infty$), Eq. (15.90) reduces exactly to the contribution of the valence electrons:

$$\chi_\infty = \frac{3 N_a \alpha}{3 - N_a \alpha}. \tag{15.91}$$

Defining the susceptibility of a classical damped oscillator without retardation as

$$\underline{\chi}_0 := \frac{\omega_p^2}{\omega_0^2} \cdot \frac{\omega_0^2}{\omega_0^2 - \omega^2 - i\omega\gamma_{TO}}, \tag{15.92}$$

Eq. (15.90) becomes

$$\underline{\chi}_e = \chi_\infty + \frac{9\underline{\chi}_0}{\left(3 - N_a\alpha - \underline{\chi}_0\right)(3 - N_a\alpha)}, \tag{15.93}$$

where the second term represents the contribution of the lattice vibrations. From Eq. (15.93) it is clear that the contribution of the phonons is not only given by the distortion of the lattice, represented by $\underline{\chi}_0$, but is also considerably influenced by the displacement of the valence electrons, given by ($N_a \alpha$).

Equation (15.90) can be rewritten in the form of a classical damped oscillator dispersion relation:

$$\underline{\chi}_e = \chi_\infty + \frac{\omega_p^{*2}}{\omega_{TO}^2} \cdot \frac{\omega_{TO}^2}{\omega_{TO}^2 - \omega^2 - i\omega\gamma_{TO}}, \tag{15.94}$$

with the characteristic quantities:

$$\begin{aligned}\omega_p^{*2} &= \omega_p^2 \left(\frac{\chi_\infty + 3}{3}\right)^2, \\ \omega_{TO}^2 &= \omega_0^2 - \frac{1}{3}\omega_p^2 \cdot \frac{\chi_\infty + 3}{3},\end{aligned} \tag{15.95}$$

where ω_{TO} and γ_{TO} denote the eigenfrequency and damping of the transverse phonon mode, respectively. The term $\left(\omega_p^{*2}/\omega_{TO}^2\right)$ equals the strength of the transverse phonon mode. Solving the differential equation by insertion of the derived formula (Eq. (15.94)) for the electrical susceptibility of an ionic crystal in which polar optical phonons can exist yields the dispersion relation. The respective frequencies of the transversal and longitudinal optical phonons are given by Eqs. (15.95) and (15.96), respectively:

$$\omega_{LO}^2 = \omega_0^2 + \frac{2}{3}\omega_p^2 \cdot \frac{\chi_\infty + 3}{3\chi_\infty + 3}. \tag{15.96}$$

Even for the simplest case of rigid ions ($\chi_\infty = 0$), a hardening of the longitudinal modes (Eq. (15.96)) and a mode softening of the low-frequency transverse mode (Eq. (15.95)) is found:

$$\omega_{LO}^2 = \omega_0^2 + \frac{2}{3}\omega_p^2 \quad \text{and} \quad \omega_{TO}^2 = \omega_0^2 - \frac{1}{3}\omega_p^2. \tag{15.97}$$

Here, ω_0 defines the resonance frequency of the phonons and ω_p defines the mode plasma frequency (see also Ref. [19]). In the case of the longitudinal optical mode, the polarization field enhances the mechanical restoring force. In contrast, the low-frequency transverse optical mode is characterized by a partial compensation of short-range lattice forces and long-range electrical fields; this mode becomes *soft*. Under certain temperature and pressure conditions, the restoring forces for the transverse optical mode are very weak and a phase transition is induced.

15.6.3
Characteristic Oscillations in Perovskite-Type Oxides

In the previous sections, we described the lattice vibrations of an ionic crystal with cubic structure and two atoms per elementary cell. We have seen that the dielectric properties of these crystals in the infrared frequency range are derived from the dispersion of the optical phonon modes. In addition, we learned (see Figure 15.16, for example) that the low-frequency tail of the imaginary part of the susceptibility caused by ionic polarization is responsible for an inherent contribution to the dielectric losses in the frequency range between 1 and 100 GHz, thus the ionic polarization losses limit the quality factor of microwave dielectrics. This is important for the selection of dielectric materials for microwave applications.

In the following, some basic properties of the technically important alkaline earth titanates are shown, which can be explained by the model of optical phonons. The alkaline earth titanates exhibit a perovskite crystal structure, which is shown in Figure 15.27 for the case of strontium titanate as an example.

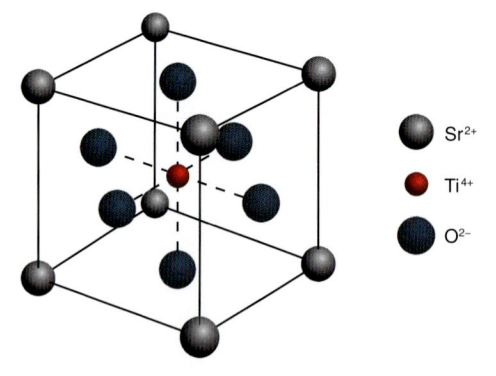

Figure 15.27 Perovskite crystal structure of strontium titanate.

Upon cooling, the materials undergo a phase transition from the high-temperature cubic phase to the distorted or tetragonal crystal structure at approximately 105 K for $SrTiO_3$ and 396 K for $BaTiO_3$. The latter shows a distortion in the tetragonal lattice cell by a displacement of cations and anions, which gives rise to the ferroelectricity of the material. Here some aspects on the dielectric properties of the room-temperature cubic phase of $SrTiO_3$ are discussed. Excellent studies are published on this topic, for example, the work of Hlinka et al. [19,20] or Fennie [21]. An analogue treatment can be performed for tetragonal phase of $BaTiO_3$, see Ref. [22] or [23], for example. For the isotropic cubic structure of $SrTiO_3$, we can distinguish between three different infrared active modes (see Figure 15.28). Here, infrared active means that the crystal exhibits a dipole moment induced by the displacement of the ions, which can interact with the light wave (see, for example, Ref. [24]).

At the highest frequency of about 540 cm^{-1} is the Axe mode, which is an oscillation of the linear O^{2-}–Ti^{4+}–O^{2-} chain against the remaining sublattice (Figure 15.28a). As a second is the Last mode where the O^{2-} octahedron together with the Ti^{4+} ion move against the other ions, as shown in Figure 15.28b. The excitation with the lowest frequency is the Slater mode, which is also called *soft mode*. Here the O^{2-} octahedron oscillates against the Ti^{4+} and Sr^{2+} ions, respectively, so all negative ions are displaced against all positive ions (Figure 15.28c). The softening of the 87 cm^{-1} TO mode describes the fact that the mode's frequency ω_{TO} decreases with decreasing temperature when approaching the phase transition temperature coming from high temperatures. For $\omega_{TO}=0$, the phase becomes unstable due to the vanishing restoring force.

15.6.4
Temperature Dependence of the Permittivity in Titanates

Taking $SrTiO_3$ as a representative example, the temperature dependence of the low-frequency dielectric constant $\varepsilon'_r(T)$ of this material in the cubic phase obeys

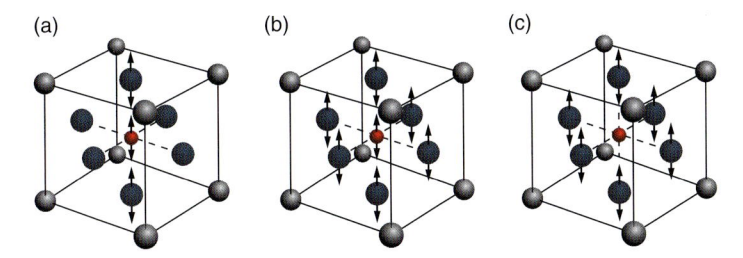

Figure 15.28 Polar optical vibrations in cubic $SrTiO_3$. (a) The Axe mode (also called displacive mode) is characterized by an oscillation of the linear O^{2-}–Ti^{4+}–O^{2-} chain against the remaining lattice at about 540 cm^{-1}. (b) The Last mode describes an oscillation of the Ti^{4+} and O^{2-} ions against the Sr^{2+} sublattice at about 180 cm^{-1}. (c) The Slater mode is characterized by an oscillation of the oxygen octahedron against the sublattice constituted by the Sr^{2+} and Ti^{4+} ions. This oscillation takes place at about 87 cm^{-1} and is called the *soft mode*.

the empirical Curie–Weiss law

$$\varepsilon'_r(T) \approx \chi'_e(T) = \frac{C}{T - \Theta}, \tag{15.98}$$

where C is the Curie constant and Θ is the Curie temperature, which in general is smaller than the critical temperature T_C of the phase transition:

$$\Theta < T_C. \tag{15.99}$$

According to the Lyddane–Sachs–Teller relation (Eq. (15.89)), the Curie–Weiss law, that is, the increase of the static dielectric constant with decreasing temperature (in the cubic phase), is caused by the decreasing frequency of the transverse optical phonon, which obeys a square root dependence as long as ω_{LO} can be assumed to be independent of the temperature:

$$\omega_{TO} \propto \sqrt{T - \Theta} \quad \leftrightarrow \quad \omega_{TO}^2 \propto (T - \Theta). \tag{15.100}$$

The correlation between the temperature dependencies of the relative permittivity (Eq. (15.99)) and the frequency of the soft phonon mode (Eq. (15.100)) has been confirmed for SrTiO$_3$, as shown in Figure 15.29 (taken from Ref. [16]).

15.6.5
Voltage Dependence of the Permittivity in Ionic Crystals

So far we have discussed an idealized ionic lattice built from harmonic oscillators. A harmonic potential means that the restoring force is a linear function of the displacement. In the case of a real ionic lattice, the local field generated by the neighboring atoms leads to an anharmonic potential for each ion, and thus to a nonlinear restoring force. The consequence of this is that the linear

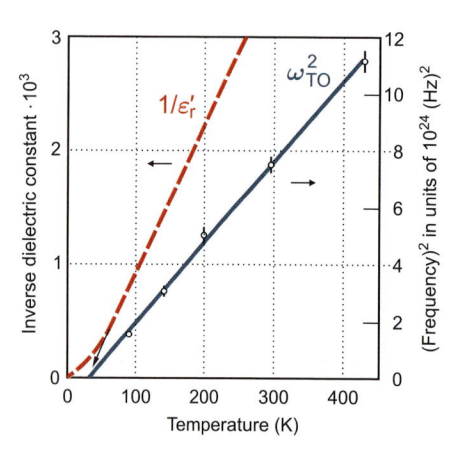

Figure 15.29 Comparison of the temperature dependencies of the reciprocal of the relative permittivity $(1/\varepsilon'_r)$ (dotted line) and of the square of the frequency of the transverse optical oscillation (ω_{TO}^2) with $k=0$ derived from neutron scattering experiments, (solid line) in SrTiO$_3$. (Reproduced with permission from Ref. [16]. Copyright 1964, American Chemical Society.)

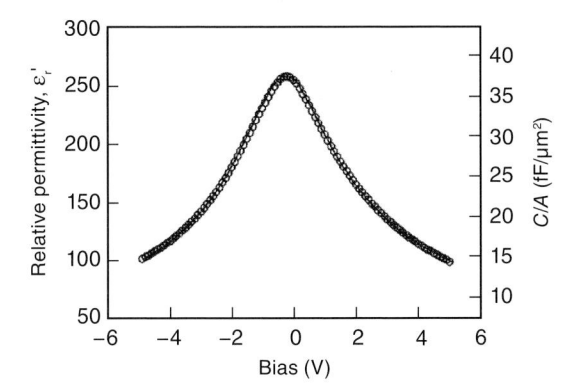

Figure 15.30 Nonlinearity of relative permittivity and capacitance density (C/A), respectively, of a $(Ba_{0.7}Sr_{0.3})TiO_3$ thin film prepared by MOCVD [25].

dependence between the dielectric displacement D and the electric field E (see Eq. (15.4))) no longer holds. Instead, we have to introduce a field-dependent permittivity $\underline{\varepsilon}_r(E)$. The effect of this nonlinearity becomes significant at high electric fields. Hence, the effect is more frequently observed in thin films than in bulk dielectrics because high electric fields are more easily reached in films at moderate voltages. In addition, for a given electric field, the effect is more pronounced for higher permittivity values. A high permittivity corresponds to a smaller restoring force between the ions of the lattice and a large atomic displacement at a given field. An example of the field-dependent permittivity is shown in Figure 15.30. The nonlinear electrical permittivity $\underline{\varepsilon}_r(E)$ is exploited, for example, for voltage-tunable microwave devices.

15.7 Summary

In this chapter, a short review of the basic concepts of the interactions between electromagnetic fields and dielectric matter is presented focusing on the spectrum of technical frequencies relevant for electronic applications. The deeper understanding of the response of the dielectric material in an electric field on the scale of the individual vibrating entity permits the design of dielectric devices with increased or variable capacitance. The DC and the low-frequency AC responses are dominated by displacing charged ions within the material. As a consequence, lattice distortions, vibrations, and the phonon dynamics play an important role and sound waves are easily coupled to alternating electrical fields. The retardation of the microscopic dipole response to the accelerating electric field is understood from the theory of the local electric field and the exchange reaction between the vibrating dipoles on the scale of the lattice dimensions. Utilizing this understanding allows a comprehensive description of dielectric materials and their behavior in alternating electric fields.

References

1. Ashcroft, N.W., Mermin, N.D., and Mermin, D. (1976) *Solid State Physics*, Holt, Rinehart and Winston, New York.
2. Ibach, H. and Lüth, H. (1996) *Solid-State Physics: An Introduction to Principles of Materials Science*, Springer, Berlin.
3. Kittel, C. (1996) *Introduction to Solid State Physics*, John Wiley & Sons, Inc., New York.
4. Madelung, O. (1978) *Introduction to Solid State Theory*, Springer Series in Solid State Sciences, vol. **2**, Springer, Berlin.
5. Bergmann, L. and Schaefer, C. (1997) *Constituents of Matter: Atoms, Molecules, Nuclei and Particles* (ed. W. Wilhelm Raith), de Gruyter, Berlin.
6. Feynman, R.P. (1989) The Feynman lectures on physics, in *Mainly Electromagnetism and Matter*, Addison-Wesley, Redwood City, CA.
7. Sze, S.M. (1981) *Physics of Semiconductor Devices*, John Wiley & Sons, Inc., New York.
8. Moulson, A.J. and Herbert, J.M. (1990) *Electroceramics: Materials, Properties, Applications*, Chapman and Hall, London.
9. Buchanan, R.C. (1991) *Ceramic Materials for Electronics: Processing, Properties, and Applications*, Marcel Dekker, New York.
10. Jaffe, B., Cook, W.R., Jr., and Jaffe, H. (1971) *Piezoelectric Ceramics*, Academic Press, London.
11. Fröhlich, H. (1986) *Theory of Dielectrics: Dielectric Constant and Dielectric Loss*, Clarendon Press, Oxford.
12. Jonscher, A.K. (1983) *Dielectric Relaxation in Solids*, Chelsea Dielectrics Press, London.
13. Born, M. and Huang, K. (1988) *Dynamical Theory of Crystal Lattices*, Clarendon Press, Oxford.
14. Singh, R.K. and Sanyal, S.P. (1990) *Phonons in Condensed Matter Physics*, John Wiley & Sons, Inc., New York.
15. Streiffer, S.K., Basceri, C., Kingon, A.I., Lipa, S., Bilodeau, S., Carl, R., and van Buskirk, P.C. (1996) *Mater. Res. Soc. Symp. Proc.*, **415**, 219.
16. Cowley, R.A. (1964) *Phys. Rev.*, **134**, A981.
17. Lyddane, R.H., Sachs, R.G., and Teller, E. (1941) *Phys. Rev.*, **59**, 673.
18. Schubert, M. (2016) *Phys. Rev. Lett.*, **117**, 215502.
19. Hlinka, J., Petzelt, J., Kamba, S., Noujni, D., and Ostapchuk, T. (2006) *Phase Transit.*, **79**, 41.
20. Ostapchuk, T., Petzelt, J., Železný, V., Pashkin, A., Pokorný, J., Drbohlav, I., Kužel, R., Rafaja, D., Gorshunov, B.P., Dressel, M., Ohly, Ch., Hoffmann-Eifert, S., and Waser, R. (2002) *Phys. Rev. B*, **66**, 235406.
21. Fennie, C.J. and Rabe, K.M. (2003) *Phys. Rev. B*, **68**, 184111.
22. Liu, S., Huang, L., Li, J., and O'Brien, S. (2012) *J. Appl. Phys.*, **112**, 014108.
23. Hlinka, J., Ostapchuk, T., Nuzhnyy, D., Petzelt, J., Kuzel, P., Kadlec, C., Vanek, P., Ponomareva, I., and Bellaiche, L. (2008) *Phys. Rev. Lett.*, **101**, 167402.
24. Kanehara, K., Hoshina, T., Takeda, H., and Tsurumi, T. (2014) *Appl. Phys. Lett.*, **105**, 042901.
25. Basceri, C., Streiffer, S.K., Kingon, A.I., and Waser, R. (1997) *J. Appl. Phys.*, **82**, 2497.

16
Mechanical Properties

Volker Schnabel,[1] Moritz to Baben,[1,2] Denis Music,[1] William J. Clegg,[3] and Jochen M. Schneider[1]

[1]RWTH Aachen University, Materials Chemistry, Kopernikusstraße 10, 52074 Aachen, Germany
[2]GTT-Technologies, Kaiserstraße 103, 52134 Herzogenrath, Germany
[3]Cambridge University, Department of Materials Science and Metallurgy, 27 Charles Babbage Rd, CB3 0FS Cambridge, UK

16.1
Introduction into Mechanical Properties

In 1638, Galileo Galilei discussed the strength of materials as one of the new sciences in his last book "Discourses and Mathematical Demonstrations Relating to Two New Sciences" [1]. Cahn suggests that Materials Science and Engineering emerged by integration of (sub)disciplines rather than by splitting [2]. While remaining firmly anchored in physics, chemistry, and mechanics, it addressed previously unanswered questions, on elastic and plastic deformation as well as fracture that now constitute the discipline of Materials Science and Engineering. The mechanical properties of materials are now a recognized research field within modern Materials Science and Engineering.

Here we seek to introduce mechanical properties of materials in a way accessible to the solid-state chemist. Therefore, elastic and plastic properties, fracture of materials, as well as friction and wear are discussed on the atomic scale, the length scale most familiar to the solid-state chemist.

In crystalline materials, the forces between atoms are repulsive at very small distances, but at longer distances become attractive, as shown in Figure 16.1.

The balance between these repulsive and attractive forces gives rise to an equilibrium spacing between the atoms at an equilibrium bond energy. Here, the equilibrium bond energy is set to zero for comparison purposes. If the atoms are displaced by a small amount from this spacing by applying a small force, there is an increase in energy, giving rise to a force in the material that causes the atoms to return to their equilibrium spacing if the force is removed. If the distance between the atoms is increased to a critical value these atoms can no longer

Handbook of Solid State Chemistry, First Edition. Edited by Richard Dronskowski, Shinichi Kikkawa, and Andreas Stein.
© 2017 Wiley-VCH Verlag GmbH & Co. KGaA. Published 2017 by Wiley-VCH Verlag GmbH & Co. KGaA.

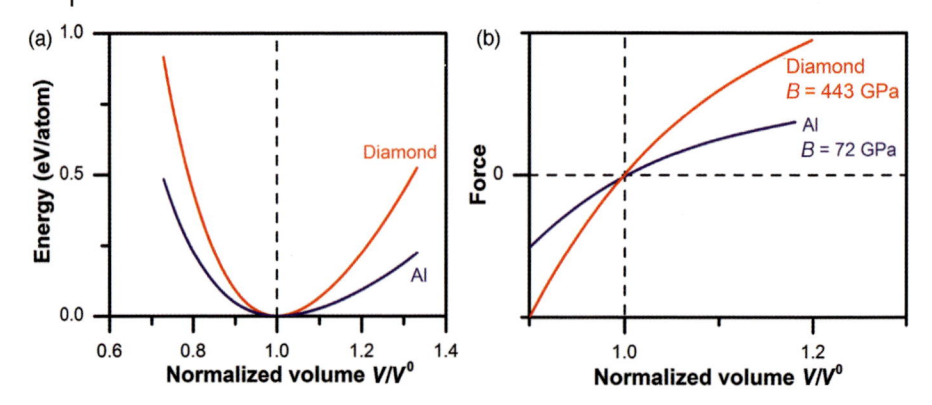

Figure 16.1 Energy per atom (a) and force (b) as a function of volume extension for aluminum and diamond, calculated by density functional theory [3,4]. Source: Adapted from Hohenberg 1964 and Kresse 1996.

return to their equilibrium position as bond breaking is predicted to occur and the material behavior is nonelastic giving an upper bound or theoretical strength of a material.

The discussion of mechanical properties on the atomic scale presented in this chapter is structured in three sections: Section 16.2 addresses elastic behavior while Section 16.3 contains information on nonelastic behavior. Finally, experimental techniques enabling the characterization of mechanical properties are discussed in Section 16.4.

16.2
Elastic Behavior

The concept of force was introduced in the seventeenth century by Isaac Newton and used by Robert Hooke in 1678, 40 years after Galileo Galilei, to describe elasticity. He wrote: *ut tensio, sic vis* (as the extension, so the force). Under these conditions, the behavior of the material is reversible and is known as *elastic*. Hooke found that under these conditions the force, F, was linearly related to the displacement, x,

$$F = -kx, \tag{16.1}$$

where k is a constant. If the bonds are treated as springs, then k is equal to the spring constant (Figure 16.2).

If a force, F, is applied parallel to the axis of a body with a cross-sectional area A, so that it is stretched the body will extend by an amount Δl. If its original length was l_0, then we can rewrite Eq. (16.1) as

$$\frac{F}{A} = k\frac{\Delta l}{l_0}, \tag{16.2}$$

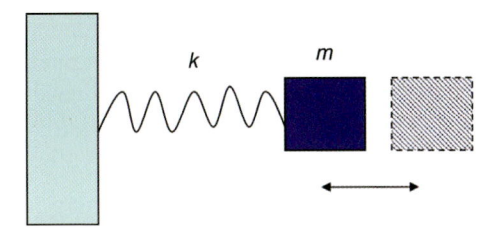

Figure 16.2 Visualization of Hooke's law via the spring model. A body with mass m oscillates as a result of loading of a spring.

or alternatively,

$$\sigma = E\varepsilon, \tag{16.3}$$

where σ is known as the stress, and ε the strain and the constant of proportionality in Eq. (16.1) is replaced by E, known as the Young modulus. This elastic modulus is thus a measure of a material's stiffness in tension. When a body is deformed in tension, it becomes thinner as well as longer. This lateral movement can also be expressed as a strain. The ratio of the transverse strain to the longitudinal strain is known as the Poisson ratio, ν. Other elastic constants also exist, giving the stiffness of the body, for instance in shear, or in hydrostatic compression, and known as the shear modulus, G, or as the bulk modulus, B, respectively.

These elastic moduli can all be related to one another, if it is assumed that the strains are small, as is the case for elastic deformation.

$$B = \frac{E}{3(1-2\nu)} \quad G = \frac{E}{2(1+\nu)} \quad E = \frac{9GB}{G+3B} \tag{16.4}$$

For an isotropic crystal, only two elastic moduli are required to completely define the elastic behavior. This is different from a cubic material, which needs not be elastically isotropic, where three elastic constants are needed. Note that there are differences in the type of deformation that each elastic modulus describes. G describes the ease with which the angle of a bond, or the sample shape may be changed, and, tends to be greater in a covalent material. B describes the ease with which bonds are elongated or compressed, while E is some mixture of the two.

An elastic body may deform in a direction other than that in which a force is applied. If the stress is acting on the i-plane and in the j-direction, then the stresses may be written as

$$\sigma_{ij} = \begin{bmatrix} \sigma_{11} & \sigma_{12} & \sigma_{13} \\ \sigma_{21} & \sigma_{22} & \sigma_{23} \\ \sigma_{31} & \sigma_{32} & \sigma_{33} \end{bmatrix} \tag{16.5}$$

With nine stresses and nine strains, a full description of elastic behavior requires a total of 81 different elastic constants. However, the symmetry of most crystals often substantially reduces the number required, which is why a cubic crystal only requires three different elastic constants to completely specify its elastic behavior [5]. In general, maximum number of independent elastic constants is 21 due to symmetry.

As both stress and strain are second-order tensors, the elastic modulus is a fourth-order tensor, written as C_{ijkl}, or S_{ijkl}, for, rather confusingly, stiffness and compliance respectively. C_{ijkl} can be considered as relating the elastic deformation when stress is acting on the i-plane and being oriented in the j-direction.

The bulk modulus, B, expresses a material's resistance to hydrostatic compression, which can be written as

$$B = -V \frac{\partial p}{\partial V}, \tag{16.6}$$

where V and p are pressure and volume, respectively.

The bulk modulus is dependent primarily on the density of bonding electrons. It is for this reason, that diamond with four valence electrons and a very small unit cell has the highest bulk modulus of 443 GPa [6] of any material. It is known [7] that the bulk modulus increases with increasing pressure.

This arises because the repulsive forces (Figure 16.1) between atoms increase rapidly as the atoms get closer together, with increasing pressure. This is known as the Birch–Murnaghan equation, after improvements by Birch [8], and has the following form:

$$P(V) = \frac{3B_0}{2} \left[\left(\frac{V_0}{V}\right)^{1/3} - \left(\frac{V_0}{V}\right)^{5/3} \right] \left\{ 1 + \frac{3}{4}(B'_0 - 4)\left[\left(\frac{V_0}{V}\right)^{2/3} - 1\right] \right\}, \tag{16.7}$$

where B_0 is the bulk modulus, Eq. (16.6), V is the deformed volume, V_0 is the reference (equilibrium) volume, B'_0 is the first derivative of the bulk modulus with respect to pressure.

Elastic properties depend on the chemical bonding. If one takes transition metals into account, the notion of band filling becomes very useful. The interatomic forces between d states in a periodic crystal lead to a band structure. The lower half of the band is commonly associated with the bonding states, while the upper portion defines anti-bonding. Hence, as the d band is being filled, electrons first occupy the bonding orbitals so that the strongest bonds as well as the largest stiffness can be found in the middle of the transition metal series. For instance, the bulk modulus of yttrium is only 37 GPa, while ruthenium reaches 321 GPa [6]. Similar relationships occur for other moduli discussed above.

16.3
Nonelastic Behavior

16.3.1
Plasticity

If the stress applied to an elastically loaded material is increased, a point will be reached, the *yield point*, above which the deformation is no longer reversible. From observations of deformed surfaces it can be seen that deformation occurred by the apparent movement of crystal planes over one another. However, the stress to achieve this can be estimated to be some orders of magnitude greater than that observed in common metals, such as Cu. Independently, Taylor [9], Polanyi [10], and Orowan [11], realized that this difficulty might be overcome, if instead of considering the simultaneous sliding of two crystal planes over one another, deformation occurred by the movement of defects, such as that shown in Figure 16.3. This so-called edge dislocation is a defect consisting of an extra half-plane of atoms inserted into the crystal lattice [12].

Different plastic deformation mechanisms exist and can be related to defects. An overview of different irreversible deformation mechanisms and the associated defects can be found in Table 16.1. This chapter focuses on dislocation movement since it is the most common plastic deformation mechanism [13].

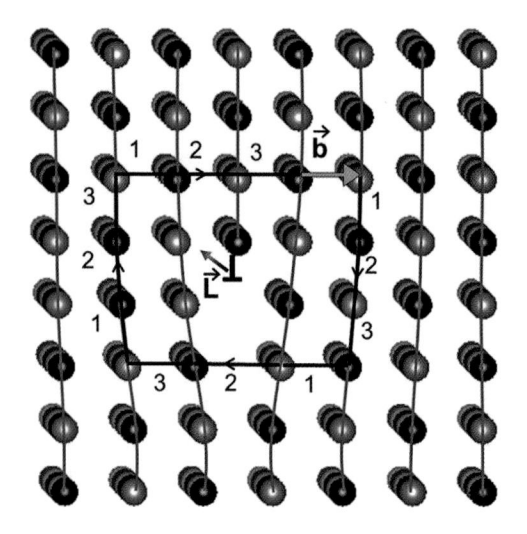

Figure 16.3 Schematic representation of an edge dislocation. Note the distortions (i.e., atomic distances greater or less than the equilibrium value) at the end of the extra half-plane. The line vector L perpendicular to the plane of projection and the Burgers vector b are indicated.

Table 16.1 Defects associated with irreversible deformation.

Mechanism	Example	Comments
Lattice diffusion (0D defects)	i. Diffusional creep; ii. Dislocation climb, (predominant time-dependent deformation (=*creep*) mechanism).	i. Atoms move from regions under compression to those under tension. Occurs at $T/T_M > 0.9$ in metals. ii. Point defects diffuse to dislocation. Occurs where diffusion is significant, that is, $T/T_M \sim 0.5$
Dislocation movement (1D defect)	Dislocation glide (predominant deformation mechanism in metals, crystalline polymers and even ceramics)	In materials with a higher lattice resistance, for example, Fe, yield stress increases as temperature is decreased; In materials with a low lattice resistance, for example, Cu, deformation is relatively temperature independent.
Grain boundary diffusion (2D defect)	Coble creep	Atoms move along grain boundaries from regions under compression to those under tension. Activation energy for diffusion down grain boundaries is ~0.6 that through the lattice.
Mechanical twinning (2D defect)	TWIP: Twinning-induced plasticity of steels	Twinning is a diffusionless transformation in which the deformed part of the crystal is a mirror image of the undeformed part. Twinning tends to occur where dislocation motion is more difficult, for example, at lower temperatures and higher strain rates.
Phase transformation (3D defect)	i. Formation of body-centered tetragonal iron (martensite) from face-centered cubic iron (austenite); ii. Transformation of tetragonal zirconia to the monoclinic form; iii. Diamond-structured Si becomes amorphous.	i. Requires sufficiently high cooling rates ii. Caused by the tetragonal form becoming thermodynamically more stable than the monoclinic form as the temperature is decreased iii. Observed when Si is indented.
Chain sliding of polymers (3D defect)	Disentanglement of polymeric chains in thermoplastics	Observed at $T \geq T_g$.
Localized shear (3D defect)	Shear bands in glasses	

An edge dislocation is characterized by a line vector L describing the edge of the inserted half-plane and the Burgers vector b describing the lattice distortion. The Burgers vector is constructed by following a continuous path around the dislocation with arbitrary distance from the dislocation core. In Figure 16.3, the starting point is the third from the right in the third row. From there, three atomic distances are counted in all directions (down, left, up, right) to the finishing point. The vector from the finishing to the starting point is the Burgers vector b. In the simple cubic crystal structure shown in Figure 16.3, its length is equal to the lattice spacing a. Additionally, there are *screw dislocations* whose line vector and Burgers vectors are parallel. Dislocations can be seen in transmission electron microscopes [14] or can be made observable at the surface by etching [15].

Unlike vacancies, there is no equilibrium concentration of dislocations. For vacancies, the equilibrium concentration exists due to the balance between the decrease in energy due to the configurational entropy and the increase in energy associated with the dangling bonds of atoms surrounding the vacancy. The highly ordered structure of the dislocation causes no increase in entropy while the structural distortion and broken bonds cost energy [16].

Dislocations normally form during solidification from the melt, or, in a very perfect crystal, they can be nucleated at the surface, often at defects such as steps, Figure 16.4. However, by far the most common is by a dislocation on a glide plane bowing between obstacles, such as particles or other dislocations. The generation of dislocations in this way takes place whenever a material is plastically deformed [17].

If the shear stress τ is small, the crystal undergoes only elastic deformation that is reversed if the crystal is unloaded (see Figure 16.4a and b). When the shear stress is increased, a dislocation pair is formed at the surface by bond breaking (Figure 16.4c). With only slight increase in shear stress the lower dislocation in Figure 16.4c moves through the crystal. giving rise to a permanent shape change (Figure 16.4d).

The crystal lattice is also likely to contain pre-existing dislocations. In this case, the energy of a dislocation increases as it moves from its initial low energy

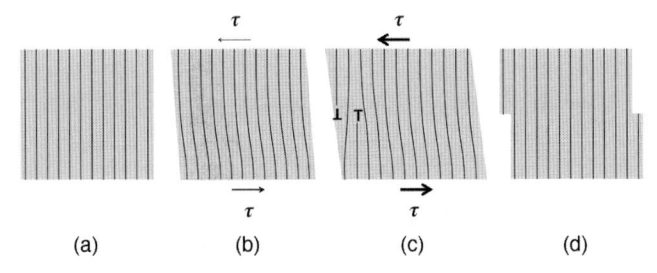

Figure 16.4 A dislocation is nucleated at a surface when the shear stress, τ, reaches a sufficient high value. The dislocation then travels across the crystal generally at a much lower load-giving rise to permanent shape change.

position, so that there is a resistance to its motion. As it arises from the atoms being displaced from their equilibrium positions it is known as the *lattice resistance*. The maximum value of the stress associated with this is known as the *Peierls stress*, τ_P, and can be estimated as [18]

$$\tau_P = \frac{2G}{(1-\nu)} \exp\left(-\frac{2\pi}{(1-\nu)}\frac{d}{b}\right), \tag{16.8}$$

where G is the shear modulus, ν the Poisson ratio (see Section 16.2), d the atom spacing normal to the slip plane, and b the Burgers vector length. The magnitude of this effect as an obstacle to dislocation motion is of the order of $10\ kT$, where k is the Boltzmann constant, so that dislocation motion becomes thermally activated. Using simple kinetics it can be shown that the yield stress of a material should decrease approximately linearly with temperature, as is observed. Here dislocation motion occurs by small segments of dislocation, called *kinks*, moving to the next low energy position. Importantly, the magnitude of the lattice resistance varies with the exponential of the negative ratio of atom spacing normal to the slip plane to that parallel to it. Materials such as diamond have a very low value of d/b, 0.354, so they are very hard, whereas metals such as copper exhibit a larger value, 1–1.4. Although the difference appears small, it enters an exponential term so the resulting effect is very large, giving rise to a difference in Peierls stress of ~5 orders of magnitude.

The most mobile dislocations are likely to be these with lower values of Peierls stress. For a given material, these are the dislocations that have the maximum lattice spacing and minimum Burgers vector. The Burgers vector is smallest in closed packed-directions, for example, <110>-direction for a face-centered cubic (fcc)-metal, where the lattice spacing is greatest for the (111)-plane. Therefore, slip tends to occur in the <110> direction and on the (111) plane in fcc metals.

If the material is assumed to be linear elastic, the energy of a dislocation per unit length L (or line energy) is given by

$$E/L = 1/2 Gb^2. \tag{16.9}$$

The energy therefore depends only on shear modulus and Burgers vector.

It is of great significance that a dislocation has a line tension. If there are obstacles on the slip plane, the applied shear stress will cause the dislocation to bow between the obstacles, in a similar way to paint flowing between obstacles. This causes an increase in the length of the dislocation, and since it has a line energy, there must be an increase in the overall energy, giving rise to a resistive force. In other words the material is strengthened by the presence of obstacles on the slip plane.

Typical obstacles include grain boundaries, solute atoms, and precipitates. As might be expected, the distortions associated with each dislocation also interact with one another and cause the material to become harder. This phenomenon, *work hardening*, is familiar to anyone who has deformed a piece of metal backwards and forwards. Above, it was shown that the high value of d/b of materials such as aluminum, copper, and nickel leads to a very low lattice resistance. It is only by modifying the structure of these materials in these ways that they become useful. A number of strengthening mechanisms exist [19], depending on the type of the obstacle and its interaction with the stress-field of the dislocation.

Most materials are not a single crystal, but made up of many crystals, that is, they are *polycrystalline*, Figure 16.5. If such an array of grains is to deform, each grain must be able to undergo a general shape change, so that it can deform with its neighbor. Clearly, one slip system is not enough. The condition for a material to undergo a general shape change is that it should have five independent slip systems [20]. A slip system is independent if the strain it produces cannot be produced by some combination of the others. In the case of a face-centered cubic metal, such as Cu, there are 12 crystallographically different slip systems. However, there are only five independent slip systems [21].

In many multicomponent compounds (intermetallics or ceramics) there is a significant difference in the stress required for yielding on different slip systems. It is therefore often the case that five slip systems do not operate. These materials have a low strain to failure.

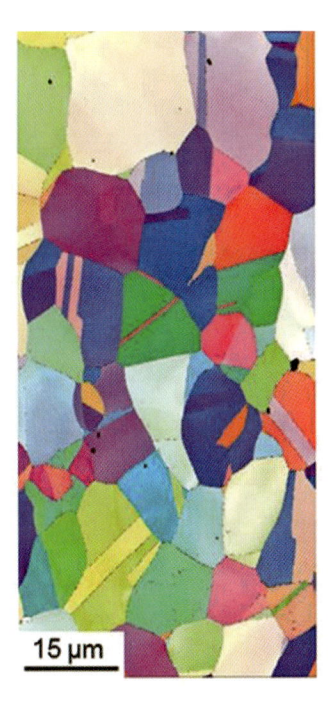

Figure 16.5 Inverse pole figure map of a high-manganese stainless Fe–Cr–Mn–C–N steel obtained by electron back-scatter diffraction. (Reprinted from Ref. [22], with permission from Elsevier.)

Ordered materials, where unlike atoms prefer to bond to one another, also tend to be stronger. Consider an AB compound that crystallizes in NaCl structure, that is, two fcc lattices where the A element occupies the first and the B element the second that is shifted by (0.5, 0, 0) with respect to the first. Such an ordered compound forms if A–B bonds are more favorable than the average of A–A and B–B bonds. When looking carefully at Figure 16.3, gray and black atoms can be distinguished, indicating the NaCl structure. The dislocation shown has a Burgers vector $b = 1/2a$, where a is the lattice vector of the NaCl lattice being considered here. In Figure 16.3 it can be seen that as the dislocation moves to the right, it leaves behind energetically less favorable A—A and B—B bonds. Therefore, if a dislocation passes through an ordered material, the bonding is disrupted so that like atoms are brought into close proximity. The energy of the system increases, so, again, there is a resistive force to the motion of the dislocation.

As dislocations are produced by plastic flow, a deformed material will have a higher energy than one that has been deformed to a lower strain. In the deformed materials, the dislocations will be arranged randomly. However, if the material is heated so that diffusion of atoms can occur, the dislocations may rearrange themselves into lower energy structures, in which the dislocations are arranged in walls. This process is known as *recovery*.

However, if the initial material is rather more heavily deformed then diffusion can give rise to regions of dislocation-free material being formed that then grow rapidly into the material with a high dislocation density. New crystals are formed. This is known as *recrystallization* and gives rise to material that contains a greatly reduced number of dislocations, expressed as a line length per unit volume, or *dislocation density*.

16.3.2
Fracture

Most manufacturing processes such as casting, machining, and welding give rise to flaws in materials. The question, therefore, is under what conditions can such flaws grow. Initial concepts were based on the idea that bonds would fail at some critical stress. Inglis [23] predicted the sharpness of a crack to be the controlling factor, whereas anyone, who has ever opened a bag of crisps knows that it is the length of the crack that is important. The first theory able to predict this effect was introduced by Griffith [24]. He treated crack growth as a reversible, thermodynamic phenomenon.

The Griffith idea is best understood by considering a stress state such as wedging. Here a wedge of the thickness h between a thin layer of material with thickness d, which is still attached to the parent block as schematically depicted in Figure 16.6. The action of the wedge can be considered to be equivalent to a force F acting on the thin layer at the point of contact with the wedge [25].

Three terms contribute to the total energy of the system U_T. There is the elastic strain energy U_E, the work done by the force U_F, and the work required to

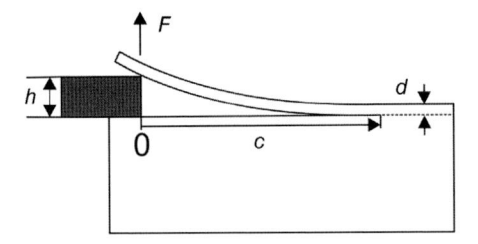

Figure 16.6 Schematic of wedging. (Reproduced from Ref. [26].)

create new surfaces U_S.

$$U_T = U_E + U_F + U_S. \tag{16.10}$$

Equilibrium occurs at

$$\frac{dU_T}{dc} = 0, \tag{16.11}$$

where c denotes the crack length. From simple beam theory, the elastic strain energy can be expressed as

$$U_E = \frac{Ed^3h^2}{8c^3}, \tag{16.12}$$

where c is the crack length. The work done by the force F is equal to the force multiplied by the distance moved in the direction of the cleavage. However, the cleavage direction is perpendicular to the force. Hence

$$U_F = 0. \tag{16.13}$$

To create new surfaces on the thin film and the parent block, respectively, the energy of

$$U_S = 2c\gamma \tag{16.14}$$

is required, where γ is the surface energy.

The idea that cracking could be treated as reversible thermodynamic phenomenon was first investigated by Obreimoff [27]. In Figure 16.7, it can be observed that the sum of the strain energy and the energy of the two new surfaces as a function of crack length exhibit a minimum. In this case the equilibrium is stable. Combining Eq. 16.12, 16.13 and 16.14 and differentiating to find the equilibrium condition, the value of the crack length at equilibrium, c_e, is given by

$$c_e = \left[\frac{3Ed^3h^2}{16\gamma}\right]^{1/4}. \tag{16.15}$$

In case of a truly thermodynamically reversible system the crack should heal, if the wedge is pulled out. When Obreimoff performed his experiments in air, exposing the surfaces to the environment no crack healing was observed. However, in vacuum, the crack did heal, which is schematically shown by the red

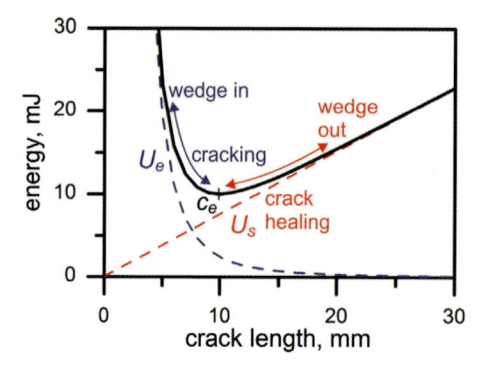

Figure 16.7 Energy as a function of crack length for the data used in the original Obreimoff experiment: $\gamma = 0.38$ J/m^2, $E = 200$ GPa, $h = 0.48$ mm, $d = 75$ μm.

arrows in Figure 16.7. This experiments show that cracking can be treated thermodynamically as Griffith suggested.

Although it is often simpler to visualize cracking in terms of overall energy changes in the cracking body, the approach becomes more difficult to use in complex stress states. An alternative way of calculating the energy change was formulated by Irwin [28]. Inglis showed that stresses were greater ahead of a crack. Irwin characterized the rate at which the stresses increase ahead of a crack using a variable known as the stress intensity factor, K. As the crack grows, these stresses are relaxed as the crack faces open and the energy release rate, G, can be expressed as

$$G = -\frac{d_{UE}}{dc} \tag{16.16}$$

The two approaches: that of Griffith and of Irwin are entirely equivalent. Both are based on energy changes during crack growth. The criterion for a crack to grow is that the stress intensity factor, K, must reach a critical value, K_C, where K_C represents the behavior of the material.

Hence, the stress intensity factor can be expressed as

$$K = \sqrt{EG} \tag{16.17}$$

Irwin showed that any stress-state could be obtained by a combination of an opening, shearing and twisting mode, schematically depicted in Figure 16.8.

In mode I the crack is opened symmetrically along the x, y-plane, whereas mode II is an asymmetric separation of the crack surface through a relative displacement in x-direction. Mode III is a separation through a relative displacement in z [29]. In real components, a mixture between the separation modes can be present. In the following part we will focus on the stress state at a crack tip in mode I, because it is the most critical stress state in application.

The stress state in a two dimensional sample at a crack tip can be expressed in terms of distance from the crack tip r and the angle φ, as depicted in Figure 16.9.

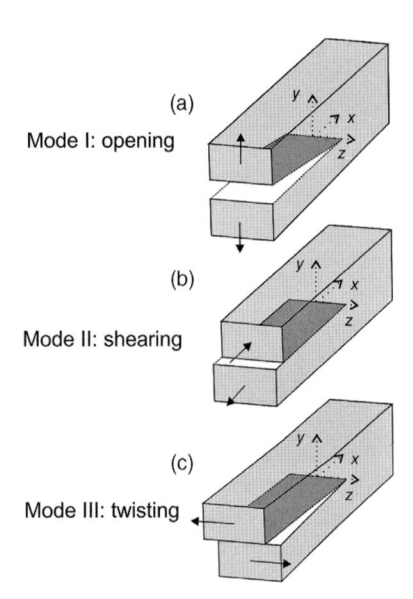

Figure 16.8 Crack opening modes.

The solution [29] for the stress state in a region around the crack tip can be expressed as

$$\begin{Bmatrix} \sigma_x \\ \sigma_y \end{Bmatrix} = \frac{K_I}{\sqrt{2\pi r}} \cos(\varphi/2) \begin{Bmatrix} 1 - \sin(\varphi/2)\sin(3\varphi/2) \\ 1 + \sin(\varphi/2)\sin(3\varphi/2) \end{Bmatrix}. \qquad (16.18)$$

In this region, though not at the crack-tip itself, the local stress is given by the stress intensity factor, K_I. Within linear elastic fracture mechanics, it is presumed that the whole body is linear elastic. Furthermore, it is assumed that any processes within the zone close to the crack tip are negligible from a macroscopic point of view. However, microscopically the processes within the zone close to the crack tip are not negligible at all. From Eq. (16.18), it can be learned that the stress in front of the crack tip at $\varphi = 0$ the stress increases with $1/\sqrt{r}$. Hence, for distances close to the crack tip, the stress becomes infinite, suggesting the material will have yielded, Figure 16.10, giving rise to a plastic zone ahead of the crack

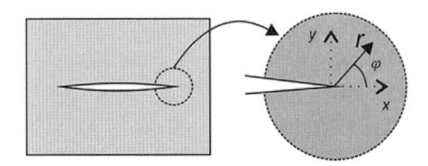

Figure 16.9 Schematic drawing of a two-dimensional crack.

16 Mechanical Properties

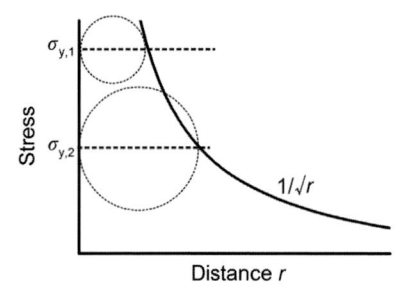

Figure 16.10 Plastic-zone size as a function of yield stress.

tip. The outer edge of this zone will occur where the stress ahead of the crack is equal to the yield stress, so that the plastic zone size, r_Y, can be written as

$$r_Y = \frac{K^2}{2\pi\sigma_Y^2}, \tag{16.19}$$

where σ_Y is the yield stress of the material. Within the plastic zone, the stress in front of the crack tip reaches the yield stress of the material. This is schematically depicted in Figure 16.10, where the $1/\sqrt{r}$ stress dependence can be observed. The vertical line represents the yield stress of the material. It can easily be understood that with decreasing yield stress from $\sigma_{y,1}$ to $\sigma_{y,2}$ the plastic zone size increases.

The total resistance to fracture R is composed of an elastic resistance, R_E, and a plastic material's resistance ΔR_P.

$$R = R_E + \Delta R_P. \tag{16.20}$$

The elastic resistance is equal to twice the surface energy of the crack, for two new surfaces are created, whereas the plastic energy term is equal to the deformation energy dissipated within the plastic zone

$$R = 2\gamma + 2\sigma_Y \bar{\varepsilon} r_Y, \tag{16.21}$$

where $\bar{\varepsilon}$ is the average strain. Substituting r_Y from Eq. (16.19) yields to

$$R = 2\gamma + 2\sigma_Y \bar{\varepsilon} \frac{K^2}{2\pi\sigma_Y^2}. \tag{16.22}$$

With Eq. (16.17), the total material's resistance can be expressed as

$$R = 2\gamma + 2\left[\frac{\bar{\varepsilon}}{2\pi}\right]\left[\frac{E}{\sigma_Y}\right]G. \tag{16.23}$$

Hence, the fracture energy should scale approximately with $1/\sigma_Y$. The conclusion, which can be drawn is that soft materials, such as Al or Cu, tend to be tough, whereas strong materials as ceramics tend to be brittle. In other words, an increase in yield strength leads to a decrease in plastic zone size (Eq. (16.23)), which goes in hand with a decrease in material's resistance to fracture. To apply linear elastic fracture mechanics, the size of the plastic process zone needs to be small compared to the whole stress field at the crack tip.

Fracture occurs, if the stress intensity factor, K, reaches a critical value K_C. The stress intensity at which fracture occurs in opening mode I, is called fracture toughness K_{IC}. The fracture toughness is a material's specific property. Common values for steels and ceramics are 14–280 MPa·$\sqrt{m}$ and 0.8–4.8 MPa·$\sqrt{m}$ [30], respectively. Static loading of a component will not lead to fracture, if the present cracks are smaller than a certain critical size in reference to the load applied. However, upon cycling loading crack growth can be observed even though the applied load is below the critical value. This phenomenon is called fatigue, which is very commonly the cause for material's failure. Furthermore, as observed from Obreimoff's experiment environmental conditions and chemical reactions at the crack tip can dramatically alter the material and hence affect fracture energy.

16.3.3
Friction and Wear

Friction and wear result from the contact that takes place between two moving surfaces: Wear is defined as the removal of material from a solid surface while friction is the tangential resistance to motion experienced as solid surfaces move relative to one another [31].

The friction coefficient, μ, is the frictional force or tangential force, F, divided by the normal force, w, on the contact [31].

$$\mu = F/w. \tag{16.24}$$

Wear can occur in a variety of ways. Holmberg and Matthews differentiate between abrasive-, adhesive-, fatigue-, and chemical wear mechanisms [31]. Abrasion describes material loss due to movement of a hard material over a soft surface. If asperities from both surfaces adhere to each other and shearing is observed upon motion, adhesive wear is observed while fatigue wear is generated by the cracking caused by repeated loading. Chemical wear implies that material is removed due to chemical reactions. Independent of the actual wear mechanism or combination of wear mechanisms, the rate of material loss from a solid surface in a sliding contact is expressed by the wear rate. An often employed unit for the wear rate is volume lost per sliding distance and per normal force applied ($mm^3/(m\ N)$).

Friction and wear are of fundamental importance in science and technology. It is immediately obvious that for the assembly of pyramids, as well as for harnessing of wind power with wind turbines friction and wear are of crucial importance. Fall et al. proposed in a 2014 paper in Physical Review Letters entitled Sliding Friction on Wet and Dry Sand that the formation of capillary water bridges increases the shear modulus of the sand, which facilitates the sliding of pyramid blocks as the friction coefficient is decreased [32]. Today in many automotive applications, antiwear tribofilms are grown in situ as solid surfaces move relative to one another. At sliding interfaces reactions between additives, for

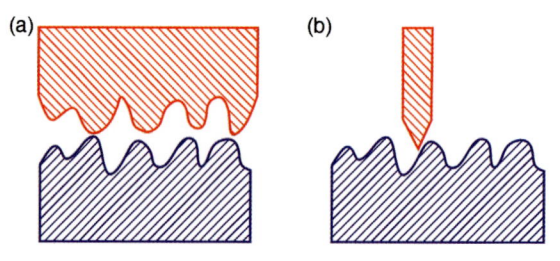

Figure 16.11 (a) Engineering tribological contact exhibiting multiple asperity contacts. (b) Single asperity nanotribological contact.

example, zinc dialkyldithiophosphates (ZDDPs), supplied in lubricants and the solid surface can result in the formation of antiwear tribofilms [33].

While friction and wear phenomena have traditionally been studied on various length scales, it is evident from the examples cited above that atomic scale mechanisms at interfaces appear to govern the tribological contact processes that in turn are relevant for the performance of machines containing moving parts. In technologically relevant sliding interface, contact occurs at multiple asperities (see Figure 16.11a). In nanotribology – the study of friction and wear at the atomic scale – typically one asperity contact is characterized theoretically and/or experimentally (see Figure 16.11b).

Scanning probe microscopy as well as the surface force apparatus in combination with computational approaches describing atomic interaction across interfaces have enabled atomic scale understanding of nanotribological phenomena.

Until about 20 years ago, most theoretical approaches to model tribological contacts were continuum based. Only recently, quantum mechanical calculations have been used to estimate mechanical properties of single asperity contacts. Only some of the *ab initio* predictions are consistent with macroscopic simulations [34].

Landman *et al.* showed in 1990 that friction and wear on the nanoscale can be understood based on atomistic simulations [35]. Joint molecular dynamics (MD) simulations and atomic force microscopy studies revealed atomic scale adhesion and contact formation between a nickel tip and a gold surface [35]. As the tip approaches the surface the calculated and measured attractive forces are monotonically increasing. Experimental and theoretical evidence for a jump-to-contact phenomenon is presented and also the calculated hysteresis upon tip retraction is verified experimentally. While the forces distance data are not shown here, the atomic configurations from the MD simulations are shown in Figure 16.12 [35].

In Figure 16.12a, evidence of inelastic deformation of the Au substrate can be seen as well as partial wetting of the tip surface by Au. Upon separation of the tip from the surface, the adhesive interaction results in the formation of an atomically thin connective neck. Furthermore, the authors analyze also indentation scenarios and present evidence for yielding and slip and fracture [35].

The AFM measurements were done in a dry box in a dry nitrogen ambient. Often technologically relevant sliding contacts operate in air while nanotribological studies are often conducted in ultrahigh vacuum. Atmosphere exposure is

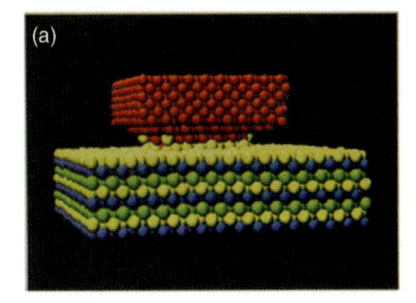

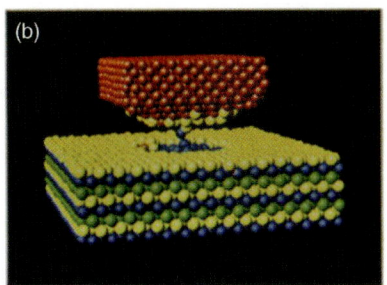

Figure 16.12 Atomic configurations generated by the MD simulations. (a) After jump-to-contact. Note bulging of the Au substrate under the Ni tip and partial wetting of the tip edges. (b) Separation after contact illustrating adherence of the top Au layer to the Ni tip and the formation of an atomically thin connective neck. (Figure and Figure captions from Ref. [35]. Copyright 1990, with permission from Elsevier.)

known to modify surface chemistry [36,37]. Physically meaningful models have to take this into account.

While the physics of single-asperity friction is a complex topic and was reviewed by Szlufarska *et al.* [38], lateral force images such as of the muscovite mica (0001) surface presented in Figure 16.13 show clear evidence for the so-called stick-slip behavior where the lateral force increases as the tip is displaced laterally until a well-defined maximum is reached (stick). Further tip displacement results in a reduction of the lateral force (slip) until the maximum in lateral force is reached again. The temporal evolution of stick and slip occurrences mirrors the periodicity of the atomic lattice.

As technologically relevant tribological contacts are multiasperity contacts, the study of single asperity contacts has been recognized as important for solid–solid interfaces also at the micro scale [40] and it is very reasonable to support Urbakh and Meyer's proposal [41] to integrate modeling contact dynamics at all length scales. However, the integration of the nanoscale to the macroscopic scale provides formidable challenges. Equally reasonable is the assumption that computational methods based on continuum mechanics are not fully applicable to describe a tribological interfaces containing a limited number of asperity contacts and are not applicable for a single asperity contact as atomistic scale mechanisms such as adhesive interactions or necking.

However, Krim *et al.* [42] study the relationship between grain size and wear by experimental and theoretical methods. Based on a continuum approach using Voronoi tessellations to represent the nanocomposite structure, the authors are able to show that nanocomposites with small grains exhibit lower wear rates and that this can be understood by the computational local stress data. While such models are useful in relating the effect of grain size and stress distribution to wear phenomena explored experimentally they are clearly not able to predict wear. For such predictions, atomistic simulation techniques would need to be employed: Representing a tribological contact with a state of the art

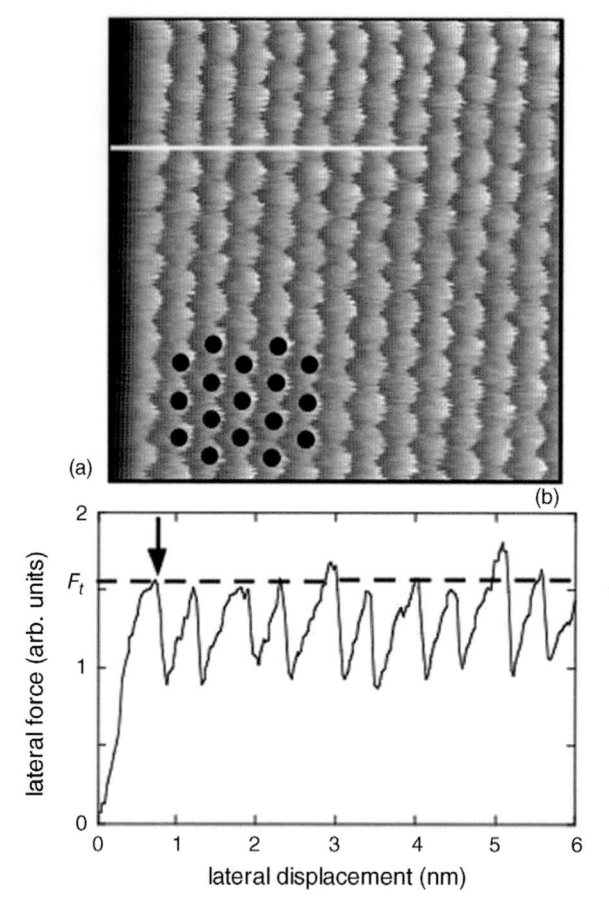

Figure 16.13 Lateral force images of the muscovite mica (0001) surface showing stick-slip behavior. (From Ref. [39], with permission of Springer.)

nanocomposite material such as adaptive solid lubricant coatings (Chameleon coatings) [43] exhibiting, for example, nanometer size Au, MoS_2, and yttria-stabilized zirconia grains in a diamond like carbon matrix is today not feasible. Connecting nanotribology with friction and wear phenomena encountered on the macroscale remains a challenge to be addressed.

16.4
Characterization Techniques

16.4.1
Diamond Anvil Cell

To measure the bulk modulus, as defined in Eq. (16.6), diamond anvil cell can be used. A simple setup is shown in Figure 16.14. Typically, high pressures

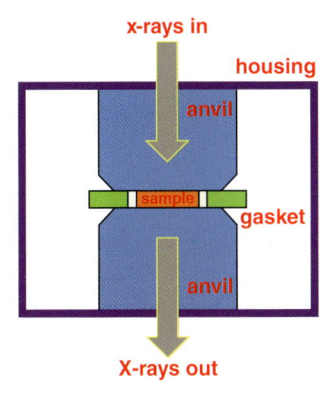

Figure 16.14 Schematics of diamond anvil cell setup.

exceeding 500 GPa can be obtained. A diamond anvil cell is made of two opposing diamonds. This uniaxial pressure loading can be transformed into uniform hydrostatic pressure using a pressure transmitting media, such as noble gases or oil. The sample is typically irradiated with X-rays to monitor the volume changes. Using Eq. (16.6), one can directly calculate the bulk modulus.

There are also modified diamond anvil cell setups with shock wave compression so that extremely larger pressures up to 5000 GPa can be achieved. It is also possible to heat the specimen in a diamond anvil cell to obtain the mechanical properties at elevated temperatures. A common approach is to use resistive heaters, but cells with lasers are also available.

Recently, the diamond anvil cell has been used to probe the sustainability of life under high pressures in astrobiology and exobiology studies. To transfer life from one planet to another, any life form must survive the high pressure upon impact, as well as the extremely high pressures that may exist on many planets with the potential for life.

Commonly, diamond anvil cell setups can also exhibit nonhydrostatic pressure components, which produce changes of shape rather than volume. These components can cause unwanted plastic deformations and are the reason for using X-rays when measuring the bulk modulus.

16.4.2
Scattering Techniques for Elasticity Measurements

Sound waves propagate through solids due to the vibrations or oscillatory motions of constituting atoms. Such a wave may be visualized as an infinite number of oscillating masses connected by elastic springs, as illustrated in Figure 16.2 and provided by Eq. (16.25). It is also clear that sound travels at different speeds in different materials. This is because the atomic masses and spring constants (bond strengths) are different for different materials. The general relationship between the speed of sound (ν) in a solid and its density (ρ), and

Young's modulus (E) is given by

$$v = E\rho. \tag{16.25}$$

Equation (16.25) is valid for longitudinal waves, while shear waves can be described if the Young's modulus is replaced by the shear modulus. Brillouin light scattering can be employed to determine acoustic velocities and elastic properties of materials based on Eq. (16.25). It is most commonly performed on transparent single crystals where all elastic constants, as defined by Eq. (16.6), can be derived. In a polycrystalline material, sound waves scatter against grain and other boundaries, which is difficult to account for in the analysis of acquired data so that the scattering techniques are normally limited to single crystals or perfectly isotropic materials, such as glasses. Another scattering technique is the surface acoustic wave method. Generally, a surface acoustic wave is an acoustic wave traveling along the surface of a material, specified by the elastic properties thereof, with an amplitude that typically decays exponentially with the depth (depth is about equal to the wavelength). In thin films, the velocity of a surface acoustic wave depends on frequency, which is known as a dispersion spectrum. A standard setup is provided in Figure 16.15. Wide-band surface wave impulses are typically generated by a laser pulse and detected with a piezoelectric transducer. Fourier transforming the acoustic signal provides the dispersion spectrum. These have a form that depends on the elastic constants and density of film and substrate as well as the film thickness.

16.4.3
Tensile Test

In a tensile test, a tensile force is applied to a specimen (see Figure 16.16) while the elongation is determined. From the applied force F and the initial cross-section area A_0 the stress can be calculated as $\sigma = F/A_0$. The strain $\varepsilon = e/L_0$ is the ratio of elongation e to initial length L_0.

During elastic loading, stress is directly proportional to strain. Therefore, the initial slope of the stress–strain diagram is given by the elastic (Young's)

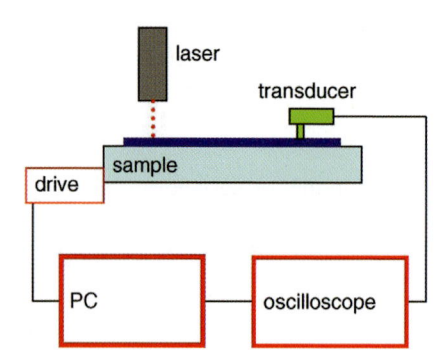

Figure 16.15 Schematics of surface acoustic waves setup.

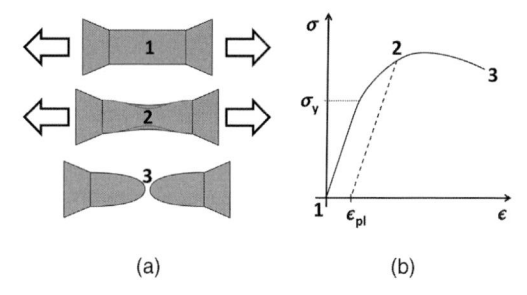

Figure 16.16 (a) Schematics of deformation of a metal under tensile load. Large stresses lead to plastic deformation and hence necking of the specimen before fracture. (b) Typical stress–strain curve of a metal. The initial conditions (1), necking (2), and fracture (3) are marked.

modulus. Once the shear stress is large enough for dislocation movement, see Section 16.3.1, plastic deformation occurs at the yield stress σ_y. Simultaneously, the shape of the specimen changes: plastic deformation leads to elongation of the specimen, however, mass conservation forces the cross-section area to decrease as shown in Figure 16.16. Since a smaller cross-section area increases stress for the same applied force, further plastic deformation will occur. This process is known as necking. Once necking has started the measured stress will decrease, see Figure 16.16. Finally, the specimen fractures, see Figure 16.16, depending on the material fracture can happen immediately after the elastic region (brittle materials) or after several percent of plastic elongation (ductile materials).

16.4.4
Nanoindentation

Nanoindentation is a method for the characterization of the elastic and plastic behavior of a material, on a very small scale. Features less than 100 nm across and thin films less than 100 nm thick can be studied with modern setups. The major breakthrough in this technique occurred in the 1990s when analysis was developed to allow the measurement of mechanical properties under load. The Oliver and Pharr method [44] -based measurements are usually carried out by the simple three-stage procedure including loading, hold region, and unloading of a tip, typically a diamond pyramid. Figure 16.17 illustrates a schematic of a load versus displacement curve for fused silica, which is usually used as a standard to measure the tip area function (A) accounting for deviations from a perfect tip geometry. The unloading part of the obtained load (P) versus displacement (h) curve is assumed to represent purely elastic behavior and the reduced elastic modulus (E_r) is obtained from Eq. (16.26)

$$S = \frac{dP}{dh} = \frac{2}{\sqrt{\pi}} E_r \sqrt{A}, \tag{16.26}$$

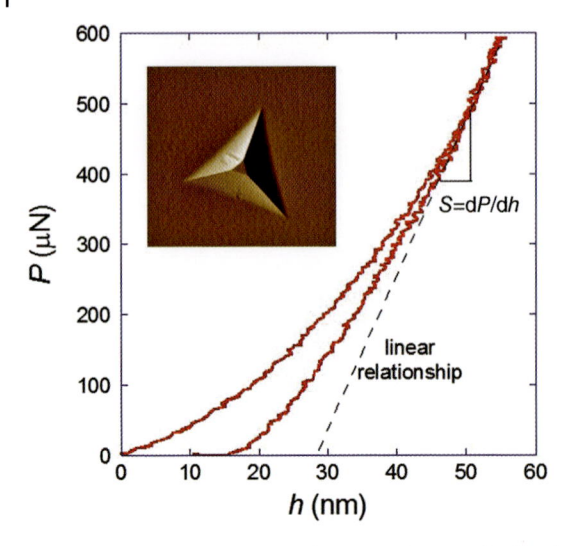

Figure 16.17 Load (P) versus displacement (h) curve for fused silica standard with a residual indent. The unloading curve can be related to the elastic modulus.

where the reduced elastic modulus is used to account for the elastic deformation of the sample and the tip

$$\frac{1}{E_r} = \left(\frac{1-\nu^2}{E}\right)_{\text{specimen}} + \left(\frac{1-\nu^2}{E}\right)_{\text{indenter}} \tag{16.27}$$

16.4.5
Fracture Properties

In order to apply linear elastic fracture mechanics, the behavior needs to be predominantly elastic. Hence, three conditions have to be met to experimentally determine the fracture toughness through linear-elastic fracture mechanics. First of all, the sample dimensions need to be large compared to the plastic zone size. Second, one needs to be able to exactly determine the load on the sample at the moment of instable fracture. Third, the relationship between the stress intensity factor and the boundary conditions as, for example, load and crack length need to be known. To determine the fracture toughness, it is common to apply tests according to ASTM-Standard E399-90 [45]. Here different sample geometries are applied, from which the single-edge notch bend test is displayed in Figure 16.18.

The first condition that the sample dimensions need to be large compared to the plastic zone size can be expressed in Eq. (16.28)

$$a, W - a, B \geq 2.5 \left(\frac{K_{IC}}{\sigma_y}\right), \tag{16.28}$$

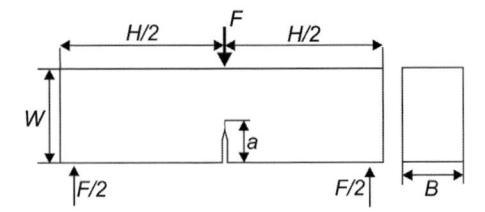

Figure 16.18 Schematics of a single edge notch bend test.

where W and B are the sample width and thickness, respectively, and a is the crack length [46]. The dependence of the stress intensity as a function of sample thickness is depicted in Figure 16.19. Only for large sample dimensions as expressed by Eq. (16.28) and illustrated in Figure 16.19, the geometry independent fracture toughness (K_{IC}) can be obtained. In case the sample thickness is chosen to small according to Eq. (16.28), the body is in plane stress conditions and nonconservative fracture intensities are obtained. With increasing sample dimensions, the body imposes an increase in resistance to plastic flow at the crack tip. Hence, the sample is closer to plane strain conditions, which results in a small process zone and hence a conservative fracture toughness measurement.

Prior to quasi static loading a notch with a known length is introduced into the sample. Afterwards, the sample is exposed to cyclic loading to induce a sharp crack tip, which also leads to a conservative evaluation of fracture toughness. The stress intensity factor and hence the fracture toughness for the sample geometry displayed in Figure 16.18 can be evaluated by Eq. (16.29)

$$K_I = \frac{F}{BW^{0.5}} f(a/W) \tag{16.29}$$

where F is the force applied and, the value of $f(a/W)$ depends on the sample geometry [46].

16.4.6
Friction and Wear Test

Friction and wear tests exist in a wide variety with some standardization [31].

On the macroscale friction and wear is often investigated by pin-on-disc experiments. A typical test configuration is shown in Figure 16.20 from Holmberg and Matthews [31].

A spherically ended pin is loaded typically with 10 N normal force and the sample disc is moved with a sliding speed of typically 0.1 m/s in a temperature and humidity-controlled environment. As the friction coefficient, μ, is the frictional force or tangential force, F, divided by the normal force, w, on the contact [31] and the normal force is known and the tangential force is measured, the friction coefficient can readily be computed and displayed versus the sliding distance. The wear rate at various sliding distances is measured by optical or

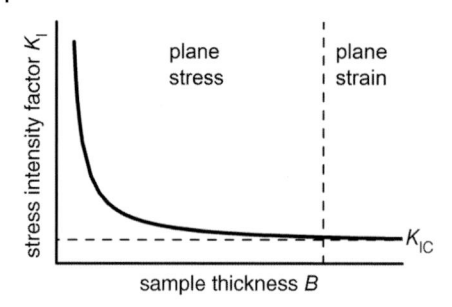

Figure 16.19 Stress intensity factor as a function of sample thickness.

topographical inspection of the wear track with confocal microscopy or with a surface stylus technique, respectively.

Holmberg and Matthews point out that in the literature reported, friction and wear data show considerable variability as they critically depend not only on the disc and pin materials but also on physical and chemical interactions in the actual pin-on-disc tester that may vary from laboratory to laboratory [31].

Friction and wear tests on the nanoscale are often conducted in a scanning tunneling microscope or in an atomic force microscope. A friction force microscope (FFM) is a variation of an atomic force microscope. [47] In FFM a few atoms in a sharp tip are brought into repulsive contact with the sliding sample

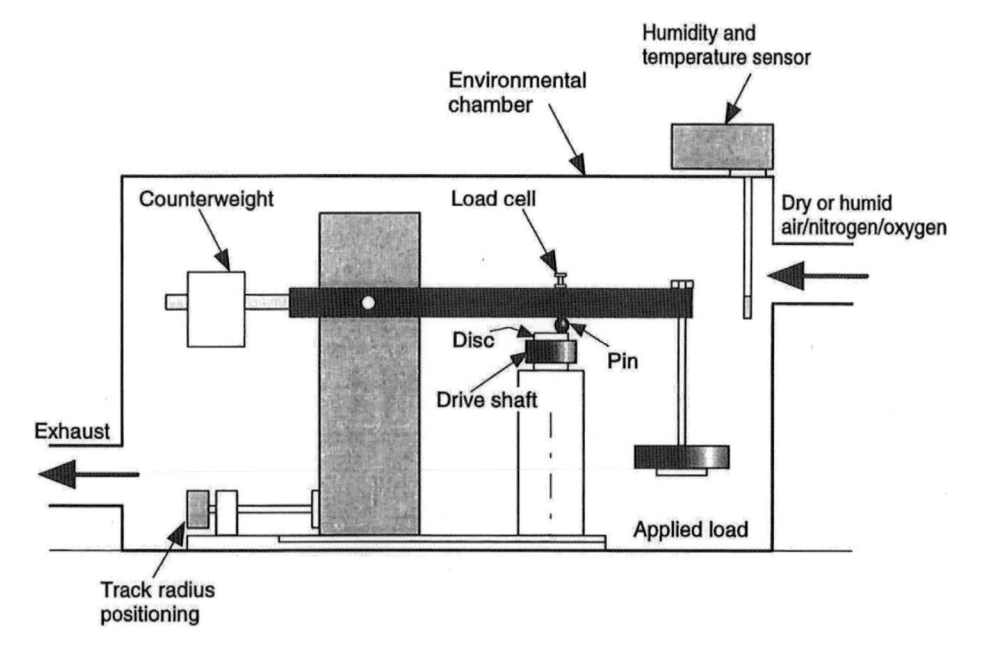

Figure 16.20 Equipment layout from a pin-on-disc tester with humidity and temperature control. (From Ref. [31], Copyright 1994, with permission from Elsevier.)

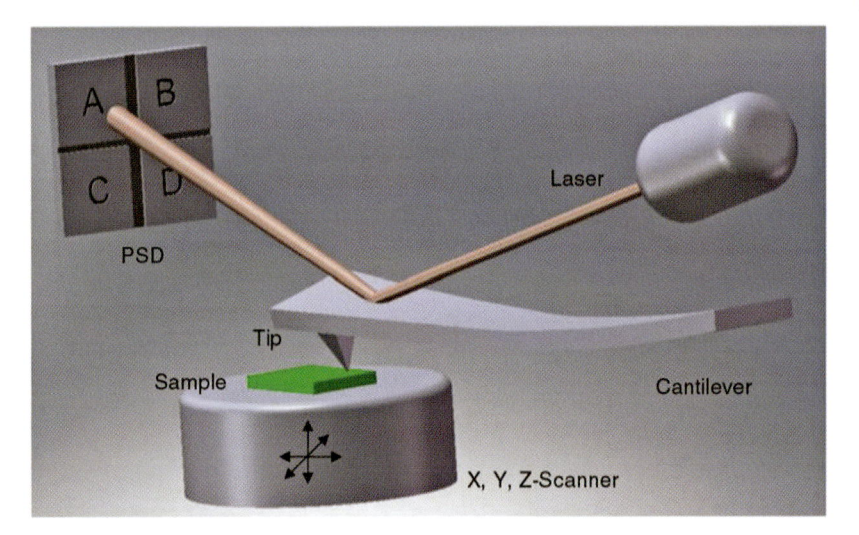

Figure 16.21 Principle of the friction force microscopy. Bending and torsion of the cantilever are measured simultaneously by detecting the lateral and vertical deflection of a laser beam while the sample is scanned in the x–y plane. The laser beam deflection is determined using the four-quadrant position sensitive detectors (PSD): $(A+B)-(C+D)$ is a measure for the bending and $(A+C)-(B+D)$ a measure for the torsion of the cantilever, if A, B, C, and D are proportional to the intensity of the incident light of the corresponding quadrant. (From Ref. [48], Copyright 2013, with permission from Springer.)

surface. The normal force and the tangential force resulting from the relative motion of the sample with respect to the tip bend and twist the tip respectively. With the help of a laser beam, the bending and twisting of the tip can be measured and the atomic scale friction coefficient calculated. Figure 16.21 shows the schematic of a friction force microscope.

Acknowledgments

The authors acknowledge support by the German National Science Foundation (DFG) within the SFB TR 87. WJC acknowledges the EPSRC/Rolls-Royce Strategic Partnership (EP/M005607/1).

References

1 Galilei, G. (1638) *Discorsi e Dimostrazioni Matematiche Intorno a Due Nuove Scienze*.
2 Cahn, R.W. (2001) *The Coming of Materials Science*, Pergamon Materials Series, Elsevier Science Ltd.
3 Hohenberg, P. and Kohn, W. (1964) Inhomogeneous electron gas. *Phys. Rev.*, **136**, 864–871.
4 Kresse, G. and Fürthmüller, J. (1996) Efficient iterative schemes for *ab initio*

total-energy calculations using a plane-wave basis set. *Phys. Rev. B*, **54**, 11169.
5. Wallace, D.C. (1998) *Thermodynamics of Crystals*, Dover Publications, Inc., Mineola, NYo, USA.
6. Kittel, C. (ed.) (1996) *Introduction to Solid State Physics*, 7th edn, John Wiley and Sons, Inc., New York.
7. Murnaghan, F.D. (1944) The compressibility of media under extreme pressures. *Proc. Natl. Acad. Sci. USA*, **30** (9), 244–247.
8. Birch, F. (1947) Finite elastic strain of cubic crystals. *Phys. Rev. A*, **71** (11), 809–824.
9. Taylor, G.I. (1934) The mechanism of plastic deformation of crystals. Part I. Theoretical. *Proc. R. Soc. Lond. A Mat.*, **145** (855), 362–387.
10. Polanyi, M. (1934) Über eine art gitterstörung, die einen kristall plastisch machen könnte. *Zeitschrift Physik*, **89** (9), 660–664.
11. Orowan, E. (1934) Zur Kristallplastizität. III. *Zeitschrift Physik*, **89** (9), 634–659.
12. Gottstein, G. (2014) *Materialwissenschaft und Werkstofftechnik*, 4th edn, Springer Vieweg, Berlin Heidelberg.
13. Frost, H.J. and Ashby, M.F. (1982) *Deformation Mechanism Maps*, Pergamon Press, Oxford.
14. Hirsch, P.B., Horne, R.W., and Whelan., M.J. (1956) Direct observations of the arrangement and motion of dislocations in aluminium. *Philos. Mag.*, **1** (7), 677–684.
15. Gilman, J.J. (2003) *Electronic Basis of the Strength of Materials*, Cambridge University Press, Cambridge, United Kingdom.
16. Cahn, R.W. and Haasen., P. (1996) *Physical Metallurgy*, 4th edn, vol. **3**, North-Holland.
17. Frank, F.C. and Read, W.T. (1950) Multiplication processes for slow moving dislocations. *Phys. Rev.*, **79** (4), 722–723.
18. Peierls, R. (1940) The size of a dislocation. *Proc. Phys. Soc. Lond.*, **52** (1), 34.
19. Hull, D. and Bacon, D.J. (2011) *Introduction to Dislocations*, 5th edn, Butterworth-Heinemann.
20. von Mises, R.E. (1928) Mechanik der plastischen Formänderung von Kristallen. *Z. Angew. Math. Mech.*, **8**, 161–185.
21. Groves, G.W. and Kelly, A. (1963) Independent slip systems in crystals. *Philos. Mag.*, **8** (89), 877–887.
22. Mosecker, L., Pierce, D.T., Schwedt, A., Beighmohamadi, M., Mayer, J., Bleck, W., Wittig, J.E. (2015) Temperature effect on deformation mechanisms and mechanical properties of a high manganese C + N alloyed austenitic stainless steel. *Mater. Sci. Eng. A*, **642**, 71–83.
23. Inglis, C.E. (1913) Stresses in plates due to the presence of cracks and sharp corners. *Trans. Inst. Naval Architects*, **55**, 219–241.
24. Griffith, A.A. (1921) The phenomena of rupture and flow in solids. *Philos. Trans. R. Soc. Lond. A*, **221**, 163–198.
25. Maugin, G.A. (1992) *The Thermomechanics of Plasticity and Fracture, Cambridge Texts in Applied Mathematics*, Cambridge University Press.
26. Lawn, B.R. (1993) *Fracture of Brittle Solids - Second Edition*, 2nd edn, Cambridge Solid State Science Series, vol **1**, Cambridge University Press, Cambridge.
27. Obreimoff, J.W. (1930) The splitting strength of mica. *Proc. R. Soc. Lond. A Mat.*, **127** (805), 290–297.
28. Irwin, G.R. (1957) Analysis of stresses and strains near the end of a crack traversing a plate. *J. Appl. Mech.*, **24**, 361–364.
29. Gross, D. and Seelig, T. (2011) *Bruchmechanik*, 5th edn, Springer-Verlag, Berlin Heidelberg.
30. Ashby, M.F. (2005) *Materials Selection in Mechanical Design*, 3rd edn, Elsevier Butterworth-Heinemann, Oxford.
31. Holmberg, K. and Matthews, A. (1994) *Coatings Tribology*, Elsevier.
32. Fall, A., Weber, B., Pakpour, M., Lenoir, N., Shahidzadeh, N., Fiscina, J., Wagner, C., and Bonn, D. (2014) Sliding friction on wet and dry sand. *Phys. Rev. Lett.*, **112** (17), 175502.
33. Gosvami, N.N., Bares, J.A., Mangolini, F., Konicek, A.R., Yablon, D.G., and Carpick., R.W. (2015) Mechanisms of antiwear tribofilm growth revealed in situ by single-asperity sliding contacts. *Supramol. Sci.*, **348** (6230), 102–106.
34. Bhushan, B., Israelachvili, J.N., and Landman., U. (1995) Nanotribology: friction, wear and lubrication at the atomic scale. *Nature*, **374** (6523), 607–616.

35 Landman, U., Luedtke, W.D., Burnham, N.A., and Colton., R.J. (1990) Atomic mechanisms and dynamics of adhesion, nanoindentation, and fracture. *Supramol. Sci.*, **248** (4954), 454–461.

36 Kunze, C., Music, D., Baben, M., Schneider, J.M., and Grundmeier., G. (2014) Temporal evolution of oxygen chemisorption on TiAlN. *Appl. Surf. Sci.*, **290**, 504–508.

37 Music, D. and Schneider., J.M. (2013) *Ab initio* study of $Ti_{0.5}Al_{0.5}N(001)$—residual and environmental gas interactions. *New J. Phys.*, **15**, 073004.

38 Szlufarska, I., Chandross, M., Carpick, W., and Robert, W. (2008) Recent advances in single-asperity nanotribology. *J. Phys. D Appl. Phys.*, **41** (12), 123001.

39 Carpick, R.W., Flater, E.E., Sridharan, K., Ogletree D.F., and Salmeron, M. (2004) Atomic-scale friction and its connection to fracture mechanics, *JOM* **56** (48–52).

40 Bhushan, B. (2005) Nanotribology and nanomechanics. *Wear*, **259** (7–12), 1507–1531.

41 Urbakh, M. and Meyer, E. (2010) Nanotribology: the renaissance of friction. *Nat. Mater.*, **9** (1), 8–10.

42 Krim, J., Dawsona, B.D., Barefoota, K., Pana, L., Pearsonb, J., Zikryb, M., Bakerc, C., and Voevodind, A. (2009) Nanoscale design of adaptive tribological coatings for gold–ytrium based nanocomposites. *Tribology*, **3** (4), 145–150.

43 Muratore, C. and Voevodin, A.A. (2009) Chameleon coatings: adaptive surfaces to reduce friction and wear in extreme environments. *Annu. Rev. Mater. Res.*, **39** (1), 297–324.

44 Oliver, W.C. and Pharr, G.M. (1992) An improved technique for determining hardness and elastic modulus using load and displacement sensing indentation experiments. *J. Mater. Res.*, **7** (6), 1564–1583.

45 ASTM (2001) Standard test method for measurement of fracture toughness.

46 Blumenauer, H. and Pusch., G. (1993) *Technische Bruchmechanik*, 3rd edn, Dt. Verl. Für Grundstoffindustry, Leipzig.

47 Bennewitz, R. (2005) Friction force microscopy. *Mater. Today*, **8** (5), 42–48.

48 Qian, L. and Yu, J. (2013) *Friction Force Microscopy*, Springer.

17
Calorimetry

Hitoshi Kawaji

Tokyo Institute of Technology, Institute of Innovative Research, Laboratory for Materials and Structures, 4259-R3-8, Nagatsuta-cho, Midori-ku, Yokohama, 226-8503, Japan

17.1 Introduction

The stability of a phase of a material can be understood essentially by the chemical thermodynamics, except for the appearance of a metastable phase due to the kinetics. The phase having the minimum Gibbs energy at the given condition (temperature, pressure, and composition of multicomponent system) should be realized as the stable phase. Thus, calorimetry is a very powerful tool for characterizing the materials from the viewpoint of chemical thermodynamics, because the Gibbs energy can be determined experimentally by calorimetry. The Gibbs energy of a material has two terms:

$$G = H - TS, \tag{17.1}$$

where H is the enthalpy of the material, T is the temperature, and S is the entropy of the material. In the chemical thermodynamics, the enthalpy term relates to the enthalpy of formation of the material, and the value can be determined using Hess's law from the reaction enthalpies of the related materials measured by a reaction calorimetry. The chemical reactions include dehydration, decomposition, neutralization, dissolving, solid-state reaction, and so on. The adiabatic calorimeter, bomb calorimeter, Calvet-type calorimeter, solution calorimeter, isothermal titration calorimeter, and differential scanning calorimeter (DSC) have been used for this purpose. The measurement of the enthalpy of reaction should be combined with the heat capacity measurement to know the enthalpy at a certain temperature T, because the reaction calorimetry usually can measure only at the temperature of calorimeter, T_1. The enthalpy at the temperature T is obtained by the following equation:

$$H(T) = H_1 + \int_{T_1}^{T} C_p dT, \tag{17.2}$$

Handbook of Solid State Chemistry, First Edition. Edited by Richard Dronskowski, Shinichi Kikkawa, and Andreas Stein.
© 2017 Wiley-VCH Verlag GmbH & Co. KGaA. Published 2017 by Wiley-VCH Verlag GmbH & Co. KGaA.

where H_1 is the enthalpy measured at T_1 and C_p is the heat capacity of the material measured from T_1 to T. The standard enthalpies of formation of various compounds are found in the thermodynamic tables [1–5], which list the values under standard state pressure (SSP) ($p^0 = 100$ or 101.325 kPa). Old tables published before 1982 use 101.325 kPa as SSP. IUPAC recommended 100 kPa for SSP in 1982, although 101.325 kPa is still used as SSP in many papers and databases on chemical thermodynamics. The tables usually give the values of the standard enthalpies of formation at $T = 298.15$ K, and the values at other temperatures should be derived using Eq. (17.2).

The entropy terms (reaction entropy) usually cannot be determined by reaction calorimetry, because most of the reactions are irreversible processes. Thus, the entropy of a phase of material should be determined by the integration of heat capacity divided by temperature using the following equation:

$$S(T) = S_1 + \int_{T_1}^{T} \frac{C_p}{T} dT, \tag{17.3}$$

where S_1 is the enthalpy at T_1. The third law of thermodynamics predicts that the entropy becomes zero at absolute zero in the absence of degeneracy in the ground state, and thus the absolute value of entropy of the materials can be determined only from the heat capacity rate from absolute zero to the target temperature using the following equation:

$$S(T) = S_0 + \int_{0}^{T} \frac{C_p}{T} dT = \int_{0}^{T} \frac{C_p}{T} dT. \tag{17.4}$$

However, it is impossible to decrease the temperature to absolute zero and to measure the heat capacity from absolute zero. Thus, the measured heat capacity to absolute zero must extrapolate using some kinds of theoretical models to obtain the entropy. Fortunately, the heat capacity is known to vanish at absolute zero, and it is enough to measure the heat capacity to a sufficiently low temperature from where it may be safely extrapolated to zero. In addition, dielectric materials usually have a low-temperature heat capacity proportional to T^3, while metals have an additional term proportional to T. These variations permit a ready extrapolation of the measured data to absolute zero. However, the detailed extrapolation method depends on the materials, for example, dielectric materials, metals, magnetic materials, superconductors, and so on.

The third law predicts $S_0 = 0$, which means the absence of degeneracy in the ground state of martials. However, there are some cases of the presence of degeneracy at very low temperatures as in glassy state or frozen-in state of materials. The entropy of gases at certain temperature can be calculated using Eq. (17.3) from the heat capacity of the solid, liquid, and gas phases as well as the latent heats of the phase transitions. It is also possible to calculate the entropy by statistical mechanics using spectroscopic observations. Both values must agree if the third law of thermodynamics and the statistical mechanics are correct. The agreement is very good in most cases, but not so well in some cases, for example, CO, H_2O, N_2O, CH_3D, and so on. The discrepancies are removed in that the solid states of these materials are not in internal equilibrium at very low temperatures, and contains frozen-in configurationally disorder not revealed in the calorimetric measurements of heat capacity and hence in the evaluation of the entropy. In the case of CO, the orientation of the molecule may be either CO or OC in the solid state. The molecule rotates freely at high temperatures, and the solid should become completely ordered at absolute zero. However, the high-temperature disordered state cannot be ordered, but becomes frozen at a temperature due to the large potential barriers opposing the molecular rotation in the solid. In these cases, many disordered configurations are degenerated at absolute zero and thus the solid has a residual entropy. All of the materials with nonequilibrium states like glasses have residual entropies.

Calorimetry is the way to measure the heat generated or absorbed during the change of the physical or the chemical state of matter. The physical changes include the temperature change of the materials and the phase transitions, that is, melting, evaporation, and many types of solid–solid phase transition, such as ferromagnetic transition, ferroelectric transition, superconducting transition, metal–insulator transition, and so on. The heat capacity of material is the heat needed to raise the temperature of the material by one degree, and relates to the amount of thermal excitation during the rise of temperature. Thus, the heat capacity measurements enable us to know the thermal behavior of lattice vibrations, electronic excitations, magnetic excitations, and so on. The calorimetry is also very powerful in detecting the presence of phase transitions, because all types of phase transitions can be detected by calorimetry, for example, a ferroelectric phase transition can be detected by calorimetry, but not by magnetic susceptibility measurements. The various types of calorimetry will be shown in this chapter.

17.2
Nonreaction Calorimetry for Heat Capacity Measurements

Nonreaction calorimetry is used for measuring the heat capacity and the heat of physical change of materials. The obtained values are related to the temperature coefficient of the enthalpy change and the phase transition enthalpy if the measurements are carried out under the constant pressure (normal condition). The phase transitions are detected as an anomalous peak in the temperature dependence of the heat capacity.

17.2.1
Adiabatic Calorimetry for Heat Capacity Measurements

The adiabatic calorimeter has been used to obtain the absolute values of heat capacity of materials in a wide temperature range for determining accurately the entropy. The absolute value of heat capacity measured by adiabatic calorimetry is most accurate from about 10 K up to about 800 K. There are few commercial apparatus of adiabatic calorimeter, and thus the almost reliable data have been reported by homemade calorimeters [6,7]. The principle of the adiabatic calorimeter for heat capacity measurements is very simple, as shown in Figure 17.1 [7,8]. If the sample is maintained under adiabatic condition, the sample comes in a thermodynamic equilibrium state at an initial temperature T_i. While keeping the adiabatic condition, a known energy Q is input to the sample. After the energy input, the next equilibrium state is attained at a higher final temperature T_f. The heat capacity of sample is simply determined by Eq. (17.5):

$$C = \frac{Q}{\Delta T} = \frac{Q}{T_f - T_i}, \qquad (17.5)$$

because all of the input energy is used for increasing the temperature of the sample under the adiabatic condition. The obtained value by Eq. (17.5) is an averaged heat capacity between the initial and final temperatures, and usually can be considered as the heat capacity at the mean temperature of the initial and final temperatures. However, the strict definition of heat capacity at T is the limiting value of Eq. (17.6) as

$$C(T) = \frac{Q}{\lim_{\Delta T \to 0} \Delta T} = \frac{Q}{\lim_{T_f \to T_i} (T_f - T_i)}. \qquad (17.6)$$

Thus, a curvature correction may be necessary to determine the accurate value at the temperature from the measured average heat capacity in the case when the temperature dependence of heat capacity becomes strongly nonlinear as observed near the phase transition temperature. Ordinarily, such curvature correction can be ignored if the temperature increment ΔT is smaller than about 1% of the mean temperature. The small breaking of adiabatic condition can be detected by the small temperature drift at the initial and final states, and the

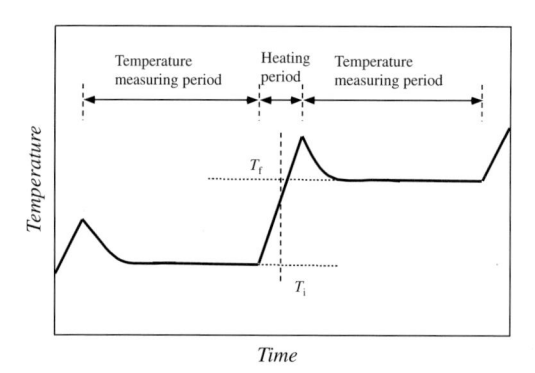

Figure 17.1 Temperature variation of adiabatic heat capacity calorimetry for heat capacity measurement. After the attainment of a thermodynamic equilibrium state at T_i, a known amount of heat is added to the sample. After finishing the heating, the next equilibrium state is attained at a higher final temperature T_f.

effect can be corrected from the drift rates. This monitoring and the correction for the small breaking of the adiabatic condition enable the high accurate measurements by the stepwise heating method. The sample is usually loaded in a calorimeter vessel equipped with a thermometer and a heater, and thus the obtained heat capacity includes the contribution of the vessel. The heat capacity of sample is calculated by subtracting the heat capacity of empty vessel, which is measured before the sample measurements, from the measured total heat capacity.

The sample vessel is surrounded by adiabatic shields to avoid heat transfer by thermal radiation, and kept in high vacuum condition below 10^{-4} Pa to prevent the heat conduction by gas. The temperature difference between the sample vessel and the adiabatic shields is detected by thermocouples, and the heater wounded on the adiabatic shields is controlled to maintain the zero temperature difference between them. The electrical lead wires for the thermometer and the heater of the sample vessel should be thin to avoid heat conduction. However, the perfect adiabatic condition is not realized by the small heat leak through the nonuniformity of temperature of the sample vessel and the adiabatic shields, and so on. Some extra shields can be set up at the outer side of the adiabatic shield to reduce the heat leak.

Figure 17.2 shows an example of cryostat for adiabatic calorimeter. Liquid nitrogen and liquid helium are used for refrigeration. The sample vessel can be cooled down to the lowest measuring temperature by introducing a small amount of helium gas for heat exchange in the vacuum space, and the helium gas is evacuated after the cooling. The heat capacity measurements of heating direction can be made using this calorimeter. The precise thermometry with a precision of 10^{-4} K leads to the accuracy of heat capacity measurement of about 0.1%. Although the stepwise heating method described is more accurate than the continuous heating method, the continuous heating may be employed to measure the precise temperature dependence of heat capacity near phase transitions.

Figure 17.3 shows an example of results of heat capacity measurement made by an adiabatic calorimeter [9]. In this figure, the heat capacity of the sample is indicated by C_p, which means heat capacity at constant pressure. In these measurements, the samples were sealed in a vessel with a small amount of helium gas for heat exchange (about 10 Pa). The pressure of the sample changes very little during the each heat capacity measurement; thus, the obtained heat capacity can be considered as the heat capacity at constant pressure without a serious error. However, this approximation works well only for solids and liquids without the large vapor pressure of materials. If the measurement is carried out near the boiling point of a liquid, the obtained value relates to the heat capacity at saturated pressure rather than the C_p. It is very difficult to measure directly the heat capacity at constant volume C_V of solids and liquids, because there is no good container keeping the volume of sample constant against the thermal expansion of the sample during the heat capacity measurement. Thus, the C_V is usually calculated from the C_p, the volume thermal expansion coefficient α, and

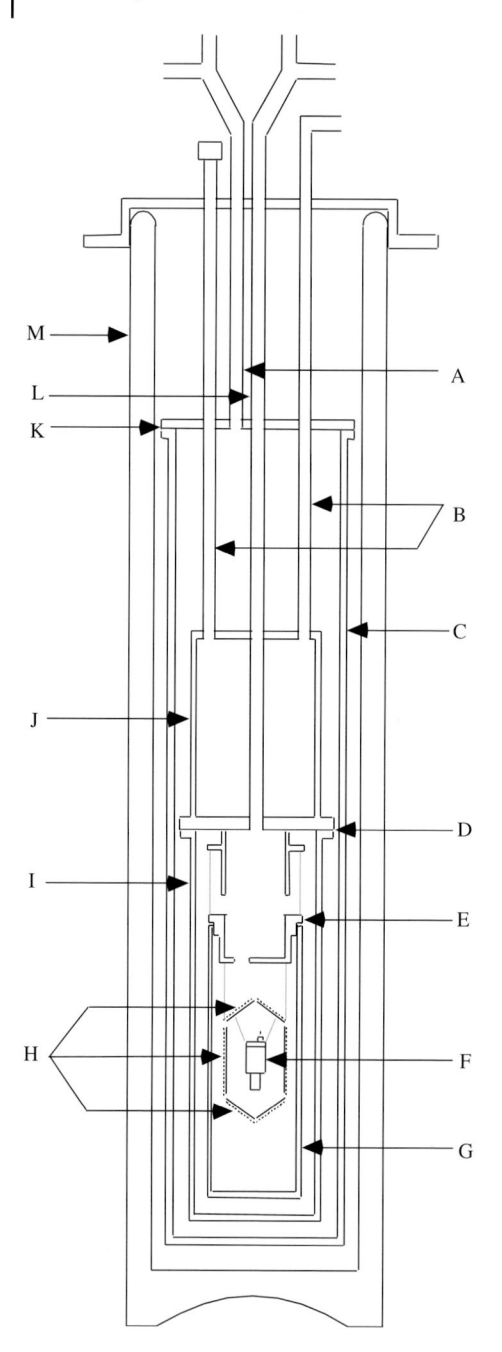

Figure 17.2 Cryostat of an adiabatic calorimeter. A: evacuation tube; B: refrigerant transfer tubes; C: outer jacket; D: indium seal; E: thermal sink; F: calorimeter vessel; G: outer shield; H: adiabatic shields; I: inner jacket; J: liquid helium can; K: indium seal; L: evacuation tube; M: liquid nitrogen Dewar vessel.

the isothermal compressibility k_T of the sample using the thermodynamic equation:

$$C_p - C_V = \frac{TV\alpha^2}{k_T}. \qquad (17.7)$$

The C_V is about a few percent smaller than C_p in many solids at room temperature, and the difference between C_V and C_p decreases rapidly as the temperature decreases.

Figure 17.3 shows the typical temperature dependence of the heat capacity of dielectric solids on the whole. The main contribution to the heat capacity solid is the thermal excitation of lattice vibrations, that is, thermal motions of atoms and ions in solids. At elevated temperatures, the C_V should approach the value expected by the classical statistical mechanics, that is, $3k_B$ for 1 atom. In the case for PMN and PMT, the heat capacity indeed approaches the expected value of $3k_B \times 5N_A = 15R \approx 125 \, \text{J/(K mol)}$. On the other hand, the heat capacity decreases rapidly with decreasing the temperature below room temperature. The decrease of heat capacity toward zero is required by the third law of thermodynamics and explained by the quantum statistical mechanics of lattice vibrations. The heat capacity of dielectric materials shows a T^3 variation at very low temperatures, which is explained theoretically by Debye's model for lattice

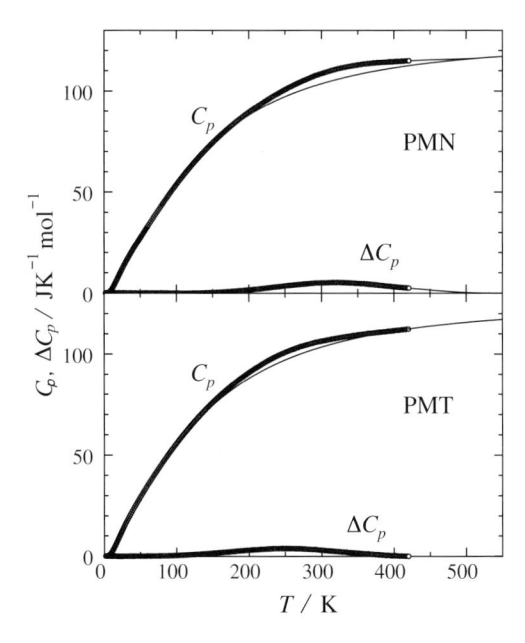

Figure 17.3 Molar heat capacity at constant pressure C_p of relaxors PMN (PbMg$_{1/3}$Nb$_{2/3}$O$_3$) and PMT (PbMg$_{1/3}$Ta$_{2/3}$O$_3$). ΔC_p calculated by subtracting the contribution of normal lattice vibrations from the measured C_p corresponds to the excess heat capacity due to the formation of polar nanoregion (PNR) in the relaxors.

vibrations in solids, and an additional T-linear term may be present for metals due to the contribution of conduction electrons.

PMN (PbMg$_{1/3}$Nb$_{2/3}$O$_3$) and PMT (PbMg$_{1/3}$Ta$_{2/3}$O$_3$) are the relaxors found very early and studied most extensively. Relaxor shows a very large peak in the temperature dependence of the dielectric constant as in ferroelectrics. However, the peak depends strongly on the measuring frequency, and thus such material is called relaxor form of the relaxation property of the dielectric constant peak. The averaged crystal structure of the relaxors belongs to the cubic space group Pm-3 m in the whole temperature range and no structural phase transition is detected near the peak temperature. On the other hand, very small polarized region, which is called polar nanoregion (PNR), is reported in the cubic matrix of the compounds. The volume fraction of PNR increases with the decreasing temperature, and reaches about 20% at about 200 K. The growth of PNR freezes around 200 K and the size of the PNR is limited to about 10 nm. The formation of PNR is reported to start below about 600 K in PMN by Burns and Dacol [10]. Figure 17.3 shows that no remarkable heat capacity anomaly is observed, but an excess heat capacity spread over a wide temperature range from about 100 K to above 500 K can be detected by the precise measurements. The excess heat capacity corresponds to the formation of PNR in the relaxors. The entropy calculated from the excess heat capacity can be estimated to 3.3 J/(K mol) for PMN and 2.9 J/(K mol) for PMT. The large values show the formation of PNR related to order–disorder mechanism in the materials as expected from the disordered crystal structure. The small entropy value obtained from the excess heat capacity corresponds to the fact that the growth of PNR does not fully progress in the crystal [9].

17.2.2
Relaxation Method for Heat Capacity Measurements

The relaxation method is originally developed to measure the heat capacity of small samples in helium temperature region, where the realization of adiabatic condition is difficult [11]. The commercial apparatus has recently become available and is widely used for studying the solid–solid phase transitions below room temperature [12,13]. The schematic simple model of the method is shown in Figure 17.4. A sample, a thermometer, and a heater are attached tightly to a sample stage, and the sample stage is connected thermally to a heat bath with a thermal conductor. If a power $\dot{Q}$ is generated from the heater, the temperature of the sample state is increased following the energy balance equation:

$$C\frac{dT}{dt} = \dot{Q} - k(T - T_{\text{bath}}), \tag{17.8}$$

where T and T_{bath} are the temperatures of sample stage and heat bath, respectively, C is the total heat capacity of the sample and sample stage, and k is the thermal conductance of the thermal resistance. The continuous heating results in a steady state, as shown in Figure 17.5. The temperature of sample stage

17.2 Nonreaction Calorimetry for Heat Capacity Measurements

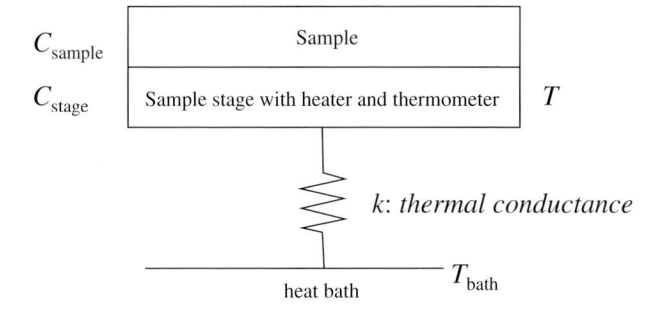

Figure 17.4 Thermal model of a simple relaxation calorimeter. The sample with the heat capacity C_{sample} is tightly connected to the sample stage with C_{stage} equipped with the heater and the thermometer, and is weakly connected with a thermal conductance k to the heat bath at T_{bath}.

becomes constant at a temperature slightly higher than T_{bath}, and the condition of $dT/dt = 0$ leads to the following equation:

$$T - T_{bath} = \Delta T = \frac{\dot{Q}}{k}, \tag{17.9}$$

where ΔT is the temperature difference between the sample stage and the heat bath at the steady state. The value of thermal conductance k at that temperature

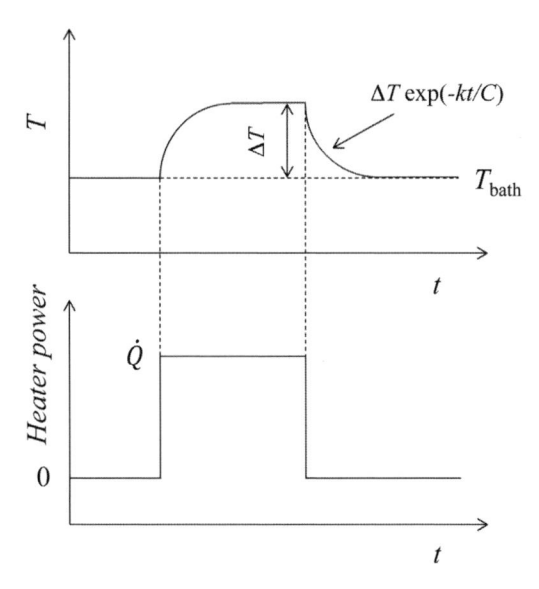

Figure 17.5 Time variations of heater power and the temperature at the thermometer through the process of heat capacity measurement by the relaxation method. If a power $\dot{Q}$ is generated from the heater, the temperature of the sample state is increased and reaches a steady state determined by the energy balance equation. After turning off the heater, the temperature of sample stage starts to relax to the bath temperature T_{bath}.

can be determined by measuring the temperature difference between ΔT and $\dot{Q}$. After the attainment to the steady state, the heater is turned off. Then the temperature of the sample stage starts to relax to the bath temperature, as shown in Figure 17.5. According to Eq. (17.8), the temperature of sample stage follows Eq. (17.10):

$$T = T_{\text{bath}} + \Delta T e^{-kt/C}. \tag{17.10}$$

The relaxation time of temperature $\tau = C/k$ can be determined by analyzing the temperature variation curve, and then the total heat capacity C is obtained. The typical time constant is about 10^{-1}–100 s. The obtained heat capacity includes the contribution of the sample stage, and the heat capacity of sample is obtained by subtracting the heat capacity of sample stage, which is measured before the sample measurements. The sample needed for the measurements is from about 0.1 to 100 mg, which is two to five orders of magnitude lower than that required for adiabatic calorimetry. By using this method, the heat capacity is measured with good accuracy and precision (better than a few percent) using a very small amount of sample. It should be noted that the detailed information about first-order phase transition is difficult to obtain by this method, because a large thermal hysteresis appears at the first-order phase transition.

The thermal uniformity of the sample and the sample stage is essential for the precise measurements by the relaxation method. Thus, the careful check for the thermal uniformity is required especially for the samples with low thermal conductivity, and for the thermal contact between the sample and the sample stage. For the case of imperfect thermal contact between the sample and the sample stage, the model for analyzing the relaxation curve has been developed [14] and the analyzing software is equipped in commercial apparatuses.

Figure 17.6 shows an example of the results of heat capacity measurements carried out by a relaxation method using the commercial apparatus PPMS (Quantum Design, Inc.) [15]. $CoCr_2O_4$ is known as a multiferroic material with spontaneous magnetization and magnetically induced ferroelectricity in the conical spin state. The ferrimagnetic transition is found as a λ-like peak at $T_C = 93$ K. The most striking feature is a very sharp peak at $T_S = 26$ K due to a transition to the conical spin state. The extremely sharp peak may indicate the first-order nature of this transition. The lock-in transition is observed as a small change in the heat capacity curve at $T_L - 13$ K. The thermal hysteresis as seen in the *inset* of Figure 17.6 suggests a first-order nature of the transition.

17.2.3
AC Calorimetry for Heat Capacity Measurements

If a sinusoidal power is applied to a sample, the temperature of the sample will vary sinusoidally and the amplitude of the oscillation will be inversely proportional to the heat capacity of the sample. The measurement of heat capacity of a small sample of about a few milligrams with high temperature resolution of 1–10 mK is possible using AC calorimetry. To take an appropriate measuring

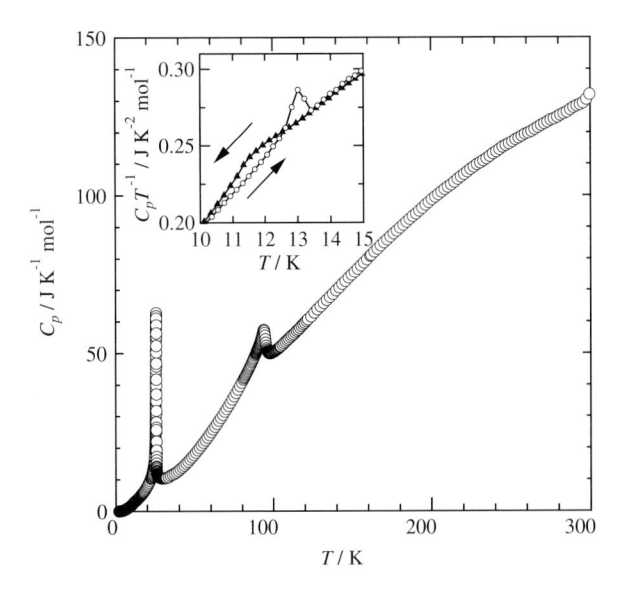

Figure 17.6 Heat capacity of the multiferroic material CoCr$_2$O$_4$ by a relaxation method using about 10 mg of sample. Sharp peaks are clearly observed at $T_C = 93$ K and $T_S = 26$ K. The *inset* shows C_p/T of the vicinity of the lock-in transition at $T_L = 13$ K.

frequency, the measurements at high temperatures [16] and at high pressures [17] are possible.

The schematic model of a typical method of AC calorimetry is shown in Figure 17.7 [18]. The side at $x = L$ of a sample with thickness L is connected to a heat bath with a thermal conductance k. If an AC heat power $\dot{Q}\sin(\omega t)$ is applied to the opposite side of the sample as shown in Figure 17.7, the amplitude of the temperature oscillation at the $x = L$ is

$$T_{ac} = \frac{\dot{Q}}{\omega C}\left[1 + (\omega\tau_{int})^2 + (\omega\tau_{ext})^{-2} + \sqrt{40\frac{\tau_{int}}{\tau_{ext}}}\right]^{-1/2}, \quad (17.11)$$

where τ_{int} is the internal relaxation time in the sample to become uniform temperature and $\tau_{ext} = C/k$ is the external relaxation time for heat dissipating from

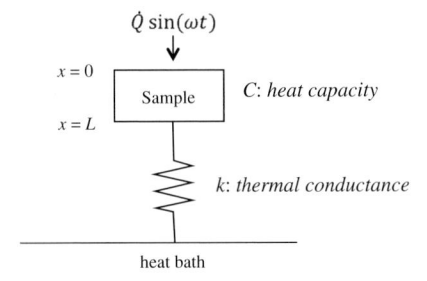

Figure 17.7 One-dimensional thermal model of an AC calorimeter. A slab-like sample with thickness L is connected to a heat bath with a thermal conductance k.

the sample to heat bath, and ω is the angular frequency. For the thermal uniformity inside the sample, the measurement frequency should be low enough to satisfy the condition of $\omega\tau_{int} \ll 1$, which indicates that the higher frequency limit is governed by the thermal conductivity and the thickness of sample. If the frequency is chosen to satisfy also the condition $\omega\tau_{ext} \gg 1$, Eq. (17.11) becomes simple to obtain the heat capacity:

$$C = \frac{\dot{Q}}{\omega T_{ac}}. \tag{17.12}$$

One of the advantages of the AC method is that the signal averaging techniques to improve signal-to-noise ratio can be applied by using a lock-in amplifier or a boxcar integrator. It allows the detection of small changes in heat capacity. Another advantage is that the condition of $\omega\tau_{ext} \gg 1$ can be realized even in nonadiabatic conditions. It allows the measurements under high pressures, where the sample is mechanically and thermally in contact with the pressure-transmitting medium, and it is very hard to realize the adiabatic condition for the sample. It also makes possible to measure the heat capacity at very high temperatures, where the thermal radiation effect is too large to realize the adiabatic condition.

There are several methods to apply the AC heating for the sample. One is the light irradiation method, in which the chopped light with a frequency ω is irradiated to the surface of sample. In this case, the very small sample can be measured because of nonnecessity of an electric heater. However, the precise estimation of heat power absorbed is difficult. Joule heating with a separated electric heater can be used for the AC heating. In this case, the absolute value of the sample heat capacity can be obtained by subtracting the heat capacity of addenda. New methods using Peltier heating [19] and modulated bath heating [20] are also developed. It should be noted that the information about the first-order phase transition and latent heat is difficult to obtain by this method due to the same reason as with relaxation method.

Figure 17.8 shows an example of the results of heat capacity measurements carried out on a small quartz single crystal with 1 mm thickness by an AC calorimetry [21]. The intermediate phase (IC phase: incommensurate phase) between the α- and β-phases of quartz is clearly detected by the heat capacity measurements. The heat capacity jump observed at the α-IC transition indicates the first-order nature of the phase transition. The contribution of latent heat and a thermal hysteresis of 0.9 K were also observed at the α-IC transition. At the IC-β transition, the heat capacity shows a sharp peak with a critical behavior. No thermal hysteresis confirms the second-order nature of the IC-β transition.

17.2.4
Heat Pulse Method for Heat Capacity Measurements

There are several calorimeters for heat capacity measurements using a heat pulse. The measurement should be done within a short period to avoid the heat

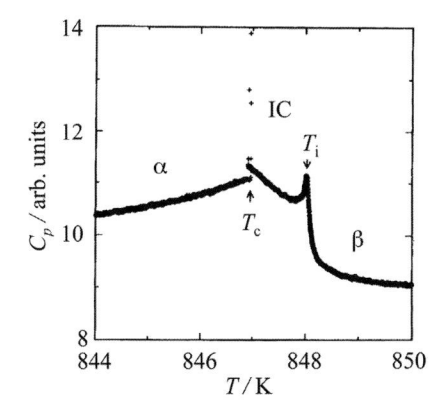

Figure 17.8 Heat capacity of a quartz single crystal near the α–β phase transition temperature measured by AC calorimetry [16]. The presence of an intermediate phase (IC) is clearly seen between the two heat capacity anomalies at T_c and T_i.

loss during the measurement, and the corrections for the heat loss should be done. For electric conductors, electric pulse for a short duration time of 0.1–1.2 s can be used for heating [22]. For insulators, laser-flash is used to apply heat to the sample (laser-flash calorimetry) [23]. The schematic view of the apparatus is shown in Figure 17.9. The laser-flash technique is mainly used for

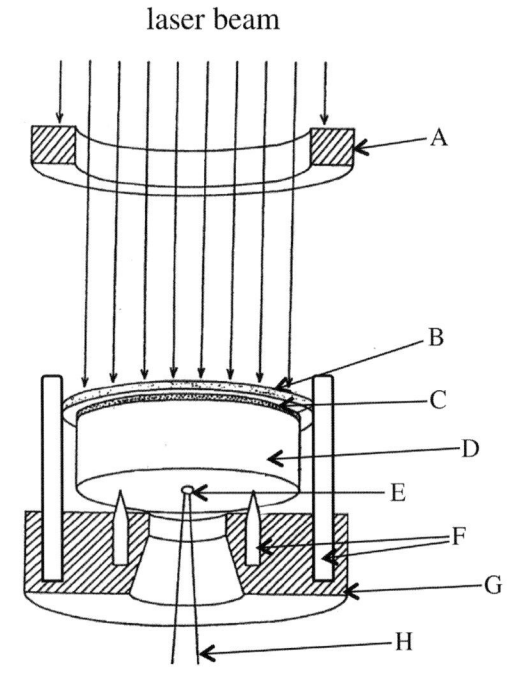

Figure 17.9 Sample holder of a laser-flash calorimeter [18]. A: laser beam slit; B: light absorbing disk; C: grease; D: sample; E: silver paste; F: fixing pins; G: holder; H: thermocouple. (Reproduced with eprmission from [18], Copyright 1968, American Physical Society.)

thermal diffusivity measurements. In the thermal diffusivity measurements, a laser pulse heats one side of a sample, and the time-dependent temperature rise on the backside is measured. The faster energy reaches to the backside that indicates the higher thermal diffusivity of the sample. In addition, the heat capacity can be determined by the maximum temperature rise of the sample after correcting the heat loss during the measurement. The amount of energy absorbed by the sample is usually estimated using a standard material of known heat capacity. To maintain the same amount of energy absorbed, the same absorbing disk is attached on the irradiation side of samples. The high accuracy data with 0.8% at room temperature and 2% at 1100 K are reported for metals and ceramics [24].

Figure 17.10 shows an example of the results of heat capacity measurements carried out by a laser-flash calorimetry [25]. The heat capacity of vanadium is accurately measured in a wide temperature range from 80 to 1000 K. No heat capacity anomalies are observed in the temperature range investigated.

17.2.5
Drop Calorimetry for Heat Capacity Measurements

Drop calorimeter, which is also called temperature jump calorimeter, determines the enthalpy difference between the two temperatures before and after the temperature jump. Usually, room temperature is used for the lower temperature. The sample kept at a high temperature T is dropped into a calorimeter assembly at room temperature, and the heat measured corresponds to the enthalpy

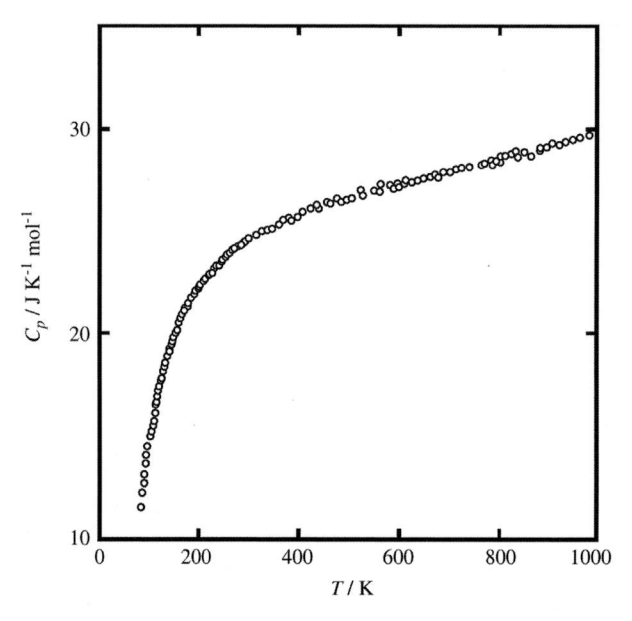

Figure 17.10 Heat capacity of vanadium measured by laser-flash calorimetry (○) [20]. (Rights managed by AIP Publishing LLC.)

difference between T and room temperature. To determine the heat capacity, the temperature-dependent enthalpy must be measured by varying the starting temperature T and differentiated with respect to the temperature. The temperature jump calorimeter is used for the accurate heat capacity measurement above 1000 K. The heat transfer by radiation and conduction into the surrounding during the drop of the sample into the calorimeter should be precisely evaluated and corrected. To avid the uncertainty of the knowledge of thermal radiation properties of the surface of a sample, a sample holder is used to keep the heat transfer into the surrounding constant and make possible the accurate correction.

17.2.6
Differential Scanning Calorimetry for Heat Capacity Measurements

Differential scanning calorimetry (DSC) is the most commonly known technique of thermal analysis, and is widely used for studying the thermal phenomena occurring in the sample. Phase transitions, chemical reactions, and so on are detected and the enthalpy change are determined. There are many commercial apparatus of different types for a wide temperature range. The schematic views of basic constructions of two major types are shown in Figure 17.11. Figure 17.11a shows a heat flux DSC. A sample is placed in the furnace with a reference material, which shows no thermal phenomena during the experiments. The sample must be kept under the same condition with the reference material, and thermally connected equally with the reference to the furnace. The basic principle of the heat flux DSC is explained as follows through Figure 17.12 [26]. If the furnace is heated at a constant rate dT_f/dt, the temperature of the furnace T_f, reference T_r, and sample T_s varies with time t at the same constant heating rate dT_f/dt. The heat transferred from the furnace to the sample (heat flow rate dQ_s/dt) should be written according to the Newton's law as

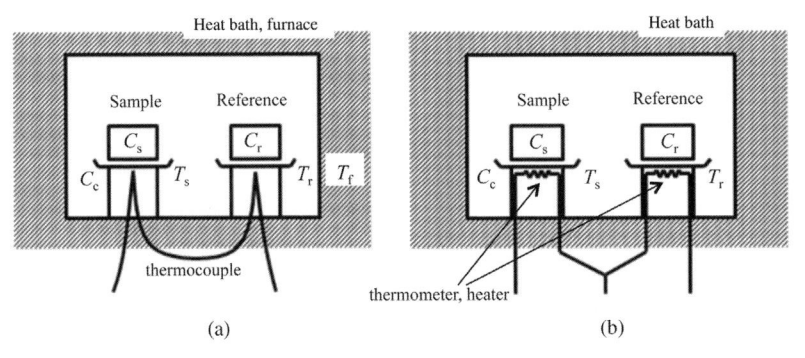

$$\frac{dQ_s}{dt} = k(T_f - T_s), \tag{17.13}$$

Figure 17.11 Basic constructions of two types of DSC. (a) Heat flux DSC. (b) Power-compensated DSC.

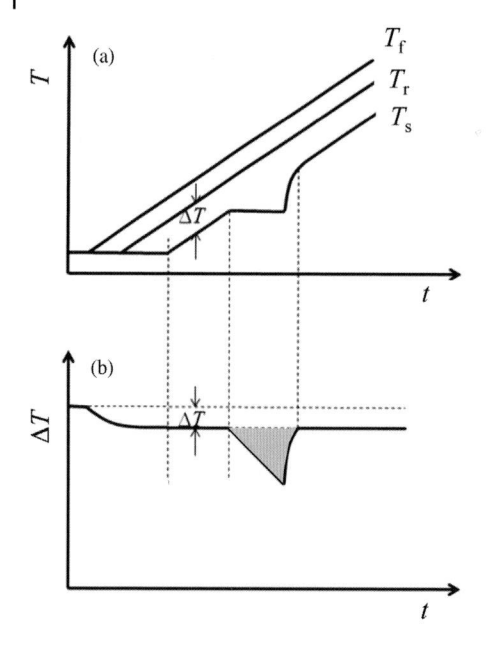

Figure 17.12 The basic principle of the heat flux DSC. (a) A typical temperature change of furnace T_f, reference T_r, and sample T_s. (b) Signal output of the differential temperature ΔT. The hatched part corresponds to a thermal phenomenon and the excess enthalpy due to the thermal phenomenon can be calculated from the area.

where k is the thermal conductance between the sample and the furnace. The heat transferred from the furnace is used for the heating of the sample, and then

$$\frac{dQ_s}{dt} = (C_s + C_c)\frac{dT_f}{dt}, \qquad (17.14)$$

where C_s, and C_c are the heat capacity of the sample and the sample cell, respectively. The similar equation should be written for the reference:

$$\frac{dQ_r}{dt} = k(T_f - T_r) = (C_r + C_c)\frac{dT_f}{dt}, \qquad (17.15)$$

where C_r is the heat capacity of the reference. From these equations, the following equation can be obtained if the thermal conductance k and the heat capacity of the cell C_c are both the same for the sample and the reference:

$$\frac{dQ_s}{dt} - \frac{dQ_r}{dt} = -k(T_s - T_r) = k\Delta T = (C_s - C_r)\frac{dT_f}{dt}. \qquad (17.16)$$

Thus, the heat capacity difference between the sample and the reference can be obtained by measuring the temperature difference ΔT, if k is determined separately. The essence of the qualitative measurement by heat flux DSC is the heat transfer via a well-defined thermal conductance. In the case of power-compensated DSC as shown in Figure 17.11b, heaters are equipped on both the sample cell and the reference cell and are controlled in order to keep the temperature difference between the sample cell and the reference cell zero. The heat capacity difference can be determined from Eq. (17.16) by measuring the difference of energy input rate calculated from the voltages and the currents for the heaters. In this case, we may less care about the thermal conductance k.

If thermal phenomena occur as shown in Figure 17.11, the enthalpy changes due to thermal phenomena can be obtained from the hatched area for heat flux DSC and from the excess input heat for power-compensated DSC. The simple explanation described above shows that the heat capacity of sample could be obtained by Eq. (17.16). However, there are small difference in the heat capacity between the sample and reference cells and in the thermal conductance of the sample and the reference cells. Thus, the heat capacity measurements are usually carried out in three runs, as shown in Figure 17.13: first run for empty cell measurement, second run for the standard material with known heat capacity, and third run for the sample. The heat capacity of the sample is calculated from Eq. (17.17):

$$C_s = -\frac{d\Delta Q_s}{d\Delta Q_{sm}} C_{sm}(T), \qquad (17.17)$$

where $C_{sm}(T)$ is the heat capacity of the standard material. In this method, the heat capacity of the reference must be known at the measuring temperatures, and α-Al_2O_3 is widely used as the reference material. In some cases, the integrations of the DSC curves instead of the heat flow rates themselves are used for calculating the average heat capacity between the scanning temperature range.

For using the commercial apparatus, some calibration procedures are very important to get the accurate data. The temperature calibration should be done for measuring the scanning rate by comparison with the known transition and/or melting temperatures of reference materials, because the DSC is a dynamic method and the temperature indicated by the apparatus may change depending

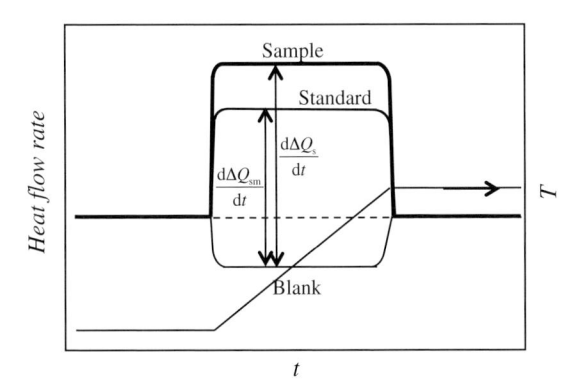

Figure 17.13 Schematic diagram of three DSC curves for heat capacity measurement. "Blank" is the DSC curve for the measurement of empty cell, "Standard" for the standard material with known heat capacity, and "Sample" for the sample.

on the scanning rate. The enthalpy calibration should be done frequently, especially for the heat flux DSC. The thermal conductance k may change depending on the condition of apparatus. In addition, it should be noted that the thermal conductance k is essentially a temperature-dependent property and the enthalpy calibration should be done near the temperature at which you want to measure the enthalpy change.

Recently, temperature-modulated DSC is also used for determining the heat capacity of materials [27]. Periodic temperature moderation is superimposed to a temperature scanning of conventional DSC. A dynamic heat capacity can be determined from the response of the sample for the temperature modulation as a complex quantity, and it becomes possible to analyze the kinetics of glass transitions, phase transitions, chemical reactions, and so on.

Figure 17.14 shows an example of the results of heat capacity measurements carried out by DSC [28]. The ferroelectric or antiferroelectric phase transition properties of lead perovskites are studied by the heat capacity measurements. In $PbTiO_3$ (PT), a sharp peak due to the ferroelectric phase transition is observed at 764 K with a relatively large phase transition entropy of $\Delta S = 7.3$ J/(K mol). In $PbZrO_3$ (PZ), a sharp peak due to the antiferroelectric phase transition is observed at 504 K with $\Delta S = 9.9$ J/(K mol). Two peaks related to the antiferroelectric phase transitions are observed at 433 and 474 K in $PbHfO_3$ (PH).

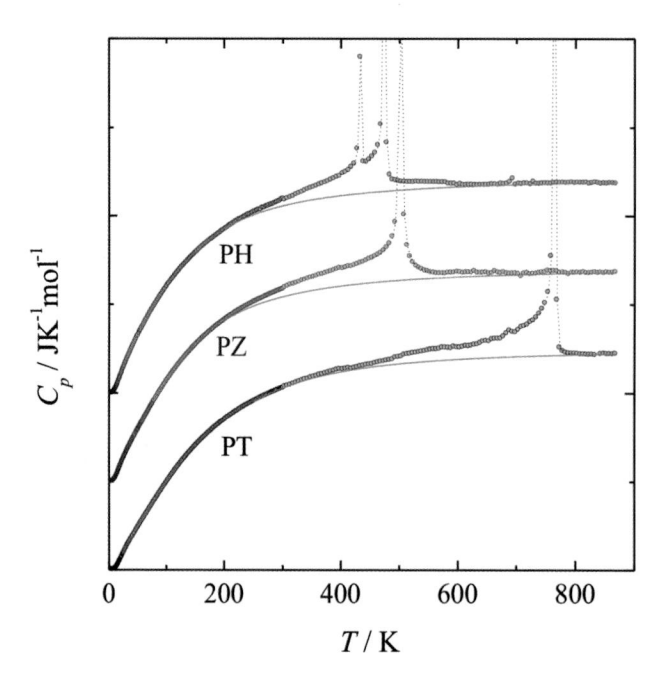

Figure 17.14 Heat capacity of $PbTiO_3$ (PT), $PbZrO_3$ (PZ), and $PbHfO_3$ (PH) by adiabatic calorimetry (below 300 K) and by DSC (above 300 K). Ferroelectric phase transition for PT and antiferroelectric phase transitions for PZ and PH are clearly observed in the heat capacity measured by DSC.

17.2.7
Heat Capacity Spectroscopy [29,30]

In general, heat capacity is defined as a static property at an equilibrium state. However, a dynamic heat capacity can be defined by a time-dependent property in the case of a system containing a relaxation process. If the temperature is changed stepwise at $t=0$ and the heat absorbed varies with time as shown in Figure 17.15, the time-dependent heat capacity can be defined by Eq. (17.18):

$$C(t) = \frac{Q(t)}{\Delta T}. \tag{17.18}$$

The equilibrium heat capacity $C_{p\infty}$ is given as $C_{p\infty} = Q_\infty/\Delta T$. On the other hand, the instant heat capacity can be expressed as $C_{p0} = Q_0/\Delta T$, which corresponds to a process with a very short relaxation time. Such time-dependent properties can be transformed to the corresponding frequency-dependent properties by Fourier transformation. Therefore, the frequency-dependent heat capacity is obtained as a complex quantity and expressed as $C_p(\omega) = C_p'(\omega) + iC_p''(\omega)$. The real part $C_p'(\omega)$ and the imaginary part $C_p''(\omega)$ should obey Kramers–Kronig relation.

A typical relaxation phenomenon is a glass transition, and many efforts have been made to measure the frequency-dependent heat capacity near the glass transition temperature. AC calorimetry may be applicable, but the measureable frequency range is limited to be narrow (<10 Hz) because of the low thermal conductivity of

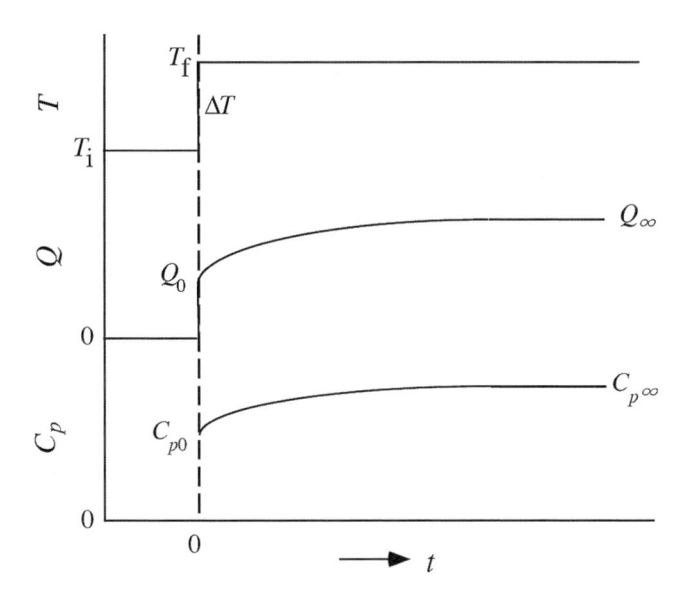

Figure 17.15 Basic concept of time-dependent heat capacity. If the heat Q is absorbed by the system of which the temperature changes stepwise at $t=0$ depends on the time, the heat capacity should be a time-dependent quantity. The instant heat capacity $C_{p0} = Q_0/\Delta T$ and the equilibrium heat capacity $C_{p\infty} = Q_\infty/\Delta T$.

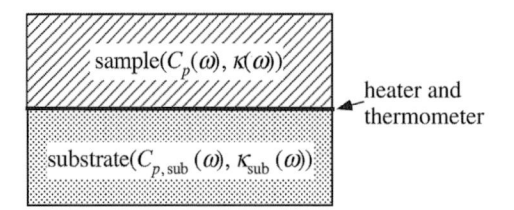

Figure 17.16 Schematic view of a typical geometry of heat capacity spectroscopy.

samples. The method of heat capacity spectroscopy has been developed by Birge and Nagel using a 3ω method [29]. Figure 17.16 shows the schematic view of the experimental geometry. The very thin metal film working as a heater is sandwiched by the sample and the substrate. If the thickness of a heater can be assumed to be zero and spread to be infinite two-dimensionally, the temperature oscillation due to the sinusoidal heat flow with the frequency ω is expressed as

$$T(\omega) = \frac{J\exp(i\pi/4)}{\left[(\varpi C_p(\varpi)\kappa(\varpi))^{1/2} + (\varpi C_{p,\text{sub}}(\varpi)\kappa_{\text{sub}}(\varpi))^{1/2}\right]}, \tag{17.19}$$

where $T(\omega)$ is the amplitude of temperature oscillation at the heater and J is the heat flow rate from the heater. The metal thin film can also work as a resistance thermometer to measure the $T(\omega)$. By this method, the product of $C_p\kappa$ of the sample can be measured for a wide frequency range up to several kilohertz.

Figure 17.17 shows an example of the results of heat capacity spectroscopy measured near the glass transition temperature of 2-n-butoxyethanol, which is known as a typical glass-forming molecule [31]. The real and imaginary parts of the thermal effusivity ($C_p\kappa$) depend on the measuring frequency near the glass transition temperature, as shown clearly in the figure. The real part of the thermal effusivity ($C_p'\kappa'$) measured at 2222 Hz starts to decrease with decreasing the temperature to about 170 K in 2-n-butoxyethanol. This indicates that the relaxation time of some degrees of freedom in the supercooled liquids increase with decreasing temperature, and the freedom becomes not to be able to response the sinusoidal heating with 2222 Hz. It should be noted that the remarkable change in $C_p\kappa$ mainly corresponds to the heat capacity C_p. The imaginary part ($C_p''\kappa''$) shows a peak at the temperature at which the slope of the real part shows the maximum as required by the Kramers–Kronig relation. The similar relaxation phenomena are also observed in the dielectric constant measurements of 2-n-butoxyethanol.

17.3
Reaction Calorimetry for Enthalpy Change Measurements

Reaction calorimetry includes the measurements of heat of various chemical reactions, dissolution, adsorption, immersion, and so on. The measurements of heat of combustion of organic materials have been carried out with oxygen or

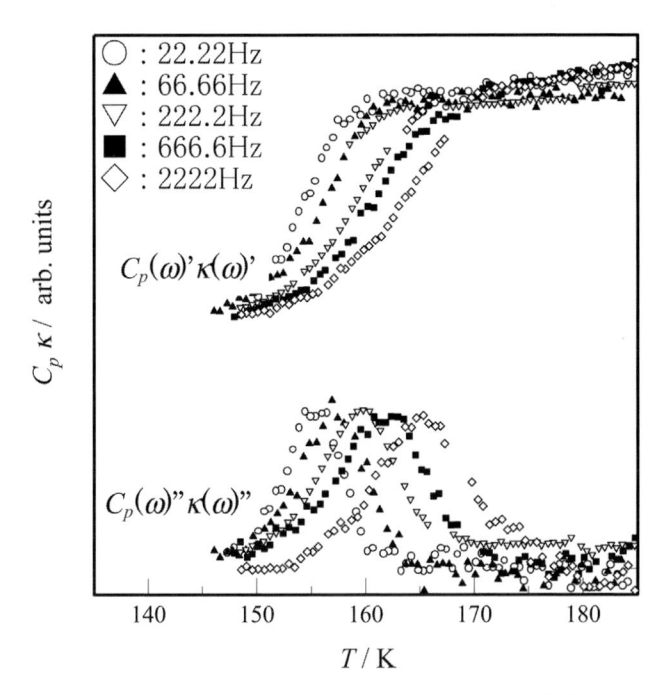

Figure 17.17 Frequency-dependent thermal effusivity ($C_p\kappa$) of 2-n-butoxyethanol near the glass transition temperature [26]. Real part $C_p(\omega)'\kappa(\omega)'$ and imaginary part $C_p(\omega)''\kappa(\omega)''$ of the thermal effusivity. ○: measured at 22.22 Hz; ▲: measured at 66.66 Hz; ▽: measured at 222.2 Hz; ■: measured at 666.6 Hz; ◇: measured at 2222 Hz.

fluorine bomb calorimeters based on isoperibol calorimetry or on adiabatic calorimetry to determine the enthalpy of formation of various organic compounds [32]. The heat of dissolution of inorganic materials in molten salts or acids and those of metals in metals or acids have been measured mainly by conduction calorimeters to obtain the enthalpies of the reactions and thus the formation entropy of the materials. The heats of adsorption and immersion are used for the characterization of the surface of materials.

17.3.1
Calvet Calorimeter for High-Temperature Reaction Calorimetry

Calvet calorimeter is a conduction-type twin calorimeter, designed by Tian and Calvet, and is widely used to measure the heat of various types of chemical reactions of inorganic compounds at high temperatures, including the dissolving of compounds in a solvent. There are some commercially available apparatuses [33]. The calorimeter has two symmetrical sample cells in a massive metal block as shown in Figure 17.18, which is maintained constant and uniform by the outside furnaces. The thermopile, in which hundreds of thermocouples

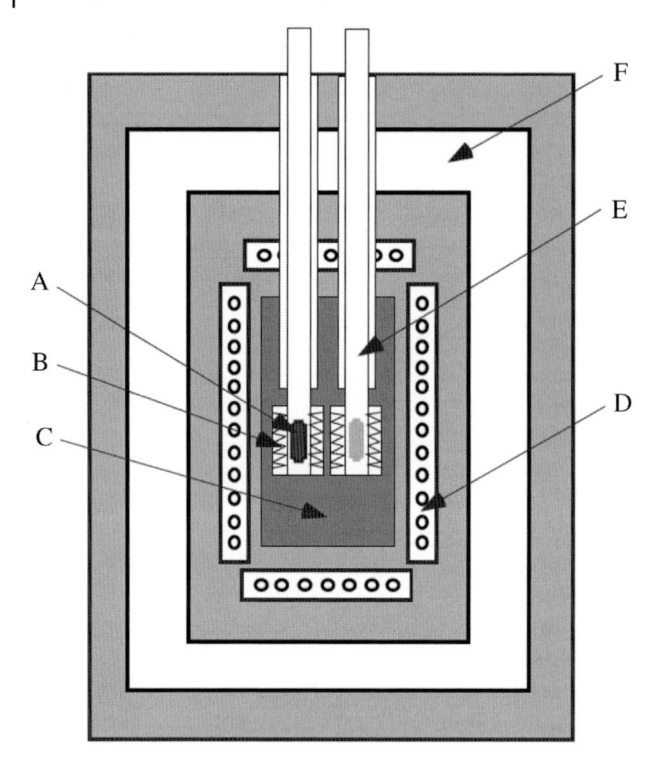

Figure 17.18 Schematic view of a high-temperature Calvet-type calorimeter. A: sample chamber; B: thermopile; C: heat block; D: heater; E: tube inductor; F: thermal insulator.

between the each cell and the block are connected in series, covers the cell uniformly. The difference in electromotive force between the two cells is measured to obtain the heat flux generated from or absorbed in one cell using another cell as a reference.

The enthalpy of solution of a sample dissolving in a solvent can be measured using a Calvet calorimeter as follows. The sample and the solvent are kept separately at the same temperature. Then the sample is inserted into the solvent and solved completely. The schematic temperature change of the sample cell is shown in Figure 17.19. The total heat generated is obtained by integrating the signal curve. The calibration of apparatus for the heat must be carried out separately using Joule heating or reference materials. The measured heat corresponds to the enthalpy of solution. If the sample can easily be solved to a solvent at room temperature, the measurements can be done at room temperature. However, high temperature is usually required to solve samples completely. For various oxides, the measurements are carried out around 1000 K using lead borate as a solvent. This method is used to determine the enthalpy of reaction at the calorimeter temperature. When the compound C is formed by the reaction of compounds A and B, the enthalpy of the reaction can be obtained by measuring the

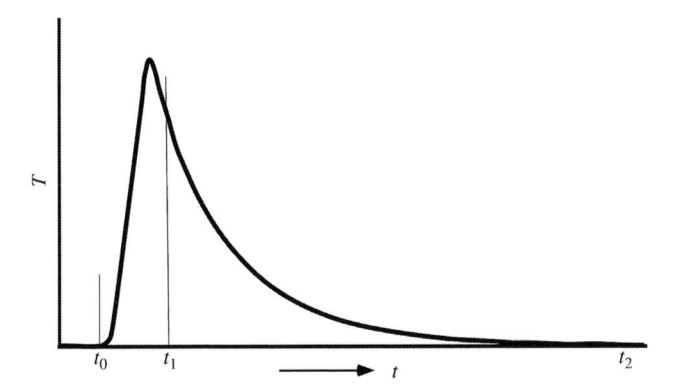

Figure 17.19 A typical temperature change of the sample cell in Calvet reaction calorimeter. The reaction started at t_0 and completed at t_1. Total reaction heat is obtained by integrating the curve from t_0 to t_2.

enthalpies of the solution to the same solvent for the samples A, B, and C. The enthalpy of the reaction is obtained using Hess's law:

$$\Delta_r H^0(A + B \rightarrow C) = \Delta_{sol} H^0(C) - \left(\Delta_{sol} H^0(A) + \Delta_{sol} H^0(B)\right). \tag{17.20}$$

It is also possible to calculate the enthalpy of formation of compound C, if the values of the enthalpy of formation of compounds A and B are available at that temperature.

The drop-solution calorimetry is also possible using a Calvet calorimeter. In this method, a sample kept at room temperature is dropped and solved completely in a solvent at the calorimeter temperature T. The measured heat corresponds to the sum of the enthalpy difference of the sample between T and room temperature and the enthalpy of solution at T. This method is very widely used for calorimetric studies of many oxides [34,35].

References

1 Wagman, D.D., Evans, W.H., Parker, I.B. et al. (1982) The NBS tables of chemical thermodynamics properties. *J. Phys. Chem. Ref. Data*, **11** (Suppl. 2).
2 Pedley, J.B., Naylor, R.D., and Kirby, S.P. (1986) *Thermochemical Data of Organic Compounds*, 2nd edn, Chapman and Hall, London.
3 Lias, S.G., Bartmess, J.E., Liebman, J.F. et al. (1988) Gas-phase ion and neutral thermochemistry. *J. Phys. Chem. Ref. Data*, **17** (Suppl. 1).
4 MALT Group (2004) https://www.kagaku.com/malt/index.html (accessed 2012).
5 National Institute of Standards and Technology (1996) https://webbook.nist.gov/chemistry/ (accessed 2015).
6 Ho, C.Y. (ed.) (1988) *Specific Heat of Solids: CINDAS Data Series on Material Properties*, vol. **I–2**, Hemisphere Publishing Corporation.
7 Atake, T., Kawaji, H., Hamano, A., and Saito, Y. (1990) Construction of an adiabatic calorimeter for heat capacity measurements from liquid helium temperature to 330K. *Report Res. Lab. Eng. Mater. Tokyo Inst. Technol.*, **15**, 23.

8 Atake, T. (2004) Adiabatic heat capacity calorimetry, in *Comprehensive Handbook of Calorimetry and Thermal Analysis* (ed. The Japan Society of Calorimetry and Thermal Analysis), John Wiley & Sons, Ltd, Chichester, pp. 68–71.

9 Moriya, Y., Kawaji, H., Tojo, T., and Atake, T. (2003) Specific-heat anomaly caused by ferroelectric nanoregions in $Pb(Mg_{1/3}Nb_{2/3})O_3$ and $Pb(Mg_{1/3}Ta_{2/3})O_3$ relaxors. *Phys. Rev. Lett.*, **90**, 205901.

10 Burns, G. and Dacol, F.H. (1983) Glassy polarization behavior in ferroelectric compounds $Pb(Mg_{1/3}Nb_{2/3})O_3$ and $Pb(Zn_{1/3}Nb_{2/3})O_3$. *Solid State Commun.*, **48**, 853–856.

11 Stewart, G.R. (1983) Measurement of low-temperature specific heat. *Rev. Sci. Instrum.*, **54**, 1–11.

12 Quantum Design, Inc. (2016) https://www.qdusa.com/products/ppms.html (accessed March 31).

13 Lashley, J.C., Hundley, M.F., Migliori, A. et al. (2003) Critical examination of heat capacity measurements made on a quantum design physical property measurement system. *Cryogenics*, **43** (6), 369–378.

14 Hwang, J.S., Lin, K.J., and Tien, C. (1997) Measurement of heat capacity by fitting the whole temperature response of a heat-pulse calorimeter. *Rev. Sci. Instrum.*, **68** (1), 94–101.

15 Kitani, S., Tachibana, M., Taira, N., and Kawaji, H. (2013) Thermal study of the interplay between spin and lattice in $CoCr_2O_4$ and $Cd Cr_2O_4$. *Phys. Rev.*, **B87**, 064402.

16 Kraftmaker, Y.A. and Strelkov, P.G. Energy of formation and concentration of vacancies in tungsten. *Sov. Phys. Solid State*, **4**, 1662–1664.

17 Bonilla, A. and Garland, C.W. (1974) High-pressure heat capacity of chromium near the Néel line. *J. Phys. Chem. Solids*, **35**, 871–877.

18 Sullivan, P.F. and Seidel, G. (1968) Steady-state, AC-temperature calorimetry. *Phys. Rev.*, **173**, 679–685.

19 Jung, D.H., Moon, I.K., and Jeong, Y.H. (2002) Peltier AC calorimeter. *Thermochim. Acta*, **391**, 7–12.

20 Graebner, J.E. (1989) Modulated-bath calorimetry. *Rev. Sci. Instrum.*, **60**, 1123–1128.

21 Yao, H. and Hatta, I. (1995) Phase transitions of quartz studied by A.C. calorimetry. *Thermochim. Acta*, **266**, 301–308.

22 Matsui, T., Arita, Y., and Watanabe, K. (2000) Development of two types of high temperature calorimeters. *Thermochim. Acta*, **352–353**, 285–290.

23 Takahashi, Y., Yokokawa, H., Kadokura, H. et al. (1979) Laser flash calorimetry I: calibration and test on alumina heat capacity. *J. Chem. Thermodyn.*, **11**, 379–394.

24 Takahashi, Y. (2004) Laser flash calorimetry, in *Comprehensive Handbook of Calorimetry and Thermal Analysis* (ed. The Japan Society of Calorimetry and Thermal Analysis), John Wiley & Sons, Ltd, Chichester, pp. 77–79.

25 Takahashi, Y., Nakamura, J., and Smith, J.F. (1982) Laser-flash calorimetry III: heat capacity of vanadium from 80 to 1000K. *J. Chem. Thermodyn.*, **14**, 977–982.

26 Mraw, S.C. (1988) Differential scanning calorimetry, in *Specific Heat of Solids: CINDAS Data Series on Material Properties*, vol. **I–2** (ed. C.Y. Ho), Hemisphere Publishing Corporation, pp. 395–436.

27 Gill, P.S., Sauerbrunn, S.R., and Reading, M. (1993) Modulated differential scanning calorimetry. *J. Therm. Anal. Calorim.*, **40**, 931–939.

28 Yoshida, T., Moriya, Y., Tojo, T. et al. (2009) Heat capacity at constant pressure and thermodynamic properties of phase transitions in $PbMO_3$ (M=Ti, Zr and Hf). *J. Therm. Anal. Calorim.*, **95**, 675–683.

29 Birge, N.O. and Nagel, S.R. (1985) Specific-heat spectroscopy of the glass transition. *Phys. Rev. Lett.*, **54** (25), 2674–2677.

30 Inada, T., Kawaji, H., and Atake, T. (1990) Construction of a heat capacity spectrometer and application to some molecular glass forming substances. *Thermochim. Acta*, **163**, 219–224.

31 Kawaji, H., Tojo, T., and Atake, T. (2001) 3ω method: simultaneous measurements of heat capacity spectroscopy and dielectric spectroscopy. *Netsu Sokutei (Calor. Therm. Anal.)*, **29**, 16–20.

32 Cox, J.D. and Pilcher, G. (1970) *Thermochemistry of Organic and Organometallic Compounds*, Academic Press, London.
33 SETARAM Instrumentation (2016) www.setaram.com/ (accessed March 31).
34 Akaogi, M. (1990) Thermodynamics and stability relations of mantle minerals, in *Dynamic Processes of Material Transport and Transformation in the Earth's Interior* (ed. F. Marumo), Terra Scientific Publishing Co., Tokyo, pp. 239–251.
35 Navrotsky, A. (1997) Progress and new directions in high temperature calorimetry revisited. *Phys. Chem. Miner.*, **24**, 222–241.

Index

a

ABF. *See* annular bright field (ABF)
ab initio structure 246, 267
absorbed photons 362
absorber temperature 451
absorption 444
– coefficient 365, 366, 370
– – μ(E) 378
– energies 455
– measurements 352
– range 362
– spectra 371, 377, 380, 384
– – angular dependence of 380
– – measurement of 380
– spectroscopy 384
– spectrum 361, 369
– – calculation 369
– – energy range of 363
– – regions 362
AC calorimeter 599
acceleration voltage 156
accelerator-driven emission spectroscopy 456
AC heating
– application 600
acoustic modes 402
acoustic phonons
– self-correlation function for 428
acoustic ringing 257
acoustic velocities 580
active pharmaceutical ingredient (API) 56
adamantane 268
adiabatic calorimeter 589, 591, 593
– cryostat of 594
– principle of 592
adiabatic calorimetry 606, 609
adiabatic condition 596, 600
adiabatic emission process 325
adiabatic shields 593

ADP. *See* atomic displacement parameter (ADP)
adsorbed atoms 313
– oxygen atoms 348
– Xe atoms 320
adsorption 317
– induced iron segregation 356
advanced electron paramagnetic resonance techniques 284
AEY. *See* Auger electron yield (AEY)
AFM. *See* atomic force microscopy (AFM)
AFMR. *See* antiferromagnetic resonance (AFMR)
agreement factors 19
ALCHEMI. *See* atom-location channeling enhanced microanalysis (ALCHEMI)
$Al_{71}Co_{13}Ni_{16}$, 141
AlK_α
– excited $Fe2p_{3/2}$, 324
– excited survey 323
– excited valence band spectra 344
alkali-doped fullerides 282
alkaline earth titanates 558
Al oxide
– double-layer
– – Moirè-type superstructure of 214
– films 213, 215
– – noncontact AFM images of the Moiré structure 216
aluminophosphate cloverite
– fluorinated 270
aluminum housing shields 195
aluminum oxide film 214
Ammann tiling 147
amorphous oxides 523
analyzer 316
Anderson impurity 380
angle- or energy-dependent measurements 352

Handbook of Solid State Chemistry, First Edition. Edited by Richard Dronskowski, Shinichi Kikkawa, and Andreas Stein.
© 2017 Wiley-VCH Verlag GmbH & Co. KGaA. Published 2017 by Wiley-VCH Verlag GmbH & Co. KGaA.

angular efficiency 350
angular frequency 545
angular momentum 326, 487
– quantum mechanical treatment of 248
– quantum number 319, 333
anharmonicities 394, 423
anharmonic thermal motion 113
anisotropic displacement 19, 22
– parameter 8, 24, 103
anisotropic extinction effects 84
anisotropic interaction 247
anisotropy exchange (AE) 281
anisotropy parameters 281
annealing procedure 54
annular bright field (ABF) 135
– scanning transmission electron microscopy 139
annular detector 139
antibonding 564
antiferromagnetic materials 383
antiferromagnetic metals chromium 516
antiferromagnetic order 470
antiferromagnetic phases 470
antiferromagnetic resonance (AFMR) 295
– in magnetically ordered phases 295
– modes 300
– response 297
– study 298
antimony-121, 458
anti-Stokes scattering 410
anti-Stokes spectrum 410
antiwear tribofilms 576
APD. See avalanche photodiodes (APD)
aperiodic crystals 111, 115, 131
aperiodicity 112
Ar dimers 386
Argand diagram 10
aromatic hydroxylation mechanism 377
aromatic ring opening 387
Arrhenius law 471
Arrhenius plot, for calculation of energy barrier 548
asperity 577
asymmetric unit 2
atomic configurations, generated by MD simulations 577
atomic displacement 447
atomic displacement parameter (ADP) 113, 134
atomic emitter 312
atomic force microscopy (AFM) 140, 184, 199, 201, 576
– cantilever
– – parabolic potential 205
– – working in noncontact mode 205
– contact mode 201
– contact-mode AFM 203
– length-extension resonator 207
– noncontact mode 203
– physical background 199
– tuning fork sensor 207
atomic form factor 80
atomic interactions 134
atomic magnetic moments 113
atomic magnetic saturation moment 490
atomic mass 311
atomic number 162, 318
atomic resolution 18
atomic scattering 8
– amplitudes 376
– factor 6, 7
atomic sensitivity factor 350
atomic surface 134, 143
– modeling method 147
atom interaction 396
atom-location channeling enhanced microanalysis (ALCHEMI) 149
atoms
– and molecules, chemical interactions between 343
– periodic arrangement 122
Au(111)
– in $CuSO_4$-containing sulfuric acid solution
– – cyclic voltammograms of 235
– electrode surface
– – cyclic voltammogram 228
Auger electron 333, 366, 368
– emission 382
Auger electron spectroscopy (AES) 210, 365
Auger electron yield (AEY) 366
Auger mechanism 331
Auger spectra 326, 334
avalanche photodiodes (APD) 462
average Hamiltonian theory 266
Avogadro number 460
azimuthal intensity
– anisotropy 347
– distribution 346
Azimuth angle-dependent O1s intensity 347

b

backscattered electron 162, 164
backscattering geometry 459
back-to-back (BABA) sequence 268
$BaCu_3V_2O_8(OH)_2$, 292
$Ba_3NbFe_3Si_2O_{14}$

– 2D arrangement of Fe^{3+}, 300
bandgap 364, 366, 368, 387
bandwidth 381
barium strontium titanate 554
barycenter rule 497, 499
barycenter splitting 497, 499
$(Ba_{0.7}Sr_{0.3})TiO_3$ thin films, dielectric response 553
$BaTiSiO_3$
– Raman spectra of 414
bcc-hafnium, for wave vectors
– phonon dispersion of 399
beamline optics 362
beam shower 163
Beer–Lambert law 361
Bethe–Salpeter equation 375
BF. See bright field (BF)
bias voltage 187
binary PtRh alloy 210
binding energy 312, 316, 318, 322, 323, 326, 361, 363, 364, 369, 374, 378
Bi_2O_5 125
biocrystallography 12
biomineralization 472
Birch–Murnaghan equation 564
Black Forest headdress 65
Bleaney–Bowers equation 493
block wave 117
Bohr magnetons 487, 489, 491
Bohr–van Leeuwen theorem 488
Boltzmann constant 81, 287, 448, 534
Boltzmann criterion 53
Boltzmann distribution 252, 533
bomb calorimeter 589
bond valence method 125
Bose factor 403
boxcar integrator 600
Bragg angles 85
Bragg–Brentano geometry 41, 56
Bragg–Brentano powder diffractometer 41
Bragg contrast 167, 170
Bragg diffraction 167
Bragg equation 9, 34, 82, 86
– constructive interference 9
– visualization of 31
Bragg intensities 84, 90, 92
Bragg peaks 104
Bragg position 86
Bragg reflection 51, 83, 84, 86, 104, 110, 112, 113, 148, 415, 419
– flipping ratios 107
– self-correlation function for 428

– splitting of 427
Bragg scattering 83
Bragg's law 87
braking radiation 175
Bremsstrahlung 33
bright field (BF) 161
– BF-TEM
– – mode, contrast generation in 166
– – – mode, contrast generation in 166
– TEM 160
Brillouin function 489
Brillouin light scattering 580
Brillouin zone 394, 422
BTA. See 1,3,5-tris(2,2-dimethylpropionyl amino)benzene (BTA)
bulk modulus 563, 564, 579
Burgers vector 570

c
cable stitch 210
cadmium-containing sulfuric acid 337
calorimetry 589, 591
– nonreaction 591
– – for heat capacity measurements 591–608
– – – AC calorimetry 598–600
– – – adiabatic calorimetry 591–596
– – – differential scanning calorimetry 603–606
– – – drop calorimetry 602
– – – heat capacity spectroscopy 606–608
– – – heat pulse method 600
– – – relaxation method 596–598
– overview 591
– reaction 589, 590
– – for enthalpy change measurements 608–611
Calvet calorimeter 589, 609, 610, 611
– for high-temperature 609–611
canonical quantization 250
capacitance 524
carbodiimide 472
Carr–Purcell–Meiboom–Gill 258
cartesian directions 498
case studies 56
– codeine phosphate sesquihydrate, crystal structure determination 56–59
– corrosion product, crystal structure determination 64
– symmetry, modes for investigation of high-pressure phase transitions 67
cathodoluminescence 175
CBED. See convergent beam electron diffraction (CBED)

C_2D_2 molecule 221
CDW. See charge density wave (CDW)
centrosymmetric structures 17
CF. See charge flipping (CF)
C_2H_2
– d^2I/dV^2 tunneling spectra of 220
characteristic X-ray 175
– scheme of generation 176
charge density 26
charge density wave (CDW) 111
charge flipping (CF) 18, 53, 144
– technique 52
charge order and related phenomena 124
charge potential model 338
Chebyshev polynomials 58
chemical bonding 113, 564
chemical connectivity 8
chemical elements 311
chemical etching 162
chemical mapping 318
chemical shift 326
– frequency 249
chemical thermodynamics 589, 590
chemical wear 575
chromatic aberration 158
chromium
– normalized form factors 79
circular dichroism 380
circular polarization 383
Clausius–Mosotti equation 541
Clebsch–Gordan coefficient 446
closed cells 180
closed cycle cryostat
– optical furnace 92
clusters 133
– embedding approach 147
CMA. See cylindrical mirror analyzer (CMA)
cobalt–olivine 102, 103
$CoCr_2O_4$
– heat capacity of 599
codeine molecule 59
codeine phosphate sesquihydrate 56, 62
coercivity 512
$Co_{89}Hf_{11}$
– alloy deformation 204
– surface 203
Cole–Cole diagram 551
Cole–Cole plot 551
combined rotation and multiple-pulse sequences (CRAMPS) 263

complex dielectric function 544
compton scattering 44, 364
computational modelling 246
concentration depth profiling 354
conducting electron spin resonance (CESR) 282
conduction band 371, 372
– energy 375
– state 365
constant energy 422
constant height mode 202
constant wavelength diffractometer
– schematic representation 85
constraints 20
– and restraints 20
continuous wave (CW) 279
– irradiation 262
conventional R factor 19
convergent beam electron diffraction (CBED) 135, 138, 170
coordinate systems 4
coplanarity 123
copper dissolution reaction (CDR) 227
copper phthalocyanine (CuPc)
– molecule, STM images 223
COP-S powder sample 57
core hole 365, 371, 373, 376, 386
– spectra 375
Co_2SiO_4
– allowed Bragg reflections 101
– CoO_6 and SiO_4 polyhedra, clinographic view 99
– crystal structure and magnetic order 102
– electron density distribution 103
– magnetic structure, graphical representation of 102
– magnetization distribution 104
– neutron powder diffraction pattern 101
– Rietveld fit 101
– x coordinate of Co_2, values and error bars 99
– x-N-difference Fourier map, deformation density from 103
Co_2SiO_4 olivine
– structure of 99
Coulomb force 164, 531
Coulomb interaction 164, 443, 449
Coulomb repulsion 200
coupled dinuclear system
– magnetic behavior per metal ion of 494

Index | 619

coupling reflection 148
coupling schemes 491
CP. See cross-polarization (CP)
crack opening modes 573
CRAMPS. See combined rotation and multiple-pulse sequences (CRAMPS)
crenel function 117
cross-polarization (CP) 245, 258
crystal axis system (CAS) 248, 254. See crystal axis system (CAS)
crystal chemical analysis, t-plots 119
crystal classes 4
crystal engineering 24
crystal field
– energies 504
– splitting 345, 495, 504
– – impact of lower symmetry 501
– of symmetries
– – splitting patterns 500
crystal field theory (CFT) 495, 509
crystallization 12
– induced phonon softening 474
crystallographers 13
crystallographic information file 25
crystallography 12, 50, 466
– open database 25
– phase problem 80, 143
crystal quality 26
crystal structure 375
– refinement 29
crystal structures
– with displacement modulations 114
$Cs_2TiSi_6O_{15}$
– Raman spectra of 414
Cu(111)
– cyclic voltammogram 232
– electrode surface
– – STM images of 230
– in iodide-containing H_2SO_4 solution
– – cyclic voltammogram 231
– sulfate-induced Moiré structure 233
– surface
– – adsorbed iodobenzene molecules 225
cubic reciprocal lattice 37
cubic $SrTiO_3$, polar optical vibrations 557
cubic symmetries 500
Cu(100) electrode
– cyclic voltammograms 238
Cu L_3 absorption 383
CuO_4 plaquettes 298
Cu_2OSeO_3, 285
cuprates 386
– high-T_c superconductors 97

Curie law 489, 494
Curie paramagnetism 488
Curie temperatures 513
Curie–von Schweidler law 553
Curie–Weiss law 493, 517, 559
Curie–Weiss temperature 299
$CuSe_2O_5$
– crystal structure 299
$CuSe_2O_5$ spin-chain compound 294
cyanopyrazine (PzCN) 385
cyclic voltammetry (CV) 227
cyclotron 457
cylindrical mirror analyzer (CMA) 317

d

damped oscillation 548
damping factor 350
dark-field (DF) 160
DARR. See dipolar-assisted rotational resonance (DARR)
DAS-model 209
data acquisition 460
1D detector 87
2D detector 87
de Broglie equation 156
Debye energy 445, 448
Debye forces 199
Debye model 404
Debye relaxation 548, 550
Debye scattering equation 33
Debye–Scherrer cones 36, 87
Debye–Scherrer geometry 42, 45, 65
– angular aberration for 45
Debye–Scherrer powder diffractometer 41
Debye–Scherrer rings 45
Debye's model
– for lattice vibrations 595
Debye temperature 448, 451
Debye-type process, bistable relaxation model 547
Debye-type relaxation 551
– equivalent circuit 548
Debye–Waller factor 83, 100, 103, 415
decagonal $Al_{72}Co_8Ni_{20}$
– structure model 146
deformation density 107
deformation energy 574
deformed volume 564
degree of retardation 535
delocalization 174
demagnetization curve 514
demagnetizing factor 454
demixing reactions 426

denominator of fraction 20
density functional theory (DFT) 198, 375
density of electronic states 370, 375
density of state (DOS) 177, 328
depolarizing factor 538
depolarizing field 538
detection region 350
detectors 13, 362
– used to obtain STEM images 161
– – ADF detector 161
– – BF detector 161
– – HAADF detector 162
deuterated naphthalene
– phonon dispersion 401
Deutsches Elektronen-Synchrotron (DESY) 43
2D exchange techniques 259
DF. See dark-field (DF)
DFT. See density functional theory (DFT)
DFT simulations 215
diamagnetic susceptibility 488
diamagnetism 488, 515
diamagnets 518
diamond anvil cell setup 579
dibenzylviologen (DBV) 238
dichloromethane 537
dielectric displacement 525, 543, 550, 560
dielectric losses 544
dielectric materials 523, 538, 590
– heat capacity of 595
– insertion with polarization 525
– temperature coefficient and dielectric permittivity 541
dielectric permittivity 526, 527, 541, 544, 549
dielectric polarization 524, 525
dielectrics
– in alternating electrical fields 543
– applications of 523
– in electrostatic field 525
– phenomena 524
dielectric solids 595
difference Fourier synthesis 15
differential pumping 180
differential scanning calorimeter 589
differential scanning calorimetry (DSC) 603
– curves 605
– heat flux 603
– – principle of 604
– power-compensated 603
diffraction 1, 23, 110, 468
– angles 22
– basic theory 77
– data, interpretation of 23
– gratings 373

– of light 30
– patterns 22, 23, 113, 159
– rotational symmetry 112
diffractometer 86
– HEiDi 81
dinuclear molecule 493
dipolar-assisted rotational resonance (DARR) 267
dipolar coupling 262, 268
dipolar recoupling enhanced by amplitude modulation (DREAM) 267
dipolar recoupling with windowless sequence (DRAWS) 267
dipolar recovery at magic angle (DRAMA) 267
dipole–dipole interactions 246, 254
– electron-nuclear 281
dipole moment 527
dipole reorientation 544
dipoles
– density 535
– and polarization 527
dislocation density 570
dispersion spectrum 580
displacement current 549, 550, 551
displacement ellipsoid 8
displacement parameters 8, 24
displacive modulation 112
displacive–structural phase transition 52
dissociation energy 342
distortion modes 50
distortion vectors 51
DM anisotropy 291, 292
DM-vector pattern 291
Doniach–Sunjic function 349
Doppler broadening 371
Doppler effect 456
Doppler velocity 461
DOS. See density of electronic states; density of state (DOS)
double-quantum (DQ) schemes 267
2D projection of lattice 8
DRAM. See dynamic random access memories (DRAM)
DRAMA. See dipolar recovery at magic angle (DRAMA)
DRAWS. See dipolar recoupling with windowless sequence (DRAWS)
DREAM. See dipolar recoupling enhanced by amplitude modulation (DREAM)
drop calorimeter 602
drop-solution calorimetry 611
Drude–Lorentz model 407

DSC. *See* differential scanning calorimetry (DSC)
3D translational symmetry 6, 109, 114
dual-space strategy 17, 58
dynamic force microscopy 203
dynamic random access memories (DRAM) 523
Dysonian lineshape 282
dysprosium 473
Dzyaloshinskii–Moriya (DM)
– interaction 281, 300
– model 291
– and symmetric exchange anisotropies 298
– vector 298

e
edge dislocation 565
EDXS. *See* energy-dispersive X-ray spectrometer (EDXS)
EELS. *See* electron-energy-loss spectroscopy (EELS)
EFG. *See* electric field gradient (EFG)
EFTEM. *See* energy-filtered transmission electron microscopy (EFTEM)
Einstein modes 430
elastic constants 418, 563
elastic deformation 563, 582
elastic fracture 573
elastic modulus 563, 564, 582
elastic scattering 164
elastic strain energy 570
electrical susceptibility 526, 544
electric conductors 601
electric dipole moment 532
electric field 156, 537, 545
– analyzers 317
– in matter
– – Clausius–Mosotti Equation 541
– – dielectric permittivity, temperature coefficient 542
– – macroscopic description 538
– – microscopic description and Lorentz field 539
electric field gradient (EFG) 246, 250
electric flux 538
electric polarization 412
electric quadrupolar 452
electrochemical atomic layer epitaxy 335
electrochemical interfaces 384
electrochemical scanning tunneling microscope (EC-STM) 195
electrocompressibility 232
electrodepositions 384

electromagnetic disturbances 195
electromagnetic field 443, 523
electromagnetic perturbations 192
electromagnetic radiation 175, 362
– and neutrons, frequency–wavelength relationship 405
electromagnetic wave 524
electromotive force 610
electron acceptance angle 318
electron analyzer 313
electron beam, diffraction 167
electron beam energy 165
electron binding energies 362
electron cloud 80
electron configurations 443
– multiple transitions for 502
electron crowding 505
electron density 7, 10, 15, 16, 18, 26, 79, 117
– distribution 134, 143
electron detection cell 460
electron detector 366
electron diffraction 77, 168, 174
– patterns 168
electron dropping 175
electronegative atoms 338
electron–electron repulsion 505
electron emission 158, 325, 365, 367
electron energy loss near edge structure (ELNES) 177
electron-energy-loss spectroscopy (EELS) 149
electron excitation 318, 320
electron excitations, photon- and electron-induced 312
electron hole spinning 327
electronic conductivity 184
electronic density 443, 448
electronic excitation 311
electronic polarization 554
electronic resonance frequency 531
electronic transitions 505
electron–matter interactions 156, 163
electron micrographs 136
electron microprobe analysis (EMPA) 149
electron microscopy 134, 155, 163
electron multiplier 318
electron nuclear double resonance (ENDOR) 281
electron, orbital motion 487
electron paramagnetic resonance
– single spin detection 284
electron paramagnetic resonance (EPR) 279
– (anti)ferromagnetic resonance in magnetically ordered phases 295

- $Ba_3NbFe_3Si_2O_{14}$ compound 299
- $CuSe_2O_5$ system 298
- exchange coupled systems 286
- high-frequency EPR spectroscopy 285
- Kubo–Tomita model 292
- Kubo–Tomita/Nagata–Tazuke theories 289
- Kubo–Tomita theory 287
- linear spin-wave theory 301
- molecular field approach to real materials 298
- moments approach 294
- Nagata–Tazuke theory 288
- Oshikawa–Affleck theory 293
- prime fields of application 282
- relevant interactions 280
electron–phonon coupling 394, 436
electrons
- elastic and inelastic tunneling 189
- energy–wavelength diagram 78
- properties 156
electron scattering 165
electron shell 529
electron spectroscopy for chemical analysis (ESCA) 312
electron spectrum 320
electron spin 313
electron spin echo envelope modulation (ESEEM) 281
electrons tunnel elastically 187
electron transitions 369
electron wavelength 156
electron waves 137
electron yield 377
electron Zeeman interaction 280
electropolishing 162
electrostatic Coulomb interaction 545
electrostatic field energy 526
electrostatic force microscopy (EFM) 208
electrostatic interaction 540
electrostatic restoring force 543
electrostriction effects 526
Elliott–Yafet spin-relaxation mechanism 282
elliptical polarization 383
ELNES. See electron energy loss near edge structure (ELNES)
emitted electron 325
emitted radiation energy 456
emitting electrons 367
EMPA. See electron microprobe analysis (EMPA)
empirical crystal field 374
energy as function of crack length 572

energy balance equation 596
energy barrier
- quantum mechanical behavior of 186
energy conservation 325
energy density 526
energy dependence 329
energy dispersive detector 366
energy-dispersive spectroscopy (EDS) 149
energy-dispersive X-ray spectrometer (EDXS) 175
energy-dispersive X-ray spectrum 176
energy-filtered transmission electron microscopy (EFTEM) 149
energy quantization 248
energy ranges 363
energy release rate 572
energy-separated electrons 318
energy shifts 255
energy-spectroscopic imaging 179
- elemental map 178
energy splittings 490
engineering tribological contact 576
enthalpy 589, 604
- measurement 589
- temperature-dependent 603
enthalpy change 591
entropy 596
- of gases 590
EPR shifts
- temperature-dependent 292
EPR spectrum 281
- linear-response theory 287
equilibrium bond energy 561
ESCA. See electron spectroscopy for chemical analysis (ESCA)
ESR line width 288
estimated isotropic displacement 19
ethyl-trifluoroacetate 337, 338
Euler angles 247, 248, 253
Eulerian cradle 89, 90
Euler transformations 247, 248
European Spallation Source (ESS) 82, 108
European Synchrotron Radiation Facility (ESRF) 63, 67, 462
europium 445, 449, 473
Ewald construction 34, 148
Ewald sphere 34, 35, 148
EXAFS. See extended X-ray absorption fine structure (EXAFS)
EXAFS oscillation 362
exchange coupled systems 280, 294
excitation energy 311, 361

excitation energy range 373
excited atom decays 367
extended X-ray absorption fine structure (EXAFS) 135, 149, 362
external field 525
extinction effect 84
extrapolation method 590
extrinsic satellites 330

f

faceting 109
Faradaic current 197
Faraday cup 318
far-infrared regime (FIR) 406
fast Fourier transformation 52
Fe atoms
– quantum corral 223
– – by atomic manipulation 224
Fe film 383
FEG. See field emission gun (FEG)
FEL. See free electron laser (FEL)
Fe_4O_5 periodic structure 126
$Fe2p_{3/2}$ spectra 353
Fermi contact 455
Fermi function 187
Fermi level 187, 215, 317, 343
Fermi's Golden Rule 327, 369
ferrimagnetic transition 598
ferrocene 268
ferroelectricity 558, 598
ferroelectric materials 111
ferroelectric properties 124
ferroelectrics 596
ferromagnetic materials
– classification 513
– hysteresis loop 511
– properties and application 514
ferromagnetics 383
ferromagnetism 380
ferromagnets
– coercivity 512
– Curie temperature 513
– differential permeability 513
– hysteresis 510
– magnetic properties of 510
– permeability 510
– remanence 512
– saturation magnetization 511, 512
FETs. See field effect transistors (FETs)
FFM. See friction force microscope (FFM)
FIB. See focused ion beam (FIB)
Fibonacci sequence (FS) 142, 143
FID. See free induction decay (FID)

field-dependent permittivity 560
field effect transistors (FETs) 523
field emission gun (FEG) 158
field emission microscopy (FEM) 183
field ion microscopy (FIM) 183
FIR radiation 407
fission process 81
flipping ratio method 107
Floquet theory 266
fluorescence detector 368
fluorescence energies 369
fluorescence photons 368, 382
fluorescence radiation 315
fluorescence yields 377, 382, 384
focused ion beam (FIB) 162
force constant matrix 396
forced damped vibration 543
force–distance curve 202
form factor fall-off 80
forward scattering 329, 345, 347
four-circle diffractometer 102
Fourier map 116, 117
Fourier series expansion 115
Fourier space 131
Fourier spectrum 131, 142
Fourier transform (Fourier synthesis) 105, 145, 544
Fourier transformation 229, 362, 607
Fourier transform IR spectroscopy 406
Fourier transform NMR spectroscopy 245
Fourier transform Raman spectrometers 411
Fourier transform spectroscopy 406
four-spin triplets 502
fractional coordinates 23
fracture toughness 575, 582
Fraissard model 259
Franck–Hertz experiment 313
free electron laser (FEL) 286, 381
free induction decay (FID) 247
free ion energy level 506
free ion, magnetic moment of 489
Frenkel–Kontorova model 111
Fresnel zone plates 382
friction coefficient 575
friction force microscope 585
Friedel's law 11, 139
frozen orbital 325, 328
FS. See Fibonacci sequence (FS)
fullerene systems 430
functional material 393
– physical and chemical 393

g

GaAs
– phonon dispersion and density of states 404
GaPO$_4$ quartz 264
gas–solid interfaces 384
Gaussian decay 287
– of spin correlation 293
Gauss' law 530
Gedanken experiment 539, 540
γ-Ge$_3$N$_4$
– band structure of 372
General Theory of Magnetic Resonance Absorption 287
Gibbs energy 589
Gittergeister 110
glass transitions 606, 607, 609
Goldanskyi–Karyagin effect 453
gold electrodes
– role of water on 384
gold–water interface 384
goniometer 13
Gram–Charlier parameters 113
graphdyine 375
graphene 282, 376
– carbon atoms in 387
– functionalization and doping of 386
graphene oxide
– electronic structure of 386
graphene rippling 386
graphene sheets 385
gravitational redshift 449
Grüneisen parameters 426
group oscillations 330
g-shift
– with magnetic susceptibility 292
Guest–Host Systems 430
gyromagnetic ratios 250, 260

h

HAADF. *See* high-angle annular dark field (HAADF)
Haeberlen–Mehring–Spiess convention 249
Hamiltonian parameters 253, 254, 263, 264, 266, 280, 283
hard X-rays 362
Harker vectors 15
harmonic phonon system 403
Hartmann–Hahn condition 258, 260
Hartree–Fock values 374
Haüy's model 109
HCl solution
– cyclic voltammogram of Cu(111) 227
HDA. *See* hemispherical deflection analyzer (HDA)
H/D-isotope effect 96
heat capacity 589, 590, 591, 593, 598, 605
– frequency-dependent 607
– molar 595
– temperature dependence of 592
– time-dependent 607
– of vanadium 602
heat capacity calorimetry, adiabatic 592
heat capacity spectroscopy 608
heat conductivity 364
heat flow rates 605
heat flux DSC 605, 606
– principle of 603
HEiDi
– closed cycle cryostat 92
– monochromator stage 91
Heisenberg exchange 281
Heisenberg spin chains CsO$_2$, 293
helium gas 593
– for heat exchange 593
Hellmann–Feynman forces 396, 417
hemispherical deflection analyzer (HDA) 317
herbertsmithite EPR spectra 291
Hermann–Mauguin symbol 4
Herzfeld–Berger method 256
Hess's law 589, 611
HetDR. *See* heteronuclear dipolar recoupling (HetDR)
heterogeneous broadening 321
heteronuclear decoupling 262
heteronuclear dipolar interaction 258, 262, 271
heteronuclear dipolar recoupling (HetDR) 268
heteronuclear dipole–dipole interactions 266
hexamethylbenzene 268
HHG, high-harmonic generation (HHG)
high-angle annular dark field (HAADF) 135, 139
higher dimensional (nD) approach 141
higher order Laue zone (HOLZ) 168
high-frequency spectra 311
high-harmonic generation (HHG) 381
high oxidation state materials 377
high-permittivity dielectrics 523
high-resolution core-level spectra 342
high-resolution powder diffraction 29
high-resolution powder patterns 43
high-resolution transmission electron microscopy (HRTEM) 135, 137, 155, 160
high-resolution XRPD pattern 57

high-temperature superconductors 386
hole formalism 500
HOLZ. *See* higher order Laue zone (HOLZ)
HomDR. *See* homonuclear dipolar recoupling (HomDR)
homemade calorimeters 592
homonuclear dipolar coupling interaction 258
homonuclear dipolar interaction 267
homonuclear dipolar recoupling (HomDR) 266
homonuclear dipole–dipole interactions 266
homonuclear rotary resonance (HORROR) 267
Hooke's law visualization 563
hopping process 95
HORROR. *See* homonuclear rotary resonance (HORROR)
HRTEM. *See* high-resolution transmission electron microscopy (HRTEM)
HRTEM principles 136
HT superconductors
– static disorder and displacements 97
Hubbard band 386
Hund's rules 502, 506
hydration energy 229
hydrogen atoms 15
hydrogen bonds 24, 123
– formation 404
hydrogen evolution reaction (HER) 227
hydrostatic compression 563, 564
hydrostatic pressure 579
hyperfine coupling 260
hyperfine coupling tensor 281
hyperfine magnetic field 471
hyperfine splitting energy 455

i

icosahedral QCs (IQCs) 132
IMS. *See* incommensurately modulated structure (IMS)
incident angle 168
incident photon energy 333, 361
incoherent scattering 415
incommensurately modulated structure (IMS) 131
inductance 543
inelastic channel 190
inelastic collisions 328
inelastic deformation 576
inelastic light scattering 409. *See also* Raman spectroscopy
inelastic neutron scattering 414
– arbitrary Brillouin zones 414

inelastic scattering data 465
inelastic structure factor 416
inelastic tunneling spectra (IETS) 221
inelastic X-ray scattering (INXS) 434
– probing superconductivity 435
infrared absorption spectroscopy 383
infrared characterization 554
infrared radiation spectroscopy 406
– electric dipole moment 406
– electromagnetic wave 406
– permittivity and the refractive index 408
– reflection coefficient 408
– refractive index 408
– transmission coefficient 408
inner shell ionization 174
In_2O_3
– rhombohedral polymorph of 376
in situ X-ray powder diffraction, application of 60–64
INS spectra 432
interatomic exchange interactions 491
interelectronic repulsion 504, 506
intermetallic compounds
– magnetic properties 97
intermolecular interactions 24, 122
internal conversion coefficient 458
International Tables for Crystallography 4
International Union of Crystallography 25
intraband transitions 363
intramolecular torsion 123
intrinsic coercivity 512
inverse partial fluorescence yield (IPFY) 377, 378
inverse susceptibility 518
ionic bonds 537
ionic conductivity 125
ionic lattice 531
ionic polarization 532, 545, 546
ionization 7
ionization limit 316
ion scattering spectroscopy (ISS) 210
IPFY. *See* inverse partial fluorescence yield (IPFY)
iron-based superconductors 465
iron oxoborate Fe_2OBO_3, 469
iron pyrazolyl 468
iron-specific inelastic scattering function 473
irrational system 485
irreversible deformation, defects associated with 566
ISODISTORT 52
isomer shift 449, 461

isomorphism class 143
isothermal titration calorimeter 589
isotopic labeling 458
isotropic materials 580
isotropic resonance 256
iteration cycles 145

j

Jahn–Teller distortions 453
J-coupling 254, 266
Joule heating 610

k

Kapton shielding 196
Kapton tube 196
KDP-type compounds
– phase transition temperatures 95
K-edge spectra 365, 369, 374
Keesom forces 199
Kelvin force microscopy (KFM) 208
KH_2PO_4 (KDP) 95
kinematical approximation 83
kinematical theory 77
kinetic energy 156, 312, 313, 315, 316, 320, 322, 354, 451
kinks 568
KND_2 salts 259
Knight shift 257
KolibriSensor^{TM}, 207
Kondo effects 282
Koopmans' theorem 326
Kramers–Kronig analysis 407
Kramers–Kronig relations 281, 408, 607, 608
Kramers' theorem 492
K_2SeO_4
– softmode dispersion 421
K-shells 364
$K_2TiSi_3O_9$
– Raman spectra of 414
K_α X-ray emission lines 311

l

$LaAlO_3$–$SrTiO_3$ interface 386
laboratory axis system (LAS) 247
La_2CuO_4
– orthorhombic LTO phase 97
$LaFeO_3$
– cell volume of 70
– Rietveld plots of 69
$LaFe_4Sb_{12}$
– measured neutron spectra 433
Lambert–Beer law 80, 350

Lamb–Mössbauer factor 447, 451, 455
– temperature dependence 444
Langevin diamagnetism 488
Langevin function 534
Laplace's equation 452
Larmor frequency 249
LAS. *See* laboratory axis system (LAS)
laser-flash calorimetry 601, 602
laser-induced collisions 381
latent semantic indexing (LSI) 57
lattice anomalies, in multiferroics 411
lattice distortion 567
lattice dynamics, fundamentals of 394
lattice expansion 542
lattice heat capacity 403
lattice magnetization 454
lattice resistance 568
lattice rigidity 447
lattice symmetry 143
lattice vibrations 595
– amplitudes 402
– – displacements of atom 402
– – normal coordinates 402
– thermal behavior of 591
– thermal excitation of 595
Laue diffractions 82
Laue equations 33
– visualization of 31
Laue groups 139
law of average environment 507
LDE. *See* low density elimination (LDE)
Le Bail method 50
L-edges 365
– of copper 378
$L_{2,3}$-edge spectra 368
LEED. *See* low energy electron diffraction (LEED)
Lee–Goldburg cross-polarization (LG-CP) 268
Lee–Goldburg (LG) sequence 263
Legendre polynom 254, 264
length-extension resonator (LER) 207
Lennard-Jones potential 200, 204
Lenz's law 488
$LiCoO_2$ battery materials
– MAS spectra of 260
ligand effect 211, 344
light irradiation method 600
light scattering experiments
– physical properties of 30
Li-ion batteries 385
linear absorption coefficient $\mu(E)$ 361, 378
linear accelerator (LINAC) 43

linear dielectric material 525
linear spin-wave theory 301
lineshapes 254
linewidths 371, 386
liquid crystalline phases 259
liquid helium
– for refrigeration 593
liquid microjets 384
liquid nitrogen 594
– for refrigeration 593
liquid–solid interfaces 384
load *vs.* displacement curve, for fused silica standard 582
local chemistry 454
local density of states (LDOS) 188, 189
local electric field 539
– feedback loop 539
local magnetic ordering 493
lock-in amplifier 600
long-range order (lro) 134
Lorentz factor 86
Lorentz field 540
Lorentz force 157
Lorentzian absorption lines 453
Lorentzian distribution 370
Lorentzian function 349, 371, 446
Lorentzian shape 288
Lorentzian spectrum 287
loss tangent 544
LO–TO splitting 555
low density elimination (LDE) 144, 145
low energy electron diffraction (LEED) 140, 342
low-energy electrons 329
lower symmetry (LS) 50
low-frequency dielectric constant 559
low loss region 177
low-spin systems 127, 491
lro. *See* long-range order (lro) 134
L-shells 364
LST. *See* Lyddane–Sachs–Teller (LST)
Luggin capillary 195
luminescence emission 387
luminescence photons 380
Lyddane–Sachs–Teller (LST) 555, 559

m

macroscopic crystal 4
macroscopic electric field 538
macroscopic field 540
macroscopic susceptibility 528
magic angle 254, 262, 263, 264
magic-angle spinning (MAS) 245, 253, 254, 255, 260, 265
magnesium ultraphosphate (MgP_4O_{11}) 268
magnetically condensed systems 493
magnetically dilute systems 490–492
magnetic anisotropy
– on Kagome lattice 290
magnetic Bragg peaks 104
magnetic chirality 80
magnetic dipole moment 454, 487
magnetic energy 301
magnetic exchange interaction 108
magnetic fields 157, 368, 487
magnetic field vector 247
magnetic force microscopy (MFM) 208
magnetic hyperfine interactions 449, 455
magnetic imaging 380
magnetic induction 454, 487
magnetic lenses 158
– schematic cross section of 157
– – copper wire 157
– – object 157
magnetic materials 510
– classification of 490
magnetic modulation vector 297
magnetic moments 77
magnetic neutron diffraction 88, 105
magnetic neutron scattering 79, 80
magnetic ordering phase transition 90
magnetic prism 178
magnetic quantities 486, 487
magnetic quantum numbers 248, 321
magnetic semiconductors 375
magnetic susceptibility 487
magnetic X-ray diffraction 80
magnetism 443, 468, 470
magnetite 341
magnetization 463, 471, 487
– definition 487
– densities 77, 107
– – three-dimensional map 107
magnetoelectric multiferroic $BiFeO_3$ 470
magnetometry 466, 468
magnetoresistive random access memory (MRAM) 523
magnetostrictive phase transition 470
Maier-Leibnitz Zentrum (MLZ) 81
main reflection 113
Markovian random processes 287, 292
MAS. *See* magic-angle spinning (MAS)
Masakova tilings 147
materials
– characterization techniques 575–578

– – diamond anvil 579
– – fracture properties 582
– – friction and wear test 583
– – nanoindentation 581
– – scattering techniques for elasticity measurements 580
– – tensile test 581
– mechanical properties 561–562
– – elastic behavior 562–564
– – nonelastic behavior 564–578
– – – fracture 570–575
– – – plasticity 565–570
– – nonelastic behavior friction and wear 561–562
matrix effects 348
matrix element 328
matter, dielectric properties 524
maxima 14
Maxwell's law 546
Maxwell's theory 539
Maxwell–Wagner polarization 536, 552
McMaille approach 40
MD. *See* molecular dynamic (MD)
mean free path length 328, 354
mean kinetic energy 451
mean polarizability 535
mean square displacement (MSD) 7, 83
mechanical anisotropy 418
mechanical stress 525
M-edges 378
Meissner effect 471
Meissner shielding 472
merohedral 22
metal compounds
– magnetic properties of 491
metal deformation, under tensile load 581
metal–electrolyte interfaces 227
metallic aluminum 331
metallic analyzer 316
metal–ligand (M–L) axis 498
metal oxidation states 509
metal spectra 345
metal, stress–strain curve of 581
MgB_2
– scattering geometry in 436
MgK_α- excited photoemission 319
microcrystalline powders 246
microdiffraction 12
microscopic dipole moment 530
microscopic polarizability 528
microspectroscopy 382
microwave dielectrics 558
microwave field 281

microwaves yields 279
Miller indices 31, 48
millielectronvolts 164
mirror plane *vs.* glide plane 3
mixed Ba–Ce oxide, TEM image 178
$(Mn_{1-x}Cr_x)_{1+\delta}Sb$
– crystal structure and site occupation 96
$(Mn_{1-x}Cr_x)_{1+\delta}Sb$ structure 95
mode Grüneisen parameters 423
model of crystal, according to Haüy 110
modern electron microscopy 77
modern synchrotron 44
modulated atomic positions 117
modulated crystals, structural chemistry 121
modulated materials, crystallography 111
modulated molecular crystals 122
modulated structures, refinement of 23
modulation functions 114
– and Fourier maps 116
modulation, incommensurability 119
modulation wavevectors 23
modulus 10
Moiré-type superstructure 214, 234
molecular crystals 429
molecular dynamics (MD) 576
molecular field approach 296
molecular field (mf) model 494
molecular vibrations 364
monochromatic powder diffraction experiments 86
monochromatic X-rays 12
monochromatization 465, 473
monochromator 85, 177, 362, 465
monochromator crystals
– wavelengths and Q ranges 91
monochromator optics 368
monochromators
– resolution function of HEiDi 91
morpholinium–BF_4, diffraction pattern 112
MRAM. *See* magnetoresistive random access memory (MRAM)
M-shells 364
Mössbauer active nuclides 446
Mössbauer data acquisition 472
Mössbauer effect data center 449
Mössbauer nuclides 460
Mössbauer, Rudolf 444
Mössbauer spectroscopy 443
– area and recoil-free fraction 447
– calibration and data treatment 461
– detectors 459
– examples and applications
– – charge order and relaxation 468

– – in situ, operando, remote operation 472
– – magnetic phase transitions 470
– – nuclear resonance scattering 473
– – site occupancy/distribution 466
– – spin crossover 468
– – superconductivity 471
– – superparamagnetism 471
– – transferred field 471
– – valence determination 466
– experimental implementation 456
– isomer shift 448
– magnetic hyperfine splitting 454
– nuclear resonance techniques 443
– quadrupole splitting 451
– radiation sources 456
– relaxation phenomena 455
– sample thickness and sample environment 460
– second-order Doppler shift 449
– spectral parameters 446
– synchrotron radiation 462
Mössbauer spectrum 452, 456, 472
Mössbauer temperature 451
multichannel plate 317
multicomponent compounds 569
multidomain samples
– Raman spectra of 412
multielectron effects 320
multiferroic $Bi_2Mn_4O_{10}$ 418
multiferroic phase transition 412
multiferroic systems
– IR reflectivity of 407
multiple diffraction 148, 169
multiple scattering effects 83, 84, 147
multiplet coupling 374
multiplet splitting 327
muon spectroscopy (μSR) 286
muscovite mica, lateral force images 578

n
NaCl film, on Cu(111)
– AFM image of a pentacene molecule adsorbed 222
Na_2CO_3, 121
Na_2CO_3, crystal structure 111, 117
Na_2CO_3 in D superspace, modulated structure of 118
Na_2CO_3, interatomic distances 121
Na_2CO_3, phase diagram 122
Nagata and Tazuke (NT) theory 288
nanocomposite material 578
nanocomposite structure 577

nanoindentation 581
nanomaterials 364
nanoparticles 162
nanoscopic photocatalysts 246
nanotribology 576
nanotubes 259
narrowband emission 387
2-n-butoxyethanol 608, 609
Nd_2CuO_4
– phonon dispersion of 400
nD hyperatoms 134
nD space groups 132
near-edge X-ray absorption fine structure (NEXAFS) 362
nearest-neighboring spins 296
Néel ordered phase 291
Néel temperature 413
neodyme–iron–boron magnets energy 514, 515
nephelauxetic effect 508
Nernst potential 235
neutral atoms 320
neutron detectors 90
neutron diffraction 8, 82–92, 83, 470
– application, examples 92–108
– concepts 85
– corrections 82
– diffracted intensities 82
– instrumentation components 85
– magnetic structures analysis, role in 103
– magnetization densities determination 107
neutron diffractometers 81
neutron filters 85
neutron flux 81
neutron powder diffraction pattern 47
– thermal evolution 101
neutron powder diffractometer SPODI 87
neutron Rietveld analysis 88
neutrons
– characteristic properties 78
– classification 78
– continuous source 82
– diffraction
– – electron densities determination 105
– energy–wavelength diagram 78
– generation 81
– magnetic moment 80
– reactor sources 81
– scattering 77
– scattering amplitudes 79
– scattering power 78

– sources 81
– spallation sources 82
– stress analysis 79
– as structural probes 78–82
neutron single-crystal diffraction 100
neutrons scatter isotope 83
neutron tomography 79
Newton's law 603
NEXAFS. *See* near-edge X-ray absorption fine structure (NEXAFS)
NFS. *See* nuclear forward scattering (NFS)
Ni_3Al alloy 197
$Ni_3Al(111)$ surface oxidization
– STM images 213
NiAs structure 95
nickel film 352
nickel oxide 344
niobium–tin superconductors 472
$Ni2p_{3/2}$ and $Fe2p_{3/2}$ XPS spectra 355
NIS. *See* nuclear inelastic scattering (NIS)
nitrogen oxide (NO) 283
NMR. *See* nuclear magnetic resonance (NMR)
noncontact atomic force microscopy (nc-AFM) 206
noncrystallographic point 7
nondispersive detectors 382
nonequilibrium studies 426
nonhydrostatic pressure 579
nonlinear electrical permittivity 560
non-negative electron density 16
nonstoichiometric compounds 124
normal coordinates, amplitudes 402
normal emission 355
normalized structure factors 16, 17
NRIXS. *See* nuclear resonant inelastic X-ray scattering (NRIXS)
nuclear absorption 447
nuclear fluorescence 464
nuclear inelastic scattering (NIS) 465
nuclear magnetic moment 455
nuclear magnetic resonance (NMR) 150, 286, 443
– crystallography 246
– experiments, simulation of 253
– spectra 257
– spectroscopy 245
nuclear magneton 454
nuclear quadrupole moment 452
nuclear reflections 104
nuclear resonance absorption 465
nuclear resonance fluorescence 465

nuclear forward scattering (NFS) 462, 472
nuclear resonant absorption 458
nuclear resonant detectors 459
nuclear resonant inelastic X-ray scattering (NRIXS) 465
nuclear resonant photons 463
nuclear scattering 96
nuclear spin interactions 245, 246, 249, 253, 260
– chemical shift 245, 246, 249
– dipolar interactions 250
– – direct dipole–dipole interaction 250
– – homonuclear dipole–dipole interaction 250
– – magnetic dipole–dipole interactions 250
– direct dipole–dipole 246
– hyperfine couplings 246
– hyperfine interaction 249
– quadrupolar coupling 245, 246, 256
– quadrupole interaction 250
– spin rotation 246
– Zeeman interaction 246, 248, 252
nuclear transition 458
nucleus with surrounding electronic environment yield, Coulomb interactions 447
numerical simulations 248

o

O_{apical}
– MSD 98
OA theory 294
oblate ellipsoids 24
observed intensities 20
occupancy merge procedure 55
octahedral crystal fields
– estimated magnetic behavior of d ions 516
octahedral d^9 complex
– electronic transition of 501
octahedral symmetry 499, 500, 506
octahedral transition metal complexes 507
OFET. *See* organic field-effect transistor (OFET)
Ohmic contact 188
$O_{inplane}$
– MSD 98
O K-edge spectrum 384
O K-edge XAS 386
OLED. *See* organic light-emitting diode (OLED)

Oliver and Pharr method 581
one-electron wave functions 498, 504
optical detection 285
optic axis 158
OPV. *See* organic photovoltaic (OPV)
orbital angular momentum 361, 487, 489
orbital points 496
orbital splitting 501
order–disorder phase transition 100
order without translational symmetry 109
organic field-effect transistor (OFET) 383
organic light-emitting diode (OLED) 383
organic photovoltaic (OPV) 383
organic thin films 364
orientation polarization 532, 535, 548
orthoenstatite 264
– structure of 265
orthorhombic crystal fields 492
osmium 463
overdamped oscillator 548
overpotential deposition (OPD) 236
oxidation state 374
oxygen adsorption 213, 348
oxygen gas and water vapor, photoemission lines 327

p

Pake doublet 256
palladium foil 459
paraelectric RDP
– difference-Fourier plot 94
paramagnetic properties 488
paramagnetic susceptibility
– temperature dependence of 516
paramagnetism 488, 515
paramagnets 515
– applications of 518
parametric Rietveld refinements 49
partial electron yield (PEY) 366
partial fluorescence yield (PFY) 366
PAS. *See* principal axis system (PAS)
pass energy 317
Patterson method 14
Pauli paramagnetism 488
Pauli principle 186
Pauli repulsion 200
Pauli's exclusion principle 503
Pawley method 50
pc-LED. *See* phosphor-converted light emitting diodes (pc-LED)
PCTF. *See* phase contrast transfer function (PCTF)

PDSD. *See* proton-driven spin diffusion (PDSD)
peak shape parameters 50
Peierls stress 568
Peltier heating 600
pentacene molecule
– structure model of 222
periodic crystals, chemistry 122
periodic distortion 110
periodicity 115
periodic table 311
permanent dipoles 532
– in electric field 532
permanent magnets 514
permittivity in ionic crystals, voltage dependence 560
permittivity in titanates, temperature dependence 559
perovskites 606
perovskite-type oxide, characteristic oscillations 558
perturbations 254
perturbative spinon 294
PES. *See* photoelectron spectroscopy (PES)
PEY. *See* partial electron yield (PEY)
PFY. *See* partial fluorescence yield (PFY)
phase-contrast electron microscopy 135
phase-contrast transfer function (PCTF) 171
phase problem in crystallography 13
phase problem, solution of 13
phase transformation 385
phase transition 592, 593, 598, 600, 601, 603, 606
– antiferroelectric 606
– enthalpy 591
– ferroelectric 606
– temperature 559
phenazine–chloranilic acid 123
phenomenological treatments 417
(phenylazophenyl)palladium(II) 60
2-phenylbenzimidazole 123
– molecular structure 123
PHONON 417
phonon bands, contribution of 426
phonon density-of-states 403, 431, 437
phonon dispersion 399
– spectrum 134
phonon-enhanced diffusion process 434
phonon-enhanced ionic transport 434
phonon frequencies 424
phonon glass 432
phonons 175, 364, 402
– in harmonic approximation

– – Brillouin zone boundary values 399
– – Fourier-transformed force constant matrix 397
– – polarization vectors 397
– – primitive cell 395
– – square root masses of the atoms 398
– – vibrational states 395
– scattering 164
– spectroscopy 472, 532
– theory 119
phosphate glasses 268
phosphor-converted light emitting diodes (pc-LED) 387
photoabsorption 334
photocurrent 327
photodissociation 356
photoelectric effect 311
photoelectron current 349
photoelectron diffraction pattern 348
photoelectron excitation, schematic energy diagram 316
photoelectrons 313, 366, 368
– scattering mechanisms of 362
photoelectrons, angular distribution 331
photoelectron spectroscopy (PES) 313, 324
– analytical applications 334–356, 335
– – chemical shift 337
– – core-level angular dependence 345
– – depth profiling 352
– – quantitative analysis 348
– Auger spectroscopy 331
– chemical shift
– – influence of ionicity 337
– – standard spectra 339
– – surface core-level shifts 341
– – valence band emission 343
– excitation, relaxation 324
– extrinsic satellites 330
– intensity 327
– photoemission experiment 313–324
– – electron analyzer and detector 315
– – photon source 315
– – sample 314
– – typical spectral features 318
– photoemission from solids 328
– photoemission process 324–334
– possible misinterpretations 356–357
– spin–orbit and multiplet splitting 326
photoemission 315, 325, 357
– experiment
– – instrumental setup 313
– – parameters 314

photoemission intensity 348
photoemission peak 332
photoemission process 324
photoemission signals 357
photoemission spectrum 322, 357
photoemitted electrons 313
photoexcitation 328, 364, 365, 368
photo-hole 349
photoionization 364
photomultiplier-based detectors 460
photon absorption 364
photon detector 377
photon emission 311
photon emitted, by free atom/nucleus 444
photon energy 315, 319, 354, 365, 366, 368, 373, 378
photons 361
photons (X-rays)
– energy–wavelength diagram 78
photon source 313
Phz–H_2ca, t-plots of selected interatomic distances 124
picoammeter 366
piezoelectric 523
pin-on-disc tester, equipment layout 584
Planck constant 81, 156, 280, 287
plasmons 175
plastic deformations 565, 579
plastic-zone size, as function of yield stress 574
PNR. See polar nanoregion (PNR)
point charge model 260, 509
Poisson equation 525
Poisson ratio 563, 568
polar angle 250
polarizability 534, 539, 546, 549
polarization 410, 524
– mechanisms 536
– – cutoff characteristics 537
– – electronic polarization 528
– – ionic polarization 531
– – material classification by 537
– – Maxwell–Wagner polarization 536
– – orientation polarization 532
– vector 51
polar nanoregion (PNR) 595, 596
polar optical phonons 554
poly(arylene vinylene) copolymers 271
polychromatic or white radiation 12
polycrystalline ceramic material, microstructure 535
polycrystalline material 580
polyethylene 537
polyhedron 122

polystyrene 537
polytetrafluoroethylene 537
polyurethane 552
porphyrin molecules
– in situ STM images 237
position-sensitive detector (PSD) 45, 57
position-sensitive photodetector 201
powder data analysis 57
powder diffraction 1, 29, 246
– atomic form factor 32
– Bragg reflection 32
– Bragg equation 32
– data, structure determination 52–55
– experiment, sketch of 87
– Fourier transform 32
– halo effect 36
– history and basics of 30
– indexing equation 39
– Laue equation 31
– primary beam 31
– real-space lattice 32
– reciprocal lattice points 36
– scattering angle 31
– scattering vector 34
– single-crystal diffractogram 30
– unit vectors 31
– X-ray diffraction experiment 30
powder diffractometers 41
– Debye–Scherrer geometry 42
– slow point detectors 42
powder diffractometry 88, 90
powder neutron diffraction 87, 88
– instrumentation 88
powder pattern 256
powder X-ray diffraction 139
primary electrons 329
primary modulation wave 124
primary photocurrent 349
primitive cell 18
– notation, of position vector of atom 395
principal axis system (PAS) 247
principal axis values 247
prompt radiation 463
protein crystallography 117
proton decoupling 245
proton-driven spin diffusion (PDSD) 267
pseudomerohedral twins 22
PtCo alloy 211
– STM image 211
PtCo alloy surface
– STM image 212
PtRh alloy
– STM image 211

Pt(111) surface
– STM image 226
pull-off-contact 203
pulsed ENDOR 284
pulsed neutron sources 90
pulsed synchrotron radiation beam, bandwidth 464
pulsed vs. c.w. electron paramagnetic resonance techniques 284
pyrazinamide (PzCONH$_2$) 385
pyroelectric 523
pyrrhotite 125

q

QC. See quasicrystal (QC)
quadrature detection 253
quadrupolar coupling 250, 255, 264
– constant 260
quadrupolar interaction 257, 259, 264, 454
– second-order 263
quadrupolar interactions 254, 262, 266
quadrupolar nuclei 264
quadrupole splitting 453, 455
quality factor 544
quantitative crystal structure analysis 135
quantitative Fe(II) and Fe(III) content 467
quantitative-phase Rietveld (QPR) 65
quantization method, for harmonic oscillators 402
quantum beat 463
quantum chemical modeling 267
quantum kagome antiferromagnets (QKA) 290
quantum numbers 487, 502
quantum sine-Gordon spin-chain model 294
quantum spin-liquid (QSL) 290
quantum statistical methods 403
quantum theory of atoms 26
quartz
– α-phases of 600
– β-phases of 600
quartz single crystal 601
quasi-continuous phase transformation 50, 51
quasicrystal (QC) 18, 111, 131, 134
– structure analysis 134
– surface structure analysis 140
quasicrystallographic space groups 132
quasicrystal structures analysis
– diffraction methods 140–148
– transmission electron microscopy 135
quasiharmonic approximation 423
quasilattices 134
quasiperiodic crystal structures 131, 132

r

Racah parameters 508
radiation damage 356, 367, 384, 387, 388
radio frequency 246
radio-frequency-driven dipolar recoupling (RFDR) 266
– finite-pulse 266
radio frequency pulses 251
Raman process
– electronic states 411
Raman scattering
– local distortions 412
Raman spectrometers 406
Raman spectroscopy 394, 409, 412
– dipole radiation 410
– polarizability tensor 410
– scattered intensity 410
– techniques 411
Raman tensor 410
rational system 485
rattlers 432
Rayleigh scattering 364, 410
RbH_2PO_4 (RDP)
– H/D ordering in ferroelectric 95
RbH_2PO_4 (RDP) paraelectric phase
– crystal structure 93
– two PO_4-tetrahedra, local configuration of 94
reaction entropy 590
real/direct space lattice parameters 40
real-space unit cell parameters 39
reciprocal lattice vectors 114
reciprocal space lattice 35, 40
reciprocal space methods 58
recoil energy 445
recoil-free fraction 447, 460
recoupling schemes 265
recoupling sequences 263, 266
recoupling techniques 245
recrystallization 570
REDOR. *See* rotational-echo double resonance (REDOR)
reference signal 356
refinement techniques 19
$\Sigma 2$ relationship 17
relative dielectric permittivity 526, 551, 560
relative displacement 530
relative (Doppler) velocity 449
relaxation calorimeter
– thermal model of 597
relaxation currents, nonexponential decay 551
relaxation curve 598
relaxation effects 325, 357, 544
relaxation phenomena 547, 597
– Debye relaxation 548
– dielectrics with distribution of relaxation times 552
– Maxwell–Wagner relaxation 552
relaxation time 542, 543, 549, 552
relaxors 595, 596
remanent magnetization 512
remanent polarization 523
Renninger effect 83
residual entropy 591
residual leakage current 544
resonance circular frequency 543
resonance effects 544
resonance energy 451, 473
resonance frequency 254, 546
resonance Raman spectra 410, 411
resonant absorption 445
resonant inelastic X-ray scattering (RIXS)
– excitations, schematic presentation of 438
– probing electron–phonon coupling 436
resonant photon energy 449
resonant X-ray scattering 80
restoring force 545, 559
restraints 20
retardation voltage 317
reversible hydrogen electrode (RHE) 195
RFDR. *See* radio-frequency-driven dipolar recoupling (RFDR)
rhodium 458
Rietveld plots 63
Rietveld refinements 29, 46, 88, 105
– agreement factors 48
– least-squares algorithm 46, 48
– mathematical description 47
– powder diffraction patterns 47
rotary resonance
– homonuclear 267
– recoupling (R^3) 268
rotational-echo double resonance (REDOR) 268
rotor axis system (RAS) 254
Runge-Ito-de Wolff method 40

s

SAED. *See* selected area electron diffraction (SAED)
SAM. *See* self-assembled monolayer (SAM)
samarium 473
samarium–cobalt permanent magnets 515
satellite reflections 110
scaling factor 263

scanning electron microscope (SEM) 56, 149, 160
scanning probe microscopy
– phase transitions and chemical reactions 183
scanning probes
– types of 208
scanning probe techniques (SPM) 184
scanning transmission electron microscopy 135, 158
scanning transmission electron microscopy (STEM) 135, 158
– schematic 161
scanning tunneling microscopy 184, 185, 342
– advantages 195
– electrochemical double-layer 229
– electrochemical STM setup 195
– at high gas pressure 226
– physical background 185
– in solution 227
– STM images 197
– STM setup in UHV 193
– surface structure 227
– tip position control 190
– tip preparation 196
scanning tunneling microscopy (STM) 140, 184, 383
– channel 217
– constant current mode imaging 190
– controller 195
– controller electronics 198
– images 198
scanning tunneling spectroscopy (STS) 189
scattered radiation 80
Scattering
– coherent 364
– incoherent 364
scattering angle
– deviation of 38
scattering effects 137
scattering experiment, layout of 415
scattering power 96
scattering vector 415
Scherzer defocus 171, 172
Schrödinger's equation 252, 417
secondary electrons 162, 175, 329
– energy distribution 330
second-order Doppler (SOD) shift 451
second-order harmonics 118
selected area electron diffraction (SAED) 136, 138, 160, 169
self-absorption 368
self-assembled monolayer (SAM) 383

self-charge mechanism 257
semiconductors 175, 316
semiconductor technology 523
sequential parametric refinements 49
shape factor 538
shear modulus 563, 568
shear stress 567
shock wave compression 579
signal to noise ratio 368
signal-to-noise-ratio (SNR) 459, 600
silicon drift detectors 459
simulated annealing (SA) 53, 58
single asperity nanotribological contact 576
single-crystal diffraction 12, 35, 90, 246
– instrumentation 90
– sample environment 92
single-crystalline oxide 354
single-crystal neutron diffractometer 90
single-crystal reciprocal space techniques 52
single-crystal X-ray (synchrotron) experiment 105
single edge notch bend test, schematics of 583
single-ion (SI) 280
single-molecule magnets (SMM) 282
single phonon mode 439
single-photon noise-free counting pixel-detector 140
Slater mode 559
small-angle neutron scattering 471
SO_4 anions
– high-resolution STM image 233
sodium chloride 537
sodium diphosphates 264
sodium iodide scintillator 459
sodium metal halide batteries 108
softmode frequency 420
softmode phase transitions 419
– antiferrodistortive mode 419
– atomic displacements 419
– Bragg peak 423
– Landau theory 419
– lock-in transition 421
– phonon mode 419
– scenarios for 421
soft X-ray
– absorption spectra 367
– – detectors 382
– – electron yield 367
– – fluorescence yield 367
– – polarization effects 382
– – technical developments
– – – inverse partial fluorescence yield (IPFY) 377
– – – measuring in transmission mode 379

– – transmission 367
– fluorescence yields in 364
soft X-ray absorption spectroscopy 361, 384, 386
soft X-ray emission spectroscopy (SXES) 150
soft X-rays 362, 364
soft X-rays absorption 364
sole electronic polarization 537
solid–liquid interfaces 184, 198, 385
solid–solid phase transitions 591, 596
– ferroelectric transition 591
– ferromagnetic transition 591
– metal–insulator transition 591
– superconducting transition 591
solid-state chemistry 30, 155, 456, 561
solid-state detector (SSD) 382, 382
solid-state NMR spectroscopy (ssNMR) 245
solid-state physics 524
solution calorimeter 589
source–absorber resonance 449
space charge polarization 536
specific adsorption 229
spectrochemical series 505–509, 507
spectromicroscopy 382
spectroscopic methods 149
spectroscopic studies 404
– energy conservation law 404
– optical spectroscopy/inelastic neutron 406
– vibrational properties, in solids 404
– X-ray scattering 406
spherical aberration 158, 179
spin canting 463
spin-crossover compounds 127
spin decoherence 285
spin density formalism 252
spin density maps 107
spin density matrix 252, 253
– size of 253
spin diffusion exchange spectra
– of triphenyl phosphite 261
spin-down electron 503
spin-echo double resonance (SEDOR) 268
spin Hamiltonian 282, 283
spin interactions
– energy levels, influence on 255
spin–lattice relaxation 253, 258
spin magnetic moment 487
spinodal decomposition 427
spin–orbit coupling 374, 492
spin–orbit splitting 319, 326
spin-paramagnets 517
spin–spin relaxation 258

spin systems 248, 249, 252, 253, 263, 266, 267, 268, 269
spin systems $SrCu_2(BO_3)_2$, 286
spin-up electron 503
spin vector 247
spin-wave theory 301
spiral magnetic order 110
spring constants 580
spring force 531
sputter profiling 352
squared structure factors 50
$SrCu_2(BO_3)_2$, 290
– 2D network of orthogonal Cu^{2+} dimers 289
– spin Hamiltonian 290
$SrTiO_3$
– Raman spectra of 414
stainless steel (SS) 193
standard deviation 461
standard spectra 339, 340, 341
standard state pressure (SSP) 590
state-of-the-art high-resolution powder 30
STEM-coupled spectroscopy 178
Sternheimer antishielding factor 453
STM. See scanning tunneling microscopy (STM)
– beetle-type setup of 191
– constant current mode 201
STMAS method 264
Stokes scattering 410
Stokes signal 410
strain energy 571
stray magnetic fields 460
stress intensity factor 572
– as function of sample thickness 584
strong-field approximation 506
strong-field limit 505
strontium titanate 558
– perovskite crystal structure 556
structure factor equation 11
subsurface carbon atoms
– STM images 219
supercell 375
supercluster constituting decagonal 137
superconducting quantum computing systems 523
superconductivity 127, 472
superconductors 471, 519
supercooled liquids 608
superspace structure model 119
surface acoustic wave 580
surface core-level shift 341

surface energy 571
surface oxidation 368
surface plasmons 330
surface reaction processes 317
surface reconstruction 209
surface refinement 49
surface science approach 209
surface segregation 209
surface-sensitive diffraction techniques 140
surface topography 175
s-wave model 188
symmetry modes, concept 50–52
synchrotron 85
synchrotron beam 381
synchrotron diffraction data 64
synchrotron light source 361
synchrotron radiation 1, 12, 44, 46, 140, 315, 335, 382, 456, 463, 465, 473
synchrotron sources 85, 380
synchrotron X-ray powder diffraction (SXPD) 43
synthetic chemistry 466
systematic absences 14

t

tangential force 575, 584
tangential resistance 575
Taylor series 301, 395
tellurium 473
temperature dependence, of reflectance 409
temperature-dependent phase transitions 111
temperature-independent paramagnetism 488
temperature jump calorimeter 602
temperature variation curve 598
tensile force 581
tensors 247, 248
– chemical shielding 249
Tesla 487
tetragonal HTT phase 97
tetragonal tungsten bronze (TTB) 172, 173
tetrahedral symmetry 498, 500
thermal analysis 603
thermal conductance 597, 604, 606
thermal decomposition reaction 473
thermal diffuse scattering (TDS) 164
thermal diffusivity 602
thermal drift 229
thermal effusivity 608
– frequency-dependent 609
thermal hysteresis 598, 600
thermal motion energy 445

thermal neutron powder diffractometer 88
thermal neutrons 81
thermal radiation 593, 600, 603
thermodynamic equation 595
thermodynamic equilibrium 592
thermodynamics
– third law 590
thermoelectrics 465
thermopile 609, 610
thermosalient (TS) 60
thin films vs. bulk samples 406
three-window method 179
tiling-decoration method 132
time-of-flight (TOF) 82
time-of-flight spectroscopy 417
Ti nanoribbons 382
TiO_2-derived nanoparticles 283
tip etching
– setup 196
Tomonaga–Luttinger-liquid (TLL) 293
TOPAS program 66
torsion angles 123, 267
total angular momentum quantum number 333
total electron yield (TEY) 366
total fluorescence yield (TFY) 366
total polarization 533
traditional single crystal diffractometer
– scheme of 89
transferred-echo double resonance (TEDOR) 268
transferred hyperfine field 471
transferred magnetic field 454
transferred moments 107
transition energy 465
transition matrix elements 375
transition metal 374
transition metal interaction 377
transition metal ion 498
transition metal oxides 383
– heterostructures 364
transition probability 327
transition rates 370
transmission electron microscope (TEM) 155
– operating modes 159
transmission integral approach 460
transmission Mössbauer spectroscopy, schematic setup 457
transmission probability 188
transverse acoustic modes 425
transverse acoustic phonon 428, 429
transverse magnetization 259

transverse optical phonon, softening 555
triphenyl phosphite 259
1,3,5-tris(2,2-dimethylpropionyl amino)
  benzene (BTA) 268, 269
tuning fork sensor (TFS) 206, 207
tunneling contact 188
tunneling gap 189
tunneling junction
– energy diagram of 187
tunneling matrix element 188
twinning 22
twin refinement 22
two-dimensional crack, schematic drawing
  of 573
two-dimensional diffraction patterns 34, 36
two-sublattice model 295, 298

u
ultrahigh vacuum (UHV) 184, 209, 314, 366
– amorphous structures 215
– anion adsorption on Cu(111) 229
– atomic manipulation 223
– dissociative oxygen adsorption 217
– gas–solid interactions 226
– investigations 208
– molecular adsorbates 219
– molecular self-assembly at metal–electrolyte
  interfaces 236
– subsurface atoms/impurities 219
– surface composition
– – alloys 210
– surface oxidation 213
– surface structure
– – reconstruction 209
– tip-induced surface reaction 224
– underpotential metal deposition 235
ultralow-permittivity dielectrics 523
ultrathin single-crystalline 213
Umweg-Anregung 83, 84
unaccounted electron density 21
unconventional coupling 435
underpotential deposition (UPD) 235
undisturbed energy 328
uniaxial single-ion anisotropy 298
unit cell 2, 82
universal curve 356
unreconstructed Si(111)
– hard sphere models 209

v
vacuum capacitor 524
vacuum permittivity 525
valence band 316

valence electrons 556
valence separation 468
van der Waals forces 199
Van Vleck paramagnetism 488
vibrating cantilever
– disadvantage 207
Voigt function 349
voltage-step response 553
volume thermal expansion coefficient 593
von Neumann equation 252

w
water bonding structure 384
water splitting
– photocatalysts for 262
wavelength-dispersive spectroscopy
  (WDS) 149
wave vector 148, 545
weak-field approximation 506
weak-field limit 505
whole powder pattern decomposition 49
– lattice parameters 50
– peak intensities 50
whole powder pattern fitting (WPPF) 42, 46
Wyckoff positions 134

x
XAS. *See* X-ray absorption spectroscopy (XAS)
X-band 279
^{129}Xe gas
– chemical shift 259
Xe lattice dynamics, fundamentals of
– phonons, in harmonic approximation 394
Xe phonons, in harmonic approximation
  394
XMCD. *See* X-ray magnetic circular dichroism
  (XMCD)
X-N-synthesis 105
X-ray 1
– absorption 366
– discovery 361
– emission 311, 333
– fluorescence 459
– fluorescence spectra 318
– form factor 79
– penetration depths 80
– photoemission 312
– scatterers 88
– scattering 365, 448
– scattering cross section 79
– scattering power 78
– spectroscopy 177
– structure analysis 93

– tubes 85
X-ray 30
X-ray absorption near-edge structure (XANES) 149, 362
X-ray absorption spectroscopy (XAS) 361, 381
X-ray data
– Fourier synthesis of 103
X-ray diffraction (XRD) 1, 7, 77, 109, 169, 375
– electron densities determination 105
X-ray excited optical luminescence (XEOL) 379
X-ray linear dichroism (XLD) 383
X-ray magnetic circular dichroism (XMCD) 383
X-ray magnetic linear dichroism (XMLD) 383
X-ray photoelectron spectroscopy (XPS) 150, 210, 312

y
Young modulus 563
ytterbium 464
yttrium 564
yttrium aluminum garnets 459

z
Zachariasen's scheme 217
Zaremba–Conroy–Wolfsberg 253
Z-contrast electron microscopy 139

ZEBRA® cell
– charge process, various phases evolution 106
ZEBRA® power cell 105, 108
– cell design 106
– reaction process 106
Zeeman interactions 284, 446
Zeeman splitting 251
ZEKE. *See* zero electron kinetic energy (ZEKE)
zeolites 259, 260, 268
zero-dimensional point detector 44
zero displacement 118
zero electron kinetic energy (ZEKE) 313
zero-field splitting 286
zero Kelvin values 188
zero loss (ZL) peak 177
zero-quantum (ZQ) transfer 266
zero-time correlation function 287
zigzag function 117
zinc dialkyldithiophosphates 576
zirconium filter 33
$Zn(CN)_2$ structure 425
$Zn_2P_2O_7$, Fourier map 116
ZrO_2 microcrystals 170
ZrO_2 microcrystals
– BF-TEM 170
– DF-TEM 170